Engineering Fundamentals

An Introduction to Engineering

Fifth Edition

Fifth Edition

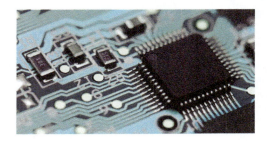

Engineering Fundamentals

An Introduction to Engineering

Saeed Moaveni

Minnesota State University, Mankato

CENGAGE
Learning

Australia • Brazil • Japan • Korea • Mexico • Singapore • Spain • United Kingdom • United States

Engineering Fundamentals:
An Introduction to Engineering, Fifth Edition
Saeed Moaveni

Publisher, Global Engineering:
  Timothy L. Anderson

Senior Development Editors: Hilda Gowans
  and Mona Zeftel

Media Assistant: Ashley Kaupert

Team Assistant: Sam Roth

Marketing Manager: Kristin Stine

Content Project Manager:
  D. Jean Buttrom

Director, Content and Media Production:
  Sharon L. Smith

Production Service: RPK Editorial Services, Inc.

Copyeditor: Shelly Gerger-Knechtl

Proofreader: Lori Martinsek

Indexer: RPK Editorial Services, Inc.

Compositor: Integra Software Services, Pvt. Ltd.

Senior Art Director: Michelle Kunkler

Cover and Internal Designer: © miller design

Cover Image: 77studio/E+/Getty Images

Intellectual Property
  Analyst: Christine Myaskovsky
  Project Manager: Sarah Shainwald

Text and Image Permissions Researcher:
  Kristiina Paul

Senior Manufacturing Planner: Doug Wilke

Library of Congress Control Number: 2014947848

ISBN: 978-1-305-08476-6

Cengage Learning
20 Channel Center Street
Boston, MA 02210
USA

Cengage Learning is a leading provider of customized learning solutions with office locations around the globe, including Singapore, the United Kingdom, Australia, Mexico, Brazil, and Japan. Locate your local office at: **www.cengage.com/global**.

Cengage Learning products are represented in Canada by Nelson Education Ltd.

To learn more about Cengage Learning Solutions, visit **www.cengage.com/engineering**.

Purchase any of our products at your local college store or at our preferred online store **www.cengagebrain.com**.

Unless otherwise noted, all items © Cengage Learning.

MATLAB is a registered trademark of The MathWorks, 3 Apple Hill Drive, Natick, MA.

Printed in the United States of America
Print Number: 03   Print Year: 2016

Contents

Preface xiii

PART 1
ENGINEERING 2

1   Introduction to the Engineering Profession 4

1.1   Engineering Work Is All Around You 6
1.2   Engineering as a Profession 9
        Before You Go On 12
1.3   Common Traits of Good Engineers 13
1.4   Engineering Disciplines 14
        Professional Profile 25
        Before You Go On 26
        Summary 26
        Key Terms 27
        Apply What You Have Learned 28
        Problems 28
        Impromptu Design I 30

2   Preparing for an Engineering Career 32

2.1   Making the Transition from High School to College 34
2.2   Budgeting Your Time 34
2.3   Study Habits and Strategies 36
        Before You Go On 42
2.4   Getting Involved with an Engineering Organization 42
2.5   Your Graduation Plan 43
        Before You Go On 44
        Professional Profile 45
        Student Profile 46
        Summary 46
        Key Terms 47
        Apply What You Have Learned 47

3   Introduction to Engineering Design 50

3.1   Engineering Design Process 52
3.2   Additional Design Considerations 67
        Before You Go On 74
3.3   Teamwork 74
3.4   Project Scheduling and the Task Chart 76
        Before You Go On 78

v

3.5   Engineering Standards and Codes 78
3.6   Water and Air Standards in the United States 84
        Before You Go On 89
        Professional Profile 90
        Summary 91
        Key Terms 92
        Apply What You Have Learned 92
        Problems 92
        Impromptu Design II 95
        Civil Engineering Design Process, A Case Study 95
        Mechanical/Electrical Engineering Design Process, A Case Study 99

4   Engineering Communication 102

4.1   Communication Skills and Presentation of Engineering Work 103
4.2   Basic Steps Involved in the Solution of Engineering Problems 104
        Before You Go On 108
4.3   Written Communication 108
        Before You Go On 111
4.4   Oral Communication 111
4.5   Graphical Communication 113
        Before You Go On 115
        Summary 116
        Key Terms 116
        Apply What You Have Learned 116
        Problems 117
        Professional Profile 122

5   Engineering Ethics 124

5.1   Engineering Ethics 125
5.2   The Code of Ethics of the National Society of Professional Engineers 126
5.3   Engineer's Creed 132
        Before You Go On 137
        Summary 138
        Key Terms 139
        Apply What You Have Learned 139
        Problems 139
        Engineering Ethics 141

PART 2
ENGINEERING FUNDAMENTALS 144

6   Fundamental Dimensions and Units 146

6.1   Fundamental Dimensions and Units 148
        Before You Go On 151
6.2   Systems of Units 151
        Before You Go On 160

6.3   Unit Conversion and Dimensional Homogeneity 160
       Before You Go On 169
6.4   Significant Digits (Figures) 169
6.5   Components and Systems 171
       Before You Go On 173
6.6   Physical Laws and Observations 173
       Before You Go On 176
       Summary 178
       Key Terms 179
       Apply What You Have Learned 180
       Problems 180

7   Length and Length-Related Variables in Engineering 188

7.1   Length as a Fundamental Dimension 191
7.2   Ratio of Two Lengths—Radians and Strain 201
       Before You Go On 202
7.3   Area 203
7.4   Volume 212
7.5   Second Moment of Area 217
       Before You Go On 222
       Summary 223
       Key Terms 223
       Apply What You Have Learned 224
       Problems 224
       Impromptu Design III 229
       An Engineering Marvel 231
       Problem 235

8   Time and Time-Related Variables in Engineering 236

8.1   Time as a Fundamental Dimension 238
8.2   Measurement of Time 241
       Before You Go On 244
8.3   Periods and Frequencies 244
8.4   Flow of Traffic 246
       Before You Go On 248
8.5   Engineering Variables Involving Length and Time 248
       Professional Profile 257
       Before You Go On 258
       Summary 258
       Key Terms 259
       Apply What You Have Learned 259
       Problems 260

9   Mass and Mass-Related Variables in Engineering 266

9.1   Mass as a Fundamental Dimension 268
9.2   Density, Specific Weight, Specific Gravity, and Specific Volume 272

9.3   Mass Flow Rate 274
       Before You Go On 275
9.4   Mass Moment of Inertia 275
9.5   Momentum 278
9.6   Conservation of Mass 280
       Before You Go On 283
       Summary 284
       Key Terms 284
       Apply What You Have Learned 285
       Problems 285
       Impromptu Design IV 291

10 Force and Force-Related Variables in Engineering 294

10.1   Force 296
10.2   Newton's Laws in Mechanics 303
       Before You Go On 306
10.3   Moment, Torque—Force Acting at a Distance 307
10.4   Work—Force Acting Over a Distance 312
       Before You Go On 313
10.5   Pressure and Stress—Force Acting Over an Area 314
10.6   Linear Impulse—Force Acting Over Time 336
       Before You Go On 338
       Summary 338
       Key Terms 339
       Apply What You Have Learned 340
       Problems 340
       Impromptu Design V 346
       An Engineering Marvel 347

11 Temperature and Temperature-Related Variables in Engineering 350

11.1   Temperature as a Fundamental Dimension 353
11.2   Temperature Difference and Heat Transfer 363
       Before You Go On 376
11.3   Thermal Comfort 377
11.4   Heating Values of Fuels 379
       Before You Go On 381
11.5   Degree-Days and Energy Estimation 381
11.6   Additional Temperature-Related Material Properties 382
       Before You Go On 386
       Summary 387
       Key Terms 388
       Apply What You Have Learned 388
       Professional Profile 388
       Problems 389

12   Electric Current and Related Variables in Engineering 396

12.1   Electric Current, Voltage, and Electric Power 399
Before You Go On 406

12.2   Electrical Circuits and Components 407
Before You Go On 416

12.3   Electric Motors 416

12.4   Lighting Systems 420
Before You Go On 425
Professional Profile 426
Summary 426
Key Terms 427
Apply What You Have Learned 428
Problems 428
Student Profile 431

13   Energy and Power 432

13.1   Work, Mechanical Energy, and Thermal Energy 434

13.2   Conservation of Energy 439
Before You Go On 442

13.3   Power 443

13.4   Efficiency 447
Before You Go On 454

13.5   Energy Sources, Generation, and Consumption 455
Before You Go On 470
Student Profile 471
Professional Profile 471
Summary 472
Key Terms 473
Apply What You Have Learned 474
Problems 474
Impromptu Design VI 477
An Engineering Marvel 477

PART 3

COMPUTATIONAL ENGINEERING TOOLS 480

14   Computational Engineering Tools Electronic Spreadsheets 482

14.1   Microsoft Excel Basics 484

14.2   Excel Functions 493
Before You Go On 498

14.3   Plotting with Excel 499

14.4   Matrix Computation with Excel 509
Before You Go On 515

14.5 An Introduction to Excel's Visual Basic for Applications 516
Before You Go On 531
Summary 532
Key Terms 532
Apply What You Have Learned 533
Problems 533

15 Computational Engineering Tools MATLAB 540

15.1 MATLAB Basics 542
Before You Go On 552
15.2 MATLAB Functions, Loop Control, and Conditional Statements 552
Before You Go On 560
15.3 Plotting with MATLAB 561
15.4 Matrix Computations with MATLAB 569
15.5 Symbolic Mathematics with MATLAB 573
Before You Go On 575
Professional Profile 576
Summary 576
Key Terms 577
Apply What You Have Learned 577
Problems 578

PART 4
ENGINEERING GRAPHICAL COMMUNICATION 586

16 Engineering Drawings and Symbols 588

16.1 Mechanical Drawings 590
Before You Go On 603
16.2 Civil, Electrical, and Electronic Drawings 603
16.3 Solid Modeling 603
Before You Go On 611
16.4 Engineering Symbols 611
Before You Go On 615
Professional Profile 616
Summary 617
Key Terms 618
Apply What You Have Learned 618
Problems 619
An Engineering Marvel 628

PART 5
ENGINEERING MATERIAL SELECTION 632

17 Engineering Materials 634

17.1 Material Selection and Origin 636
Before You Go On 641

17.2   The Properties of Materials 641
        Before You Go On 648
17.3   Metals 648
        Before You Go On 653
17.4   Concrete 654
        Before You Go On 655
17.5   Wood, Plastics, Silicon, Glass, and Composites 656
        Before You Go On 661
17.6   Fluid Materials: Air and Water 661
        Before You Go On 666
        Summary 666
        Key Terms 669
        Apply What You Have Learned 669
        Problems 669
        Impromptu Design VII 672
        Professional Profile 674
        An Engineering Marvel 674

PART 6
MATHEMATICS, STATISTICS, AND ENGINEERING ECONOMICS 676

18   Mathematics in Engineering 678

18.1   Mathematical Symbols and Greek Alphabet 680
18.2   Linear Models 682
        Before You Go On 688
18.3   Nonlinear Models 689
        Before You Go On 694
18.4   Exponential and Logarithmic Models 694
        Before You Go On 699
18.5   Matrix Algebra 700
        Before You Go On 711
18.6   Calculus 711
        Before You Go On 719
18.7   Differential Equations 720
        Before You Go On 722
        Summary 722
        Key Terms 723
        Apply What You Have Learned 723
        Problems 724

19   Probability and Statistics in Engineering 730

19.1   Probability—Basic Ideas 732
19.2   Statistics—Basic Ideas 733
19.3   Frequency Distributions 734
        Before You Go On 736
19.4   Measures of Central Tendency and Variation—Mean, Median, and Standard Deviation 737

19.5   Normal Distribution 741
        Before You Go On 750
        Summary 751
        Key Terms 751
        Apply What You Have Learned 752
        Problems 752

20   Engineering Economics 758

20.1   Cash Flow Diagrams 760
20.2   Simple and Compound Interest 761
20.3   Future Worth of a Present Amount and Present Worth of a Future Amount 762
20.4   Effective Interest Rate 765
        Before You Go On 767
20.5   Present and Future Worth of Series Payment 768
        Before You Go On 773
20.6   Interest–Time Factors 773
20.7   Choosing the Best Alternatives—Decision Making 778
20.8   Excel Financial Functions 780
        Before You Go On 784
        Summary 784
        Key Terms 785
        Apply What You Have Learned 785
        Problems 786

        Appendix 793
        Index 805

Changes in the Fifth Edition

The Fifth Edition, consisting of twenty chapters, includes a number of new features, additions, and changes that were incorporated in response to pedagogical advances, suggestions, and requests made by professors and students using the Fourth Edition of the book. The major changes include:

New Features

To promote active learning, we have added eight new features in the fifth edition of this book. These features include: (1) Learning Objectives (LO), (2) Discussion Starter—What Do You Think? (3) Before You Go On, (4) Highlighted Key Concepts, (5) Summary, (6) Key Terms, (7) Apply What You Have Learned, and (8) Life-long Learning Exercises.

1. **Learning Objectives (LO)**

   Each chapter begins by stating the learning objectives (LO).

2. **Discussion Starter**

   Pertinent articles serve as chapter openers to engage students and promote active learning. The discussion starters provide a current context for why the content that the students are about to learn is important. An instructor can start class by asking students to read the Discussion Starter and then ask the students for their thoughts and reactions.

3. **Before You Go On**

   This feature encourages students to test their comprehension and understanding of the material discussed in section(s) by answering questions, before they continue to the next section(s).

   Vocabulary—It is important for students to understand that they need to develop a comprehensive vocabulary to communicate effectively as well educated engineers and intelligent citizens. This feature promotes growing vocabulary by asking students to state the meaning of new words that are covered in section(s).

4. **Highlighted Key Concepts**

   Key concepts are highlighted in blue boxes and displayed throughout the book.

5. **Summary**

   Each chapter concludes by summarizing what the student should have gained from studying the chapter. Moreover, the learning objectives and the summary are tied together as a refresher for the students.

6. **Key Terms**

   The key terms are indexed at the end of each chapter so that students may return to them for review.

7. **Apply What You Have Learned**

   This feature encourages students to apply what they have learned to an interesting problem or a situation. To emphasize the importance of teamwork and to encourage group participation, many of these problems require group work.

8. **Life-long Learning Exercises**

   Problems that promote life-long learning are denoted by ⚷.

MindTap

This textbook is also available as a course or supplement to the textbook through Cengage Learning's MindTap, a personalized learning program. Students who purchase the MindTap version have access to the book's MindTap Reader and are able to complete homework and assessment material online, through their desktop, laptop, or iPad. If you are using a Learning Management System (such as Blackboard or Moodle) for tracking course content, assignments, and grading, you can seamlessly access the MindTap suite of content and assessments for this course.

In MindTap, instructors can:

- Personalize the Learning Path to match the course syllabus by rearranging content or appending original material to the online content.
- Connect a Learning Management System portal to the online course and Reader
- Customize online assessments and assignments
- Track student progress and comprehension
- Promote student engagement through interactivity and exercises

Additionally, students can listen to the text through ReadSpeaker, take notes, create their own flashcards, highlight content for easy reference, and check their understanding of the material through practice quizzes and homework.

Additional Content in the Fifth Edition

- A new section on Visual Basic for Applications (VBA). Excel's VBA is a programming language that allows students to use Excel more effectively and use the capabilities of VBA to solve a wide range of engineering problems. In Section 14.5, we explain how to input and retrieve data, display results, create a subroutine, and how to use Excel's Built-in functions in a VBA program. We also explain how to create a loop and the use of arrays. Students learn how to create a custom dialogue box.

- Over fifty new problems have been added throughout the book.

- Instructor resources include new Lecture Note PowerPoint slides and Test Banks for each chapter.

Organization

This book is organized into six parts and twenty chapters. Each chapter begins by stating its objectives and concludes by summarizing what the student should have gained from studying that chapter. I have included enough material for two semester-long courses. The reason for this approach is to give the instructor sufficient materials and the flexibility to choose specific topics to meet his or her needs. Relevant, everyday examples with which students can easily associate are provided in every chapter. Each chapter includes many hands-on problems, requiring the student to gather and analyze information. Moreover, information collection and proper use of information are encouraged in this book by asking students to complete a number of assignments that require information gathering by using the Internet as well as employing traditional methods. Many of the problems require students to make brief reports so that they learn that successful engineers need to have good written and oral communication skills. To emphasize the importance of teamwork in engineering and to encourage group participation, many of the assignment problems require group work; some require the participation of the entire class.

The main parts of the book are:

Part One: Engineering—An Exciting Profession

In Chapters 1 through 5, we introduce the students to the engineering profession, how to prepare for an exciting engineering career, the design process, engineering communication, and ethics. Chapter 1 provides a comprehensive introduction to the engineering profession and its branches. It explains some of the common traits of good engineers. Various engineering disciplines and engineering organizations are discussed. In Chapter 1, we also emphasize that engineers are problem solvers. Engineers have a good grasp of fundamental physical and chemical laws and mathematics, and apply these fundamental laws and principles to design, develop, test, and supervise the manufacture of millions of products and services. The examples, demonstrate the many satisfying and challenging jobs for engineers. We point out that although the activities of engineers can be quite varied, there are some personality traits and work habits that typify most of today's successful engineers:

- Engineers are problem solvers.
- Good engineers have a firm grasp of the fundamental principles that can be used to solve many different problems.
- Good engineers are analytical, detailed oriented, and creative.
- Good engineers have a desire to be life-long learners. For example, they take continuing education classes, seminars, and workshops to stay abreast of new innovations and technologies.
- Good engineers have written and oral communication skills that equip them to work well with their colleagues and to convey their expertise to a wide range of clients.
- Good engineers have time management skills that enable them to work productively and efficiently.
- Good engineers have good "people skills" that allow them to interact and communicate effectively with various people in their organization.
- Engineers are required to write reports. These reports might be lengthy, detailed, and technical, containing graphs, charts, and engineering drawings. Or they may take the form of a brief memorandum or an executive summary.
- Engineers are adept at using computers in many different ways to model and analyze various practical problems.
- Good engineers actively participate in local and national discipline-specific organizations by attending seminars, workshops, and meetings. Many even make presentations at professional meetings.
- Engineers generally work in a team environment where they consult each other to solve complex problems. Good interpersonal and communication skills have become increasingly important now because of the global market.

Chapter 1 explains the difference between an *engineer* and an *engineering technologist,* and the difference in their career options. In Chapter 2, the transition from high school to college is explained in terms of the need to form good study habits and suggestions are provided on how to budget time effectively. Chapter 3 provides an introduction to engineering design, sustainability, teamwork, and standards and codes. We show that engineers, regardless of their background, follow certain steps when designing products and services. Chapter 4 shows that presentations are an integral part of any engineering project. Depending on the size of the project, presentations might be brief, lengthy, frequent, and may follow a certain format requiring calculations, graphs, charts, and engineering drawings. In Chapter 4, various forms of engineering communication, including homework presentation, brief technical memos, progress reports, detailed technical reports, and research papers are explained. Chapter 5 emphasizes engineering ethics by noting that engineers design many products and provide many services that affect our quality of life and safety. Therefore, engineers must perform under a standard of professional behavior that requires adherence to the highest principles of ethical conduct. A large number of engineering ethics-related case studies are presented in this chapter.

Part Two: Engineering Fundamentals—Concepts Every Engineer Should Know

Chapters 6 through 13 focus on engineering fundamentals and introduce students to the basic principles and physical laws that they will encounter repeatedly during the next four years. Successful engineers have a good grasp of the Fundamentals, which they can use to understand and solve many different problems. These are concepts that every engineer, regardless of his or her area of specialization, should know.

In these chapters, we emphasize that we need only a few physical quantities to fully describe events and our surroundings. These are length, time, mass, force, temperature, mole, and electric current. We also explain that we need not only physical dimensions to describe our surroundings, but also some way to scale or divide these physical dimensions. For example, time is considered a physical dimension, but it can be divided into both small and large portions, such as seconds, minutes, hours, days, years, decades, centuries, and millennia.

We discuss common systems of units and emphasize that engineers must know how to convert from one system of units to another and always show the appropriate units that go with their calculations. We also explain that the physical laws and formulas that engineers use are based on observations of their surroundings. We show that we use mathematics and basic physical quantities to express our observations.

In these chapters, we also explain that there are many engineering design variables that are related to the fundamental dimensions (quantities). To become a successful engineer a student must fully understand these fundamental and related variables and the pertaining governing laws and formulas. Then it is important for the student to know how these variables are measured, approximated, calculated, or used in practice.

Chapter 6 explains the role and importance of fundamental dimension and units in analysis of engineering problems. Basic steps in the analysis of any engineering problem are discussed in detail.

Chapter 7 introduces length and length-related variables and explains their importance in engineering work. For example, we discuss the role of area in heat transfer, aerodynamics, load distribution, and stress analysis. Measurement of length, area, and volume, along with numerical estimation (such as trapezoidal rule) of these values, are presented.

Chapter 8 considers time and time-related engineering variables. Periods, frequencies, linear and angular velocities and accelerations, volumetric flow rates and flow of traffic are also discussed in Chapter 8.

Chapter 9 covers mass and mass-related variables such as density, specific weight, mass flow rate, and mass moment of inertia, and their role in engineering analysis.

Chapter 10 discusses the importance of force and force-related variables in engineering. The important concepts in mechanics are explained conceptually. What is meant by force, internal force, reaction, pressure, modulus of elasticity, impulsive force (force acting over time), work (force acting over a distance) and moment (force acting at a distance) are covered in detail.

Chapter 11 presents temperature and temperature-related variables. Concepts such as temperature difference and heat transfer, specific heat, and thermal conductivity also are covered. As future engineers, it is important for students to understand some simple-energy-estimation procedures given current energy and sustainability concerns. Because of this fact, we have a section on Degree-Days and Energy Estimation.

Chapter 12 considers topics such as direct and alternating current, electricity, basic circuit components, power sources, and the tremendous role of electric motors in our everyday life. Lighting systems account for a major portion of electricity use in buildings and have received much attention lately. Section 12.4 introduces the basic terminology and concepts in lighting systems. All future engineers regardless of their area of expertise need to understand these basic concepts.

Chapter 13 presents energy and power and explains the distinction between these two topics. The importance of understanding what is meant by work, energy, power, watts, horsepower, and efficiency is emphasized. Energy sources, generation, and consumption in the United States are also discussed in this chapter. With the world's growing demand for energy being among the most difficult challenges that we face today, as future engineers, students need to understand two problems: energy sources and emission. Section 13.6 introduces conventional and renewable energy sources, generation, and consumption patterns.

Part Three: Computational Engineering Tools—Using Available Software to Solve Engineering Problems

In Chapters 14 and 15, we introduce Microsoft Excel™ and MatLab™—two computational tools that are used commonly by engineers to solve engineering problems. These computational tools are used to record, organize, analyze data using formulas, and present the results of an analysis in chart forms. MatLab is also versatile enough that students can use it to write their own programs to solve complex problems.

Part Four: Engineering Graphical Communication—Conveying Information to Other Engineers, Machinists, Technicians, and Managers

Chapter 16 introduces students to the principles and rules of engineering graphical communication and engineering symbols. A good grasp of these principles will enable students to convey and understand information effectively. We explain that engineers use technical drawings to convey useful information to others in a standard manner. An engineering drawing provides information, such as the shape of a product, its dimensions, materials from which to fabricate the product, and the assembly steps. Some engineering drawings are specific to a particular discipline. For example, civil engineers deal with land or boundary, topographic, construction, and route survey drawings. Electrical and electronic engineers, on the other hand, could deal with printed circuit board assembly drawings, printed circuit board drill plans, and wiring diagrams. We also show that engineers use special symbols and signs to convey their ideas, analyses, and solutions to problems.

Part Five: Engineering Material Selection—An Important Design Decision

As engineers, whether you are designing a machine part, a toy, a frame of a car, or a structure, the selection of materials is an important design decision. Chapter 17 looks more closely at materials such as metals and their alloys, plastics, glass, wood, composites, and concrete that commonly are used in various engineering applications. We also discuss some of the basic characteristics of the materials that are considered in design.

Part Six: Mathematics, Statistics, and Engineering Economics—Why Are They Important?

Chapters 18 through 20 introduce students to important mathematical, statistical, and economical concepts. We explain that engineering problems are mathematical models of physical situations. Some engineering problems lead to linear models, whereas others result in nonlinear models. Some engineering problems are formulated in the form of differential equations and some in the form of integrals. Therefore, a good understanding of mathematical concepts is essential in the formulation and solution of many engineering problems.

Moreover, statistical models are becoming common tools in the hands of practicing engineers to solve quality control and reliability issues, and to perform failure analyses. Civil engineers use statistical models to study the reliability of construction materials and structures, and to design for flood control, for example. Electrical engineers use statistical models for signal processing and for developing voice-recognition software. Manufacturing engineers use statistics for quality control assurance of the products they produce. Mechanical engineers use statistics to study the failure of materials and machine parts.

Economic factors also play important roles in engineering design decision making. If you design a product that is too expensive to manufacture, then it cannot be sold at a price that consumers can afford and still be profitable to your company.

Case Studies—Engineering Marvels

To emphasize that engineers are problem solvers and that engineers apply physical and chemical laws and principles, along with mathematics, to *design* products and services that we use in our everyday lives, we include case studies throughout the book. Each case study is followed by assigned problems. The solutions to these problems incorporate the engineering concepts and laws that are discussed in the preceding chapters. There are also a number of engineering ethics case studies, from the National Society of Professional Engineers, in Chapter 5, to promote the discussion on engineering ethics.

Impromptu Designs

I have included seven inexpensive impromptu designs that could be developed during class times. The basic ideas behind some of the impromptu designs have come from the ASME.

References

In writing this book, several engineering books, Web pages, and other materials were consulted. Rather than giving you a list that contains hundreds of resources, I cite some of the sources that I believe to be useful to you. All freshmen engineering students should own a reference handbook in their chosen field. Currently, there are many engineering handbooks available in print or electronic format, including chemical engineering handbooks, civil engineering handbooks, electrical and electronic engineering handbooks, and mechanical engineering handbooks. I also believe all engineering students should own chemistry, physics, and mathematics handbooks. These texts can serve as supplementary resources for all your classes. Many engineers may find useful the ASHRAE handbook, the *Fundamental Volume*, by the American Society of Heating, Refrigerating, and Air Conditioning Engineers.

In this book, some data and diagrams were adapted with permission from the following sources:

- Baumeister, T., et al., *Mark's Handbook*, 8th ed., McGraw Hill, 1978.
- *Electrical Wiring*, 2nd ed., AA VIM, 1981.
- *Electric Motors*, 5th ed., AA VIM, 1982.
- Gere, J. M., *Mechanics of Materials*, 6th ed., Thomson, 2004.
- Hibbler, R. C., *Mechanics of Materials*, 6th ed., Pearson Prentice Hall.
- *U.S. Standard Atmosphere*, Washington D.C., U.S. Government Printing Office, 1962.
- Weston, K. C., *Energy Conversion*, West Publishing, 1992.

Acknowledgments

I would like to express my sincere gratitude to the editorial and production team at Cengage, especially Hilda Gowans, Mona Zeftel, and Kristiina Paul. I am also grateful to Rose Kernan of RPK Editorial Services, Inc. I would also like to thank Dr. Karen Chou of Northwestern University, Mr. James Panko, and Paulsen Architects, who provided the section on civil engineering design process and the related design case study, and Mr. Pete Kjeer and Johnson Outdoors who provided the mechanical/electrical engineering case study. I am also thankful to all the reviewers who offered general and specific comments including Charles Duvall, Southern Polytechnic State University, Eddie Jacobs, University of Memphis, Thaddeus Roppel, Auburn University, and Steve Warner, University of Massachusetts—Dartmouth.

I would also like to thank the following individuals for graciously providing their insights for our Student and Professional Profiles sections: Nahid Afsari, Jerry Antonio, Celeste Baine, Suzelle Barrington, Steve Chapman, Karen Chou, Ming Dong, Duncan Glover, Dominique Green, Lauren Heine, John Mann, Katie McCullough, Susan Thomas, and Nika Zolfaghari.

Thank you for considering this book, and I hope you enjoy the Fifth Edition.

Saeed Moaveni

Engineering Fundamentals

An Introduction to Engineering

Fifth Edition

Engineering

An Exciting Profession

violetkaipa/Shutterstock.com

In Part One of this book, we will introduce you to the engineering profession. Engineers are problem solvers. They have a good grasp of fundamental physical and chemical laws and mathematics and apply these laws and principles to design, develop, test, and supervise the manufacture of millions of products and services. Engineers, regardless of their background, follow certain steps when designing the products and services we use in our everyday lives. Successful engineers possess good communication skills and are team players. Ethics plays a very important role in engineering. As eloquently stated by the National Society of Professional Engineers (NSPE) code of ethics, "Engineering is an important and learned profession. As members of this profession, engineers are expected to exhibit the highest standards of honesty and integrity. Engineering has a direct and vital impact on the quality of life for all people. Accordingly, the services provided by engineers require honesty, impartiality, fairness and equity, and must be dedicated to the protection of the public health, safety, and welfare. Engineers must perform under a standard of professional behavior which requires adherence to the highest principles of ethical conduct." In the next five chapters, we will introduce you to the engineering profession, how to prepare for an exciting engineering career, the design process, engineering communication, and ethics.

> Good engineers are problem solvers and have a firm grasp of mathematical, physical, and chemical laws and principles. They apply these laws and principles to design products and services that we use in our everyday lives. They also have good written and oral communication skills.

CHAPTER 1   INTRODUCTION TO THE ENGINEERING PROFESSION

CHAPTER 2   PREPARING FOR AN ENGINEERING CAREER

CHAPTER 3   INTRODUCTION TO ENGINEERING DESIGN

CHAPTER 4   ENGINEERING COMMUNICATION

CHAPTER 5   ENGINEERING ETHICS

Introduction to the Engineering Profession

carlosseller/Shutterstock.com

Golden Pixels LLC/Shutterstock.com

NASA

Engineers are problem solvers. Successful engineers possess good communication skills and are team players. They have a good grasp of fundamental physical laws and mathematics. Engineers apply physical and chemical laws and mathematics to design, develop, test, and supervise the manufacture of millions of products and services. They consider important factors such as efficiency, cost, reliability, and safety when designing products. Good engineers are dedicated to lifelong learning and service to others.

LEARNING OBJECTIVES

LO¹ **Engineering Work is All Around You:** give examples of products and services that engineers design that make our lives better

LO² **Engineering as a Profession:** describe what engineers do and give examples of common careers for engineers

LO³ **Common Traits of Good Engineers:** describe the important traits of successful engineers

LO⁴ **Engineering Disciplines:** give examples of common engineering disciplines and how they contribute to the comfort and betterment of our everyday lives

WHO ARE ENGINEERS?

Engineers are problem solvers. They have a good grasp of fundamental physical and chemical laws and mathematics and apply these laws and principles to design, develop, test, and supervise the manufacture of millions of products and services. **Engineers**, regardless of their background, follow certain steps when designing the products and services we use in our everyday lives. Successful engineers possess good communication skills and are team players.

Ethics plays a very important role in engineering. As eloquently stated by the National Society of Professional Engineers (NSPE) code of ethics, "Engineering is an important and learned profession. As members of this profession, engineers are expected to exhibit the highest standards of honesty and integrity. Engineering has a direct and vital impact on the quality of life for all people. Accordingly, the services provided by engineers require honesty, impartiality, fairness and equity, and must be dedicated to the protection of the public health, safety and welfare. Engineers must perform under a standard of professional behavior which requires adherence to the highest principles of ethical conduct."

zentilia/Shutterstock.com

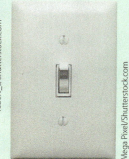

robert_s/Shutterstock.com

Mega Pixel/Shutterstock.com

Luis Santos/Shutterstock.com

MO_SES Premium/Shutterstock.com

tkemot/Shutterstock.com

To the students: What do you think engineers do? Why do you want to study engineering? Name at least two products or services that are not available now that you think will be readily available in the next 20 years. Which engineering disciplines do you think will be involved in design and development of these products and services?

Possibly some of you are not yet certain you want to study engineering during the next four years in college and may have questions similar to the following:

Do I really want to study engineering?

What is engineering and what do engineers do?

What are some of the areas of specialization in engineering?

How many different engineering disciplines are there?

Do I want to become a mechanical engineer, or should I pursue civil engineering? Or would I be happier becoming an electrical engineer?

How will I know that I have picked the best field for me?

Will the demand for my area of specialization be high when I graduate, and beyond that?

The main objectives of this chapter are to provide some answers to these and other questions you may have and to introduce you to the engineering profession and its various branches.

LO¹ 1.1   Engineering Work Is All Around You

Engineers make products and provide services that make our lives better (see Figure 1.1). To see how engineers contribute to the comfort and the betterment of our everyday lives, tomorrow morning when you get up, just look around you more carefully. During the night, your bedroom was kept at the right temperature thanks to some mechanical engineers who designed the heating, air-conditioning, and ventilating systems in your home. When you get up in the morning and turn on the lights, be assured that thousands of mechanical and electrical engineers and technicians at power plants and power stations around the country are making certain the flow of electricity remains uninterrupted so that you have enough power to turn the lights on or turn on your TV to watch the morning news and weather report for the day. The TV you are using—to get your morning news or to see how your favorite team did—was designed by electrical and electronic engineers. There are, of course, engineers from other disciplines involved in creating the final product; for example, manufacturing and industrial engineers. When you are getting ready to take your morning shower, the clean water you are about to use is coming to your home thanks to civil and mechanical

arka38/Shutterstock.com

vovan/Shutterstock.com

Alexandru Nika/Shutterstock.com

You can more/Shutterstock.com

FIGURE 1.1   Examples of products and services designed by engineers.

engineers. Even if you live out in the country on a farm, the pump you use to bring water from the well to your home was designed by mechanical and civil engineers. The water could be heated by natural gas that is brought to your home thanks to the work and effort of chemical, mechanical, civil, and petroleum engineers. After your morning shower, when you get ready to dry yourself with a towel, think about what types of engineers worked behind the scenes to produce the towels. Yes, the cotton towel was made with the help of agricultural, industrial, manufacturing, chemical, petroleum, civil, and mechanical engineers. Think about the machines that were used to pick the cotton, transport the cotton to a factory, clean it, and dye it to a pretty color that is pleasing to your eyes. Then other machines were used to weave the fabric and send it to sewing machines that were designed by mechanical engineers. The same is true of the clothing you are about to wear. Your clothing may contain some polyester, which was made possible with the aid of petroleum and chemical engineers. "Well," you may say, "I can at least sit down and eat my breakfast and not wonder whether some engineers made this possible as well." But the food you are about to eat was made with the help and collaboration of various engineering disciplines, from agricultural to mechanical. Let's say you are about to have some cereal. The milk was kept fresh in your refrigerator thanks to the efforts and work of mechanical engineers who designed the refrigerator components and chemical engineers who investigated alternative refrigerant fluids with appropriate thermal properties and other environmentally friendly properties that can be used in your refrigerator. Furthermore, electrical engineers designed the control and the electrical power units.

Now you are ready to get into your car or take the bus to go to school. The car you are about to drive was made possible with the help and collaboration of automotive, mechanical, electrical, electronic, industrial, material, chemical, and petroleum engineers. So, you see there is not much that you do in your daily life that has not involved the work of engineers. Be proud of the decision you have made to become an engineer. Soon you will become one of those whose behind-the-scenes efforts will be taken for granted by billions of people around the world. But you will accept that fact gladly, knowing that what you do will make people's lives better.

Engineers Deal with an Increasing World Population and Sustainability Concerns

We as people, regardless of where we live, need the following things: food, clothing, shelter, clear air, and water. In addition, we need various modes of transportation to get to different places, because we may live and work in different cities or wish to visit friends and relatives who may live elsewhere. We also like to have some sense of security, to be able to relax and be entertained. We need to be liked and appreciated by our friends and family, as well.

Increasingly, because of worldwide socioeconomic population trends, environmental concerns, and the earth's finite resources, more is expected of engineers. Future engineers are expected to provide goods and services that increase the standard of living and advance health care while also addressing serious environmental and sustainability concerns. At the turn of the 21st century, there were approximately six billion of us inhabiting the earth. As a means of comparison, it is important to note that the world population about 115 years ago, at the start of the 20th century, was one billion. Think about it. It took us since the beginning of human existence to reach a population of one billion. It only took 115 years to increase the population by fivefold. Some of us have a good standard of living, but some of us living in developing countries do not. You will probably agree that our world would be a better place if every one of us had enough to eat, a comfortable and safe place to live, meaningful work to do, and some time for relaxation.

According to the latest estimates and projections of the U.S. Census Bureau, the world population will reach 9.3 billion people by the year 2050. Not only will the number of people inhabiting the earth continue to rise but the age structure of the world population will also change. The world's elderly population—the people at least 65 years of age—will more than double in the next 25 years (see Figure 1.2).

How is this information relevant? Well, now that you have decided to study to become an engineer, you need to realize that what you do in a few years after your graduation is very important to all of us. The world's

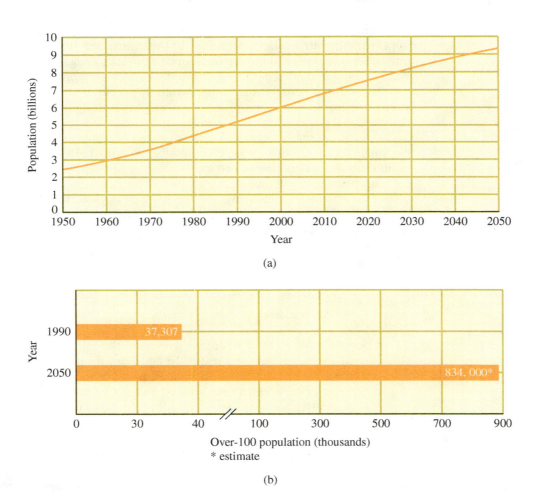

(a)

(b)

| FIGURE 1.2 | (a) The latest projection of world population growth. (b) The latest estimate of U.S. elderly population growth. |

Data from the U.S. Census Bureau

current economic development is not sustainable—the world population already uses approximately 20% more of the world's resources than the planet can sustain. (United Nations *Millenium Ecosystem Assessment Synthesis Report,* 2005.) You will design products and provide services especially suited to the needs and demands of an increasing elderly population as well as increased numbers of people of all ages. So prepare well to become a good engineer and be proud that you have chosen the engineering profession in order to contribute to raising the living standard for everyone and at the same time addressing environmental and sustainability concerns. Today's world economy is very dynamic. Corporations continually employ new technologies to maximize efficiency and profits. Because of this ongoing change and emerging technologies, new jobs are created and others are eliminated. Computers and smart electronic devices are continuously reshaping our way of life. Such devices influence the way we do things and help us provide the necessities of our lives—clean water, clean air, food, and shelter. You need to become a lifelong learner so that you can make informed decisions and anticipate as well as react to the global changes caused by technological innovations as well as population and environmental changes. According to the Bureau of Labor Statistics, U.S. Department of Labor, among the fastest-growing occupations are engineers, computer specialists, and systems analysts.

LO² 1.2   Engineering as a Profession

In the following sections, we will first discuss **engineering** in a broad sense, and then we will focus on selected aspects of engineering. We will also look at the traits and characteristics common to many engineers. Next we will discuss some specific engineering disciplines. As we said earlier in this chapter, perhaps some of you have not yet decided what you want to study during your college years and consequently may have many questions, including: What is engineering and what do engineers do? What are some of the areas of specialization in engineering? Do I really want to study engineering? How will I know that I have picked the best field for me? Will the demand for my area of specialization be high when I graduate, and beyond that?

The following sections are intended to help you make a decision that you will be happy with; don't worry about finding answers to all these questions right now. You have some time to ponder them because most of the coursework during the first year of engineering is similar for all engineering students, regardless of their specific discipline. So you have at least a year to consider various possibilities. This is true at most educational institutions. Even so, you should talk to your advisor early to determine how soon you must choose an area of specialization. Don't be concerned about your chosen profession changing in a way that makes your education obsolete. Most companies assist their engineers in acquiring further training and education to keep up with changing technologies. A good engineering education will enable you to become a good problem solver throughout your life. You may wonder during the next few years of school why you need to be learning some of the material you are studying. Sometimes your homework may seem irrelevant, trivial, or out-of-date. Rest assured that you are learning both content information and strategies of thinking and analysis that will equip you to face future challenges, ones that do not even exist yet.

What Is Engineering and What Do Engineers Do?

Engineers apply physical and chemical laws and principles and mathematics to design millions of products and services that we use in our everyday lives. These products include cars, computers, aircraft, clothing, toys, home appliances, surgical equipment, heating and cooling equipment, health care devices, tools and machines that makes various products, and so on (see Figure 1.3). Engineers consider important factors such as cost, efficiency, reliability, sustainability, and safety when designing these products. Engineers perform tests to make certain that the products they design withstand various loads and conditions. They are continuously searching for ways to improve already existing products as well. They also design and supervise the construction of buildings, dams, highways, and mass transit systems and the construction of power plants that supply power to manufacturing companies, homes, and offices. Engineers play a significant role in the design and maintenance of a nation's infrastructure, including communication systems, public utilities, and transportation. Engineers continuously develop new, advanced materials to make products lighter and stronger for different applications. They are also responsible for finding suitable ways to extract petroleum, natural gas, and raw materials from the earth, and they are involved in coming up with ways of increasing crop, fruit, and vegetable yields along with improving the safety of our food products.

> Engineers consider important factors such as cost, efficiency, reliability, sustainability, and safety when designing products.

The following represent some common careers for engineers. In addition to design, some engineers work as sales representatives for products, while others provide technical support and troubleshooting for customers of their products. Many engineers decide to become involved in sales and customer support, because their engineering background enables them to explain and discuss technical information and to assist with installation, operation, and maintenance of various products and machines. Not all engineers work for private industries; some work for federal, state, and local governments in various

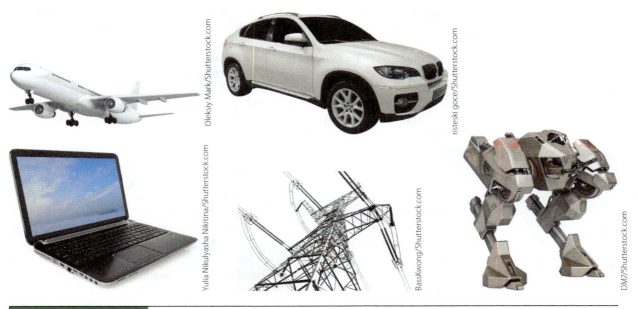

FIGURE 1.3   As an engineer you will apply physical and chemical laws and principles and mathematics to design various products and services.

capacities. Engineers work in departments of agriculture, defense, energy, and transportation. Some engineers work for the National Aeronautics and Space Administration (NASA). As you can see, there are many satisfying and challenging jobs for engineers.

These are some other facts about engineering that are worth noting.

- For almost all entry-level engineering jobs, a bachelor's degree in engineering is required.

According to the U.S. Bureau of Labor Statistics:

- The starting salaries of engineers are significantly higher than those of Bachelor's-degree graduates in other fields. The outlook for engineering is very good. Good employment opportunities are expected for new engineering graduates during 2015–2025.
- Most engineering degrees are granted in electrical, mechanical, and civil engineering, the parents of all other engineering branches.
- In the year 2013, engineers held approximately 1.6 million jobs (see Table 1.1).

TABLE 1.1	Engineering Employment by Disciplines-Data from U.S. Bureau of Labor Statistics
Total, All Engineers	**1,547,590**
Civil	262,170
Mechanical	258,630
Industrial	230,580
Electrical	168,100
Electronics, except computer	135,350
Computer hardware	77,670
Aerospace	71,500
Environmental	53,020
Petroleum	34,910
Chemical	33,300
Materials	24,190
Health and safety, except mining safety	23,850
Biomedical	19,890
Nuclear	16,400
Mining and geological, including mining safety	7,990
Marine engineers and naval architects	6,640
Agricultural	2,590
All other engineers	120,810

Data from U.S. Bureau of Labor Statistics

The distribution of employment by disciplines is shown in Table 1.1.

As mentioned previously, engineers earn some of the highest salaries among those holding bachelors degrees. The average starting salary for engineers is shown in Table 1.2. The data shown in Table 1.2 is the result of the May 2013 survey conducted by the U.S. Bureau of Labor Statistics.

TABLE 1.2	The Mean Salary Offer for Engineers (2013)–Data from U.S. Bureau of Labor Statistics
Disciplines or Specialties	**Mean Salary Offer**
Aerospace/aeronautical/astronautical	$103,870
Agricultural	74,450
Bioengineering and biomedical	88,670
Chemical	95,730
Civil	80,770
Computer	104,250
Electrical /electronics and communications	89,180
Industrial /manufacturing	80,300
Materials	87,330
Mechanical	82,100
Mining and mineral	86,870
Nuclear	101,600

Data from U.S. Bureau of Labor Statistics

According to the U.S. Bureau of Labor Statistics, mean annual salaries for engineers ranged from $104,250 in computer engineering to $74,450 in agricultural engineering in March 2009.

Before You Go On

Answer the following questions to test your understanding of the preceding sections.

1. What are the essential needs of people?

2. Give examples of products and services that make our everyday lives better.

3. What is the latest projection of the U.S. Census Bureau for the world population?

4.   Give examples of the changes that are continuously reshaping our way of life.

5.   What do engineers do?

LO³ 1.3 Common Traits of Good Engineers

Although the activities of engineers are quite varied, there are some personality traits and work habits that typify most of today's successful engineers.

- Engineers are problem solvers.
- Good engineers have a firm grasp of the fundamental principles of engineering, which they can use to solve many different problems.
- Good engineers are analytical, detailed oriented, and creative.
- Good engineers have a desire to be lifelong learners. For example, they take continuing education classes, seminars, and workshops to stay abreast of innovations and new technologies. This is particularly important in today's world because the rapid changes in technology will require you as an engineer to keep pace with new technologies. Moreover, you will risk being laid off or denied promotion if you are not continually improving your engineering education.
- Good engineers, regardless of their area of specialization, have a core knowledge that can be applied to many areas. Therefore, well-trained engineers are able to work outside their area of specialization in other related fields. For example, a good mechanical engineer with a well-rounded knowledge base can work as an automotive engineer, an aerospace engineer, or as a chemical engineer.
- Good engineers have written and oral communication skills that equip them to work well with their colleagues and to convey their expertise to a wide range of clients.
- Good engineers have time-management skills that enable them to work productively and efficiently.
- Good engineers have good "people skills" that allow them to interact and communicate effectively with various people in their organization. For example, they are able to communicate equally well with the sales and marketing experts and their own colleagues.
- Engineers are required to write reports. These reports might be lengthy, detailed technical reports containing graphs, charts, and engineering drawings, or they may take the form of brief memoranda or executive summaries.
- Engineers are adept at using computers in many different ways to model and analyze various practical problems.
- Good engineers actively participate in local and national discipline-specific organizations by attending seminars, workshops, and meetings. Many even make presentations at professional meetings.
- Engineers generally work in a team environment where they consult each other to solve complex problems. They divide up the task into smaller, manageable problems among themselves; consequently,

productive engineers must be good team players. Good interpersonal and communication skills are increasingly important now because of the global market. For example, various parts of a car could be made by different companies located in different countries. In order to ensure that all components fit and work well together, cooperation and coordination are essential, which demands strong communication skills.

Clearly, an interest in building things or taking things apart or solving puzzles is not all that is required to become an engineer. In addition to having a dedication to learning and a desire to find solutions, an engineer needs to foster certain attitudes and personality traits.

LO⁴ 1.4   Engineering Disciplines

Now that you have a general sense of what engineers do, you may be wondering about the various branches or specialties in engineering. Good places to learn more about areas of specialization in engineering are the Web sites of various engineering organizations. We will explain in Chapter 2 that as you spend a little time reading about these organizations, you will discover many share common interests and provide some overlapping services that could be used by engineers of various disciplines. Following is a list of a few Web sites that you may find useful when searching for information about various engineering disciplines.

American Academy of Environmental Engineers
www.aaees.org

American Institute of Aeronautics and Astronautics
www.aiaa.org

American Institute of Chemical Engineers
www.aiche.org

The American Society of Agricultural and Biological Engineers
www.asabe.org

American Society of Civil Engineers
www.asce.org

American Nuclear Society
www.ans.org

American Society for Engineering Education
www.asee.org

American Society of Heating, Refrigerating, and Air-Conditioning Engineers
www.ashrae.org

American Society of Mechanical Engineers
www.asme.org

Biomedical Engineering Society
www.bmes.org

Institute of Electrical and Electronics Engineers
www.ieee.org

Institute of Industrial Engineers
The Global Association of Productivity & Efficiency Professionals
www.iienet2.org

National Academy of Engineering
www.nae.edu

National Science Foundation
www.nsf.gov

National Society of Black Engineers
www.nsbe.org

National Society of Professional Engineers
www.nspe.org

Society of Automotive Engineers
www.sae.org

Society of Hispanic Professional Engineers
www.shpe.org

Society of Manufacturing Engineering
www.sme.org

Society of Women Engineers
www.swe.org

Tau Beta Pi
(All-Engineering Honor Society)
www.tbp.org

NASA Centers
Ames Research Center
www.arc.nasa.gov

Dryden Flight Research Center
www.dfrc.nasa.gov

Goddard Space Flight Center
www.gsfc.nasa.gov

Jet Propulsion Laboratory
www.jpl.nasa.gov

Johnson Space Center
www.jsc.nasa.gov

Kennedy Space Center
www.ksc.nasa.gov

Langley Research Center
www.larc.nasa.gov

Marshall Space Flight Center
www.msfc.nasa.gov

Glenn Research Center
www.grc.nasa.gov

Inventors Hall of Fame
www.invent.org

U.S. Patent and Trademark Office
www.uspto.gov

For an additional listing of engineering-related Web sites, please see this book's companion Web site.

What Are Some Areas of Engineering Specialization?

There are over 20 major disciplines or specialties that are recognized by professional engineering societies. Moreover, within each discipline there exist a number of branches. For example, the mechanical engineering program can be traditionally divided into two broad areas: (1) thermal/fluid systems and (2) structural /solid systems. In most mechanical engineering programs, during your senior year you can take elective classes that allow you to pursue your interest and broaden your knowledge base in these areas. So, for example, if you are interested in learning more about how buildings are heated during the winter or cooled during the summer, you will take a heating, ventilating, and air-conditioning class. To give you additional ideas about the various branches within specific engineering disciplines, consider civil engineering. The main branches of a civil engineering program normally are environmental, geotechnical, water resources, transportation, and structural. The branches of electrical engineering may include power generation and transmission, communications, control, electronics, and integrated circuits.

Not all engineering disciplines are discussed here, but you are encouraged to visit the Web sites of appropriate engineering societies to learn more about a particular engineering discipline.

> Civil engineers design and supervise the construction of buildings, roads and highways, bridges, dams, tunnels, mass transit systems, airports, municipal water supplies, and sewage systems.

Accreditation Board for Engineering and Technology

Over 300 colleges and universities in the United States offer bachelor's-degree programs in engineering that are accredited by the Accreditation Board for Engineering and Technology (**ABET**). ABET examines the credentials of the engineering programs faculty, curricular content, facilities, and admissions standards before granting accreditation. It may be wise for you to find out the accreditation status of the engineering program you are planning to attend. ABET maintains a Web site with a list of all accredited programs; visit www.abet.org for more information. According to ABET, accredited engineering programs must demonstrate that their graduates, by the time of graduations, have

- an ability to apply knowledge of mathematics, science, and engineering;
- an ability to design and conduct experiments, as well as to analyze and interpret data;
- an ability to design a system, component, or process to meet desired needs;
- an ability to function on multidisciplinary teams;
- an ability to identify, formulate, and solve engineering problems;
- an understanding of professional and ethical responsibility;
- an ability to communicate effectively;
- the broad education necessary to understand the impact of engineering solutions in a global and societal context;
- a recognition of the need for and an ability to engage in lifelong learning;
- a knowledge of contemporary issues; and
- an ability to use the techniques, skills, and modern engineering tools (Figure 1.4) necessary for engineering practice.

Therefore, these are the educational outcomes that are expected of you when you graduate from your engineering program. Bachelor's-degree programs in engineering are typically designed to last four years; however, many students take five years to acquire their engineering degrees. In a typical engineering program, you will spend the first two years studying mathematics, English, physics, chemistry, introductory engineering, computer science, humanities, and social sciences. These first two years are often referred to as *pre-engineering*. In the last two years, most courses are in engineering, usually with a concentration in one branch. For example, in a typical mechanical engineering program, during the last two years of your studies, you will take courses such as thermodynamics, mechanics of materials, fluid mechanics, heat transfer, applied thermodynamics, and design. During the last two years of your civil engineering studies, you can expect to take courses in fluid mechanics, transportation, geotechnical engineering, hydraulics, hydrology, and steel or concrete design. Some programs offer a general engineering curriculum; students then specialize in graduate school or on the job.

Many community colleges around the country offer the first two years of engineering programs, which are normally accepted by the engineering schools. Some engineering schools offer five-year master's-degree programs. Some engineering schools, in order to provide hands-on experience, have a cooperative plan whereby students take classes during the first three years and

> Mechanical engineers are involved in the design, development, testing, and manufacturing of machines, robots, tools, power-generating equipment (such as steam and gas turbines), heating, cooling, and refrigerating equipment, and internal combustion engines.

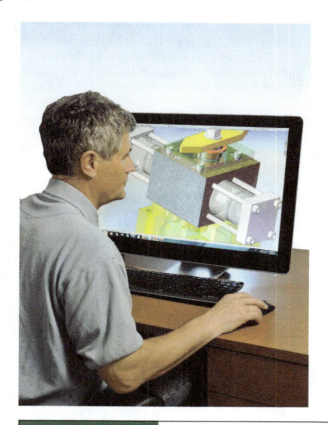

FIGURE 1.4   Engineers are adept at using computers in many different ways to model and analyze various practical problems.
Chuck Rausin/Shutterstock.com

then may take a semester off from studying to work for an engineering company. Of course, after a semester or two, students return to school to finish their education. Schools that offer cooperative programs generally offer full complements of classes every semester so that students can graduate in four years if they desire.

Professional Engineer

All 50 states and the District of Columbia require registration for engineers whose work may affect the safety of the public. As a first step in becoming a registered **professional engineer** (PE), you must have a degree from an ABET-accredited engineering program. You also need to take your **Fundamentals of Engineering Exam** (FE).

The FE exam is designed for recent graduates and students who are close to finishing an undergraduate engineering degree. It is a computer-based exam and is administered year-round at National Council of Examiners for Engineering and Surveying (NCEES) approved testing centers. The FE exam is six hours long, which includes a tutorial, a break, the exam, and a brief survey at the end. It is a closed-book exam with electronic references and contains 110 multiple-choice questions. Moreover, it has questions in both SI and U.S. Customary systems of units, so study Chapter 6 of this book carefully. The exam is offered in seven disciplines: chemical, civil, electrical and computer, environmental, industrial, mechanical, and "other." After you pass your FE exam, you need to gain four years of relevant engineering work experience and pass another eight-hour exam (the Principles and Practice of Engineering Exam) given by the state. Candidates choose an exam from one of 16 engineering disciplines. Some engineers are registered in several states. Normally, civil, mechanical, chemical, and electrical engineers seek professional registrations.

As a recent engineering graduate, you should expect to work under the supervision of a more experienced engineer. Based on your assigned duties, some companies may have you attend workshops (short courses that could last for a week) or a day-long seminar to obtain additional training in communication skills, time management, or a specific engineering method. As you gain more knowledge and experience, you will be given more freedom to make engineering decisions. Once you have many years of experience, you may then elect to become a manager in charge of a team of engineers and technicians. Some engineers fresh out of college begin their careers not in a specific area of engineering, but in sales or marketing related to engineering products and services.

As already mentioned, there are more than 20 engineering disciplines recognized by the professional societies. However, most engineering degrees are granted in civil, electrical, and mechanical engineering. Therefore, these disciplines are discussed here first.

Civil Engineering   Civil engineering is perhaps the oldest engineering discipline. As the name implies, civil engineering is concerned with providing public infrastructure and services. Civil engineers design and supervise the construction of buildings, roads and highways, bridges, dams, tunnels, mass transit systems, and airports. They are also involved in the design and supervision of municipal water supplies and sewage systems. The major branches within the civil engineering discipline include structural, environmental, transportation, water resources, and geotechnical. Civil engineers (Figure 1.5) work as consultants, construction supervisors, city engineers, and public utility and transportation engineers. According to the Bureau of Labor Statistics, the job outlook for graduates of civil engineering is good because as population grows, more civil engineers are needed to design and supervise the construction of new buildings, roads, and water supply and sewage systems. They are also needed to oversee the maintenance and renovation of existing public structures, roads, bridges, and airports.

Electrical and Electronic Engineering   Electrical and electronic engineering is the largest engineering discipline. Electrical engineers design, develop, test, and supervise the manufacturing of electrical equipment, including lighting and wiring for buildings, cars, buses, trains, ships, and aircrafts; power generation and transmission equipment for utility companies; electric motors found in various products; control devices; and radar equipment. The major branches of electrical engineering include power generation, power transmission and distribution, and controls. Electronic engineers (Figure 1.6) design, develop, test, and supervise the production of electronic equipment, including computer hardware; computer network hardware; communication devices such as cellular phones, television, and audio and

Electrical engineers design, develop, test, and supervise the manufacturing of electrical equipment. This includes lighting and wiring for buildings, cars, buses, trains, ships, and aircraft; power generation and transmission equipment for utility companies; electric motors found in various products; control devices; and radar equipment.

| FIGURE 1.5 | A civil engineer at work. |

Charles Thatcher/Stone/Getty Images

| FIGURE 1.6 | An electrical engineer at work. |

© Virgo Productions/zefa/Corbis

video equipment; as well as measuring instruments. Growing branches of electronic engineering include computer and communication electronics. The job outlook for electrical and electronic engineers is good because businesses and government need faster computers and better communication systems. Of course, consumer electronic devices will play a significant role in job growth for electrical and electronic engineers as well.

Mechanical Engineering The mechanical engineering discipline, which has evolved over the years as new technologies have emerged, is one of the broadest engineering disciplines. Mechanical engineers are involved in the design, development, testing, and manufacturing of machines, robots, tools, power generating equipment such as steam and gas turbines, heating, cooling, and refrigerating equipment, and internal combustion engines. The major branches of mechanical engineering include thermal /fluid systems and structural /solid systems. The job outlook for mechanical engineers is also good, as more efficient machines and power generating equipment and alternative energy-producing devices are needed. You will find mechanical engineers working for the federal government, consulting firms, various manufacturing sectors, the automotive industry, and other transportation companies.

The other common disciplines in engineering include aerospace engineering, biomedical, chemical engineering, environmental engineering, petroleum engineering, nuclear engineering, and materials engineering.

> Aerospace engineers design, develop, test, and supervise the manufacture of commercial and military aircraft, helicopters, spacecraft, and missiles. They also may work on projects dealing with research and development of guidance, navigation, and control systems.

Aerospace Engineering Aerospace engineers design, develop, test, and supervise the manufacture of commercial and military aircraft, helicopters, spacecraft, and missiles. They may work on projects dealing with research and development of guidance, navigation, and control systems. Most aerospace engineers work for aircraft and missile manufacturers, the Department of Defense, and NASA. If you decide to pursue an aerospace engineering career, you should expect to live in California, Washington, Texas, or Florida, because these are the states with large aerospace manufacturing companies. According to the Bureau of Labor Statistics, the job outlook for aerospace engineers is also good. Because of population growth and the need to meet the demand for more passenger air traffic, commercial airplane manufacturers are expected to do well.

Biomedical Engineering Biomedical engineering is a new discipline that combines biology, chemistry, medicine, and engineering to solve a wide range of medical and health-related problems. They apply the laws and the principles of chemistry, biology, medicine, and engineering to design artificial limbs, organs, imaging systems, and devices used in medical procedures. They also perform research alongside of medical doctors, chemists, and biologists to better understand various aspects of biological systems and the human body. In addition to their training in biology and chemistry, biomedical engineers have a strong background in either mechanical or electrical engineering.

There are a number of specializations within biomedical engineering, including: biomechanics, biomaterials, tissue engineering, medical imaging, and rehabilitation. Computer-assisted surgery and tissue engineering are among the fastest growing areas of research in biomedical engineering.

According to the Bureau of Labor Statistics, the job outlook for graduates of biomedical engineering is very good, because of the focus on health issues and the aging population.

Chemical Engineering As the name implies, chemical engineers use the principles of chemistry and basic engineering sciences to solve a variety of problems related to the production of chemicals and their use in various industries, including the pharmaceutical, electronic, and photographic industries. Most chemical engineers are employed by chemical, petroleum refining, film, paper, plastic, paint, and other related industries. Chemical engineers also work in metallurgical, food processing, biotechnology and fermentation industries. They usually specialize in certain areas such as polymers, oxidation, fertilizers, or pollution control. To meet the needs of the growing population, the job outlook for chemical engineers is also good, according to the Bureau of Labor Statistics.

> Chemical engineers use the principles of chemistry and basic engineering sciences to solve a variety of problems related to the production of chemicals and their use in various industries, including the pharmaceutical, electronic, paint, paper, and plastic industries.

Environmental Engineering Environmental engineering is another new discipline that has grown out of our concern for the environment. As the name implies, environmental engineering is concerned with solving problems related to the environment. They apply the laws and the principles of chemistry, biology, and engineering to address issues related to water and air pollution control, hazardous waste, waste disposal, and recycling. These issues, if not addressed properly, will affect public health. Many environmental engineers get involved with the development of local, national, and international environmental policies and regulations. They study the effects of industrial emissions and the automobile emissions that lead to acid rain and ozone depletion. They also work on problems dealing with cleaning up existing hazardous waste. Environmental engineers work as consultants or work for local, State, or Federal agencies.

> Environmental engineers apply the laws and the principles of chemistry, biology, and engineering to address issues related to water and air pollution control, hazardous waste, waste disposal, and recycling.

According to the Bureau of Labor Statistics, the job outlook for graduates of environmental engineering is very good, because environmental engineers will be needed in greater numbers to address and control the environmental issues discussed above. It is important to note that the job outlook for environmental engineers, more than engineers in other disciplines, is affected by politics. For example, looser environmental policies could lead to a fewer jobs, whereas stricter policies could lead to a greater number of jobs.

Manufacturing Engineering Manufacturing engineers develop, coordinate, and supervise the process of manufacturing all types of products. They are concerned with making products efficiently and at minimum cost. Manufacturing engineers are involved in all aspects of production, including scheduling and materials handling and the design, development, supervision, and control of assembly lines.

Manufacturing engineers employ robots and machine-vision technologies for production purposes. To demonstrate concepts for new products, and to save time and money, manufacturing engineers create prototypes of products before proceeding to manufacture actual products. This approach is called *prototyping*. Manufacturing engineers are employed by all types of industries, including automotive, aerospace, and food processing and packaging. The job outlook for manufacturing engineers is expected to be good.

Petroleum Engineering Petroleum engineers specialize in the discovery and production of oil and natural gas. In collaboration with geologists, petroleum engineers search the world for underground oil or natural gas reservoirs. Geologists have a good understanding of the properties of the rocks that make up the earth's crust. After geologists evaluate the properties of the rock formations around oil and gas reservoirs, they work with petroleum engineers to determine the best drilling methods to use. Petroleum engineers are also involved in monitoring and supervising drilling and oil extraction operations. In collaboration with other specialized engineers, petroleum engineers design equipment and processes to achieve the maximum profitable recovery of oil and gas. They use computer models to simulate reservoir performance as they experiment with different recovery techniques. If you decide to pursue petroleum engineering, you are most likely to work for one of the major oil companies or one of the hundreds of smaller, independent companies involved in oil exploration, production, and service. Engineering consulting firms, government agencies, oil field services, and equipment suppliers also employ petroleum engineers. According to the U.S. Department of Labor, large numbers of petroleum engineers are employed in Texas, Oklahoma, Louisiana, Colorado, and California, including offshore sites. Many American petroleum engineers also work overseas in oil-producing regions of the world such as Russia, the Middle East, South America, or Africa.

The job outlook for petroleum engineers depends on oil and gas prices. In spite of this fact, if you do decide to study petroleum engineering, employment opportunities for petroleum engineers should be favorable because the number of degrees granted in petroleum engineering has traditionally been low. Also, petroleum engineers work around the globe, and many employers seek U.S.-trained petroleum engineers for jobs in other countries.

Nuclear Engineering Only a few engineering colleges around the country offer a nuclear engineering program. Nuclear engineers design, develop, monitor, and operate nuclear power equipment that derives its power from nuclear energy. Nuclear engineers (Figure 1.7) are involved in the design, development, and operation of nuclear power plants to generate electricity or to power Navy ships and submarines. They may also work in such areas as the production and handling of nuclear fuel and the safe disposal of its waste products. Some nuclear engineers are involved in the design and development of industrial and diagnostic medical equipment. Nuclear engineers work for the U.S. Navy, nuclear power utility companies, and the Nuclear Regulatory Commission of the Department of Energy. Because of the high cost and numerous safety concerns on the part of the public, there are only a few nuclear power plants under construction. Even so, the job outlook for nuclear engineers is not too bad, because currently there are not many graduates in this field. Other job opportunities exist for nuclear engineers in the departments of Defense and Energy, nuclear medical technology, and nuclear waste management.

Mining Engineering There are only a few mining engineering schools around the country. Mining engineers, in collaboration with geologists and metallurgical engineers, find, extract, and prepare coal for use by utility companies; they also look for metals and minerals to extract from the earth

FIGURE 1.7 A nuclear engineer at work.
© Picture Contact / Alamy

for use by various manufacturing industries. Mining engineers design and supervise the construction of aboveground and underground mines. Mining engineers could also be involved in the development of new mining equipment for extraction and separation of minerals from other materials mixed in with the desired minerals.

Most mining engineers work in the mining industry, some work for government agencies, and some work for manufacturing industries. The job outlook for mining engineers is not as good as for other disciplines. The mining industry is somewhat similar to the oil industry in that the job opportunities are closely tied to the price of metals and minerals. If the price of these products is low, then the mining companies will not want to invest in new mining equipment and new mines. Similar to petroleum engineers, U.S. mining engineers may find good opportunities outside the United States.

Materials Engineering There are only a few engineering colleges that offer a formal program in materials engineering, ceramic engineering, or metallurgical engineering. Materials engineers research, develop, and test new materials for various products and engineering applications. These new materials could be in the form of metal alloys, ceramics, plastics, or composites. Materials engineers study the nature, atomic structure, and thermo-physical properties of materials. They manipulate the atomic and molecular structure of materials in order to create materials that are lighter, stronger, and more durable. They create materials with specific mechanical, electrical, magnetic, chemical, and heat-transfer properties for use in specific applications: for example, graphite tennis racquets that are much lighter and stronger than the old wooden racquets; the composite materials used in stealth military planes with specific electromagnetic properties; and the ceramic tiles on the space shuttle that protected the shuttle during reentry into the atmosphere (ceramics are nonmetallic materials that can withstand high temperatures).

Materials engineering may be further divided into metallurgical, ceramics, plastics, and other specialties. You can find materials engineers working in aircraft manufacturing; various research and testing labs; and electrical, stone, and glass products manufacturers. Because of the low number of current graduates, the job opportunities are good for materials engineers.

Engineering Technology

In the preceding text, we introduced you to the engineering profession and its various areas of specialization. Let us now say a few words about engineering technology. For those of you who tend to be more hands-on and less interested in theory and mathematics, engineering technology might be the right choice for you. Engineering technology programs typically require the knowledge of basic mathematics up to integral and differential calculus level, and focus more on the application of technologies and processes. Although to a lesser degree than engineers, engineering technologists use the same principles of science, engineering, and mathematics to assist engineers in solving problems in manufacturing, construction, product development, inspection, maintenance, sales, and research. They may also assist engineers or scientists in setting up experiments, conducting tests, collecting data, and calculating some results. In general, the scope of an engineering technologist's work is more application-oriented and requires less understanding of mathematics, engineering theories, and scientific concepts that are used in complex designs.

Engineering technology programs usually offer the same type of disciplines as engineering programs. For example, you may obtain your degree in Civil Engineering Technology, Mechanical Engineering Technology, Electronics Engineering Technology, or Industrial Engineering Technology. However, if you decide to pursue an engineering technology degree, note that graduate studies in engineering technology are limited and registration as a professional engineer might be more difficult in some states.

The engineering technology programs are also accredited by the Accreditation Board for Engineering and Technology (ABET). According to ABET, the engineering technology curriculum must effectively develop the following subject areas in support of student outcomes and program educational objectives: mathematics, technical content, physical and natural science, and a capstone or integrating experience. The technical content of a particular engineering technology program is focused on the applied side of science and engineering, and is intended to develop the skills, knowledge, methods, procedures, and techniques associated with that particular technical discipline.

According to ABET, an accredited engineering technology program must demonstrate that their students at the time of graduations have

a. an ability to select and apply the knowledge, techniques, skills, and modern tools of the discipline to broadly defined engineering technology activities;

Katie McCullough

I have enjoyed math, science, and general problem solving since middle and high school. Engineering sparked my interest as a way to engage in all of these disciplines. Pursuing my Bachelor of Science degree in Chemical Engineering led me through many challenges. Besides keeping up with reading assignments and learning class materials, one of my greatest challenges was deciding what industry to enter and what type of work I wanted to perform upon graduation. Two actions that helped me in this decision were taking a variety of elective classes in the engineering department and taking advantage of an engineering internship.

Today, I am in the oil and gas industry, supporting facilities on a refined products pipeline system. One of the things I have enjoyed most about engineering is the diversity of work experiences I have had. In my current position, I participate in incident investigations and risk analyses that evaluate hazards in a facility. I drive risk-reduction projects that impact communities, employees, and the environment by minimizing leak risks and safety hazards. In past positions, I have managed a variety of commercial projects, as well as developed training programs for engineers new to a facility. I look forward to continuing to be challenged by new roles and to exploring different opportunities in engineering as my career continues.

Courtesy of Katie McCullough

b. an ability to select and apply a knowledge of mathematics, science, engineering, and technology to engineering technology problems that require the application of principles and applied procedures or methodologies;

c. an ability to conduct standard tests and measurements; to conduct, analyze, and interpret experiments; and to apply experimental results to improve processes;

d. an ability to design systems, components, or processes for broadly defined engineering technology problems appropriate to program educational objectives;

e. an ability to function effectively as a member or leader on a technical team;

f. an ability to identify, analyze, and solve broadly defined engineering technology problems;

g. an ability to apply written, oral, and graphical communication in both technical and nontechnical environments; and an ability to identify and use appropriate technical literature;

h. an understanding of the need for and an ability to engage in self-directed continuing professional development;

i. an understanding of and a commitment to address professional and ethical responsibilities, including a respect for diversity;

j. a knowledge of the impact of engineering technology solutions in a societal and global context; and

k. a commitment to quality, timeliness, and continuous improvement.

Therefore, these are the educational outcomes that are expected of you when you graduate from an engineering technology program.

Before You Go On

Answer the following to test your understanding of the preceding sections.

1. Give examples of common traits of successful engineers.

2. Why is lifelong learning important in engineering?

3. Why are good oral and written communication skills important in engineering?

4. What do nuclear engineers do?

5. What do petroleum engineers do?

6. What is the difference between engineering and engineering technology?

Vocabulary—As a well-educated engineer and intelligent citizen, it is important for you to understand that you need to develop a comprehensive vocabulary to communicate effectively. State the meaning of the following terms:

Professional Engineer _____

Mechanical Engineer _____

Electrical Engineer _____

Civil Engineer _____

Aerospace Engineer _____

Manufacturing Engineer _____

Chemical Engineer _____

Environmental Engineer _____

SUMMARY

LO¹   Engineering Work Is All Around You

By now, you should understand how engineers contribute to the comfort and betterment of our everyday lives. Engineers apply physical and chemical laws and principles and mathematics to design millions of products and services that we use in our everyday lives. These products include cars, computers, aircraft, clothing, toys, home appliances, surgical equipment, heating and cooling equipment, health-care devices, tools, and machines that make various products. Engineers also play a significant role in the design and maintenance of a nation's infrastructure, including communication systems, public utilities, and transportation. They design and supervise the construction of buildings, dams, highways, mass-transit systems, and the construction of power plants that supply power to manufacturing companies, homes, and offices.

LO²   Engineering as a Profession

Engineers consider important factors such as cost, efficiency, reliability, and safety when designing products and services. Engineers perform tests to make certain that the products they design withstand various conditions and situations. They are continuously searching for ways to improve already existing products as well. Engineers continuously develop new, advanced materials to make products lighter and stronger for different applications. They are also responsible for finding suitable ways to extract petroleum, natural gas, and raw materials from the earth, and they are involved in coming up with ways of increasing crop, fruit, and vegetable yields along with improving the safety of our food products. In addition to design, some engineers work as sales representatives for products, while others provide technical support and troubleshooting for the customers of their products. Many engineers decide to become involved in sales and customer support, because their engineering background enables them to explain and discuss technical information and to assist with installation, operation, and maintenance of various products and machines. Not all engineers work for private industries; some work for federal, state, and local governments in various capacities. Engineers work in departments of agriculture, defense, energy, and transportation. Some engineers work for the National Aeronautics and Space Administration (NASA).

LO³   Common Traits of Good Engineers

Although the activities of engineers are quite varied, there are some personality traits and work habits that typify most of today's successful engineers. Good engineers are problem solvers and have a firm grasp of the fundamental principles of engineering, which they can use to solve many different problems. They have a desire to be lifelong learners. They also have written and oral communication skills that allow them to work well with their colleagues and to convey

their expertise to a wide range of clients. Engineers are required to write reports. These reports might be lengthy, detailed technical reports containing graphs, charts, and engineering drawings, or they may take the form of brief memoranda or executive summaries. Good engineers have time-management skills that enable them to work productively and efficiently. They also actively participate in local and national discipline-specific organizations by attending seminars, workshops, and meetings.

LO⁴   Engineering Disciplines

There are more than 20 engineering disciplines recognized by the professional societies. Most engineering degrees are granted in civil, electrical, and mechanical engineering. Civil engineering is the oldest engineering discipline, and as the name implies, civil engineering is concerned with providing public infrastructure and services. Electrical and electronic engineering is the largest engineering discipline. Electrical engineers design, develop, test, and supervise the manufacturing of electrical equipment, including lighting and wiring for buildings, cars, buses, trains, ships, and aircraft; power generation and transmission equipment for utility companies; electric motors found in various products; control devices; and radar equipment. The mechanical engineering discipline is perhaps the broadest engineering discipline. Mechanical engineers are involved in the design, development, testing, and manufacturing of machines, robots, tools, power-generating equipment (such as steam and gas turbines), heating, cooling, and refrigerating equipment, and internal combustion engines. The other common disciplines in engineering include aerospace engineering, biomedical, chemical engineering, environmental engineering, petroleum engineering, nuclear engineering, and materials engineering.

KEY TERMS

ABET 16

Aerospace Engineering 20

Biomedical Engineering 20

Chemical Engineering 21

Civil Engineering 18

Electrical Engineering 18

Engineer 5

Engineering 9

Environmental Engineering 21

Fundamentals of Engineering
    Exam 17

Materials Engineering 23

Mechanical Engineering 20

Nuclear Engineering 22

Petroleum Engineering 22

Professional Engineer 17

APPLY WHAT YOU HAVE LEARNED

This is a class project. Each of you is to ask his or her parents/grandparents to think back to when they graduated from high school or college and to create a list of products and services that are available in their everyday lives now that were not available to them then. Ask them if they ever imagined that these products and services would be available today. To get your parents started, here are few examples: cellular phones, ATM cards, personal computers, airbags in cars, price scanners at the supermarket, E-Z passes for tolls, and so on. Ask your parents/grandparents to explain how these products have made their lives better (or worse).

wavebreakmedia/Shutterstock.com

Next, each of you is to compile a list of products and services that are not available now that you think will be readily available in the next 20 years. Discuss which engineering disciplines will be involved in the design and development of these products and services. Per your instructor's guidelines, you may work in a group. Present your results to the entire class. As a future engineer, how do you think you could contribute to the comfort and betterment of our everyday lives?

Dmitry Kalinovsky/Shutterstock.com

PROBLEMS

Problems that promote life-long learning are denoted by 🔑

1.1 Observe your own surroundings. What are some of the engineering achievements that you couldn't do without today?

1.2 Using the Internet, find the appropriate organization for the following list of engineering disciplines. Depending on your personal interests, prepare a brief two-page report about the goals and missions of the organization you have selected.

Bioengineering
Ceramic engineering
Chemical engineering
Civil engineering
Computer engineering
Electrical engineering
Electronic engineering
Environmental engineering
Industrial engineering
Manufacturing engineering
Materials engineering
Mechanical engineering

1.3 To increase public awareness about the importance of engineering and to promote engineering education and careers among the younger generation, prepare and give a 15-minute Web-based presentation at a mall in your town. You need to do some planning ahead of time and ask permission from the proper authorities.

1.4 If your introduction to engineering class has a term project, present your final work at a mall at the date set by your instructor. If the project has a competitive component, hold the design competition at the mall.

1.5 Prepare a 15-minute oral presentation about engineering and its various disciplines, and the next time you go home present it to the juniors in your high school. Ask your college engineering department and engineering organizations on your campus to provide engineering-related brochures to take along.

1.6 This is a class project. Prepare a Web site for engineering and its various branches. Elect a group leader, then divide up the tasks among yourselves. As you work on the project, take note of both the pleasures and problems that arise from working in a team environment. Write a brief report about your experiences regarding this project. What are your recommendations for others who may work on a similar project?

1.7 This is a team project. Prepare a Web-based presentation of the history and future of engineering. Collect pictures, short videos, graphs, and so on. Provide links to major engineering societies as well as to major research and development centers.

1.8 Perform a Web search to obtain information about the number of engineers employed by specific area and their mean salaries in recent years. Present your findings to your instructor.

1.9 If you are planning to study chemical engineering, investigate what is meant by each of the following terms: *polymers, plastics, thermoplastics,* and *thermosetting*. Give at least ten examples of plastic products that are consumed every day. Write a brief report explaining your findings.

1.10 If you are planning to study electrical engineering, investigate how electricity is generated and distributed. Write a brief report explaining your findings.

1.11 Electric motors are found in many appliances and devices around your home. Identify at least ten products at home that use electric motors.

1.12 Identify at least 20 different materials that are used in various products at home.

1.13 If you are planning to study civil engineering, investigate what is meant by *dead load, live load, impact load, wind load,* and *snow load* in the design of structures. Write a brief memo to your instructor discussing your findings.

1.14 This is a group project. As you can see from our discussion of the engineering profession in this chapter, people rely quite heavily on engineers to provide them with safe and reliable goods and services. Moreover, you realize that there is no room for mistakes or dishonesty in engineering! Mistakes made by engineers could cost not only money but also more importantly lives. Think about the following: An incompetent and unethical surgeon could cause the death of at most one person at one time, whereas an incompetent and unethical engineer could cause the deaths of hundreds of people at one time. If in order to save money an unethical engineer designs a bridge or a part for an airplane that does not meet the safety requirements, hundreds of peoples' lives are at risk!

Visit the Web site of the National Society of Professional Engineers and research engineering ethics. Discuss why engineering ethics is so important, and explain why engineers are expected to practice engineering using the highest standards of honesty and integrity. Give examples of engineering codes of ethics. Write a brief report to your instructor explaining your findings.

1.15 The Fundamentals of Engineering exam is offered in seven disciplines: chemical, civil, electrical and computer, environmental, industrial, mechanical, and "other disciplines." Visit the National Council of Examiners for Engineering and Surveying (NCEES) website at www.ncees.org and look up the number of questions that you should expect from mathematics, probability and statistics, chemistry, ethics, engineering economics, statics, and so forth in the discipline that you are planning to study. In the future, as you take classes in these topics, remind yourself as to their importance for your preparation for the FE exam and becoming a successful engineer.

Impromptu Design I

Objective: To design a vehicle from the materials listed below and adhere to the following rules.

- You must use all the items provided.
- The vehicle is to be dropped from a height of 10 ft.
- The vehicle must land in the marked area (4 ft × 4 ft).
- Each design is allowed one practice drop.
- The vehicle design with the slowest drop time wins.
- 30 minutes will be allowed for preparation.
- Explain the rationale behind your design.

Provided Materials: 2 paper plates: 1 paper cup;
2 balloons; 3 rubber bands; 1 straw; 12 self-adhesive labels

Saeed Moavni

Stocksnapper/Shutterstock.com

"I find that the harder I work, the more luck I seem to have."

— THOMAS JEFFERSON (1743—1826)

CHAPTER

2

Preparing for an Engineering Career

Ammentorp Photography/Shutterstock.com

Making the transition from high school to college requires extra effort. In order to have a rewarding education, you should realize that you must start studying and preparing from the first day of class, attend class regularly, get help right away, take good notes, select a good study place, and form study groups. You should also consider the time-management ideas discussed in this chapter to arrive at a reasonable weekly schedule. Your education is an expensive investment. Invest wisely.

LEARNING OBJECTIVES

LO¹   **Making the Transition from High School to College:** realize that in high school most of your learning took place in class, whereas in college most of your education takes place outside the classroom

LO²   **Budgeting Your Time:** describe effective ways to manage your time so that you can have a fulfilling college experience and have adequate time for studying, social events, and work

LO³   **Study Habits and Strategies:** explain study habits and strategies that would lead to good academic performance

LO⁴   **Getting Involved with an Engineering Organization:** describe why it is important to join an engineering organization

LO⁵   **Your Graduation Plan:** realize the importance of having a graduation plan

PRACTICE COLLEGE STUDY SKILLS DURING HIGH SCHOOL

The changes students will face as they make the transition from high school to college academics are inevitable. While high schoolers may not be able to wrap their heads around the idea of professors instead of teachers or lecture halls filled with hundreds of students, they can begin adapting to the necessary study methods before college orientation.

Assignments and tests will undoubtedly become more challenging once students enter college, so it's critical to instill disciplined study habits sooner rather than later. Some current college students say that identifying these study strategies in high school paves the path to success in university academics.

"High school students should try a variety of study methods to figure out what works best for them," said Lexie Swift, a senior at the University of Iowa. "That way, when they are in college, they will know what is most effective and won't hurt their grades trying to figure it out." It is pivotal to understand the importance of actually completing a reading assignment and not skipping it altogether. Tiffany Sorensen, a 2013 graduate of Stony Brook University (SUNY) feels this is a particularly important piece of advice.

When preparing for exams, high school students should get into the habit of creating their own comprehensive study guides. Caleb Zimmerman, a senior at King's College in New York City, advises taking all of the information you need for the test and organizing it in one spot. He said it has helped him immensely in studying

for his college exams. "Whenever I have a big test, I get everything I need to know together in one place. Usually, a Word document does the trick," he said. "Sometimes that document will run 10 to 20 pages, but at least your mind won't feel scattered." However you choose to create the study guide is up to you, but having one is a very important step to succeeding on a college exam.

Another element that cannot be underestimated is getting to know your professors. Some students say it is one of the best ways to ultimately support your academic goals and better understand what material you need to study. High schoolers can practice this now by reaching out to their teachers for extra help in order to develop stronger relationships with them — some of them might even become prime candidates for writing those letters of recommendation later on. Zimmerman says he regrets not taking this approach when he was a college freshman. "I wish I had gotten to know my professors better," he said. "Professors are surprisingly approachable in most cases, and getting to know them on a personal level will give you very helpful hints as to how they think."

If high school students truly apply themselves to experiment with various study methods, fully process assignments, consolidate information for tests, and understand exactly what relationships with professors can bring to the table, success in college will be on the horizon.

Source: Cathryn Sloane, U.S. News, Oct. 14, 2013

To the Students: How many hours did you study in high school? How many hours do you think you need to study now that you are in college? What do you think are the major differences between high school and college teaching and learning structures? Your instructor may invite a few seniors to class to share their thoughts as well.

I n this chapter, you will be introduced to some very important suggestions and ideas that could, if followed, make your engineering education more rewarding. Read this section very carefully, and think about how you can adapt the strategies offered here to get the optimum benefit from your college years. If you encounter difficulty in your studies, reread this chapter for ideas to help you maintain some self-discipline.

LO¹ 2.1   Making the Transition from High School to College

You belong to an elite group of students now because you are studying to be an engineer. According to the *Chronicle of Higher Education,* approximately only 5% of students who graduate with a B.S. degree from universities and colleges across the United States are engineers. You will be taught to look at your surroundings differently than other people do. You will learn how to ask questions to find out how things are made, how things work, how to improve things, how to design something from scratch, and how to take an idea from paper to reality and actually build something.

Some of you may be on your own for the very first time. Making the transition from high school to college may be a big step for you. Keep in mind that what you do for the next four or five years will affect you for the rest of your life. Remember that how successful and happy you are will depend primarily on you. You must take the responsibility for learning; nobody can make you learn. Depending on which high school you attended, you may not have had to study much to get good grades. In high school, most of your learning took place in class. In contrast, in college most of your education takes place outside the class. Therefore, you may need to develop some new habits and get rid of some of your old habits in order to thrive as an engineering student. The rest of this chapter presents suggestions and ideas that will help you make your college experience successful. Consider these suggestions and try to adapt them to your own unique situation.

LO² 2.2   Budgeting Your Time

Each of us has the same 24 hours in a day, and there is only so much that a person can do on an average daily basis to accomplish certain things. Many of us need approximately 8 hours of sleep every night. In addition, we all need to have some time for work, friends and family, studying, relaxation and recreation, and just goofing around.

> In high school, most of your learning took place in class. In contrast, in college most of your education takes place outside the class. So, you need to develop some new study habits and get rid of some of the old ones!

Suppose you were given a million dollars when you reached your adulthood and were told that is all the money that you would have for the rest of your life for clothing, food, entertainment, leisure activities, and so on. How would you go about spending the money? Of course, you would make reasonable efforts to spend and invest it wisely. You would carefully budget for various needs, trying to get the most for your money. You would look for good sales and plan to buy only what was necessary, and you would attempt not to waste any money. Think of your education in a similar manner. Don't just pay your tuition and plan to sit in class and daydream. Your education is an expensive investment, one that requires your responsible management. A student at a private university went to her instructor to drop a class because she was not getting the grade she wanted. The instructor asked her how much she had paid for the class. She said that she had spent approximately $2000 for the four-credit class. The instructor happened to have a laptop computer on his desk and asked her the following question: If you bought a laptop computer from a computer store, took it home to install some software on it, and had some difficulty making the computer work, would you throw it in the trash? The student looked at her instructor as if he had asked a stupid question. He explained to her that her dropping a class she had already paid tuition for is similar in many ways to throwing away a computer the first time she has trouble with some software. Try to learn from this example. Generally speaking, for most of us learning is a lot of work at the beginning, and it's not much fun. But often, after even a short period of time, learning will become a joy, something you work at that raises your own self-esteem. Learning and understanding new things can be downright exciting. Let us examine what you can do to enhance your learning during the next few years to make the engineering education you are about to receive a fulfilling and rewarding experience.

Let us begin by performing some simple arithmetic to see how efficient we might be in using our time. With 24 hours in a given day, we have, for a one-week period, 168 hours available. Let's allocate liberal time periods to some activities common to most students. When following this example, refer to Table 2.1. Notice that the time periods allocated to various activities in this table are very generous and you don't have to deprive yourself of sleep or relaxation or socializing with your friends. These numbers are meant only to give you a reasonable starting point to help you budget your time on a weekly basis. You may prefer to spend an hour a day relaxing during the week and use the additional social hours on weekends. Even with generous relaxation and social time, this sample allows 68 hours a week to devote to your education. A typical engineering student takes 16 semester credits, which simply means about 16 hours a week are spent in the classroom. You still have 52 hours a week to study. A good rule of thumb is to spend at least 2 to 3 hours of studying for each hour of class time, which amounts to at least 32 hours and at most 48 hours a week of studying. Of course, some classes are more demanding than others and will require more time for preparation and

TABLE 2.1	An Example of Weekly Activities
Activity	**Required Time per Week**
Sleeping: (7 days/week) × (8 hours/day)	= 56 (hours/week)
Cooking and eating: (7 days/week) × (3 hours/day)	= 21 (hours/week)
Grocery shopping	= 2 (hours/week)
Personal grooming: (7 days/week) × (1 hour/day)	= 7 (hours/week)
Spending time with family, (girl/boy) friends, relaxing, playing sports, exercising, watching TV: (7 days/week) × (2 hours/day)	= 14 (hours/week)
Total	**= 100 (hours/week)**

homework, projects, and lab work. You still have from 4 to 20 hours a week in your budget to allocate at your own discretion.

You may not be an 18-year-old freshman whose parents are paying most of your tuition. You may be an older student who is changing careers. Or you may be married and have children, so you must have at least a part-time job. In this case, obviously you will have to cut back in some areas. For example, you may want to consider not taking as many credits in a given semester and follow a five-year plan instead of a four-year plan. Depending on how many hours a week you need to work, you can rebudget your time. The purpose of this time budget example is mainly to emphasize the fact that you need to learn to manage your time wisely if you want to be successful in life. Every individual, just like any good organization, monitors his or her resources. No one wants you to turn into a robot and time yourself to the second. These examples are provided to give you an idea of how much time is available to you and to urge you to consider how efficiently and wisely you are allocating and using your time. The point is that budgeting your time is very important.

With the exception of a few courses, most classes that you will take are scheduled for 50-minute periods, with a 10-minute break between classes to allow students to attend several classes in a row. The other important reason for having a 10-minute break is to allow time to clear your head. Most of us have a limited attention span and cannot concentrate on a certain topic for a long period of time without a break. Taking a break is healthy; it keeps your mind and body working well.

Typically, as a first-term freshman in engineering you may have a course load similar to the one shown here:

Chemistry (3)
Chemistry Lab (1)
Introduction to Engineering (2)
Calculus (4)
English Composition (3)
Humanities/Social Science electives (3)

Table 2.2 is an example of a schedule for a freshman engineering student. You already know your strengths and weaknesses; you may have to make several attempts to arrive at a good schedule that will fit your needs the best. You may also need to modify the example schedule shown to allow for any variability in the number of credits or other engineering program requirements at your particular school. Maintain a daily logbook to keep track of how closely you are following the schedule and where time is being used inefficiently, and modify your schedule accordingly.

LO³ 2.3   Study Habits and Strategies

You start studying and preparing from the very first day of class! It is always a good idea to read the material that your professor is planning to cover in class ahead of time. This practice will improve both your understanding and retention of the lecture materials. It is also important to go over the material that was discussed in class again later the same day after the lecture was given. When you are reading the material ahead of a lecture, you are familiarizing yourself with the information that the instructor will present to you in class. Don't worry if you don't fully understand everything you are reading

TABLE 2.2		An Example of Weekly Schedule for a Freshman Engineering Student				

Hour	Monday	Tuesday	Wednesday	Thursday	Friday	Saturday	Sunday
7–8	Shower/ Dress/ Breakfast	Shower/ Dress/ Breakfast	Shower/ Dress/ Breakfast	Shower/ Dress/ Breakfast	Shower/ Dress/ Breakfast	Extra sleep	Extra sleep
8–9	**CALCULUS CLASS**	**CALCULUS CLASS**	**Study English**	**CALCULUS CLASS**	**CALCULUS CLASS**	Shower/ Dress/ Breakfast	Shower/ Dress/ Breakfast
9–10	**ENGLISH CLASS**	**Study Intro. to Eng.**	**ENGLISH CLASS**	**Study Intro. to Eng.**	**ENGLISH CLASS**	Grocery shopping	**OPEN HOUR**
10–11	**Study Calculus**	**INTRO. TO ENG. CLASS**	**Study H/SS**	**INTRO. TO ENG. CLASS**	**Study Calculus**	Grocery shopping	**OPEN HOUR**
11–12	**H/SS CLASS**	**Study Chemistry**	**H/SS CLASS**	**Study Chemistry**	**H/SS CLASS**	Exercise	Relax
12–1	Lunch	Lunch	Lunch	Lunch	Lunch	Lunch	Lunch
1–2	**CHEM. CLASS**	**Study Calculus**	**CHEM. CLASS**	**Study Intro. to Eng.**	**CHEM. CLASS**	Relax	Relax
2–3	**Study H/SS**	**CHEM. LAB**	**Study H/SS**	**Study Intro. to Eng.**	**Study H/SS**	**Study Chemistry**	**Study English**
3–4	**Study Calculus**	**CHEM. LAB**	**Study H/SS**	**Study Calculus**	**Study Calculus**	**Study Chemistry**	**Study H/SS**
4–5	Exercise	**CHEM. LAB**	Exercise	Exercise	Exercise	**Study Chemistry**	**Study Chemistry**
5–6	Dinner	Dinner	Dinner	Dinner	Dinner	Dinner	Relax
6–7	**Study Chemistry**	**Study Calculus**	**Study Calculus**	**Study Calculus**	Relax	Relax	Dinner
7–8	**Study Chemistry**	**Study Calculus**	**Study Calculus**	**Study Chemistry**	**Study Calculus**	**Study Intro. to Eng.**	**Study Calculus**
8–9	**Study Intro. to Eng.**	**Study Chemistry**	**Study Intro. to Eng.**	**Study Chemistry**	Recreation	Recreation	**Study Calculus**
9–10	**Study English**	**Study Chemistry**	**Study English**	**Study English**	Recreation	Recreation	**Study English**
10–11	Relax/Get ready for bed	Relax/Get ready for bed	Relax/Get ready for bed	Relax/Get ready for bed	Recreation	Recreation	Relax/Get ready for bed

| FIGURE 2.1 | Read the material your professor is planning to cover in class ahead of time. |

© Golden Pixels LLC / Alamy

Allowing for 8 hours a night for sleeping and 6 hours every day for personal activities such as cooking, eating, relaxing, and grooming, you still have 10 hours a day for other activities such as school and/ or work. How do you spend your 10 hours?

at that time. During the lecture, you can focus on the material that you did not fully understand and ask questions (Figure 2.1). When you go over the material after the lecture, everything then should come together. Remember to read before the class and study the material after the class on the same day!

Attend Your Classes Regularly Yes, even if your professor is a bore, you can still learn a great deal from attending the class. Your professor may offer additional explanations and discussion of some material that may not be well presented in your textbook. Moreover, you can ask questions in class. If you have read the material before class and have made some notes about the concepts that you do not fully understand, during the lecture you can ask questions to clarify any misunderstanding. If you need more help, then go to your professor's office and ask for additional assistance.

Get Help Right Away When you need some help, don't wait till the last minute to ask! Your professor should have his or her office hours posted on the office door or be available on the Web. The office hours are generally stated in the course syllabus. If for some reason you cannot see your professor during the designated office hours, ask for an appointment. Almost all professors are glad to sit down with you and help you out if you make an appointment with them (Figure 2.2). After you have made an appointment, be on time and have your questions written down so that you remember what to ask. Once again, remember that most professors do not want you to wait until the last minute to get help!

Any professor can tell you some stories about experiences with students who procrastinate. Recently, I had a student who sent an e-mail to me on

> Remember to read before the class and study the material after the class on the same day, attend class regularly, and get help away!

Sunday night at 10:05 P.M. asking for an extension on a homework assignment that was due on the following day. I asked the student the next morning, "When did you start to do the assignment?" He replied, "at 10:00 P.M. Sunday night." On another occasion, I had a student who came to my office and introduced himself to me for the first time. He asked me to write him a recommendation letter for a summer job he was applying for. Like most professors, I do not write recommendation letters for students, or anyone, whom I do not actually know. Get to know your professors and visit them often!

Take Good Notes Everyone knows that it is a good idea to take notes during lecture, but some students may not realize that they should also take notes when reading the textbook. Try to listen carefully during the lectures so you can identify and record the important ideas and concepts. If you have read ahead of time the text materials that your professor is planning to cover in class, then you are prepared to write down notes that complement what is already in your textbook. You don't need to write down everything that your professor says, writes, or projects onto a screen. The point is to listen very carefully and write down only notes regarding the important concepts that you did not understand when you read the book.

Use wirebound notebooks for your notes. Don't use loose papers, because it is too easy to lose some of your notes that way. Keeping a notebook is a good habit to develop now. As an engineer you will need to keep records of meetings, calculations, measurements, and so forth with time and date recorded so you can refer back to them if the need arises. Thus it is best that you keep the notes in a wirebound notepad or a notebook with the pages sewn into the binding so you won't lose any pages. Study your notes for at least an hour or two the same day you take them. Make sure you understand all the concepts and ideas that were discussed in class before you attempt to do your homework assignment. This approach will save you a great deal of time in the long run! Don't be among those students who spend as little time as possible on understanding the underlying concepts and try to take a shortcut by finding an example problem in the book similar to the assigned homework problem. You may be able to do the homework problem, but you won't develop an understanding of the material. Without a firm grasp of the basic concepts, you will not do well on the exams, and you will be at a disadvantage later in your other classes and in life when you practice engineering.

Take good legible notes so that you can go back to them later if you need to refresh your memory before exam time. Most of today's engineering textbooks provide a blank margin on the left and right side of each sheet. Don't be afraid to write in these margins as you study your book. Keep all your engineering books; don't sell them back to the bookstore—some day you may need them. If you have a computer, you may want to type up a summary sheet of all important concepts. Later, you can use the *Find* command to look up selected terms and concepts. You may wish to insert links between related concepts in your notes. Digital notes may take some extra time to type, but they can save you time in the long run when you search for information.

Select a Good Study Place   You may already know that you should study in a comfortable place with good lighting. You do not want any distractions while you are studying (Figure 2.3). For example, you do not want to study in front of a TV while watching your favorite situation comedy. A library is certainly a good place for studying, but you can make your dorm room or your apartment room into a good place for studying (Figure 2.4). Talk to your roommate(s) about your study habits and study time. Explain to them that you prefer to study in your own room and appreciate not being disturbed while studying. If possible, find a roommate with a declared engineering major, who is likely to be more understanding of your study needs. Remember

FIGURE 2.3   Not a good way to study!

FIGURE 2.4	A good way to study!
	lightpoet/Shutterstock.com

that a bad place for studying is in your girlfriend's or boyfriend's lap or arms (or any other acceptable engineering configuration). Another useful idea is to keep your desk clean and avoid having a picture of your sweetheart in front of you. You don't want to daydream as you are studying. There is plenty of time for that later.

Form Study Groups Your professor will be the first person to tell you that the best way to learn something is to teach it. In order to teach something, though, you have to first understand the basic concepts. You need to study on your own first, and then get together with your classmates to discuss and explain key ideas and concepts to each other. Everyone in your study group should agree that they need to come prepared to discuss appropriate materials and that they all need to contribute to the discussions. It should be understood that the study groups serve a different purpose than that of a tutoring session. However, if another student in your class asks for assistance, help if you can by explaining new concepts you have learned. If you have difficulty explaining the material to someone else, that could be an indication you don't fully understand the concept yourself and you need to study the material in more detail. So remember that a good way to learn something is to form discussion groups where you explain ideas and concepts in your own words to others in the group. *Be an active learner, not a passive learner!*

Prepare for Examinations If you study and prepare from the first day of class, then you should perform admirably on your exams. Keep reminding yourself that there is absolutely no substitute for daily studying. Don't wait until the night before the exam to study! That is not the best time for learning new concepts and ideas. The night before the exam is the time for review only. Just before an exam, spend a few hours reviewing your notes and sample problems. Make sure that you get a good night's rest so you can be fresh and think clearly when you take the exam. It may be a good idea to ask your instructor ahead of time what type of exam it will be, how many questions there will be, or what suggestions she or he has to help you prepare better for the exam. As with any test, be sure you understand what the questions are asking. Read the questions carefully before you proceed with answering them. If there is some ambiguity in the exam questions, ask the instructor for clarification. After you have looked over the exam, you may want to answer the easy questions first and then come back to the more complicated questions. Finally, some of you may have experienced test anxiety when taking an exam. To reduce the anxiety, prepare well and consider timing yourself when doing your homework problems.

> Select a good study place! Remember that a bad place for studying is in your girlfriend's or boyfriend's lap or arms.

Before You Go On

Answer the following questions to test your understanding of the preceding section.

1. What is the main difference between learning in high school and college?

2. Why is it important to create a weekly schedule for your activities?

3. Give examples of practices that you need to follow to be successful in college.

4. Give examples of good note taking.

LO⁴ 2.4   Getting Involved with an Engineering Organization

There are many good reasons to join an **engineering organization**. Networking, participating in plant tours, listening to technical guest speakers, participating in design competitions, attending social events, taking advantage of learning opportunities through short courses, seminars, and conferences, and obtaining student loans and scholarships are a few of the common benefits of belonging to an engineering organization. Moreover, good places to learn more about areas of specialization in engineering are the Web sites of various engineering organizations. As you spend a little time reading about these organizations, you will discover that many share common interests and provide some overlapping services that could be used by engineers of various disciplines. You will also note that the primary purpose of these professional engineering organizations is to offer the following benefits:

1. They conduct conferences and meetings to share new ideas and findings in research and development.

2. They publish technical journals, books, reports, and magazines to help engineers in particular specialties keep up-to-date.

3. They offer short courses on current technical developments to keep practicing engineers abreast of the new developments in their respective fields.

4. They advise the federal and state governments on technology-related public policies.

5. They create, maintain, and distribute codes and standards that deal with correct engineering design practices to ensure public safety.

6. They provide a networking mechanism through which you get to know people from different companies and institutions. This is important for two reasons: (1) If there is a problem that you feel requires assistance from outside your organization, you have a pool of colleagues whom you have met at the meetings to help you solve the problem. (2) When you know people in other companies who are looking for good engineers to hire and you are thinking about something different to do, then you may be able to find a good match.

> Join an engineering organization! Networking, participating in design competitions, and obtaining student loans and scholarships are a few of the common benefits of belonging to an engineering organization.

Find out about the local student chapters of national engineering organizations on your campus. Attend the first meetings. After collecting information, choose an organization, join, and become an active participant. As you will see for yourself, the benefits of being a member of an engineering organization are great!

LO⁵ 2.5   Your Graduation Plan

At most schools, there are three levels of admissions. First, you get admitted to the university. For that to happen, you must meet certain requirements. For example, you need to rank in the top x % of your high school class, have a certain ACT or SAT score, and have so many years of English, mathematics, sciences, and social studies. After you complete your freshman year, you may need to apply to the college in which the engineering program of your interest resides. To be accepted to, say, the college of engineering at your university, you need to meet additional requirements. Finally, at the end of the second year, upon successful completion of math, chemistry, physics, and basic engineering classes, then you need to apply and gain admission to a specific engineering program, for example, civil, electrical, mechanical, and so on. Make sure you meet with your advisor so that you understand what the requirements are for admission to the college and the specific program, because at many universities, admission to an engineering program is highly selective.

It is also a good idea to sit down with your advisor and plan your graduation. List all of the classes that you need to take in order to obtain your degree in four or five years. You can always modify this plan later as your interests change. Make sure you understand the prerequisites for each class and in which semester a class typically is offered; a program flowchart will be quite useful. In order to make you aware of your social responsibilities as an engineer, you also are required to take a certain number of classes in social sciences and humanities. Think about your current interests and plan your social science and humanity electives as well. Again, don't worry about your interests changing; you can always modify your plan. For those of you who currently are studying at a community college and planning to transfer to a university later, you should contact the university, learn about their engineering course transfer policies and requirements, and prepare your graduation plan accordingly.

Other Considerations

Doing Volunteer Work If your study schedule allows, volunteer for a few hours a week to help those in need in your community. The rewards are unbelievable! Not only will you feel good about yourself, but you will gain a sense of satisfaction and feel connected to your community. Volunteering could also help develop communication, management, or supervisory skills that you may not develop by just attending school.

Vote in Local and National Elections Most of you are 18 or older. Take your civic duties seriously. Exercise your right to vote, and try to play an active role in your local, state, or federal government. Remember that *freedom is not free*. Be a good, responsible citizen.

Get to Know Your Classmates There are many good reasons for getting to know one or two other students in your classes. You may want to study with someone from class, or if you are absent from class, you have someone to contact to find out what the assignment is or find out what was covered in class. Record the following information on the course syllabi for all your classes: the name of a student sitting next to you, his or her telephone number, and his or her e-mail address.

Get to Know an Upper-Division Engineering Student Becoming acquainted with junior and senior engineering students can provide you with valuable information about their engineering education experience and campus social issues. Ask your instructor to introduce you to a junior or a senior engineering student, and record the following information on your introduction to engineering class syllabus: the name of an upper-division engineering student, his or her telephone number, and his or her e-mail address.

Before You Go On

Answer the following questions to test your understanding of the preceding section.

1. Why is it important to get involved with an engineering organization?
2. Explain why it is important to have a graduation plan.
3. Give examples of rewards that one gains from doing volunteer work.
4. Why is it important to get to know upper-division engineering students?

Suzelle Barrington

Courtesy of Suzelle Thauvette Barrington

Today, job aspiration is challenging with so many opportunities and new branches sprouting regularly. And yet, will there be work in this branch when I graduate three or four years from now? Engineering became an interesting career for me when I discovered what I could do something interesting with the mathematics and sciences which I loved. I could apply them to solve all kinds of issues in a manner far simpler that if I had remained in physics, for example. After my first university year in General Sciences, I thus decided to change over to an engineering faculty. I still believe that, as a career choice today, engineering opens wide doors for those interested in design and conception. Furthermore, if the engineering branch you choose does not quite fit your career opportunities, it is easy to use your basic knowledge to cross over into a neighboring field of engineering.

Some 40 years ago, I started my career as an employee of the Quebec Ministry of Agriculture, Food, and Fisheries managing rural drainage programs and designing agri-food buildings, such as grain centers and specialized fruit and vegetable storage. At that time, there were few women working as engineers—about 1% as a matter of fact. But I did not spend a lot of time thinking about this, as I was busy developing my skills and starting a family. Today, the percentage of women working as professional engineers has climbed to 10%. This higher percentage—not only in the workplace but also at university—means that both genders expect to work with each other. In another generation, the number of women entering the engineering profession will likely double to 20%.

While I was working for the Quebec Ministry of Agriculture, Food, and Fisheries, I decided to go back to school on a part-time basis to finish an M. Sc. Degree in engineering. Things went so well that, by the second year, I decided to jump into the Ph. D. program. Five years later, with a brand new degree, I had outgrown my job and joined the university life as a professor. Well, you are not a professor right from the start, as you must climb from Assistant to Associate and then

to full Professor. This new job opened once more a whole new category of opportunities.

As a university professor, I had a chance to share my passion for engineering. Engineering for me is letting your imagination run wild in creating and solving. Every day, engineers use their imagination to solve problems and introduce innovative technologies that makes life more enjoyable for everyone. Furthermore, the modern engineer must develop social skills, as the solutions developed must not only be integrated to the needs of society but must be acceptable to society in general. For example, I helped a Montreal community develop a composting center. Developing the composter was the easy job; selecting the site and organizing the operation of the composting center was the challenge, or no one would have used the facility.

As a university professor, I was also able to show many science students that engineering was not an out-of-reach profession. As compared to other scientists, engineers must be able to resolve issues quickly, and with a fairly good degree of accuracy. So, engineers have developed coefficients for every possible application and have applied safety factors to cover for the unknown events.

To end my university career, I accepted an International Research Chair with l'Université européenne de Bretagne, Rennes, France, where I spent one year coercing my four university partners into working together by introducing new blood and ideas for research projects. Then after 26 years of university research and teaching, I decided to join a consulting company owned by one of my former university students, and where two M. Sc. students whom I supervised were already working. Why? Because once more I was in search of brand new opportunities. Yes, there is no end to engineering and its opportunities. So why not join our ranks?

Susan Thomas

As a young girl from Utah with medical school aspirations, my high-school education was a whir of biology, math, and chemistry. My first years of college were spent at the University of Utah as a biology student, where my interest in the complex world of medicine took a turn that I was not expecting. I realized that I was not in love with the practice of medicine, as a physician should be, but with the *technology* of medicine; I was fascinated by CAT scans, surgical lasers, and the artificial heart. Add to this a long-standing love of math and the physical sciences and the influence of a successful "engineer" grandfather, and the decision was made. I switched my major to mechanical engineering.

One of the most challenging things about my engineering studies also turned out to be one of my greatest assets. As there were very few women in the program, I initially felt like somewhat of a misfit. However, I discovered that being one of a few gave me the chance to stand out and to enjoy a very personable, close relationship with my peers and professors. Engineering appealed to me the most because it was, as Herbert Hoover once said, "the job of clothing the bare bones of science with life, comfort and hope." I even found my medical niche in the field of micro-technology, which has led to the development of painless micro-needles, and the future possibility of tiny, injectable, cancer-zapping robots. When a professor approached me with the opportunity to complete my doctorate in this area, I jumped at the chance. My greatest aspiration is to participate in the research and development of innovative, medical (micro?) devices that will make a difference.

Courtesy of Susan Thomas

SUMMARY

LO¹   Making the Transition from High School to College

Making the transition from high school to college requires extra effort. You must realize that in high school most of your learning took place in class, whereas in college most of your education takes place outside the classroom. You must start studying and preparing from the first day of class, attend class regularly, get help right away, take good notes, select a good study place, and form study groups.

LO²   Budgeting Your Time

Your education is an expensive investment. Invest wisely! You need to learn to manage your time wisely if you want to be successful in college and life. If you budget your time appropriately and create a weekly schedule for your activities, you could then have a fulfilling college experience and have adequate time for studying, social events, and work.

LO³   Study Habits and Strategies

The study strategies that would lead to good academic performance include studying and preparing from the first day of class, attending class regularly, getting help right away, taking good notes, selecting a good study place, and forming study groups.

LO⁴   Getting Involved with an Engineering Organization

There are many good reasons to join an engineering organization. Networking, participating in plant tours, listening to technical guest speakers, participating in design competitions, attending social events, taking advantage of learning opportunities through short courses, seminars, and conferences, and obtaining student loans and scholarships are a few of the common benefits of belonging to an engineering organization.

LO⁵ Your Graduation Plan

It is also a good idea to sit down with your advisor and plan your graduation. List all of the classes that you need to take in order to obtain your degree in four or five years. You can always modify this plan later as your interests change. For those of you who currently are studying at a community college and planning to transfer to a university later, you should contact the university, learn about engineering course transfer policies and requirements, and prepare your graduation plan accordingly.

KEY TERMS

Daily Preparation 36	Graduation Plan 43	Transition from High School 34
Engineering Organization 42	Study Group 41	Volunteer Work 44
Exam Preparation 41	Time Management 34	

APPLY WHAT YOU HAVE LEARNED

1. Prepare a schedule for the current semester; also prepare two additional alternative schedules. Discuss the pros and cons of each schedule. Select what you think is the best schedule, and discuss it with your instructor or advisor. Consider his or her suggestions and modify the schedule if necessary; then present the final schedule to your instructor. Maintain a daily logbook to keep track of how closely you are following the schedule and where time is being used inefficiently. Write a one-page summary discussing each week's activities that deviated from the planned schedule, and come up with ways to improve or modify its shortcomings. Turn in a biweekly summary report to your instructor or advisor. Think of this exercise as an ongoing test similar to other tests that engineers perform regularly to understand and improve things.
2. Also, meet with your advisor and prepare your graduation plan, as discussed in Section 2.5.

Hour	Monday	Tuesday	Wednesday	Thursday	Friday	Saturday	Sunday
7–8							
8–9							
9–10							
10–11							
11–12							
12–1							
1–2							
2–3							
3–4							

Hour	Monday	Tuesday	Wednesday	Thursday	Friday	Saturday	Sunday
4–5							
5–6							
6–7							
7–8							
8–9							
9–10							
10–11							

"I have never let my schooling interfere with my education."
—MARK TWAIN (1835–1910)

Introduction to Engineering Design

Nucleartist/Shutterstock.com

1971yes/Shutterstock.com

Engineers design millions of products and services that we use in our everyday lives. To arrive at solutions, engineers, regardless of their backgrounds, follow certain steps including understanding the problem, conceptualizing ideas for possible solutions, evaluating good ideas in more detail, and presenting the final solution.

LEARNING OBJECTIVES

LO¹ Engineering Design Process: explain the basic steps that engineers follow to design something and to arrive at a solution to a problem

LO² Additional Design Considerations: describe what is meant by sustainability and its role in design; also explain the roles of engineering economics and material in engineering design

LO³ Teamwork: explain what is meant by a design team and describe the common traits of good teams; also explain how good teams manage conflicts

LO⁴ Project Scheduling and the Task Chart: describe the process that engineering managers use to ensure that a project is completed on time and within the allocated budget

LO⁵ Engineering Standards and Codes: describe why we need standards and codes and give examples of standards and codes organizations in the United States and abroad

LO⁶ Water and Air Standards in the United States: describe the drinking water, indoor and outdoor air sources of pollutants, and the water and air quality standards in the United States

DESIGN

Engineers, regardless of their backgrounds, follow certain steps when designing the products and services we use in our everyday lives. These steps are (1) recognizing the need for a product or service, (2) defining and understanding the problem (the need) thoroughly, (3) doing the preliminary research and preparation, (4) conceptualizing ideas for possible solutions, (5) synthesizing the results, (6) evaluating good ideas in more detail, (7) optimizing the result to arrive at the best possible solution, and (8) presenting the solution.

Economics also plays an important role in engineering decision making. The fact is that companies design products and provide services not only to make our lives better but also to make money! As design engineers, whether you are designing a machine part, a toy, or a frame for a car or a structure, the selection of material is also an important design decision. Engineers also work as a team to solve a problem. A design team may be defined as a group of individuals with complementary expertise, problem solving skills, and talent who are working together to solve the problem. When a group of people work together, conflicts sometimes arise. Managing conflicts is an important part of a team dynamic. Engineers also follow a project schedule to ensure that the project is completed on time and within the allocated budget. They also make certain that the product or service adheres to established national and international standards and codes.

To the students: Have you ever designed something? Did you follow steps similar to the ones mentioned here? If so, explain in greater detail what they mean. Have you ever worked with others on a project? How would you describe your experience working on a team?

Engineers are problem solvers. In this chapter, we will introduce you to the **engineering design** process. As we discussed in Chapter 1, engineers apply physical and chemical laws and principles and mathematics to design millions of products and services that we use in our everyday lives. Here, we will look more closely at what the term design means and learn more about how engineers go about designing these products and services. We will discuss the basic steps that most engineers follow when designing something. We will also introduce you to the economic considerations, material selection, team work, project scheduling, and engineering standards and codes—all an integral part of the design process and product and service development.

LO¹ 3.1   Engineering Design Process

Let us begin by emphasizing what we said in Chapter 1 about what engineers do. Engineers apply physical laws, chemical laws and principles, and mathematics to *design* millions of products and services that we use in our everyday lives. These products include cars, computers, aircrafts, clothing, toys, home appliances, surgical equipment, heating and cooling equipment, health care devices, tools and machines that make various products, and so on. Engineers consider important factors such as cost, efficiency, reliability, sustainability, and safety when designing the products, and they perform tests to make certain that the products they design withstand various loads and conditions. Engineers are continuously searching for ways to improve already existing products as well. Engineers also *design* and supervise the construction of buildings, dams, highways, and mass transit systems. They also *design* and supervise the construction of power plants that supply power to manufacturing companies, homes, and offices. Engineers play a significant role in the *design* and maintenance of nations' infrastructures, including communication systems, utilities, and transportation. They continuously develop new advanced materials to make products lighter and stronger for different applications. Engineers are also responsible for finding *suitable ways* and *designing* the necessary equipment to extract petroleum, natural gas, and raw materials from the earth.

Engineers follow certain steps when designing products and services. These steps include understanding the need for the product or service, doing the preliminary research, generating ideas for a solution, selecting the best idea, evaluating it in more detail and testing it, optimizing it if necessary, and presenting the final design.

Let us now look more closely at what constitutes the **design process**. These are the basic steps that engineers, regardless of their background, follow to arrive at solutions to problems. The steps include (1) recognizing the need for a product or a service, (2) defining and understanding the problem (the need) completely, (3) doing preliminary research and preparation, (4) conceptualizing ideas for possible solutions, (5) synthesizing the findings, (6) evaluating good ideas in more detail, (7) optimizing solutions to arrive at the best possible solution, (8) and presenting the final solution.

Keep in mind that these steps, which we will discuss soon, are not independent of one another and do not necessarily follow one another in the order in which they are presented here. In fact, engineers often need to return to Steps 1 and 2 when clients decide to change design parameters. Quite often, engineers are also required to give oral and written progress reports on a regular time basis. Therefore, be aware of the fact that even though we listed presentation of the design process as Step 8, it could well be an integral part of many other **design steps**. Let us now take a closer look at each step, starting with the need for a product or a service.

Step 1: Recognizing the Need for a Product or a Service

All you have to do is look around to realize the large number of products and services—designed by engineers—that you use every day. Most often, we take these products and services for granted until, for some reason, there is an interruption in the services they provide. Some of these existing products are constantly being modified to take advantage of new technologies. For example, cars and home appliances are constantly being redesigned to incorporate new technologies. In addition to the products and the services already in use, new products are being developed every day for the purpose of making our lives more comfortable, more pleasurable, and less laborious. There is also that old saying that "every time someone complains about a situation, or about a task, or curses a product, right there, there is an opportunity for a product or a service." As you can tell, the need for products and services exists; what one has to do is to identify it. The need may be identified by you, the company that you may eventually work for, or by a third-party client who needs a solution to a problem or a new product to make what it does easier and more efficient.

Step 2: Problem Definition and Understanding

One of the first things you need to do as a design engineer is to fully understand the problem. *This is the most important step in any design process.* If you do not have a good grasp of what the problem is or of what the client wants, you will not come up with a solution that is relevant to the need of the client. The best way to fully understand a problem is by asking many questions.

You may ask the client questions such as: How much money are you willing to spend on this project? Are there restrictions on the size or the type of materials that can be used? When do you need the product or the service? How many of these products do you need? Questions often lead to more questions that will better define the problem. Moreover, keep in mind that engineers generally work in a team environment where they consult each other to solve complex problems. They divide up the task into smaller, more manageable problems among themselves; consequently, productive engineers must be good team players. Good interpersonal and communication skills are increasingly important now because of the global market. You need to make sure you clearly understand your portion of the problem and how it fits with the other problems. For example, various parts of a product could be made by different companies located in different states or countries. In order to ensure that all components fit and work well together, cooperation and coordination are essential, which demands good teamwork and strong communication skills. Make sure you understand the problem, and make sure that the problem is well defined before you move on to the next step. *This point cannot be emphasized enough.* Good problems solvers are those who first fully understand what the problem is.

Step 3: Research and Preparation

Once you fully understand the problem, as a next step you need to collect useful information. Generally speaking, a good place to start is by searching to determine if a product already exists that closely meets the need of your client. Perhaps a product or components of a product already have been developed by your company that you could modify to meet the need. You do not want to "reinvent the wheel!" As mentioned earlier, depending on the scope, some projects require collaboration with other companies, so you need to find out what is available through these other companies as well. Try to collect as much information as you can. This is where you spend lots of time not only with the client but also with other engineers and technicians. Internet search engines are becoming increasingly important tools to gather such information. Once you have collected all pertinent information, you must then review it and organize it in a suitable manner.

Step 4: Conceptualization

During this phase of design, you need to generate some ideas or concepts that could offer reasonable solutions to your problem. In other words, without performing any detailed analysis, you need to come up with some possible ways of solving the problem. You need to be creative and perhaps develop several alternative solutions. At this stage of design, you do not need to rule out any reasonable working concept. If the problem consists of a complex system, you need to identify the components of the system. You do not need to look at details of each possible solution yet, but you need to perform enough analysis to see whether the concepts that you are proposing have merit. Simply stated, you need to ask yourself the following question: Would the concepts be likely to work if they were pursued further? Throughout the design process, you must also learn to budget your time. Good engineers have time-management skills that enable them to work productively and efficiently. You must learn to create a milestone chart detailing your time plan for completing the project. You need to show the time periods and the corresponding tasks that are to be performed during these time periods.

Evaluating Alternatives Once you have narrowed down your design to a few workable concepts, it is customary to use an evaluation table similar to the one shown in Table 3.1 to evaluate alternative concepts in more detail. You start by assigning a level of importance (I) to each design criterion. For example, you may use a scale of 1 to 5, with $I = 1$ indicating little importance, and $I = 5$ signifying extremely important. Next you will rate (R) each workable concept in terms of how well it meets each design criterion. You may use a scale of $R = 3$, $R = 2$, and $R = 1$ for high, medium, and low, respectively.

| TABLE 3.1 | Table Used to Evaluate Alternative Concepts |

Design Criterion	I	Design I		Design II	
		R	R × I	R	R × I
Positive					
Originality					
Practicability					
Manufacturability					
Reliability					
Performance					
Durability					
Appearance					
Profitability					
Other					
Negative					
Production cost					
Operating cost					
Maintenance cost					
Time to complete the project					
Environmental impact					
Other					
Net score					

Note that the design criteria that we have listed in Table 3.1 are to serve as an example, not as absolute design criteria. The design criteria vary depending on a project. For your class project, you should list the design criteria that you feel are important. Moreover, note that it is customary to divide the design criteria into positive and negative criteria. After you assign the I and R values to your design options, you add the $R \times I$ scores for each design and select the design with the highest overall rating. An example demonstrating how to evaluate alternatives is shown in Table 3.2.

Step 5: Synthesis

Recall from our discussion in Chapter 1 that good engineers have a firm grasp of the fundamental principles of engineering, which they can use to solve many different problems. Good engineers are analytical, detailed oriented,

TABLE 3.2   Comparing Two Alternative Designs

Design Criterion	I	R	Design I $R \times I$	R	Design II $R \times I$
Positive					
Originality	4	2	8	3	12
Practicability	5	3	15	2	10
Manufacturability	5	3	15	2	10
Reliability	5	3	15	3	15
Performance	5	3	15	2	10
Durability	4	2	8	2	8
Appearance	4	2	8	3	12
Profitability	5	3	15	2	10
Other					
			99		87
Negative					
Production cost	5	2	10	3	15
Operating cost	4	2	8	2	8
Maintenance cost	3	2	6	3	9
Time to complete the project	5	3	15	3	15
Environmental impact	5	2	10	3	15
Other					
			49		62
Net score			50		25

and creative. During this stage of design, you begin to consider details. You need to perform calculations, run computer models, narrow down the type of materials to be used, size the components of the system, and answer questions about how the product is going to be fabricated. You will consult pertinent codes and standards and make sure that your design will be in compliance with these codes and standards. We will discuss engineering codes and standards in Section 3.5.

Step 6: Evaluation

Analyze the problem in more detail (Figure 3.1). You may have to identify critical design parameters and consider their influence in your final design. At this stage, you need to make sure that all calculations are performed correctly. If there are some uncertainties in your analysis, you must perform experimental investigation. When possible, working models must be created and tested. At this stage of the design procedure, the best solution must be identified from alternatives. Details of how the product is to be fabricated must be worked out fully.

Step 7: Optimization

Optimization means minimization or maximization. There are two broad types of design: a functional design and an optimized design. A functional design is one that meets all of the preestablished design requirements but allows for improvement to be made in certain areas. To better understand the concept

| FIGURE 3.1 | Two engineers considering details during the design process. |

Courtesy of DOE/NREL

of a functional design, we will consider an example. Let us assume that we are to design a 3-meter-tall (10 ft) ladder to support a person who weighs 1335 newtons (300 pounds) with a certain factor of safety. We will come up with a design that consists of a steel ladder that is 3 meter tall (10 ft) and can safely support the load of 1335 N (300 lb) at each step. The ladder would cost a certain amount of money. This design would satisfy all of the requirements, including those of strength and size, and thus constitutes a functional design. Before we can consider improving our design, we need to ask ourselves what criterion we should use to optimize the design. Design optimization is always based on some particular criterion, such as cost, strength, size, weight, reliability, noise, or performance. If we use the weight as an optimization criterion, then the problem becomes one of minimizing the weight of the ladder without jeopardizing its strength. For example, we may consider making the ladder from aluminum. We would also perform stress analysis on the new ladder to see if we could remove material from certain sections of the ladder without compromising the loading and safety requirements.

Another important fact to keep in mind is that optimizing individual components of an engineering system does not necessarily lead to an optimized system. For example, consider a thermal-fluid system such as a refrigerator. Optimizing the individual components independently—such as the compressor, the evaporator, or the condenser—with respect to some criterion does not lead to an optimized overall system (refrigerator).

Traditionally, improvements in a design come from the process of starting with an initial design, performing an analysis, looking at results, and deciding whether or not we can improve the initial design. This procedure is shown in Figure 3.2. In the past few decades, the optimization process has grown into a discipline that

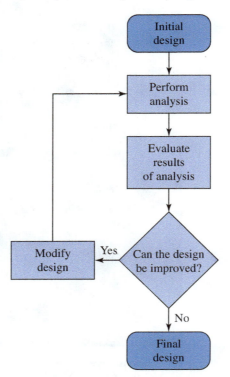

FIGURE 3.2   An optimization procedure.

ranges from linear to nonlinear programming techniques. As is the case with any discipline, the optimization field has its own terminology. There are advanced classes that you can take to learn more about the design optimization process.

Step 8: Presentation

Now that you have a final solution, you need to communicate your solution to the client, who may be your boss, another group within your company, or an outside customer. You may have to prepare not only an oral presentation (Figure 3.3) but also a written report. As we said in Chapter 1, engineers are required to write reports. Depending on the size of the project, these reports might be lengthy, detailed technical reports containing graphs, charts, and engineering drawings, or they may take the form of a brief memorandum or executive summaries.

A reminder again that, although we have listed the presentation as Step 8 of the design process, quite often engineers are required to give oral and written progress reports on a regular time basis to various groups. Consequently, presentation could well be an integral part of many other design steps. Because of the importance of communication, we have devoted an entire chapter to engineering communication (see Chapter 4).

Finally, recall that in our discussion in Chapter 1 regarding the attributes of good engineers, we said that good engineers have written and oral communication skills that equip them to work well with their colleagues and to convey their expertise to a wide range of clients. Moreover, engineers have good "people skills" that allow them to interact and communicate effectively with various people in their organization. For example, they are able to communicate equally well with the sales and marketing experts and with their own colleagues in engineering.

FIGURE 3.3 Many oral presentations also require the engineer to create written materials.
Bloomberg via Getty Images

In Step 7 of the design process, we discussed optimization. Let us now use a simple example to introduce you to some of the fundamental concepts of optimization and its terminology.

EXAMPLE 3.1

Assume that you have been asked to look into purchasing some storage tanks for your company, and for the purchase of these tanks, you are given a budget of $1680. After some research, you find two tank manufacturers that meet your requirements. From Manufacturer A, you can purchase 16-ft³-capacity tanks that cost $120 each. Moreover, the type of tank requires a floor space of 7.5 ft². Manufacturer B makes 24-ft³-capacity tanks that cost $240 each and that require a floor space of 10 ft². The tanks will be placed in a section of a lab that has 90 ft² of floor space available for storage. You are looking for the greatest storage capacity within the budgetary and floor-space limitations. How many of each tank must you purchase?

First, we need to define the *objective function,* which is the function that we will attempt to minimize or maximize. In this example, we want to maximize storage capacity. We can represent this requirement mathematically as

$$\text{maximize } Z = 16x_1 + 24x_2 \qquad \text{3.1}$$

subject to the following constraints:

$$120x_1 + 240x_2 \leq 1680 \qquad \text{3.2}$$

$$7.5x_1 + 10x_2 \leq 90 \qquad \text{3.3}$$

$$x_1 \geq 0 \qquad \text{3.4}$$

$$x_2 \geq 0 \qquad \text{3.5}$$

In Equation (3.1), Z is the objective function, while the variables x_1 and x_2 are called *design variables,* and represent the number of 16-ft³-capacity tanks and the number of 24-ft³-capacity tanks, respectively. The limitations imposed by the inequalities (3.2) through (3.5) are referred to as a set of *constraints.* Although there are specific techniques that deal with solving linear programming problems (the objective function and constraints are linear), we will solve this problem graphically to illustrate some additional concepts.

Let us first review how you would plot the regions given by the inequalities. For example, to plot the region as given by the linear inequality $120x_1 + 240x_2 \leq 1680$, we must first plot the line $120x_1 + 240x_2 = 1680$ and then determine which side of the line represents the region. For example, after plotting the line $120x_1 + 240x_2 = 1680$, we can test points $x_1 = 0$ and $x_2 = 0$ to see if they fall inside the inequality region; because substitution of these points into the inequality satisfies the inequality, that is, $(120)(0) + (240)(0) \leq 1680$, the shaded region represents the given inequality (see Figure 3.4(a)). Note that if we were to substitute a set of points outside the region, such as $x_1 = 15$ and $x_2 = 0$, into the inequality, we would find that the inequality is not satisfied. The inequalities (3.2) through (3.5) are plotted in Figure 3.4(b).

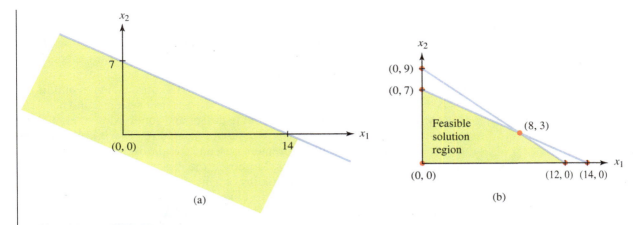

FIGURE 3.4   (a) The region as given by the linear inequality $120x_1 + 240x_2 \leq 1680$.
(b) The feasible solution for Example 3.1.

The shaded region shown in Figure 3.4(b) is called a *feasible solution region*. Every point within this region satisfies the constraints. However, our goal is to maximize the objective function given by Equation (3.1). Therefore, we need to move the objective function over the feasible region and determine where its value is maximized. It can be shown that the maximum value of the objective function will occur at one of the corner points of the feasible region. By evaluating the objective function at the corner points of the feasible region, we see that the maximum value occurs at $x_1 = 8$ and $x_2 = 3$. This evaluation is shown in Table 3.3.

Thus, we should purchase eight of the 16-ft^3-capacity tanks from Manufacturer A and three of the 24-ft^3-capacity tanks from Manufacturer B to maximize the storage capacity within the given constraints.

It is worth noting here that most of you will take specific design classes during the next four years. In fact, most of you will work on a relatively comprehensive design project during your senior year. Therefore, you will learn more in depth about design process and its application to your specific discipline. For now our intent has been to introduce you to the design process, but keep in mind that more design is coming your way.

TABLE 3.3   Values of the Objective Function at the Corner Points of the Feasible Region

Corner Points (x_1, x_2)	Value of $Z = 16x_1 + 24x_2$
(0, 0)	0
(0, 7)	168
(12, 0)	192
(8, 3)	200 (max.)

Civil Engineering Design Process

The civil engineering design process is slightly different from other disciplines such as mechanical, electrical, or chemical engineering. As we explained in Chapter 1, civil engineering is concerned with providing public infrastructure and services. Civil engineers design and supervise the construction of buildings, roads and highways, bridges, dams, tunnels, mass transit systems, and airports (Figure 3.5). They are also involved in the design and supervision of municipal water supplies and sewage systems. Because of the nature of their projects, they must follow specific procedures, regulations, and standards that are established by local, state, or federal agencies. Moreover, the design procedure will be different for a bridge than say for a building or a mass transit system. To shed some light on how a civil engineering project may be carried out, we will next cover the basic steps that are followed to design a building. The design process for buildings such as schools, offices, shopping malls, medical clinics, and hospitals usually includes the following steps. Note that the description in parentheses corresponds to the design steps we discussed in the previous section.

1. Recognizing the need for a building (*Step 1: recognizing the need for a product or a service*)
2. Define the usage of the building (*Step 2: problem definition and understanding*)

FIGURE 3.5   Civil engineers design and supervise the design of buildings, roads and highways, bridges, dams, tunnels, mass transit systems, and airports.
Elena Rooraid/Photo Edit

3. Project planning (*Step 3: research and preparation*)
4. Schematic design phase (*Steps 4 & 8: conceptualization and presentation*)
5. Design development phase (*Steps 5, 6, & 8: synthesis, evaluation, and presentation*)
6. Construction documentation phase (*Steps 5 & 7: synthesis and optimization*)
7. Construction administration phase

Step 1: Recognizing the Need for a Building There could be a number of reasons for wanting a new building. For example, due to demographic change in a district, a new elementary school may be needed, or an existing building may need to be expanded to accommodate the increase in population of children between 6 and 12 years old; or a new medical clinic is needed to address the growing need of an aging population; or a factory could need to be expanded to increase production to meet the demand.

In the private sector, the need is usually determined by the owner of a business or real estate. On the other hand, in the public sector the need is usually identified by others, such as a school principal, a city engineer, or a district engineer. Moreover, the need must be approved by a corresponding oversight body, such as a school board, city council, or the department of transportation and state legislatures.

Step 2: Define the Usage of the Building Before any design work is done, the person (client) who has identified the need for a building determines the types of activities that would take place in the building. For example, in the case of a new elementary school, the principal forecasts the number of students expected to enroll in the near future. The enrollment projection data then allows the principal to determine the number of classrooms and computer labs, and the need for a library or a cafeteria. For a medical clinic, other activities are considered, such as the number of examination rooms, x-ray labs, reception areas, record rooms, and so on. The usage and activity data will help the architect determine the amount of area (square footage) needed to accommodate the client's projected need (Figure 3.6).

Step 3: Project Planning During this stage, the person who has requested the project (the client) selects potential sites for the new building. In addition to the cost and whether the location is ideal for the proposed building, other factors are considered. These factors include zoning, environmental impact, achaeological impact, and traffic flow. Although detailed study or design is not necessary at this stage, it is important for the client to recognize all the factors that would affect the project's cost and feasibility. For example, if the potential site is zoned as residential, it would be extremely difficult if not impossible to rezone the area as commercial. A change of zoning would require public hearings and approval by the city council with recommendation from the zoning committee. The most common environmental impact issues may include noise level created by additional traffic drawn by the new building, wet land remediation, effect on wildlife, and so on. If a site contains achaeological artifacts, disturbance may not be permitted. The design may have to change before construction can start. Historical knowledge of the site and construction information from projects near the potential site would help civil engineers assess the potential extent of achaeological artifacts in the potential project site.

FIGURE 3.6	Engineers must take the usage of the building into account when completing the project plan.

Courtesy of DOE/NREL

Also in this stage, the client selects an architectural firm or a contractor to initiate the design phase. If the client selects a contractor, then the architectural firm is hired by the contractor. It is also important to note that the architectural firm does not perform any design; instead, they collaborate with structural engineers, mechanical engineers, electrical engineers, interior designers, drafters, and project managers to complete the project.

Step 4: Schematic Design Phase During this phase, the architect consults with the client to fully understand the intended usage for the building and to obtain an approximate budget for the project. The architect then prepares multiple schematic designs for the building. Through continuous communication with the client, the architect then narrows down the options to one or two designs. The design would show the layout of the space and rooms. The material type and the framing system also would be proposed at this stage. The architect then presents the schematic design(s) to the client and gets feedback from the client for the next design phase.

Step 5: Design Development (DD) Phase In the design development (DD) phase, the architect continues to finalize the layout of the building. The architect consults with the structural engineer at this phase so that limits on column size and beam size can be determined. The limitation on beam sizes is influenced by the ceiling height and the overall height of the building. The limitation on column sizes would affect how the rooms are arranged. This is to avoid having many columns in the middle of a room or a space.

Once the architect has finalized the layout of the rooms and spaces, the structural engineer then begins to perform a preliminary design for the building. Preliminary designs of the HVAC (heating, ventilation, and air conditioning) system will be done by the mechanical engineer, while the electric design is done by the electrical engineer. Next, the interior designer provides a preliminary design for the interior of the building. Based on the preliminary design, the contractor will then provide a cost estimate for the project.

At the end of this phase, the architect who represents the engineers, the interior designer, and the contractor would then meet with the client to present the preliminary design and seek feedback. The client may request the rearrangement of rooms or change in the interior design based on cost or other factors. After receiving clients input, the next phase of design is initiated.

Step 6: Construction Documentation (CD) Phase During this phase of the project, all of the detail work is done. The construction document includes design specifications and drawings from the architect, civil, structural, mechanical, and electrical engineers, and the interior designers. In some projects, the work of the landscape architect is also included. While the engineers work on their own design components, the architect continues with fine tuning the layout of the building, from roof top to the bottom floor and everything in between.

The civil engineer also provides the site plan design, which includes the grading of the ground from the perimeter of the building to the sidewalk, the grading of the parking area, and drainage for surface runoff. If there are existing structures and power lines within the construction site, the civil engineer also needs to include a demolition plan and the relocation of powerlines.

The structural engineer provides all the design details for structural components, including the foundation, beams and columns, interior and exterior walls, connections, additional support for openings such as windows, doors, roof supports, floor supports, canopies, and so on. The engineers must bear in mind all the design specifications required by the building codes as established by local government. In most cases, the local government would adopt the state's building code. The building code does not only specify engineering design standards, but also specifies information such as the width of corridors, number of emergency exits in a building, and so on.

Before construction can commence, the construction document must be reviewed and approved by the building inspectors. If the client has not selected a contractor, as it is common for publicly funded projects, interested contractors would then purchase a hard copy of the construction document or download it from the architect's web site for bid preparation.

Step 7: Construction Administration Phase Once the construction document is approved and a general contractor is selected, construction will commence. The general contractor will have a superintendent on site to manage the construction and its progress and coordinates all the subcontractors (Figure 3.7). In large-scale projects such as bridges, highways, or power plants, a construction manager, who is a registered professional engineer, would be responsible for the site supervision.

A project manager representing the architect would then meet with the site superintendent and the client on a regular basis to review the construction progress and to respond to any issues that require further attention. Many projects would require adjustment in design due to unforeseen matters.

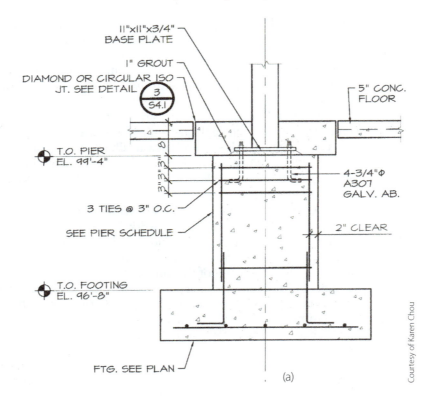

11"x11"x3/4"
BASE PLATE

1" GROUT

DIAMOND OR CIRCULAR ISO
JT. SEE DETAIL
3
S4.1

5" CONC.
FLOOR

T.O. PIER
EL. 99'-4"

3" 3" 3"

4-3/4"Φ
A307
GALV. AB.

3 TIES @ 3" O.C.

SEE PIER SCHEDULE

2" CLEAR

T.O. FOOTING
EL. 96'-8"

FTG. SEE PLAN

(a)

Courtesy of Karen Chou

(b)

FIGURE 3.7   Professional engineers are responsible for overseeing plans as well as on-site
supervision.
Courtesy of Karen Chou

The structural engineer also visits the construction site periodically to observe the progress of the project. These visits are particularly important, as the foundation is being built and as the skeleton of the building is being erected. Even though the site superintendent would make sure the workers are building per given construction drawings and the project manager would walk through the site frequently to ensure the building is constructed as designed, sometimes it takes the eyes of the structural engineer to see if the building is being properly built per design specifications. Besides visiting the construction site, the structural engineer is also responsible for reviewing the shop drawings submitted by the fabricators through the general contractor. The shop drawings (Figure 3.7(a)) show the details of structural components. For example, the steel fabricator must draw to scale the exact beam length to be delivered to the site. It also includes the number of holes and hole sizes at each end of the beam for connection purposes. Each beam in the project will be assigned a unique identification number so that the steel worker will know exactly the location of each beam in the project.

When the project is completed, the project manager will walk through the building with the client and the superintendent to go through a "punch" list. The punch list identifies areas that need to be completed certain ways or adjusted. Finally, the building inspector must approve the building, before it can be occupied.

LO² 3.2 Additional Design Considerations

Sustainability in Design*

In the past few years, you have been hearing or reading a great deal about **sustainability**. What does sustainability mean and why is important for you, as future engineers, to get a good grasp of it? To start with, it is important to know that there is no universal definition for sustainability and sustainable engineering. It means different things to different professions. However, one of the generally accepted definitions is: *"design and development that meets the needs of the present without compromising the ability of future generations to meet their own needs."*

As you know by now, engineers contribute to both private and public sectors of our society. In the private sector, they design and produce the goods and services that we use in our daily lives; the same goods and services that have allowed us to enjoy a high standard of living. We have also explained the role of engineer in the public sector. Engineers support local, state, and federal missions, such as meeting our infrastructure needs, energy and food security, and national defense. Increasingly, because of worldwide socioeconomic trends, environmental concerns, and the earth's finite resources, more is expected of engineers.

As future engineers, you will be expected to design and provide goods and services that increase the standard of living and advance health care, while addressing serious environmental and sustainability concerns. In other words, when you design products and services, you must consider

*Based on Board of Direction Views Sustainability Strategy as Key Priority, ASCE News, January 2009 Volume 34, Number 1, http://www.asce.org/Content.aspx?id=2147484152

Sustainability, economics, and material selection also play significant roles in design decision making processes.

the link among earth's finite resources, environmental, social, ethical, technical, and economical factors. Moreover, there is an international competition for engineers who can come up with solutions that address energy and food security and simultaneously address the sustainability issues. The potential shortage of engineers with training in sustainability—engineers who can apply the sustainability concepts, methods, and tools to their problem-solving and decision-making processes—could have serious consequences for our future. Because of this fact, in recent years, the engineering organizations including the American Society of Civil Engineers (ASCE), the American Society for Engineering Education (ASEE), the American Society of Mechanical Engineers (ASME), and the Institute of Electrical and Electronics Engineers (IEEE) all have come out in support of sustainability education in engineering curricula.

The civil engineers play an increasing important role in addressing the climate change and sustainability issues that are being discussed nationally and internationally among policy makers and politicians. The following American Society of Civil Engineers (ASCE) sustainability statement is a testament to this fact: *"The public's growing awareness that it is possible to achieve a sustainable built environment, while addressing such challenges as natural and man-made disasters, adaptation to climate change, and global water supply, is reinforcing the civil engineer's changing role from designer/constructor to policy leader and life-cycle planner, designer, constructor, operator, and maintainer (sustainer). Civil engineers are not perceived to be a significant contributors to sustainable world"**.

On November 4, 2008, ASCE Board of Direction adopted sustainability as the fourth ASCE priorities. The other three are renewing the nation's infrastructure, raising the bar on civil engineering education, and addressing the role of the civil engineers in today's changing professional environment. Moreover, in the January 8, 2009 article of the ASCE News, entitled: "Board of Direction Views Sustainability Strategy as Key Priority," William Wallace—founder and president of the Wallace Futures Group of Steamboat Springs, Colorado, and the author of "Becoming Part of the Solution: The Engineer's Guide to Sustainable Development," Washington, D.C.: American Council of Engineering Companies, 2005—offers five issues that must be understood if engineers are to assume new responsibilities in sustainability. These are:

1. The world's current economic development is not sustainable—the world population already uses approximately 20% more of the world's resources than the planet can sustain. (UN Millennium Ecosystem Assessment Synthesis Report, 2005.)

2. The effects of outpacing the earth's carrying capacity have now reached crisis proportions—spiking energy costs, extreme weather events causing huge losses, and the prospect of rising sea levels threatening coastal cities. Global population increase outstrips the capacity of institutions to address it.

3. An enormous amount of work will be required if the world is to shift to sustainable development—a complete overhaul of the world's processes, systems, and infrastructure will be needed.

4. The engineering community should be leading the way toward sustainable development but has not yet assumed that responsibility. Civil

engineers have few incentives to change. Most civil engineers deliver conventional engineering designs that meet building codes and protect the status quo.

5. People outside the engineering community are capitalizing on this new opportunity—accounting firms and architects are examples cited by Wallace. The architects bring their practices into conformity with the U.S. Green Building Council's Leadership in Energy and Environmental Design (LEED) Green Building Rating System.

As mentioned previously, other organizations also have come to realize the importance of sustainability in engineering education as well. For example, in January 2009, the Institute of Electrical and Electronics Engineers (IEEE) formed the Sustainability Ad Hoc Committee to map and coordinate sustainability related issues across IEEE. The IEEE also studied sustainability activities of other organizations to determine areas of collaboration, and took an active role in creating a worldwide Earth-monitoring network to "take the pulse of the planet." The project, known as the Global Earth Observation System of Systems (GEOSS), involves collecting data from thousands sensors, gauges, buoys, and weather stations across the globe. The goal of GEOSS is to help foster sustainable development. The IEEE defines sustainable development as *"development that meets the needs of the present without compromising the ability of future generations to meet their own needs."* As evident from the approaches taken and statements made by different engineering organizations, sustainability has to be a major part of your engineering education and any engineering design. Let us now define the key sustainability concepts, methods, and tools. These terms are self explanatory; think about them for a while and then using your own words explain what they mean to you.

Key sustainability concepts—understanding the earth's finite resources and environmental issues; socioeconomic issues related to sustainability; ethical aspects of sustainability; sustainable development.

Key sustainability methods—life-cycle based analysis; resource and waste management (material, energy); environmental impact analysis.

Key sustainability tools—life-cycle assessment; environmental assessment; use of sustainable-development indicators; U.S. Green Building Council (USGBC) Leadership in Energy and Environmental Design (LEED) rating system.

As stated on their website, "LEED is an internationally recognized green building certification system, providing third-party verification that a building or community was designed and built using strategies aimed at improving performance across all the metrics that matter most: energy savings, water efficiency, CO_2 emissions reduction, improved indoor environmental quality, and stewardship of resources and sensitivity to their impacts. Developed by the U.S. Green Building Council (USGBC), LEED provides building owners and operators a concise framework for identifying and implementing practical and measurable green building design, construction, operations and maintenance solutions." You can learn more about LEED by visiting www. usgbc.org/LEED.

As you take additional courses in engineering and design, gradually you will learn in more detail about these concepts, methods, and tools, and will apply them to the solutions of engineering problems and design.

Engineering Economics

Economic factors always play important roles in engineering design decision making. If you design a product that is too expensive to manufacture, then it can not be sold at a price that consumers can afford and still be profitable to your company. The fact is that companies design products and provide services not only to make our lives better but also to make money! In Chapter 20, we will discuss the basics of engineering economics. The information provided in Chapter 20 not only applies to engineering projects but can also be applied to financing a car or a house or borrowing from or investing money in banks.

Material Selection

As design engineers, whether you are designing a machine part, a toy, or a frame for a car or a structure, the selection of material is an important design decision. There are a number of factors that engineers consider when selecting a material for a specific application. For example, they consider the properties of material such as density, ultimate strength, flexibility, machinability, durability, thermal expansion, electrical and thermal conductivity, and resistance to corrosion. They also consider the cost of the material and how easily it can be repaired. Engineers are always searching for ways to use advanced materials to make products lighter and stronger for different applications.

In Chapter 17, we will look more closely at materials that commonly are used in various engineering applications. We will also discuss some of the basic physical characteristics of materials that are considered in design. We will examine the application and properties of common solid materials; such as metals and their alloys, plastics, glass, and wood and those that solidify over time; such as concrete. We will also investigate in more detail basic fluids; such as air and water.

By now, it should be clear that material properties and material cost are important design factors. In general, the properties of a material may be divided into three groups: electrical, mechanical, and thermal. In electrical and electronic applications, for example, the electrical resistivity of materials is important. How much resistance to flow of electricity does the material offer? In many mechanical, civil, and aerospace engineering applications, the mechanical properties of materials are important. These properties include the modulus of elasticity, modulus of rigidity, tensile strength, compression strength, strength-to-weight ratio, modulus of resilience, and modulus of toughness. In applications dealing with fluids (liquids and gases), thermophysical properties such as density, thermal conductivity, heat capacity, viscosity, vapor pressure, and compressibility are important properties. Thermal expansion of a material, whether solid or fluid, is also an important design factor. Resistance to corrosion is another important factor that must be considered when selecting materials.

Material properties depend on many factors, including how the material was processed, its age, its exact chemical composition, and any nonhomogenity or defect within the material. Material properties also change with temperature and time as the material ages. Most companies that sell materials will provide, upon request, information on the important properties of their manufactured materials. Keep in mind that when practicing as an engineer, you should use the manufacturers' material property values in your design calculations. The property values given in this and other textbooks should be used as typical values—not as exact values.

In the upcoming chapters, we will explain in detail the properties of materials and what they mean. For the sake of continuity of presentation, a summary of important material properties follows.

Electrical Resistivity—The value of electrical resistivity is a measure of resistance of material to flow of electricity. For example, plastics and ceramics typically have high resistivity, whereas metals typically have low resistivity, and among the best conductors of electricity are silver and copper.

Density—Density is defined as mass per unit volume; it is a measure of how compact the material is for a given volume. For example, the average density of aluminum alloys is 2700 kg/m^3; when compared to steel's density of 7850 kg/m^3, aluminum has a density that is approximately one third of the density of steel.

Modulus of Elasticity (Young's Modulus)—Modulus of elasticity is a measure of how easily a material will stretch when pulled (subject to a tensile force) or how well the material will shorten when pushed (subject to a compressive force). The larger the value of the modulus of elasticity is, the larger the required force would be to stretch or shorten the material. For example, the modulus of elasticity of aluminum alloy is in the range of 70 to 79 GPa (gigapascal, giga $= 10^9$), whereas steel has a modulus of elasticity in the range of 190 to 210 GPa; therefore, steel is approximately three times stiffer than aluminum alloys.

Modulus of Rigidity (Shear Modulus)—Modulus of rigidity is a measure of how easily a material can be twisted or sheared. The value of the modulus of rigidity, also called shear modulus, shows the resistance of a given material to shear deformation. Engineers consider the value of shear modulus when selecting materials for shafts, which are rods that are subjected to twisting torques. For example, the modulus of rigidity or shear modulus for aluminum alloys is in the range of 26 to 36 GPa, whereas the shear modulus for steel is in the range of 75 to 80 GPa. Therefore, steel is approximately three times more rigid in shear than aluminum is.

Tensile Strength—The tensile strength of a piece of material is determined by measuring the maximum tensile load a material specimen in the shape of a rectangular bar or cylinder can carry without failure. The tensile strength or ultimate strength of a material is expressed as the maximum tensile force per unit cross-sectional area of the specimen. When a material specimen is tested for its strength, the applied tensile load is increased slowly. In the very beginning of the test, the material will deform elastically, meaning that if the load is removed, the material will return to its original size and shape without any permanent deformation. The point at which the material ceases to exhibit this elastic behavior is called the yield point. The yield strength represents the maximum load that the material can carry without any permanent deformation. In certain engineering design applications, the yield strength is used as the tensile strength.

Compression Strength—Some materials are stronger in compression than they are in tension; concrete is a good example. The compression strength of a piece of material is determined by measuring the maximum compressive load a material specimen in the shape of rectangular bar, cylinder, or cube can carry without failure. The ultimate compressive strength of a material is expressed as the maximum compressive force per unit cross-sectional area of the specimen. Concrete has a compressive strength in the range of 10 to 70 MPa (megapascal, mega $= 10^6$).

Modulus of Resilience—Modulus of resilience is a mechanical property of a material that shows how effective the material is in absorbing mechanical energy without going through any permanent damage.

Modulus of Toughness—Modulus of toughness is a mechanical property of a material that indicates the ability of the material to handle overloading before it fractures.

Strength-to-Weight Ratio—As the term implies, it is the ratio of strength of the material to its specific weight (weight of the material per unit volume). Based on the application, engineers use either the yield or the ultimate strength of the material when determining the strength-to-weight ratio of a material.

Thermal Expansion—The coefficient of linear expansion can be used to determine the change in the length (per original length) of a material that would occur if the temperature of the material were changed. This is an important material property to consider when designing products and structures that are expected to experience a relatively large temperature swing during their service lives.

Thermal Conductivity—Thermal conductivity is a property of materials that shows how good the material is in transferring thermal energy (heat) from a high-temperature region to a low-temperature region within the material.

Heat Capacity—Some materials are better than others in storing thermal energy. The value of heat capacity represents the amount of thermal energy required to raise the temperature one kilogram mass of a material by one degree Celsius, or, using U.S. Customary units, the amount of thermal energy required to raise one pound mass of a material by one degree Fahrenheit. Materials with large heat capacity values are good at storing thermal energy.

Viscosity, vapor pressure, and bulk modulus of compressibility are additional fluid properties that engineers consider in design.

Viscosity—The value of viscosity of a fluid represents a measure of how easily the given fluid can flow. The higher the viscosity value is, the more resistance the fluid offers to flow. For example, it would require less energy to transport water in a pipe than it does to transport motor oil or glycerin.

Vapor Pressure—Under the same conditions, fluids with low vapor-pressure values will not evaporate as quickly as those with high values of vapor pressure. For example, if you were to leave a pan of water and a pan of glycerin side by side in a room, the water will evaporate and leave the pan long before you would notice any changes in the level of glycerin.

Bulk Modulus of Compressibility—A fluid bulk modulus represents how compressible the fluid is. How easily can one reduce the volume of the fluid when the fluid pressure is increased? For example, as we will discuss in Chapter 10, it would take a pressure of 2.24×10^9 N/m^2 to reduce 1 m^3 volume of water by 1%, or said another way, to a final volume of 0.99 m^3.

Patent, Trademark, and Copyright

In the early days, trade information and invention were kept in the family and passed on from one generation to the next. For example, when a plow maker came up with a better design, he kept the details of the design to himself and shared the specifications of the new invention only with his family, including son(s), brother(s), and so on. The new designs and inventions stayed in the family to protect the business and to prevent others from duplicating the inventor's design. However, new designs and inventions need to be shared if they are to bring about improvements in everyone's lives. At the same time, the person(s) who comes up with a new idea should benefit from it. Traded information and invention, if not protected, can be stolen. So you can see that, in order for a government to promote new ideas and inventions, it must also provide means for protecting others from stealing someone's new ideas and inventions, which are considered *intellectual property.*

Patents, trademarks, service marks, and copyrights are examples of means by which intellectual property is protected by United States and other countries laws.

Patent When you come up with an invention, in order to prevent others from making, using, or selling your invention, you may file for a patent with the U.S. Patent and Trademark Office. The right given by the patent in the language of the statute and of the grant itself includes "the right to exclude others from making, using, offering for sale, or selling" the invention in the United States or "importing" the invention into the United States.

It is important to understand that the patent does not grant the inventor the right to make, use, or sell the invention, but it excludes others from making, using, or selling the invention. For a new patent, an invention is protected for a term of 20 years from the date on which the application for the patent was filed.

The U.S. Patent and Trademark Office recommends that all prospective applicants retain the services of a registered patent attorney or patent agent to prepare and execute their applications. A design patent is good for 14 years from the time it was granted. A utility patent lasts for either 17 years from the time it was granted or 20 years from the earliest filing date, whichever is longer. A utility patent is issued for the way an item works and is used, whereas a design patent protects the way an item looks.

Trademark and Service Mark A trademark is a name, word, or symbol that a company uses to distinguish its products from others. It is important to note that the trademark right issued to a company excludes others from using the same or similar mark, but it does not prevent other companies from making the same or similar products. Coke® and Pepsi® are examples of similar products with different trademarks. A service mark is a name, word, or symbol that a company uses to distinguish its services from others. A service mark is the same as a trademark, except that it applies to a service rather than a product.

Copyright Copyright is a form of protection provided by the laws of the United States to the authors of "original works of authorship." The copyright laws cover literary, dramatic, musical, artistic, and other types of intellectual works and is obtainable for both published and unpublished work. The copyright laws protect the form of expression used by the authors, not the content or the subject matter of their work. For example, an author can write a book about the fundamentals of physics. The copyright law protects the author's work from others copying the exact way things were explained or described. It does not prevent others from writing another book about the same fundamentals of physics, nor does it prevent others from using the fundamental laws of physics. For a work created after January 1, 1978, the copyright laws protect it for a term that is equal to the author's life plus 70 years after the author's death. For a piece of work that has two or more authors, the term extends to the last surviving author's life plus 70 years. It is also important to note that currently there are no international copyright laws that would protect an author's work throughout the world.

Before You Go On

Answer the following questions to test your understanding of the preceding sections:

1. Describe the basic steps that engineers follow to design something.

2. Describe the process by which engineers evaluate alternatives.

3. In your own words, explain sustainability.

4. Why do economics and material selection play important roles in the design decision making process?

5. How is intellectual property protected in the United States?

Vocabulary—As a well-educated engineer and intelligent citizen, it is important for you to understand that you need to develop a comprehensive vocabulary to communicate effectively. State the meaning of the following terms:

Optimization _____

Sustainability _____

LEED _____

Trademark _____

Patent _____

Copyright _____

LO³ 3.3   Teamwork

A **design team** may be defined as a group of individuals with complementary expertise, problem solving skills, and talent who are working together to solve a problem or achieve a common goal. The goal might be providing a service; designing, developing, and manufacturing a product; or improving an existing service or product.

A good team is one that gets the best out of each other. The individuals making up a good team know when to compromise for the good of the team and its common goal. Communication is an essential part of successful teamwork. The individuals making up the team need also to clearly understand the role of each team member and how each task fits together.

> A design team may be defined as a group of individuals with complementary expertise, problem solving skills, and talent who are working together to solve a problem.

Common Traits of Good Teams

More and more, employers are looking for individuals who not only have a good grasp of engineering fundamentals but who can also work well with others in a team environment. Successful teams have the following components:

1. The project that is assigned to a team must have clear and realistic goals. These goals must be understood and accepted by all members of the team.
2. The team should be made up of individuals with complementary expertise, problem solving skills, background, and talent.
3. The team must have a good leader.
4. The team leadership and the environment in which discussions take place should promote openness, respect, and honesty.
5. Team needs and goals should come before individual needs and goals.

In addition to these characteristics that make up a good team, Dr. R. Meredith Belbin, in his book *Management Teams: Why They Succeed or Fail,* identifies additional roles for good team members. A team with members who represent the following secondary roles tends to be very successful.

The **organizer** is someone who is experienced and confident. This person is trusted by members of the team and serves as a coordinator for the entire project. The organizer does not have to be the smartest or most creative member of the team; however, he or she needs to be good at clarifying goals and advancing decision making.

The **creator** is someone who is good at coming up with new ideas, sharing them with other team members, and letting the team develop the ideas further. The creator is also good at solving difficult problems, but may have problems with following certain protocols.

The **gatherer** is someone who is enthusiastic and good at obtaining things, looking for possibilities, and developing contacts.

The **motivator** is someone who is energetic, confident, and outgoing. The motivator is good at finding ways around obstacles. Because the motivator is logical and doesn't like vagueness, he or she is good at making objective decisions.

The **evaluator** is someone who is intelligent and capable of understanding the complete scope of the project. The evaluator is also good at judging outcomes correctly.

The **team worker** is someone who tries to get everyone to come together, because he or she does not like friction or problems among team members.

The **solver** is someone who is reliable and decisive and can turn concepts into practical solutions.

The **finisher** is someone who can be counted on to finish his or her assigned task on time. The finisher is detail orientated and may worry about the team's progress toward finishing the assignment.

There are many other factors that influence team performance, including:

- the way a company is organized;
- how projects are assigned;
- what resources are available to a team to perform their tasks, and
- the corporate culture: whether openness, honesty, and respect are promoted.

These factors are considered external to the team environment, meaning that the team members do not have much control over them. However, there are some factors that are internal to the team environment, factors which the team can control. Communication, the decision making process, and the level of collaboration are examples of factors that are controllable at the team level.

Conflict Resolution

When a group of people work together, conflicts sometimes arise. Conflicts could be the result of miscommunication, personality differences, or the way events and actions are interpreted by a member of a team. Managing conflicts is an important part of a team dynamic. When it comes to managing conflicts, a person's response may be categorized in one of the following ways. There are those in a team environment who try to avoid conflicts. Although this may seem like a good approach, it demonstrates low assertiveness and a low level of cooperation. Under these conditions, the person who is assertive will dominate, making progress as whole difficult. *Accommodating team members* are highly cooperative, but their low assertiveness could result in poor team decisions. This is because the ideas of the most assertive person in the group may not necessarily reflect the best solution. *Compromising team members* demonstrate a moderate level of assertiveness and cooperation. Compromised solutions should be considered as a last resort. Again, by compromising, the team may have sacrificed the best solution for the sake of group unity. A better approach is the *collaborative* "conflict resolution" approach, which demonstrates a high level of assertiveness and cooperation by the team. With this approach, instead of pointing a finger at someone and blaming an individual for the problem, the conflict is treated as a problem to be solved by the team. The team proposes solutions, means of evaluation, and perhaps combines solutions to reach an ideal solution.

However, in order to reach a resolution to a problem, a plan with clear steps must be laid out. Good communication is an integral part of any conflict resolution. One of the most important rules in communication is to make sure that the message sent is the message received—without misunderstanding. Team members must listen to each other. Good listeners do not interrupt; they allow the speaker to feel at ease and do not get angry or criticize. You may want to ask relevant questions to let the speaker know that you really are listening.

Now you have some idea about teamwork and what makes successful teams; next we will discuss project scheduling.

LO⁴ 3.4   Project Scheduling and the Task Chart

Project scheduling is a process that engineering managers use to ensure that a project is completed on time and within the allocated budget. A good schedule will assign an adequate amount of time for various project activities. It will also make use of personnel and the available resources for planning, organizing, and controlling the completion of the project. A well-planned schedule could also improve the efficiency of the operation and eliminate redundancy in task assignments. An example of a simple project schedule and task assignment is shown in Table 3.4. This schedule was used for a small design project for an introduction to engineering class.

> Project scheduling is a process that engineering managers use to ensure that a project is completed on time and within the allocated budget.

TABLE 3.4 Example of a Task Chart

Task	Personnel	Week												
		1	2	3	4	5	6	7	8	9	10	11	12	13
		9/9	9/16	9/23	9/30	10/7	10/14	10/21	10/28	11/4	11/11	11/18	11/25	12/2
Research and preparation	Jim and Julie													
Progress reports	Lisa													
Concept development	Jim, Julie, and Lisa													
Synthesis and evaluation	Jim, Julie, and Lisa													
Fabrication	Mr. Machinist													
Testing	Julie and Lisa													
Optimization	Julie													
Preparing the written and oral reports	Lisa													
Final presentation	Jim, Julie, and Lisa													

Before You Go On

Answer the following questions to test your understanding of the preceding sections.

1. Explain what is meant by a design team and describe the common traits of good teams.

2. Describe how good teams manage conflicts.

3. What is the process that engineering managers use to ensure that a project is completed on time and within budget?

Vocabulary—As a well-educated engineer and intelligent citizen, it is important for you to understand that you need to develop a comprehensive vocabulary to communicate effectively. State the meaning of the following terms:

A Design Team _____

Conflict Resolution _____

A Task Chart _____

LO⁵ 3.5   Engineering Standards and Codes

Today's existing standards and codes ensure that we have safe structures, safe transportation systems, safe drinking water, safe indoor/outdoor air quality, safe products, and reliable services. Standards also encourage uniformity in the size of parts and components that are made by various manufacturers around the world.

Why Do We Need Standards and Codes?

Standards and codes have been developed over the years by various organizations to ensure product safety and reliability in services. The standardization organizations set the authoritative standards for safe food supplies, safe structures, safe water systems, safe and reliable electrical systems, safe and reliable transportation systems, safe and reliable communication systems, and so on. In addition, standards and codes ensure uniformity in the size of parts and components that are made by various manufacturers around the world (Figure 3.8). In today's globally driven economy where parts for a product are made in one place and assembled somewhere else, there exists an even greater need than ever before for uniformity and consistency in parts and components and in the way they are made. These standards ensure that parts manufactured in one place can easily be combined with parts made in other places on an assembly line. An automobile is a good example of this concept. It has literally thousands of parts that are manufactured by various companies in different parts of the United States and the world, and all of these parts must fit together properly.

Standards and codes have been developed over the years by various organizations to ensure product safety and reliability in services.

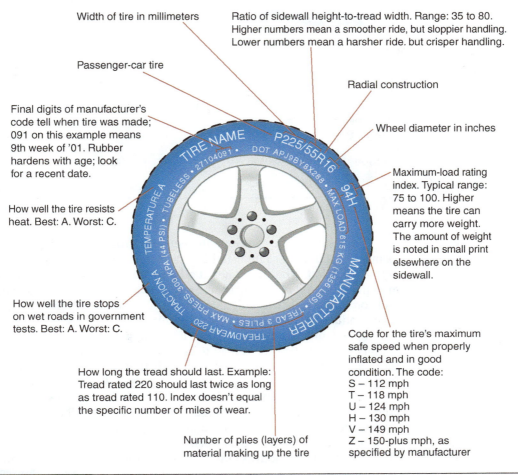

Width of tire in millimeters

Ratio of sidewall height-to-tread width. Range: 35 to 80. Higher numbers mean a smoother ride, but sloppier handling. Lower numbers mean a harsher ride. but crisper handling.

Passenger-car tire

Final digits of manufacturer's code tell when tire was made; 091 on this example means 9th week of '01. Rubber hardens with age; look for a recent date.

Radial construction

Wheel diameter in inches

Maximum-load rating index. Typical range: 75 to 100. Higher means the tire can carry more weight. The amount of weight is noted in small print elsewhere on the sidewall.

How well the tire resists heat. Best: A. Worst: C.

How well the tire stops on wet roads in government tests. Best: A. Worst: C.

Code for the tire's maximum safe speed when properly inflated and in good condition. The code:
S – 112 mph
T – 118 mph
U – 124 mph
H – 130 mph
V – 149 mph
Z – 150-plus mph, as specified by manufacturer

How long the tread should last. Example: Tread rated 220 should last twice as long as tread rated 110. Index doesn't equal the specific number of miles of wear.

Number of plies (layers) of material making up the tire

FIGURE 3.8 Standards and codes have been developed over the years to ensure that we have safe structures, safe transportation systems, safe electrical systems, safe drinking water, and safe indoor/outdoor air quality. A tire is a good example of an engineered product that adheres to such standards. There are many national and international standardization organizations that set these authoritative standards. Standards and codes also ensure uniformity in the size of parts and components made by various manufacturers around the world.
Based on Rubber Manufacturer's Association

To shed more light on why we need standards and codes, let us consider products that we all are familiar with, for example, shoes or shirts. In the United States, you are familiar with shoe sizes of 9, 10, or 11 and so on, as shown in Table 3.5. In Europe, the standard shoe sizes are 43, 44, or 45 and so on. Similarly, the standard shirt sizes in the United States are 15, $15\frac{1}{2}$, or 16 and so on, whereas in Europe the standard shirt sizes are 38, 39, or 41 and so on. If a shirt manufacturer in Europe wants to sell shirts in the United States, it has to label them such that people understand the sizes so that they can choose a shirt of the correct size. Conversely, if a shoe manufacturer from the United States wants to sell shoes in Europe, it has to label them such that the shoe sizes are understood by European customers. Would it not be easier if every shirt or shoe manufacturer in the world used uniform

TABLE 3.5	Standard Shoe and Shirt Sizes in the United States and Europe						
Men's Shirts							
Europe	36	37	38	39	41	42	43
U.S.	14	$14\frac{1}{2}$	15	$15\frac{1}{2}$	16	$16\frac{1}{2}$	17
Men's Shoes							
Europe	38	39	41	42	43	44	45	46
U.S.	5	6	7	8	9	10	11	12

size identifications to eliminate the need for cross referencing? These simple examples demonstrate the need for uniformity in the size and the way products are labeled. Now, think about all possible parts and components that are manufactured every day by thousands of companies around the world: parts and components such as bolts, screws, nuts, cables, tubes, pipes, beams, gears, paints, adhesives, springs, wires, tools, lumber, fasteners, and so on. You see that if every manufacturer built products using its own standards and specifications, this practice could lead to chaos and many misfit parts! Fortunately, there are existing international standards that are followed by many manufacturers around the world.

A good example of a product that uses international standards is your credit card or your bankcard (Figure 3.9). It works in all the ATM machines or store credit card readers in the world. The size of the card and the format of information on the card conform to the standard set by the International Organization of Standards (ISO), thus allowing the card to be read by ATM machines everywhere. The 35-mm camera film speed (e.g., 100, 200, 400) is another example of ISO standards being used by film manufacturers. As another example, warning and functional symbols based on ISO standards on the instrument panel of your car have become commonplace. The ISO standards are being implemented by more and more companies around the world every day.

FIGURE 3.9   Example of product conforming to an ISO standard.
ded pixto/Shutterstock.com

There are many standardization organizations in the world, among them various engineering organizations. Recall from our discussion in Chapter 2 that most national/international engineering organizations create, maintain, and distribute codes and standards that deal with uniformity in size of parts and correct engineering design practices so that public safety is ensured. In fact, the American Society of Mechanical Engineers (ASME) discussed at its first meeting in 1880 the need for standardized sizes for screws. Here we will focus on some of the larger standardization organizations in the United States, Canada, Europe, and Asia. We will briefly describe the role of these organizations and how they may interact. Among the more well-known and internationally recognized organizations are the American National Standards Institute (ANSI), the American Society for Testing and Materials (ASTM), the Canadian Standards Association (CSA), the British Standards Institute (BSI), the German *Deutsches Institut für Normung* (DIN), the French *Association Française de normalisation* (AFNOR), the Swedish *Standardiserigen I Sverige* (SIS), the China State Bureau of Quality and Technical Supervision (CSBTS), the International Organization for Standardization (ISO), and the European Union C€ marking. We will briefly describe these organizations in the following sections.

Examples of Standards and Codes Organizations in the United States

The American National Standards Institute The American National Standards Institute (**ANSI**) was founded in 1918 by five engineering societies and three government agencies to administer and coordinate standards in the United States. The ANSI is a not-for-profit organization, which is supported by various public and private organizations. The institute itself does not develop the standards, but instead it assists qualified groups, such as various engineering organizations, with the development of the standards and sets the procedures to be followed. Today, the American National Standards Institute represents the interests of well over a thousand companies and other members. According to the American National Standards Institute, there are over 13,000 approved ANSI standards in use today, and more standards are being developed.

> Among the more well-known and internationally recognized organizations are the American National Standards Institute (ANSI), the American Society for Testing and Materials (ASTM), the International Organization for Standardization (ISO), and the European Union C€ standards.

The American Society for Testing and Materials (ASTM) Founded in 1898, the American Society for Testing and Materials (**ASTM**) is another not-for-profit organization. It publishes standards and test procedures that are considered authoritative technical guidelines for product safety, reliability, and uniformity. The testing is performed by its national and international member laboratories. The ASTM collects and publishes the work of over 100 standards-writing committees dealing with material test methods. For example, ASTM sets the standard procedures for tests and practices to determine elastic properties of materials, impact testing, fatigue testing, shear and torsion properties, residual stress, bend and flexure testing, compression, ductility, and linear thermal expansion. The ASTM also sets the standards for medical devices and equipment, including bone cements, screws, bolts, pins, prostheses, and plates, and specifications for alloys used in surgical implants.

Electrical insulation and electronics-related standards are also set by ASTM. Additional examples of ASTM work include the following:

- Test guidelines for evaluating mechanical properties of silicon or other procedures for testing semiconductors, such as germanium dioxide.
- Testing methods for trace metallic impurities in electronic-grade aluminum-copper and aluminum-silicon.
- Standards related to the chemical analysis of paints or detection of lead in paint, along with tests to measure the physical properties of applied paint films, such as film thickness, physical strength, and resistance to environmental or chemical surroundings.
- Standard procedures for evaluating properties of motor, diesel, and aviation fuels, crude petroleum, hydraulic fluids, and electric insulating oils.
- Test procedures for measurements of insulation properties of materials.
- Standard procedures for soil testing, such as density characteristics, soil texture, and moisture content.
- Building-construction-related tests and procedures, such as measuring the structural performance of sheet metal roofs.
- Tests for evaluating the properties of textile fibers, including cotton and wool.
- Standards for steel piping, tubing, and fittings.

You can find specific standards dealing with any of the materials in these examples in the *Annual Book* of ASTM, which includes large volumes of standards and specifications in the following areas:

Iron and steel products
Nonferrous metal products
Metals test methods and analytical procedures
Construction
Petroleum products, lubricants, and fossil fuels
Paints, related coatings, and aromatics
Textiles
Plastics
Rubber
Electric insulation and electronics
Water and environmental technology
Nuclear, solar, and geothermal energy
Medical devices and services
General methods and instrumentation
General products, chemical specialities, and end-use products

ASTM Standards are now available on CD-ROM and online. ASTM also publishes a number of journals:

Cement, Concrete & Aggregates
Geotechnical Testing Journal
Journal of Composites Technology and Research
Journal of Forensic Sciences
Journal of Testing and Evaluation

National Fire Protection Association (NFPA) Losses from fires total billions of dollars per year. Fire, formally defined as a process during which rapid oxidization of a material occurs, gives off radiant energy that can not

FIGURE 3.10

Courtesy of Underwriters
Laboratories Inc.

only be felt but also seen. Fires can be caused by malfunctioning electrical systems, hot surfaces, and overheated materials. The National Fire Protection Association (NFPA) is a not-for-profit organization that was established in 1896 to provide codes and standards to reduce the burden of fire. The NFPA publishes the *National Electrical Code*®, the *Life Safety Code*®, the *Fire Prevention Code*™, the *National Fuel Gas Code*®, and the *National Fire Alarm Code*®. It also provides training and education.

Underwriters Laboratories (UL) The Underwriters Laboratories Inc. (UL) is a nonprofit organization that performs product safety tests and certifications. Founded in 1894, today Underwriters Laboratories has laboratories in the United States, England, Denmark, Hong Kong, India, Italy, Japan, Singapore, and Taiwan. Its certification mark, shown in Figure 3.10, is one of the most recognizable marks on products.

Examples of International Standards and Codes

The International Organization for Standardization (ISO) As the name implies, the International Organization for Standardization, established in 1947, consists of a federation of national standards from various countries. The International Organization for Standardization promotes and develops standards that can be used by all countries in the world, with the objective of facilitating standards that allow for free, safe exchange of goods, products, and services among countries. It is recognized by its abbreviation, or short form, ISO, which is derived from *isos*, a Greek word meaning "equal." As you take more engineering classes, you will see the prefix *iso* in many engineering terms; for example, *iso*bar, meaning equal pressure, or *iso*therm, meaning equal temperature. ISO was adopted instead of International Organization of Standards (IOS) so that there would not be any nonuniformity in the way the abbreviation is presented in other languages.

CE Standards All products sold in Europe must now comply with CE standards. Before the formation of the European Union and the utilization of CE standards, manufacturers in Europe and those exporting to Europe had to comply with different standards based on the requirements that were dictated by a specific country. CE provides a single set of safety and environmental standards that is used throughout Europe. The CE marking on a product ensures conformity to European standards. The letters CE come from the French words Conformité Européenne.

Other Internationally Recognized Standardization Organizations The British Standards Institute (BSI) is another internationally known organization that deals with standardization. In fact, BSI, founded in 1901, is one of the oldest standardization bodies in the world. It is a nonprofit organization that organizes and distributes British, European, and International standards. Other internationally recognized standardization organizations include the German *Deutsches Institut für Normung* (DIN), the French *Association Française de normalisation* (AFNOR), the Swedish *Standardiserigen I Sverige* (SIS), and the China State Bureau of Quality and Technical Supervision (CSBTS). Visit the Web sites of these organizations to obtain more information about them.

LO⁶ 3.6   Water and Air Standards in the United States

Drinking Water

The U.S. Environmental Protection Agency (EPA) sets the standards for the maximum level of contaminants that can be in our drinking water and still be considered safe to drink (Figure 3.11). Basically, the EPA sets two standards for the level of water contaminants: (1) the maximum contaminant level goal (MCLG) and (2) the maximum contaminant level (MCL). The MCLG represents the maximum level of a given contaminant in the water that causes no known harmful health effects. On the other hand, the MCL, which may represent slightly higher levels of contaminants in the water, are the levels of contaminants that are legally enforceable. The EPA attempts to set MCL close to MCLG, but this goal may not be attainable because of economic or technical reasons. Examples of drinking water standards are shown in Table 3.6.

See Problem 3.27 for other standards set by the EPA for the Surface Water Treatment Rule (SWTR).

Outdoor Air

The source of outdoor air pollution (Figure 3.12) may be classified into three broad categories: the *stationary sources, the mobile sources, and the natural sources.* Examples of stationary sources include power plants, factories, and dry cleaners. The mobile sources of air pollution consist of cars, buses, trucks, planes, and trains. As the name implies, the sources of natural air pollution could include windblown dust, volcanic eruptions, and forest fires. The Clean Air Act, which sets the standard for six major air pollutants, was signed into

| FIGURE 3.11 | The EPA sets the standard for MCL in our drinking water. |

windu/Shutterstock.com

TABLE 3.6			Examples of Drinking Water Standards
Contaminant	**MCLG**	**MCL**	**Source of Contaminant by Industries**
Antimony	6 ppb	6 ppb	copper smelting, refining, porcelain plumbing fixtures, petroleum refining, plastics, resins, storage batteries
Asbestos	7 M.L. (million fibers per liter)	7 M.L.	asbestos products, chlorine, asphalt felts and coating, auto parts, petroleum refining, plastic pipes
Barium	2 ppm	2 ppm	copper smelting, car parts, inorganic pigments, gray ductile iron, steel works, furnaces, paper mills
Beryllium	4 ppb	4 ppb	copper rolling and drawing, nonferrous metal smelting, aluminum foundries, blast furnaces, petroleum refining
Cadmium	5 ppb	5 ppb	zinc and lead smelting, copper smelting, inorganic pigments
Chromium	0.1 ppm	0.1 ppm	pulp mills, inorganic pigments, copper smelting, steel works
Copper	1.3 ppm	1.3 ppm	primary copper smelting, plastic materials, poultry slaughtering, prepared feeds
Cyanide	0.2 ppm	0.2 ppm	metal heat treating, plating, and polishing
Lead	zero	15 ppb	lead smelting, steel works and blast furnaces, storage batteries, china plumbing fixtures
Mercury	2 ppb	2 ppb	electric lamps, paper mills
Nickel	0.1 ppm	0.1 ppm	petroleum refining, gray iron foundries, primary copper, blast furnaces, steel
Nitrate	10 ppm	10 ppm	nitrogenous fertilizer, fertilizing mixing, paper mills, canned foods, phosphate fertilizers
Nitrite	1 ppm	1 ppm	Nitrates in fertilizers, once taken into the body, are converted to nitrites
Selenium	0.05 ppm	0.05 ppm	metal coatings, petroleum refining
Thallium	0.5 ppb	2 ppb	primary copper smelting, petroleum refining, steel works, blast furnaces

law in 1970. The Environmental Protection Agency is responsible for setting standards for these six major air pollutants: carbon monoxide (CO), lead (Pb), nitrogen dioxide $\left(NO_2\right)$, ozone $\left(O_3\right)$, sulfur dioxide $\left(SO_2\right)$, and particulate matter (PM). The EPA measures the concentration levels of these pollutants in many urban areas and collects air quality information by actual measurement of pollutants from thousands of monitoring sites located throughout the country. According to a study performed by the EPA (1997), between 1970 and 1997, the U.S. population increased by 31% and the vehicle miles traveled increased by 127%. During this period, the total emission of air pollutants

(a)

ssuaphotos/Shutterstock.com

(b)

Hung Chung Chih/Shutterstock.com

(c)

M. Shcherbyna/Shutterstock.com

FIGURE 3.12   Outdoor pollution!

from stationary and mobile sources decreased by 31% because of improvements made in the efficiency of cars and in industrial practices, along with the enforcement of the Clean Air Act regulations.

However, there are still approximately 107 million people who live in areas with unhealthy air quality. The EPA is continuously working to set standards and monitor the emission of pollutants that cause acid rain and damage to bodies of water and fish (currently there are over 2000 bodies of water in the United States that are under fish consumption advisories), damage to the stratospheric ozone layer, and damage to our buildings and our national parks. The unhealthy air has more pronounced adverse health effects on children and elderly people. The human health problems associated with poor air quality include various respiratory illnesses and heart or lung diseases. Congress passed amendments to the Clean Air Act in 1990, which required the EPA to address the effect of many toxic air pollutants by setting new standards. Since 1997, the EPA has issued 27 air standards that are to be fully implemented in the coming years. The EPA is currently working with the individual states in this country to reduce the amount of sulfur in fuels and setting more stringent emission standards for cars, buses, trucks, and power plants.

The U.S Environmental Protection Agency (EPA) sets the standards for the maximum level of contaminants that can be in our drinking water and still be considered safe to drink. It is also responsible for setting standards for air pollutants such as carbon monoxide (CO), lead (Pb), nitrogen dioxide (NO_2), ozone (O_3), sulfur dioxide (SO_2), and particulate matter (PM).

We all need to understand that air pollution is a global concern that can affect not only our health, but also affect our climate. It may have triggered the onset of global warming that could lead to unpleasant natural events. Because we all contribute to this problem, we need to be aware of the consequences of our lifestyles and find ways to reduce pollution. We could carpool or take public transportation when going to work or school. We should not leave our cars running idle for long periods of time, and we can remind others to consume less energy. We should conserve energy around home and school. For example, turn off the light in a room that is not in use. When at home, in winter, set the thermostat at 65° F or slightly lower and wear a sweater to feel warm. During summer, at home set the air-conditioning thermostat at 78° F or slightly higher. By consuming less energy and driving less, we can help our environment and reduce air pollution.

Indoor Air

In the previous section, we discussed outdoor air pollution and the related health effects. Indoor air pollution can also create health risks. According to EPA studies of human exposure to air pollutants, the indoor levels of pollutants may be two to five times higher than outdoor levels. Indoor air quality is important in homes, schools, and workplaces. Because most of us spend approximately 90% of our time indoors, the indoor air quality is very important to our short-term and long-term health. Moreover, lack of good indoor air quality can reduce productivity at the workplace or create an unfavorable learning environment at school, causing sickness or discomfort to building occupants. Failure to monitor indoor air quality (IAQ) or failure to prevent indoor air pollution can also have adverse effects on equipment and the physical appearance of buildings. In recent years, liability issues related to people who suffer dizziness or headaches or other illness related to "sick buildings" are becoming a concern for building managers. According to the EPA, some common health symptoms caused by poor indoor air quality are

- headache, fatigue, and shortness of breath;
- sinus congestion, coughing, and sneezing;
- eye, nose, throat, and skin irritation; and
- dizziness and nausea.

As you know, some of these symptoms may also be caused by other factors and are not necessarily caused by poor air quality. Stress at school, work, or at home could also create health problems with symptoms similar to the ones mentioned. Moreover, individuals react differently to similar problems in their surroundings.

The factors that influence air quality can be classified into several categories: the heating, ventilation, and air-conditioning (HVAC) system; sources of indoor air pollutants; and occupants. In recent years, we have been exposed to more indoor air pollutants for the following reasons. (1) In order to save energy, we are building tight houses that have lower air infiltration or exfiltration compared to older structures. In addition, the ventilation rates have also been reduced to save additional energy. (2) We are using more synthetic building materials in newly built homes that could give off harmful vapors. (3) We are using more chemical pollutants, such as pesticides and household cleaners.

As shown in Table 3.7, indoor pollutants can be created by sources within the building or they can be brought in from the outdoors. It is important to keep in mind that the level of contaminants within a building can vary with time.

TABLE 3.7	Typical Sources of Indoor Air Pollutants		
Outside Sources	**Building Equipment**	**Component/ Furnishings**	**Other Indoor Sources**
Polluted outdoor air Pollen, dust, fungal spores, industrial emissions, and vehicle emissions	**HVAC equipment** Microbiological growth in drip pans, ductwork, coils, and humidifiers; improper venting of combustion products; and dust or debris in ductwork	**Components** Microbiological growth on soiled or water-damaged materials, dry traps that allow the passage of sewer gas, materials containing volatile organic compounds, inorganic compounds, damaged asbestos, and materials that produce particles (dust)	Science laboratories; copy and print areas; food preparation areas; smoking lounges; cleaning materials, emission from trash; pesticides; odors and volatile organic compounds from paint, chalk, and adhesives; occupants with communicable diseases; dry-erase markers and similar pens; insects and other pests; personal hygiene products
Nearby sources Loading docks, odors from dumpsters, unsanitary debris or building exhausts near outdoor air intakes	**Non-HVAC equipment** Emissions from office equipment, and emissions from shops, labs and cleaning processes		
Underground sources Radon, pesticides, and leakage from underground storage tanks		**Furnishings** Emissions from new furnishings and floorings and microbiological growth on or in soiled or water-damaged furnishings	

EPA Fact Sheets, EPA-402-F-96-004, October 1996

For example, in order to protect floor surfaces from wear and tear, it is customary to wax them. During the period when waxing is taking place, it is possible, based on the type of chemical used, for anyone near the area to be exposed to harmful vapors. Of course, one simple remedy to this indoor air problem is to wax the floor late on Friday afternoons to avoid exposing too many occupants to harmful vapors. Moreover, this approach will provide some time for the vapor to be exhausted out of the building by the ventilation system over the weekend when the building is not occupied.

The primary purpose of a well-designed heating, ventilation, and air-conditioning (HVAC) system is to provide thermal comfort to its occupants. Based on the building's heating or cooling load, the air that is circulated through the building is conditioned by heating, cooling, humidifying, or dehumidifying. The other important role of a well-designed HVAC system is to filter out the contaminants or to provide adequate ventilation to dilute air-contaminant levels.

The air flow patterns in and around a building will also affect the indoor air quality. The air flow pattern inside the building is normally created by the HVAC system. However, the outside air flow around a building envelope that is dictated by wind patterns could also affect the air flow pattern within the building as well. When looking at air flow patterns, the important concept to keep in mind is that air will always move from a high-pressure region to a low-pressure region.

Methods to Manage Contaminants

There are several ways to control the level of contaminants: (1) source elimination or removal, (2) source substitution, (3) proper ventilation, (4) exposure control, and (5) air cleaning.

A good example of source elimination is not allowing people to smoke inside the building or not allowing a car engine to run idle near a building's outdoor air intake. In other words, eliminate the source before it spreads out! It is important for engineers to keep that idea in mind when designing the HVAC systems for a building—avoid placing the outdoor air intakes near loading docks or dumpsters. A good example of source substitution is to use a gentle cleaning product rather than a product that gives off harmful vapors when cleaning bathrooms and kitchens. Local exhaust control means removing the sources of pollutants before they can be spread through the air distribution system into other areas of a building. Everyday examples include use of an exhaust fan in restrooms to force out harmful contaminants. Fume hoods are another example of local exhaust removal in many laboratories. Clean outdoor air can also be mixed with the inside air to dilute the contaminated air. The American Society of Heating, Refrigerating and Air Conditioning Engineers (ASHRAE) has established a set of codes and standards for how much fresh outside air must be introduced for various applications. Air cleaning means removing harmful particulate and gases from the air as it passes through some cleaning system. There are various methods that deal with air contaminant removal, including absorption, catalysis, and use of air filters.

Finally, you can bring the indoor air quality issues to the attention of friends, classmates, and family. We all need to be aware and try to do our part to create and maintain healthy indoor air quality.

Before You Go On

Answer the following questions to test your understanding of the preceding sections:

1.   Explain why we need standards and codes.

2.   Give examples of organizations that set standards and codes in the United States.

3.   Give examples of international organizations that set standards.

4.   Give examples of sources of water pollutions.

5.   What are the major outdoor air pollutants?

Vocabulary—As a well-educated engineer and intelligent citizen, it is important for you to understand that you need to develop a comprehensive vocabulary to communicate effectively. State the meaning of the following terms:

ASTM _____

NFPA _____

ISO _____

Maximum Contaminant Level _____

EPA _____

Lauren Heine, Ph.D

Courtesy of Lauren Heine

I became an Environmental Engineer because I was interested in waste management and in preserving the environment. I earned a Ph.D. in Civil and Environmental Engineering at Duke University. My advisor encouraged me to explore some non-traditional coursework, including ethics and environmental chemistry and toxicology. I quickly became hooked on solving interdisciplinary problems—a natural extension of environmental engineering.

After graduating, I became a Fellow with the American Association for the Advancement of Science at the U.S. Environmental Production Agency (USEPA). I chose to work at the Green Chemistry Program in the Office of Pollution Prevention and Toxics. On my first day in the office, one of the Branch Chiefs at USEPA asked me how it felt to know that my degree was obsolete. I must have looked pretty puzzled. He continued, explaining that environmental engineering is about cleaning up waste at the end of the pipe, **but the future is about designing problems out from the start**. While of course my mind was trying to figure out how he intended to design out human biological waste, his words made a lot sense to me.

While in the Green Chemistry Program at USEPA, I had the opportunity to work with one of the visionary co-authors of the 12 Principles of Green Chemistry. I also became intrigued with the work of the founders of Cradle-to-Cradle design. I felt inspired and challenged to bring these big ideas into practice in the world. My role was to find ways to translate the big vision and principles of Green Chemistry and Cradle-to-Cradle design into real products and processes. Both approaches advocate using design to avoid problems of waste and toxics in the first place and to create synergistic benefits by designing sustainable material and product systems.

I started by helping to create a new non-profit organization in Portland, Oregon, that focused on working with organizations to eliminate waste and toxics through engineering solutions. It was great fun to work with a range of individuals from commercial printers to manufacturers of silicon wafers, lighting fixtures, storm water-treatment technology, and anaerobic digesters on dairy farms. I was at the forefront of the sustainability wave. I am amazed now as I look back and remember how (at first) people were very resistant to the idea that industry could strive to be simultaneously profitable and benign to the environment. People assumed it would be too expensive but instead, many found that it could save money by

1. avoiding costs to manage waste and toxics, and
2. driving innovation for the development of new products.

Since then, I have worked with other environmental organizations, including Green Blue Institute and Clean Production Action. Finding like-minded colleagues who enjoy creating change is fun and challenging. I led the creation of CleanGredients, a database of green chemicals for use in designing environmentally safer cleaning products. With Clean Production Action, I co-authored the Green Screen for Safer Chemicals, a method to help organizations identify safer chemical alternatives.

Along the way, I learned conflict mediation. I did not foresee how valuable that training would become. It taught me to appreciate and engage stakeholders from very different sectors and to focus on solutions. Using conflict mediation and facilitation skills along with environmental engineering and chemistry, I found wonderful opportunities to create solutions with other scientists and engineers from government, environmental organizations, and industry leading to products and processes that are more benign for human health and the environment.

While in the past, environmental organizations like Clean Production Action worked in opposition to industry, now there are opportunities to partner with proactive companies ranging from manufacturers of cleaning products to major electronics firms and retailers to create the positive change we want to see in the world. Examples include Apple, IBM, Hewlett Packard, Wal-Mart, Staples, Method Home, Seventh Generation, and more.

Engineers are designers. While environmental engineers may not typically work as product designers, no matter where we work in the value chain, we need to design solutions not only with consideration of cost and performance but with the entire environmental, economic, and social system in mind.

Courtesy of Lauren Heine

SUMMARY

LO¹ Engineering Design Process

By now you should know the basic design steps that all engineers follow to design products and services. These steps are (1) recognizing the need for a product or a service, (2) defining and understanding the problem (the need) completely, (3) doing the preliminary research and preparation, (4) conceptualizing ideas for possible solutions, (5) synthesizing the results, (6) evaluating good ideas in more detail, (7) optimizing the solution to arrive at the best possible solution, and (8) presenting the solution.

LO² Additional Design Considerations

You should also realize that economics and material selection play important roles in engineering decision making. Also, engineers should design and develop products and services in a way that meets the needs of the present without compromising the ability of future generations to meet their own needs (sustainability).

LO³ Teamwork

A design team may be defined as a group of individuals with complementary expertise, problem-solving skills, and talent who are working together to solve a problem or achieve a common goal. A good team is one that gets the best out of each other. Moreover, the goals of a project must be understood and accepted by all members of the team, and they should put the team needs and goals before individual needs and goals. A team must have a good leader as well. Moreover, when a group of people work together, conflicts are bound to happen. Managing conflicts is an essential part of a team dynamic.

LO⁴ Project Scheduling and Task Chart

Project scheduling is a process that engineering managers use to ensure that a project is completed on time and within the allocated budget. A well-planned schedule will assign adequate time for various activities and make proper use of personnel to eliminate redundancy in task assignments.

LO⁵ Engineering Standards and Codes

Standards and codes have been developed to ensure product safety and reliability in services. There are many organizations that set the standards for safe food supplies, safe structures, safe water systems, safe and reliable electrical systems, safe and reliable transportation systems, safe and reliable communication systems, and so on. Among the more well-known and internationally recognized organizations are the American National Standards Institute (ANSI), the American Society for Testing and Materials (ASTM), the National Fire Protection Association (NFPA), the Underwriters Laboratories Inc. (UL), the International Organization for Standardization (ISO), and the European Union C€ Standards.

LO⁶ Water and Air Standards in the United States

The U.S. Environmental Protection Agency (EPA) sets the standards for the maximum contaminant level (MCL) such as lead, copper, cyanide, cadmium, chromium that can be in our drinking water and still be considered safe to drink. The EPA is also responsible for setting standards for major air pollutants such as carbon monoxide (CO), lead (Pb), nitrogen dioxide $\left(NO_2\right)$, ozone $\left(O_3\right)$, sulfur dioxide $\left(SO_2\right)$, and particulate matter (PM).

KEY TERMS

ANSI 81
ASTM 81
Conflict Resolution 76
Design Steps 53
Design Team 74

Engineering Design 52
Engineering Economics 70
ISO 83
Material Selection 70
NFPA 83

Project Scheduling 76
Standards and Codes 78
Sustainability 67
Task Chart 77
UL 83

APPLY WHAT YOU HAVE LEARNED

Engineering organizations such ASME, ASCE, and IEEE have annual student design competitions that encourage a team of students to work together and present their solutions to a particular design problem. Typically, you and your team members are required to read a problem statement and then design, build, and operate a prototype that meets the requirements that are stated in the problem statement. Moreover, you will have the opportunity to compete with teams from other universities.

Contact the student section of the ASME, ASCE, or IEEE on your campus and get involved with this year's annual design competition. At this stage of your education, perhaps, it is best that you join a team that has juniors and seniors. That way you can also gain from their knowledge and experience as well.

© ZUMA Press, Inc. / Alamy

PROBLEMS

Problems that promote life-long learning are denoted by 🔑

3.1 List at least ten products that already exist, which you use, that are constantly being modified to take advantage of new technologies

3.2 List five products that are not currently on the market, which could be useful to us and will most likely be designed by engineers and others.

3.3 List five sports-related products that you think should be designed to make playing sports more fun.

3.4 List five Internet-based services that are not currently available, but that you think will eventually become realities.

3.5 You have seen bottle and container caps in use all around you. Investigate the design of bottle caps used in the following products: Pepsi or Coke bottles, aspirin bottles, shampoo bottles, mouthwash bottles, liquid cleaning containers, hand lotion containers, aftershave bottles, ketchup or mustard bottles. Discuss what you think are important design parameters. Discuss the advantages and disadvantages associated with each design.

3.6 Mechanical clips are used to close bags and keep things together. Investigate the design of various paper clips, hair clips, and potato chip bag clips. Discuss what you think are important design parameters. Discuss the advantages and disadvantages associated with each of three designs.

3.7 Discuss in detail at least two concepts (for example, activities, processes, or methods) that can be employed during busy hours to better serve customers at grocery stores.

3.8 Discuss in detail at least two methods or procedures that can be employed by airline companies to pick up your luggage at your home and drop it off at your final destination.

3.9 In the near future, NASA is planning to send a spaceship with humans to Mars. Discuss important concerns and issues that must be planned for on this trip. Investigate and discuss issues such as how long it would take to go to Mars and when the spaceship should be launched, considering that the distance between the earth and Mars changes based on where they are in their respective orbits around the sun. What type and how much food reserves are needed for this trip? What type of exercise equipment should be on board so muscles won't atrophy on this long trip? What should be done with the waste? What is the energy requirement for such a trip? What do you think are the important design parameters for such a trip? Write a report discussing your findings.

3.10 You have been using pens and mechanical pencils for many years now. Investigate the designs of at least five different pens and mechanical pencils. Discuss what you think are important design parameters. Discuss the advantage and disadvantage associated with each design. Write a brief report explaining your findings.

3.11 This is an in-class design project. Given a 12in.×12in. aluminum sheet, design a boat that can hold as many pennies as possible. What are some of the important design parameters for this problem? Discuss them among yourselves. You may want to choose a day in advance on which to hold a competition to determine the good designs.

3.12 This is an in-class design project. Given a bundle of drinking straws and paper clips, design a bridge between two chairs that are 18 in. apart. The bridge should be designed to hold at least 3 lb. You may want to choose a day in advance on which to hold a competition to determine the design that carries maximum load. Discuss some of the important design parameters for this problem.

3.13 Identify and make a list of at least ten products around your home that are certified by the Underwriters Laboratories.

3.14 Create a table showing hat sizes in the United States and Europe.

Continental	51	52	53	.	.	.
Europe	$6\frac{1}{4}$	$6\frac{3}{8}$	$6\frac{1}{2}$	.	.	.

3.15 Create a table showing wrench sizes in metric and U.S. units.

3.16 Obtain information about what the colors on an electrical resistor mean. Create a table showing the electrical resistor codes. Your table should have a column with the colors Black, Brown, Red, Orange, Yellow, Green, Blue, Violet, Gray, White, Gold, and Silver, and a column showing the values. Imagine that you are making this table for others to use; therefore, include at least two examples of how to read the codes on electrical resistors at the bottom of your table.

3.17 Collect information on U.S. standard steel pipes ($\frac{1}{4}$ in. to 20 in.). Create a table showing nominal size, schedule number, inside diameter, outside diameter, wall thickness, and cross-sectional area.

3.18 Collect information on the American Wire Gage (AWG) standards. Create a table for annealed copper wires showing the gage number, diameter in mils, cross-sectional area, and resistance per 1000 ft.

3.19 Write a brief memo to your instructor explaining the role and function of the U.S. Department of Transportation (DOT).

3.20 Obtain information on the standard sign typefaces used for highway signs in the United States. The Federal Highway Administration publishes a set of standards called *Standard Alphabets for Highway Signs*. Write a brief memo to your instructor explaining your findings.

3.21   Obtain information about the Nuclear Regulatory Commission (NRC), which sets the standards for handling and other activities dealing with radioactive materials. Write a brief report to your instructor regarding your findings.

3.22   Obtain information about the classification of fire extinguishers. Write a brief report explaining what is meant by Class A, Class B, Class C, and Class D fires.

3.23   Write a brief report detailing the development of safety belts in cars. When was the first safety belt designed? Which was the first car manufacturer to incorporate safety belts as standard items in its cars?

3.24   Investigate the mission of each of the following standards organizations. For each of the organizations listed, write a one-page memo to your instructor about its mission and role.

   a.   The European Committee for Electrotechnical Standardization (CENELEC)

   b.   European Telecommunication Standards Institute (ETSI)

   c.   Pan American Standards Commission (COPANT)

   d.   Bureau of Indian Standards (BID)

   e.   Hong Kong Standards and Testing Centre Ltd. (HKSTC)

   f.   Korea Academy of Industrial Technology (KAITECH)

   g.   Singapore Academy of Industrial Technology (PSB)

   h.   Standards New Zealand (SNZ)

3.25   Write a brief report explaining what is meant by ISO 9001 and ISO 14001 certification.

3.26   Ask your local city water supplier to give you a list of the chemicals that it tests for in your water. Also ask how your city water is being treated. You may want to contact your state department of health/environment to get additional information. For help in locating state and local agencies, or for information on drinking water in general, you can call the EPA's Safe Drinking Water Hotline: (800)

426-4791 or by visiting the EPA Web site at www.epa.org. You can also obtain information about the uses and releases of chemicals in your state by contacting the Community Right-to-Know Hotline: (800) 535-0202.

3.27   Collect information on the Surface Water Treatment Rule (SWTR) standards set by the EPA. Write a brief memo to your instructor explaining your findings.

3.28   Obtain the EPA's consumer fact sheets (they are now available on the Web) on antimony, barium, beryllium, cadmium, cyanide, and mercury. After reading the fact sheets, prepare a brief report explaining what they are, how they are used, and what health effects are associated with them.

3.29   In 1970 the U.S. Congress passed the Occupational Safety and Health Act. The following is a duplicate of the act, which reads:

Public Law 91-596
91st Congress, S. 219
December 29, 1970
As Amended by Public Law 101-552,
Section 3101, November 5, 1990
As Amended by Public Law 105-198,
July 16, 1998
As Amended by Public Law 105-241
September 29, 1998

An Act

To assure safe and healthful working conditions for working men and women; by authorizing enforcement of the standards developed under the Act; by assisting and encouraging the States in their efforts to assure safe and healthful working conditions; by providing for research, information, education, and training in the field of occupational safety and health; and for other purposes.

Be it enacted by the Senate and House of Representatives of the United States of America in Congress assembled, that this Act may be cited as the "Occupational Safety and Health Administration Compliance Assistance Authorization Act of 1998."

Visit the Department of Labor's Occupational Safety and Health Act (OSHA) home page at www.osha.gov, and write a brief report describing the type of safety and health standards that are covered by OSHA.

For Problems 3.30 through 3.40, you are asked to look up some examples of specific standards and codes that are used in various engineering products in the United States. The publisher and reference are given in parentheses.

3.30 General requirements for helicopter air conditioning (SAE)

3.31 Safety codes for elevators and escalators (ASE, ANSI)

3.32 Chimneys and fireplaces (ANSI/NFPA)

3.33 Refrigerated vending machines (ANSI/UL)

3.34 Drinking water coolers (ANSI/UL)

3.35 Household refrigerators and freezers (ANSI/UL)

3.36 Mobile homes (CAN/CSA)

3.37 Electric motors (ANSI/UL)

3.38 Centrifugal pumps (ASME)

3.39 Electric air heaters (ANSI/UL)

3.40 Recreational vehicles (NFPA)

Impromptu Design II

Objective: To build the tallest tower from an $8\frac{1}{2}$" × 11" sheet of paper and 20 in. of tape that will stand for at least one minute. Thirty minutes will be allowed for preparation.

Brainstorming Session

Define the purpose of the brainstorming session and the ground rules.
Examples of ground rules: no criticism of ideas as they are being presented, one person talks at one time and for an agreed-upon period.
Choose a facilitator to keep track of ground rules.
Record all ideas where everyone can see them.
Don't worry about looking foolish; record all your ideas.
It is a good practice not to associate a person with an idea. Think of each idea as a group idea.

After Brainstorming

Identify the promising ideas. Don't evaluate the ideas in detail yet!
Discuss ways to improve a promising idea or promising ideas.
Choose and list the ideas for detail evaluation.
Evaluate the most promising ideas!

Saeed Moaveni

Civil Engineering Design Process

A Case Study: Health Clinic

The Board of Directors of a health clinic recognized that in order to enhance its health service to meet the increasing needs of their city and its surrounding communities, they needed to expand the existing facilities adjacent to the hospital. The health service expansion consisted of a physician office building and a clinic. The physician office building (POB) was to attach to the existing hospital with the clinic connecting to the POB. The structures were treated as separate projects with two different

Karen Chou

design teams working on them. The focus of this case study is the clinic.

Step 1: Recognizing the Need for a Building

As stated previously, the Board of Directors of the clinic recognized the need for expansion to meet the increasing demand of health service in their city and its surrounding communities. To better serve the people in these communities, the Board of Directors decided to build a new clinic.

Step 2: Define the Usage of the Building

After the Board of Directors recognized that there was a need for expansion to meet the increasing heath services demand, the Board had to define in detail the types of building usage. Parameters such as number of examination room and reception areas, laboratory facilities such as X-ray and MRI rooms, staff rooms, meeting rooms, and managerial and maintenance facilities were considered. Anticipated number of patients, visitors, and staff was also included during this decision making process. The Board also considered future expansions; future expansion potentials, regardless how far in the future they may occur, could impact the planning and design of the current structure.

Step 3: Project Planning

The owner also needed to identify possible building sites. The selection criteria for the site are usually based on economical, zoning, environmental, and other factors. In the case of the clinic, the proximity of the hospital and the future physician office building were the major factors that led to the building site.

Since the clinic was a privately funded structure, the owner could have selected an architect or contractor to initiate the design phase, or requested bids from architects or contractors to lead the project.

Step 4: Schematic Design Phase

During this phase of the design process, the architect designer met with the staff of the clinic to learn more about how the new clinic was to be used. The architect designer and the contractor also learned about the estimated budget. For the clinic, additional coordination with the architect of the physician office building was warranted since both buildings, shared some columns and foundations.

The clinic was designed as a steel frame structure. The primary supporting components of the building are made of structural steel. The bricks, masonry, wood, etc. are to provide closure and esthetics to the building. When the architect designer prepared the schematic design, usually with multiple alternatives, the designer consulted with the structural engineer for information such as maximum span length of steel beams. This information would help the architect, contractor, and owner to determine a good design and estimated project cost.

Step 5: Design Development (DD) Phase

In the design development (DD) phase, the architect designer laid out the locations, sizes, and orientations of the reception areas, examination rooms, laboratories, business administration offices, maintenance facilities, entrances to the physician office building, and the street. The arrangement of these rooms along with the maximum span length of steel beams were used to determine the locations of supporting columns. These column locations are the basis the architect used to set up the gridlines. Gridlines are a set of lines running in two directions. As a convention, one set of lines is named in alphabetical order.

The second set of lines is named in numerical sequence. When a new gridline is inserted between two existing lines, the new line would either be named C.# or 3.#, depending on the direction of this new gridline. The "#" sign represents a number between 1 and 9 depending on the relative location of the new line with respect to the two existing ones. Gridlines are used by the design and construction teams to reference the location of all components in the project.

The structural engineer provided the size of major support components of the building such as beams, columns, and foundations. Non-load bearing components would be neglected in this phase. However, the contractor would include them in the cost estimate. The mechanical and electrical engineers then provided their preliminary mechanical and electrical designs.

A set of architectural drawings with superimposed structural, mechanical, and electrical information was then provided to the clinic from the contractor. Through multiple review and revisions, the final layout of user space and estimated project cost was approved, the project moved to the construction documentation phase.

Step 6: Construction Documentation (CD) Phase

All the detailed comprehensive designs: architectural, structural, civil, interior, mechanical, electrical, plumbing, etc. were performed during the construction documentation (CD) phase. The project manager who represented the architect during all the construction meetings was responsible for overseeing the completion of the design and document produced in this phase. The project manager also compiled a set of specifications for the project and checked that the design conformed to the current building codes. Each engineering group provided the specifications pertinent to the group. Some building codes requirements included the number of handicap parking spaces, number of exits and their locations, and minimum dimensions of public areas and corridors, in addition to safety requirements set in the engineering specifications.

During this phase, the civil engineer was responsible for the grading of the surface outside the building (such as the parking lot and sidewalk), handicap parking signs and other signs, drainage of the paved surfaces to the storm water line, connections from the clinic to the city water line and sewer line.

The structural engineer was responsible for the design of all the load bearing and non-load-bearing components and connections. Some of the designs included the sizing of steel beams, steel columns, isolated reinforced concrete footings, bracing necessary to support wind load, steel joists to support the roof, and snow loads. In addition, the structural engineers also provided additional design details to support the roof top unit (mechanical system for heating and air conditioning) and X-ray equipment. The documentation of the structural design included a set of very detailed drawings of the layout of the beams, columns, and their sizes; steel joist sizes and spacing; connections between beams and columns, joists and beams, columns and footings, etc; steel reinforcing details of the footings; masonry wall sizes and steel reinforcements, metal studs spacing. The structural drawings also included special details to support door and window openings, and other architectural components such as canopy at entrances.

Since the clinic and the physician office building shared a common gridline and the beams from the clinic at this common gridline were supported by the

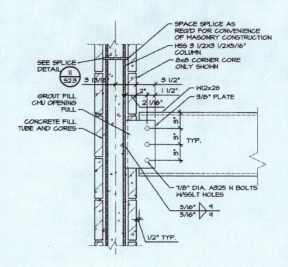

columns designed by the engineers of the physician office building, the structural engineer at the architects provided design information to the engineer for the physician office building.

Step 7: Construction Administration (CA) Phase

During the construction administration (CA) phase, there were weekly meetings between the site superintendent (from the contractor), the project manager (from the architects), representatives from different subcontractors such as electricians, plumbers, steel erectors, etc. The site superintendent was responsible for the logistic of the construction process and all the communications among all the subcontractors, project manager, and the clinic. His primary responsibilities were to ensure that the construction progressed as scheduled, that supplies were available when needed, and that the project manager was informed when concerns or issues arose during construction. Minutes from each construction meeting were recorded by the project manager and distributed to all parties.

Karen Chou

Periodically, the project manager and site superintendent also met with the owner to report the progress of the construction and to address the owner's concern. The structural engineer, though not required, was strongly recommended to visit the site to observe the construction process especially during foundation construction and framing of the building and to attend the construction meeting periodically during that time. The main purpose of the site observations was to verify that the structure was constructed as designed and the design was proper.

Beside visiting construction site, the structural engineer was required to review all the shop drawings of structural components such as beam sizes and length, connection details, etc. that were submitted by the fabricators directly or through the general contractor.

When the framing was done, other contractors went on site to do the wiring, plumbing, roofing, installing equipment. The interior designers began their part of the project when the interior part of the building was ready such as walls, floors, and ceilings. When the project was completed to a point where the building inspector issued the permit of occupancy, the clinic staff could start using the new clinic. The project manager, site superintendent, and the clinic performed a walk-through checking everything to make sure they were acceptable. This walkthrough is also called "punch list." The contractor and project manager took notes of all the fixes needed and items remained to be finished such as touch up on paint, cleaning, missing cover plate on light switches, etc. Generally, the owner would hold back the last 5 to 10% of the payment until he or she is completely satisfied with the construction.

Case study courtesy of Karen Chou

Mechanical/Electrical Engineering Design Process

A Case Study: Minnkota Electric Outboard Drive, Johnson Outdoors, Mankato, MN*

The marketing department at Johnson Outdoors recognized the growing interest in environmentally friendly power sources for their boating industry. Johnson Outdoors is the leading manufacturer of outdoor recreational equipment in the world. The engineering research and development and manufacturing facilities are located in Mankato, Minnesota, and the headquarters is located in Racine, Wisconsin.

*By Peter M. Kjeer.

Step 1: Recognizing the Need for a Product or a Service

As stated above, the marketing department at Johnson Outdoors recognized the growing interest in environmentally friendly power sources for their boating industry. To better serve the consumers and environment, the marketing department contacted the engineering department to discuss the feasibility of developing new generation of motors that are environmentally friendly. Increasingly, more states were enacting regulations banning the use of gasoline boat motors in public water ways including lakes and rivers.

Step 2: Problem Definition and Understanding

After the marketing people met with the engineers, the details of the project requirements were defined. The design specifications included—the motor had to move a 17 feet long pontoon at a speed of 5 mph minimum. The motor had to run at least for 2 hours on full battery charge. In addition, the boat operator had to have the capability to trim and tilt (raise the motor out of the water) from a remote console. The motor also had to be compatible with the industry-standard steering wheel mechanism.

Step 3: Research and Preparation

The engineers checked their existing design inventory to determine if a motor already existed that would meet some or all requirements. Moreover, a mechanical engineering student intern was commissioned to look at state regulations concerning the use of gasoline vs. electric boat motors.

Step 4: Conceptualization

During this phase of the design process, the engineering designers (twelve of them) met on weekly basis to brain storm and bounce ideas off each other. They also reviewed the information that was gathered in Step 3, and developed few concepts to pursue further. An additional idea that surfaced was the use of an electric linear actuator in place of a hydraulic actuator. This idea was pursued further because the potential leaks associated with hydraulic actuators.

Step 5: Synthesis

During this phase of design, the design engineers began to consider details. They consulted pertinent codes and standards to make sure that their design was in compliance. Most of the design work was done in ProE®, and prototypes were built in machine and electrical labs. Both technicians and engineers were involved in the fabrication of the prototypes. An interesting result of this project was that the unique design of the propeller required the use of a manufacturing process known as investment casting.

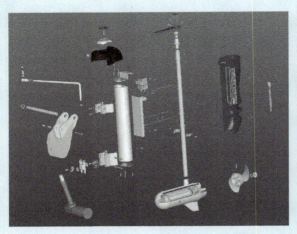

Exploded diagram of motor.

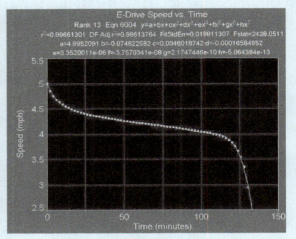

Engineers used ProMechanica® to conduct numerical experiments on the motor.

Step 6: Evaluation

Numerical experiments were conducted using ProMechanica®. Finite element techniques were used to look at stresses in critical components of the motor itself and the mounting bracket and the lifting mechanism. Numerical experiments also were performed to study the hydrodynamics of propeller designs including thrust, cavitation, speed, and drag. Using a GPS unit, the speed of the boat was measured over a period of several hours. This test was performed to quantify the motors' speed as a function of time. From the collected data the acceleration and position time functions were determined mathematically and compared to competitors' motors.

Step 7: Optimization

Based on results obtained from Step 6, modifications were made to the design and additional analyses performed. The finite element numerical experiments helped reshape the mounting bracket significantly to better withstand loading conditions. The results of numerical experiments dealing with propeller performance were also used to optimize the shape of the propeller for the final design. Many hours of testing also helped with the optimization of the final design. The testing included actual field testing in water and simulated life testing in labs.

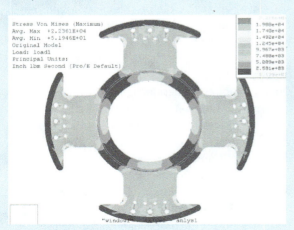

Actual testing of the system in a lake.
Peter M. Kjeer

Testing of the system in a laboratory setting.
Peter M. Kjeer

Step 8: Presentation

The product development process took approximately two years. During this period, the design engineers gave weekly progress reports to the rest of design group; quarterly status oral and written reports were given to marketing department and the group vice president in the headquarters. At the end of the project, a final presentation was given to the Board of Directors by the lead mechanical and electrical design engineers. The presentation addressed several issues, including development cost, unit cost, market outlook, performance characteristics, testing results, and environmental impact. The duration of presentations ranged from 15 minutes to an entire afternoon.
Case prepared by Peter M. Kjeer

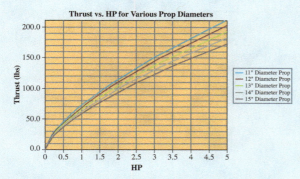

The thrust vs. horsepower results for different diameter propellers.

Engineering Communication

Sean Prior/Shutterstock.com

Presentations are an integral part of any engineering project. Depending on the size of the project, the presentations might be brief, lengthy, frequent, and may follow a certain format requiring calculations, graphs, charts, and engineering drawings.

LEARNING OBJECTIVES

LO¹ **Communication Skills and Presentation of Engineering Work:** explain why you should have good written and oral communication skills as an engineer

LO² **Basic Steps Involved in the Solution of Engineering Problems:** describe the basic steps that you need to follow to solve an engineering problem

LO³ **Written Communication:** explain different modes of written communication in engineering and their purpose

LO⁴ **Oral Communication:** describe the key concepts that must be followed when giving an oral presentation

LO⁵ **Graphical Communication:** realize the importance of graphical communication (drawings) in conveying ideas and design information

DISCUSSION STARTER

COMMUNICATION: THE ENGINEER'S RESPONSIBILITY

That communication is the engineer's responsibility is inherent in the nature of engineering. The engineer's first order of responsibility is to the organization for which he or she works. Engineers fill specific roles in business, industry, public institutions, and government agencies, and their responsibilities are to get things done for those organizations. The engineer's responsibility, therefore, is to communicate effectively so that these changes occur.

The engineer's second order of responsibility is to society, because he or she belongs to a profession whose objective is to improve the conditions of human life by changing the physical environment and systems. Science and technology are considered the basis for the transition to a postindustrial society and for the postindustrial society itself. The engineer will have to communicate effectively to establish and maintain relationships between the sphere of technology and production and the spheres of social services and political institutions. The engineer's responsibility to communicate, therefore, is a commitment the engineer must accept. Those who fail to assume the responsibility to communicate—to interact with the community—ultimately fail in their responsibilities to themselves.

Based on Mathes, J. C., ERIC, 1980

To the students: Do you agree with what is said here? What are your thoughts? Electronic communication (e.g., text messages, e-mail(s), electronic file marking and sharing, etc.) is becoming increasingly important. What type of communication changes do you anticipate to occur within next 30 years?

Engineers are problem solvers. Once they have obtained a solution to a problem, they need to communicate effectively their solution to various people inside or outside their organization. Presentations are an integral part of any engineering project. As an engineering student, you would be asked to present your solutions to assigned homework problems, write a technical report, or give an oral presentation to your class, student organization, or an audience at a student conference. Later, as an engineer, you could be asked to give presentations to your boss, colleagues in your design group, sales and marketing people, the public, or an outside customer. Depending on the size of the project, the presentations might be brief, lengthy, or frequent, and they may follow a certain format. You might be asked to give detailed technical presentations containing graphs, charts, and engineering drawings, or they may take the form of brief project updates. In this chapter, we will explain some of the common engineering presentation formats.

LO¹ 4.1   Communication Skills and Presentation of Engineering Work

As an engineering student, you need to develop good written and oral communication skills. During the next four or five years, you will learn how to express your thoughts, present a concept for a product or a service and an engineering analysis of a problem and its solution, or show your findings from experimental work. Moreover, you will learn how to communicate design ideas by means of engineering drawings or computer-aided modeling techniques. Starting right now, it is important to understand that the ability to communicate your solution to a problem is as important as the solution itself. You may spend months on a project, but if you cannot effectively communicate to others, the results of all your efforts may not be understood and appreciated. Most engineers are required to write reports. These reports might be lengthy, detailed, technical reports containing charts, graphs, and engineering drawings, or they may take the form of a brief memorandum or executive summary. Some of the more common forms of engineering communication are explained briefly next.

LO² 4.2   Basic Steps Involved in the Solution of Engineering Problems

> To analyze an engineering problem: (1) define the problem, (2) simplify the problem by making assumptions and estimations, (3) perform the analysis, and (4) verify the results

Before we discuss some of the common engineering presentation formats, let us say a few words about the basic steps involved in the solution of an engineering problem. There are four basic steps that must be followed when analyzing an engineering problem: (1) defining the problem, (2) simplifying the problem by assumptions and estimations, (3) performing the solution or analysis, and (4) verifying the results.

Step 1: Defining the Problem

Before you can obtain an appropriate solution to a problem, you must first thoroughly understand the problem itself. There are many questions that you need to ask before proceeding to determining a solution. What is it exactly that you want to analyze? What do you really *know* about the problem, or what are some of the things *known* about the problem? What are you looking for? What exactly are you trying to find a solution to?

Taking time to understand the problem completely at the beginning will save lots of time later and help to avoid a great deal of frustration. Once you understand the problem, you should be able to divide any given problem into two basic questions: What is known? and What is to be found?

Step 2: Simplifying the Problem

Before you can proceed with the analysis of the problem, you may first need to simplify it.

Assumptions and Estimations   Once you have a good understanding of the problem, you should then ask yourself this question: Can I simplify the problem by making some reasonable and logical assumptions and yet obtain an appropriate solution? Understanding the physical laws and the fundamental concepts, as well as where and when to apply them and their limitations, will benefit you greatly in making assumptions and solving the problem. It is very important as you take different engineering classes in the next few years that you develop a good grasp of the fundamental concepts in each class that you take.

Step 3: Performing the Solution or Analysis

Once you have carefully studied the problem, you can proceed with obtaining an appropriate solution. You will begin by applying the physical laws and fundamental concepts that govern the behavior of engineering systems to solve the problem. Among the engineering tools in your toolbox you will find mathematical tools. It is always a good practice to set up the problem in symbolic or *parametric* form, that is, in terms of the variables involved. You should wait until the very end to substitute for the given values. This approach will allow you to change the value of a given variable and see its influence on the final result. The difference between numerical and symbolic solutions is explained in more detail in Section 6.5.

Step 4: Verifying the Results

The final step of any engineering analysis should be the verification of results. Various sources of error can contribute to wrong results. Misunderstanding a given problem, making incorrect assumptions to simplify the problem, applying a physical law that does not truly fit the given problem, and incorporating inappropriate physical properties are common sources of error. Before you present your solution or the results to your instructor or, later in your career, to your manager, you need to learn to think about the calculated results. You need to ask yourself the following question: Do the results make sense? A good engineer must always find ways to check results. Ask yourself this additional question: What if I change one of the given parameters? How would that change the result? Then consider if the outcome seems reasonable. If you formulate the problem such that the final result is left in parametric (symbolic) form, then you can experiment by substituting different values for various parameters and look at the final result. In some engineering work, actual physical experiments must be carried out to verify one's findings. Starting today, get into the habit of asking yourself if your solution to a problem makes sense. Asking your instructor if you have come up with the right answer and checking the back of your textbook to match answers are not good approaches in the long run. You need to develop the means to check your results by asking yourself the appropriate questions. Remember, once you start working for hire, there are no answer books. You will not want to run to your boss to ask if you did the problem right!

Homework Presentation

Engineering paper is specially formatted for use by engineers and engineering students. The paper has three cells on the top that may be used to convey such information as course number, assignment due date, and your name. A given problem may be divided into a "Given" section, a "Find" section, and a "Solution" section. It is a good practice to draw horizontal lines to separate the known information (Given section) from the information that is to be found (Find section) and the analysis (Solution section), as shown in Figure 4.1. Do not write anything on the back of the paper. The grid lines on the back provide scale and an outline for freehand sketches, tables, or plotting data by hand. The grid lines, which can be seen from the front of the paper, are there to assist you in drawing things or presenting tables and

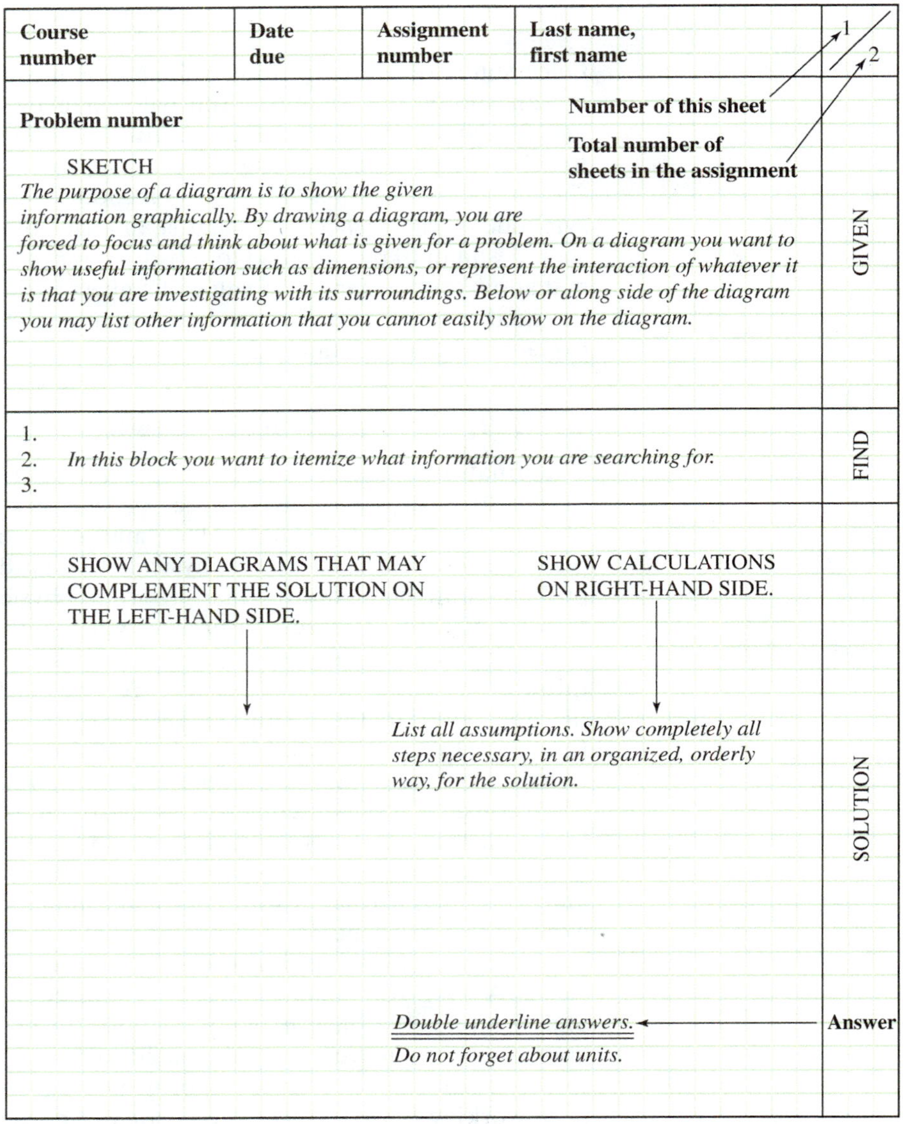

Course number	Date due	Assignment number	Last name, first name	

Problem number

Number of this sheet

Total number of sheets in the assignment

SKETCH

GIVEN

The purpose of a diagram is to show the given information graphically. By drawing a diagram, you are forced to focus and think about what is given for a problem. On a diagram you want to show useful information such as dimensions, or represent the interaction of whatever it is that you are investigating with its surroundings. Below or along side of the diagram you may list other information that you cannot easily show on the diagram.

1.
2.   *In this block you want to itemize what information you are searching for.*
3.

FIND

SHOW ANY DIAGRAMS THAT MAY COMPLEMENT THE SOLUTION ON THE LEFT-HAND SIDE.

SHOW CALCULATIONS ON RIGHT-HAND SIDE.

List all assumptions. Show completely all steps necessary, in an organized, orderly way, for the solution.

SOLUTION

Double underline answers. ◄———— **Answer**

Do not forget about units.

FIGURE 4.1   An example of engineering problem presentation.

graphical information on the front of the page neatly. These grid lines also allow you to present a freehand engineering drawing with its dimensions. Your engineering assignments will usually consist of many problems, thus you will present your work on many sheets; which should be stapled together. Professors do not generally like loose papers, and some may even deduct points from your assignment's total score if the assignment sheets are not stapled together. The steps for presenting an engineering problem are demonstrated in Example 4.1. If you are presenting solutions to simple problems and you think you can show the complete solution to more than one problem on one page, then separate the two problems by a relatively thick line or a double line, whichever is more convenient for you.

EXAMPLE 4.1

Determine the mass of compressed air in a scuba diving tank, given the following information. The internal volume of the tank is 10 L and the absolute air pressure inside the tank is 20.8 MPa. The temperature of the air inside the tank is 20° C. Use the ideal gas law to analyze this problem. The ideal gas law is given by

$$PV = mRT$$

where

P = absolute pressure of the gas, Pa

V = volume of the gas, m^3

m = mass, kg

R = gas constant, $\dfrac{J}{kg \cdot K}$

T = absolute temperature, Kelvin, K

The gas constant R for air is 287 J/kg · K.

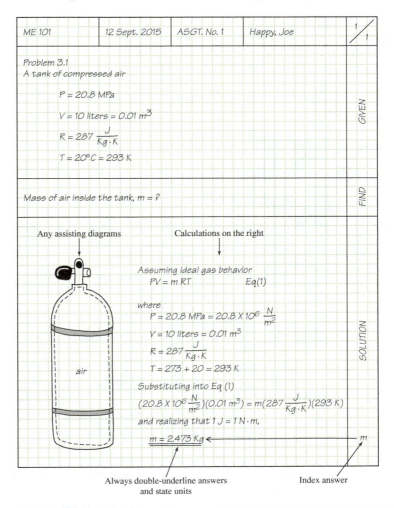

ME 101 | 12 Sept. 2015 | ASGT. No. 1 | Happy, Joe | 1/1

Problem 3.1
A tank of compressed air

$P = 20.8$ *MPa*

$V = 10$ *liters* $= 0.01$ *m^3*

$R = 287 \dfrac{J}{Kg \cdot K}$

$T = 20°C = 293$ *K*

GIVEN

Mass of air inside the tank, m = ?

FIND

Any assisting diagrams   Calculations on the right

air

Assuming ideal gas behavior
$PV = m\,RT$ *Eq(1)*

where
$P = 20.8$ *MPa* $= 20.8 \times 10^6 \dfrac{N}{m^2}$

$V = 10$ *liters* $= 0.01$ *m^3*

$R = 287 \dfrac{J}{Kg \cdot K}$

$T = 273 + 20 = 293$ *K*

Substituting into Eq (1)

$(20.8 \times 10^6 \dfrac{N}{m^2})(0.01\ m^3) = m(287 \dfrac{J}{Kg \cdot K})(293\ K)$

and realizing that 1 J = 1 N·m,

$\underline{\underline{m = 2.473\ Kg}}$ ← m

SOLUTION

Always double-underline answers   Index answer
and state units

FIGURE 4.2   An example of engineering homework presentation for Example 4.1.

At this time, do not worry about understanding the ideal gas law. This law will be explained to you in detail in Chapter 11. The purpose of this example is to demonstrate how a solution to an engineering problem is presented. Make sure you understand and follow the steps shown in Figure 4.2.

Before You Go On

Answer the following questions to test your understanding of the preceding section.

1. Describe the basic steps that are involved in the solution of engineering problems.

2. Explain how you should present your engineering homework problems.

LO³ 4.3   Written Communication

Written and oral presentations are important parts of engineering. Written communications might be brief, as in progress reports or short memos, or longer and follow a certain format requiring calculations, graphs, charts, and engineering drawings.

Progress Report

Progress reports are means of communicating to others in an organization or to the sponsors of a project how much progress has been made and which of the main objectives of the project have been achieved to date. Based on the total time period required for a project, progress reports may be written for a period of a week, a month, several months, or a year. The format of the progress report may be dictated by a manager in an organization or by the project's sponsors.

Executive Summary

Executive summaries are means of communicating to people in top management positions, such as a vice president of a company, the findings of a detailed study or a proposal. The executive summary, as the name implies, must be brief and concise. It is generally no more than a few pages long. In the executive summary, references may be made to more comprehensive reports so that readers can obtain additional information if they so desire.

Short Memos

Short memos are yet another way of conveying information in a brief way to interested individuals. Generally, short memos are under two pages in length. A general format for a short memo follows. The header of the memo contains information such as the date, who the memo is from, to whom it is being sent, and a subject (Re:) line. This is followed by the main body of the memo.

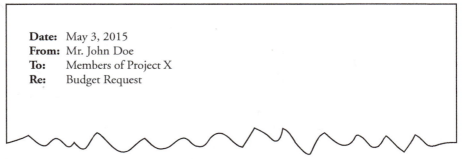

Date: May 3, 2015
From: Mr. John Doe
To: Members of Project X
Re: Budget Request

Short Memos

Detailed Technical Report

Detailed technical reports dealing with experimental investigations generally contain the following items:

Title The title of a report should be a brief informative description of the report contents. A sample of an acceptable title (cover) sheet is shown in Figure 4.3. If the report is long, a table of contents should follow the title page.

Abstract This is a very important part of a report because readers often read this part first to decide if they should read the report in detail. In the abstract, in complete but concise sentences you state the precise objective, emphasize significant findings, and present conclusions and/or recommendations.

Objectives The purpose of the objectives section is to state what is to be investigated through the performance of the experiment. Be sure to list your objectives explicitly (e.g., 1., 2.,).

Theory and Analysis There are several purposes of the theory and analysis section:

- To state pertinent principles, laws, and equations (equations should be numbered);
- To present analytical models that will be used in the experiment;
- To define any unfamiliar terms or symbols; and
- To list important assumptions associated with the experimental design.

Apparatus and Experimental Procedures There are two main purposes of the apparatus and experimental procedures section:

1. To present a list of apparatus and instrumentation to be used, including the instrument ranges, least count, and identification numbers.
2. To describe how you performed the experiment. The procedure should be itemized (step 1., 2., etc.) and a schematic or diagram of the instrument setup should be included.

All State University
Department of Mechanical Engineering

Course Title

Experiment No. _____

Experiment Title _____

_____

_____

Date Experiment Completed _____

Students' Names _____

_____

_____

Example of title (cover) sheet.

Data and Results The purpose of this section is to present the results of the experiment, as described in the stated objective, in a tabular and/or graphical form. These tables and graphs show the results of all your efforts. Include descriptive information such as titles, column or row headings, units, axis labels, and data points (data points should be marked by $\odot$, $\boxdot$, $\triangle$, etc.). All figures must be numbered and have a descriptive title. The figure number and the title should be placed below the figure. All tables must be numbered and have a descriptive title as well. However, for tables, the table number and the title should be placed above the table. It is sometimes necessary to note in this section that you have included the original data sheets in the appendix to your report.

Discussion of the Results The purpose of the results section is to emphasize and explain to the reader the important results of the experiment and point out their significance. When applicable, be sure to compare experimental results with theoretical calculations.

Conclusions and Recommendations The conclusions and recommendations section compares your objectives with your experimental results. Support your conclusions with appropriate reference materials. Be sure to state recommendations based on the conclusions.

Appendix The appendix serves several purposes:

- To provide the reader with copies of all original data sheets, diagrams, and supplementary notes.
- To display sample calculations used in processing the data. The sample calculations should contain the following parts:

  A title of the calculation
  A statement of mathematical equation
  Calculation using one sample of data

References A list of references that have been numbered in the text must be included in the report. Use the following format examples:

  For Books: Author, title, publisher, place of publication, date (year), and page(s).
  For Journal Articles: Author, title of article (enclosed in quotation marks), name of journal, volume number, issue number, year, and page(s).
  For Internet Materials: Author, title, date, and URL address.

Before You Go On

Answer the following questions to test your understanding of the preceding section.

1. What is the purpose of an executive summary?

2. What is the purpose of a progress report?

3. Describe the main components of a detailed technical report.

LO⁴ 4.4 Oral Communication

We communicate orally to each other all the time. Informal communication is part of our everyday life. We may talk about sports, weather, what is happening around the world, or a homework assignment. Some people are better at expressing themselves than others are. Sometimes we say things that are misunderstood and the consequences could be unpleasant. When it comes to formal presentations, there are certain rules and strategies that you need to follow. Your oral presentation may show the results of all your efforts regarding a project that you may have spent months or a year to develop. If the listener cannot follow how a product was designed or how the analysis was performed, then all your efforts become insignificant. It is very important that all information be conveyed in a manner easily understood by the listener.

The oral technical presentation in many ways is similar to a written one. You need to be well organized and have an outline of your presentation ready, similar to the format for a written report. It may be a good idea to write down what you are planning to present. Remember it is harder to erase or correct what you say after you have said it than to write it down on a piece of paper and correct it before you say it. You want to make every effort to ensure that what is said (or sent) is what is understood (or received) by the listener.

Rehearse your presentation before you deliver it to a live audience. You may want to ask a friend to listen and provide helpful suggestions about your style of presentation, delivery, content of the talk, and so on.

> Rehearse your oral presentation before you deliver it. Ask a friend to listen and provide suggestions about your style of presentation, delivery, and content of the talk. Make sure to present the information in a way that is easily understood by your audience.

Present the information in a way that is easily understood by your audience. Avoid using terminology or phrases that may be unfamiliar to listeners. You should plan so that you won't overexplain key concepts and ideas, because those who are really interested in a specific area of your talk can always ask questions later.

Try to keep your talk to about half an hour or less, because the attention span of most people is about 20 to 30 minutes. If you have to give a longer talk, then you may want to mix your presentation with some humor or tell some interesting related story to keep your audience's attention. Maintain eye contact with all of your audience, not just one or two people. Don't ever have your back to the audience for too long! Use good visual aids. Use presentation software such as PowerPoint to prepare your talk. When possible, incorporate charts, graphs, animated drawings, short videos, or a model. When it is available, you can use prototyping technology to demonstrate concepts for new products and have a prototype of the product on hand as part of your presentation. You may also want to have copies of the outline, along with notes on the important concepts and findings, ready to hand out to interested audience members. In summary, be organized, be well prepared, get right to the point when giving an oral presentation, and consider the needs and expectation of your listeners.

Now here are a few words about Microsoft PowerPoint presentations. You can use PowerPoint to generate and organize your slides showing text, charts, graphs, and video clips. With PowerPoint, you can also create supplementary materials (such as handouts) for your audience and preparatory notes for your presentation. PowerPoint offers a number of attractive *templates* and *layout options*. Moreover, for formal presentations, you may want to include your university or organization logo and the date on the bottom of every slide in your presentation. As you may know, to do this, you need to create a *slide master* first. *Animation* is yet another way of adding visual and sound effects to your slides. For example, you can show a bulleted list of items one at a time, or you can have each item fade away as you move to the next item. *Slide transition* options can also give your slide presentations sound and visual effects as one slide disappears and the next slide appears. You may also consider using *action buttons,* which allow you to move to a specific slide in another PowerPoint file, Word document, or Excel file without having to leave your current presentation to access the file. The action button is linked to the file, and upon clicking it, you automatically will go to this file containing the desired slide or document. Finally, as we said earlier, after you prepare your presentation, you need to rehearse your presentation before you deliver it to a live audience. Another reason for rehearsing your presentation is that, typically, engineering presentations must be made within an allotted time period. PowerPoint offers a *rehearse timing* option that allows you to time your presentation.

LO⁵ 4.5 Graphical Communication

In the previous sections, we showed you how to present your homework solutions and to write technical and progress reports, an executive summary, and short memos. Now, we discuss engineering **graphical communication**. Engineers use special kinds of drawings, called engineering drawings, to convey their ideas and design information about products. These drawings portray vital information, such as the shape of the product, its size, type of material used, and assembly steps. Moreover, machinists use the information provided by engineers or draftspersons on the engineering drawings to make the parts. For complicated systems made of various parts, the drawings also serve as a how-to-assemble guide, showing how the various

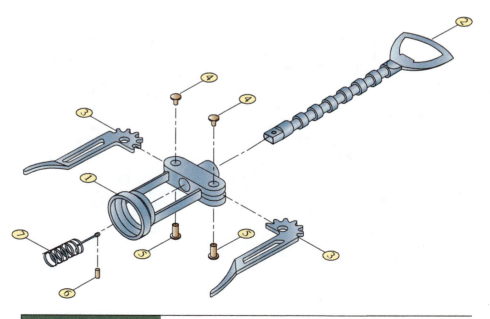

FIGURE 4.4	Assembly drawing of corkscrew.

Based on Kasey Cassell.

parts fit together. Examples of these types of drawings are shown in Figures 4.4 through 4.6. In Chapter 16, we provide an introduction to engineering graphical-communication principles. We will discuss why engineering drawings are important, how they are drawn, and what rules must be followed to create such drawings. Engineering symbols and signs also provide valuable information. These symbols are a "language" used by engineers to convey their ideas, solutions to problems, or analyses of certain situations. In Chapter 16, we will also discuss the need for conventional engineering symbols and will show some common symbols used in civil, electrical, and mechanical engineering.

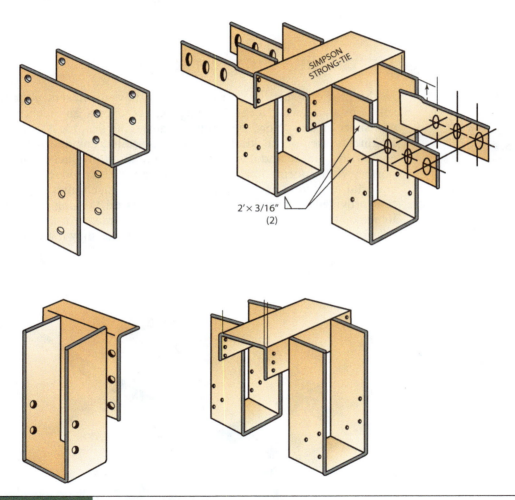

FIGURE 4.5	Common manufactured metal beam connectors.
	Based on Simpson Strong-Tie Company, Inc.

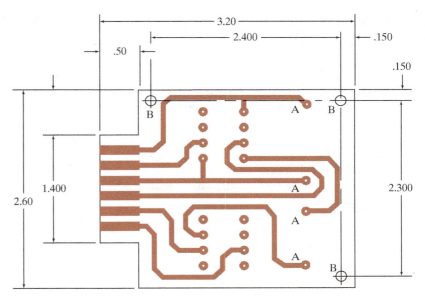

(a) A printed circuit board drill plan

LTR	QTY	SIZE	REMARKS
NONE	16	.040	PLATED THRU
A	4	.063	PLATED THRU
B	3	.125	

Engineers use drawings to convey their ideas and design information. These drawings provide information, such as the shape of a product, its size, type of material used, and assembly steps. For complicated systems made of various parts, the drawings also show how various parts of a product fit together.

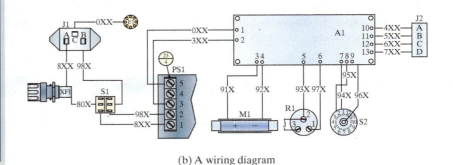

(b) A wiring diagram

FIGURE 4.6   Examples of drawings used in electrical and electronic engineering.

Before You Go On

Answer the following questions to test your understanding of the preceding section.

1. What are some of the important concepts that you should consider when preparing for an oral presentation?

2. Explain what we mean by engineering graphical communication and why it is important.

SUMMARY

LO¹   Communication Skills and Presentation of Engineering Work

As an engineer, you should have good written and oral communication skills. You are expected to express your thoughts, present a concept for a product, provide an engineering analysis of a problem and its solution, or show your findings from experimental work. It is also very important to understand that the ability to communicate your solution to a problem is as important as the solution itself.

LO²   Basic Steps Involved in the Solution of Engineering Problem

When you analyze an engineering problem, there are four steps that you must follow: (1) define the problem, (2) simplify the problem by making assumptions and estimations, (3) perform the analysis, and (4) verify the results.

LO³   Written Communication

Writing reports are an integral part of engineering tasks. Depending on the size of a project, you could be expected to write brief and frequent progress reports, memos, an executive summary, or a lengthy report with calculations, graphs, charts, and engineering drawings.

LO⁴   Oral Communication

The oral technical presentation in many ways is similar to a written report. It needs to be well organized and presented in a way that is easily understood by the audience. Terminology or phrases that may be unfamiliar to listeners should be avoided. The presentation also should be rehearsed for a smooth delivery.

LO⁵   Graphical Communication

Engineers use drawings to convey their ideas and design information. These drawings provide important information such as shape of a product, size, type of material used, and assembly steps. For complicated systems made of various parts, the drawings also show how the various parts fit together. Moreover, it is important to realize that machinists use engineering drawings to make the parts. Therefore, the drawings have to be complete and convey proper information so that a product is fabricated correctly.

KEY TERMS

Analysis 105
Assumptions 104
Defining the Problem 104
Executive Summary 108

Graphical Communication 113
Homework Presentation 105
Oral Communication 111
Progress Report 108

Short Memo 108
Verifying the Results 105
Written Communication 108

APPLY WHAT YOU HAVE LEARNED

Lighting systems account for a major portion of electricity used in buildings and have received much attention recently due to energy and sustainability concerns. Prepare a 15-minute PowerPoint presentation about new lighting systems. Begin your presentation by providing some background. For example, explain what we mean by illumination, efficacy, and color rendition index. Then discuss various types of lights (e.g. compact fluorescent, LED, etc.). Also, prepare a table that shows in comparison efficacy, life (hours), and color rendition index for various types of lights. Share your findings with your class.

Roman Samokhin/Shutterstock.com   Dmitriy Raykin/Shutterstock.com

PROBLEMS

Problems that promote lifelong learning are denoted by 🔑

4.1 Investigate the operation of various turbines. Write a brief report explaining the operation of steam turbines, hydraulic turbines, gas turbines, and wind turbines.

4.2 In a brief report, discuss why we need various modes of transportation. How did they evolve? Discuss the role of public transportation, water transportation, highway transportation, railroad transportation, and air transportation.

4.3 Identify the major components of a computer, and briefly explain the function or the role of each component.

4.4 Electronic communication is becoming increasingly important. In your own words, identify the various situations under which you should write a letter, send an e-mail, make a telephone call, or talk to someone in person. Explain why one particular form of communication is preferable to the others available.

You may have seen examples of emoticons (derived from emotion and icons)—simple printable characters used in e-mails to convey human facial expressions. Here are some examples of emoticons:

:) or: -)	smiling
:-D	laughing
: (or: -(	sad
:,(or: .(	crying
:-O	surprised, shocked
;) or; -)	wink
>: -O	angry/yelling
>: -(	angry/grumpy
: -*	kiss
: -**	returning a kiss
<3	heart (e.g. i <3 u)
</3	broken heart

The following assignments could be done as group projects.

4.5 Prepare a 15-minute PowerPoint presentation about engineering and its various disciplines, and the next time you go home present it to the juniors in your high school.

4.6 Prepare a 15-minute PowerPoint presentation about your plans to receive a rewarding education and the preparation that it takes to have a successful career in engineering. When preparing your presentation, consider a four or five year detailed plan of study, involvement with extra curricular activities, an internship, volunteer activities, and so on. Share your plans with your classmates.

4.7 From the subjects presented in this book, choose a topic, prepare a 15-minute PowerPoint presentation, and deliver it to your class.

4.8 If your introduction to engineering class has a term project, prepare a PowerPoint presentation of a length specified by your instructor and deliver it to your class at a date specified by your instructor.

4.9 Prepare a 20-minute PowerPoint presentation about the history and future of engineering. Incorporate pictures, short video clips, graphs, and so on in your presentation.

4.10 Prepare a 15-minute PowerPoint presentation about an engineering topic, such as alternative energy or an environmental issue that interests you and that you would like to learn more about. Present it to your class.

4.11 Visit the Web site of the National Society of Professional Engineers and research engineering ethics. Prepare a PowerPoint presentation showing why engineering ethics is so important and explaining why honesty and integrity in engineering are essential. Give examples of engineering codes of ethics. Present an ethics-related case and involve the class by discussing it during your presentation.

4.12 Visit the Web site of an engineering organization, such as ASME, IEEE, or ASCE, and prepare a PowerPoint presentation about this year's student design competitions. Deliver your presentation at one of your engineering organization's meetings.

4.13 Prepare a 10-minute PowerPoint presentation about engineering professional registration. Explain why professional registration is important and what the requirements are.

Problems 4.14–4.20

As described in the Data and Results part of Section 4.3, all tables and graphs must have descriptive information such as titles, column or row headings, units, and axis labels, and data points must be clearly marked. All figures must be numbered and have a descriptive title. The figure number and the title should be placed below the figure. All tables must be numbered and have a descriptive title as well. However, for tables, the table number and the title should be placed above the table.

4.14 Plot the following data. Use two different y-axes. Use a scale of zero to 100° F for temperature, and zero to 12 mph for wind speed. Present your work using the ideas discussed in this chapter and engineering papers.

Time (p.m.)	Temperature (°F)	Wind Speed (mph)
1	75	4
2	80	5
3	82	8
4	82	5
5	78	5
6	75	4
7	70	3
8	68	3

4.15 Create a table that shows the relationship between the units of temperature in degrees Celsius and Fahrenheit in the range of −50° C to 50° C. Use increments of 10° C. Present your work incorporating the ideas discussed in this chapter and using engineering paper.

4.16 Create a table that shows the relationship between the units of mass in kilograms and pound mass in the range of 50 kg to 120 kg. Use increments of 10 kg. Present your work incorporating the ideas discussed in this chapter and using engineering paper.

4.17 The given data show the result of a model known as *stopping sight distance,* used by civil engineers to design roadways. This simple model estimates the distance a driver needs in order to stop his or her car, traveling at a certain speed, after detecting a hazard. Plot the data using engineering paper and incorporating the ideas discussed in this chapter.

Speed (mph)	Stopping Sight Distance (ft)
0	0
5	21
10	47
15	78
20	114
25	155
30	201
35	252
40	309
45	370
50	436
55	508
60	584
65	666
70	753
75	844
80	941

4.18 The given data represent the velocity distribution for a flow of a fluid inside a circular pipe with a radius of 0.1 m. Plot the data using engineering paper and incorporating the ideas discussed in this chapter.

Radial distance, r (m) r = 0 corresponds to the center of pipe	U(r) Velocity (m/s)
−0.1	0
−0.09	0.095
−0.08	0.18
−0.07	0.255

Radial distance, r (m) r = 0 corresponds to the center of pipe	U(r) Velocity (m/s)
−0.06	0.32
−0.05	0.375
−0.04	0.42
−0.03	0.455
−0.02	0.48
−0.01	0.495
0	0.5
0.01	0.495
0.02	0.48
0.03	0.455
0.04	0.42
0.05	0.375
0.06	0.32
0.07	0.255
0.08	0.18
0.09	0.095
0.1	0

1	308
1.2	252
1.4	207
1.6	172
1.8	143
2	121
2.2	103
2.4	89
2.6	78
2.8	69
3	62
3.2	57
3.4	52
3.6	49
3.8	46
4	44
4.2	42
4.4	40
4.6	39
4.8	38
5	38

4.19 In an annealing process—a process wherein materials such as glass and metal are heated to high temperatures and then cooled slowly to toughen them — thin steel plates are heated to temperatures of 900° C and then cooled in an environment with temperature of 35° C. The results of an annealing process for a thin plate is shown below. Plot the data using engineering paper incorporating the ideas discussed in this chapter.

Time (hr)	Temperature (°C)
0	900
0.2	722
0.4	580
0.6	468
0.8	379

4.20 The relationship between a spring force and its deflection is given in the table at the top of the next column. Plot the results using engineering paper and incorporating the ideas discussed in this chapter.

Deflection, X (mm)	Spring force, F (N)
0	0
5	10
10	20
15	30
20	40

4.21 Present Example 6.1 in Chapter 6 using the format discussed in Section 4.2. Divide the example problem into "Given," "Find," and "Solution" sections.

4.22 Present Example 6.3 in Chapter 6 using the format discussed in Section 4.2. Divide the example problem into "Given," "Find," and "Solution" sections.

4.23 Present Example 7.1 in Chapter 7 using the format discussed in Section 4.2. Divide the example problem into "Given," "Find," and "Solution" sections.

4.24 Present Example 7.4 in Chapter 7 using the format discussed in Section 4.2. Divide the example problem into "Given," "Find," and "Solution" sections.

4.25 Present Example 8.3 in Chapter 8 using the format discussed in Section 4.2. Divide the example problem into "Given," "Find," and "Solution" sections.

4.26 Present Example 8.4 in Chapter 8 using the format discussed in Section 4.2. Divide the example problem into "Given," "Find," and "Solution" sections.

4.27 Present Example 9.3 in Chapter 9 using the format discussed in Section 4.2. Divide the example problem into "Given," "Find," and "Solution" sections.

4.28 Present Example 9.4 in Chapter 9 using the format discussed in Section 4.2. Divide the

example problem into "Given," "Find," and "Solution" sections.

4.29 Present Example 10.7 in Chapter 10 using the format discussed in Section 4.2. Divide the example problem into "Given," "Find," and "Solution" sections.

4.30 Present Example 10.14 in Chapter 10 using the format discussed in Section 4.2. Divide the example problem into "Given," "Find," and "Solution" sections.

4.31 Present Example 11.5 in Chapter 11 using the format discussed in Section 4.2. Divide the example problem into "Given," "Find," and "Solution" sections.

4.32 Present Example 12.4 in Chapter 12 using the format discussed in Section 4.2. Divide the example problem into "Given," "Find," and "Solution" sections.

4.33 Present Example 13.1 in Chapter 13 using the format discussed in Section 4.2. Divide the example problem into "Given," "Find," and "Solution" sections.

4.34 Present Example 13.6 in Chapter 13 using the format discussed in Section 4.2. Divide the example problem into "Given," "Find," and "Solution" sections.

4.35 Present Example 13.9 in Chapter 13 using the format discussed in Section 4.2. Divide the example problem into "Given," "Find," and "Solution" sections.

"Who has never tasted what is bitter does not know what is sweet." – GERMAN PROVERB

Dr. Karen Chou

I was born and raised in Hong Kong until I was 14 years old. I have three (two older and one younger) brothers and an older sister with an age span of over 20 years. My parents and my older siblings came from a village in Guangzhou (Canton), China. My mother, two brothers and I immigrated to the United States in 1970. Shortly afterwards, my father passed away in Hong Kong. Raised by a single parent in the land of opportunity (uncertainty at that time), I was also the first generation to attend college and the only one holds a Ph.D. in my family tree.

Choosing engineering as my career was an accident. It was a decision that I never regret. When I applied to colleges, my interest was in mathematics. My high school teacher had suggested I consider civil engineering. Unfortunately, I had no idea what it was beyond building bridges! I promised my teacher to consider that option and selected a college that had engineering programs just in case. My choice was Tufts University. During the freshmen orientation, a faculty from engineering college informed us the option of double major in engineering and mathematics. That sealed my choice – double major in civil engineering and mathematics (I had never heard of other engineering disciplines anyway). I thought if I couldn't make it in engineering, I could always fall back to mathematics. After receiving a B.S.C.E. with dual majors in 1978, I went on to graduate study in structural engineering at Northwestern University. After completing my M.S. program in 1979, I decided to enter the workforce to obtain some engineering experience before I pursued my Ph.D.

Educated in structural engineering with the traditional topics of concrete and steel design of buildings, my first full time engineering position was structural engineer at Mongometry-Harza (formerly Harza) Engineering in Chicago. I worked with engineers, mostly with advanced structural engineering degrees in designing structures associated with hydropower plants such as spillways, powerhouse, intakes, double curvature arch dams, gravity dam, piers supporting gates, etc. It was a learning experience especially we were rarely taught any of these structures in schools. I didn't just learn how to design, I had to learn new vocabulary – components of hydropower projects. I was concern of my inadequate preparation to be a structural engineer. My section head told me that as long as I understand the fundamental concepts of engineering mechanics and designs, I would be fine.

Courtesy of Karen Chou

While I was working full time as an engineer, I also had a desire to obtain a Ph.D. Hence, I returned to the classroom part time and eventually returning for full time study and received a Ph.D. in structural engineering in 1983. Instead of returning to practicing engineering, I pursued a career in academic. However, the desire to gain more practical experience so that I can incorporate it in the classroom has never diminished. Over the past 30 years, I have taught at Syracuse University (10 years), University of Tennessee at Knoxville (8 years), and Minnesota State University, Mankato (9 years), and now back in my alma mater, Northwestern University. In addition, I was also a visiting faculty and adjunct faculty at the University of Minnesota. The experience at each university was different in teaching, research, and interaction with students. The reward of seeing the students succeed is immense. Between 2006 and 2010, I also fulfilled my desire to gain practical engineering experience and brought them into the classroom. I am a registered professional engineer in seven states. Few years ago, I even passed the first part of the structural engineering exam. In engineering, life-long learning is crucial, even at my age!

It would be a disservice to the potential female engineers who are pondering the decision of majoring in engineering and not to mention some of the obstacles many female engineers faced in this, still, male dominated field. For most part, the engineering profession and the country had placed more emphases in equal opportunity and had improved on their attitudes towards having female colleagues. I have seen more female engineers and students. In academic, female civil engineering faculty is still very few (I continue to find myself to be the only or one of two female engineers or civil engineering faculty) within a department or institution. The loneliness is not too different from 30 years ago. The lack of interaction with or inclusiveness from the male colleagues has improved but not sufficient. This oftentimes hampers the effectiveness of doing collaborative research in professional discussion. The attitude of "women don't belong in engineering" has reduced and is less obvious.

Nevertheless, it is still there. With the climate still being cool towards female engineers, would I discourage any of my female students consider engineering as her career. The answer is ABSOLUTELY NOT!!! I believed 30 years ago and still do, we should all choose a career that we would enjoy regardless of gender and ethnical background. So, if you enjoy science and math and willing to learn and work hard, consider engineering. The challenges and rewards are tremendous.

Source: Courtesy of Karen Chou

Engineering Ethics

Stockbyte/Thinkstock

Engineers design many products and provide many services that affect our quality of life and safety. They supervise the construction of buildings, dams, highways, bridges, mass-transit systems, and power plants. Engineers must perform under a certain standard of professional behavior that requires adherence to the highest principles of ethical conduct.

LEARNING OBJECTIVES

LO¹ **Engineering Ethics:** explain what is meant by engineering ethics

LO² **The Code of Ethics of the National Society of Professional Engineers:** give examples of fundamental canons and rules of practice

LO³ **Engineer's Creed:** explain what is meant by the Engineer's Creed and give examples

LO⁴ **Academic Dishonesty, Conflict of Interest, Professional Responsibility:** explain what each of these terms mean

USE OF CD-ROM FOR HIGHWAY DESIGN: CASE NO. 98-3

Facts: Engineer A, a chemical engineer with no facilities design and construction experience, receives a solicitation in the mail with the following information:

"Engineers today cannot afford to pass up a single job that comes by, including construction projects that may be new or unfamiliar.

Now—thanks to a revolutionary new CD-ROM—specifying designing and costing out any construction project is as easy as pointing and clicking your mouse—no matter your design experience. For instance, never designed a highway before? No problem. Just point to the 'Highways' window and click.

Simply sign and return this letter today and you'll be among the first engineers to see how this full-featured interactive library of standard design can help you work faster than ever and increase your firm's profits."

Engineer A orders the CD-ROM and begins to offer facilities design and construction services.

_____
NSPE Cases No. 98-3

To students: What is your definition for ethics and ethical conduct? Was it ethical for Engineer A to offer facilities design and construction services under the facts presented?

As eloquently stated in the National Society of Professional Engineers (NSPE) code of ethics, "Engineering is an important and learned profession. As members of this profession, engineers are expected to exhibit the highest standards of honesty and integrity. Engineering has a direct and vital impact on the quality of life for all people. Accordingly, the services provided by engineers require honesty, impartiality, fairness, and equity; and must be dedicated to the protection of the public health, safety, and welfare. Engineers must perform under a standard of professional behavior which requires adherence to the highest principles of ethical conduct." In this chapter, we will discuss the importance of engineering ethics and will present the National Society of Professional Engineers code of ethics in detail. We will also provide two case studies that you may want to discuss in class.

LO¹ 5.1 Engineering Ethics

Ethics refers to the study of morality and the moral choices that we all have to make in our lives. Professional societies, such as medical and engineering, have long established guidelines, standards, and rules that govern the conduct of their members. These rules are also used by the members of the board of ethics of the professional organization to interpret ethical dilemmas that are submitted by a complainant.

As we discussed in Chapter 1, engineers design many products, including cars, computers, aircraft, clothing, toys, home appliances, surgical equipment, heating and cooling equipment, health care devices, tools and machines that make various products. Engineers also design and supervise the construction

> Ethics refers to the study of morality and the moral choices that we all have to make in our lives. Professional societies, such as the NSPE and ASME, have long established guidelines, standards, and rules that govern the conduct of their members.

"A man who has committed a mistake and doesn't correct it is committing another mistake." — CONFUCIUS

of buildings, dams, highways, and mass transit systems. They also design and supervise the construction of power plants that supply power to manufacturing companies, homes, and offices. Engineers play a significant role in the design and maintenance of nations' infrastructures, including communication systems, utilities, and transportation. Engineers are involved in coming up with ways of increasing crop, fruit, and vegetable yields along with improving the safety of our food products.

As you can see, people rely quite heavily on engineers to provide them with safe and reliable goods and services. There is no room for mistakes or dishonesty in engineering! Mistakes made by engineers could cost not only money but also more importantly lives. Think about the following: An incompetent and unethical surgeon could cause at most the death of one person at one time (when a pregnant woman dies on the operating table, two deaths may result), whereas an incompetent and unethical engineer could cause the deaths of hundreds of people at one time. If an unethical engineer in order to save money designs a bridge or a part for an airplane that does not meet the safety requirements, hundreds of people's lives are at risk!

You realize that there are jobs where a person's mistake could be tolerated. For example, if a waiter brings you Coke instead of the Pepsi that you ordered, or instead of french fries brings you onion rings, you can live with that mistake. These are mistakes that usually can be corrected without any harm to anyone. But if an incompetent or unethical engineer incorrectly designs a bridge, or a building, or a plane, he or she could be responsible for killing hundreds of people. Therefore, you must realize why it is so important that as future practicing engineers you are expected to hold to the highest standards of honesty and integrity.

In the section that follows, we will look at an example of a code of ethics, namely, the National Society of Professional Engineers code. The American Society of Mechanical Engineers, the American Society of Civil Engineers, and the Institute of Electrical and Electronics Engineers also have codes of ethics. They are typically posted at their Web sites.

LO² 5.2   The Code of Ethics of the National Society of Professional Engineers

The National Society of Professional Engineers (NSPE) ethics code is very detailed. The NSPE ethical code of conduct is used in making judgments about engineering ethic-related cases that are brought before the NSPE's Board of Ethics Review. The NSPE ethical code of conduct follows.

Code of Ethics for Engineers*

Preamble Engineering is an important and learned profession. As members of this profession, engineers are expected to exhibit the highest standards

*From Code of Ethics for Engineers by National Society of Professional Engineers, Copyright © 2001 National Society of Professional Engineers. Reprinted by permission.

of honesty and integrity. Engineering has a direct and vital impact on the quality of life for all people. Accordingly, the services provided by engineers require honesty, impartiality, fairness, and equity, and must be dedicated to the protection of the public health, safety and welfare. Engineers must perform under a standard of professional behavior that requires adherence to the highest principles of ethical conduct.

I. Fundamental Canons Engineers, in the fulfillment of their professional duties, shall:

1. Hold paramount the safety, health and welfare of the public.
2. Perform services only in areas of their competence.
3. Issue public statements only in an objective and truthful manner.
4. Act for each employer or client as faithful agents or trustees.
5. Avoid deceptive acts.
6. Conduct themselves honorably, responsibly, ethically, and lawfully so as to enhance the honor, reputation and usefulness of the profession.

> Engineers, in the fulfillment of their professional duties, shall hold paramount the safety, health, and welfare of the public.

II. Rules of Practice

1. Engineers shall hold paramount the safety, health, and welfare of the public.

    a. If engineers' judgment is overruled under circumstances that endanger life or property, they shall notify their employer or client and such other authority as may be appropriate.

    b. Engineers shall approve only those engineering documents which are in conformity with applicable standards.

    c. Engineers shall not reveal facts, data, or information without the prior consent of the client or employer except as authorized or required by law or this Code.

    d. Engineers shall not permit the use of their name or associate in business ventures with any person or firm which they believe is engaged in fraudulent or dishonest enterprise.

    e. Engineers having knowledge of any alleged violation of this Code shall report thereon to appropriate professional bodies and, when relevant, also to public authorities, and cooperate with the proper authorities in furnishing such information or assistance as may be required.

2. Engineers shall perform services only in the areas of their competence.

    a. Engineers shall undertake assignments only when qualified by education or experience in the specific technical fields involved.

    b. Engineers shall not affix their signatures to any plans or documents dealing with subject matter in which they lack competence, nor to any plan or document not prepared under their direction and control.

    c. Engineers may accept assignments and assume responsibility for coordination of an entire project and sign and seal the engineering documents for the entire project, provided that each technical segment is signed and sealed only by the qualified engineers who prepared the segment.

> Engineers, in the fulfillment of their professional duties, shall perform services only in areas of their competence.

3. Engineers shall issue public statements only in an objective and truthful manner.

   a. Engineers shall be objective and truthful in professional reports, statements, or testimony. They shall include all relevant and pertinent information in such reports, statements, or testimony, which should bear the date indicating when it was current.

   b. Engineers may express publicly technical opinions that are founded upon knowledge of the facts and competence in the subject matter.

   c. Engineers shall issue no statements, criticisms, or arguments on technical matters which are inspired or paid for by interested parties, unless they have prefaced their comments by explicitly identifying the interested parties on whose behalf they are speaking, and by revealing the existence of any interest the engineers may have in the matters.

4. Engineers shall act for each employer or client as faithful agents or trustees.

   a. Engineers shall disclose all known or potential conflicts of interest which could influence or appear to influence their judgment or the quality of their services.

> Engineers, in the fulfillment of their professional duties, shall act for each employer or client as faithful agents or trustees.

   b. Engineers shall not accept compensation, financial or otherwise, from more than one party for services on the same project, or for services pertaining to the same project, unless the circumstances are fully disclosed and agreed to by all interested parties.

   c. Engineers shall not solicit or accept financial or other valuable consideration, directly or indirectly, from outside agents in connection with the work for which they are responsible.

   d. Engineers in public service as members, advisors, or employees of a governmental or quasi-governmental body or department shall not participate in decisions with respect to services solicited or provided by them or their organizations in private or public engineering practice.

   e. Engineers shall not solicit or accept a contract from a governmental body on which a principal or officer of their organization serves as a member.

5. Engineers shall avoid deceptive acts.

   a. Engineers shall not falsify their qualifications or permit misrepresentation of their or their associates' qualifications. They shall not misrepresent or exaggerate their responsibility in or for the subject matter of prior assignments. Brochures or other presentations incident to the solicitation of employment shall not misrepresent pertinent facts concerning employers, employees, associates, joint venturers or past accomplishments.

   b. Engineers shall not offer, give, solicit, or receive, either directly or indirectly, any contribution to influence the award of a contract by public authority, or which may be reasonably construed by the public as having the effect of intent to influence the awarding of a contract. They shall not offer any gift or other valuable consideration in order to secure work. They shall not pay a commission, percentage, or brokerage fee in order to secure work, except to a bona fide employee or bona fide established commercial or marketing agencies retained by them.

III. Professional Obligations

1. Engineers shall be guided in all their relations by the highest standards of honesty and integrity.

   a. Engineers shall acknowledge their errors and shall not distort or alter the facts.

   b. Engineers shall advise their clients or employers when they believe a project will not be successful.

   c. Engineers shall not accept outside employment to the detriment of their regular work or interest. Before accepting any outside engineering employment, they will notify their employers.

   d. Engineers shall not attempt to attract an engineer from another employer by false or misleading pretenses.

   e. Engineers shall not promote their own interest at the expense of the dignity and integrity of the profession.

2. Engineers shall at all times strive to serve the public interest.

   a. Engineers shall seek opportunities to participate in civic affairs; career guidance for youths; and work for the advancement of the safety, health, and well-being of their community.

   b. Engineers shall not complete, sign, or seal plans and/or specifications that are not in conformity with applicable engineering standards. If the client or employer insists on such unprofessional conduct, they shall notify the proper authorities and withdraw from further service on the project.

   c. Engineers shall endeavor to extend public knowledge and appreciation of engineering and its achievements.

> Engineers, in the fulfillment of their professional duties, shall avoid deceptive acts.

3. Engineers shall avoid all conduct or practice that deceives the public.

   a. Engineers shall avoid the use of statements containing a material misrepresentation of fact or omitting a material fact.

   b. Consistent with the foregoing, engineers may advertise for recruitment of personnel.

   c. Consistent with the foregoing, engineers may prepare articles for the lay or technical press, but such articles shall not imply credit to the author for work performed by others.

4. Engineers shall not disclose, without consent, confidential information concerning the business affairs or technical processes of any present or former client or employer, or public body on which they serve.

   a. Engineers shall not, without the consent of all interested parties, promote or arrange for new employment or practice in connection with a specific project for which the engineer has gained particular and specialized knowledge.

   b. Engineers shall not, without the consent of all interested parties, participate in or represent an adversary interest in connection with a specific project or proceeding in which the engineer has gained particular specialized knowledge on behalf of a former client or employer.

5. Engineers shall not be influenced in their professional duties by conflicting interests.

   a. Engineers shall not accept financial or other considerations, including free engineering designs, from material or equipment suppliers for specifying their product.

   b. Engineers shall not accept commissions or allowances, directly or indirectly, from contractors or other parties dealing with clients or employers of the engineer in connection with work for which the engineer is responsible.

6. Engineers shall not attempt to obtain employment or advancement or professional engagements by untruthfully criticizing other engineers, or by other improper or questionable methods.

   a. Engineers shall not request, propose, or accept a commission on a contingent basis under circumstances in which their judgment may be compromised.

   b. Engineers in salaried positions shall accept part-time engineering work only to the extent consistent with policies of the employer and in accordance with ethical considerations.

   c. Engineers shall not, without consent, use equipment, supplies, laboratory, or office facilities of an employer to carry on outside private practice.

> Engineers, in the fulfillment of their professional duties, shall issue public statements only in an objective and truthful manner.

7. Engineers shall not attempt to injure, maliciously or falsely, directly or indirectly, the professional reputation, prospects, practice, or employment of other engineers. Engineers who believe others are guilty of unethical or illegal practice shall present such information to the proper authority for action.

   a. Engineers in private practice shall not review the work of another engineer for the same client, except with the knowledge of such engineer, or unless the connection of such engineer with the work has been terminated.

   b. Engineers in governmental, industrial, or educational employ are entitled to review and evaluate the work of other engineers when so required by their employment duties.

   c. Engineers in sales or industrial employ are entitled to make engineering comparisons of represented products with products of other suppliers.

8. Engineers shall accept personal responsibility for their professional activities, provided, however, that engineers may seek indemnification for services arising out of their practice for other than gross negligence, where the engineer's interests cannot otherwise be protected.

   a. Engineers shall conform with state registration laws in the practice of engineering.

   b. Engineers shall not use association with a nonengineer, a corporation, or partnership as a "cloak" for unethical acts.

9. Engineers shall give credit for engineering work to those to whom credit is due, and will recognize the proprietary interests of others.

   a. Engineers shall, whenever possible, name the person or persons who may be individually responsible for designs, inventions, writings, or other accomplishments.

b. Engineers using designs supplied by a client recognize that the designs remain the property of the client and may not be duplicated by the engineer for others without express permission.

c. Engineers, before undertaking work for others in connection with which the engineer may make improvements, plans, designs, inventions, or other records that may justify copyrights or patents, should enter into a positive agreement regarding ownership.

d. Engineers' designs, data, records, and notes referring exclusively to an employer's work are the employer's property. Employer should indemnify the engineer for use of the information for any purpose other than the original purpose.

As Revised February 2001 "By order of the United States District Court for the District of Columbia, former Section 11(c) of the NSPE Code of Ethics prohibiting competitive bidding, and all policy statements, opinions, rulings or other guidelines interpreting its scope, have been rescinded as unlawfully interfering with the legal right of engineers, protected under the antitrust laws, to provide price information to prospective clients; accordingly, nothing contained in the NSPE Code of Ethics, policy statements, opinions, rulings or other guidelines prohibits the submission of price quotations or competitive bids for engineering services at any time or in any amount."

Statement by NSPE Executive Committee In order to correct misunderstandings which have been indicated in some instances since the issuance of the Supreme Court decision and the entry of the Final Judgment, it is noted that in its decision of April 25, 1978, the Supreme Court of the United States declared: "The Sherman Act does not require competitive bidding."

It is further noted that as made clear in the Supreme Court decision:

1. Engineers and firms may individually refuse to bid for engineering services.

2. Clients are not required to seek bids for engineering services.

3. Federal, state, and local laws governing procedures to procure engineering services are not affected, and remain in full force and effect.

4. State societies and local chapters are free to actively and aggressively seek legislation for professional selection and negotiation procedures by public agencies.

5. State registration board rules of professional conduct, including rules prohibiting competitive bidding for engineering services, are not affected and remain in full force and effect. State registration boards with authority to adopt rules of professional conduct may adopt rules governing procedures to obtain engineering services.

6. As noted by the Supreme Court, "nothing in the judgment prevents NSPE and its members from attempting to influence governmental action …".

---

[1]"Sustainable development" is the challenge of meeting human needs for natural resources, industrial products, energy, food, transportation, shelter, and effective waste management while conserving and protecting environmental quality and the natural resource base essential for future development. - See more at: http://www.nspe.org/resources/ethics/code-ethics#sthash.P8yBvwSc.dpuf

Note: In regard to the question of application of the Code to corporations vis-a-vis real persons, business form or type should not negate nor influence conformance of individuals to the Code. The Code deals with professional services, which services must be performed by real persons. Real persons in turn establish and implement policies within business structures. The Code is clearly written to apply to the engineer and items incumbent on members of NSPE to endeavor to live up to its provisions. This applies to all pertinent sections of the Code.

LO³ 5.3   Engineer's Creed

The engineer's creed, which was adopted by NSPE in 1954, is a statement of belief, similar to the Hippocratic oath taken by medical practitioners. It was developed to state the engineering philosophy of service in a brief way. The NSPE **Engineer's Creed** is:

- To give the utmost of performance;
- To participate in none but honest enterprise;
- To live and work according to the laws of man and the highest standards of professional conduct; and
- To place
  - service before profit,
  - the honor and standing of the profession before personal advantage, and
  - the public welfare above all other considerations.

In humility and with need for Divine guidance, I make this pledge.

The engineer's creed is typically used in graduation ceremonies or licensure certificate presentations.

These are additional definitions that should be studied carefully.

Academic Dishonesty—Honesty is very important in all aspects of life. Academic dishonesty refers to behavior that includes cheating on tests, homework assignments, lab reports; plagiarism; lying about being sick and not taking a test because of it; signing the attendance sheet for another student, or asking another student to sign the sheet for you in your absence. Universities have different policies that deal with academic dishonesty, including giving the dishonest student a failing grade for the course or requiring the student to drop the class or placing a student on probation.

Plagiarism—Plagiarism refers to presenting someone else's work as your own. You may use or cite the work of others including information from journal articles, books, online sources, TV or radio, but make sure that you cite where you obtain the information from. In Chapter 4, we discussed in detail how you should give proper reference in your oral and written communications.

Conflict of Interest—A conflict between the individual's personal interests and the individual's obligations because of the position he or she holds.

Contract—Contract is an agreement among two or more parties, which they entered into freely. A legal contract is one that is legally binding, meaning if not fulfilled it could have legal consequences.

Professional Responsibility—It is the responsibility associated with the mastery of special kind of knowledge that a person possesses and the use of knowledge for well-being and benefit of the society.

Read the cases—from the following list—assigned to you by your instructor before class and be prepared to discuss them in class.

2012 Milton F. Lunch Ethics Contest

Facts Engineer A works for Company X, which is owned by Engineer B. Company X is currently experiencing financial problems and Engineer B recently created another company, Company Y. Engineer A has learned that Engineer B recently advised clients of Company X to remit payments for work performed by Company X and its employees to Company Y.

Question What are Engineer A's ethical obligations under the circumstances?

2013 Milton F. Lunch Ethics Contest

Facts A marketing company establishes a Web portal and offers a service to customers whereby customers type in questions on various topics (e.g., law, medicine, accounting, engineering, etc.) and, following the receipt of the responses, which are generally fairly detailed responses, the customer pays the marketing company what the customer believes the service is worth, plus an access fee for the Web portal. Following receipt of the payment, the marketing company passes along the customer payment to the service provider (lawyer, physician, accountant, engineer, etc.). Engineer A, a structural engineer, wants to know if it would be ethical for him to participate in this type of business.

Question Would it be ethical for Engineer A, a structural engineer, to participate in this type of business?

The following are ethics-related cases that were brought before NSPE's Board of Ethical Review. These cases were adapted with permission from the National Society of Professional Engineers.*

Confidentiality of Engineering Report: Case No. 82-2

Facts Engineer A offers a home owner inspection service, whereby he undertakes to perform an engineering inspection of residences by prospective purchasers. Following the inspection, Engineer A renders a written report to the prospective purchaser. Engineer A performed this service for a client (husband and wife) for a fee and prepared a one-page written report, concluding that the residence under consideration was in generally good condition requiring no major repairs, but noting several minor items needing attention. Engineer A submitted his report to the client showing that a carbon copy was sent to the real estate firm handling the sale of the residence. The client objected that

Source: Reprinted by Permission of the National Society of Professional Engineers (NSPE) www.nspe.org

such action prejudiced their interests by lessening their bargaining position with the owners of the residence. They also complained that Engineer A acted unethically in submitting a copy of the report to any others who had not been a party to the agreement for the inspection services.

Question Did Engineer A act unethically in submitting a copy of the home inspection report to the real estate firm representing the owners?

Gift Sharing of Hotel Suite: Case No. 87-4

Facts Engineer B is director of engineering with a large governmental agency that uses many engineering consultants. Engineer A is a principal in a large engineering firm that performs services for that agency. Both are members of an engineering society that is conducting a two-day seminar in a distant city. Both plan to attend the seminar, and they agree to share costs of a two-bedroom hotel suite in order to have better accommodations.

Question Was it ethical for Engineer A and B to share the hotel suite?

Credit for Engineering Work-Design Competition: Case No. 92-1

Facts Engineer A is retained by a city to design a bridge as part of an elevated highway system. Engineer A then retains the services of Engineer B, a structural engineer with expertise in horizontal geometry, superstructure design, and elevations to perform certain aspects of the design services. Engineer B designs the bridge's three curved welded plate girder spans, which were critical elements of the bridge design.

Several months following completion of the bridge, Engineer A enters the bridge design into a national organization's bridge design competition. The bridge design wins a prize. However, the entry fails to credit Engineer B for his part of the design.

Question Was it ethical for Engineer A to fail to give credit to Engineer B for his part in the design?

Services-Same Services for Different Clients: Case No. 00-3

Facts Engineer A, a professional engineer, performs a traffic study for Client X as part of the client's permit application for traffic flow for the development of a store. Engineer A invoices Client X for a complete traffic study.

Later, Client X learns that part of the traffic study provided by Engineer A to Client X was earlier developed by Engineer A for a developer, Client Y, at a nearby location and that Engineer A invoiced Client Y for the complete traffic study. The second study on a new project for Client X utilized some of the same raw data as was in the report prepared for Client Y. The final conclusion of the engineering study was essentially the same in both studies.

Question Was it ethical for Engineer A to charge Client X for the complete traffic study?

Use of Alleged Hazardous Material in a Processing Facility: Case No. 99-11

Facts Engineer A is a graduate engineer in a company's manufacturing facility that uses toxic chemicals in its processing operations. Engineer A's job has nothing to do with the use and control of these materials.

A chemical called "MegaX" is used at the site. Recent stories in the news have reported alleged immediate and long-term human genetic hazards from inhalation of or other contact with MegaX. The news items are based on findings from laboratory experiments, which were done on mice, by a graduate student at a well-respected university's physiology department. Other scientists have neither confirmed nor refuted the experimental findings. Federal and local governments have not made official pronouncements on the subject.

Several colleagues outside of the company have approached Engineer A on the subject and ask Engineer A to "do something" to eliminate the use of MegaX at the processing facility. Engineer A mentions this concern to her manager who tells Engineer A, "Don't worry, we have an Industrial Safety Specialist who handles that."

Two months elapse and MegaX is still used in the factory. The controversy in the press continues, but since there is no further scientific evidence pro or con in the matter, the issues remain unresolved. The use of the chemical in the processing facility has increased and now more workers are exposed daily to the substance than was the case two months ago.

Question Does Engineer A have an obligation to take further action under the facts and circumstances?

Software Design Testing: NSPEBER Case No. 96-4

Facts Engineer A is employed by a software company and is involved in the design of specialized software in connection with the operations of facilities affecting the public health and safety (i.e., nuclear, air quality control, water quality control). As part of the design of a particular software system, Engineer A conducts extensive testing, and although the tests demonstrate that the software is safe to use under existing standards, Engineer A is aware of new draft standards that are about to be released by a standard setting organization— standards that the newly designed software may not meet. Testing is extremely costly, and the company's clients are eager to begin to move forward. The software company is eager to satisfy its clients, protect the software company's finances, and protect existing jobs; but at the same time, the management of the software company wants to be sure that the software is safe to use. A series of tests proposed by Engineer A will likely result in a decision whether to move forward with the use of the software. The tests are costly and will delay the use of the software at least six months, which will put the company at a competitive disadvantage and cost the company a significant amount of money. Also, delaying implementation will mean the state public service commission utility rates will rise significantly during this time. The company requests Engineer A's recommendation concerning the need for additional software testing.

Question Under the Code of Ethics, does Engineer A have a professional obligation to inform his company of the reasons for needed additional testing and his recommendations that it be undertaken?

Whistleblowing: Case No. 82-5

Facts Engineer A is employed by a large industrial company that engages in substantial work on defense projects. Engineer A's assigned duties relate to the work of subcontractors, including review of the adequacy and acceptability of the plans for material provided by subcontractors. In the course of this work, Engineer A advised his superiors by memoranda of problems he found with certain submissions of one of the subcontractors, and urged management to reject such work and require the subcontractors to correct the deficiencies he outlined. Management rejected the comments of Engineer A, particularly his proposal that the work of a particular subcontractor be redesigned because of Engineer A's claim that the subcontractor's submission represented excessive cost and time delays. After the exchange of further memoranda between Engineer A and his management superiors and continued disagreement between Engineer A and management on the issues he raised, management placed a critical memorandum in his personnel file and subsequently placed him on three months' probation, with the further notation that if his job performance did not improve, he would be terminated. Engineer A has continued to insist that his employer had an obligation to ensure that subcontractors deliver equipment according to the specifications, as he interprets same, and thereby save substantial defense expenditures. He has requested an ethical review and determination of the propriety of his course of action and the degree of ethical responsibility of engineers in such circumstances.

Question Does Engineer A have an ethical obligation, or an ethical right, to continue his efforts to secure change in the policy of his employer under these circumstances, or to report his concerns to proper authority?

Academic Qualifications: Case No. 79-5

Facts Engineer A received a Bachelor of Science degree in 1940 from a recognized engineering curriculum and subsequently was registered as a professional engineer in two states. Later, he was awarded an earned "Professional Degree" from the same institution. In 1960 he received a Ph.D. degree from an organization that awards degrees on the basis of correspondence without requiring any form of personal attendance or study at the institution and is regarded by state authorities as a "diploma mill." Engineer A has since listed his Ph.D. degree among his academic qualifications in brochures, correspondence, and otherwise, without indicating its nature.

Question Was Engineer A ethical in citing his Ph.D. degree as an academic qualification under these circumstances?

Advertising—Misstating Credentials: Case No. 92-2

Facts Engineer A is an EIT who is employed by a medium-sized consulting engineering firm in a small city. Engineer A has a degree in mechanical engineering and has performed services almost exclusively in the field of mechanical engineering. Engineer A learns that the firm has begun a marketing campaign and in its literature lists Engineer A as an electrical engineer. There are other electrical engineers in the firm. Engineer A alerts the marketing director, also an engineer, to the error in the promotional literature, and the marketing director indicates that the error will be corrected. However, after a period of six months, the error is not corrected.

Question Under the circumstances, what actions, if any, should Engineer A take?

Advertising–Statement of Project Success: Case No. 79-6

Facts Engineer A published an advertisement in the classified section of a daily newspaper under the heading, "Business Services," which read in full: "Consulting Engineer for Industry. Can reduce present process heating fuel consumption by 30% to 70% while doubling capacity in same floor space. For more information contact Engineer A, telephone 123-456-7890."

Question Was Engineer A's advertisement ethical?

Using Technical Proposal of Another Without Consent: Case No. 83-3

"Keep company with good men, and you'll increase their number." —ITALIAN PROVERB

Facts Engineer B submitted a proposal to a county council following an interview concerning a project. The proposal included technical information and data that the council requested as a basis for the selection. Smith, a staff member of the council, made Engineer B's proposal available to Engineer A. Engineer A used Engineer B's proposal without Engineer B's consent in developing another proposal, which was subsequently submitted to the council. The extent to which Engineer A used Engineer B's information and data is in dispute between the parties.

Question Was it unethical for Engineer A to use Engineer B's proposal without Engineer B's consent in order for Engineer A to develop a proposal that Engineer A subsequently submitted to the council?

Before You Go On

Answer the following questions to test your understanding of the preceding sections.

1. In your own words, explain what is meant by ethics.

2. What is engineering ethics and why is important to have established guidelines, standards, and rules?

3. Give two examples of the fundamental canons of the NSPE's Code of Ethics.

4. Give two examples of the NSPE's Professional Obligations.

5. What is the NSPE's Engineer's Creed?

6. Give two examples of Engineer's creed.

Vocabulary—State the meaning of the following terms:

Ethics _____

Conflict of Interest _____

Academic Dishonesty _____

Plagiarism _____

Contract _____

SUMMARY

LO¹   Engineering Ethics

By now you should have learned that ethics refers to the study of morality and the moral choices that we all have to make in our lives. Moreover, professional engineering societies have established guidelines, standards, and rules that govern the conduct of their members. These rules are also used by the members of the board of ethics of professional organizations to interpret ethical dilemmas that are submitted by a complainant.

LO²   The Code of Ethics of the National Society of Professional Engineers

The NSPE's Code of Ethics for Engineers includes a preamble, fundamental canons, rules of practice, and professional obligations. The fundamental canons states that "engineers, in the fulfillment of their professional duties, shall:

- Hold paramount the safety, health and welfare of the public.
- Perform services only in areas of their competence.
- Issue public statements only in an objective and truthful manner.
- Act for each employer or client as faithful agents or trustees.

- Avoid deceptive acts.
- Conduct themselves honorably, responsibly, ethically, and lawfully so as to enhance the honor, reputation and usefulness of the profession."

LO³   Engineer's Creed

The Engineer's Creed is a statement of belief that is similar to the Hippocratic Oath taken by physicians. It was developed to state the engineering philosophy of service in a brief way. Examples include: to give the utmost of performance; to participate in none but honest enterprise; and to place service before profit.

LO⁴   Academic Dishonesty, Conflict of Interest, Professional Responsibility

Academic dishonesty refers to behavior that includes cheating on tests, homework assignments, and lying about being sick (and not taking a test because of it). Conflict of interest refers to a conflict between the individual's personal interests and the individual's obligations because of the position he or she holds. Professional responsibility refers to the responsibility associated with the mastery of special kind of knowledge that a person possesses and the use of knowledge for the well-being and benefit of society.

KEY TERMS

Academic Dishonesty 132
Conflict of Interest 132
Contract 132
Engineer's Creed 132

Engineering Ethics 125
Ethics 125
Fundamental Canons 127
Plagiarism 132

Professional Obligations 129
Professional Responsibility 133
Rules of Practice 127

APPLY WHAT YOU HAVE LEARNED

Visit the National Society of Engineers site at www.nspe.org and look up the Milton F. Lunch Ethics Contest for the current year. Download the contest rules, and read them carefully. All entries must be 750 words or less (Discussion and Conclusion Sections only) and received by a certain date. Good Luck!

PROBLEMS

Problems that promote lifelong learning are denoted by. 🔑

5.1 The following is a series of questions pertaining to the NSPE Code of Ethics. Please indicate whether the statements are true or false. These questions are provided by the NSPE.

Note: This ethics test is intended solely to test individual knowledge of the specific language contained in the NSPE Code of Ethics and is not intended to measure individual knowledge of engineering ethics or the ethics of individual engineers or engineering students.

1. Engineers in the fulfillment of their professional duties must carefully consider the safety, health, and welfare of the public.
2. Engineers may perform services outside of their areas of competence as long as they inform their employer or client of this fact.
3. Engineers may issue subjective and partial statements if such statements are in writing and consistent with the best interests of their employer, client or the public.
4. Engineers shall act for each employer or client as faithful agents or trustees.

5. Engineers shall not be required to engage in truthful acts when required to protect the public health, safety, and welfare.
6. Engineers may not be required to follow the provisions of state or federal law when such actions could endanger or compromise their employer or their client's interests.
7. If engineers' judgment is overruled under circumstances that endanger life or property, they shall notify their employer or client and such other authority as may be appropriate.
8. Engineers may review but shall not approve those engineering documents that are in conformity with applicable standards.
9. Engineers shall not reveal facts, data or information without the prior consent of the client or employer except as authorized or required by law or this Code.
10. Engineers shall not permit the use of their name or their associate's name in business ventures with any person

or firm that they believe is engaged in fraudulent or dishonest enterprise, unless such enterprise or activity is deemed consistent with applicable state or federal law.

11. Engineers having knowledge of any alleged violation of this Code, following a period of thirty days during which the violation is not corrected, shall report thereon to appropriate professional bodies and, when relevant, also to public authorities, and cooperate with the proper authorities in furnishing such information or assistance as may be required.

12. Engineers shall undertake assignments only when qualified by education or experience in the specific technical fields involved.

13. Engineers shall not affix their signatures to plans or documents dealing with subject matter in which they lack competence, but may affix their signatures to plans or documents not prepared under their direction and control where the engineer has a good faith belief that such plans or documents were competently prepared by another designated party.

14. Engineers may accept assignments and assume responsibility for coordination of an entire project and shall sign and seal the engineering documents for the entire project, including each technical segment of the plans and documents.

15. Engineers shall strive to be objective and truthful in professional reports, statements or testimony, with primary consideration for the best interests of the engineer's client or employer. The engineer's reports shall include all relevant and pertinent information in such reports, statements or testimony, which shall bear the date on which the engineer was retained by the client to prepare the reports.

16. Engineers may express publicly technical opinions that are founded upon knowledge of the facts and competence in the subject matter.

17. Engineers shall issue no statements, criticisms, or arguments on technical matters that are inspired or paid for by interested parties, unless they have prefaced their comments by explicitly identifying the interested parties on whose behalf they are speaking, and by revealing the existence of any interest the engineers may have in the matters.

18. Engineers may not participate in any matter involving a conflict of interest if it could influence or appear to influence their judgment or the quality of their services.

19. Engineers shall not accept compensation, financial or otherwise, from more than one party for services on the same project, or for services pertaining to the same project, unless the circumstances are fully disclosed and agreed to by all interested parties.

20. Engineers shall not solicit but may accept financial or other valuable consideration, directly or indirectly, from outside agents in connection with the work for which they are responsible, if such compensation is fully disclosed.

21. Engineers in public service as members, advisors or employees of a governmental or quasi-governmental body or department may participate in decisions with respect to services solicited or provided by them or their organizations in private or public engineering practice as long as such decisions do not involve technical engineering matters for which they do not possess professional competence.

22. Engineers shall not solicit or accept a contract from a governmental body on which a principal or officer of their organization serves as a member.

23. Engineers shall not intentionally falsify their qualifications or actively permit written misrepresentation of their or their associate's qualifications. Engineers may accept credit for previous work performed where the work was performed during the period the engineer was employed by the previous employer. Brochures or other presentations incident to the solicitation of employment shall specifically indicate the work performed and the dates the engineer was employed by the firm.

24. Engineers shall not offer, give, solicit, or receive, either directly or indirectly, any contribution to influence the award of a contract by a public authority, or which may be reasonably construed by the public as having the effect or intent of influencing the award of a contract unless such contribution is made in accordance with applicable federal or state election campaign finance laws and regulations.

25. Engineers shall acknowledge their errors after consulting with their employer or client.

Engineering Ethics

A Case Study From NSPE*

The following is an ethics-related case that was brought before NSPE's Board of ethics review.

Facts

Engineer A is a licensed professional engineer and a principal in a large-sized engineering firm. Engineer B is a graduate engineer who works in industry and has also worked as a student in Engineer A's firm during one summer. Although Engineer B was employed in Engineer A's firm, Engineer A did not have direct knowledge of Engineer B's work. Engineer B is applying for licensure as a professional engineer and requests that Engineer A provide him with a letter of reference testifying as to Engineer B's engineering experience and that the engineer (Engineer A) was in direct charge of Engineer B. Engineer B was under the assumption that Engineer A had personal knowledge of Engineer B's work. Engineer A inquired about Engineer B's experience from someone who had direct knowledge of Engineer B's experience. Based on the inquiry, Engineer A provides the letter of reference explaining the professional relationship between Engineer A and Engineer B.

Question

Was it ethical for Engineer A to provide the letter of reference for Engineer B attesting as to Engineer B's engineering experience even though Engineer A did not have direct control of Engineer B's engineering work?

References

Section II.3.— Code of Ethics: Engineers shall issue public statements only in an objective and truthful manner.

Section II.3.a.— Code of Ethics: Engineers shall be objective and truthful in professional reports, statements or testimony. They shall include all relevant and pertinent information in such reports, statements or testimony, which should bear the date indicating when it was current.

Section II.5.a.— Code of Ethics: Engineers shall not falsify their qualifications or permit misrepresentation of their, or their associates' qualifications. They shall not misrepresent or exaggerate their responsibility in or for the subject matter of prior assignments. Brochures or other presentations incident to the solicitation of employment shall not misrepresent pertinent facts concerning employers, employees, associates, joint venturers or past accomplishments.

Section III.1.— Code of Ethics: Engineers shall be guided in all their relations by the highest standards of honesty and integrity.

Section III.8.a.— Code of Ethics: Engineers shall conform with state registration laws in the practice of engineering.

Discussion

The Board has, on prior occasions, considered cases involving the misstatement of credentials of an engineer employed in a firm. In BER Case No. 92-1,

*Reprinted by Permission of the National Society of Professional Engineers (NSPE) www.nspe.org

Engineer A was an EIT who was employed by a medium-sized consulting engineering firm in a small city. Engineer A had a degree in mechanical engineering and had performed services almost exclusively in the field of mechanical engineering. Engineer A learned that the firm had begun a marketing campaign and in its literature listed Engineer A as an electrical engineer. There were other electrical engineers in the firm. Engineer A alerted the marketing director, also an engineer, to the error in the promotional literature and the marketing director indicated that the error would be corrected. However, after a period of six months, the error was not corrected. In ruling that the firm should take actions to correct the error, the Board noted that the firm's marketing director had been informed by the engineer in question that the firm's marketing brochure contained inaccurate information that could mislead and deceive a client or potential client. Under earlier BER Case No. 90-4, the marketing director had an ethical obligation to take expeditious action to correct the error. The Board noted that the marketing director, a professional engineer, had an ethical obligation both to the clients and potential clients, as well as to Engineer A, to expeditiously correct the misimpression which may have been created.

The Board of Ethical Review can certainly understand in the present case the desire of Engineer A to assist another engineer (Engineer B) in enhancing career opportunities and becoming licensed as a professional engineer. Obviously such assistance should not come under misleading or deceptive circumstances. Engineers have an ethical obligation to be honest and objective in their professional reports, and such reports include written assessments of the qualifications and abilities of engineers and others under their direct supervision. Engineers that are not in a position to offer an evaluation of the qualifications and abilities of other individuals should not provide such evaluations or prepare reports that imply that they are providing such evaluations. Claiming to be in responsible charge of another engineer without actually having direct control or personal supervision over that engineer is inconsistent with the letter and the spirit of the NSPE Code.

By providing the report in the manner described, the Board believes Engineer A is sending the right message to Engineer B about what will be expected of Engineer B and his colleagues as professional engineers. Clearly, Engineer B desired the letter of reference from Engineer A, a principal in a consulting firm, in order to improve his chances to become licensed as a professional engineer, and Engineer B is taking conscientious action to address the request. Professional engineers must always be mindful that their conduct and actions as professional engineers set an example for other engineers, particularly those that are beginning their professional careers and who are looking for models and mentors upon which to build their professional identities. A professional engineer providing such a letter of reference should demonstrate that the author has obtained sufficient information about the candidate to write a letter of substance and detail the individuals technical abilities as well as the individual's character. A letter of recommendation for engineering licensure generally requires the recommending professional engineer to state in detail that the candidate possesses legitimate and progressive engineering work experience.

The Board is of the view that an alternative approach could have been for Engineer A to refer Engineer B back to the engineer in the firm that was in responsible charge of engineering for the letter of recommendation. However, the letter provided by Engineer A was just as adequate and ethical.

Conclusion

It was ethical for Engineer A to provide the letter of reference for Engineer B testifying as to Engineer B's engineering experience.

Board of Ethical Review

Lorry T. Bannes, P.E., NSPE
E. Dave Dorchester, P.E., NSPE
John W. Gregorits, P.E., NSPE
Paul E. Pritzker, P.E., NSPE
Richard Simberg, P.E., NSPE
Harold E. Williamson, P.E., NSPE
C. Allen Wortley, P.E., NSPE, Chair

PART 2

Engineering Fundamentals

Concepts Every Engineer Should Know

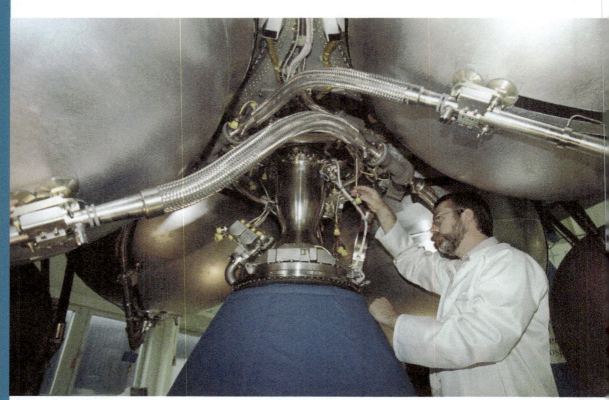

Engineers are problem solvers. Successful engineers possess good communication skills and are team players. Successful engineers have a good grasp of fundamentals, which they can use to understand and solve many different problems. In Part Two of this book, we will focus on these engineering fundamentals. These are concepts that every engineer, regardless of his or her area of specialization, should know. From our observation of our surroundings, we have learned that we need only a few physical quantities to describe events and our surroundings. These quantities are length, time, mass, force, temperature, mole, and electric current. There are also many design variables that are related to these fundamental quantities. For example, dimension of length is needed to describe how tall, how long, or how wide something is. The fundamental dimension of length and length-related variables, such as area and volume, play important roles in engineering design. To become a successful engineer, you need to first fully understand these fundamental and related variables. Then, it is important for you to know how these variables are measured, approximated, calculated, or used in engineering formulas.

CHAPTER 6	FUNDAMENTAL DIMENSIONS AND UNITS
CHAPTER 7	LENGTH AND LENGTH-RELATED VARIABLES
CHAPTER 8	TIME AND TIME-RELATED VARIABLES
CHAPTER 9	MASS AND MASS-RELATED VARIABLES
CHAPTER 10	FORCE AND FORCE-RELATED VARIABLES
CHAPTER 11	TEMPERATURE AND TEMPERATURE-RELATED VARIABLES
CHAPTER 12	ELECTRIC CURRENT AND RELATED VARIABLES
CHAPTER 13	ENERGY AND POWER

Fundamental Dimensions and Units

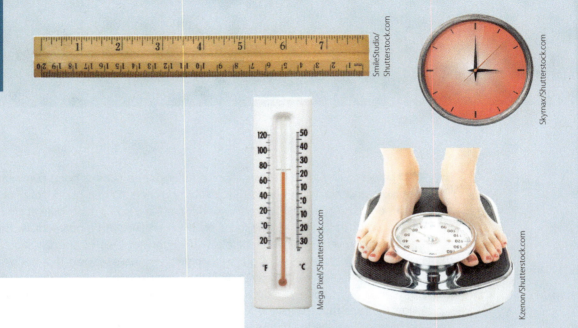

SmileStudio/Shutterstock.com

Skymax/Shutterstock.com

Mega Pixel/Shutterstock.com

Kzenon/Shutterstock.com

Dimension is a physical quantity, such as length, time, mass, or temperature, that makes it possible for us to communicate. For example, length is needed to describe how tall we are, or how wide, or how long a room is. To describe how cold or hot something is, we need a physical quantity or dimension that we refer to as temperature. Time is another physical dimension that allows us to explain our surroundings and answer questions such as: "How old are you?" We also have learned that some things are more massive than other things, so there is a need for another physical dimension to describe that observation.

LEARNING OBJECTIVES

LO¹ **Fundamental Dimensions and Units:** explain what they mean and give examples

LO² **Systems of Units:** describe what system of units represent and give examples of SI (metric), British, and U.S. Customary units for length, time, mass, force, and temperature

LO³ **Unit Conversion and Dimensional Homogeneity:** know how to convert data from the SI units to British and U.S. Customary units (and vice versa) and check for dimensional homogeneity in formulas

LO⁴ **Significant Digits (Figures):** explain the extent to which recoded or computed data is dependable

LO⁵ **Components and Systems:** describe what they mean and give examples

LO⁶ **Physical Laws and Observations:** state what they mean and give examples

WHAT DO YOU THINK?

CHICAGO—William Holdorf's amusingly vehement reaction against the metric system ("A foolish U.S. push to go metric," *Voice*, Jan. 3) illustrates the very American habit of flaring our individualism at the slightest provocation. People are understandably put off by the argument that we must go metric because everyone else has, but the truth is that we do benefit from adopting certain universal conventions. Like Arabic numerals and the English language, the metric system is convenient and powerful just by being so widespread.

But that's only a circumstantial advantage. As Mr. Holdorf correctly implies, a foot is still exactly as accurate as a meter. Accuracy depends on our instruments and technique, not on our choice of units. What's the big deal, then? Is our government just trying to appease those foreigners? No. The metric system actually is better. Its units are systematically, uniformly, and therefore predictably divided in multiples of 10. In order to convert inches to feet to yards to rods to furlongs to miles, you must divide by 12, 3, 5.5, 40, and 8, respectively. To go from centimeters to decimeters to meters to decameters to hectometers to kilometers, you divide, respectively, by 10, 10, 10, 10, and 10.

Which list of numbers would you rather memorize in school? Which would you prefer if your work required frequent measurements and conversions? In the metric system, all you ever do is add or delete zeros or move the decimal point around, whereas conversions with the English system usually need long division or a calculator.

The metric system was tailored to our good old decimal-number system, which, for those who find comfort in the anatomical implications of feet and inches, is in turn a monument to that accident of evolution that gave us 10 fingers. For better or for worse, this has given us a familiarity with the number 10 and its multiples, which is why we use them as yardsticks (sorry!) in phrases like "parts per million," "mortality rate per 1,000 live births," and "30 percent." Familiarity is the only thing that makes our system seem easier. The metric system looks hard to fathom (yikes!) only because we never get to pour a liter or lift a kilogram. In many countries where the commercial influence of the United States ensures frequent contact with our units, people have no problem buying sugar by the pound while using kilos for official and technical matters.

I agree with Mr. Holdorf: Scientifically, we are still in a league (oops!) of our own, and not going metric won't send us back to the Stone Age. I also agree that the public cannot—and should not—be forced to use it, which is why the Metric Conversion Act of 1975 merely encourages voluntary conversion. But I think it is perfectly appropriate for the government to gently prod its contractors and agencies into becoming familiar with the metric system.

The American heart and soul are not at stake here. This is an issue of tidiness. The U.S. will inch (ouch!) toward the metric system through gradual exposure. The government might as well have some fun and adopt as a slogan my favorite reason for going metric: It builds character.

Manuel Sanchez, Chicago Tribune, February 17, 1996

To the students: As you can see, not much has changed since 1996. What are your thoughts? Do you think we should switch to metric units, and if so, what are the benefits?

I n this chapter, we will explain fundamental engineering dimensions, such as length and time, and their units, such as meter and second, and their role in engineering analysis and design. As an engineering student and later as practicing engineer, when performing an analysis, you will find a need to convert from one system of units to another. We will explain the steps necessary to convert information from one system of units to another correctly.

In this chapter, we will also emphasize the fact that you must always show the appropriate units that go with your calculations. Moreover, we will explain what is meant by an engineering system and an engineering component. Finally, we will explain that physical laws are based on observations, and we will use mathematics to express our observations in the form of useful equations.

LO¹ 6.1   Fundamental Dimensions and Units

In this section, we will introduce you to the concepts of dimension and unit. You have been using these concepts most of your life; here, in this chapter, we will define them in a formal way. For example, when asked how tall you are, you may respond with "I am 6 feet (183 centimeters) tall." Or when asked what is the expected outside air temperature today, you could answer with something like: "Today is going to be hot and reach 100° F $\left(38° C\right)$."

The evolution of the human intellect has taken shape over a period of thousands of years. Men and women all over the world observed and learned from their surroundings. They used the knowledge gained from their observations of nature to design, develop, test, and fabricate tools, shelter, weapons, water transportation, and the means to cultivate and produce more food.

> Dimension is a physical quantity such as length, time, mass, and temperature that makes it possible for us to describe our surroundings and events.

Moreover, they realized that they needed only a few physical quantities (*dimensions*) to fully describe natural events and their surroundings. *Dimension* is a physical quantity, such as *length, time, mass*, or *temperature,* that makes it possible for us to communicate. For example, the dimension of length was needed to describe how tall, how long, or how wide something was. They also learned that some things are heavier than other things, so there was a need for another physical quantity (dimension) to describe that observation: the concept of **mass** and *weight*. Early humans did not fully understand the concept of gravity; consequently, the correct distinction between mass and weight (which is a force) was made later.

What is **force**? The simplest form of a force that represents the interaction of two objects is a push or a pull. When you push or pull on a vacuum cleaner, that interaction between your hand and the vacuum cleaner is called *force*. In this example, the force is exerted by one body (your hand) on another body (vacuum cleaner) by direct contact. Not all forces result from direct contact. For example, gravitational force is not exerted by direct contact. If you hold this book, say, 3 feet above the ground and let it go, what happens? It falls; that is due to gravitational force that is exerted by the earth on the book. The gravitational attractive forces act at a distance. The weight of an object is the force that is exerted on the mass of the object by the earth's gravity. We will discuss the concepts of mass and force in much greater detail in Chapters 9 and 10.

Time was another physical dimension that humans needed to understand in order to be able to explain their surroundings and answer questions such

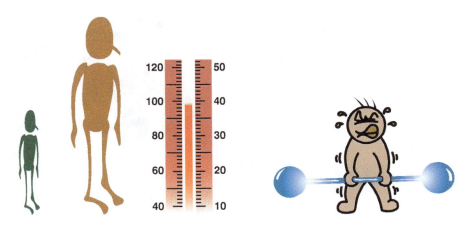

as: "How old are you?" "How long does it take to go from here to there?" The response to these questions in those early days may have been something like this: "I am many moons old," or "It takes a couple of moons to go from our village to the other village on the other side of the mountains." Moreover, to describe how cold or hot something was, humans needed yet another physical quantity, or physical dimension, that we now refer to as *temperature*. Think about the important role of temperature in your everyday life in describing various states of things. Do you know the answer to some of these questions: What is your body temperature? What is your bedroom air temperature? What is the temperature of the water that you used this morning to take a shower? What is the temperature of the air inside your refrigerator that kept the milk cold overnight? What is the temperature of the air coming out of your hair dryer? Once you start thinking about the role of temperature in quantifying what goes on in our surroundings, you realize that you could ask hundreds of similar questions. Temperature represents the level of molecular activity of a substance. The molecules of a substance at a high temperature are more active than at a lower temperature.

Early humans relied on the sense of touch or vision to measure how cold or how warm something was. In fact, we still rely on touch today. When you are planning to take a bath, you first turn the hot and cold water on and let the bathtub fill with water. Before you enter the tub, however, you first touch the water to feel how warm it is. Basically, you are using your sense of touch to get an indication of the temperature. Of course, using touch alone, you can't quantify the temperature of water accurately. You cannot say, for example, that the water is at 70° F (21° C). We will discuss the importance of temperature, its measurement, and engineering concepts related to temperature in Chapter 11.

Today, based on what we know about our physical world, we need **seven** *fundamental dimensions* to correctly express ourselves in our surroundings. They are *length, mass, time, temperature, electric current, amount of substance*, and *luminous intensity*. With the help of these base dimensions, we can derive all other necessary physical quantities that would allow us to describe how things work.

By now, you understand why we need to formally define physical variables using fundamental dimensions. The other important fact you need to realize is that early humans needed not only physical dimensions to describe their surroundings but also *some way to scale or divide these physical dimensions*. This realization led to the concept of **units**. For example, time is considered a physical dimension, but it can be divided into both small and large portions, such as seconds, minutes, hours, days, months, years, decades, centuries, millennia, and so on. Today, when someone asks you how old you are, you reply by saying, "I am 19 years old." You don't say that you are approximately 6939 days, 170,000 hours, or 612,000,000 seconds old—even though these statements may very well be true at that instant! Or to describe the distance between two cities, we may say that they are 100 miles (~ 161 kilometers) apart; we don't say the cities are 528,000 feet (~ 161,000 meters) apart. The point of these examples is that we use appropriate divisions of physical dimensions to keep numbers manageable. We have learned to create an appropriate scale for these fundamental dimensions and divide them properly so that we can describe particular events, the size of an object, the thermal state of an object, or its interaction with the surroundings correctly, and to do so without much difficulty.

I am 612,000,000 seconds old, have a mass of 80,000 grams, and am 0.00185 kilometers tall!

Answer the following questions to test your understanding of the preceding section.

1.   Name at least four fundamental dimensions.

2.   What is the difference between a dimension and unit?

3.   Name at least two units that you use every day.

4.   What is the difference between mass and weight?

Vocabulary—State the meaning of the following terms:

Dimension _____

Unit _____

Mass _____

Temperature _____

LO² 6.2   Systems of Units

In the previous section, we explained that a dimension or physical quantity such as time can be divided into small and large portions such as seconds, hours, and days. Today, throughout the world, there are several systems of units in use, among which are the *International System* (abbreviated as *SI*, from French *Systéme international d'unités*) or sometimes called *metric units*, the *British Gravitational (BG)*, and the *U.S. Customary units*. Let us now examine these systems of units in greater detail.

International System (SI) of Units

We begin our discussion of systems of units with the International System (SI) of units, because SI is the most common system of units used in the world. The origin of the present day International System of units can be traced back to 1799 with *meter* and *kilogram* as the first two *base units*. By promoting the use of the *second* as a base unit of time in 1832, Carl Friedrich Gauss (1777–1855), who was an important figure in mathematics and physics, including magnetism and astronomy, had a great impact in many areas of science and engineering.

It was not until 1946 that the proposal for the *ampere* as a base unit for electric current was approved by the General Conference on Weights and Measures (CGPM). In 1954, the CGPM included ampere, *kelvin*, and *candela* as base units. The *mole* was added as a base unit by the 14th CGPM in 1971. A list of SI base units is given in Table 6.1.

Meter, kilogram, second, kelvin (or degree Celsius), ampere, mole, and candela are units of length, mass, time, temperature, electric current, amount of substance, and luminous intensity in the SI system.

| TABLE 6.1 | The SI Base Units | | |

Physical Quantity (Dimension)		Name of SI Base Unit	SI Symbol
Length	1.6 m–2.0 m Range of height for most adults	Meter	m
Mass	50 kg–120 kg Range of mass for most adults	Kilogram	kg
Time	Fastest person can run 100 meters in approximately 10 seconds	Second	s
Thermodynamic temperature	Ice water: 0° C or 273 K Comfortable room temperature: 22° C or 295 K	Kelvin	°C or K
Electric current	27 watts 120 volts 0.225 amps	Ampere	A
Amount of substance	Uranium 238 ← One of the heaviest atoms known Gold 197 Silver 108 Copper 64 Calcium 40 Aluminum 27 Carbon 12 ← Common carbon is Helium 4 used as a standard. Hydrogen 1 ← Lightest atom	Mole	mol
Luminous intensity	A candle has luminous intensity of approximately 1 candela	Candela	cd

Listed here are formal definitions of base units as provided by the Bureau International des Poids et Mesures.

The **meter** is the length of the path traveled by light in a vacuum during a time interval of 1/299,792,458 of a second.

The **kilogram** is the unit of mass; it is equal to the mass of the international prototype of the kilogram.

The **second** is the duration of 9,192,631,770 periods of the radiation corresponding to the transition between the two hyperfine levels of the ground state of the cesium 133 atom.

The **ampere** is that constant current that, if maintained in two straight parallel conductors of infinite length, of negligible circular cross section, and placed 1 meter apart in a vacuum, would produce between these conductors a force equal to 2×10^7 newton per meter of length.

The **kelvin,** a unit of thermodynamic temperature, is the fraction 1/273.16 of the thermodynamic temperature of the triple point of water (a point at which ice, liquid water, and water vapor coexist). The unit of Kelvin is related to the degree Celsius, according to $K = {}^\circ C + 273.16$.

The **mole** is the amount of substance of a system that contains as many elementary entities as there are atoms in 0.012 kilogram of carbon 12. When the mole is used, the elementary entities must be specified and may be atoms, molecules, ions, electrons, other particles, or specified groups of such particles.

The **candela** is the luminous intensity, in a given direction, of a source that emits monochromatic radiation of frequency 540×10^{12} hertz and has a radiant intensity in that direction of 1/683 watt per steradian.

You need not memorize the formal definitions of base units as provided by the Bureau International des Poids et Mesures. From your everyday life experiences, you have a pretty good idea about some of them. For example, you know how short a time period a second is or how long a period a year is. However, you may need to develop a "feel" for some of the other base units. For example, how long is a meter? How tall are you? Under 2 meters or perhaps above 1.5 meters? Most adult people's height is approximately between 1.6 meters and 2 meters. There are exceptions of course. What is your mass in kilograms? Developing a "feel" for units will make you a better engineer. For example, assume you are designing and sizing a new type of hand-held tool, and based on your stress calculation, you arrive at an average thickness of 1 meter. Having a "feel" for these units, you will be alarmed by the value of the thickness and realize that somewhere in your calculations you must have made a mistake. Also, when you travel abroad, the knowledge of these units could be quite useful to you. We will discuss in detail the role of the base units and other derived units in the upcoming chapters in this book.

Let us now turn our attention to the SI units for temperature. Most of you have seen a thermometer, a graduated glass rod that is filled with mercury. On the **Celsius** scale, which is a SI unit, under standard atmospheric conditions, the value zero was arbitrarily assigned to the temperature at which water freezes, and the value of 100 was assigned to the temperature at which water boils. It is important for you to understand that the numbers were assigned arbitrarily; had someone decided to assign a value of 100 to the ice water temperature and a value of 1000 to boiling water, we would have had a very different type of temperature scale today! In fact, as you see in Figure 6.1, in the British Gravitation system of units and U.S. Customary system of units, on a **Fahrenheit** temperature scale under standard atmospheric conditions, the temperature at which water freezes is assigned a value of 32, and the temperature at which the water boils is assigned a value of 212.

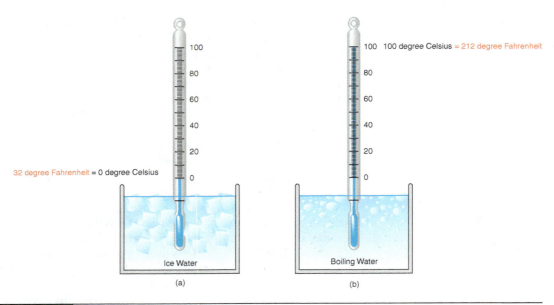

32 degree Fahrenheit = 0 degree Celsius

100 degree Celsius = 212 degree Fahrenheit

Ice Water
(a)

Boiling Water
(b)

FIGURE 6.1

Because both the Celsius and the Fahrenheit scales are arbitrarily defined, scientists recognized a need for a better temperature scale. This need led to the definition of an absolute scale, the *Kelvin* and *Rankine scales*, which are based on the behavior of an ideal gas, and that at zero absolute temperature all molecular activities stop. We will discuss this concept in much greater detail in Chapter 11. For now, it is important for you to know that in SI, the unit of temperature is degree Celsius (°C) or in terms of absolute temperature kelvin (K). The relationship between the degree Celsius and kelvin is given by:

$$\text{temperature (K)} = \text{temperature (°C)} + 273 \qquad \boxed{6.1}$$

The General Conference on Weights and Measures in 1960 also adapted the first series of *prefixes* and symbols of *decimal multiples* of SI units. Over the years, the list has been extended to include those listed in Table 6.2. When studying Table 6.2, note that **nano** (10^{-9}), **micro** (10^{-6}), **centi** (10^{-2}), **kilo** (10^3), **mega** (10^6), **giga** (10^9), and **tera** (10^{12}) are examples of decimal multiples and prefixes used with SI units. You already use some of these multiples and prefixes in your daily conversations. Examples include *milli*meter, *centi*meter, *kilo*meter, *milli*gram, *mega*bytes, *giga*bytes, and *tera*bytes.

The units for other physical quantities such as speed, force, pressure, energy, or power used in engineering calculations can be derived from the base (fundamental) units. For example, the unit for force is the *newton*. It is derived from Newton's second law of motion. One newton is defined as a magnitude of a force that, when applied to 1 kilogram of mass, will accelerate the mass at a rate of 1 meter per second squared (m/s^2). That is: $1\,\text{N} = (1\,\text{kg})(1\,\text{m/s}^2)$. As a well-educated engineer, it is important to know the difference between mass and weight. As we mentioned previously, the weight of an object is the force that is exerted on the mass of the object by the earth's gravity and is based on the universal law of gravitational

Nano (10^{-9}), micro (10^{-6}), centi (10^{-2}), kilo (10^3), mega (10^6), giga (10^9), and tera (10^{12}) are examples of decimal multiples and prefixes used with SI units.

TABLE 6.2	The List of Decimal Multiples and Prefixes Used with SI Base Units	
Multiplication Factors	**Prefix**	**SI Symbol**
$1,000,000,000,000,000,000,000,000 = 10^{24}$	yotta	Y
$1,000,000,000,000,000,000,000 = 10^{21}$	zetta	Z
$1,000,000,000,000,000,000 = 10^{18}$	exa	E
$1,000,000,000,000,000 = 10^{15}$	peta	P
$1,000,000,000,000 = 10^{12}$	tera	T
$1,000,000,000 = 10^{9}$	giga	G
$1,000,000 = 10^{6}$	mega	M
$1000 = 10^{3}$	kilo	k
$100 = 10^{2}$	hecto	h
$10 = 10^{1}$	deka	da
$0.1 = 10^{-1}$	deci	d
$0.01 = 10^{-2}$	centi	c
$0.001 = 10^{-3}$	milli	m
$0.000,001 = 10^{-6}$	micro	μ
$0.000,000,001 = 10^{-9}$	nano	n
$0.000,000,000,001 = 10^{-12}$	pico	p
$0.000,000,000,000,001 = 10^{-15}$	femto	f
$0.000,000,000,000,000,001 = 10^{-18}$	atto	a
$0.000,000,000,000,000,000,001 = 10^{-21}$	zepto	z
$0.000,000,000,000,000,000,000,001 = 10^{-24}$	yocto	y

attraction. The following mathematical relationship shows the relationship among the weight of an object, its mass, and the acceleration due to gravity.

$$\text{weight} = (\text{mass})(\text{accelaration due to gravity})$$

  6.2

I wonder how long it will take for this watermelon to hit the ground!

This is a good place to say a few words about acceleration due to Earth's gravity, which has an approximate value of 9.8 m/s^2. To better understand what this value represents, consider a situation where you let go of something from the rooftop of a tall building. If you were to express your observation, you will note the following. At the instant the object is released, it has a zero speed. The speed of the object will then increase by 9.8 m/s each second after you release it, resulting in speeds of 9.8 m/s after 1 second, 19.6 m/s after 2 seconds, and 29.4 m/s after 3 seconds, and so on. Moreover, when an object changes speed, we say it is accelerating. Weight is equal to an equivalent force that we must exert to prevent the object from accelerating. You know this from your daily experiences. For example, when you are holding on to a suitcase above ground, you feel the force that your hand has to apply to prevent the suitcase from falling to the ground and thus prevent it from accelerating.

Examples of commonly derived SI units used by engineers are shown in Table 6.3. The physical quantities shown in Table 6.3 will be discussed in detail in the following chapters of this book. Starting in Chapter 7, we will discuss their physical meaning, their significance and relevance in engineering, and their use in engineering analysis.

TABLE 6.3   Examples of Derived Units in Engineering

Physical Quantity	Name of SI Unit	Symbol for SI Unit	Expression in Terms of Base Units
Acceleration			m/s^2
Angle	radian	rad	
Angular acceleration			rad/s^2
Angular velocity			rad/s
Area			m^2
Density			kg/m^3
Energy, work, heat	joule	J	$N \cdot m$ or $kg \cdot m^2/s^2$
Force	newton	N	$kg \cdot m/s^2$
Frequency	hertz	Hz	s^{-1}
Impulse			$N \cdot S$ or $kg \cdot m/s$
Moment or torque			$N \cdot m$ or $kg \cdot m^2/s^2$
Momentum			$kg \cdot m/s$
Power	watt	W	J/s or $N \cdot m/s$ or $kg \cdot m^2/s^3$
Pressure, stress	pascal	Pa	N/m^2 or $kg/m \cdot s^2$
Velocity			m/s
Volume			m^3
Electric charge	coulomb	C	$A \cdot s$
Electric potential	volt	V	J/C or $m^2 \cdot kg/(s^3 \cdot A^2)$
Electric resistance	ohm	Ω	V/A or $m^2 \cdot kg/(s^3 \cdot A^2)$
Electric conductance	siemens	S	$1/\Omega$ or $s^3 \cdot A^2/(m^2 \cdot kg)$
Electric capacitance	farad	F	C/V or $s^4 \cdot A^2/(m^2 \cdot kg)$
Magnetic flux density	tesla	T	$V \cdot s/m^2$ or $kg/(s^2 \cdot A)$
Magnetic flux	weber	Wb	$V \cdot s$ or $m^2 \cdot kg/(s^2 \cdot A)$
Inductance	henry	H	$V \cdot s/A$
Absorbed dose of radiation	gray	Gy	J/kg or m^2/s^2

Michael Ransburg/Shutterstock.com

EXAMPLE 6.1

Consider a situation wherein an exploration vehicle that has a mass of 250 kilogram on Earth (gravity$_{\text{Earth}}$ $= 9.8$ m/s^2) is sent to the Moon and planet Mars to explore their surfaces. What is the mass of the vehicle on the Moon, where acceleration due to gravity is 1.6 m/s^2, and the planet Mars, where the acceleration due to gravity is 3.7 m/s^2? What is the weight of the vehicle on the Earth, Moon, and Mars?

The mass of the vehicle is 250 kg on the Moon and on the planet Mars as well. The mass of the vehicle is always 250 kg, regardless of where it is located. The mass represents the matter that makes up the vehicle, and since that does not change, the mass remains constant. However, the weight of the vehicle varies depending on the gravitational pull of the location. On Earth, the vehicle will have a weight of

$$\text{weight}_{\text{on Earth}} = (250 \text{ kg})\left(9.8\,\frac{\text{m}}{\text{s}^2}\right) = 2450 \text{ N}$$

Whereas on the Moon and Mars, the weights of the vehicle are

$$\text{weight}_{\text{on Moon}} = (250 \text{ kg})\left(1.6\,\frac{\text{m}}{\text{s}^2}\right) = 400 \text{ N}$$

$$\text{weight}_{\text{on Mars}} = (250 \text{ kg})\left(3.7\,\frac{\text{m}}{\text{s}^2}\right) = 925 \text{ N}$$

So as you can see, the vehicle will weigh the least on the surface of the moon and would require the least amount of effort to lift it off the ground when necessary.

British Gravitational System of Units

In the British Gravitational (BG) system of units, the unit of length is a **foot** (ft), which is equal to 0.3048 meter; the unit of time is a *second* (s); and the unit of force is a pound (lb), which is equal to 4.448 newton. Note that in the BG system, a **pound force** is considered a base or primary unit and the unit of mass, the *slug* is derived from Newton's second law. When one slug is subjected to one pound force, it will accelerate at a rate of 1 foot per second squared (ft/s^2). That is, 1 lb $= (1$ slug$)(1$ ft/s$^2)$. In the British Gravitational system, the unit of temperature (T) is expressed in degree Fahrenheit (°F) or in terms of absolute temperature degree Rankine (°R). The relationship between the degree Fahrenheit and degree Rankine is given by

$$T(°R) = T(°F) + 460 \qquad \boxed{6.3}$$

The relationship between degree Fahrenheit and degree Celsius is given by

$$T(°C) = \frac{5}{9}\left[T(°F) - 32\right] \qquad \boxed{6.4}$$

$$T(°F) = \frac{9}{5}T(°C) + 32 \qquad \boxed{6.5}$$

And the relationship between the degree Rankine and the Kelvin is

$$T(°R) = \frac{9}{5}T(K) \qquad \boxed{6.6}$$

We will explore these relationships further in Chapter 11.

U.S. Customary System of Units

In the United States, most engineers still use the U.S. Customary system of units. The unit of length is a *foot* (ft), which is equal to 0.3048 meter; the unit of mass is a **pound mass** (lbm), which is equal to 0.453592 kg; and the unit of time is a *second* (s). In the U.S. Customary system, the unit of force is *pound force* (lbf) and 1 lbf is defined as the weight of an object having a mass of 1 lbm at a certain location, where acceleration due to gravity is 32.2 ft/s². One pound force is equal to 4.448 newton (N). Because the pound force is not defined formally using Newton's second law and instead is defined at a specific location, a correction factor must be used in formulas when using U.S. Customary units. The units of temperature in the U.S. Customary system are identical to the BG system, which we discussed earlier, that is *degree Fahrenheit* (°F) or in terms of absolute temperature *degree Rankine* (°R).

Finally, by comparing the units of mass in the BG system (that is one slug) to the U.S. system pound mass, we note 1 slug ≈ 32.2 lbm. For those of us who might be slightly massive (overweight), it might be tempting to express our mass in slugs rather than in pound mass or kilogram. For example, a person who has a mass of 150 lbm or 68 kg would sound skinny if he were instead to express his mass as 4.6 slugs. Note that 150 lbm = 68 kg = 4.6 slugs, and therefore, he is telling the truth about his mass! So you don't have to lie about your mass; knowledge of units can bring about instant results without any exercise or diet. (See Example 6.4.)

The relationships among magnitudes of various SI and U.S. Customary units are depicted in Figure 6.2. When examining Figure 6.2, note that *1 meter is slightly larger than 3 feet, 1 kilogram is slightly larger than 2 pounds, and every 10° C difference is equal to 18 °F difference.*

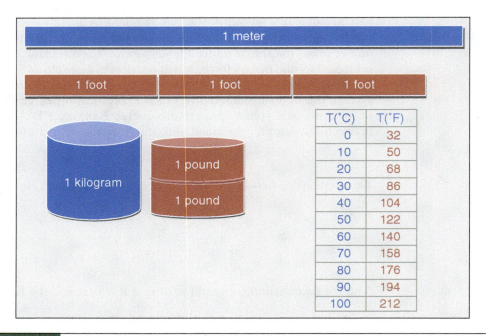

T(°C)	T(°F)
0	32
10	50
20	68
30	86
40	104
50	122
60	140
70	158
80	176
90	194
100	212

FIGURE 6.2   The relationships among magnitudes of various SI and U.S. Customary units.

Examples of SI and U.S. Customary units used in our everyday lives are shown in Tables 6.4 and 6.5, respectively.

TABLE 6.4	Examples of SI Units in Everyday use
Examples of SI Unit Usage	**SI Units Used**
Camera film	35 mm
Medication dose such as pills	100 mg, 250 mg, or 500 mg
Sports:	
Swimming	100 m breaststroke or butterfly stroke
Running	100 m, 200 m, 400 m, 5000 m, and so on
Automobile engine capacity	2.2 L (liter), 3.8 L, and so on
Light bulbs	60 W, 100 W, or 150 W
Electric consumption	kWh (kilowatt-hour)
Radio broadcasting signal frequencies	88–108 MHz (FM broadcast band)
	0.54–1.6 MHz (AM broadcast band)
Police, fire signal frequencies	153–159 MHz
Global positioning system signals	1575.42 MHz and 1227.60 MHz

TABLE 6.5	Examples of U.S. Customary Units in Everyday use
Examples of U.S. Customary Unit Usage	**U.S. Customary Units Used**
Fuel tank capacity	20 gallons or 2.67ft^3 (1 ft^3 = 7.48 gallons)
Sports (length of a football field)	100 yd or 300 ft
Power capacity of an automobile	150 hp or 82500 lb · ft/s (1 hp = 550 lb · ft/s)
Distance between two cities	100 miles (1 mile = 5280 ft)

As shown in Table 6.4, a common SI unit for volume is *liter*, which is equal to 1000 cm^3 or 0.264 gallon (1 liter $\approx$ ¼ gallon), and 1000 liters is equal to 1 cubic meter (i.e., 1000 liters = 1 m^3). Also, note that 1 cubic foot is 7.48 gallons (1 ft^3 $\approx$ 7.5 gallons). It is also worth noting that a liter of water has a mass of 1 kilogram, and a gallon of water has a mass of approximately 8.3 pounds. These are good numbers to remember!

The watt (W) and horsepower (hp) are units of power in SI and U.S., respectively, and kilowatt-hour (kWh) is a SI unit of energy. We will discuss these units in Chapter 13 after we explain the different forms of energy and power. The units of frequency are commonly expressed in kilohertz (kHz), megahertz (MHz), or gigahertz (GHz). Frequency represents the number of cycles per seconds. For example, the alternating electric current in your home is 60 cycles per second (60 hertz). *Alternating current (ac)* is the flow of electric charge that periodically reverses. We will discuss this concept in greater detail in Chapter 12 when we discuss electricity.

Before You Go On

Answer the following questions to test your understanding of the preceding section.

1. What are the two most common systems of units?

2. What are the base SI units?

3. Name at least three prefixes and symbols of decimal multiple of SI units.

4. What are the units of mass and weight in U.S.?

5. What do we mean by absolute zero temperature?

6. What are the units of mass and weight in BGS?

Vocabulary—State the meaning of the following terms:

Absolute Zero Temperature _____

Rankine Temperature Scale _____

Kelvin Temperature Scale _____

One Slug _____

LO³ 6.3   Unit Conversion and Dimensional Homogeneity

Some of you may know that not too long ago NASA lost a spacecraft called Mars Climate Orbiter because two groups of engineers working on the project neglected to communicate correctly their calculations with appropriate units. According to an internal review conducted by NASA's Jet Propulsion Laboratory, "a failure to recognize and correct an error in a transfer of information between the Mars Climate Orbiter spacecraft team in Colorado and the mission navigation team in California led to the loss of the spacecraft." The peer review findings indicated that one team used U.S. Customary units (e.g., foot and pound) while the other used SI units (e.g., meter and kilogram) for a key spacecraft operation. According to NASA, the information exchanged between the teams was critical to the maneuvers required to place the spacecraft in the proper Mars orbit. An overview of the Mars polar lander mission is given in Figure 6.3.

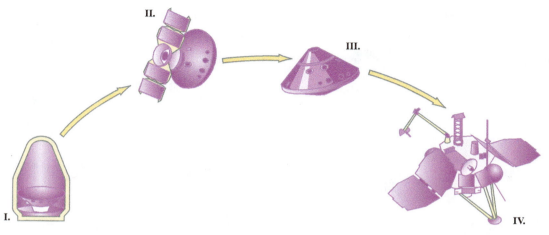

I. Launch

- Delta II 7425
- Launched January 3, 1999
- Launch mass: 574 kilograms

II. Cruise

- Thruster attitude control
- Four trajectory-correction maneuvers (TCM); site-adjustment maneuver September 1, 1999; contingency 5th TCM at entry −24 hours
- Eleven month cruise
- Near-simultaneous tracking with Mars Climate Orbiter or Mars Global Surveyor during approach

III. Entry, Descent, Landing

- Arrival: December 3, 1999
- Jettison cruise stage; microprobes separate from cruise stage
- Hypersonic entry
- Parachute descent; propulsive landing
- Descent imaging of landing site

IV. Landed Operations

- Lands in Martian spring at 76 degrees South latitude, 195 degrees West longitude (76S, 195W)
- 90-day landed mission
- Meteorology, imaging, soil analysis, trenching
- Data relay via Mars Climate Orbiter, Mars Global Surveyor, or direct-to-Earth high-gain antenna

FIGURE 6.3   Mars polar lander mission overview.
Courtesy of NASA

Unit Conversion

As you can see, as engineering students and later as practicing engineers, when performing analysis, you will find a need to convert from one system of units to another. It is very important for you at this stage in your education to learn to convert information from one system of units to another correctly. It is also important for you to understand and to remember to show all your calculations with proper units. This point cannot be emphasized enough! Always show the appropriate units that go with your calculations. The conversion factors for the fundamental and derived dimensions commonly encountered in engineering are shown in Table 6.6. Detailed conversion tables are also given on the back endpapers of this book. Examples 6.2, 6.3, 6.4, and 6.5 show the steps that you need to take to convert from one system of units to another.

TABLE 6.6	Systems of Units and Conversion Factors			
	System of Units			
Dimension	**SI**	**BG**	**U.S. Customary**	**Conversion Factors**
Length	meter (m)	foot (ft)	foot (ft)	1 ft = 0.3048 m 1 m = 3.2808 ft
Time	second (s)	second (s)	second (s)	
Mass	kilogram (kg)	slugs*	pound mass (lbm)	1 lbm = 0.4536 kg 1 kg = 2.2046 lbm 1 slug = 32.2 lbm
Force	newton (N) $1\,\text{N} = (1\,\text{kg})\left(1\dfrac{\text{m}}{\text{s}^2}\right)$	$1\,\text{lbf}^{**} = (1\,\text{slug})\left(1\dfrac{\text{ft}}{\text{s}^2}\right)$	One pound mass*** weighs one pound force at sea level	1 N = 224.809E−3 lbf 1 lbf = 4.448 N
Temperature	degree Celsius (°C) or kelvin (K)**** K = °C + 273.15	degree Fahrenheit (°F) or degree Rankine (°R) °R = °F + 459.67	degree Fahrenheit (°F) or degree Rankine (°R) °R = °F + 459.67	$°C = \dfrac{5}{9}[°F - 32]$ $°F = \dfrac{9}{5}°C + 32$ $K = \dfrac{5}{9}°R$ $°R = \dfrac{9}{5}K$
Work, Energy	joule (J) = (1 N)(1 m)	lbf ft = (1 lbf)(1 ft) Commonly written as ft · lbf	lbf · ft = (1 lbf)(1 ft)	1 J = 0.7375 ft · lbf 1 ft · lbf = 1.3558 J 1 Btu = 778.17 ft · lbf
Power	watt (W) = $\dfrac{1\,\text{Joule}}{1\,\text{second}}$ kW = 1000 W	$\dfrac{\text{lbf·ft}}{\text{second}} = \dfrac{(1\,\text{lbf})(1\,\text{ft})}{1\,\text{second}}$	$\dfrac{\text{lbf·ft}}{\text{second}} = \dfrac{(1\,\text{lbf})(1\,\text{ft})}{1\,\text{second}}$	$1\,\text{W} = 0.7375\dfrac{\text{ft·lbf}}{\text{s}}$ $1\,\text{hp} = 550\dfrac{\text{ft·lbf}}{\text{s}}$ 1 hp = 0.7457 kW

*Derived or secondary dimension
**Fundamental dimension
***Note unlike SI and BG systems, the relationship between pound force and pound mass is not defined using Newton's second law
****Note a temperature value expressed in K reads Kelvin not degree Kelvin

EXAMPLE 6.2	What is the equivalent value of T = 50° C in degrees Fahrenheit, degree Rankine, and Kelvin?

To convert the value of temperature from Celsius to Fahrenheit, we use Equation (6.4) and substitute the value of 50 for the temperature (°C) variable as

$$T(°F) = \frac{9}{5}T(°C) + 32 = \frac{9}{5}(50) + 32 = 122°\,F$$

And to convert the result to degree Rankine, we use Equation (6.3), so

$$T(°R) = T(°F) + 460 = 122 + 460 = 582°\,R$$

Finally, to covert the value of $T = 50°$ C to kelvin, we use Equation (6.1) or Table 6.6:

$$T(\mathrm{K}) = T(°\mathrm{C}) + 273 = (50) + 273 = 323\ \mathrm{K}$$

EXAMPLE 6.3

A person who is 6 feet and 3 inches tall and weighs 185 pound force (lbf) is driving a car at a speed of 65 miles per hour over a distance of 25 miles between two cities. The outside air temperature is 80° F. Let us convert all of the values given in this example from U.S. Customary units to SI units.

The steps to convert the person's height from feet and inches to meters and centimeters are explained next.

$$\text{height} = \underbrace{\left[6\ \text{ft} + \underbrace{\overbrace{(3\ \text{in.})\left(\frac{1\ \text{ft}}{12\ \text{in.}}\right)}^{\text{step 1}}}_{\text{step 2}} \right]\left(\frac{0.3048\ \text{m}}{1\ \text{ft}}\right)}_{\text{step 3}} = 1.905\ \text{m}$$

or

$$\text{height} = (1.905\ \text{m})\overbrace{\left(\frac{100\ \text{cm}}{1\ \text{m}}\right)}^{\text{step 4}} = 190.5\ \text{cm}$$

1. Start by converting the inch value into feet by realizing that 1 foot is equal to 12 inches. The expression $\left(\frac{1\ \text{ft}}{12\ \text{in}}\right)$ conveys the same fact, except when you write in fraction form and multiply it by the "3 in." as shown: $(3\ \cancel{\text{in.}})\left(\frac{1\ \text{ft}}{12\ \cancel{\text{in.}}}\right)$, the inch units in the numerator and denominator cancel out, and the 3 in. value is now represented in feet.

2. Add the results of step 1 to 6 ft.

3. Multiply the results of step 2 by $\left(\frac{0.3048\ \text{m}}{1\ \text{ft}}\right)$, because 1 ft is equal to 0.3048 m, and the foot units in the numerator and denominator cancel out. This step leads to the person's height, expressed in meters, as $\left[6\,\cancel{\text{ft}} + (1\,\cancel{\text{in}})\left(\frac{1\ \cancel{\text{ft}}}{12\ \cancel{\text{in}}}\right)\right]\left(\frac{0.3048\ \text{m}}{1\ \cancel{\text{ft}}}\right)$.

4. To convert the result from meters to centimeters, we multiply 1.905 $\cancel{\text{m}}$ $\left(\frac{100\ \text{cm}}{1\ \cancel{\text{m}}}\right)$, because 1 meter is equal to 100 cm, and this step cancels out the meter in the numerator and denominator.

To convert the person's weight from pound force to newtons is

$$\text{weight} = (185\ \text{lbf})\overbrace{\left(\frac{4.448\ \text{N}}{1\ \text{lbf}}\right)}^{\text{step 5}} = 822.8\ \text{N}$$

5. To convert the person's weight, we multiply the 185 lbf value by (4.448 N/1 lbf), because 1 lbf is equal to 4.448 newtons (N). This leads to pound force units in the numerator and denominator canceling out and the person's weight being expressed in newtons, as (185 lbf)(4.448 N/1 lbf).

To convert the speed of the car from miles per hour to kilometers per hour, use

$$\text{speed} = \overbrace{\overbrace{\overbrace{\left(65\frac{\text{miles}}{\text{hr}}\right)\left(\frac{5280\text{ ft}}{1\text{ mile}}\right)}^{\text{step6}}\left(\frac{0.3048\text{ m}}{1\text{ ft}}\right)}^{\text{step7}}\left(\frac{1\text{ km}}{1000\text{ m}}\right)}^{\text{step 8}} = 104.6\frac{\text{km}}{\text{hr}}$$

6. To convert the speed of the car from 65 miles/h to km/h, we start by converting the 65 miles value to feet; knowing that 1 mile is equal to 5280 feet, we multiply the 65 miles by 5280. Thus, $\left(65\frac{\text{miles}}{\text{hr}}\right)\left(\frac{5280\text{ ft}}{1\text{ mile}}\right) = \left((65)(5280)\frac{\text{ft}}{\text{hr}}\right)$. This step cancels out the miles units in the numerator and denominator and result in the speed value being represented in feet per hour (ft/h).

7. Next, multiply the results of step 6 by $(0.3048\text{ m}/1\text{ ft})$, because 1 foot is equal to 0.3048 meters. This step cancels out the feet units in the numerator and denominator and leads to $\left(65\frac{\text{miles}}{\text{hr}}\right)\left(\frac{5280\text{ ft}}{1\text{ mile}}\right)\left(\frac{0.3048\text{ m}}{1\text{ ft}}\right) = 104,607\text{ m/hr}$,

8. To convert the result of step 7 from m/h to km/h, we note that 1 kilometer is equal to 1000 meters, and multiply $\left(104,607\frac{\text{m}}{\text{hr}}\right)$ by $\left(\frac{1\text{ km}}{1000\text{ m}}\right)$ to cancel out the meter unit in the numerator and denominator. The speed of the car is now expressed in kilometers per hour (km/hr).

To convert the distance traveled between two cities from miles to kilometers, use steps similar to these discussed previously.

$$\text{distance} = \overbrace{\overbrace{\overbrace{(25\text{ miles})\left(\frac{5280\text{ ft}}{1\text{ mile}}\right)}^{\text{step 9}}\left(\frac{0.3048\text{ m}}{1\text{ ft}}\right)}^{\text{step 10}}\left(\frac{1\text{ km}}{1000\text{ m}}\right)}^{\text{step 11}} = 40.2\text{ km}$$

9. Convert the miles to feet by multiplying $(25\text{ miles})\left(\frac{5280\text{ ft}}{1\text{ mile}}\right)$.

10. Convert the feet to meters by $(25\text{ miles})\left(\frac{5280\text{ ft}}{1\text{ mile}}\right)\left(\frac{0.3048\text{ m}}{1\text{ ft}}\right)$.

11. Convert the meters to kilometers by $(25\text{ miles})\left(\frac{5280\text{ ft}}{1\text{ mile}}\right)\left(\frac{0.3048\text{ m}}{1\text{ ft}}\right)\left(\frac{1\text{ km}}{1000\text{ m}}\right)$.

To convert the air temperature from degree Fahrenheit to Celsius, we substitute for $T(°F)$ the value 80 in Equation (6.4). Thus,

$$T(°C) = \frac{5}{9}[T(°F) - 32]$$

$$T(°C) = \frac{5}{9}[80 - 32] = 26.7° \text{ C}$$

EXAMPLE 6.4

You don't have to lie about your weight! For those of us who might be slightly massive or even overweight, it might be more appealing to express our mass in kilogram rather than in pound mass. For example, a person who has a mass of 150 pound-mass (lbm) would sound skinny if he/she were instead to convert this value and express his/her mass in kilograms (kg).

$$(150 \text{ lbm})\left(\frac{1 \text{ kg}}{2.2 \text{ lbm}}\right) = (150 \text{ \sout{lbm}})\left(\frac{1 \text{ kg}}{2.2 \text{ \sout{lbm}}}\right) = 68 \text{ kg}$$

To convert the mass from pound mass (lbm) to kilogram (kg), we note that 1 kg is equal to 2.2 lbm. To obtain the result in kilogram we multiplied the 150 lbm by the conversion factor $\dfrac{1 \text{ kg}}{2.2 \text{ lbm}}$, which reads 1 kg is equal to 2.2 lbm. This step cancels out the pound-mass units in the numerator and denominator as shown. As you can see from the result, 150 lbm is equal to 68 kg; therefore, he or she is telling the truth about his or her mass! So you don't have to lie about your mass; knowledge of units can bring about instant results without any exercise or diet!

EXAMPLE 6.5

For the following problems, use the conversion factors given on the front and back end covers of this book.

(a) Convert the given value of area A from cm^2 to m^2. Note $A = 100$ cm^2.

$$A = 100 \text{ cm}^2$$

$$A = (100 \text{ cm}^2)\left(\frac{1 \text{ m}}{100 \text{ cm}}\right)^2 = 0.01 \text{m}^2$$

(b) Convert the given value of volume V from mm^3 to m^3. Note $V = 1000$ mm^3.

$$V = 100 \text{ mm}^3$$

$$V = (100 \text{ mm}^3)\left(\frac{1 \text{ m}}{1000 \text{ mm}}\right)^3 = 10^{-6} \text{m}^3$$

(c) Convert the given value of atmospheric pressure P from N/m^2 to lbf/in.^{2}

$$P = 10^5 \text{ N/m}^2$$

$$P = \left(10^5 \frac{\text{N}}{\text{m}^2}\right)\left(\frac{1 \text{ lbf}}{4.448 \text{ N}}\right)\left(\frac{0.0254 \text{ m}}{1 \text{ in.}}\right)^2 = 14.5 \text{ lbf/in.}^2$$

(d) Convert the given value of the density of water ρ from kg/m^3 to lbm/ft^3.

$$\rho = 1000 \text{ kg/m}^3$$

$$\rho = \left(1000 \frac{\text{kg}}{\text{m}^3}\right)\left(\frac{1 \text{ lbm}}{0.4536 \text{ kg}}\right)\left(\frac{1 \text{ m}}{3.28 \text{ ft}}\right)^3 = 62.5 \text{ lbm/ft}^3$$

Dimensional Homogeneity

Another important concept that you need to understand is that all formulas used in engineering analysis must be *dimensionally homogeneous*. What do we mean by dimensionally homogeneous? Can you, say, add someone's height who is 6 feet tall to his weight of 285 lbf and his body temperature of 98° F; that is, 6 + 285 + 98 = 389? Of course not! What would be the result of such a calculation?

Therefore, if we were to use the formula $L = a + b + c$, in which the variable L on the left-hand side of the equation has a dimension of length, then the variables a, b, and c on the right-hand side of equation should also have dimensions of length. Otherwise, if variables a, b, and c had dimensions such as length, weight, and temperature, respectively, the given formula would not be homogeneous, which would be like adding someone's height to his weight and body temperature (Figure 6.4)! Example 6.6 shows how to check for homogeneity of dimensions in an engineering formula.

> All formulas used in engineering analysis must be dimensionally homogeneous.

PeJo/Shutterstock.com

| FIGURE 6.4 | Example of elements that are not homogeneous. |

EXAMPLE 6.6

(a) When a constant load is applied to a bar with a constant cross section, as shown in Figure 6.5, the amount by which the end of the bar will deflect can be determined from the relationship

$$d = \frac{PL}{AE} \qquad \text{6.7}$$

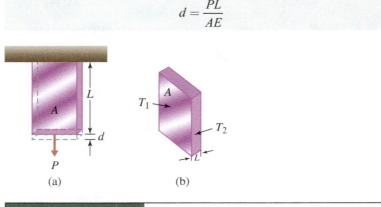

(a)   (b)

| FIGURE 6.5 | (a) The bar in Example 6.6 and (b) heat transfer through a solid material. |

where

d = end deflection of the bar in meter (m)
P = applied load in newton (N)
L = length of the bar in meter (m)
A = cross-sectional area of the bar in m^2
E = modulus of elasticity of the material

What are the units for modulus of elasticity?

For Equation (6.7) to be dimensionally homogeneous, the units on the left-hand side of the equation must equal the units on the right-hand side. This equality requires the modulus of elasticity to have the units of N/m^2, as

$$d = \frac{PL}{AE} \Rightarrow m = \frac{(N)(m)}{(m^2)E}$$

Solving for the units of E leads to N/m^2 (called newton per squared meter or force per unit area).

(b)   The heat transfer rate through a solid material is governed by Fourier's law:

$$q = kA\frac{T_1 - T_2}{L} \qquad \boxed{6.8}$$

where

q = heat transfer rate
k = thermal conductivity of the solid material in watts per meter degree Celsius, W/m·°C
A = area in m^2
$T_1 - T_2$ = temperature difference, °C
L = thickness of the material, m

What is the appropriate unit for the heat transfer rate q?

Substituting for the units of k, A, T_1, T_2, and L in Equation 6.8, we have

$$q = kA\frac{T_1 - T_2}{L} = \left(\frac{W}{m \cdot °C}\right)(m^2)\left(\frac{°C}{m}\right) = W$$

From this, you can see that the appropriate SI unit for the heat transfer rate is the watt.

Numerical versus Symbolic Solutions

When you take your engineering classes, you need to be aware of two important things: (1) understanding the basic concepts and principles associated with that class and (2) how to apply them to solve real physical problems (situations). In order to gain an understanding of the basic concepts, you need to study carefully the statements of governing laws and the derivations of engineering formulas and their limitations. After you have studied the underlying concepts, you then need to apply them to physical situations by solving problems. After studying a certain concept initially, you may think that you completely understand the concept, but it is through the application of the concept (by doing the homework problems) that you really can test your understanding.

Moreover, homework problems in engineering typically require either a numerical or a symbolic solution. For problems that require numerical solution, data is given. In contrast, in the symbolic solution, the steps and the final answer are presented with variables that could be substituted with data—if necessary. The following example will demonstrate the difference between numerical and symbolic solutions.

EXAMPLE 6.7

Determine the load that can be lifted by the hydraulic system shown. All of the necessary information is shown in Figure 6.6.

The general relationship among force, pressure, and area is explained in detail in Chapter 10. At this time, don't worry about understanding these relationships. The purpose of this example is to demonstrate the difference between a numerical and a symbolic solution. The concepts that are used to solve this problem are $F_1 = m_1 g$, $F_2 = m_2 g$, and $F_2 = (A_2/A_1) F_1$, where F denotes force, m is mass, g is acceleration due to gravity ($g = 9.81$ m/s^2), and A represents area.

Numerical Solution

We start by making use of the given data and substituting them into appropriate equations.

$$F_1 = m_1 g = (100 \text{ kg})(9.81 \text{ m/s}^2) = 981 \text{ N}$$

$$F_2 = \frac{A_2}{A_1} F_1 = \frac{\pi (0.15 \text{ m})^2}{\pi (0.05 \text{ m})^2}(981 \text{ N}) = 8829 \text{ N}$$

$$F_2 = 8829 \text{ N} = (m_2 \text{ kg})(9.81 \text{ m/s}^2) \Rightarrow m_2 = 900 \text{ kg}$$

Symbolic Solution

For this problem, we could start with the equation that relates F_2 to F_1 and then simplify the similar quantities such as π and g in the manner:

$$F_2 = \frac{A_2}{A_1} F_1 = m_2 g = \frac{\pi (R_2)^2}{\pi (R_1)^2}(m_1 g)$$

$$m_2 = \frac{(R_2)^2}{(R_1)^2} m_1 \Rightarrow m_2 = \frac{(15 \text{ cm})^2}{(5 \text{ cm})^2}(100 \text{ kg}) = 900 \text{ kg}$$

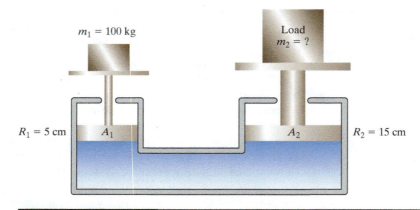

$m_1 = 100$ kg

Load
$m_2 = ?$

$R_1 = 5$ cm A_1 A_2 $R_2 = 15$ cm

FIGURE 6.6   The hydraulic system of Example 6.7.

Often, this approach is preferred over the direct substitution of values into the equation right away, because it allows us to change a value of a variable such as m_1 or the areas and see what happens to the result. For example, using the symbolic approach, we can see clearly that if m_1 is increased to a value of 200 kg, then m_2 changes to 1800 kg.

Before You Go On

Answer the following questions to test your understanding of the preceding section.

1. Why is it important to know how to convert from one system of units to another?

2. What do we mean by dimensional homogeneity? Give an example.

3. Show the steps that you would take to convert your height from feet and inches to meters and centimeters.

4. Show the steps that you would take to convert your weight from pound-force to Newtons.

Vocabulary—State the meaning of the following terms:

Dimensional Homogeneity _____

Unit Conversion _____

LO⁴ 6.4   Significant Digits (Figures)

Engineers make measurements and carry out calculations. Engineers then record the results of measurements and calculations using numbers. Significant digits (figures) represent and convey the extent to which recorded or computed data is dependable. For example, consider the instruments shown in Figure 6.7. We are interested in measuring the temperature of room air using a thermometer, the dimensions of a credit card using an engineering ruler, and the pressure of a fluid in a line using the pressure gage shown. As you can see from these examples, the measurement readings fall between the smallest scale division of each instrument. In order to take the guess work out of the reading and for consistency, we record the measurement to one half of the smallest scale division of the measuring instrument. One half of the smallest scale division commonly is called the *least count* of the measuring instrument. For example, referring to Figure 6.7, it should be clear that the least count for the thermometer is 1° F (the smallest division is 2° F), for the ruler is 0.05 in., and for the pressure gage is 0.5 in. of water. Therefore, using the given thermometer, it would be incorrect to record the air temperature as 71.25° F and later use this value to carry out other calculations. Instead, it should be recorded as 71 ± 1° F. This way, you are telling the reader or the user of your measurement that the temperature reading falls between 70° F and 72° F. Note the correct way of recording the dependability of a measurement using the ± sign and the least count value.

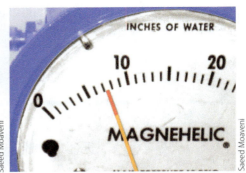

	FIGURE 6.7	Examples of recorded measurements.

As stated earlier, significant digits (figures) represent and convey the extent to which recorded or computed data is dependable. Significant digits are numbers zero through nine. However, when zeros are used to show the position of a decimal point, they are not considered significant digits. For example, each of the following numbers 175, 25.5, 1.85, and 0.00125 has three significant digits. Note the zeros in number 0.00125 are not considered to be significant digits, since they are used to show the position of the decimal point. As another example, the number of significant digits for the number 1500 is not clear. It could be interpreted as having two, three, or four significant digits based on what the role of the zeros is. In this case, if the number 1500 was expressed by 1.5×10^3, 15×10^2, or 0.015×10^5, it would be clear that it has two significant digits. By expressing the number using the power of ten, we can make its accuracy more clear. However, if the number was initially expressed as 1500.0, then it has four significant digits and, this would imply that the accuracy of the number is known to 1/10000.

> Significant digits (figures) represent and convey the extent to which recorded or computed data is dependable.

Addition and Subtraction Rules When adding or subtracting numbers, the result of the calculation should be recorded such that the last significant digit in the result is determined by the position of the last column of digits common to all of the numbers being added or subtracted. For example,

152.47 +	or	132.853 −	
3.9		5	
156.37		127.853	(Your calculator will display.)
156.3		127	(However, the result should be recorded this way.)

The numbers 152.47 and 3.9 have five and two significant digits, respectively. When we add these two numbers, the calculator will display 156.37; however, since the first column after the decimal point is common to these numbers, the result should be recorded as 156.3.

Multiplication and Division Rules When multiplying or dividing numbers, the result of the calculation should be recorded with the least number of significant digits given by any of the numbers used in the calculation. For example,

| 152.47 × or | 152.47 ÷ | |
3.9	3.9	
594.633	39.0948717949	(Your calculator will display.)
5.9×10^2	39	(However, the result should be recorded this way.)

In this example, the number 152.47 has five significant digits, and the number 3.9 has two significant digits. Therefore, the result of the calculations should be recorded with two significant digits, because the number 3.9 used in the calculations has the least significant digits.

Finally, it is worth noting that in many engineering calculations it may be sufficient to record the results of a calculation to a fewer number of significant digits than obtained following the rules explained previously. In this book, we present the results of example problems with two or three decimal points.

LO⁵ 6.5   Components and Systems

Every engineered product is made of components. Let us start with a simple example to demonstrate what we mean by an engineering system and its components. Most of us own a winter coat, which can be looked at as a system. First, the coat serves a purpose. Its primary function is to offer additional insulation for our bodies so that our body heat does not escape as quickly and as freely as it would without protective covering. The coat may be divided into smaller components: the fabric comprising the main body of the coat, insulating material, a liner, threads, zipper(s), and buttons. Moreover, each component may be further subdivided into smaller components. For example, the main body of the jacket may be divided into sleeves, collar, pockets, the chest section, and the back section (see Figure 6.8). Each component serves a purpose: The pockets were designed to hold things, the sleeves cover our arms, and so on. The main function of the zipper is to allow us to open and close the front of the jacket freely. It too consists of smaller components. Think once more about the overall purpose of the coat and the function of each component. A well-designed coat not only looks appealing to the eyes but also has functional pockets and keeps us warm during the winter.

Engineering systems are similar to a winter coat. Any given engineered product or engineering system can be divided into smaller, more manageable subsystems, and each subsystem can be further divided into smaller and smaller components. The components of a well-designed engineering system should function and fit well together so that the primary purpose of the product is attained. Let us consider another common example. The primary function of a car is to move us from one place to another in a reasonable amount of time. The car must provide a comfortable area for us to sit within. Furthermore, it must shelter us and provide some protection from the outside elements, such as harsh weather and harmful objects outside. The automobile consists of thousands of parts. When viewed in its entirety, it is a complicated system. Thousands of engineers have contributed to the design, development, testing, and supervision of the manufacture of an automobile. These include electrical engineers, electronic engineers,

> Every product is considered a system that serves a purpose. A system is made up of smaller parts called components.

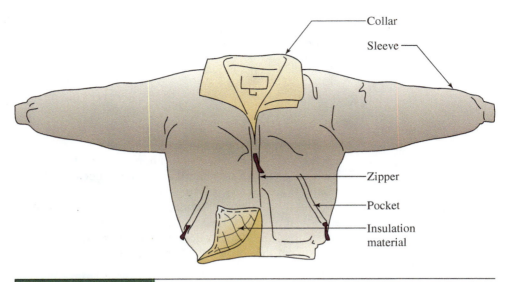

Collar

Sleeve

Zipper

Pocket

Insulation material

| FIGURE 6.8 | A simple system and its components. |

combustion engineers, materials engineers, aerodynamics experts, vibration and control experts, air-conditioning specialists, manufacturing engineers, and industrial engineers.

When viewed as a system, the car may be divided into major subsystems or units, such as electrical, body, chassis, power train, and air conditioning (see Figure 6.9). Each major **component** can be further subdivided into smaller subsystems and their components. For example, the main body of the car consists of doors, hinges, locks, windows, and so on. The windows are controlled by mechanisms that are activated by hand or motors. And the electrical system of a car consists of a battery, a starter, an alternator, wiring, lights, switches, radio, microprocessors, and so on. The car's air-conditioning system consists of components such as a fan, ducts, diffusers, a compressor, an evaporator, and a condenser. Again, each of these components can be

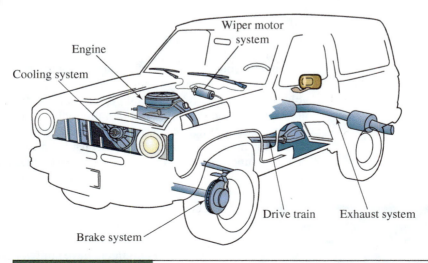

Wiper motor system

Engine

Cooling system

Drive train

Exhaust system

Brake system

| FIGURE 6.9 | An engineering system and its main components. |

further divided into yet smaller components. For example, the fan consists of an impeller, a motor, and a casing. From these examples, it should be clear that, in order to understand a system, we must first fully understand the role and function of its components.

During the next four or five years, you will take a number of engineering classes that will focus on specific topics. You may take a statics class, which deals with the equilibrium of objects at rest. You will learn about the role of external forces, internal forces, reaction forces, and their interactions. Later, you will learn the underlying concepts and equilibrium conditions for designing parts. You will also learn about other physical laws, principles, mathematics, and correlations that will allow you to analyze, design, develop, and test various components that make up a system. It is imperative that during the next four or five years you fully understand these laws and principles so that you can design components that fit well together and work in harmony to fulfill the ultimate goal of a given system. Thus, you can see the importance of learning the fundamentals. If you don't, you are likely to design poor components that, when put together, will result in an even poorer system!

Before You Go On

Answer the following questions to test your understanding of the preceding section.

1. What is the difference between a component and a system?

2. What are the major components of a building?

3. How would you define major components for a supermarket?

4. What are the major components of a bicycle?

Vocabulary—State the meaning of the following terms:

A component _____

A system _____

LO⁶ 6.6   Physical Laws and Observations

As stated earlier, engineers apply physical and chemical laws and principles along with mathematics to design, develop, test, and mass-produce products and services that we use in our everyday lives. The key concepts that you need to keep in the back of your mind are the physical and chemical laws and principles and mathematics.

Having had a high school education, you have a pretty good idea of what we mean by mathematics. But what do we mean by physical laws? Well, the universe, including the earth that we live on, was created a certain way. There are differing opinions as to the origin of the universe. Was it put together by God, or did it start with a big bang? We won't get into that discussion here. But we have learned through observation and by the collective effort of those before us

that things work a certain way in nature. For example, if you let go of something that you are holding in your hand, it will fall to the ground. That is an observation that we all agree upon. We can use words to explain our observations or use another language, such as mathematics, to express our findings. Sir Isaac Newton (1642–1727) formulated that observation into a useful mathematical expression that we know as the universal law of gravitational attraction.

An important point to remember is that the physical laws are based on observations. Moreover, we use mathematics and basic physical quantities to express our observations in the form of a law. Even so, to this day, we may not fully understand why nature works the way it does. We just know it works. There are physicists who spend their lives trying to understand on a more fundamental basis why nature behaves the way it does. Some engineers may focus on investigating the fundamentals, but most engineers use fundamental laws to design things.

As another example, when you place some hot object in contact with a cold object, the hot object cools down while the cold object warms up until they both reach an equilibrium temperature somewhere between the two initial temperatures. From your everyday experience, you know that the cold object does not get colder while the hot object gets hotter! Why is that? Well, it is just the way things work in nature! The second law of thermodynamics, which is based on this observation, simply states that heat flows spontaneously from a high-temperature region to a low-temperature region. The object with the higher temperature (more energetic) transfers some of its energy to the low-temperature (less energetic) object. When you put some ice cubes in a glass of warm soda, the soda cools down while the ice warms up and eventually melts away. You may call this "sharing resources." Unfortunately, we as people do not follow this law closely when it comes to social issues.

To better understand the second law of thermodynamics, consider another example. Some of you may have young children or young brothers and sisters. If you placed the child with some toys in a room that is tidied up and orderly, let the child play with the toys for a while, and then came back in a few minutes, you would find toys scattered all over the room in a disorderly way. Why won't you find toys put away nicely? Well, that is because things work spontaneously in a certain direction in nature. These two examples demonstrate the second law of thermodynamics. Things in nature work in a certain direction by themselves.

Engineers are also good bookkeepers. What do we mean by this? Any of us with a checking account knows the importance of accurate record keeping. In order to avoid problems, most of us keep track of the transactions in terms of payments (debits) and deposits (credits). Good bookkeepers can tell you instantly what the balance in their account is. They know they need to add to the recorded balance whenever they deposit some money and subtract from the balance with every withdrawal from the account. Engineers, like everyone else, need to keep track of their accounts. Moreover, similar to bookkeeping a checking account, engineers keep track of (bookkeep) physical quantities when analyzing an engineering problem.

To better understand this concept, consider the air inside a car tire. If there are no leaks, the mass of air inside the tire remains constant. This is a statement expressing the *conservation of mass,* which is based on our observations. If the tire develops a leak, then you know from your experience that the amount of air within the tire will decrease until you have a flat tire. Furthermore, you know the air that escaped from the tire was not destroyed; it simply became part of the surrounding atmosphere. The conservation of mass statement is similar to a bookkeeping method that allows us to account for what happens to the mass in an engineering problem. What happens if we try to pump some air into the tire that has a hole? Well, it all depends on the size of the hole and the pressure and flow rate of the pressurized air available to us. If the hole is small, we may be able to inflate the tire temporarily. Or the hole may be so large that the same amount of air that we put into the tire comes right back out. To completely describe all possible situations pertaining to this tire problem, we can express the conservation of mass as the rate at which air enters the tire minus the rate at which the air leaves the tire should be equal to the rate of accumulation or depletion of air inside the tire. Of course, we will use the physical quantity mass along with mathematics to express this statement. We will discuss the conservation of mass in more detail in Chapter 9.

There are other physical laws based on our observations that we use to analyze engineering problems. *Conservation of energy* is another good example. It is again similar to a bookkeeping method that allows us to keep track of various forms of energy and how they may change from one form to another. We will spend more time discussing the conservation of energy in Chapter 13.

> Physical laws are based on observation and experimentation and expressed using mathematical formulae.

Another important law that all of you have heard about is *Newton's second law of motion.* If you place a book on a smooth table and push it hard enough, it will move. This is simply the way things work. Newton observed this and formulated his observation into what we call Newton's second law of motion. This is not to say that other people had not made this simple observation before, but Newton took it a few steps further. He noticed that as he increased the mass of the object being pushed, while keeping the magnitude of the force constant (pushing with the same effort), the object did not move as quickly. Moreover, he noticed that there was a direct relationship between the push, the mass of the object being pushed, and the acceleration of the object. He also noticed that there was a direct relationship between the direction of the force and the direction of the acceleration. Newton expressed his observations using mathematics, but simply expressed, this law states that unbalanced force is equal to mass times acceleration. You will have the opportunity to take physics classes that will allow you to study and explore Newton's second law of motion further. Some of you may even take a dynamics class that will focus in greater detail on motion and forces and their relationship. Don't lose sight of the main idea: Physical laws are based on observations.

Another important idea to keep in mind is that a physical law may not fully describe all possible situations. Statements of physical laws have limitations because we may not fully understand how nature works; thus, we may fail to account for all variables that can affect the behavior of things within our natural world. Some natural laws are stated in a particular way to keep the mathematical expressions describing the observations simple. Often, we resort to experimental work dealing with specific engineering applications. For example, to better understand the aerodynamics of a car, we place it inside a wind tunnel to measure the drag force acting on the car. We may represent our experimental findings in the form of a chart or a correlation that can be used for design purposes over a predetermined range. The main difference between laws and other forms of experimental findings is that the laws represent the results of a much broader observation of nature, and almost everything that we know in our physical world obeys these laws. The engineering correlations, on the other hand, apply over a very limited and specific range of variables.

Before You Go On

Answer the following questions to test your understanding of the preceding section.

1. What do we mean by a physical law and what are they based on?

2. Give two examples of physical laws.

3. In your own words, describe the conservation of mass.

Vocabulary—State the meaning of the following terms:

Physical Law _____

Correlation _____

Learning Engineering Fundamental Concepts and Design Variables from Fundamental Dimensions

A Note to Students In Chapters 7 through 12, we will focus on teaching you some of the engineering fundamentals that you will see over and over in some form or other during your college years. Please try to study these concepts carefully and understand them completely. Unfortunately, today, many students graduate without a good grasp of these fundamental concepts—concepts that every engineer, regardless of his or her area of specialization, should know. We will focus on an innovative way to teach some of the engineering fundamental concepts using fundamental dimensions. Moreover, we will explain them in a way that could be easily grasped at a freshman level.

As we explained previously, from the observation of our surroundings, we have learned that we need only a few physical quantities (fundamental dimensions) to describe events and our surroundings. With the help of these fundamental dimensions, we can then define or derive engineering variables that are commonly used in analysis and design. As you will see in the following chapters, there are many engineering design variables that are related to these fundamental dimensions (quantities). As we also discussed and emphasized previously, we do need not only physical dimensions to describe our surroundings, but we also need some way to scale or divide these physical dimensions. For example, time is considered to be a physical dimension, but it can be divided into both small and large portions (such as seconds, minutes, hours, and so on). To become a successful engineer, you must first fully understand these fundamentals. Then it is important for you to know how these variables are measured, approximated, calculated, or used in engineering analysis and design. A summary of fundamental dimensions and their relationship to engineering variables is given in Table 6.7. After you understand these concepts, we will explain the concepts of energy and power in Chapter 13. Study this table carefully.

TABLE 6.7   **Fundamental Dimensions and How They Are Used in Defining Variables that Are Used in Engineering Analysis and Design**

Chapter	Fundamental Dimension	Related Engineering Variables		
7	Length (L)	Radian (L/L) Strain (L/L)	Area (L^2)	Volume (L^3)	Area moment of inertia (L^4)
8	Time (t)	Angular speed ($1/t$) Angular acceleration ($1/t^2$) Linear speed (L/t) Linear acceleration (L/t^2)		Volume flow rate (L^3/t)	
9	Mass (M)	Mass flow rate (M/t) Momentum (ML/t) Kinetic energy (ML^2/t^2)		Density (M/L^3), Specific volume (L^3/M)	

Continued

TABLE 6.7						

Fundamental Dimensions and How They Are Used in Defining Variables that Are Used in Engineering Analysis and Design *(Continued)*

Chapter	Fundamental Dimension		Related Engineering Variables			
10	Force (F)	Moment (LF)	Pressure (F/L^2)	Specific weight (F/L^3),		
		Work, energy (FL)	Stress (F/L^2)			
		Linear impulse (Ft),	Modulus of elasticity (F/L^2)			
		Power (FL/t)	Modulus of rigidity (F/L^2)			
11	Temperature (T)	Linear thermal expansion (L/LT)			Volume thermal expansion (L^3/L^3T),	
		Specific heat (FL/MT)				
12	Electric Current (I)	Charge (It)	Current density (I/L^2)			

SUMMARY

LO¹   Fundamental Dimensions and Units

By now, you should understand the importance of fundamental dimensions in everyday life and why, as good future engineers, you should develop a good grasp of them. As people, we have realized that we need only a few physical dimensions or quantities to describe our surroundings and daily events. For example, we need a length dimension to describe how tall, how long, or how wide something is. Time is another physical dimension that we need to answer questions such as: "How old are you?" or "How long it takes to go from here to there." You should also know that today, based on what we know about our world, we need seven fundamental dimensions to correctly express our observations in our surroundings. They are length, mass, time, temperature, electric current, amount of substance, and luminous intensity. The other

important concept that you should know is that not only do we need to define these physical dimensions to describe our surroundings but that we also need to devise some way to scale or divide them into units. For example, the dimension time can be divided into both small and large portions, such as seconds, minutes, hours, days, months, years, etc.

LO²   Systems of Units

The SI (from French *Systéme international d' unités*) is the most common system of units used in the world, and you should be familiar with the units of length (meter), time (second), mass (kilogram), temperature (kelvin or degree Celsius), electric current (ampere), amount of substance (mole), and luminous intensity (candela). You should also have a good feel for what these units represent. For example, how much a kilogram is or what a meter represents. The SI units also make use of series of prefixes and symbols of decimal multiples such mega, giga, kilo, etc.

You also should be familiar with British Gravitational units such as slugs. The U.S. Customary Systems of units is used only in the United States. You should be familiar with formal definitions for the units of length (feet), time (second), mass (pound-mass), temperature (degree Rankine or degree Fahrenheit), electric current (ampere), amount of substance (mole), and luminous intensity (candela). You also should have a good feel for what these units represent.

LO³ Unit Conversion and Dimensional Homogeneity

You should know how to convert values from one system of units to another. For example, you should be able to convert SI data, such as meter, kilogram, or kelvin, to U.S. Customary units of feet, pound-mass, and Rankine, and vice versa. You should know what we mean when we say an equation must be dimensionally homogeneous. For example, you already know that you cannot add someone's height who is 6 feet tall to his weight of 285 lbf and his body temperature of 98° F; that is, $6 + 285 + 98 = 389$? What would be the result of such a calculation? Therefore, if you were to use the formula $L = a + b + c$, in which the variable L on the left-hand side of the equation has a dimension of length, then the variables a, b, and c on the right-hand side of equation should also have dimensions of length. Common sense!

LO⁴ Significant Digits (Figures)

As we explained in this chapter, engineers make measurements and carry out calculations. Engineers then record the results of measurements and calculations using numbers. Significant digits (figures) represent and convey the extent to which recorded or computed data is dependable.

LO⁵ Components and Systems

You should be able to explain what is meant by a system and its component and give examples. For example, every product that you own or will purchase some day is considered a system and is made of components. The next time you purchase a product think of it in terms of a system and its components and be mindful of the entire life cycle of the product. Could the components of the system be recycled and/or used for another purpose?

LO⁶ Physical Laws and Observations

You should realize that physical laws are based on observation and experimentation. We have learned through observation and by the collective effort of those before us that things work a certain way in nature. For example, if you let go of something that you are holding in your hand, it will fall to the ground. That is an observation that we all agree upon. We can use words to explain our observations or use another language, such as mathematics and formulas, to express our findings. Sir Isaac Newton and many other scientists formulated their observations into useful mathematical expressions that we now can use to design various things.

K E Y   T E R M S

Ampere 153	Giga 154	Nano 154
Candela 153	Kelvin 153	Physical Law 173
Celsius 153	Kilogram 152	Pound Force 157
Centi 154	Kilo 154	Pound Mass 158
Component 172	Mass 148	Rankine 154
Dimension 146	Mega 154	Second 153
Fahrenheit 153	Meter 152	System 171
Foot 157	Micro 154	Tera 154
Force 148	Mole 153	Unit 150

APPLY WHAT YOU HAVE LEARNED

You are planning a business trip to Europe, and in order to prepare yourself well for conversations that may arise during your visit, you would need to convert the following data from U.S. Customary units to SI units: your height from feet and inches to meter and centimeter; your mass from pound to kilogram; your desired room temperature thermostat setting from Fahrenheit to Celsius; a gallon of drinking water to liters; fifteen gallons of gasoline to liters; speed limits of 30, 40, 50, 60, and 70 from miles per hour to kilometers per hour. Create a useful table that you can take along with you.

PROBLEMS

Problems that promote lifelong learning are denoted by 🔑

6.1 Convert the information given in the accompanying table from SI units to U.S. Customary units. Refer to the conversion tables on the inside front and back covers of this book. Show all steps of your solutions. See Examples 6.3 and 6.4.

Convert from SI Units	To U.S. Customary Units
120 km/hr	miles/hr and ft/s
1000 W	Btu/hr and hp
100 m³	ft³
80 kg	lbm
1000 kg/m³	lbm/ft³
900 N	lbf
100 kPa	lbf/in²
9.81 m/s²	ft/s²

6.2 Convert the information given in the accompanying table from U.S. Customary units to SI units. Refer to conversion tables on the inside front and back covers of this book. Show all steps of your solutions. See Examples 6.3 and 6.4.

Convert from U.S. Customary Units	To SI Units
65 miles/hr	km/hr and m/s
60,000 Btu/hr	W
120 lbm/ft³	kg/m³
30 lbf/in²	kPa
200 lbm	kg
200 lbf	N

6.3 The angle of twist for a shaft subjected to twisting torque can be expressed by the equation:

$$\phi = \frac{TL}{JG}$$

where

ϕ = the angle of twist in radians

T = applied torque (N·m)

L = length of the shaft in meter (m)

J = shaft's polar moment of inertia (measure of resistance to twisting)

G = shear modulus of the material (N/m²)

What is the appropriate unit for J if the preceding equation is to be homogeneous in units?

6.4 Which one of the following equations is dimensionally homogeneous? Show your proof.

a. $F = ma$

b. $F = m\dfrac{V^2}{R}$

c. $F(t_2 - t_1) = m(V_2 - V_1)$

d. $F = mV$

e. $F = m\dfrac{(V_2 - V_1)}{(t_2 - t_1)}$

where

F = force (**N**)

m = mass (**kg**)

a = acceleration (**m/s²**)

V = velocity (**m/s**)

R = radius (**m**)

t = time (**s**)

6.5 Determine the number of significant digits for the following numbers.

	Number of Significant Digits
286.5	
2.2×10^2	
2200	
0.0286	

6.6 Present the results of the following operations using the proper number of significant digits.

	Your Calculator Displays	Should Be Recorded As
1.2856 + 10.1 =		
155 − 0.521 =		
155 − 0.52 =		
1558 × 12 =		
3.585 ÷ 12 =		

6.7 For the following systems, identify the major components, and briefly explain the function or the role of each component: a. a dress shirt, b. pants, c. a skirt, d. shoes, e. a bicycle, and f. roller blades.

6.8 For the following systems, identify the major components:

a. a household refrigerator

b. a computer

c. the human body

d. a building

e. a hot water heater

f. a toaster

g. an airplane

6.9 Investigate what observations the following laws describe:

a. Fourier's law

b. Darcy's law

c. Newton's law of viscosity

d. Newton's law of cooling

e. Coulomb's law

f. Ohm's law

g. the ideal gas law

h. Hooke's law

i. the first law of thermodynamics

j. Fick's law

k. Faraday's law

6.10 Investigate the operation of various turbines. Write a brief report explaining the operation of steam turbines, hydraulic turbines, gas turbines, and wind turbines.

6.11 In a brief report, discuss why we need various modes of transportation. How did they evolve? Discuss the role of public transportation, water transportation, highway transportation, railroad transportation, and air transportation.

6.12 Identify the major components of a computer, and briefly explain the function or the role of each component.

6.13 Which one of the following equations is dimensionally homogeneous? Show your proof.

a. $F(x_2 - x_1) = \dfrac{1}{2}mV_2^2 - \dfrac{1}{2}mV_1^2$

b. $F = \frac{1}{2}mV_2^2 - \frac{1}{2}mV_1^2$

c. $F(V_2 - V_1) = \frac{1}{2}mx_2^2 - \frac{1}{2}mx_1^2$

d. $F(t_2 - t_1) = mV_2 - mV_1$

where

F = force (N)

x = distance (m)

m = mass (kg)

V = velocity (m/s)

t = time (s)

6.14 A car has a mass of 1500 kg. Express the mass and the weight of the car using BG and U.S. Customary units. Show the conversion steps.

6.15 Express the kinetic energy $\frac{1}{2}(mass)(speed)^2$ of a car with a mass of 1200 kg and moving at a speed of 100 km/h using SI, BG, and U.S. Customary units. Show the conversion steps.

6.16 A machine shop has a rectangular floor shape with dimensions of 30 ft by 50 ft. Express the area of the floor in ft², m², in², and cm² . Show the conversion steps.

6.17 A trunk of a car has a listed luggage capacity of 18 ft³. Express the capacity in in³, m³, and cm³. Show the conversion steps.

6.18 The 2005 Acura RL has a listed 300 horsepower, 3.5 liter engine. Express the engine size in both kW and in³. Show the conversion steps.

6.19 The density of air at standard room conditions is 1.2 kg/m³. Express the density in BG and U.S. Customary units. Show the conversion steps.

6.20 On a summer day in Phoenix, Arizona, the inside room temperature is maintained at 68° F while the outdoor air temperature is a sizzling 110° F. What is the outdoor–indoor temperature difference in (a) degrees Fahrenheit, (b) degrees Rankine, (c) degrees Celsius, and (d) kelvin? Is one degree temperature difference in Celsius equal to one temperature difference in kelvin, and is one degree temperature difference in Fahrenheit equal to one degree temperature difference in Rankine? If so, why?

6.21 A cantilever beam shown in the accompanying figure is used to support a load acting on a balcony. The deflection of the centerline of the beam is given by the equation:

$$y = \frac{-wx^3}{24EI}(x^2 - 4Lx + 6L^2)$$

where

y = deflection at a given x location, (m)

w = distributed load

E = modulus of elasticity (N/m²)

I = second moment of area (m⁴)

x = distance from the support as shown (m)

L = length of the beam (m)

What is the appropriate unit for w if the preceding equation is to be homogeneous in units? Show all steps of your work.

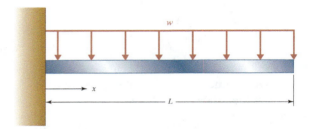

Problem 6.21   A cantilever beam.

6.22 A model known as *stopping sight distance* is used by civil engineers to design roadways. This simple model estimates the distance a driver needs in order to stop his car while traveling at a certain speed after detecting a hazard. The model proposed by the American Association of State Highway Officials (AASHO) is given by

$$S = \frac{V^2}{2g(f \pm G)} + TV$$

where

S = stopping sight distance (ft)

V = initial speed (ft/s)

g = accelaration due to gravity, 32.2 ft/s²

f = coefficient of friction between tires and roadways

G = grade of road

T = driver reaction time (s)

What are the appropriate units for f and G if the preceding equation is to be homogeneous in units? Show all steps of your work.

6.23 In an annealing process—a process wherein materials such as glass and metal are heated to high temperatures and then cooled slowly to toughen them—the following equation may be used to determine the temperature of a thin piece of material after some time t.

$$\frac{T - T_{environment}}{T_{initial} - T_{environment}} = \exp\left(-\frac{2h}{\rho c L} t\right)$$

where

T = temperature ($^{\circ}$C)

h = heat transfer coefficient

ρ = density (kg/m^3)

c = specific heat (J/kg$\cdot$K)

L = plate thickness (m)

t = time (s)

Those of you who will pursue aerospace, chemical, mechanical, or materials engineering will learn about the underlying concepts that lead to the solution in your heat-transfer class. What is the appropriate unit for h if the preceding equation is to be homogeneous in units? Show all steps of your work.

6.24 The air resistance to the motion of a vehicle is something important that engineers investigate. As you may also know, the drag force acting on a car is determined experimentally by placing the car in a wind tunnel. For a given car, the experimental data is generally represented by a single coefficient that is called the drag coefficient. It is given by the relationship:

$$C_d = \frac{F_d}{\frac{1}{2}\rho V^2 A}$$

where

C_d = drag coefficient

F_d = measured drag force (lb)

ρ = air density (slugs/ft^3)

V = air speed inside the wind tunnel (ft/s)

A = frontal area of the car (ft^2)

What is the appropriate unit for C_d if the preceding equation is to be homogeneous in units? Show all steps of your work.

6.25 Fins, or extended surfaces, commonly are used in a variety of engineering applications to enhance cooling. Common examples include a motorcycle engine head, a lawnmower engine head, heat sinks used in electronic equipment, and finned-tube heat exchangers in room heating and cooling applications. For long fins, the temperature distribution along the fin is given by the exponential relationship:

$$T - T_{ambient} = (T_{base} - T_{ambient})e^{-mx}$$

where

T = temerature (K)

$m = \sqrt{\dfrac{hp}{kA}}$

h = the transfer coeficient (W/m^2 $\cdot$ K)

p = perimeter of the fin (m)

A = cross-section area of the fin (m^2)

k = thermal onductivity of the fin material

x = distance from the base of the fin (m)

What is the appropriate unit for k if the preceding equation is to be homogeneous in units? Show all steps of your work.

6.26 A person who is 180 cm tall and weighs 750 newtons is driving a car at a speed of 90 kilometers per hour over a distance of 80 kilometers. The outside air temperature is 30° C and has a density of 1.2 kilograms per cubic meter (kg/m^3). Convert all of the values given in this example from SI units to U.S. Customary units.

6.27 Use the conversion factors given on the front and back end covers of this text to convert the given values: (a) area $A = 16$ in^2 to ft^2, (b) volume $V = 64$ in^3 to ft^3, and (c) area moment of inertia $I = 21.3$ in^4 to ft^4.

6.28 The acceleration due to gravity g is 9.81 m/s^2. Express the value g in U.S. Customary and B.G. units. Show all conversion steps.

6.29 Sir Isaac Newton discovered that any two masses m_1 and m_2 attract each other with a force that is equal in magnitude and acts in the opposite direction, according to the relationship:

$$F = \frac{Gm_1 m_2}{r^2}$$

where

F = attractive force (N)
G = Universal Gravitational Constant
m_1 = mass of particle -1 (kg)
m_2 = mass of particle -2 (kg)
r = distance between the center of each particle (m)

What is the appropriate unit for G, if the preceding equation is to be homogeneous in units?

6.30 Convert the atmospheric pressure in the given units to requested units. Show all the conversion steps. (a) 14.7 lbf/in^2 to lbf/ft^2, (b) 14.7 lbf/in^2 to Pa, (c) 14.7 lbf/in^2 to kPa, and (d) 14.7 lbf/in^2 to bar (1 bar $=100$ kPa).

6.31 For gases under certain conditions, there is a relationship between the pressure of the gas, its volume, and its temperature as given by what is commonly called the *ideal gas law*. The ideal gas law is

$PV = mRT$

where

P = absolute pressure of the gas (Pa)
V = volume of the gas (m^3)
m = mass (kg)
R = gas constant
T = absolute temperature (kelvin)

What is the appropriate unit for R if the preceding equation is to be homogeneous in units?

6.32 The amount of radiant energy emitted by a surface is given by

$$q = \varepsilon \sigma A T_s^4$$

where

q represents the rate of thermal energy (per unit time) emitted by the surface in watts;

ε = the emissivity of the surface $0 < \varepsilon < 1$ and is unitless
σ = Stefan-Boltzman constant ($\sigma = 5.67 \times 10^8$)
A represents the area of the surface in m^2
T_s = surface temperature of the object expressed in kelvin

What is the appropriate unit for σ if the equation is to be homogeneous in units?

6.33 A person's body temperature is controlled by (1) convective and radiative heat transfer to the surroundings, (2) sweating, (3) respiration by breathing surrounding air and exhaling it at approximately body temperature, (4) blood circulation near the surface of the skin, and (5) metabolic rate. Metabolic rate determines the rate of conversion of chemical to thermal energy within a person's body. The metabolic rate depends on the person's activity level. A unit commonly used to express the metabolic rate for an average person under sedentary conditions (per unit surface area) is called *met*; 1 met is equal to 58.2 W/m^2. Convert this value to Btu/hr · ft^2. Also, calculate the amount of energy dissipated by an average adult person sleeping for 8 hours if he or she generates 0.7 mets and has a body surface area of 19.6 ft^2. Express your results in Btu and Joules (1 Btu $= 1055$ Joules).

6.34 The *calorie* is defined as the amount of heat required to raise the temperature of 1 gram of water by 1° C. Also, the energy content of food is typically expressed in *Calories,* which is equal here to 1000 calories. Convert the results of the previous problem to calories (1 Btu $= 252$ calories).

6.35 Convert the strength of selected materials given in the accompanying table from MPa to ksi, where $1000 \text{ lbf/in}^2 = 1 \text{ ksi}$.

Material	Ultimate strength (MPa)	Ultimate strength (ksi)
Aluminum alloys	100–550	
Concrete (compression)	10–70	
Steel		
Machine	550–860	
Spring	700–1,900	
Stainless	400–1,000	
Tool	900	
Structural Steel	340–830	
Titanium alloys	900–1,200	
Wood (Bending)		
Douglas fir	50–80	
Oak	50–100	
Southern pine	50–100	

6.36 The density of water is 1000 kg/m^3. Express the density of water in lbm/ft^3 and $\text{lbm/gallon}(7.48 \text{ gallons} = 1\text{ft}^3)$.

6.37 A unit that is generally used to express the insulating value of clothing is called *clo*. 1 clo is equal to $0.155 \text{ m}^2 \cdot \text{°C/W}$. Express this value in U.S. Customary units $(\text{°F} \cdot \text{ft}^2 \cdot \text{hr/Btu})$.

6.38 Start with $1 \text{ lbf} \cdot \text{ft/s}$ and convert it to $\text{N} \cdot \text{m/s}$ and show that $1 \text{ lbf} \cdot \text{ft/s}$ is equal to 1.36 W. Knowing that $550 \text{ lbf} \cdot \text{ft/s}$ is equal to 1 hp (horsepower), how many kW is that?

6.39 Viscosity of fluid plays a significant role in the analyses of many fluid dynamics problems. The viscosity of water can be determined from the correlation:

$$\mu = c_1 10^{\left(\frac{c_2}{T - c_3}\right)}$$

where

$$\mu = \text{viscosity } (\text{N/s} \cdot \text{m}^2)$$
$$T = \text{temperature } (\text{K})$$
$$c_1 = 2.414 \times 10^{-5}$$
$$c_2 = 247.8 \text{ (K)}$$
$$c_3 = 140 \text{ (K)}$$

What is the appropriate unit for c_1 if the preceding equation is to be homogeneous in units?

6.40 For the ideal gas law given in Problem 6.31, if the units of P, V, m, and T are expressed in lbf/ft^2, ft^3, lbm, and degree Rankine (°R), respectively, what is the appropriate unit for gas constant R if the ideal gas law is to be homogeneous in units?

6.41 For the fin equation described in Problem 6.25, if the units of T, h, P, A, and x are expressed in degree Rankine (°R), $\text{Btu/hr} \cdot \text{ft}^2 \cdot \text{°R}$, ft, ft^2, and ft, respectively, what is the appropriate unit for thermal conductivity k if the fin equation is to be homogeneous in units? Show all steps of your work.

6.42 For Example 6.6 part (a), if the units of d, P, L, and A are given in in., lbf, inches, and in^2, respectively, what are the units for modulus of elasticity E?

6.43 For Example 6.6 part (b), if the units of k, A, $T_1 - T_2$, and L are given in $\text{Btu/hr} \cdot \text{ft} \cdot \text{°F}$, ft^2, °F, and ft, respectively, what is the appropriate unit for heat transfer rate q?

6.44 For the drag coefficient relationship given in Problem 6.24, if the units of F_d, V, A, and C_d, are expressed in N, m/s, m^2, and unitless, respectively, what is the appropriate unit for ρ?

6.45 For the cantilever beam relationship given in Problem 6.21, if the units of y, w, E, x, and L are expressed in in., lbf/in, lbf/in^2, in., and in., respectively, what is the appropriate unit for I?

6.46 The rotation of a rigid object is governed by the relationship:

$$\sum M = I\alpha$$

where

$\sum M$ = sum of the moments due to external
forces $(N \cdot m)$
I = mass moment of inertia
α = angular acceleration of the
object (rad/s^2)

What is the appropriate unit for mass moment of inertia I?

6.47   The value of viscosity of a fluid represents a measure of how easily the given fluid can flow. The higher the viscosity value is, the more resistance the fluid offers to flow. For example, it would require less energy to transport water in a pipe than it would to transport motor oil or glycerin. The viscosity of many fluids is governed by Newton's law of viscosity

$$\tau = \mu \frac{du}{dy}$$

where

τ = shear stress (N/m^2)

μ = viscosity

du = change in flow speed (m/s) over a height dy (m)

What is the appropriate unit for viscosity?

6.48   The power output of a water turbine is given by

$$P = \rho g Q h$$

where

P = power

ρ = density of water (kg/m^3)

g = acceleration due to gravity (m/s^2)

Q = water flow rate (m^3/s)

h = the available water head (m)

What is the appropriate unit for P?

6.49   The head loss due to flow of a fluid inside a pipe is calculated from

$$h_{Loss} = f \frac{L}{D} \frac{V^2}{2g}$$

where

h_{Loss} (m),

f = friction factor

L = pipe length (m)

D = pipe diameter (m)

V = average flow velocity (m/s)

g = acceleration due to gravity (m/s^2).

What is the appropriate unit for friction factor f?

6.50   For Problem 6.49, h_{Loss} is expressed in feet, L and D in inches, V in (ft/s), and g in (ft/s^2). What is the appropriate unit for the friction factor f?

6.51   Look up the metric specifications for a car of your choice (body, trunk, engine sizes, and gas consumption) and a home appliance such as an air-conditioning unit (size, cooling capacity, and energy consumption). Convert your findings to U.S. Customary units.

Georgios Kollidas/Shutterstock.com

"Nothing is too wonderful to be true if it be consistent with the laws of nature."— MICHAEL FARADAY (1791–1867)

Length and Length-Related Variables in Engineering

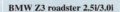

BMW Z3 roadster 2.5i/3.0i

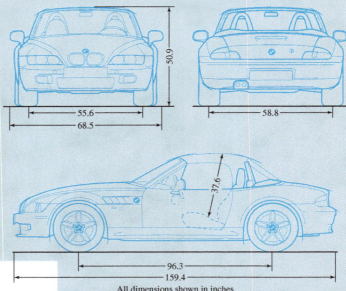

50.9

55.6

68.5

58.8

37.6

96.3

159.4

All dimensions shown in inches

Z3 Roadster 2.5i/3.0i schematics reproduced with permission of BMW AG Munich.

The important dimensions of a BMW Z3 Roadster are shown in the illustration. The fundamental dimension length and other length-related variables, such as area and volume (e.g., seating surface or trunk capacity), play important roles in engineering design. As a good future engineer, it is important for you to know how to measure, how to calculate, and how to approximate length, area, and volume.

LEARNING OBJECTIVES

LO¹ **Length as a Fundamental Dimension:** describe the role of length in engineering analysis and design, as well as its units, measurement, and calculation

LO² **Ratios of Two Lengths—Radians and Strain:** explain what is meant by radians and strain—both length-related quantities— and their role in engineering analysis and design

LO³ **Area:** describe the role of area—a length-related quantity—in engineering analysis and design, as well as its units, calculation, and measurement

LO⁴ **Volume:** describe the role of volume—a length-related quantity—in engineering analysis and design, as well as its units, calculation, and measurement

LO⁵ **Second Moment of Area:** explain what is mean by second moment of area—a length-related quantity—its role in engineering analysis and design, and its calculation

HAND GEOMETRY

Hand geometry recognition is the longest implemented biometric type, debuting in society in the late 1980s. The systems are widely implemented for their ease of use, public acceptance, and integration capabilities. One of the shortcomings of the hand geometry characteristic is that it is not highly unique, limiting the applications of the hand geometry system to verification tasks only.

The devices use a simple concept of measuring and recording the length, width, thickness, and surface area of an individual's hand when guided on a plate. Hand geometry systems use a camera to capture a silhouette image of the hand. The hand of the subject is placed on the plate, palm down, and guided by five pegs that sense when the hand is in place. The image captures both the top surface of the hand and a side image that is captured using an angled mirror. Upon capture of the silhouette image, 31,000 points are analyzed and 90 measurements are taken; the measurements range from length of the fingers, to distance between knuckles, to the height or thickness of the hand and fingers.

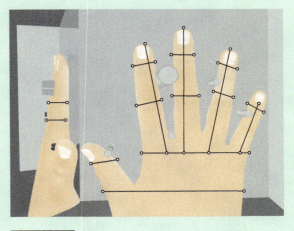

http://www.biometrics.gov

To the students: Can you give other examples of length and length-related variables in everyday life? What is the role of length and length-related quantities in engineering analysis and design? Can you give some examples?

When you become a practicing engineer, you will find out that you don't stop learning new things even after obtaining your engineering degree. For example, you may work on a project in which the noise of a machine is a concern, and you may be asked to come up with ways to reduce the level of noise. It may be the case that during the four or five years of your engineering education, you did not take a class in noise control. Considering your lack of understanding and background in noise reduction, you may first try to find someone who specializes in noise control who could solve the problem for you. But your supervisor may tell you that because of budget constraints and because this is a one-time project, you must come up with a reasonable solution yourself. Therefore, you will have to learn something new and learn fast. If you have a good grasp of underlying engineering concepts and fundamentals, the learning process could be fun and quick. The point of this story is that during the next four years you need to make sure that you learn the fundamentals well.

TABLE 6.7	Fundamental Dimensions and How They Are Used in Defining Variables that Are Used in Engineering Analysis and Design				

Chapter	Fundamental Dimension	Related Engineering Variables			
7	Length (L)	Radian (L/L), Strain (L/L)	Area (L^2)	Volume (L^3)	Area moment of inertia (L^4)
8	Time (t)	Angular speed ($1/t$), Angular acceleration ($1/t^2$) Linear speed (L/t), Linear acceleration (L/t^2)		Volume flow rate (L^3/t)	
9	Mass (M)	Mass flow rate (M/t), Momentum (ML/t), Kinetic energy (ML^2/t^2)		Density (M/L^3), Specific volume (L^3/M)	
10	Force (F)	Moment (LF), Work, energy (FL), Linear impulse (Ft), Power (FL/t)	Pressure (F/L^2), Stress (F/L^2), Modulus of elasticity (F/L^2), Modulus of rigidity (F/L^2)	Specific weight (F/L^3),	
11	Temperature (T)	Linear thermal expansion (L/LT), Specific heat (FL/MT)		Volume thermal expansion (L^3/L^3T)	
12	Electric Current (I)	Charge (It)	Current density (I/L^2)		

As an engineering student, you also need to develop a keen awareness of your surroundings. In this chapter, we will investigate the role of length, area, and volume along with other length-related variables in engineering applications. You will learn how these physical variables affect engineering design decisions. The topics introduced in this chapter are fundamental in content, so developing a good grasp of them will make you a better engineer. You may see some of the concepts and ideas introduced here in some other form in the engineering classes that you will take later. The main purpose of introducing these concepts here is to help you become aware of their importance and learn to look for their relation to other engineering parameters in your future classes when you study a specific topic in detail.

Table 6.7 is repeated here to show the relationship between the content of this chapter and other fundamental dimensions discussed in other chapters.

LO¹ 7.1   Length as a Fundamental Dimension

When you walk down a hallway, can you estimate how tall the ceiling is? Or how wide a door is? Or how long the hallway is? You should develop this ability because having a "feel" for dimensions will help you become a better engineer. If you decide to become a design engineer, you will find out that size and cost are important design parameters. Having developed a "feel" for the size of objects in your surroundings enables you to have a good idea about the acceptable range of values when you design something.

As you know, every physical object has a size. Some things are bigger than others. Some things are wider or taller than others. These are some common ways of expressing the relative size of objects. As discussed in Chapter 6, through their observation of nature, people recognized the need for a physical quantity or a physical dimension (which today we call **length**) so that they could describe their surroundings better. They also realized that having a common definition for a physical quantity, such as length, makes communication easier. Earlier humans may have used a finger, arm, stride, stick, or rope to measure the size of an object. Chapter 6 also emphasized the need for having scales or divisions for the dimension length so that numbers could be kept simple and manageable. Today, we call these divisions or scales *systems of units*. In this chapter, we will focus our attention on length and such length-related derived quantities as area and volume.

Length is one of the seven fundamental or base dimensions that we use to properly express what we know of our natural world. In today's globally driven economy, where products are made in one place and assembled somewhere else, there exists an even greater need for a uniform and consistent way of communicating information about the fundamental dimension length and other related length variables so that parts manufactured in one place can easily be combined on an assembly line with parts made in other places. An automobile is a good example of this concept. It has literally thousands of parts that are manufactured by various companies in different parts of the world.

As explained in Chapter 6, we have learned from our surroundings and formulated our observations into laws and principles. We use these laws and physical principles to design, develop, and test products and services. Are you observing your surroundings carefully? Are you learning from your everyday observations? Here are some questions to consider: Have you thought about the size of a soda can? What are its dimensions? What do you think are the important design factors? Most of you drink a soda every day, so you know that it fits in your hand. You also know that the soda can is made from aluminum, so it is lightweight. What do you think are some of its other design factors? What are important considerations when designing signs for a highway? How wide should a hallway be? When designing a supermarket, how wide should an aisle be? Most of you have been going to class for at least 12 years, but have you thought about classroom seating arrangements? For example, how far apart are the desks? Or how far above the floor is the presentation board? For those of you who may take a bus to school, how wide are the seats in a bus? How wide is a highway lane? What do you think are the important factors when determining the size of a car seat? You can also look around your home to think about the dimension length. Start with your bed: What are its dimensions; how far above the floor is it? What is a typical standard height for steps in a stairway? When you tell someone that you own a 32-inch TV, to what dimension are you referring? How high off the floor are a doorknob, showerhead, sink, light switch, and so on?

> Coordinate systems are used to locate things with respect to a known origin. Based on the location of the thing, we use different types of coordinate systems, such as rectangular, cylindrical, and spherical.

You are beginning to see that length is a very important fundamental dimension, and it is thus commonly used in engineering products. Coordinate systems are examples of another application where length plays an important role. Coordinate systems are used to locate things with respect to a known origin. In fact, you use coordinate systems every day, even though you don't think about it. When you go from your home to school or a grocery store or to meet a friend for lunch, you use coordinate systems. The use of coordinate systems is almost second nature to you. Let's say you live downtown, and your school is located on the northeast side of town. You know the exact location of the school with respect to your home. You know which streets to take for what distance and in which directions to move to get to school. You have been using a coordinate system to locate places and things most of your life, even if you did not know it. You also know the specific location of objects at home relative to other objects or to yourself. You know where your TV is located relative to your sofa or bedroom.

There are different types of coordinate systems such as rectangular, **cylindrical**, **spherical**, and so on, as shown in Figure 7.1. Based on the nature of a particular problem, we may use one or another. The most common coordinate system is the rectangular or **Cartesian coordinate system** (Figure 7.1). When you are going to school from home or meeting a friend for lunch you use the rectangular coordinate system. But you may not call it that; you may use the directions north, east, west, or south to get where you are going. You can think of the axes of a rectangular coordinate system as aligning with, for example, the east and north direction. People who are blind are expert users of rectangular coordinate systems. Because they cannot rely on their visual perception, people with a vision disability know how many steps to take and in which direction to move to go from one location to another. So to better understand coordinate systems, perform the following experiment. While at home, close your eyes for a few minutes and try to go from your bedroom to the bathroom. Note the number of steps you had to take and in which directions you had to move. Think about it!

All of us, at one time or another, have experienced not knowing where we are going. In other words, we were lost! The smarter ones among us use a map or stop and ask someone for directions and distances (x and y coordinates) to the desired place. An example of using a map to locate a place is shown in Figure 7.2. **Coordinate systems** are also integrated into software that drives computer numerically controlled (CNC) machines, such as a milling machine or a lathe that cuts materials into specific shapes.

Now that you understand the importance of the length dimension, let us look at its divisions or units. There are several systems of units in use in engineering today. We will focus on two of these systems: the International System of Units (SI) and the United States Customary units. The unit of length in SI is the meter (m). We can use the multiples and fractions of this unit according to Table 6.2. Common multipliers of the meter are micrometer (μm), millimeter (mm), centimeter (cm), and kilometer (km). Recall from our discussion of units and multiplication prefixes in Chapter 6 that we use these multiplication prefixes to keep the numbers manageable. The International System of Units

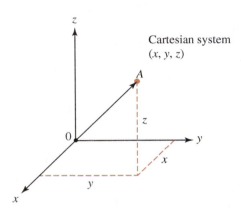

Cartesian system
(x, y, z)

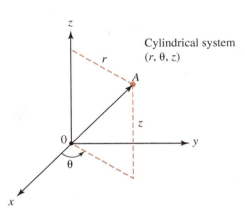

Cylindrical system
(r, θ, z)

To locate an object at point A, with respect to the origin (point 0) of the Cartesian system, you move along the x axis by x amount (or steps) and then you move along the dashed line parallel to the y axis by y amount. Finally, you move along the dashed line parallel to the z direction by z amount. How would you get to point A using the Cylindrical or Spherical system? Explain.

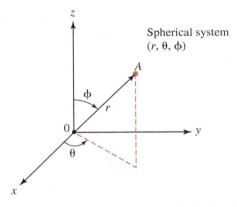

Spherical system
(r, θ, ϕ)

FIGURE 7.1   Examples of coordinate systems.

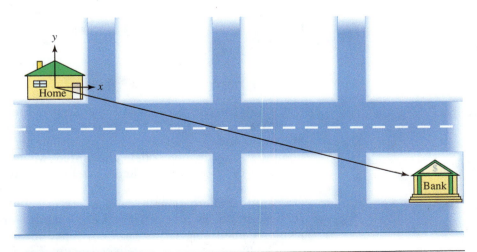

FIGURE 7.2   An example demonstrating the use of a coordinate system.

is used almost universally, except in the United States. The unit of length in the U.S. Customary system is foot (ft). The relation between foot and meter units is given by $1\,\text{ft} = 0.3048\,\text{m}$. Table 7.1 shows other commonly used units and their equivalent values in an increasing order and includes both SI and U.S. Customary units to give you a sense of their relative magnitude.

Some interesting dimensions in the natural world are

The highest mountain peak (Everest):	8848 m (29,028 feet)
Pacific Ocean:	Average depth: 4028 m (13,215 feet)
	Greatest known depth: 11,033 m (36,198 feet)

TABLE 7.1	Units of Length and their Equivalent Values
Units of Length in Increasing Order	**Equivalent Value**
1 angstrom	1×10^{-10} meter (m)
1 micrometer or 1 micron	1×10^{-6} meter (m)
1 mil	1/1000 inch $\approx 2.54 \times 10^{-5}$ meter (m)
1 point (printer's)	3.514598×10^{-4} (m)
1 millimeter	1/1000 meter (m)
1 pica (printer's)	4.217 millimeter (mm)
1 centimeter	1/100 meter (m)
1 inch	2.54 centimeter (cm)
1 foot	12 inches (in.)
1 yard	3 feet (ft)
1 meter	1.0936 yard $\approx$ 1.1 yard (yd)
1 kilometer	1000 meter (m)
1 mile	1.6093 kilometer (km) = 5280 feet (ft)

Measurement and Calculation of Length

Early humans may have used finger length, arm span, stride length (step length), a stick, rope, chains, and so on to measure the size or displacement of an object. Today, depending on how accurate the measurement needs to be and the size of the object being measured, we use other measuring devices, such as a ruler, a yardstick, and a steel tape. All of us have used a ruler or tape measure to measure a distance or the size of an object. These devices are based on internationally defined and accepted units such as millimeters, centimeters, or meters or inches, feet, or yards. For more accurate measurements of small objects, we have developed measurement tools such as the micrometer or the Vernier caliper (Figure 7.3), which allow us to measure dimensions within 1/1000 of an inch. In fact, machinists use micrometers and Vernier calipers every day.

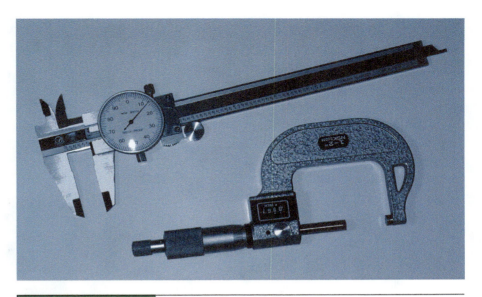

FIGURE 7.3   A Vernier caliper and a micrometer.
Saeed Moaveni

On a larger scale, you have seen the milepost markers along interstate highways. Some people actually use the mileposts to check the accuracy of their car's odometer. By measuring the time between two markers, you can also check the accuracy of the car's speedometer. In the last few decades, electronic distance measuring instruments (EDMI) have been developed that allow us to measure distances from a few feet to many miles with reasonable accuracy. These electronic distance measuring devices are used quite commonly for surveying purposes in civil engineering applications (Figure 7.4). The instrument sends out a light beam that is reflected by a system of reflectors located at the unknown distance. The instrument and the reflector system are situated such that the reflected light beam is intercepted by the instrument. The instrument then interprets the information to determine the distance between the instrument and the reflector. The Global Positioning System (GPS) is another example of recent advances in locating objects on the surface of the earth with good accuracy. As of the year 2013, radio signals were sent from approximately 64 artificial satellites orbiting the earth. Tracking stations are located around the world to receive and interpret the signals sent from the satellites. Although originally the GPS was funded and controlled by the U.S. Department of Defense for military applications, it now has hundreds of millions of users. GPS navigation receivers are now common in airplanes, automobiles, buses, cellphones, and hand-held receivers used by hikers.

Sometimes, distances or dimensions are determined indirectly using trigonometric principles. For example, let us say that we want to determine the height of a building similar to the one shown in Figure 7.5 but do not have access to accurate measuring devices. With a drinking straw, a protractor, and a steel tape we can determine a reasonable value for the height of the building by measuring the angle α and the dimensions d and h_1. The analysis is shown in Figure 7.5. Note that h_1 represents the distance from the ground to the eye of a person looking through the straw. The angle α (alpha) is the angle between the straw, which is focused on the edge of the roof, and a horizontal line. The protractor is used to measure the angle that the straw makes with the horizontal line.

FIGURE 7.4	An example of an electronic distance measuring instrument used in surveying.

Brian McEntire/Shutterstock.com

It is very likely that you have used trigonometric tools to analyze some problems in the past. It is also possible that you may have not used them recently. If that is the case, then the tools have been sitting in your mental toolbox for a while and quite possibly have collected some rust in your head! Or you may have forgotten altogether how to properly use the tools. Because of their importance, let us review some of these basic relationships and definitions. For a right triangle, the Pythagorean relation may be expressed by

$$a^2 + b^2 = c^2$$

In the *right triangle* shown in Figure 7.6(a), the angle facing side a is denoted by α (alpha), and the angle facing side b by β (beta). The sine, cosine, and the tangent of an angle are defined by

$$\sin \alpha = \frac{\text{opposite}}{\text{hypotenuse}} = \frac{a}{c}, \ \cos \alpha = \frac{\text{adjacent}}{\text{hypotenuse}} = \frac{b}{c}, \ \tan \alpha = \frac{\sin \alpha}{\cos \alpha} = \frac{\text{opposite}}{\text{adjacent}} = \frac{a}{b}$$

$$\sin \beta = \frac{\text{opposite}}{\text{hypotenuse}} = \frac{b}{c}, \ \cos \beta = \frac{\text{adjacent}}{\text{hypotenuse}} = \frac{a}{c}, \ \tan \beta = \frac{\sin \beta}{\cos \beta} = \frac{\text{opposite}}{\text{adjacent}} = \frac{b}{a}$$

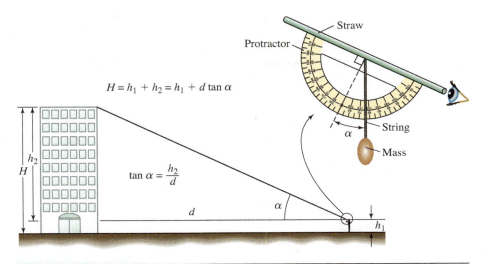

$$H = h_1 + h_2 = h_1 + d \tan \alpha$$

$$\tan \alpha = \frac{h_2}{d}$$

FIGURE 7.5   Measuring the height of a building indirectly.

The sine and the cosine law (rule) for an arbitrarily shaped triangle Figure 7.6(b) is

The sine rule: $\dfrac{a}{\sin \alpha} = \dfrac{b}{\sin \beta} = \dfrac{c}{\sin \theta}$

The cosine rule:
$$a^2 = b^2 + c^2 - 2\,bc\,(\cos \alpha)$$
$$b^2 = a^2 + c^2 - 2\,ac\,(\cos \beta)$$
$$c^2 = a^2 + b^2 - 2\,ba\,(\cos \theta)$$

or

$$\cos \alpha = \frac{b^2 + c^2 - a^2}{2\,bc}$$

$$\cos \beta = \frac{a^2 + c^2 - b^2}{2\,ac}$$

$$\cos \theta = \frac{a^2 + b^2 - c^2}{2\,ba}$$

Some other useful trigonometry identities are

$$\sin^2 \alpha + \cos^2 \alpha = 1$$
$$\sin 2\alpha = 2 \sin \alpha \cos \alpha$$
$$\cos 2\alpha = \cos^2 \alpha - \sin^2 \alpha = 2 \cos^2 \alpha - 1 = 1 - 2 \sin^2 \alpha$$
$$\sin(-\alpha) = -\sin \alpha$$
$$\cos(-\alpha) = \cos \alpha$$
$$\sin(\alpha + \beta) = \sin \alpha \cos \beta + \sin \beta \cos \alpha$$
$$\sin(\alpha - \beta) = \sin \alpha \cos \beta - \sin \beta \cos \alpha$$
$$\cos(\alpha + \beta) = \cos \alpha \cos \beta - \sin \alpha \sin \beta$$
$$\cos(\alpha - \beta) = \cos \alpha \cos \beta - \sin \alpha \cos \beta$$

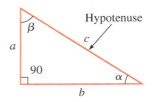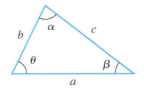

FIGURE 7.6

As you will learn later in your physics, statics, and dynamics classes, the trigonometry relations are quite useful.

Nominal Sizes versus Actual Sizes

You have all seen or used a 2×4 piece of lumber. If you were to measure the dimensions of the cross section of a 2×4, you would find that the actual width is less than 2 inches (approximately 1.5 inches) and the height is less than 4 inches (approximately 3.5 inches). Then why is it referred to as a "2 by 4"? Manufacturers of engineering parts (See Figure 7.7) use round numbers so that it is easier for people to remember the size and thus more easily refer to a specific part. The 2×4 is called the **nominal size** of the lumber. If you were to investigate other structural members, such as I-beams, you would also note that the nominal size given by the manufacturer is different from the **actual size**. You will find a similar situation for pipes, tubes, screws, wires, and many other engineering parts. Agreed-upon standards are followed by manufacturers when providing information about the size of the parts that they make. But manufacturers do provide actual sizes of parts in addition to nominal sizes. This fact is important because, as you will learn in your future engineering classes, you need the actual size of parts for various engineering calculations. Examples of nominal sizes versus actual sizes of some engineering parts are given in Table 7.2.

FIGURE 7.7   Manufacturers provide actual and nominal sizes for many parts and structural members.

| TABLE 7.2 | | | | Examples of Nominal Size versus Actual Size of Some Engineering Products |

Dimension of Copper Tubing (Type K—for high pressure—temperature use)

Nominal Size (in.)	Outside Diameter (in.)	Inside Diameter (in.)	Wall Thickness (in.)	Flow Cross Section (in²)
1/4	0.375	0.305	0.035	0.073
3/8	0.5	0.402	0.049	0.127
1/2	0.625	0.527	0.049	0.218
5/8	0.75	0.652	0.049	0.334
3/4	0.875	0.745	0.065	0.436
1	1.125	0.995	0.065	0.778
$1\frac{1}{4}$	1.375	1.245	0.065	1.22
$1\frac{1}{2}$	1.625	1.481	0.072	1.72
2	2.125	1.959	0.083	3.01
$2\frac{1}{2}$	2.625	2.435	0.095	4.66
3	3.125	2.907	0.109	6.64
$3\frac{1}{2}$	3.625	3.385	0.120	9.00
4	4.125	3.857	0.134	11.7
5	5.125	4.805	0.160	18.1

Dimension of Copper Tubing (Type L—for HVAC applications)

Nominal Size (in.)	Outside Diameter (in.)	Inside Diameter (in.)	Wall Thickness (in.)	Flow Cross Section (in²)
1/4	0.375	0.315	0.030	0.078
3/8	0.5	0.44	0.035	0.145
1/2	0.625	0.545	0.040	0.233
5/8	0.75	0.666	0.042	0.334
3/4	0.875	0.785	0.045	0.484
1	1.125	1.025	0.065	0.825
$1\frac{1}{4}$	1.375	1.265	0.050	1.26

Continued

TABLE 7.2	Examples of Nominal Size versus Actual Size of Some Engineering Products *(Continued)*

Dimension of Copper Tubing (Type L—for HVAC applications)

Nominal Size (in.)	Outside Diameter (in.)	Inside Diameter (in.)	Wall Thickness (in.)	Flow Cross Section (in²)
$1\frac{1}{2}$	1.625	1.505	0.055	1.78
2	2.125	1.985	0.060	3.09
$2\frac{1}{2}$	2.625	2.465	0.070	4.77
3	3.125	2.945	0.080	6.81
$3\frac{1}{2}$	3.625	3.425	0.090	9.21
4	4.125	3.905	0.100	11.0
5	5.125	4.875	0.110	18.1

Dimension of Copper Tubing (Type M—for HVAC and domestic water applications)

Nominal Size (in.)	Outside Diameter (in.)	Inside Diameter (in.)	Wall Thickness (in.)	Flow Cross Section (in²)
3/8	0.5	0.450	0.025	0.159
1/2	0.625	0.569	0.028	0.254
3/4	0.875	0.811	0.032	0.517
1	1.125	1.055	0.035	0.874
$1\frac{1}{4}$	1.375	1.291	0.042	1.31
$1\frac{1}{2}$	1.625	1.527	0.049	1.83
2	2.125	2.009	0.058	3.17
$2\frac{1}{2}$	2.625	2.495	0.065	4.89
3	3.125	2.981	0.072	6.98
$3\frac{1}{2}$	3.625	3.459	0.083	9.40
4	4.125	3.935	0.095	12.2
5	5.125	4.907	0.109	18.9

| TABLE 7.2 | | **Examples of Nominal Size versus Actual Size of Some Engineering Products** (Continued) | |

Examples of Standard Wide-Flange Beams (SI version)

Designation 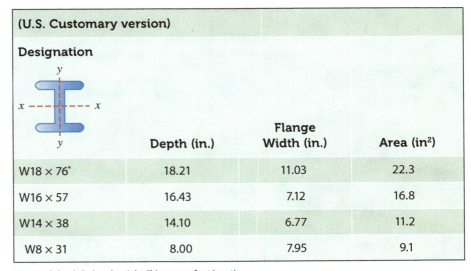	Depth (mm)	Flange Width (mm)	Area (mm²)
W460 × 113*	463	280	14,400
W410 × 85	417	181	10,800
W360 × 57	358	172	7,230
200 × 46.1	203	203	5,890

*Nominal depth (mm) and weight (kg) per one meter length.

| **(U.S. Customary version)** | | | |
Designation	Depth (in.)	Flange Width (in.)	Area (in²)
W18 × 76*	18.21	11.03	22.3
W16 × 57	16.43	7.12	16.8
W14 × 38	14.10	6.77	11.2
W8 × 31	8.00	7.95	9.1

*Nominal depth (in.) and weight (lb) per one foot length.

LO² 7.2   Ratio of Two Lengths—Radians and Strain

> Radians and strain are considered length-related variables because their values are determined from the ratio of two lengths.

Radians as a Ratio of Two Lengths

Consider the circular arc shown in Figure 7.8. The relationship among the arc length, S, radius of the arc, R, and the angle in radians, θ, is given by

$$\theta = \frac{S_1}{R_1} = \frac{S_2}{R_2}$$

7.1

Note that **radians** represents the ratio of two lengths and thus is unitless. As you will learn later in your physics and dynamics classes, you can use Equation (7.1) as the basis for establishing a relationship between the translational speed of a point on an object and its rotational speed. Also note 2π radians is equal to 360 degrees, and 1 radian is equal to 57.30 degrees.

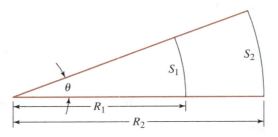

| FIGURE 7.8 | The relationship among arc length, radius, and angle. |

Strain as a Ratio of Two Lengths

When a material (e.g., in the shape of a rectangular bar) is subjected to a tensile load (pulling load), the material will deform. The deformation, ΔL, divided by the original length, L, is called *normal strain*, as shown in Figure 7.9. In Chapter 10, we will discuss the concepts of stress and strain in more detail. We will also explain how stress and strain information is used in engineering analysis. But for now, remember that strain is the ratio of deformation length to original length and thus is unitless.

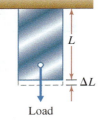

FIGURE 7.9

A bar subjected to a pulling load.

$$\text{strain} = \frac{\Delta L}{L} \qquad \boxed{7.2}$$

Note that strain is another length-related engineering variable.

Before You Go On

Answer the following questions to test your understanding of the preceding sections.

1. Give examples of the important role of length in everyday life.

2. Give examples of the units of length in SI and U.S. Customary systems.

3. What is a coordinate system? Explain.

4. Explain the difference between nominal and actual size.

5. What do radians represent?

6. What does strain represent?

Vocabulary—State the meaning of the following terms:

CNC Machine _____

Strain _____

Coordinate System _____

LO³ 7.3 Area

Area is a derived, or secondary, physical quantity. It plays a significant role in many engineering problems. For example, the rate of heat transfer from a surface is directly proportional to the exposed surface area. That is why a motorcycle engine or a lawn mower engine has extended surfaces, or fins, as shown in Figure 7.10. If you look closely inside buildings around your campus, you will also see heat exchangers or radiators with extended surfaces under windows and against some walls. As you know, these heat exchangers or radiators supply heat to rooms and hallways during the winter. For another example, have you ever thought about why crushed ice cools a drink faster than ice cubes? It is because given the same amount of ice, the crushed ice has more surface exposed to the liquid. You may have also noticed that given the same amount of meat, it takes longer for a roast of beef to cook than it takes stew. Again, it is because the stew has more surface area exposed to the liquid in which it is being cooked. So next time you are planning to make some mashed potatoes, make sure you first cut the potatoes into smaller pieces. The smaller the pieces, the sooner they will cook. Of course, the reverse is also true. That is, if you want to reduce the heat loss from something, one way of doing this would be to reduce the exposed surface area. For example, when we feel cold we naturally try to curl up, which reduces the surface area exposed to the cold surroundings. You can tell by observing nature that trees take advantage of the effect and importance of surface area. Why do trees have lots of leaves rather than one big leaf? It is because they can absorb more solar radiation that way. Surface area is also important in engineering problems related to mass transfer. Your parents or grandparents perhaps recall when they used to hang out clothes to dry. The

> Area is a length-related variable, because its value is determined from the product of two lengths. It plays an important role in analysis and design of problems in heating and cooling, aerodynamics, cutting tools, and building foundations.

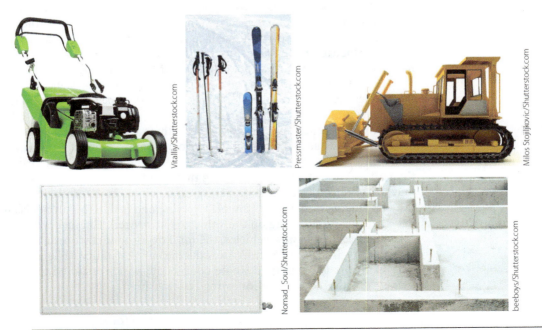

FIGURE 7.10   The important role of area in the design of various systems.

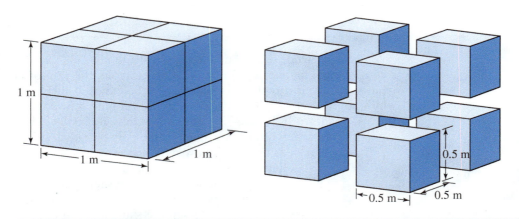

FIGURE 7.11   The relationship between the area and volume of a cube.

clothes were hung over the clothesline in such way that they had a maximum area exposed to the air. Bed-sheets, for example, were stretched out fully across the clothesline.

Let us now investigate the relationship between a given volume and exposed surface area. Consider a 1 m × 1 m × 1 m cube. What is the volume? 1 m³. What is the exposed surface area of this cube? 6 m². If we divide each dimension of this cube by half, we get 8 smaller cubes with the dimensions of 0.5 m × 0.5 m × 0.5 m, as shown in Figure 7.11. What is the total volume of the 8 smaller cubes? It is still 1 m³. What is the total exposed surface area of the cubes? Each cube now has an exposed surface area of 1.5m², which amounts to a total exposed surface area of 12 m².

Let us proceed with dividing the dimensions of our 1 m × 1 m × 1 m original cube into even smaller cubes with the dimensions of 0.25 m × 0.25 m × 0.25 m. We will now have 64 smaller cubes, and we note that the total volume of the cubes is still 1 m³. However, the surface area of each cube is 0.375 m², leading to a total surface area of 24 m². Thus, the same cube divided into 64 smaller cubes now has an exposed surface area that is four times the original surface area.

This mental exercise gives you a good idea why, given the same amount of ice, the crushed ice cools a drink faster than ice cubes do. You may not know it, but you already own the best heat and mass exchanger in the world. Your lungs! Human lungs are the best heat and mass exchangers that we know of, with an approximate area density (surface area of heat exchange per unit volume) of 20,000 m²/m³.

Area plays an important role in aerodynamics as well. Air resistance to the motion of a vehicle is something that all of you are familiar with. As you may also know, the drag force acting on a vehicle is determined experimentally by placing it in a wind tunnel. The airspeed inside the tunnel is changed, and the drag force acting on the vehicle is measured. Engineers have learned that when designing new vehicles, the total exposed surface area and the frontal area are important factors in reducing air resistance. The experimental data is normally given by a single coefficient, which is called the *drag coefficient*. It is defined by the following relationship:

$$\text{drag coefficient} = \frac{\text{drag force}}{\frac{1}{2}(\text{air density})(\text{air speed})^2 (\text{frontal area})}$$

The frontal area represents the frontal projection of the vehicle's area and could be approximated simply by calculating 0.85 times the width and the height of a rectangle that outlines the front of the vehicle. This is the area that you see when you view the car or truck from a direction normal to the front grill. Later in your engineering education, some of you may take a class in the physics of flight, fluid mechanics, or aerodynamics, where you will learn that the lift force acting on the wings of a plane is proportional to the planform area of the wing. The planform area is the area that you would see if you were to look from above at the wing from a direction normal to the wing.

Cross-sectional area also plays an important role in distributing a force over an area. Foundations of buildings, hydraulic systems, and cutting tools (see Figure 7.12) are examples of objects for which the role of area is important. For example, have you ever thought about why the edge of a sharp knife cuts well? What do we mean by a "sharp" knife? A good sharp knife is one that has a cross-sectional area as small as possible along its cutting edge. The pressure along the cutting edge of a knife is simply determined by

$$\text{pressure at the cutting surface} = \frac{\text{force}}{\substack{\text{cross-sectional area} \\ \text{at the cutting edge}}}$$

7.3

FIGURE 7.12   Area plays an important role in the design of cutting tools.

You can see from Equation (7.3) that for the same force (push) on the knife, you can increase the cutting pressure by decreasing the cross-sectional area. As you can also tell from Equation (7.3), we can also reduce the pressure by increasing the area. In skiing, we use the area to our advantage and distribute our weight over a bigger surface so we won't sink into the snow. Next time you go skiing think about this relationship. Equation (7.3) also makes clear why high-heeled shoes are designed poorly as compared to walking shoes. The purpose of these examples is to make you realize that area is an important parameter in engineering design. During your engineering education, you will learn many new concepts and laws that are either directly or inversely proportional to the area. So keep a close watch out for area as you study various engineering topics.

TABLE 7.3	Units of Area and their Equivalent Values
Units of Area in Increasing Order	**Equivalent Value**
1 mm^2	1×10^{-6} m^2
1 cm^2	1×10^{-4} m^2 = 100 mm^2
1 in^2	645.16 mm^2
1 ft^2	144 in^2
1 yd^2	9 ft^2
1 m^2	1.196 yd^2
1 acre	43,560 ft^2
1 km^2	1,000,000 m^2 = 247.1 acres
1 square mile	2.59 km^2 = 640 acres

Now that you understand the significance of area in engineering analysis, let us look at its units. The unit of area in SI is m^2. We can also use the multiples and the submultiples (see Table 6.2) of SI fundamental units to form other appropriate units for area, such as mm^2, cm^2, km^2, and so on. Remember, the reason we use these units is to keep numbers manageable. The common unit of area in the U.S. Customary system is ft^2. Table 7.3 shows other units commonly used in engineering practice today and their equivalent values.

Area Calculations and Measurement

The areas of common shapes, such as a triangle, a circle, and a rectangle, can be obtained using the area formulas shown in Table 7.4. It is a common practice to refer to these simple areas as *primitive areas*. Many composite surfaces with regular boundaries can be divided into primitive areas. To determine the total area of a composite surface, such as the one shown in Figure 7.13, we first divide the surface into the simpler primitive areas that make it up, and then we sum the values of these areas to obtain the total area of the composite surface.

Examples of the more useful area formulas are shown in Table 7.4.

Approximation of Planar Areas

There are many practical engineering problems that require calculation of planar areas of irregular shapes. If the irregularities of the boundaries are such that they will not allow for the irregular shape to be represented by a sum of primitive shapes, then we need to resort to an approximation method. For these situations, you may approximate planar areas using any of the procedures discussed next.

FIGURE 7.13   A composite surface (surface of a heat sink) that may be divided into primitive areas.
Wlg/Shutterstock.com

TABLE 7.4	Some Useful Area Formulas

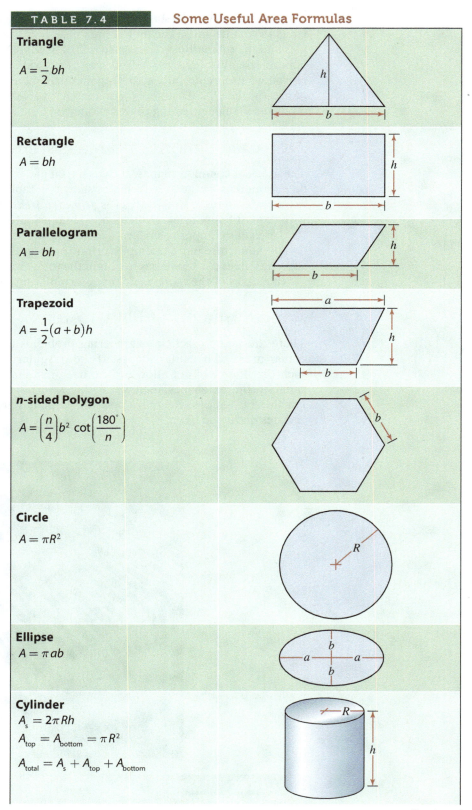

Triangle

$A = \dfrac{1}{2} bh$

Rectangle

$A = bh$

Parallelogram

$A = bh$

Trapezoid

$A = \dfrac{1}{2}(a + b)h$

n-sided Polygon

$A = \left(\dfrac{n}{4}\right)b^2 \cot\left(\dfrac{180°}{n}\right)$

Circle

$A = \pi R^2$

Ellipse

$A = \pi ab$

Cylinder

$A_s = 2\pi Rh$

$A_{top} = A_{bottom} = \pi R^2$

$A_{total} = A_s + A_{top} + A_{bottom}$

Continued

TABLE 7.4	Some Useful Area Formulas *(Continued)*

Right Circular Cone

$A_s = \pi R s = \pi R \sqrt{R^2 + h^2}$

$A_{bottom} = \pi R^2$

$A_{total} = A_s + A_{bottom}$

Sphere

$A = 4\pi R^2$

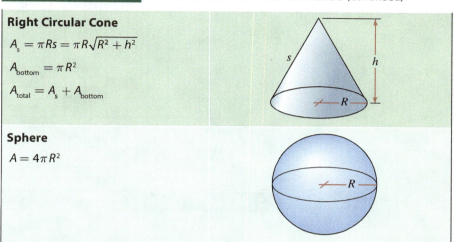

The Trapezoidal Rule You can approximate the planar areas of an irregular shape with reasonably good accuracy using the **trapezoidal rule**. Consider the planar area shown in Figure 7.14. To determine the total area of the shape shown in Figure 7.14, we use the trapezoidal approximation. We begin by dividing the total area into small trapezoids of equal height h, as depicted in Figure 7.14. We then sum the areas of the trapezoids. Thus, we begin with the equation

$$A = A_1 + A_2 + A_3 + \cdots + A_n \qquad \boxed{7.4}$$

Substituting for the values of each trapezoid,

$$A \approx \frac{h}{2}(y_0 + y_1) + \frac{h}{2}(y_1 + y_2) + \frac{h}{2}(y_2 + y_3) + \cdots + \frac{h}{2}(y_{n-1} + y_n) \qquad \boxed{7.5}$$

and simplifying Equation (7.5) leads to

$$A \approx h\left(\frac{1}{2}y_0 + y_1 + y_2 + \cdots + y_{n-2} + y_{n-1} + \frac{1}{2}y_n\right) \qquad \boxed{7.6}$$

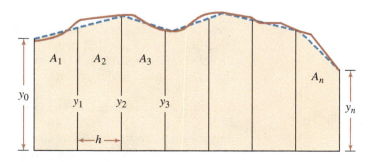

FIGURE 7.14	Approximation of a planar area by the trapezoidal rule.

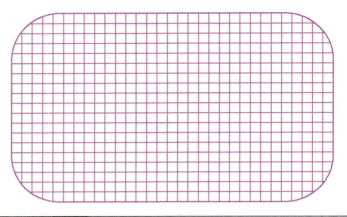

| FIGURE 7.15 | Approximation of a planar area using small squares. |

| FIGURE 7.16 | Approximation of a planar area using a rectangular primitive and small squares and triangles. |

Equation (7.6) is known as the trapezoidal rule. Also note that the accuracy of the area approximation can be improved by using more trapezoids. This approach will reduce the value of h and thus improve the accuracy of the approximation.

Counting the Squares There are other ways to approximate the surface areas of irregular shapes. One such approach is to divide a given area into small squares of known size and then count the number of squares. This approach is depicted in Figure 7.15. You then need to add to the areas of the small squares the leftover areas, which you may approximate by the areas of small triangles.

Subtracting Unwanted Areas Sometimes, it may be advantageous to first fit large primitive area(s) around the unknown shape and then approximate and subtract the unwanted smaller areas. An example of such a situation is shown in Figure 7.16. Also keep in mind that for symmetrical areas you may make use of the symmetry of the shape. Approximate only 1/2, 1/4, or 1/8, of the total area, and then multiply the answer by the appropriate factor.

Weighing the Area Another approximation procedure requires the use of an accurate analytical balance from a chemistry lab. Assuming the profile of the area to be determined can be drawn on a $8\frac{1}{2} \times 11$ sheet of paper, first weigh a blank $8\frac{1}{2} \times 11$ sheet of paper, then using the analytical balance, weigh the sheet and record its weight. Next, draw the boundaries of the unknown area on the blank paper, and then cut around the boundary of that area. Determine the weight of the piece of paper that has the area drawn on it. Now, by comparing the weights of the blank sheet of paper to the weight of the paper with the profile, you can determine the area of the given profile. In using this approximation method, we assumed that the paper has uniform thickness and density.

| EXAMPLE 7.1 | Using the trapezoidal rule, determine the total ground-contact area of the athletic shoe shown in Figure 7.17. All dimensions are shown in inches. |

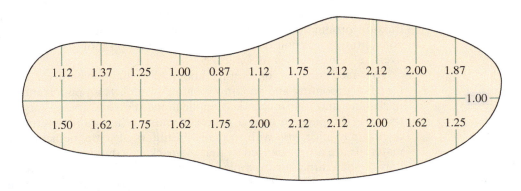

| **FIGURE 7.17** | The shoe profile for Example 7.1. |

We have divided the profile into two parts and each part into 12 trapezoids of equal heights of 1.0 in. Applying the trapezoidal rule, we have

$$A \approx h\left(\frac{1}{2}y_0 + y_1 + y_2 + \cdots + y_{n-2} + y_{n-1} + \frac{1}{2}y_n\right)$$

$$A_1 \approx (1.0)\left[\frac{1}{2}(0) + 1.12 + 1.37 + 1.25 + 1.00 + 0.87 + 1.12 + 1.75\right.$$

$$\left. + 2.12 + 2.12 + 2.00 + 1.87 + \frac{1}{2}(0)\right] \approx 16.6 \text{ in}^2$$

$$A_2 \approx (1.0)\left[\frac{1}{2}(0) + 1.50 + 1.62 + 1.75 + 2.00 + 2.12 + 2.12 + 2.12\right.$$

$$\left. + 2.00 + 1.62 + 1.25 + \frac{1}{2}(0)\right] \approx 19.3 \text{ in}^2$$

Then the total area is given by

$$A_{\text{total}} \approx 16.6 + 19.3 \approx 35.9 \text{ in}^2$$

LO⁴ 7.4 Volume

Volume is another important physical quantity, or physical variable, that does not get enough respect. We live in a three-dimensional world, so it is only natural that volume would be an important player in how things are shaped or how things work. Let us begin by considering the role of volume in our daily lives. Today you may have treated yourself to a can of soda, which on average contains 12 fluid ounces or 355 milliliters of your favorite beverage. You may have driven a car whose engine size is rated in liters. For example, if you own a BMW 3-series, its engine size is 2 liters. Depending on the size of your car, it is also safe to say that in order to fill the gas tank you need to put in about 15 to 20 gallons (57 to 77 liters) of gasoline. We also express the gas consumption rate of a car in terms of so many miles per gallon of gasoline. Doctors tell us that we need to drink at least 8 glasses of water (approximately 2.5 to 3 liters) a day. We breathe in oxygen at a rate of approximately 1.6 ft³/h (0.0453m³/h). Of course, as you would expect, the volume of oxygen consumption or carbon dioxide production depends on the level of physical activity. The oxygen consumption, carbon dioxide production, and pulmonary ventilation for an average man is shown in Table 7.5.

We each consume on average about 20 to 40 gallons of water per day for personal grooming and cooking. Volume also plays an important role in food packaging and pharmaceutical applications. For example, a large milk container is designed to hold a gallon of milk. When administering drugs, the doctor may inject you with so many milliliters of some medicine. Many other materials are also packaged so that the package contains so many liters or gallons of something, for example, a gallon can of paint. Clearly, we use volume to express quantities of various fluids that we consume. Volume also plays a significant role in many other engineering concepts. For example, the density

> We live in a three-dimensional world. Naturally, volume would play an important role in how things are shaped or how things work.

TABLE 7.5   The Oxygen Consumption, Carbon Dioxide Production, and Pulmonary Ventilation for a Man

Level of Physical Activity	Oxygen Consumption (ft³/h)	Carbon Dioxide Production (ft³/h)	Rate of Breathing (ft³/h)
Exhausting effort	6.6	5.7	146
Strenuous work or sports	4.44	3.8	97
Moderate exercise	2.96	2.5	64
Mild exercise; light work	1.84	1.55	40
Standing; desk work	1.10	0.93	24
Sedentary, at ease	0.74	0.62	16
Reclining, at rest	0.3	0.56	12

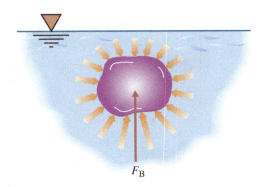

F_B

| FIGURE 7.18 | Buoyancy force acting on a submerged surface. |

of a material represents how light or heavy a material is per unit volume. We will discuss density and other related mass and volume properties in Chapter 9. Buoyancy is another engineering principle where volume plays an important role. **Buoyancy** is the force that a fluid exerts on a submerged object. The net upward buoyancy force arises from the fact that the fluid exerts a higher pressure at the bottom surfaces of the object than it does on the top surfaces of the object, as shown in Figure 7.15. Thus, the net effect of fluid pressure distribution acting over the submerged surface of an object is the buoyancy force. The magnitude of the buoyancy force is equal to the weight of the volume of the fluid displaced. It is given by

$$F_B = \rho V_g \qquad \qquad \boxed{7.7}$$

where F_B is the buoyancy force (N), ρ represents the density of the fluid (kg/m^3), and g is acceleration due to gravity (9.81 m/s^2). If you were to fully submerge an object with a volume V, as shown in Figure 7.18, you would see that an equal volume of fluid has to be displaced to make room for the volume of the object. In fact, you can use this principle to measure the unknown volume of an object. This idea is demonstrated in Example 7.2. NASA astronauts also make use of buoyancy and train underwater to prepare for the in-orbit repair of satellites. This type of training takes place in the Underwater Neutral Buoyancy Simulator, shown in the accompanying photograph of Figure 7.19. The changes in the apparent weight of the astronaut allows him or her to prepare to work under near-zero-gravity (weightless) conditions.

Now that you understand the significance of volume in the analysis of engineering problems, let us look at some of the more common units in use. These units are shown in Table 7.6. As you go over Table 7.6, try to develop a "feel" for the order of volume quantity. For example, ask yourself whether a pint or a liter is the larger quantity, and so on.

Volume Calculations

The volume of simple shapes, such as a cylinder, a cone, or a sphere, may be obtained using volume formulas as shown in Table 7.7. Indirect and direct estimation of volumes of objects are demonstrated in Examples 7.2 and 7.3.

TABLE 7.6	Units of Volume and their Equivalent Values*	
Volume Units in Increasing Capacity Order	**Equivalent Value**	
1 milliliter	1/1000 liter	
1 teaspoon (tsp)	4.928 milliliter	
1 tablespoon (tbsp)	3 tsp	
1 fluid ounce	2 tbsp $\approx$ 1/1000 ft^3	
1 cup	8 ounces = 16 tbsp	
1 pint	16 ounces = 2 cups	
1 quart	2 pints	
1 liter	1000 cm^3 $\approx$ 4.2 cups	
1 gallon	4 quarts	
1 cubic foot	7.4805 gallons	
1 cubic meter	1000 liters $\approx$ 264 gallons $\approx$ 35.3 ft^3	

*1 milliliter < 1 teaspoon < 1 tablespoon < 1 fluid ounce < 1 cup < 1 pint < 1 quart < 1 liter < 1 gallon < 1 cubic foot < 1 cubic meter.

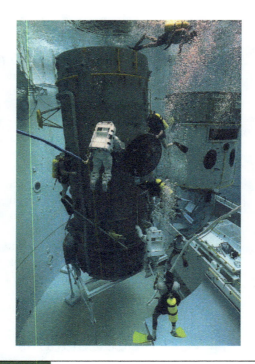

| FIGURE 7.19 | NASA Astronauts training in the Underwater Neutral Buoyancy Simulator to prepare for the in-orbit Hubbell Space Telescope repair.
NASA, Marshall Space Flight Center |

TABLE 7.7	Some Useful Volume Formulas
Cylinder $V = \pi R^2 h$	
Right Circular Cone $V = \dfrac{1}{3}\pi R^2 h$	
Section of a Cone $V = \dfrac{1}{3}\pi h\left(R_1^2 + R_2^2 + R_1 R_2\right)$	
Sphere $V = \dfrac{4}{3}\pi R^3$ **Section of a Sphere** $V = \dfrac{1}{6}\pi h\left(3a^2 + 3b^2 + h^2\right)$	

Finally, it is worth noting here that numerical solid modeling is an engineering topic that deals with computer generation of the surface areas and volumes of an actual object. Solid-modeling software programs are becoming quite common in engineering practice. Computer-generated solid models provide not only great visual images but also such information as magnitude of the area and the volume of the model. To generate numerical solid models of simple shapes, area and volume primitives are used. Other means of generating surfaces include dragging a line along a path or rotating a line about an axis, and, as with areas, you can also generate volumes by dragging or sweeping an area along a path or by rotating an area about a line. We will discuss computer solid modeling ideas in more detail in Chapter 16.

EXAMPLE 7.2

Object

FIGURE 7.20

The object in Example 7.2.

Determine the exterior volume of the object shown in Figure 7.20.

For this example, we will use the buoyancy effect to measure the exterior volume of the object shown. We will consider two procedures. First, we obtain a large container that can accommodate the object. We will then fill the container completely to its rim with water and place the container inside a dry, empty tub. We next submerge the object with the unknown volume into the container until its top surface is just below the surface of the water. This will displace some volume of water, which is equal to the volume of the object. The water that overflowed and was collected in the tub can then be poured into a graduated cylinder to measure the volume of the object. This procedure is shown in Figure 7.21.

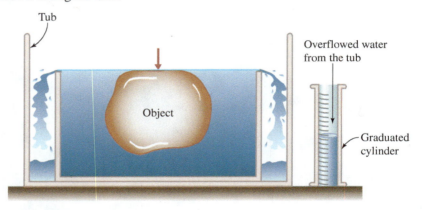

Tub

Overflowed water from the tub

Object

Graduated cylinder

FIGURE 7.21   Using a displaced volume of water to measure the volume of the given object.

The second procedure makes direct use of the buoyancy force. We first suspend the object in air from a spring scale to obtain its weight. We then place the object, still suspended from the spring, into a container filled with water. Next, we record the apparent weight of the object. The difference between the actual weight of the object and the apparent weight of the object in water is the buoyancy force. Knowing the magnitude of the buoyancy force and using Equation (7.7), we can then determine the volume of the object. This procedure is depicted in Figure 7.22.

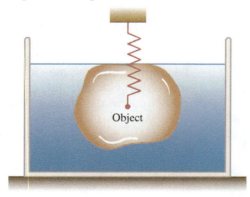

Object

FIGURE 7.22   Using apparent weight (actual weight minus the buoyancy force) to determine the volume of a given object.

EXAMPLE 7.3

Estimate the inside volume of a soda can. We have used a ruler to measure the height and the diameter of the can, as shown in Figure 7.23.

We may approximate the inside volume of the soda can using the volume of a cylinder of equal dimensions:

$$V = \pi R^2 h = (3.1415)\left(\frac{6.3 \text{ cm}}{2}\right)^2 (12.0 \text{ cm}) = 374 \text{ cm}^3 = 374 \text{ mL}$$

diameter = 6.3 cm

12 cm

FIGURE 7.23

Soda can in Example 7.3.

Compared to the 355-mL value shown on a typical soda can, the approximated value seems reasonable. The difference between the approximated value and the indicated value may be explained in a number of ways. First, the soda container does not represent a perfect cylinder. If you were to look closely at the can you would note that the diameter of the can reduces at the top. This could explain our overestimation of volume. Second, we measured the outside diameter of the can, not the inside dimensions. However, this approach will introduce smaller inaccuracies because of the small thickness of the can.

We could have measured the inside volume of the can by filling the can with water and then pouring the water into a graduated cylinder or beaker to obtain a direct reading of the volume.

LO⁵ 7.5   Second Moment of Area

In this section, we will consider a property of an area known as the **second moment of area**. The second moment of area, also known as the **area moment of inertia**, is an important property of an area that provides information on how hard it is to bend something. Next time you walk by a construction site, take a closer look at the cross-sectional area of the support beams, and notice how the beams are laid out. Pay close attention to the orientation of the cross-sectional area of an I-beam with respect to the directions of expected loads. Are the beams laid out in the orientation shown in Figure 7.24(a) or in Figure 7.24(b)?

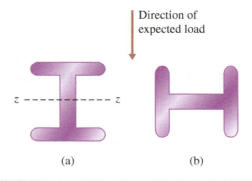

Direction of expected load

z ――――――― z

(a)   (b)

FIGURE 7.24   Which way is an I-beam oriented with respect to loading?

Steel I-beams, which are commonly used as structural members to support various loads, offer good resistance to bending, and yet they use much less material than beams with rectangular cross sections. You will find I-beams supporting guard rails and I-beams used as bridge cross members and also as roof and flooring members. The answer to the question about the orientation of I-beams is that they are oriented with respect to the loads in the configuration shown in Figure 7.24(a). The reason for having I-beams support loads in that configuration is that about the $z–z$ axis shown, the value of the area moment of inertia of the I-beam is higher for configuration (a) than it is for configuration (b).

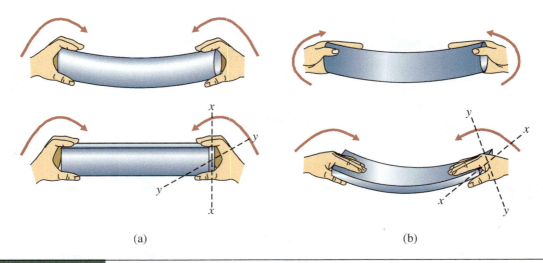

(a)　　　　　　　　　　　　(b)

FIGURE 7.25　　Bend the rod and yardstick in the directions shown.

To better understand this important property of an area and the role of the second moment of area in offering a measure of resistance to bending, try the following experiment. Obtain a thin wooden rod and a yardstick. First try to bend the rod in the directions shown in Figure 7.25.

If you were to report your findings, you would note that the circular cross section of the rod offers the same resistance to bending regardless of the direction of loading. This is because the circular cross section has the same distribution of area about an axis going through the center of the area. Note that we are concerned with bending a member, not twisting it! Now, try bending the yardstick in the directions shown in Figure 7.25. Which way is it harder to bend the yardstick? Of course, it is much harder to bend the yardstick in the direction shown in Figure 7.25(a). Again, that is because in the orientation shown in Figure 7.25(a), the second moment of area about the centroidal axis is higher.

Most of you will take a statics class, where you will learn more in depth about the formal definition and formulation of the second moment of area, or area moment of inertia, and its role in the design of structures. But for now, let us consider the simple situations shown in Figure 7.26. For a small area element A, located at a distance r from the axis $z–z$, the area moment of inertia is defined by

$$I_{z-z} = r^2 A$$

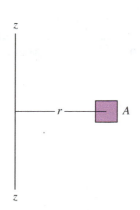

FIGURE 7.26

Small area element located at distance r from the $z–z$ axis.

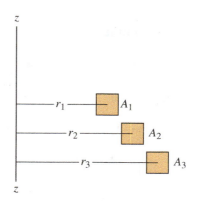

FIGURE 7.27   Second moment of area for three small area elements.

Now let us expand this problem to include more small area elements, as shown in Figure 7.27. The area moment of inertia for the system of discrete areas shown about the z–z axis is now

$$I_{z-z} = r_1^2 A_1 + r_2^2 A_2 + r_3^2 A_3 \qquad \boxed{7.9}$$

Similarly, we can obtain the second moment of area for a cross-sectional area such as a rectangle or a circle by summing the area moment of inertia of all the little area elements that make up the cross section. As you take calculus classes, you will learn that you can use integrals instead of summing the $r^2 A$ terms to evaluate the area moment of inertia of a continuous cross-sectional area. After all, the integral sign, $\int$, is nothing but a big "S" sign, indicating summation.

$$I_{z-z} = \int r^2 \, dA \qquad \boxed{7.10}$$

Also note that the reason this property of an area is called "second moment of area" is that the definition contains the product of *distance squared* and an area, hence the name "second moment of area." In Chapter 10, we will discuss the proper definition of a moment and how it is used in relation to the tendency of unbalanced forces to rotate things. As you will learn later, the magnitude of a moment of a force about a point is determined by the product of the perpendicular *distance* from the point about which the moment is taken to the line of action of the force and the magnitude of that force. You have to pay attention to what is meant by "a moment of a force about a point or an axis" and the way the term *moment* is incorporated into the name "the second moment of area" or "the area moment of inertia." Because the distance term is multiplied by another quantity (area), the word "moment" appears in the name of this property of an area.

The second moment of area, also known as the area moment of inertia, is an important property of an area that provides information on how hard it is to bend something.

You can obtain the area moment of inertia of any geometric shape by performing the integration given by Equation (7.10). You will be able to perform the integration and better understand what these terms mean in another semester or two. Keep a close watch for them in the upcoming semesters. For now, we will give you the formulas for area moment of inertia without proof. Examples of area moment of inertia formulas for some common geometric shapes are given next.

Rectangle

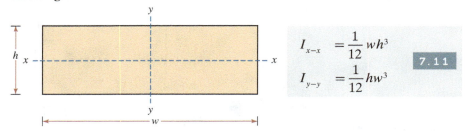

$$I_{x-x} = \frac{1}{12} wh^3$$

$$I_{y-y} = \frac{1}{12} hw^3$$

7.11

Circle

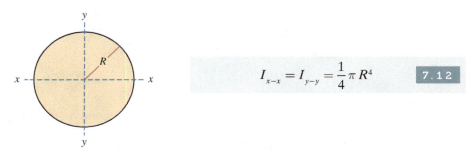

$$I_{x-x} = I_{y-y} = \frac{1}{4} \pi R^4$$

7.12

The values of the second moment of area for standard wide-flange beams are shown in Table 7.8.

In Chapter 9, we will look at another similarly defined property of an object, mass moment of inertia, which provides a measure of resistance to rotational motion.

EXAMPLE 7.4

Calculate the second moment of area for a 2×4 stud with the actual dimensions shown on Figure 7.28.

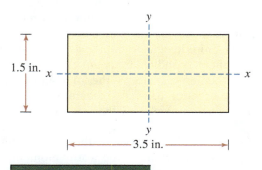

FIGURE 7.28

$$I_{x-x} = \frac{1}{12} w h^3 = \left(\frac{1}{12}\right)(3.5 \text{ in.})(1.5 \text{ in.})^3 = 0.98 \text{ in}^4$$

$$I_{y-y} = \frac{1}{12} h w^3 = \left(\frac{1}{12}\right)(1.5 \text{ in.})(3.5 \text{ in.})^3 = 5.36 \text{ in}^4$$

Note that the 2×4 lumber will show more than five times more resistance to bending about the y–y axis than it does about the x–x axis.

TABLE 7.8   **Examples of the Second Moment of Area of Standard Beams**

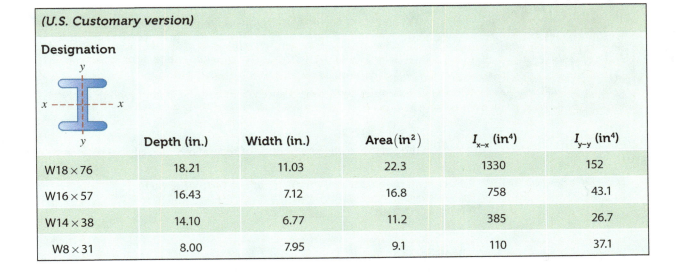

(SI Version)

Designation

	Depth (mm)	Width (mm)	Area (mm^2)	$I_{x-x}(\text{mm}^4)$	$I_{y-y}(\text{mm}^4)$
W460 $\times$ 113	463	280	14,400	554×10^6	63.3×10^6
W410 $\times$ 85	417	181	10,800	316×10^6	17.9×10^6
W360 $\times$ 57	358	172	7,230	160×10^6	11.1×10^6
W200 $\times$ 46.1	203	203	5,890	45.8×10^6	15.4×10^6

(U.S. Customary version)

Designation

	Depth (in.)	Width (in.)	Area (in^2)	I_{x-x} (in^4)	I_{y-y} (in^4)
W18 $\times$ 76	18.21	11.03	22.3	1330	152
W16 $\times$ 57	16.43	7.12	16.8	758	43.1
W14 $\times$ 38	14.10	6.77	11.2	385	26.7
W8 $\times$ 31	8.00	7.95	9.1	110	37.1

EXAMPLE 7.5

Calculate the second moment of area for a 5 cm diameter shaft about the *x–x* and the *y–y* axes shown in Figure 7.29.

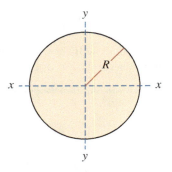

FIGURE 7.29   The shaft in Example 7.5.

$$I_{x-x} = I_{y-y} = \frac{1}{4}\pi R^4 = \frac{1}{4}(3.1415)\left(\frac{5\text{ cm}}{2}\right)^4 = 30.7\text{ cm}^4$$

Finally, it is important to emphasize the fact that all physical variables discussed in this chapter are based on the fundamental dimension of length. For example, area has a dimension of (length)^2, volume has a dimension of (length)^3, and the second moment of area has a dimension of (length)^4. In Chapter 8, we will look at time and time-and-length-related quantities in engineering analysis and design.

Before You Go On

Answer the following questions to test your understanding of the preceding sections.

1. Give examples of the important role of area in engineering analysis and design.

2. Describe two different methods that you can use to approximate an area of something.

3. Give examples of the important role of volume in engineering analysis and design.

4. What is buoyancy?

5. What does the relative magnitude of second moment of area represent?

Vocabulary—State the meaning of the following terms:

Drag Coefficient _____

Trapezoidal Rule _____

Buoyancy_____

Area Moment of Inertia_____

SUMMARY

LO¹ Length as a Fundamental Dimension

By now, you should understand that the fundamental dimension of length and other length-related variables such as area and volume play important roles in engineering analysis and design. As a good engineer, you need to know how to measure, calculate, and approximate length and length-related variables. Length is one of the seven fundamental or base dimensions that we use to properly express what we know of our natural world. The unit of length in SI is the meter (m), and in the U.S. Customary system, it is foot (ft). Coordinate systems are examples of application where length plays an important role. Coordinate systems are used to locate things with respect to a known origin. The most common coordinate system is the rectangular, or Cartesian, coordinate system.

Depending on how accurate the measurement needs to be and the size of the object being measured, we use different measuring devices, such as a ruler, a yardstick, a steel tape, micrometer, Vernier caliper, and electronic distance measuring instruments (EDMI).

LO² Ratios of Two Lengths—Radians and Strain

Radians and strain represent the ratio of two lengths. Radians represent the ratio of an arc length to radius of the arc. It is unitless. When a piece of material in a shape of rectangular bar is subjected to a tensile load, the material will deform. The ratio of deformation length to original length of the bar is called strain. Strain also is unitless.

LO³ Area

Area plays an important role in analysis and design of many engineering problems including heating and cooling, aerodynamics, foundation of structures, and cutting tools. The areas of common shapes such as a triangle, a circle, and a rectangle can be obtained using simple area formulas. For irregular shapes, we use approximation methods such as the Trapezoidal rule.

LO⁴ Volume

We live in a three-dimensional world, so it is natural that volume would play an important role in how things are shaped or how things work. The volume of simple shapes such as a cylinder, a cone, or a sphere may be obtained using simple volume formulas. We can use the buoyancy effect to measure the exterior volume of objects with irregular shapes. Solid-modeling software programs are becoming quite common in engineering practice. Computer-generated solid models provide not only great visual images but also such information as magnitude of the area and the volume of the model.

LO⁵ Second Moment of Area

The second moment of area, also known as the area moment of inertia, is an important property of an area that provides information on how hard it is to bend something. Also note that the reason this property of an area is called "second moment of area" is that its definition contains the product of distance squared and an area, hence the name "second moment of area." For common geometric shapes, area moment of inertia formulas may be used.

KEY TERMS

Actual Size 198
Area 203
Area Moment of Inertia 217
Buoyancy 213
Cartesian Coordinate System 192
Cylindrical 192

Coordinate System 192
Cylindrical Coordinate System 192
Length 191
Nominal Size 198
Radians 202
Second Moment of Area 217

Spherical Coordinate System 192
Spherical 192
Strain 202
Trapezoidal Rule 209
Volume 212

APPLY WHAT YOU HAVE LEARNED

Investigate the following dimensions and write a brief report to your instructor about your findings.

1. Measure and record the length of each finger of ten male adults in your class. Also, measure and record the length of each finger of ten female adults. Compute the averages for males and females, and compare the female results to the male results. Present the results in both SI and U.S. Customary units.

2. Measure and record the surface area of legs of ten male adults by covering their legs with paper and then measuring the area of the paper. How much plaster would be required for a plaster cast around an average adult male leg, assuming a 2-mm thickness? What is the volume of the plaster needed?

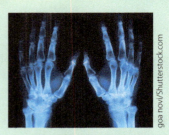

goa novi/Shutterstock.com

racorn/Shutterstock.com

PROBLEMS

Problems that promote lifelong learning are denoted by 🔑

7.1 Seasoned engineers are good at estimating physical values without using tools. Therefore, you need to begin developing a "feel" for the sizes of various physical quantities. This exercise is intended to help you develop this ability. Using the table below, first estimate the dimensions of the given objects. Next measure, or look up, the actual dimensions of the objects, and compare them to your estimated values. How close are your estimations? Do you have a "feel" for units of length yet?

Object	Estimated Values (cm)	(in.)	Measured Values (cm)	(in.)	Difference (cm)	(in.)
This book						
A pen or a pencil						
A laptop computer (closed)						
A 12-fl.-oz. soda can						
The distance from home to school Use your car's odometer to measure the actual distance	(km)	(miles)	(km)	(miles)	(km)	(miles)
A dollar bill	(cm)	(in.)				
The height of your engineering building	(m)	(ft)				
Wingspan of a Boeing 747	(m)	(ft)	(m)	(ft)	(m)	(ft)

7.2 The following exercises are designed to help you become aware of the significance of various dimensions around you. You see these dimensions every day, but perhaps you never looked at them with the eyes of an engineer. Measure and discuss the significance of the dimensions of the following items.

   a. The dimensions of your bedroom or living room

   b. The dimensions of the hallway

   c. The window dimensions

   d. The width, height, and thickness of your apartment doors or dormitory doors

   e. The distance from the floor to the doorknob

   f. The distance from the floor to the light switches

   g. The dimensions of your desk

   h. The dimensions of your bed

   i. The distance from the floor to the bathroom sink

   j. The distance from the tub surface to the showerhead

7.3 In this exercise, you are to explore the size of your classroom, the seating arrangements, the location of the chalkboard (or the whiteboard) with respect to the classroom's main entrance, and the size of the board relative to classroom size. Discuss your findings in a brief report to your instructor.

   a. What are the dimensions of the classroom?

   b. How far apart are the seats placed? Is this a comfortable arrangement? Why or why not?

   c. How far above the floor is the chalkboard (or whiteboard) placed? How wide is it? How tall is it? What is its relative placement in the classroom? Can someone sitting in the back corner of the room see the board without too much difficulty?

7.4 This is a sports-related assignment. First look up the dimensions, and then show the dimensions on a diagram. You may need to do some research to obtain the information required here.

   a. A basketball court

   b. A tennis court

   c. A football field

   d. A soccer field

   e. A volleyball court

7.5 These dimensions concern transportation systems. Look up the dimensions of the following vehicles; give the model year and the source of your information.

   a. A car of your choice

   b. BMW 760 Li

   c. Honda Accord

   d. How wide is the driver's seat in each car mentioned in (a) through (c)?

   e. A city bus

   f. How wide are the bus seats?

   g. A high-speed passenger train

   h. How wide are the train seats in the coach section?

7.6 This is a bioengineering assignment. Investigate the following dimensions and write a brief report to your instructor about your findings.

   a. What is the average diameter of a healthy red blood cell?

   b. What is the average diameter of a white blood cell?

   c. What are the lengths of human small and large intestines?

7.7 This assignment is related to civil engineering.

   a. How wide is each lane of a street in your neighborhood? Talk to your city engineer to obtain information.

   b. Visit the U.S. Department of Transportation Web site to find out how wide each lane of an interstate highway is. Do all interstate highway lanes have the same width?

   c. Find out how tall the Hoover Dam is. How wide is the Hoover Dam? What is the area of the dam exposed to upstream water? Discuss how and why the thickness of the dam varies with its height.

   d. How tall are the interstate bridges so that an average truck can go under them?

e. How far above the road are the highway signs placed?

f. What is the average height of a tunnel? What is the area of a tunnel at the entrance?

Write a brief report discussing your findings.

7.8 Investigate the size of the main waterline in your neighborhood. What are the inside and outside diameters? What is the nominal size of the pipe? What is the cross-sectional area of water flow? Investigate the size of piping used in your home. Write a brief memo to your instructor discussing your findings.

7.9 Look up the length of the Alaska pipeline. What are the inside and outside diameters of the pipes used in transporting oil? How far apart are the booster pump stations? How thick is the pipeline? What is the cross-sectional area of the pipe? What is the nominal size of the pipe? Write a brief memo to your instructor discussing your findings.

7.10 Investigate the size of pipes used in transporting natural gas to your state. What are the inside and outside diameters of the pipe? What is the distance between boosting stations? What is the cross-sectional area of the pipe? Write a brief memo to your instructor discussing your findings.

7.11 Investigate the diameter of the electrical wire used in your home. How thick are the interstate transmission lines, and what is their cross-sectional area? Write a brief memo to your instructor discussing your findings.

7.12 What is the operating wavelength range for the following items?

a. A cellular phone

b. FM radio transmissions

c. Satellite TV broadcasting

Write a brief memo to your instructor discussing your findings.

7.13 Derive the formula given for the area of a trapezoid. Start by dividing the area into two triangular areas and one rectangular area.

7.14 Trace on a white sheet of paper the boundaries of the area of the United States shown in the accompanying diagram.

a. Use the trapezoidal rule to determine the total area.

b. Approximate the total area by breaking it into small squares. Count the number of total squares and add what you think is an appropriate value for the remaining area. Compare your findings to part (a).

c. Use an analytical balance from a chemistry lab to weigh an $8\frac{1}{2} \times 11$ sheet of paper. Record the dimensions of the paper. Draw the boundaries of the area shown on the accompanying figure and cut around the boundary of the area. Determine the weight of the piece of paper that has the area drawn on it. Compare the weights and determine the area of the given profiles. What assumptions did you make to arrive at your solution? Compare the area computed in this manner to your results in parts (a) and (b). Are there any other ways that you could have determined the area of the profile? Explain.

0   500 miles

Problem 7.14

7.15 Obtain a woman's high-heeled dress shoe and a woman's athletic walking shoe. For each shoe, make imprints of the floor contact surfaces. Determine the total area for each shoe, and assuming the woman who wears these shoes weighs 120 lb, calculate the average pressure at the bottom of the shoe for each shoe style. What are your findings? What are your recommendations? Recall that

$$\text{pressure} = \frac{\text{force}}{\text{area}}$$

7.16 Obtain two different brands of cross-country skis. Make imprints of the expected contact surface of the skies with snow. Obtain a value for the average pressure for each ski, assuming the person wearing the ski weighs 180 lb.

7.17 Estimate the average pressure exerted on the road by the following.

    a. A family car

    b. A truck

    c. A bulldozer

    Discuss your findings.

7.18 Using area as your variable, suggest ways to cool freshly baked cookies faster.

7.19 Estimate how much material is needed to make 100,000 stop signs. Investigate the size of one side of a stop sign and the kind of material it is made from. Write a brief memo to your instructor discussing your findings.

7.20 Estimate how much material is needed to make 100,000 traffic yield signs. Investigate the size of one side of a yield sign and the kind of material it is made from. Write a brief memo to your instructor discussing your findings.

7.21 As explained in the chapter, the air resistance to motion of a vehicle is determined experimentally by placing it in a wind tunnel. The air speed inside the tunnel is changed, and the drag force acting on the vehicle is measured. The experimental data is normally represented by a single coefficient that is called the *drag coefficient*. It is defined by the following relationship:

$$\frac{\text{drag}}{\text{coefficient}} = \frac{\text{drag force}}{\frac{1}{2}(\text{air density})(\text{air speed})^2(\text{frontal area})}$$

or, in a mathematical form,

$$C_d = \frac{F_d}{\frac{1}{2}\rho V^2 A}$$

It was also explained in this chapter that the frontal area A is the frontal projection of the area and could be approximated simply by multiplying 0.85 times the width and the height of a rectangle that outlines the front of a vehicle when you view it from a direction normal to the front windshield. The 0.85 factor is to adjust for rounded corners, open space

below the bumper, and so on. Typical drag coefficient values for sports cars are between 0.27 and 0.38, and for sedans the values are between 0.34 and 0.5. This assignment requires you to actually measure the frontal area of a car. Tape a ruler or a yardstick to the bumper of your car. The ruler will serve as a scale. Take a photograph of the car and use any of the methods discussed in this chapter to compute the frontal area of the car.

7.22 A machinist in an engineering machine shop has ordered a sheet of plastic with dimensions of 10 ft × 12 ft × 1 in. width, length, and thickness, respectively. Can the machinist get the sheet of plastic inside the machine shop provided that the door dimensions are 6 ft × 8 ft? Give the maximum dimensions of a sheet that can be moved inside the shop and explain how.

7.23 Investigate the volume capacity of a barrel of oil in gallons, cubic feet, and cubic meters. Also, determine the volume capacity of a bushel of agricultural products in cubic inches, cubic feet, and cubic meters. Write a brief memo to your instructor discussing your findings.

7.24 Measure the outside diameter of a flagpole or a street-light pole by first wrapping a piece of string around the object to determine its circumference. Why do you think the flagpole or the streetlight pole is designed to be thicker at the bottom near the ground than at the top? Explain your answer.

7.25 Measure the width and the length of a treadmill machine. Discuss what some of the design factors are that determine the appropriateness of the values you measured.

7.26 Collect information on the standard sizes of automobile tires. Create a list of these dimensions. What do the numbers indicated on a tire mean? Write a brief memo to your instructor discussing your findings.

7.27 Visit a hardware store and obtain information on the standard sizes of the following items.

    a. Screws

    b. Plywood sheets

c. PVC pipes

d. Lumber

Create a list with both the actual and the nominal sizes.

7.28 Using just a yardstick or a measuring tape, open the classroom door 35 degrees. *You cannot use a protractor.*

7.29 This is a group project. Determine the area of each class member's right hand. You can do that by tracing the profile of everyone's fingers on a white sheet of paper and using any of the techniques discussed in this chapter to compute the area of each hand. Compile a data bank containing the areas of the right hands of each person in class. Based on some average, classify the data into small, medium, and large. Estimate how much material (outer leather and inner lining) should be ordered to make 10,000 gloves. How many spools of thread should be ordered? Describe and/or draw a diagram of the best way to cut the hand profiles from sheets of leather to minimize wasted materials.

7.30 This is a group project. Given a square kilometer of land, how many cars can be parked safely there? Determine the appropriate spacing between cars and the width of drive lanes. Prepare a diagram showing your solution, and write a brief report to your instructor discussing your assumptions and findings.

7.31 Estimate the length of tubing used to make a bicycle rack on your campus.

7.32 Determine the base area of an electric steam iron.

7.33 Determine the area of a compact fluorescent 30-W light bulb.

7.34 Obtain a trash bag and verify its listed volume.

7.35 Discuss the size of a facial tissue, a paper towel, and a sheet of toilet paper.

7.36 Estimate how much silver is needed to make a four-piece set of silverware containing teaspoons, tablespoons, forks, and knives (16 pieces in all). Discuss your assumptions and findings.

7.37 Calculate the second moment of area of a 4 in. diameter shaft about the *x–x* and *y–y* axes, as shown.

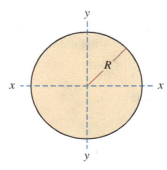

Problem 7.37

7.38 Calculate the second moment of area for a 2 × 6 piece of lumber about the *x–x* and *y–y* axes, as shown. Visit a hardware store and measure the actual dimensions of a typical 2 × 6 piece of lumber.

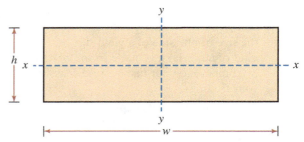

Problem 7.38

7.39 Calculate the buoyancy force acting on a partially immersed boat (V = 5.0 m³). The density of water is 1000 kg/m³. Express your answer in both SI and U.S. Customary units.

7.40 How many 5 in. diameter, 1/4-in. thick disks can be fitted into a 4 ft long, 5 1/4 in. diameter cylindrical container?

7.41 In the U.S., the fuel consumption of an automobile is expressed in *X* miles per gallon. Obtain a single factor that could be used to convert the *X* miles per gallon to km per liter.

7.42 Calculate the arc length S_1 in the accompanying figure. $R_2 = 8$ cm, $R_1 = 5$ cm, and $S_2 = 6.28$ cm.

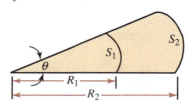

Problem 7.42

7.43 A 10 cm long rectangular bar (when subjected to a tensile load) deforms by 0.1 mm. Calculate the normal strain.

7.44 Calculate the volume of material used in making 100 ft of 4 in. type K copper tubing.

7.45 Calculate the volume of material used in making 100 ft of 4 in. type L copper tubing.

7.46 Calculate the volume of material used in making 100 ft of 4 in. type M copper tubing.

7.47 By what factor do you increase the resistance to bending (about x–x axis) if you were to use $W18 \times 76$ beam over $W8 \times 31$?

7.48 By what factor do you increase the resistance to bending (about y–y axis) if you were to use $W18 \times 76$ beam over $W8 \times 31$?

7.49 Estimate the frontal area (excluding the display panel) of the cell phone shown in the accompanying figure. Express your answer in both SI and U.S. Customary units.

7.50 Estimate the frontal area of the car shown in the accompanying figure. Express your answer in both SI and U.S. Customary units.

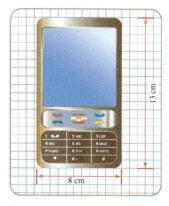

13 cm

8 cm

Problem 7.49

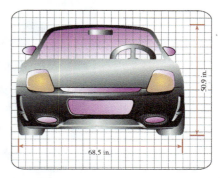

50.9 in.

68.5 in.

Problem 7.50

Impromptu | Design III

Objective: To build a boat from a $6'' \times 6''$ sheet of aluminum foil, which will carry as many pennies as possible. It should float for at least 1 minute. Thirty minutes will be allowed for preparation.

Saeed Moaveni

"Teamwork divides the task and doubles the success."—Andrew Carnegie
(1835—1919)

An Engineering Marvel

The New York City Water Tunnel No. 3*

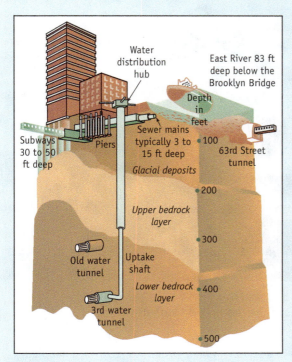

Based on Don Foley/National Geographic Image Collection

In 1954, New York City recognized the need for a third water tunnel to meet the growing demand on its more than 150-year-old water supply system. Planning for City Tunnel No. 3 began in the early 1960s, and actual construction commenced nearly a decade later in 1970. The New York City Water Tunnel No. 3, because of its size, length, controlling devices, and depth of excavation, represents one of the most complex and challenging engineering projects in the world today. The New York City Water Tunnel No. 3 also represents the largest capital improvement project in the city's history. Tunnel No. 3, when complete, is projected to cost nearly $6 billion dollars. Constructed by the New York City Department of Environmental Protection (DEP) employees, engineers, and underground construction workers (known as sandhogs), the tunnel will eventually span more than 60 miles. The tunnel is expected to be complete in 2020. Since 1970, when construction on the tunnel began, a total of 24 people have died in construction-related accidents.

Although City Tunnel No. 3 will not replace City Tunnels No. 1 and No. 2, it will enhance and improve the adequacy and dependability of the entire water supply system as well as improve service and pressure to outlying areas of the city. It will also allow the DEP to inspect and repair City Tunnels No. 1 and No. 2 for the first time since they were put into operation in 1917 and 1936, respectively.

The City Tunnel No. 3 project is to be completed in four stages. Stage 1 of the project has already been completed. Similar to City Tunnels No. 1 and 2, stage 1 of City Tunnel No. 3 begins at the Hillview Reservoir in Yonkers. It is constructed in bedrock 250 to 800 ft below the surface and runs 13 miles, extending across Central Park until about Fifth Avenue and 78th Street, then stretching eastward under the East River and Roosevelt Island into Astoria, Queens. The first stage of the tunnel, which cost $1 billion to construct, consists of a 24 ft diameter concrete-lined pressure tunnel that steps down in diameter to 20 ft. Water travels along this route and rises from the tunnel via 14 supply shafts, or "risers," and feeds into the city's water distribution system. Currently, City Tunnel No. 3 is serving the Upper East and Upper West Sides of Manhattan, Roosevelt Island, and many neighborhoods in the Bronx, west of the Bronx River.

There are four unique valve chambers that will allow stage 1 to connect to future portions of the tunnel without disrupting the flow of water service. Each of the existing valve chambers contains a series of 96 in. diameter conduits with valves and flowmeters to direct, control, and measure the flow

Materials were adapted with permission of the New York City Department of Environmental Protection.

Photo courtesy of New York City Department of Environmental Protection

Excavation of stage 1 of City Water Tunnel No. 3 looked like this in 1972. Today, water filling this portion of the tunnel is serving areas of the Bronx and Manhattan.

of water in sections of the tunnel. The largest of the valve chambers is in Van Cortlandt Park in the Bronx. Built 250 ft below the surface, this valve chamber will control the daily flow of water from the Catskill and Delaware water supply systems into Tunnel No. 3. In a design departure from the two existing tunnels, valves that control the flow of water in Tunnel No. 3 will be housed in large underground valve chambers, making them accessible for maintenance and repair. The valves for City Tunnels No. 1 and 2 are at the tunnel level and thus inaccessible when the tunnels are in service. Three of these four unique subsurface valve chambers have already been built to allow the connection of future stages of the tunnel without removing the water or taking any other stage of the tunnel out of service. These three valve chambers are located in the Bronx at Van Cortlandt Park (Shaft 2B), in Manhattan at Central Park (Shaft 13B), and Roosevelt Island (Shaft 15B).

Stage 2 of City Tunnel No. 3 construction is comprised of a two-leg Brooklyn/Queens section

and a section in Manhattan. Stage 2 was scheduled to be completed in 2009 at an approximate cost of $1.5 billion. The combination of stages 1 and 2 will provide the system with the ability to bypass one or both of City Tunnels No. 1 and 2.

Stage 3 involves the construction of a 16 mi. long section extending from the Kensico Reservoir to the Valve Chamber in the Bronx, which contains water from the Catskill and Delaware systems. When stage 3 is completed, City Tunnel No. 3 will operate at greater pressure, induced by the high elevation of Kensico Reservoir. It will also provide an additional aqueduct to supply water to the city that will parallel the Delaware and Catskill aqueducts. Stage 4, which will be 14 mi long, will travel from Van Cortlandt Park under the East River into Woodside, Queens.

Construction of the remaining three stages of City Tunnel No. 3 is being accelerated by the use of a mechanical rock excavator called a tunnel-boring machine (TBM). This machine, which is lowered in sections and assembled on the tunnel floor, chips off sections of bedrock through the continuous rotation of a series of steel cutting teeth. The TBM, which replaces the conventional drilling and blasting methods used during the construction of stage 1, allows for faster and safer excavation.

A photo summary illustrating various aspects of Tunnel No. 3 is shown in the accompanying photos.*

A form is constructed to prepare the tunnel to be lined with concrete.
Source: Photo courtesy of New York City Department of Environmental Protection

This finished section of the tunnel, which is 20 feet in diameter, is under Shaft 13B, near the Central Park Reservoir.
Source: Photo courtesy of New York City Department of Environmental Protection

Photo courtesy of New York City Department of Environmental Protection

The valve chamber in Van Cortlandt Park was completed in the early 1990s.
Photo courtesy of New York City Department of Environmental Protection

The tunnel boring machine (TBM), which was lowered in sections and assembled on the tunnel floor, chips off sections of bedrock in stage 2 of City Water Tunnel No. 3. The TBM's rotating steel cutting teeth replace the dynamite needed for earlier excavation.
Photo courtesy of New York City Department of Environmental Protection

A close-up view of the tunnel boring machine (TBM)
Photo courtesy of New York City Department of Environmental Protection

PROBLEM

1 Estimate the volume of the earth that has to be removed to make room for the 96 km, 7 m diameter, concrete-lined water tunnel. Also, investigate the capacity of typical dump trucks used in removal of earth materials.

Time and Time-Related Variables in Engineering

© Geray Sweeney/CORBIS

Time is a fundamental dimension and plays an important role in our daily life in describing events, processes, and many occurrences in our surroundings. Good engineers recognize the role of time in their lives and in calculating speed, acceleration, and flow of materials and substances as well as traffic. They know what is meant by *frequency* and *period* and understand the difference between a steady and a transient process. Good engineers have a comfortable grasp of rotational motion and understand how it differs from translational motion.

LEARNING OBJECTIVES

LO¹ **Time as a Fundamental Dimension:** realize that time is a fundamental dimension and is needed to describe many engineering problems, situations, and processes; also explain the difference between a steady and a transient process

LO² **Measurement of Time:** describe how time is measured and how its measurement has evolved

LO³ **Periods and Frequencies:** explain what we mean by a period or a frequency and how they are related to the fundamental dimension time

LO⁴ **Flow of Traffic:** describe traffic variables such as flow, density, and average speed that make use of the fundamental dimension of time

LO⁵ **Engineering Variables Involving Length and Time:** describe engineering quantities such speed, acceleration, and volume flow rate that are based on the fundamental dimensions of length and time

TIME

Before the development of modern science and for most of human existence, time was perceived as a circle or spiral: a cyclical pattern of renewal and rebirth. More familiar to us now is the Western tradition of linear time, which is a forward-moving direction or flow that represents a line between the past and the future implicit with the idea of progress. As the discoveries of evolution have come to underpin most of modern science, so the "arrow of time" has become a given in our collective consciousness.

You can't see, smell, taste, or hear it, yet it is of the physical world. We witness evidence of time all around in death and decay. Yet perception of time is largely a human phenomenon, and certainly, we are the only creatures who measure it, apply tools to it, and create tools to use it. Time is an aspect of the natural world, and the characteristics of physical time are determined by the processes of the physical world. It is the essence of

NASA; ESA; Z. Levay and R. van der Marel, STScI; T. Hallas; and A. Mellinger

cosmology, astronomy, and physics but is equally important in the disciplines of biology and geology. Certainly, it is critical to modern technology. The precision of its measurement drives our daily lives…. From Newtonian to Einsteinian physics and quantum mechanics to thermodynamics and satellite communications, time is of the essence to virtually all scientific disciplines.

US Library of Congress, http://www.loc.gov/rr/scitech/tracer-bullets/timetb.html

To the students: How would you define time? What is the role of time in your daily life? What is the role of time in technology and engineering? Give some examples.

In the previous chapter, we considered the role of length and length-related parameters such as area and volume in engineering analysis and design. In this chapter, we will investigate the role of time as a fundamental dimension and other time-related engineering parameters, such as frequency and period. We will first discuss why the time variable is needed to describe events, processes, and other occurrences in our physical surroundings. We will then explain the role of periods and frequency in recurring, or periodic, events (events that repeat themselves). A brief introduction to parameters describing traffic flow is also given in this chapter. Finally, we will look at engineering variables that involve length and time, including linear velocities and accelerations and volume flow rate of fluids. Rotational motion is also introduced at the end of this chapter. Table 6.7 is repeated here to show the relationship between the contents of this chapter and other fundamental dimensions discussed in other chapters.

| TABLE 6.7 | Fundamental Dimensions and How They Are Used in Defining Variables that Are Used in Engineering Analysis and Design |

Chapter	Fundamental Dimension	Related Engineering Variables			
7	Length (L)	Radian (L/L), Strain (L/L)	Area (L^2)	Volume (L^3)	Area moment of inertia (L^4)
8	Time (t)	Angular speed ($1/t$), Angular acceleration ($1/t^2$), Linear speed (L/t), Linear acceleration (L/t^2)		Volume flow rate (L^3/t)	
9	Mass (M)	Mass flow rate (M/t), Momentum (ML/t), Kinetic energy (ML^2/t^2)		Density (M/L^3), Specific volume (L^3/M)	
10	Force (F)	Moment (LF), Work, energy (FL), Linear impulse (Ft), Power (FL/t)	Pressure (F/L^2), Stress (F/L^2), Modulus of elasticity (F/L^2), Modulus of rigidity (F/L^2)	Specific weight (F/L^3)	
11	Temperature (T)	Linear thermal expansion (L/LT), Specific heat (FL/MT)		Volume thermal expansion (L^3/L^3T)	
12	Electric Current (I)	Charge (It)	Current density (I/L^2)		

LO¹ 8.1   Time as a Fundamental Dimension

We live in a dynamic world. Everything in the universe is in a constant state of motion. Think about it! Everything in the universe is continuously moving. The earth and everything associated with it moves around the sun. All of the solar planets, and everything that comprises them, are moving around the sun. We know that everything outside our solar system is moving too. From our everyday observation we also know that some things move faster than others. For example, people can move faster than ants, or a rabbit can move faster than a turtle. A jet plane in flight moves faster than a car on a highway.

A long time ago humans learned that, by defining a variable called time, they could use it to describe the occurrences of events in their surroundings.

Time is an important parameter in describing motion. How long does it take to cover a certain distance? A long time ago, humans learned that by defining a parameter called **time** they could use it to describe the occurrences of events in their surroundings. Think about the questions frequently asked in your everyday life: How old are you? How long does it take to go from here to there? How long does it take to cook this food? How late are you open? How long is your vacation? We have also associated time with natural occurrences in our lives; for example, to express the relative position of the earth with respect to the sun, we use day, night, 2:00 A.M. or 3:00 P.M., or May 30. The parameter time has been conveniently divided into smaller and larger intervals, such as seconds, minutes, hours, days, months, years, centuries, and millennia. We are continuously learning more and more about our surroundings in terms of how nature was put together and how it works; thus we need shorter and shorter time divisions, such as microseconds and nanoseconds. For example, with the advent of high-speed communication lines, the time that it takes electrons to move between short distances can be measured in nanoseconds.

We have also learned from our observation of the world around us that we can combine the parameter time with the parameter length to describe how fast something is moving. When we ask how fast, we should be careful to state with respect to what. Remember, everything in the universe is moving.

Before discussing the role of time in engineering analysis, let us focus on the role of time in our lives—our limited time budget. Today, we can safely assume that the average life expectancy of a person living in the Western world is around 75 years. Let us use this number and perform some simple arithmetic operations to illustrate some interesting points. Converting the 75 years to hours, we have

$$(75 \text{ years})(365.25 \text{ days/year})(24 \text{ hours/day}) = 657,450 \text{ hours}$$

On an average basis, we spend about 1/3 of our lives sleeping; this leaves us with 438,300 waking hours. Considering that traditional college freshmen are 18 years old, you have 333,108 waking hours still available to you if you live to the age of 75 years. Think about this for a while. If you were given only $333,108 for the rest of your life, would you throw away a dollar here and a dollar there as you were strolling through life? Perhaps not, especially knowing that you will not get any more money. Life is short! Make good use of your time, and at the same time enjoy your life.

Now let us look at the role of time in engineering problems and solutions. Most engineering problems may be divided into two broad areas: *steady* and *unsteady*. The problem is said to be *steady* when the value of a physical quantity under investigation does not change over time. For example, the length and width of your credit card doesn't change with time if it is not subjected to a temperature change or a load. If the value of a physical quantity changes with time, then the problem is said to be *unsteady* or *transient*.

In engineering, the problem is said to be steady when the value of a physical quantity under investigation does not change over time. If the value of the quantity changes with time, the problem is said to be unsteady or transient.

A good example of an unsteady situation that is familiar to everyone is the rate of a person's physical growth. From the time you were born until you reached your late teens or early twenties, your physical dimensions changed with time. You became

There is still time for you to do some good things in life

taller, your arms grew longer, your shoulders got wider. Of course, in this example not only the dimension length changes with time but also your mass changes with time. Another common example of an unsteady event that you are familiar with deals with the physical variable temperature. When you lay freshly baked cookies out to cool, the temperature of the cookies decreases with time until they reach the air temperature in the kitchen. There are many engineering problems that are also unsteady. You will find examples of unsteady and steady processes in the cooling of electronic equipment, biomedical applications, combustion, materials casting, materials processing, plastic forming, building heating or cooling applications, and in food processing. The transient response of a mechanical or structural system to a suddenly applied force is another instance of an unsteady engineering problem: for example, the response of a car's suspension system as you drive through a pothole or the response of a building to an earthquake.

Once I reach the age of 20, does my physical dimension really become steady? What if I were to gain mass?

LO² 8.2   Measurement of Time

Early humans relied on the relative position of the earth with respect to the sun, moon, stars, or other planets to keep track of time. The lunar calendar was used by many early civilizations. These celestial calendars were useful in keeping track of long periods of time, but humans needed to devise a means to keep track of shorter time intervals, such as what today we call an hour. This need led to the development of clocks (see Figure 8.1). Sun clocks, also known as shadow clocks or sundials, were used to divide a given day into smaller periods. The moving shadow of the dial marked the time intervals. Like other human-invented instruments, the sundial evolved over time into elaborate instruments that accounted for the shortness of the day during the winter as compared to the summer to provide for a better year-round accuracy. Sand glasses (glass containers filled with sand) and water clocks were among the first time-measuring devices that did not make use of the relative position of the earth with respect to the sun or other celestial bodies. Most of you have seen a sand glass (sometimes referred to as an hourglass); the water clocks were basically made of a graduated container with a small hole near the bottom. The container held water and was tilted so that the water would drip out of the hole slowly. Graduated cylindrical containers, into which water dripped at a constant rate, were also used to measure the passage of an hour. Over the years, the design of water clocks was also modified. The next revolution in timekeeping came during the 14th century when weight-driven mechanical clocks were used in Europe. Later, during the 16th century, came spring-loaded clocks. The spring mechanism design eventually led to smaller clocks and to watches. The oscillation of a pendulum was the next advancement in the design of clocks.

Quartz clocks eventually replaced the mechanical clocks during the second half of the 20th century. A quartz clock or watch makes use of the piezoelectric property of quartz crystal. A quartz crystal, when subjected to a mechanical pressure, creates an electric field. The inverse is also true—that is to say, the shape of the crystal changes when it is subjected to an electric field.

Dimedrol68/Shutterstock.com

Alistair Scott/Shutterstock.com

AGCuesta/Shutterstock.com

Ian 2010/Shutterstock.com

FIGURE 8.1   Are we really measuring time?

These principles are used to design clocks that make the crystal vibrate and generate an electric signal of constant frequency.

As we stated in Chapter 6, the natural frequency of the cesium atom was adopted as the new standard unit of time. The unit of a second is now formally defined as the duration of 9,192,631,770 periods of the radiation corresponding to the transition between the two hyperfine levels of the ground state of the cesium 133 atom.

The Need for Time Zones

You know that the earth rotates about an axis that runs from the South Pole to the North Pole and that it takes the earth 24 hours to complete one revolution about this axis. Moreover, from studying globes and maps, you may have noticed that the earth is divided into 360 circular arcs that are equally spaced from east to west; these arcs are called *longitudes*. The zero longitude was arbitrarily assigned to the arc that passes through Greenwich, England. Because it takes the earth 24 hours to complete one revolution about its axis, every 15 degrees longitude corresponds to 1 hour (360 degrees/24 hours = 15 degrees per one hour). For example, someone exactly 15 degrees west of Chicago will see the sun in the same exact position one hour later as it was observed by another person in Chicago one hour before. The earth is also divided into latitudes, which measure the angle formed by the line connecting the center of the earth to the specific location on the surface of the earth and the equatorial plane, as shown in Figure 8.2. The latitude varies from 0 (equatorial plane) degrees to 90 degrees north (North Pole) and from 0 degrees to 90 degrees south (South Pole).

The need for **time zones** was not realized until the latter part of the 19th century when the railroad companies were expanding. The railroad companies realized a need for standardizing their schedules. After all, 8:00 A.M. in New York City did not correspond to 8:00 A.M. in Denver, Colorado. Thus, a need for a uniform means to keep track of time and its relationship to other locations on the earth was born. It was railroad scheduling and commerce

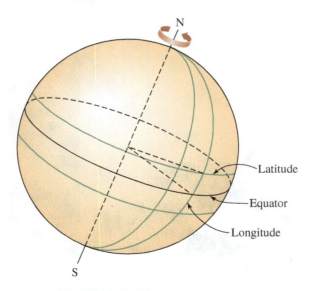

| FIGURE 8.2 | The longitudes and latitudes of the earth. |

that eventually brought nations together to define time zones. The standard time zones are shown in Figure 8.3.

Daylight Saving Time

Daylight saving time was originally put into place to save fuel (energy) during hard times such as World War I and World War II. The idea is simple; by setting the clock forward in the spring and keeping it that way during the summer and the early fall, we extend the daylight hours and, consequently, we save energy. For example, on a certain day in the spring, without daylight saving time, it will get dark at 8:00 P.M., but with the clocks set forward by one hour, it will then get dark at 9:00 P.M. So we turn our lights on an hour later. According to a U.S. Department of Transportation study, daylight saving time saves energy because we tend to spend more time outside of our homes engaging in outdoor activities. Moreover, because more people drive during the daylight hours, daylight saving can also reduce the number of automobile accidents, consequently saving many lives.

In 1966, the U.S. Congress passed the Uniform Time Act to establish a system of uniform time. Moreover, in 1986, Congress changed the initiation of daylight saving time from the last Sunday in April to the first Sunday in April. In 2005, Congress approved an amendment to have daylight saving time start on the second Sunday in March and end on the first Sunday in November, commencing in 2007. Today, most countries around the world follow some type of daylight saving schedule. In the European Union countries, daylight saving time begins on the last Sunday in March and continues through the last Sunday in October.

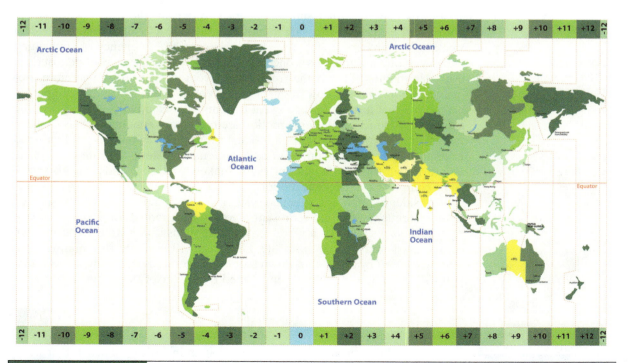

FIGURE 8.3 Standard time zones.
Designua/Shutterstock.com

LO³ 8.3 Periods and Frequencies

> For periodic events, a period is the time that it takes for the event to repeat itself. The inverse of a period is called a frequency.

For periodic events, a **period** is the time that it takes for the event to repeat itself. For example, every 365.24 days the earth lines up in exactly the same position with respect to the sun. The orbit of the earth around the sun is said to be periodic because this event repeats itself. The inverse of a period is called a **frequency**. For example, the frequency at which the earth goes around the sun is once a year.

Let us now use other simple examples to explain the difference between period and frequency. It is safe to assume that most of you do laundry once a week or buy groceries once a week. In this case, the frequency of your doing laundry or buying groceries is once a week. Or you may go see your dentist once every six months. That is the frequency of your dentist visits. Therefore, frequency is a measure of how frequently an event or a process occurs, and period is the time that it takes for that event to complete one cycle. Some engineering examples that include periodic motion are oscillatory systems such as shakers, mixers, and vibrators. The piston inside a car's engine cylinder is another good example of periodic motion. Your car's suspension system, the wings of a plane in a turbulent flight, or a building being shaken by strong winds may also show some component of periodic motion.

Consider a simple spring–mass system, as shown in Figure 8.4. The spring–mass system shown could represent a very simple model for a vibratory system, such as a shaker or a vibrator. What happens if you were to push down on the mass and then let go of it? The mass will oscillate in a manner that manifests itself by an up-and-down motion. If you study mechanical vibration, you will learn that the natural undamped frequency of the system is given by

$$f_n = \frac{1}{2\pi}\sqrt{\frac{k}{m}} \qquad \boxed{8.1}$$

where f_n is the natural frequency of the system in cycles per second, or hertz (Hz), k represents the stiffness of the spring or an elastic member (N/m), and m is the mass of the system (kg).

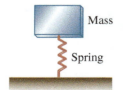

Mass

Spring

FIGURE 8.4

A spring–mass system.

The period of oscillation, T, for the given system—or in other words, the time that it takes for the mass to complete one cycle—is given by

$$T = \frac{1}{f_n}$$

8.2

Most of you have seen oscillating pendulums in clocks. The pendulum is another good example of a periodic system. The period of oscillation for a pendulum is given by

$$T = 2\pi\sqrt{\frac{L}{g}}$$

8.3

where L is the length of the pendulum (m) and g is the acceleration due to gravity (m/s^2). Note that the period of oscillation is independent of the mass of the pendulum. Not too long ago, oil companies used measured changes in the period of an oscillating pendulum to detect variations in acceleration due to gravity that could indicate an underground oil reservoir.

An understanding of periods and frequencies is also important in the design of electrical and electronic components. In general, excited mechanical systems have much lower frequencies than electrical/electronic systems. Examples of frequencies of various electrical and electronic systems are given in Table 8.1.

TABLE 8.1	Examples of Frequencies of Various Electrical and Electronic Systems
Application	**Frequency**
Alternating current (USA)	60 Hz
AM radio	540 kHz–1.6 MHz
FM radio	88–108 MHz
Emergency, fire, police	153–159 MHz
Personal computer clocks (as of year 2013)	up to 5.5 GHz
Wireless router (2013)	5.8 GHz

EXAMPLE 8.1

Determine the natural frequency of the simple spring-mass system shown in Figure 8.5.

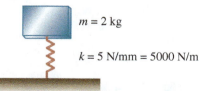

$m = 2$ kg

$k = 5$ N/mm = 5000 N/m

FIGURE 8.5   A simple spring-mass system.

Using Equation (8.1), we have

$$f_n = \frac{1}{2\pi}\sqrt{\frac{k}{m}} = \frac{1}{2\pi}\sqrt{\frac{5000 \text{ N/m}}{2 \text{ kg}}} \approx 8 \text{ Hz}$$

LO⁴ 8.4   Flow of Traffic

A branch of civil engineering deals with the design and layout of highways, roads, and streets and the location and timing of traffic control devices that move vehicles efficiently. Traffic variables such as the average speed of cars are time related.

Those of you who live in a big city know what we mean by traffic (Figure 8.6). A branch of civil engineering deals with the design and layout of highways, roads, and streets and the location and timing of traffic control devices that move vehicles efficiently. In this section, we will provide a brief overview of some elementary concepts related to traffic engineering. These variables are time related. Let us begin by defining what we mean by traffic flow. In civil engineering, **traffic flow** is formally defined by

$$q = \frac{3600\,n}{T} \qquad \text{8.4}$$

In Equation (8.4), q represents the traffic flow in terms of number of vehicles per hour, n is the number of vehicles passing a known location during a time duration T in seconds. Another useful variable of traffic information is **density**—how many cars occupy a stretch of a highway. Density is defined by

$$k = \frac{1000\,n}{d} \qquad \text{8.5}$$

where k is density and represents the number of vehicles per kilometer, and n is the number of vehicles on a stretch of highway, d, measured in meters.

Knowing the average speed of cars also provides valuable information for the design of road layouts and the location and timing of traffic control devices. The **average speed** ($\bar{u}$) of cars is determined from

$$\bar{u} = \frac{1}{n}\sum_{i=1}^{n} u_i \qquad \text{8.6}$$

| FIGURE 8.6 | A congested flow! |

Bart Everett/Shutterstock.com

In Equation (8.6), u_i is the speed of individual cars (we will discuss the definition of speed in more detail in Section 8.5.), and n represents the total number of cars. There is a relationship among the traffic parameter—namely, the flow of traffic, density, and the average speed—according to

$$q = k\,\overline{u}$$

8.7

where q, k, and $\overline{u}$ were defined earlier by Equations (8.4) through (8.6).

The relationship among flow, density, and average speed is shown in Figures 8.7 and 8.8. Figure 8.7 shows that when the average speed of moving cars is high, the value of density is near zero, implying that there are not that many cars on that stretch of highway. As you may expect, as the value of density increases (the number of vehicles per kilometer), the average speed of vehicles decreases and will eventually reach a zero value, meaning bumper-to-bumper traffic. Figure 8.8 shows the relationship between the average speed of vehicles and the flow of traffic. For bumper-to-bumper traffic, with no cars moving, the flow of traffic stops, and thus q has a value of zero. This is the beginning of the congested region shown in Figure 8.8. As the average speed of the vehicles increases, so does the flow of traffic, eventually reaching a maximum value. As the average speed of vehicles continues to increase, the flow of traffic (the number of vehicles per hour) decreases. This region is marked by an uncongested flow region.

Finally, traffic engineers use various measurement devices and techniques to obtain real-time data on the flow of traffic. They use the collected information to make improvements to move vehicles more efficiently. You may have seen examples of traffic measurement devices such as pneumatic road tubes and counters. Other common traffic measurement devices include magnetic induction loops and speed radar.

EXAMPLE 8.2   Show that Equation (8.7) is dimensionally homogeneous by carrying out the appropriate units for each term in that equation.

$$q\left(\frac{\text{vehicles}}{\text{hour}}\right) = \left[k\left(\frac{\text{vehicles}}{\text{kilometer}}\right)\right]\left[\overline{u}\left(\frac{\text{kilometer}}{\text{hour}}\right)\right]$$

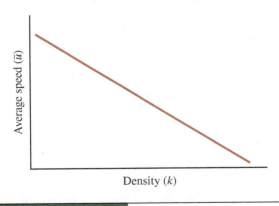

FIGURE 8.7   The relationship between speed and density

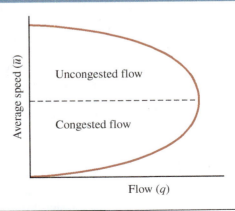

FIGURE 8.8   The relationship between speed and flow.

LO⁵ 8.5   Engineering Variables Involving Length and Time

In this section, we will consider *derived physical quantities* that are based on the fundamental dimensions of length and time. We will first discuss the concepts of linear speed and acceleration and then define volumetric flow rate.

Linear Velocities

Today, we use the fundamental dimension time in various engineering situations and in calculating frequency, linear and rotational speed and acceleration, and flow of materials and substances.

We will begin by explaining the concept of linear speed. Knowledge of linear speed and acceleration is important to engineers when designing conveyer belts that are used to load luggage into airplanes and product assembly lines, treadmills, elevators, automatic walkways, escalators, water or gas flow inside pipes, space probes, roller coasters, transportation systems (such as cars, boats, airplanes, and rockets), snow removal equipment, backup computer tape drives, and so on. Civil engineers are also concerned with velocities, particularly wind velocities, when designing structures. They need to account for the wind speed and its direction in their calculations when sizing structural members.

Let us now take a close look at what we mean by linear speed and linear velocity. All of you are familiar with a car speedometer. It shows the instantaneous speed of a car. Before we explain in more detail what we mean by the term **instantaneous speed**, let us define a physical variable that is more easily understood, the **average speed**, which is defined as

$$\text{average speed} = \frac{\text{distance traveled}}{\text{time}} \qquad \boxed{8.8}$$

Note that the fundamental (base) dimensions of length and time are used in the definition of the average speed. The average speed is called a *derived physical quantity* because its definition is based on the fundamental dimensions of length and time. The SI unit for speed is m/s, although for fast-moving objects km/h is also commonly used. In U.S. Customary units, ft/s and miles/h (mph) are used to quantify the magnitude of a moving object. Now let us look at what we mean by the instantaneous speed and see how it is related to the average speed.

To understand the difference between average and instantaneous speed, consider the following mental exercise. Imagine that you are going from New York City to Boston, a distance of 220 miles (354 km). Let us say that it took you 4.5 hours to go from the outskirts of New York City to the edge of Boston. From Equation (8.8), you can determine your average speed, which is 49 mph (79 km/h). You may have made a rest stop somewhere to get a cup of coffee. Additionally, the posted highway speed limit may have varied from 55 mph (88 km/h) to 65 mph (105 km/h), depending on the stretch of highway. Based on the posted speed limits and other road conditions, and how you felt, you may have driven the car faster during some stretches, and you may have gone slower during other stretches. These conditions led to an average speed of 49 mph (79 km/h). Let us also imagine that you recorded the speed of your car as indicated by the speedometer every second. The actual speed of the car at any given instant while you were driving it is called the **instantaneous speed**.

To better understand the difference between the average speed and the instantaneous speed, ask yourself the following question. If you needed to locate the car, would you be able to locate the car knowing just the average speed of the car? The knowledge of the average speed of the car would not be sufficient. To know where the car is at all times, you need more information, such as the instantaneous speed of the car and the direction in which it is traveling. This means you must know the *instantaneous velocity* of the car. Note that when we say velocity of a car, we not only refer to the speed of the car but also the direction in which it moves.

Physical quantities that possess both a magnitude and a direction are called *vectors*. You will learn more about vectors in your calculus, physics, and mechanics classes. For now, just remember the simple definition of a vector quantity–a quantity that has both magnitude and direction. A physical quantity that is described only by a magnitude is called a *scalar* quantity. Examples of scalar quantities include temperature, volume, and mass. Examples of the range of speed of various objects are given in Table 8.2.

> Engineering variables such as instantaneous speed and acceleration are vectors. Variables that possess both a magnitude and a direction are called vectors.

TABLE 8.2   **Examples of Some Speeds**

Situation	m/s	km/h	ft/s	mph
Average speed of a person walking	1.3	4.7	4.3	2.9
The fastest runner in the world (100 m)	10.2	36.7	33.5	22.8
Professional tennis player serving a ball	58	209	190	130
Top speed of a sports car	67	241	220	150
777 Boeing airplane (cruise speed)	248	893	814	555
Orbital speed of the space shuttle	7741	27,869	25,398	17,316
Average orbital speed of the earth around the sun	29,000	104,400	95,144	64,870

Linear Accelerations

Acceleration provides a measure of how velocity changes with time. Something that moves with a constant velocity has a zero acceleration. *Because velocity is a vector quantity and has both magnitude and direction, any change in either the direction or the magnitude of velocity can create acceleration.* For example, a car moving at a constant speed following a circular path has an acceleration component due to the change in the direction of the velocity vector, as shown in Figure 8.9. Here, let us focus on an object moving along a straight line. The **average acceleration** is defined as

$$\text{average acceleration} = \frac{\text{change in velocity}}{\text{time}} \qquad \boxed{8.9}$$

Again note that acceleration uses only the dimensions of length and time. Acceleration represents the rate at which the velocity of a moving object changes with time. Therefore, acceleration is the time rate of change of velocity. The SI unit for acceleration is m/s² and in U.S. Customary units ft/s².

The difference between **instantaneous acceleration** and average acceleration is similar to the difference between instantaneous velocity and average velocity. The instantaneous acceleration can be obtained from Equation (8.9) by making the time interval smaller and smaller. That is to say, instantaneous acceleration shows how the velocity of a moving object changes at any instant.

Let us now turn our attention to acceleration due to gravity. Acceleration due to gravity plays an important role in our everyday lives in the weight of objects and in the design of projectiles. What happens when you let go of an object in your hand? It falls to the ground. Sir Isaac Newton discovered that two masses attract each other according to

$$F = \frac{Gm_1 m_2}{r^2} \qquad \boxed{8.10}$$

where F is the attractive force between the masses (N), G represents the universal gravitational attraction and is equal to $6.7 \times 10^{-11} \, \text{m}^3/\text{kg} \cdot \text{s}^2$, m_1 and m_2 are the mass of each object (kg), and r denotes the distance between the center of each object. Using Equation (8.10), we can determine the weight of

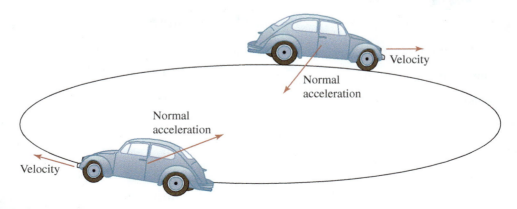

Velocity

Normal
acceleration

Normal
acceleration

Velocity

FIGURE 8.9   The acceleration of a car moving at a constant speed following a circular path.

an object having a mass m (on earth) by substituting m for m_1, substituting for m_2 the mass of the earth, and using the radius of the earth as the distance r between the center of m_1 and m_2. At the surface of the earth, the attractive force F is called the *weight* of an object, and the acceleration created by the force is referred to as the *acceleration due to gravity* (g). At the surface of the earth, the value of g is equal to 9.81 m/s^2, or 32.2 ft/s^2. In general, acceleration due to gravity is a function of latitude and longitude, but for most practical engineering applications near the surface of the earth the values of 9.81 m/s^2 or 32.2 ft/s^2 are sufficient.

Table 8.3 shows speed, acceleration, and the distance traveled by a falling object from the roof of a high-rise building. We have neglected air resistance in our calculations. Note how the distance traveled by the falling object and its velocity change with time.

TABLE 8.3 Speed, Acceleration, and Distance Traveled by a Falling Object Neglecting the Air Resistance

Time (seconds)	Acceleration of the Object (m/s²)	Speed of the Falling Object (m/s) ($V = gt$)	Distance Traveled (m) $\left(d = \dfrac{1}{2}gt^2\right)$
0	9.81	0	0
1	9.81	9.81	4.90
2	9.81	19.62	19.62
3	9.81	29.43	44.14
4	9.81	39.24	78.48
5	9.81	49.05	122.62
6	9.81	58.86	176.58

EXAMPLE 8.3

A car starts from rest and accelerates to a speed of 100 km/h in 15 seconds. The acceleration during this period is constant. For the next 30 minutes, the car moves with a constant speed of 100 km/h. At this time the driver of the car applies the brake and the car decelerates to a full stop in 10 seconds. The variation of the speed of the car with time is shown in Figure 8.10. We are interested in determining the total distance traveled by the car and its average speed over this distance.

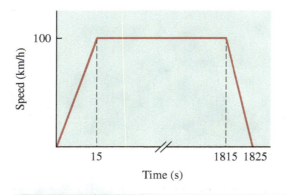

FIGURE 8.10   The variation of the speed of the car with time for Example 8.3.

During the initial 15 seconds, the speed of the car increases linearly from a value of zero to 100 km/h. Therefore, the average speed of the car is 50 km/h during this period. The distance traveled during this period is

$$d_1 = (\text{time})(\text{average speed}) = (15\,\text{s})\left(50\,\frac{\text{km}}{\text{h}}\right)\left(\frac{1\,\text{h}}{3600\,\text{s}}\right)\left(\frac{1000\,\text{m}}{1\,\text{km}}\right) = 208.3\,\text{m}$$

During the next 30 minutes (1800 s), the car moves with a constant speed of 100 km/h and the distance traveled during this period is

$$d_2 = (1800\,\text{s})\left(100\,\frac{\text{km}}{\text{h}}\right)\left(\frac{1\,\text{h}}{3600\,\text{s}}\right)\left(\frac{1000\,\text{m}}{1\,\text{km}}\right) = 50000\,\text{m}$$

Because the car decelerates at a constant rate from a speed of 100 km/h to 0, the average speed of the car during the last 10 seconds is also 50 km/h, and the distance traveled during this period is

$$d_3 = (10\,\text{s})\left(50\,\frac{\text{km}}{\text{h}}\right)\left(\frac{1\,\text{h}}{3600\,\text{s}}\right)\left(\frac{1000\,\text{m}}{1\,\text{km}}\right) = 138.9\,\text{m}$$

The total distance traveled by the car is

$$d = d_1 + d_2 + d_3 = 208.3 + 50000 + 138.9 = 50347.2\,\text{m} = 50.3472\,\text{km}$$

And finally, the average speed of the car for the entire duration of travel is

$$V_{\text{average}} = \frac{\text{distance travelled}}{\text{time}} = \frac{50347.2}{1825} = 27.56\,\text{m/s} = 99.2\,\text{km/h}$$

Volume Flow Rate

Engineers design flow-measuring devices to determine the amount of a material or a substance flowing through a pipeline in a processing plant. Volume flow rate measurements are necessary in many industrial processes to keep track of the amount of material being transported from one point to the next point in a plant. Additionally, knowing the flow rate of a material, engineers can determine the consumption rate so that they can provide the necessary supply for a steady state operation. Flow-measuring devices are also employed to determine the amount of water or natural gas being used by us in our homes during a specific period of time. City engineers need to know the daily or monthly volumetric water consumption rates in order to provide an adequate supply of water to our homes and commercial plants. Many homes in the United States and around the world use natural gas for cooking purposes or for space heating. For example, for a gas furnace to heat cold air, natural gas is burned inside a heat exchanger that transfers the heat of combustion to the cold air supply. Companies providing the natural gas need to know how much fuel–how many cubic meters of natural gas–is burned every day, or every month, by each home so that they can correctly charge their customers. In sizing the heating or cooling units for buildings, the volumetric flow of warm or cool air must be determined to adequately compensate for the heat loss or heat gain for a given building. Ventilation rates dealing with the introduction of fresh air into a building are also expressed in cubic feet per minute (cfm) or cubic meter per minute or per hour. On a smaller scale, to keep the microprocessor inside your computer operating at a safe temperature, the volumetric flow of air must also be determined.

Another example that you may be familiar with is the flow rate of a water–antifreeze mixture necessary to cool your car's engine. Engineers who design the cooling systems for a car engine need to determine the volumetric flow rate of the water–antifreeze mixture (gallons of water–antifreeze mixture per minute) through the car's engine block and through the car's radiator to keep the engine block at safe temperatures.

Now that you have an idea what we mean by the term *volumetric flow rate*, let us formally define it. The **volume flow rate** is simply defined by the volume of a given substance that flows through something per unit time.

$$\text{volume flow rate} = \frac{\text{volume}}{\text{time}} \qquad \boxed{8.11}$$

Note that in the definition given by Equation (8.11), the fundamental dimensions of length (length cubed) and time are used. Some of the more common units for volume flow rate include m^3/s, m^3/h, L/s, ml/s, ft^3/s (cfs), gal/min (gpm), or gal/day (gpd).

> The volume flow rate is defined as the volume of a substance (e.g., water) that flows through something (e.g., pipe) per unit time.

A Fun Experiment for You to Conduct Use a stopwatch and an empty Pepsi or Coke can to determine the volumetric flow rate of water coming out of a drinking fountain. Present your results in liters per second and in gallons per minute.

For fluids flowing through pipes, conduits, or nozzles, there exists a relationship between the volumetric flow rate, the average velocity of the flowing fluid, and the cross-sectional area of the flow, according to this equation.

$$\text{volumetric flow rate} = (\text{average velocity})(\text{cross-sectional area of the flow}) \qquad \boxed{8.12}$$

In fact, some flow-measuring devices make use of Equation (8.12) to determine the volume flow rate of a fluid by first measuring the fluid's average velocity and the cross-sectional area of the flow. Finally, note that in Chapter 9 we will explain another closely defined variable, mass flow rate of a flowing material, which provides a measure of time rate of mass flow through pipes or other carrying conduits.

EXAMPLE 8.4

Consider the piping system shown in Figure 8.11. The average speed of water flowing through the 12-inch-diameter section of the piping system is 5 ft/s. What is the volume flow rate of water in the piping system? Express the volume flow rate in ft³/s, gpm, and L/s. For the case of steady flow of water through the piping system, what is the average speed of water in the 6 inch diameter section of the system?

FIGURE 8.11   The piping system of Example 8.4.

We can determine the volume flow rate of water through the piping system using Equation (8.12).

$$Q = \text{volume flow rate} = (5 \text{ ft/s})\left(\frac{\pi}{4}\right)(1 \text{ ft})^2 = 3.926 \text{ ft}^3/\text{s}$$

$$Q = (3.926 \text{ ft}^3/\text{s})\left(\frac{7.48 \text{ gal}}{1 \text{ ft}^3}\right)\left(\frac{60 \text{ s}}{1 \text{ min}}\right) = 1762 \text{ gpm}$$

$$Q = (3.926 \text{ ft}^3/\text{s})\left(\frac{28.3168 \text{ L}}{1 \text{ ft}^3}\right) = 111.2 \text{ L/s}$$

For the steady flow of water through the piping system the volume flow rate is constant. This fact allows us to calculate the speed of water in the 6 inch section of the pipe.

$$Q = 3.926 \text{ ft}^3/\text{s} = (\textbf{average velocity})(\textbf{cross-sectional area of flow})$$

$$3.926 \text{ ft}^3/\text{s} = (\textbf{average speed})\left(\frac{\pi}{4}\right)(0.5 \text{ ft})^2$$

$$\text{average speed} = 20 \text{ ft/s}$$

From the results of this example you can see that when you reduce the pipe diameter by a factor of 2, the water speed in the reduced section of the pipe increases by a factor of 4.

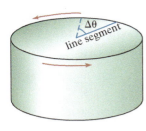

Angular Motion

In the next two sections, we will discuss angular motion, including angular speeds and accelerations of rotating objects.

Angular (Rotational) Speeds In the previous section, we explained the concept of linear speed and acceleration. Now we will consider variables that define angular motion. Rotational motion is also quite common in engineering applications. Examples of engineering components with rotational motion include shafts, wheels, gears, drills, pulleys, fan or pump impellers, helicopter blades, hard drives, CD drives, and so on.

The average **angular speed** of a line segment located on a rotating object such as a shaft is defined as the change in its angular position (angular displacement) over the time that it took the line to go through the angular displacement.

$$\omega = \frac{\Delta \theta}{\Delta t}$$

8.13

In Equation (8.13), ω represents the average angular speed in radians per second, $\Delta \theta$ is the angular displacement (radians), and Δt is the time interval in seconds. Similar to the definition of instantaneous velocity given earlier, the instantaneous angular speed is defined by making the time increment smaller and smaller. Again, when we speak of angular velocity, we not only refer to the speed of rotation but also to the direction of rotation. It is a common practice to express the angular speed of rotating objects in revolutions per minute (**rpm**) instead of radians per second (rad/s). For example, the rotational speed of a

FIGURE 8.12   (a) A stroboscope and (b) a tachometer.

pump impeller may be expressed as 1600 rpm. To convert the angular speed value from rpm to rad/s, we make the appropriate conversion substitutions.

$$1600\left(\frac{\text{revolutions}}{\text{minutes}}\right)\left(\frac{2\pi\,\text{radians}}{1\,\text{revolution}}\right)\left(\frac{1\,\text{minute}}{60\,\text{seconds}}\right) = 167.5\,\frac{\text{rad}}{\text{s}}$$

In practice, the angular speed of rotating objects is measured using a strobo-scope or a tachometer (Figure 8.12).

To get some idea how fast some common objects rotate, consider the following examples: a dentist's drill runs at 400,000 rpm; a current state-of-the-art computer hard drive runs at 7200 rpm; the earth goes through one complete revolution in 24 hours, thus the rotational speed of earth is 15 degrees per hour or 1 degree every 4 minutes.

There exists a relationship between linear and angular velocities of objects (Figure 8.13) that not only rotate but also translate as well. For example, a car wheel, when not slipping, will not only rotate but also translate. To establish the relationship between rotational speed and the translational speed we begin with

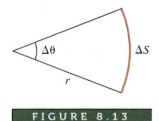

FIGURE 8.13

$$\Delta S = r\Delta\theta \qquad \text{8.14}$$

dividing both sides by time increment Δt

$$\frac{\Delta S}{\Delta t} = r\frac{\Delta\theta}{\Delta t} \qquad \text{8.15}$$

and making use of the definitions of linear and angular velocities, we have

$$V = r\omega \qquad \text{8.16}$$

For example, the actual linear velocity of a particle located 0.1 m away from the center of a shaft that is rotating at an angular speed of 1000 rpm (104.7 rad/s) is approximately 10.5 m/s.

EXAMPLE 8.5

Determine the rotational speed of a car wheel if the car is translating along at a speed of 55 mph. The radius of the wheel is 12.5 in.

Using Equation (8.16), we have

$$\omega = \frac{V}{r} = \frac{\left(55\,\dfrac{\text{miles}}{\text{h}}\right)\left(\dfrac{1\,\text{h}}{3600\,\text{s}}\right)\left(\dfrac{5280\,\text{ft}}{1\,\text{mile}}\right)}{(12.5\,\text{in.})\left(\dfrac{1\,\text{ft}}{12\,\text{in.}}\right)} = 77.4\,\text{rad/s} = 739\,\text{rpm}$$

Angular Accelerations In the previous section, we defined angular speed and its importance in engineering applications. We now define angular accel-eration in terms of the rate of change of angular velocity. Angular accelera-tion is a vector quantity. Note also that similar to the relationship between the average acceleration and instantaneous acceleration, we could first define an average **angular acceleration** as

$$\text{angular acceleration} = \frac{\text{change in angular speed}}{\text{time}}$$

8.17

and then define the instantaneous angular acceleration by making the time interval smaller and smaller in Equation (8.17).

EXAMPLE 8.6

It takes 5 s for a shaft of a motor to go from zero to 1600 rpm. Assuming constant angular acceleration, what is the value of the angular acceleration of the shaft?

First, we convert the final angular speed of the shaft from rpm to rad/s.

$$1600 \left(\frac{\text{revolutions}}{\text{minutes}} \right) \left(\frac{2\pi \text{ radians}}{1 \text{ revolution}} \right) \left(\frac{1 \text{ minute}}{60 \text{ seconds}} \right) = 167.5 \, \frac{\text{rad}}{\text{s}}$$

Then we use Equation (8.17) to calculate the angular acceleration.

$$\text{angular acceleration} = \frac{\text{change in angular speed}}{\text{time}} = \frac{(167.5 - 0) \, \text{rad/s}}{5 \, \text{s}} = 33.5 \, \frac{\text{rad}}{\text{s}^2}$$

Finally, remember that all the parameters discussed in this chapter involved time or length and time.

Nahid Afsari

Courtesy of Nahid Afsari

My interest in engineering has taken me from building a spaghetti bridge in high school to designing the reconstruction of a real "spaghetti bowl"—downtown Milwaukee's Marquette Interchange. The transition from pasting pasta to designing part of an $810 million project for the Wisconsin Department of Transportation has been rewarding. As a structural engineer at CH2M HILL assigned to the replacement of the high-profile complex interchange that links three interstate highways near Marquette University, I design bridges for the project's half-mile-long east leg. My diverse experience also includes a "low-profile" project, which calls for a highway undercrossing that enables a threatened species of snake to safely get to the other side of the road.

Ever since I can remember, I have been intrigued by puzzles, how things work, and how pieces fit together to make something whole. I enjoy problem solving and teamwork to reach a goal, and an engineering career enables that in a big way. One of the nicest things about civil engineering is seeing the finished product and knowing that people in the community are benefiting from what I have designed. Besides engineering's appeal to my basic nature, the fact that my father and my older

brother are engineers and have thoroughly enjoyed their work also influenced my career choice.

The biggest challenge I faced as an undergraduate student at the University of Wisconsin—Madison was balancing study time with the rest of my activities. On top of being a full-time student on track to finish my engineering degree in four years, I was on the women's intercollegiate soccer team and a member of the American Society of Civil Engineers student chapter (participating in several regional and national concrete canoe competitions). My busy schedule taught me a lot in terms of how to prioritize tasks and manage time effectively. Those lessons have proven to be invaluable in my professional life, especially in the fast-paced and demanding atmosphere of a major project office.

Courtesy of Nahid Afsari

SUMMARY

LO¹   Time as a Fundamental Dimension

You should realize that we live in a dynamic world. Everything in the universe is in a constant state of motion. By now, you should also understand the important role of the fundamental dimension of time in various engineering situations and in calculating frequency, linear and rotational speed and acceleration, and flow of materials and substances.

LO²   Measurement of Time

Time can be divided into both small and large portions, such as seconds, minutes, hours, days, months, and years. Early humans relied on the relative position of the earth with respect to the moon, stars, or other planets to keep track of long periods of time, but humans needed to devise a means to keep track of shorter time intervals, such as what today we call an hour. Sun clocks were used to divide a given day into smaller periods. The sundial evolved over time into elaborate instruments that accounted for the shortness of the day during the winter as compared to the summer to provide for a better year-round accuracy. The next revolution in timekeeping came during the 14th century,

when weight-driven mechanical clocks were used in Europe. Later, during the 16th century, came spring-loaded clocks. Quartz clocks eventually replaced the mechanical clocks during the second half of the 20th century.

The need for time zones was realized in the latter part of the 19th century when the railroad companies were expanding. The railroad companies realized a need for standardizing their schedules.

Daylight saving time was originally put into place to save fuel (energy) during hard times in World War I and World War II. The idea is simple; by setting the clock forward in the spring and keeping it that way during the summer and the early fall, we extend the daylight hours, and consequently, we save energy.

LO³   Periods and Frequencies

Frequency is a measure of how frequently an event or a process occurs, and period is the time that it takes for that event to complete one cycle. In other words, for periodic events, a period is the time that it takes for the event to repeat itself, and the inverse of a period is called a frequency.

LO⁴   Flow of Traffic

A branch of civil engineering deals with the design and layout of highways, roads, and streets and the location and timing of traffic control devices that move vehicles efficiently. Traffic variables such as the average speed of cars and traffic flow (number of vehicles per hour) are time related.

LO⁵   Engineering Variables Involving Length and Time

There are many engineering variables that are based on the fundamental dimensions of length and time. Among them are linear speed and acceleration and the volumetric flow rate of material or a substance. Knowledge of linear speed and acceleration is important to engineers when designing conveyer belts, assembly lines, treadmills, elevators, automatic walkways, escalators, water or gas flow inside pipes, space probes, roller coasters, transportation systems (such as cars, boats, airplanes, and rockets), and snow removal equipment. Volume flow rate measurements are necessary to keep track of the amount of material being transported or consumed. For example, engineers design flow-measuring devices to determine the volume of water or natural gas that is consumed in our homes per month. Knowing the flow rate of a material, engineers can determine the consumption rate so that they can provide the necessary supply for a steady–state operation.

KEY TERMS

Angular Acceleration 256
Angular Speed 255
Angular Acceleration 256
Average Speed 246
Daylight Saving Time 243
Frequency 244

Instantaneous Acceleration 250
Instantaneous Speed 248
Period 244
RPM 255
Time 239
Time Zones 242

Traffic Average Speed 246
Traffic Density 246
Traffic Flow 246
Volume Flow Rate 253

APPLY WHAT YOU HAVE LEARNED

This is a project to determine your class water footprint. Each of you is to estimate how much water you consume each year for taking showers and flushing toilets. Follow these steps to determine your shower and flushing water consumption.

1. Obtain a container of a known volume, and then time how long it takes to fill the container when placed under shower head.
2. Calculate the volumetric flow rate in gallons per minute. Then, measure the time that you spend on average when taking showers. Calculate the volume of the water consumed on daily basis.
3. Multiply the daily value by 365 to get the yearly value.
4. Look up the size of your toilet and estimate on average how many times per day you flush the toilet.

Telekhovskyi/Shutterstock.com

Tom Grundy/Shutterstock.com

5. Calculate the volume of water you consume on a daily basis.
6. Multiply the daily volume by 365 to get the annual value.

Compile your findings into a single report and present it to the class. Discuss annual consumption per class average. Suggest ways to conserve.

PROBLEMS

Problems that promote lifelong learning are denoted by. o—

8.1 Create a calendar showing the beginning and the end of daylight saving time for the years 2016–2022.

Year	Daylight Saving Time Begins at 2:00 A.M.	Daylight Saving Ends at 2:00 A.M.
2016		
2017		
2018		
2019		
2020		
2021		
2022		

8.2 According to the Department of Transportation analysis of energy consumption figures for 1974 and 1975, observation of daylight saving time in the month of April in the years of 1974 and 1975 saved the country an estimated energy equivalent of 10,000 barrels of oil each day. Estimate how much energy is saved in the United States or your country due to observing daylight saving time in the current year.

8.3 Is there a need for a country near the equator to observe daylight saving time? Explain.

8.4 Besides energy savings, what are other advantages that observation of daylight saving time may bring?

8.5 Every year all around the world we celebrate certain cultural events dealing with our past. For example, in Christianity, Easter is celebrated in the spring and Christmas is celebrated in December; in the Jewish calendar, Yom Kippur is celebrated in October; and Ramadan, the time of fasting, is celebrated by Muslims according to a lunar calendar. Briefly discuss the basis of the Christian, Jewish, Muslim, and Chinese calendars.

8.6 In this problem you are asked to investigate how much water a leaky faucet wastes in one week, one month, and one year. Perform an experiment by placing a container under a leaky faucet and actually measure the amount of water accumulated in an hour (you can simulate a leaky faucet by just partially closing the faucet). You are to design the experiment. Think about the parameters that you need to measure. Express and project your findings in gallons/day, gallons/week, gallons/month, and gallons/year. At this rate, how much water is wasted by 10,000,000 households with leaky faucets. Write a brief report to discuss your findings.

8.7 Next time you are putting gasoline in your car, determine the volumetric flow rate of the gasoline at the pumping station. Record the time that it takes to pump a known volume of gasoline into your car's gas tank. The flow meter at the pump will give you the volume in gallons, so all you have to do is to measure the time. Investigate the size of the storage tanks in your neighborhood gas station. Estimate how often the storage tank needs to be refilled. State your assumptions.

8.8 You are to investigate the water consumption rates in the United States. For bookkeeping purposes, it is customary to group major activities that consume water into public, domestic, irrigation, livestock, aquaculture, industrial, mining, and thermoelectric power generation. For example, the "livestock" category represents water use in dairy operations and feed lots, whereas water consumed to farm fish, shrimp, or other animals or plants that live in water is grouped into "aquaculture" category. As another example, the water that is used for industrial purposes such as making paper, chemicals, or steel is classified as "industrial." In a brief report, discuss the magnitude of the annual water consumption in each category.

8.9 Using the concepts discussed in this chapter, measure the volumetric flow rate of water out of a drinking fountain.

8.10 Convert the following speed limits from miles per hour (mph) to kilometers per hour (km/h) and from feet per second (ft/s) to meters per second (m/s). Think about the relative magnitude of values as you go from mph to ft/s and as you go from km/h to m/s. You may use Microsoft Excel to solve this problem.

Speed Limit (mph)	Speed Limit (km/h)	Speed Limit (ft/s)	Speed Limit (m/s)
15			
25			
30			
35			
40			
45			
55			
65			
70			

8.11 Most car owners drive their cars an average of 12,000 miles a year. Assuming a 20 miles/gallon gas consumption rate, determine the amount of fuel consumed by 150 million car owners on the following time bases:

   a. average daily basis
   b. average weekly basis

   c. average monthly basis
   d. average yearly basis
   e. over a period of ten years
   Express your results in gallons and liters.

8.12 Calculate the speed of sound for the U.S. standard atmosphere using $c = \sqrt{kRT}$, where c represents the speed of sound in m/s, k is the specific heat ratio for air ($k = 1.4$), and R is the gas constant for the air ($R = 286.9$ J/kg·K) and T represents the temperature of the air in Kelvin. The speed of sound in atmosphere is the speed at which sound propagates through the air. You may use Excel to solve this problem.

Altitude (m)	Air Temperature (K)	Speed of Sound (m/s)	Speed of Sound (km/h)
500	284.9		
1000	281.7		
2000	275.2		
5000	255.7		
10,000	223.3		
15,000	216.7		
20,000	216.7		
40,000	250.4		
50,000	270.7		

8.13 Express Equation (8.5), the traffic density, in terms of number of vehicles per mile.

8.14 Express the angular speed of the earth in rad/s and rpm.

8.15 What is the magnitude of the speed of a person at the equator due to rotational speed of the earth?

8.16 Calculate the average speed of the gasoline exiting a nozzle at a gas station. Next time you go to a gas station, measure the volumetric flow rate of the gas first, and then measure the diameter of the nozzle. Use Equation (8.12) to calculate the average speed of the gasoline coming out of the nozzle. *Hint:* First measure the time that it takes to put so many gallons of gasoline into the gas tank!

8.17 Measure the volumetric flow rate of water coming out of a faucet. Also determine the average velocity of water leaving the faucet.

8.18 Determine the speed of a point on the earth's surface in ft/s, m/s, mph, km/h. State your assumptions.

8.19 Into how many time zones are the United States and its territories divided?

8.20 Estimate the rotational speed of your car wheels when you are traveling at 60 mph.

8.21 Determine the natural frequency of a pendulum whose length is 15 ft.

8.22 Determine the spring constant for Example 8.1 if the system is to oscillate with a natural frequency of 5 Hz.

8.23 Determine the traffic flow if 100 cars pass a known location during 10 s.

8.24 A conveyer belt runs on 4 in. drums that are driven by a motor. If it takes 5 s for the belt to go from zero to the speed of 4 ft/s, calculate the final angular speed of the drum and its angular acceleration. Assume constant acceleration.

8.25 Chinook, a military helicopter, has two three-blade rotor systems, each turning in opposite directions. Each blade has a diameter of approximaterly 41 ft. The blades can spin at angular speeds of up to 225 rpm. Determine the translational speed of a particle located at the tip of a blade. Express your answer in ft/s, mph, m/s, and km/h.

8.26 Consider the piping system shown in the accompanying figure. The speed of water flowing through the 4 in. diameter section of the piping system is 3 ft/s. What is the volume flow rate of water in the piping system? Express the volume flow rate in ft^3/s, gpm, and L/s. For the case of steady flow of water through the piping system, what is the speed of water in the 3 in. diameter section of the system?

4 in. → 3 ft/s   →   3 in.

Problem 8.26

8.27 Consider the duct system shown in the accompanying figure. Air flows through two 8 in. by 10 in. ducts that merge into a 18 in. by 14 in. duct. The average speed of the air in each of the 8×10 ducts is 30 ft/s. What is the volume flow rate of air in the 18×14 duct? Express the volume flow rate in ft^3/s, ft^3/min, and m^3/s. What is the average speed of air in the 18×14 duct?

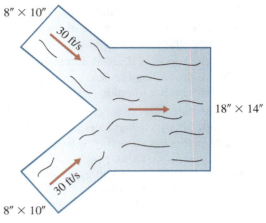

8″ × 10″

30 ft/s

18″ × 14″

8″ × 10″

30 ft/s

Problem 8.27

8.28 A car starts from rest and accelerates to a speed of 60 mph in 20 seconds. The acceleration during this period is constant. For the next 20 minutes the car moves with the constant speed of 60 mph. At this time the driver of the car applies the brake and the car decelerates to a full stop in 10 seconds. The variation of the speed of the car with time is shown in the accompanying diagram. Determine the total distance traveled by the car and the

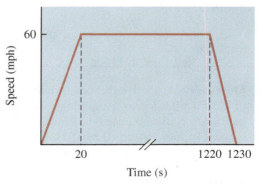

Problem 8.28

average speed of the car over this distance. Also plot the acceleration of the car as a function of time.

8.29 The drum of a clothing dryer is turning at a rate of 1 revolution every second when you suddenly open the door of the dryer. You noticed that it took 2 seconds for the drum to completely stop. Determine the deceleration of the drum. State your assumptions.

8.30 A drill bit is turning at a rate of 1200 revolutions per minute when you suddenly stop it by turning off the power. If the deceleration of the bite is 40 rad/s², how long would it take for the bit to completely stop?

8.31 On a gusty windy day, the blades of a wind turbine are turning at a rate of 200 revolutions per minute when suddenly the brakes are applied to stop the turbine to avoid failure. If the brakes cause a deceleration of 2 rad/s², how long would it take for the blades of the wind turbine to come to rest?

8.32 A plugged dishwasher sink with the dimensions of 2 ft × 1.5 ft × 1 ft is being filled with water from a faucet with an inner diameter of 1 in. If it takes 40 seconds to fill the sink to its rim, estimate the volumetric flow of water coming out of the faucet. What is the average velocity of water coming out of the faucet?

8.33 Imagine the plug in the sink described in Problem 8.32 leaks. If it now takes 45 seconds to fill the sink to its rim, estimate the volumetric flow rate of the leak.

8.34 A rectangular duct with dimensions of 12 in × 14 in delivers conditioned air to a room at a rate of 1000 ft³/min. What should be the size of a circular duct if the average air velocity inside the duct is to remain the same?

8.35 The tank shown in the accompanying figure is being filled by pipes 1 and 2. If the water level is to remain constant, what is the volumetric flow rate of water leaving the tank at 3? What is the average velocity of the water leaving the tank?

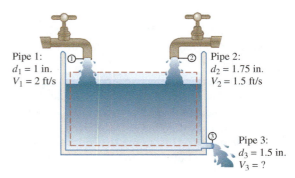

Pipe 1:
$d_1 = 1$ in.
$V_1 = 2$ ft/s

Pipe 2:
$d_2 = 1.75$ in.
$V_2 = 1.5$ ft/s

Pipe 3:
$d_3 = 1.5$ in.
$V_3 = ?$

Problem 8.35

8.36 Imagine that the water level in Problem 8.35 rises at a rate of 0.1 in/s. Knowing the diameter of the tank is 6 in., what is the average velocity of the water leaving the tank?

8.37 If it takes a bicyclist 25 seconds to reach a speed of 10 mph from rest, what is her acceleration? For the next 10 minutes, she moves at the constant speed of 10 mph, and at this time, she applies her brakes, and the bicycle decelerates to a full stop in 5 seconds. What is the total distance travelled by the bicyclist? Determine the average speed of the cyclist during the first 25 seconds, 620 seconds, and 625 seconds.

8.38 An object is dropped from the roof of a high-rise building at a distance of 450 ft. Prepare a table similar to Table 8.3 showing the speed and acceleration of the object and the distance travelled by the object as a function of time.

8.39 Solve Problem 8.38 for a situation in which the object is given an initial vertical upward velocity of 4 ft/s. Again, prepare a table similar to Table 8.3 showing the speed and acceleration of the object and the distance travelled by the object as a function of time.

8.40 Determine the natural frequency of the system given in Example 8.1 if its mass is doubled.

8.41 The period of oscillation for a pendulum on Earth is 2 seconds. If the given pendulum oscillates with a period of 4.9 seconds on the surface of the Moon, what is the acceleration due to gravity on the Moon's surface? Express your answer in both SI and U.S. Customary units.

8.42 What is the period of oscillation of the pendulum given in Problem 8.41 on Mars's surface? Given $g_{Earth} = 9.81$ m/s^2 and $g_{Mars} = 3.70$ m/s^2.

8.43 The power to an electric motor running at a constant speed of 1600 rpm is suddenly turned off. It takes 10 seconds for the motor to come to rest. What is the deceleration of the motor? How many turns does the motor make before it stops? State your assumptions.

8.44 The 2013 World Record for the 100 m sprint is 9.75 seconds and belongs to an American runner named Tyson Gay. Assuming constant acceleration, determine the speed of Mr. Gay at distances of 10 m, 20 m, 30 m, ..., 80 m, 90 m, and 100 m.

8.45 The 2013 women's pole vault World Record of 4.89 m belongs to Yelena Isinbayeva of Russia. If the pole vault mat has the dimensions of 6.0 m $\times 8.0$ m $\times 0.8$ m, what is the vertical speed of the vaulter right before she strikes the mat?

8.46 A typical household of four people consume approximately 80 gallons of water per day. Express the annual consumption rate of a city with a population of 100,000 people. Express your answer in gallons per year, ft^3 per year, liters per year, and m^3 per year.

8.47 The American Society of Heating, Ventilating, and Air Conditioning Engineers (ASHRAE) sets the standard for outdoor air requirement for ventilation purposes. For example, for a gymnasium, an outdoor air requirement of 20 cfm (cubic feet per minute) per person is required. What is the total volume of the outdoor air that must be introduced into a gym during a 12-hour period if on average 30 people are using the gym every minute? Express your answer in ft^3, liters, and m^3.

8.48 Lake Mead near the Hoover Dam, which is the largest man-made lake in the United States, contains 28,537,000 acre-foot of water (an acre-foot is the amount of water required to cover 1 acre to a depth of 1 foot). Express this water volume in gallons and m^3.

8.49 Within the next 5 to 10 years, wind turbines with rotor diameters of 180 m are anticipated to be developed and installed in Europe. If the blades of such turbines turn at a rate of 5 revolutions per minute, what is the speed of a point located at a tip of a blade? Express your answer in ft/s, m/s, km/h, and mph.

8.50 The design fluid (typically water and antifreeze) flow rate through a solar hot-water heater system is 0.02 gpm/ft^2. If a system runs continuously for 3 hours and makes use of two solar panels (each 4 ft $\times$ 8 ft), what is the total volume of the fluid that goes through the collector during this period? Express your answer in gallons, ft^3, liters, and m^3.

"If you study to remember, you will forget, but, if you study to understand, you will remember." —Unknown

Mass and Mass-Related Variables in Engineering

Ivan Chudakov/Shutterstock.com

A kayaker maneuvers his kayak in a whitewater slalom event. Mass is another important fundamental dimension that plays an important role in engineering analysis and design. Mass provides a measure of resistance to motion. Knowledge of mass is important in determining the momentum of moving objects.

LEARNING OBJECTIVES

LO[1] **Mass as Fundamental Dimension:** explain what is meant by mass, give examples of its units, and describe its important roles in engineering analysis and applications

LO[2] **Density, Specific Weight, Specific Gravity, and Specific Volume:** describe how these terms are used to show how light or heavy materials are

LO[3] **Mass Flow Rate:** explain what is meant by mass flow rate and how it is related to volume flow rate

LO[4] **Mass Moment of Inertia:** describe what is meant by mass moment of inertia and its role in rotational motion

LO[5] **Momentum:** explain what is meant by momentum

LO[6] **Conservation of Mass:** describe the conservation of mass for an engineering situation

BONE MASS MEASUREMENTS

Osteoporosis or "porous bone" is a disease of the skeletal system characterized by low bone mass and deterioration of bone tissue. Osteoporosis produces an enlargement of the pore spaces in the bone, causing increased fragility and an increased risk for fracture, typically in the wrist, hip, and spine. An estimated 10 million Americans have osteoporosis, and more than 34 million Americans have low bone mass, placing them at increased risk for osteoporosis. One out of every two women and one in four men aged 50 and older will have an osteoporosis-related fracture in their lifetime. The good news is that osteoporosis is a preventable and treatable disease. Early diagnosis and treatment can reduce or prevent fractures.

Bone density is usually studied using diagnostic bone-mass measurement techniques recognized by the FDA. Bone density can be measured at the wrist, spine, hip, or calcaneus (heel). Single and combined measurements may be required to diagnose bone disease, monitor bone changes with disease progression, or monitor bone changes with therapy.

Bone densitometry is a radiological test that can diagnose osteoporosis early enough for it to be treated. Bone densitometry is an outpatient radiological test. Most machines use x-rays, but some use ultrasound. A densitometry machine calculates the density of bone and creates a chart that compares the patient's density to what the density should be.

Department of Health and Human Services

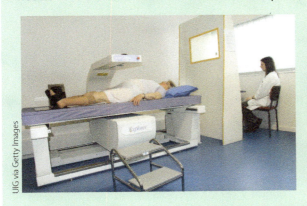

UIG via Getty Images

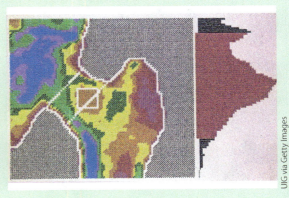

UIG via Getty Images

To the students: What do we mean by mass? What is the purpose of defining density? Give examples of situations from everyday life where mass plays an important role.

In Chapters 7 and 8, we explained the role of length, time, and length- and time-dependent quantities such as area, volume, speed, and volume flow rate in engineering analysis and design. The objective of this chapter is to introduce the concept of mass and mass-related quantities encountered in engineering. We will begin by discussing the building blocks of all matter, atoms and molecules. We will then introduce the concept of mass in terms of a quantitative measure of the amount of atoms possessed by a substance. We will then define and discuss other mass-dependent engineering quantities, such as density, specific gravity, mass moment of inertia, momentum, and mass flow rate. In this chapter, we will also consider conservation of mass and its application in engineering. Table 6.7 is repeated here to remind you of the role of fundamental dimensions and how they are combined to define mass-dependent variables such as moment of inertia, momentum, and mass flow rate.

TABLE 6.7   **Fundamental Dimensions and How They Are Used in Defining Variables that Are Used in Engineering Analysis and Design**

Chapter	Fundamental Dimension	Related Engineering Variables			
7	Length (L)	Radian (L/L), Strain (L/L)	Area (L^3)	Volume (L^3)	Area moment of inertia (L^4)
8	Time (t)	Angular speed ($1/t$), Angular acceleration ($1/t^2$), Linear speed (L/t), Linear acceleration (L/t^2)		Volume flow rate (L^3/t)	
9	Mass (M)	Mass flow rate (M/t), Momentum (ML/t), Kinetic energy (ML^2/t^2)		Density (M/L^3), Specific volume (L^3/M)	
10	Force (F)	Moment (LF), Work, energy (FL), Linear impulse (Ft), Power (FL/t)	Pressure (F/L^2), Stress (F/L^2), Modulus of elasticity (F/L^2), Modulus of rigidity (F/L^2)	Specific weight, (F/L^3),	
11	Temperature (T)	Linear thermal expansion (L/LT), Specific heat (FL/MT)		Volume thermal expansion (L^3/L^3T)	
12	Electric Current (I)	Charge (It)	Current density (I/L^2)		

LO¹ 9.1   Mass as a Fundamental Dimension

As we discussed in Chapter 6, from their day-to-day observations humans noticed that some things were heavier than others and thus recognized the need for a physical quantity to describe that observation. Early humans did not fully understand the concept of gravity; consequently, the correct distinction between mass and weight was made later. Let us now look more carefully at mass as a physical variable. Consider the following. When you look around at your surroundings, you will find that matter exists in various forms and shapes. You will also notice that matter can change shape when its condition or its surrounding conditions are changed. All objects and living things are made of matter, and matter itself is made of atoms, or chemical elements. There are 106 known chemical elements to date. Atoms of similar characteristics are grouped together in a table, which is called the *periodic table of chemical elements.* An example of the chemical periodic table is shown in Figure 9.1.

Atoms are made up of even smaller particles we call *electrons, protons,* and *neutrons.* In your first chemistry class you will study these ideas in more detail, if you have not yet done so. Some of you may decide to study chemical engineering, in which case you will spend much more time studying chemistry. But for now, remember that atoms are the basic building blocks of all matter.

Atoms are combined naturally, or in a laboratory setting, to create molecules. For example, as you already know, water molecules are made of two atoms of hydrogen and one atom of oxygen. A glass of water is made up of billions and billions of homogeneous water molecules. Molecules are the smallest portion of a given matter that still possesses its characteristic properties. Matter can exist in four states, depending on its own and the surrounding conditions: solid, liquid, gaseous, or plasma. Let us consider the water that we drink every day. As you already know, under certain conditions, water exists in a solid form that we call ice. At a standard atmospheric pressure, water exists in a solid form as long as its temperature is kept under $0°$ C. Under standard atmospheric pressure, if you were to heat the ice and consequently change its temperature, the ice would melt and change into a liquid form. Under standard pressure at sea level, the water remains liquid up to a temperature of $100°$ C as you continue heating the water. If you were to carry out this experiment further by adding more heat to the liquid water, eventually the liquid water changes its phase from a liquid into a gas. This phase of water we commonly refer to as steam. If you had the means to heat the water to even higher temperatures, temperatures exceeding $2000°$ C, you would find that you can break up the water molecules into their atoms, and eventually the atoms break up into free electrons and nuclei that we call **plasma**.

Well, what does all this have to do with mass? Mass provides a quantitative measure of how many molecules or atoms are in a given object. The matter may change its phase, but its mass remains constant. Some of you will take a class in dynamics where you will learn that on a macroscopic scale mass also serves as a measure of resistance to motion. You already know this from your daily observations. Which is harder to push, a motorcycle or a truck? As you know, it takes more effort to push a truck. When you want to rotate something, the distribution of the mass about the center of rotation also plays a significant role. The further away the mass is located from the center of rotation, the harder it will be to rotate the mass about that axis. A measure of how hard it is to rotate something with respect to center of rotation is called *mass moment of inertia.* We will discuss this in detail in Section 9.4.

IA																		VIIIA
1 **H** 1.0079	IIA											IIIA	IVA	VA	VIA	VIIA		2 **He** 4.003
3 **Li** 6.941	4 **Be** 9.012											5 **B** 10.811	6 **C** 12.011	7 **N** 14.007	8 **O** 15.999	9 **F** 18.998		10 **Ne** 20.180
11 **Na** 22.990	12 **Mg** 24.305	IIIB	IVB	VB	VIB	VIIB		VIIIB		IB	IIB	13 **Al** 26.982	14 **Si** 28.086	15 **P** 30.974	16 **S** 32.066	17 **Cl** 35.453		18 **Ar** 39.948
19 **K** 39.098	20 **Ca** 40.078	21 **Sc** 44.956	22 **Ti** 47.88	23 **V** 50.942	24 **Cr** 51.996	25 **Mn** 54.938	26 **Fe** 55.845	27 **Co** 58.933	28 **Ni** 58.69	29 **Cu** 63.546	30 **Zn** 65.39	31 **Ga** 69.723	32 **Ge** 72.61	33 **As** 74.922	34 **Se** 78.96	35 **Br** 79.904		36 **Kr** 83.8
37 **Rb** 85.468	38 **Sr** 87.62	39 **Y** 88.906	40 **Zr** 91.224	41 **Nb** 92.906	42 **Mo** 95.94	43 **Tc** 98	44 **Ru** 101.07	45 **Rh** 102.906	46 **Pd** 106.42	47 **Ag** 107.868	48 **Cd** 112.411	49 **In** 114.82	50 **Sn** 118.71	51 **Sb** 121.76	52 **Te** 127.60	53 **I** 126.905		54 **Xe** 131.29
55 **Cs** 132.905	56 **Ba** 137.327	57 **La** 138.906	72 **Hf** 178.49	73 **Ta** 180.948	74 **W** 183.84	75 **Re** 186.207	76 **Os** 190.23	77 **Ir** 192.22	78 **Pt** 195.08	79 **Au** 196.967	80 **Hg** 200.59	81 **Tl** 204.383	82 **Pb** 207.2	83 **Bi** 208.980	84 **Po** 209	85 **At** 210		86 **Rn** 222
87 **Fr** 223	88 **Ra** 226.025	89 **Ac** 227.028	104 **Rf** 261	105 **Db** 262	106 **Sg** 263	107 **Bh** 262	108 **Hs** 265	109 **Mt** 266	110 **Uun** 269	111 **Uuu** 272	112 **Uub** 277		114		116			118

Lanthanide series	58 **Ce** 140.115	59 **Pr** 140.908	60 **Nd** 144.24	61 **Pm** 145	62 **Sm** 150.36	63 **Eu** 151.964	64 **Gd** 157.25	65 **Tb** 158.925	66 **Dy** 162.5	67 **Ho** 164.93	68 **Er** 167.26	69 **Tm** 168.934	70 **Yb** 173.04	71 **Lu** 174.967
Actinide series	90 **Th** 232.038	91 **Pa** 231.036	92 **U** 238.029	93 **Np** 237.048	94 **Pu** 244	95 **Am** 243	96 **Cm** 247	97 **Bk** 247	98 **Cf** 251	99 **Es** 252	100 **Fm** 257	101 **Md** 258	102 **No** 259	103 **Lr** 262

FIGURE 9.1   The chemical elements to date (2014).

The other mass-related parameter that we will investigate in this chapter is momentum. Consider two objects with differing masses moving with the same velocity. Which object is harder to bring to a stop, the one with the small mass or the one with the larger mass? Again, you already know the answer to this question. The object with the bigger mass is harder to stop.

You see that in the game of football, too. If two players of differing mass are running at the same speed, it is harder to bring down the bigger player. These observations lead to the concept of momentum, which we will explain in Section 9.5.

Mass also plays an important role in storing thermal energy. The more massive something is, the more thermal energy you can store within it. Some materials are better than others at storing thermal energy. For example, water is better at storing thermal energy than air is. In fact, the idea of storing thermal energy within a massive medium is fully utilized in the design of passive solar houses. There are massive brick or concrete floors in the sunrooms. Some people even place big barrels of water in the sunrooms to absorb the available daily solar radiation and store the thermal energy in the water to be used overnight. We will discuss heat capacities of materials in more detail in Chapter 11.

Measurement of Mass

The *kilogram* is the unit of mass in SI; it is equal to the mass of the international prototype of the kilogram. As we explained in Chapter 6, in the British Gravitational system of units, the unit of mass is slug, and in U.S. Customary system of units, the unit of mass is lbm (1 kg = 0.0685 slugs = 2.205 lbm and 1 slug = 32.2 lbm). In practice, the mass of an object is measured indirectly by using how much something weighs. The weight of an object on earth is the force that is exerted on the mass due to the gravitational pull of the earth. You are familiar with spring scales that measure the weight of goods at a supermarket or bathroom scales at home. Force due to gravity acting on the unknown mass will make the spring stretch or compress. By measuring the deflection of the spring, one can determine the weight and consequently the mass of the object that created that deflection. Some weight scales use force transducers consisting of a metallic member that behaves like a spring, except the deflection of the metal member is measured electronically using strain gauges. You should understand the difference between

Mass serves as a measure of resistance to translational motion. When you want to rotate something, the distribution of the mass about the center of rotation also plays a significant role. The further away the mass is located from the center of rotation, the harder it will be to rotate the mass about that axis.

weight and mass and be careful how you use them in engineering analysis. We will discuss the concept of weight in more detail in Chapter 10.

LO² 9.2   Density, Specific Weight, Specific Gravity, and Specific Volume

In engineering practice, to represent how light or how heavy materials are we often define properties that are based on a unit volume; in other words, how massive something is per unit volume. Given 1 cubic foot of wood and 1 cubic foot of steel, which one has more mass? The steel of course! The **density** of any substance is defined as the ratio of the mass to the volume that it occupies, according to

$$\text{density} = \frac{\text{mass}}{\text{volume}} \qquad \boxed{9.1}$$

Density provides a measure of how compact the material is for a given volume. Materials such as mercury or gold with relatively high values of density have more mass per 1ft^3 volume or 1m^3 volume than those with lower density values, such as water. It is important to note that the density of matter changes with temperature and could also change with pressure. The SI unit for density is kg/m^3, and in BG and U.S. Customary system the density is expressed in slugs/ft^3, and lbm/ft^3, respectively.

Specific weight is another way to measure how truly heavy or light a material is for a given volume. Specific weight is defined as the ratio of the weight of the material to the volume that it occupies, according to

$$\text{specific weight} = \frac{\text{weight}}{\text{volume}} \qquad \boxed{9.2}$$

Again, the weight of something on earth is the force that is exerted on its mass due to gravity. We will discuss the concept of weight in more detail in Chapter 10. In this chapter, it is left as an exercise (see Problem 9.4) for you to show the relationship between density and specific weight:

$$\text{specific weight} = (\text{density})(\text{acceleration due to gravity})$$

Another common way to represent the heaviness or lightness of some material is by comparing its density to the density of water. This comparison is called the **specific gravity** of a material and is formally defined by

$$\text{specific gravity} = \frac{\text{density of a material}}{\text{density of water @ 4° C}} \qquad \boxed{9.3}$$

> In engineering practice, to represent how light or how heavy materials are, we define properties that are based on a unit volume. Examples include density and specific weight.

It is important to note that specific gravity is unitless because it is the ratio of the value of two densities. Therefore, it does not matter which system of units is used to compute the specific gravity of a substance, as long as consistent units are used.

Specific volume, which is the inverse of density, is defined by

$$\text{specific volume} = \frac{\text{volume}}{\text{mass}}$$

9.4

Specific volume is commonly used in the study of thermodynamics. The SI unit for specific volume is m^3/kg.

The density, specific gravity, and the specific weight of some materials are shown in Table 9.1.

TABLE 9.1   Density, Specific Gravity, and Specific Weight of Some Materials (at room temperature or at the specified temperature)

Material	Density (kg/m³)	Specific Gravity	Specific Weight (N/m³)
Aluminum	2740	2.74	26,880
Asphalt	2110	2.11	20,700
Cement	1920	1.92	18,840
Clay	1000	1.00	9810
Fireclay brick	1790@100° C	1.79	17,560
Glass (soda lime)	2470	2.47	24,230
Glass (lead)	4280	4.28	41,990
Glass (Pyrex)	2230	2.23	21,880
Iron (cast)	7210	7.21	70,730
Iron (wrought)	7700@100° C	7.70	75,540
Paper	930	0.93	9120
Steel (mild)	7830	7.83	76,810
Steel (stainless 304)	7860	7.86	77,110
Wood (ash)	690	0.69	6770
Wood (mahogany)	550	0.55	5400
Wood (oak)	750	0.75	7360
Wood (pine)	430	0.43	4220
Fluids			
Standard air	1.225	0.0012	12
Gasoline	720	0.72	7060
Glycerin	1260	1.26	12,360
Mercury	13,550	13.55	132,930
SAE 10 W oil	920	0.92	9030
Water	1000@4° C	1.0	9810

LO³ 9.3   Mass Flow Rate

In Chapter 8, we discussed the significance of volume flow rate. The mass flow rate is another closely related parameter that plays an important role in many engineering applications. There are many industrial processes that depend on the measurement of fluid flow. Mass flow rate tells engineers how much material is being used or moved over a period of time so that they can replenish the supply of the material. Engineers use flowmeters to measure volume or mass flow rates of water, oil, gas, chemical fluids, and food products. The design of any flow measuring device is based on some known engineering principles. Mass and volume flow measurements are also necessary and common in our everyday life. For example, when you go to a gasoline service station, you need to know how many gallons of gas are pumped into the tank of your car. Another example is measuring the amount of domestic water used or consumed. There are over 100 types of commercially available flowmeters to measure mass or volume flow rates. The selection of a proper flowmeter will depend on the application type and other variables, such as accuracy, cost, range, ease of use, service life, and type of fluid: gas or liquid, dirty or slurry or corrosive, for example.

The **mass flow rate** is simply defined by the amount of mass that flows through something per unit of time.

$$\text{mass flow rate} = \frac{\text{mass}}{\text{time}} \tag{9.5}$$

Some of the more common units for mass flow rate include kg/s, kg/min, kg/h or slugs/s or lbm /s. How would you measure the mass flow rate of water coming out of a faucet or a drinking fountain? Place a cup under a drinking fountain and measure the time that it takes to fill the cup. Also, measure the total mass of the cup and the water and then subtract the mass of the cup from the total to obtain the mass of the water. Divide the mass of the water by the time interval it took to fill the cup.

Mass and volume flow measurements are also necessary and common in our everyday life. For example, these determine how much water or natural gas is consumed by a home, a city, or a country over a period of month or a year.

We can relate the volume flow rate of something to its mass flow rate provided that we know the density of the flowing fluid or flowing material. The relationship between the mass flow rate and the volume flow rate is given by

$$\text{mass flow rate} = \frac{\text{mass}}{\text{time}} = \frac{(\text{density})(\text{volume})}{\text{time}} = (\text{density})\left(\frac{\text{volume}}{\text{time}}\right) \tag{9.6}$$

$$= (\text{density})(\text{volume flow rate})$$

The mass flow rate calculation is also important in excavation or tunnel-digging projects in determining how much soil can be removed in one day or one week, taking into consideration the parameter of the digging and transport machines.

Answer the following questions to test your understanding of the preceding sections.

1. Give examples of the roles that mass play in engineering applications.

2. Describe properties that represent how light or how heavy materials are per unit volume.

3. What is the relationship between mass flow rate and volume flow rate?

Vocabulary—State the meaning of the following terms:

Specific Weight _____

Specific Gravity _____

Mass Flow Rate _____

LO⁴ 9.4   Mass Moment of Inertia

As mentioned earlier in this chapter, when it comes to the rotation of objects, the distribution of mass about the center of rotation plays a significant role. The farther away the mass is located from the center of rotation, the harder it is to rotate the mass about the given center of rotation. A measure of how hard it is to rotate something with respect to center of rotation is called **mass moment of inertia**. All of you will take a class in physics, and some of you may even take a dynamics class, where you will learn in more depth about the formal definition and formulation of mass moment of inertia. But for now, let us consider the following simple situation as shown in Figure 9.2. For a single mass particle m, located at a distance r from the axis of rotation z–z, the mass moment of inertia is defined by

$$I_{z-z} = r^2\, m \qquad \boxed{9.7}$$

Now let us expand this problem to include a system of mass particles, as shown in Figure 9.3. The mass moment of inertia for the system of masses shown about the z–z axis is now

$$I_{z-z} = r_1^2\, m_1 + r_2^2\, m_2 + r_3^2\, m_3 \qquad \boxed{9.8}$$

Similarly, we can obtain the mass moment of inertia for a body, such as a wheel or a shaft, by summing the mass moment of inertia of each mass particle that makes up the body. As you take calculus classes, you will learn that you can use integrals instead of summations to evaluate the mass moment of inertia of continuous objects. After all, the integral sign $\int$ is nothing but a big "S" sign, indicating summation.

$$I_{z-z} = \int r^2\, dm \qquad \boxed{9.9}$$

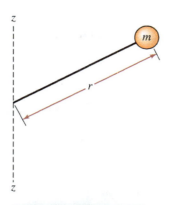

FIGURE 9.2   The mass moment of inertia of a point mass.

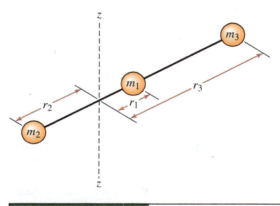

FIGURE 9.3   Mass moment of inertia of a system consisting of three point masses.

> A measure of how hard it is to rotate something with respect to center of rotation is called mass moment of inertia.

The mass moment of inertia of objects with various shapes can be determined from Equation (9.9). You will be able to perform this integration in another semester or two. Examples of mass moment of inertia formulas for some typical bodies, such as a cylinder, disk, sphere, and a thin rectangular plate, are given on the following page.

Disk

$$I_{z-z} = \frac{1}{2}mR^2 \qquad \boxed{9.10}$$

Circular Cylinder

$$I_{z-z} = \frac{1}{2}mR^2$$

9.11

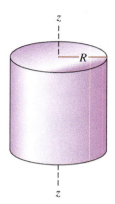

Sphere

$$I_{z-z} = \frac{2}{5}mR^2$$

9.12

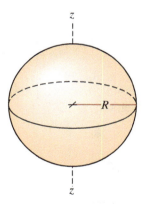

Thin Rectangular Plate

$$I_{z-z} = \frac{1}{12}mW^2$$

9.13

EXAMPLE 9.1

Determine the mass moment inertia of a steel shaft that is 2 m long and has a diameter (d) of 10 cm. The density of steel is 7860 kg/m³.

We will first calculate the mass of the shaft using Equation (9.1). The volume of a cylinder is given by

$$\text{volume} = (\pi/4)(d^2)(\text{length}) = (\pi/4)(0.1\ \text{m})^2(2\ \text{m}) = 0.01571\ \text{m}^3$$

$$\text{density} = \frac{\text{mass}}{\text{volume}}$$

$$7860\ \text{kg/m}^3 = \frac{\text{mass}}{0.01571\ \text{m}^3}$$

$$\text{mass} = 123.5\ \text{kg}$$

To calculate the mass moment of inertia of the shaft about its longitudinal axis, we use Equation (9.11).

$$I_{z-z} = \frac{1}{2}mR^2 = \frac{1}{2}(123.5\ \text{kg})(0.05\ \text{m})^2 = 0.154\ \text{kg}\cdot\text{m}^2$$

LO⁵ 9.5 Momentum

Mass also plays an important role in problems dealing with moving objects. **Momentum** is a physical variable that is defined as the product of mass and velocity.

$$\vec{L} = m\vec{V}$$

9.14

In Equation (9.14), $\vec{L}$ represents momentum vector, m is mass, and $\vec{V}$ is the velocity vector. Because the velocity of the moving object has a direction, we associate a direction with momentum as well. The momentum's direction is the same as the direction of the velocity vector or the moving object. So a 1000-kg car moving north at a rate of 20 m/s has a momentum with a magnitude of 20,000 kg · m/s in the north direction. Momentum is one of those physical concepts that are commonly abused by sports broadcasters when discussing sporting events. During a timeout period when athletes are standing still and listening to their coaches, the broadcaster may say, "Bob, clearly the momentum has shifted, and team B has more momentum now going into the fourth quarter." Using the preceding definition of momentum, now you know that relative to the earth, the momentum of an object or a person at rest is zero!

Because the magnitude of linear momentum is simply mass times velocity, something with relatively small mass could have a large momentum value, depending on its velocity. For example, a bullet with a relatively small mass shot out of a gun can do lots of harm and penetrate a surface because of its high velocity. The magnitude of the momentum associated with the bullet could be relatively large. You have seen stunt actors falling from the top of tall buildings. As the stunt performer approaches the ground, he

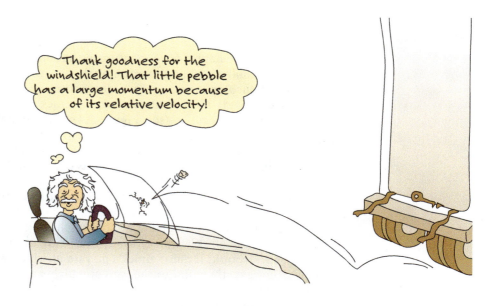

Thank goodness for the windshield! That little pebble has a large momentum because of its relative velocity!

For a moving object, momentum is defined as the product of mass and velocity. Momentum is a vector quantity: a quantity with both magnitude and direction. The momentum's direction is the same as the direction of the velocity vector.

or she has a relatively large momentum, so how are injuries avoided? Of course, the stunt performer falls onto an air bag and some soft materials that increase the time of contact to reduce the forces that act on his or her body. We will discuss the relationship between linear momentum and linear impulse (force acting over time) in Chapter 10 after we discuss the concept of force.

EXAMPLE 9.2

Determine the linear momentum of a person whose mass is 80 kg and who is running at a rate of 3 m/s. Compare it to the momentum of a car that has a mass of 2000 kg and is moving at a rate of 30 m/s in the same direction as the person.

We will use Equation (9.14) to answer the questions.

$$\text{Person:}\quad \vec{L} = m\vec{V} = (80 \text{ kg})(3 \text{ m/s}) = 240 \text{ kg·m/s}$$

$$\text{Car:}\quad \vec{L} = m\vec{V} = (2000 \text{ kg})(30 \text{ m/s}) = 60{,}000 \text{ kg·m/s}$$

Kinetic energy is another quantity that is mass dependent and is used in engineering analysis and design. An object having a mass m and moving with a speed V has a kinetic energy that is equal to $\frac{1}{2}mV^2$. In Chapter 13, we will explain the concept of kinetic energy in more detail after we define mechanical work.

LO⁶ 9.6   Conservation of Mass

Recall that in Chapter 6 we mentioned that engineers are good bookkeepers. In the analysis of engineering work, we need to keep track of physical quantities such as mass, energy, momentum, and so on. Let us now look at how engineers may go about keeping track of mass and the associated bookkeeping procedure (see Figure 9.4). Simply stated, the **conservation of mass** says that we cannot create or destroy mass. Consider the following example. You are taking a shower in your bathtub. You turn on the water and begin to wash yourself. Let us focus our attention on the tub. With the drain open and clear of any hair and dead skin tissue, the rate at which water comes to the tub from the showerhead is equal to the rate at which water leaves the tub. This is a statement of conservation of mass applied to the water inside the tub. Now what happens if the drain becomes plugged up by hair or dead skin tissue? Many of us have experienced this at one time or another. Liquid Drāno® time! The rate at which water comes to the tub now is not exactly equal to the rate at which water leaves the tub, which is why the water level in the tub begins to rise. How would you use the conservation of mass principle to describe this situation? You can express it this way: The rate at which water comes to the tub is equal to the rate at which the water leaves the tub plus the time rate of accumulation of the mass of water within the tub.

What happens if you were taking a bath and you had the tub filled with water? Let us say that after you are done taking the bath you open the drain, and the water begins to leave. Being the impatient person that you are, you slowly turn on the shower as the tub is being drained. Now you notice that the water level is going down but at a slow rate. The rate the water is coming into the tub minus the rate at which water is leaving the tub is equal to the rate of water depletion (reduction) within the tub.

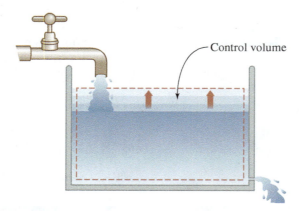

| FIGURE 9.4 | The rate at which water enters the container minus the rate at which water leaves the container should be equal to the rate of accumulation or depletion of the mass of water within the container. |

Let us now turn our attention to an engineering presentation of conservation of mass. In engineering we refer to the tub as a control volume, because we focus our attention on a specific object occupying a certain volume in space. The control volume could represent the flow boundaries in a pump, or a section of a pipe, or the inside volume of a tank, the flow passage in a compressor or water heater, or the boundaries of a river. We can use the bathtub example to formulate a general statement for conservation of mass, which states: The rate at which a fluid enters a control volume minus the rate at which the fluid leaves the control volume should be equal to the rate of accumulation or depletion of the mass of fluid within the given control volume. There is also a broad area in mathematics, operations research, engineering management, and traffic called *queuing*. It is a study of "queues" of people waiting in service lines, products waiting in assembly lines, cars waiting in lanes, or digital information waiting to move through computer networks. What happens during busy hours at banks, gas stations, or supermarket checkout counters? The lines formed by people or cars grow longer. The flow into a line is not equal to the flow out of the line, and thus the line gets longer. You can think of this in terms of conservation of mass. During busy hours, the rate at which people enter a line is not equal to the rate at which people leave the line, and thus there is a rate of accumulation at the line. This analogy is meant to make you think about other issues closely related to the flow of things in everyday life.

> Conservation of mass for a control volume states that the rate at which a fluid enters a control volume minus the rate at which the fluid leaves the control volume should be equal to the rate of accumulation or depletion of the mass of fluid within the given control volume.

The rate at which water comes into the tub is equal to the rate at which water leaves the tub plus the time rate of accumulation of the mass of water within the tub.

I think I'll Invite my class tomorrow to demonstrate the conservation of mass!

EXAMPLE 9.3

How much water is stored after 5 minutes in each of the tanks shown in Figure 9.5? How long will it take to fill the tanks completely provided that the volume of each tank is 12 m³? Assume the density of water is 1000 kg/ m³.

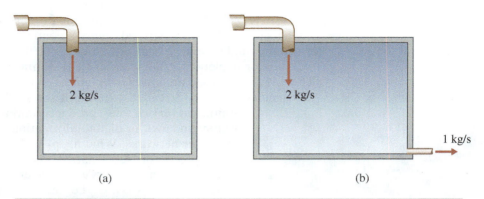

(a)   (b)

FIGURE 9.5   The tanks of Example 9.3.

We will use the conservation of mass statement to solve this problem.

$$\left(\begin{array}{l}\text{the rate at which}\\ \text{fluid enters a control}\\ \text{volume}\end{array}\right) - \left(\begin{array}{l}\text{the rate at which}\\ \text{the fluid leaves the}\\ \text{control volume}\end{array}\right) = \left(\begin{array}{l}\text{the rate of accumulation}\\ \text{or depletion of the mass}\\ \text{of fluid within the given}\\ \text{control volume}\end{array}\right)$$

Realizing that no water leaves tank (a) we have

$$(2 \text{ kg/s}) - (0) = \frac{\text{changes of mass inside the control volume}}{\text{change in time}}$$

After 5 minutes,

$$\text{change of mass inside the control volume} = (2 \text{ kg/s})(5 \text{ min}) \left(\frac{60 \text{ s}}{1 \text{ min}}\right) = 600 \text{ kg}$$

To determine how long it will take to fill the tank, first we will make use of the relationship among mass, density, and volume—mass = (density) (volume)—to compute how much mass each tank can hold.

$$\text{mass} = (1000 \text{ kg/m}^3)(12 \text{ m}^3) = 12,000 \text{ kg}$$

By rearranging terms in the conservation of mass equation, we can now solve for the time that is required to fill the tank.

$$\text{time required to fill the tank} = \frac{12,000 \text{ kg}}{(2 \text{ kg/s})} = (6000 \text{ s}) = 100 \text{ min}$$

Water enters tank (b) at 2 kg/s and leaves the tank at 1 kg/s. Applying conservation of mass to tank (b), we have

$$(2 \text{ kg/s}) - (1 \text{ kg/s}) = \frac{\text{changes of mass inside the control volume}}{\text{change in time}}$$

After 5 minutes,

$$\text{change of mass inside the control volume} = \left[(2\ \text{kg/s}) - (1\ \text{kg/s})\right](5\ \text{min})\left(\frac{60\ \text{s}}{1\ \text{min}}\right)$$

$$= 300\ \text{kg}$$

and the time that is required to fill the tank (b)

$$\text{time required to fill the tank} = \frac{12000\ \text{kg}}{\left[(2\ \text{kg/s}) - (1\ \text{kg/s})\right]} = 12000\ \text{s} = 200\ \text{min}$$

EXAMPLE 9.4

We are interested in determining the mass-flow rate of fuel from the gasoline tank of a small car to its fuel injection system. The gasoline consumption of the car is 15 kilometers per liter when the car is moving at the speed of 90 km/h. The specific gravity of gasoline is 0.72. If there were one million of these cars on the road, how many kilograms of gasoline would be burned every hour?

First, we will use Equation (9.3) to compute the density of gasoline.

$$\text{SG}_{\text{gasoline}} = 0.72 = \frac{\text{density of gasoline}}{1000\ \text{kg/m}^3} \rightarrow \text{density of gasoline} = 720\ \text{kg/m}^3$$

The volume-flow rate of fuel for a single car is determined from:

$$\text{volume-flow rate} = \frac{90\ \text{km/h}}{15\ \text{km/liter}} = 6\ \text{liter/h}$$

Next, we will use Equation (9.6) to calculate the mass-flow rate of the fuel.

$$\text{mass-flow rate} = (\text{density})(\text{volume flow rate})$$

$$\text{mass-flow rate} = \left(720\ \frac{\text{kg}}{\text{m}^3}\right)\left(6\ \frac{\text{liter}}{\text{h}}\right)\left(\frac{1\ \text{m}^3}{1000\ \text{liters}}\right) = 4.32\ \text{kg/h}$$

Each car burns 4.32 kg/h, so for one million cars in each hour 4,320,000 kg of gasoline is burned!

Before You Go On

Answer the following questions to test your understanding of the preceding sections.

1. Explain what the value of mass moment of inertia represents.

2. Explain what the value of momentum represents.

3. In your own words, explain the conservation of mass principle.

Vocabulary—State the meaning of the following terms:

Mass Moment of Inertia _____

Momentum _____

SUMMARY

LO¹   Mass as Fundamental Dimension

By now, you should have a good understanding of what is meant by mass and know about the important roles of mass in engineering applications and analysis. Mass provides a quantitative measure of how many molecules or atoms are in a given object. It also provides a measure of resistance to translational motion. It also plays a significant role in storing thermal energy. The more massive something is, the more thermal energy one can store within it. The kilogram is the unit of mass in SI units, and slugs and pound-mass are used as units of mass in British Gravitational and U.S. Customary systems, respectively.

LO²   Density, Specific Weight, Specific Gravity, and Specific Volume

In engineering, to show how light or heavy materials are, we use properties such as density, specific weight, specific gravity, and specific volume. You should know the formal definition of these properties. For example, specific weight is defined as the ratio of the weight of the material by the volume that it occupies. Another common way to represent the heaviness or lightness of some materials is by comparing its density to the density of water. This comparison is called the specific gravity of a material.

LO³   Mass Flow Rate

Mass flow rate is defined by the amount of mass that flows through something per unit of time. There are many engineering applications that depend on the measurement of fluid flow. Mass flow rate tells engineers how much material or fluid, such as water or natural gas, is being used over a period of time so that they can replenish the supply of the material or the fluid. Moreover, the mass flow rate is related to volume flow rate, provided that the density of the flowing material or fluid is known. The mass flow rate is equal to the product of volume flow rate and the density of the material or fluid.

LO⁴   Mass Moment of Inertia

Whereas mass provides a measure of resistance to translational motion, the mass moment of inertia provides a measure of resistance to rotational motion. If you want to rotate something, the distribution of mass about the center of rotation also plays an important role. The further the mass from a center of rotation, the more resistance mass offers to rotational motion.

LO⁵   Momentum

By now you should know how we define momentum for a moving object. Momentum is defined as the product of mass and velocity. Because the velocity of the moving object has a direction, we associate a direction with momentum as well. The momentum's direction is the same as the direction of the velocity vector or the moving object. You should also know that something with a relatively small mass could have a relatively large momentum if it has a large velocity.

LO⁶   Conservation of Mass

In the analysis of engineering work, we often need to keep track of a physical quantity such as mass. Conservation of mass simply states that we cannot create or destroy mass. In order to keep track of mass, we define a control volume (i.e., the volume under investigation through which the mass flows). Then the rate at which a fluid enters a control volume minus the rate at which the fluid leaves the control volume should be equal to the rate of accumulation or depletion of the mass of fluid within the given control volume.

KEY TERMS

Conservation of Mass 280
Density 272
Mass 271
Mass Flow Rate 274

Mass Moment of Inertia 275
Momentum 278
Plasma 268
Specific Gravity 272

Specific Volume 273
Specific Weight 272

APPLY WHAT YOU HAVE LEARNED

We discussed buoyancy in Chapter 7. The buoyancy principle is often used to determine a person's percentage of body fat. Bone, muscle, and lean tissue have a specific gravity that is greater than 1 (heavier than water), whereas fat tissue has a specific gravity of less than 1 (lighter than water). Athletic departments at many universities have relatively large hydro-static weighting tanks that allow for determining a person's percentage of body fat from measuring the individual's weight when fully immersed under water. You are to design a similar and a simple set-up to measure the densities of lean beef and fat. You need to purchase a quarter pound of lean beef and a quarter pound of fat to test your apparatus. Write a brief report per your instructor's guidelines. Can you determine the densities of lean beef and fat using other engineering principles?

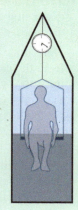

PROBLEMS

Problems that promote lifelong learning are denoted by 🔑

9.1 Look up the mass of the following objects:
   a. recent model of an automobile of your choice
   b. the earth
   c. a fully loaded Boeing 777
   d. an ant

9.2 The density of standard air is a function of temperature and may be approximated using the ideal gas law, according to

$$\rho = \frac{P}{RT}$$

   where

   ρ = density (kg/m^3)

   P = standard atmospheric pressure $(101.3\ kPa)$

   R = gas constant; its value for air is 286.9

   $$\left(\frac{J}{kg \cdot K}\right)$$

   T = air temperature in kelvin

   Create a table that shows the density of air as a function of temperature in the range of $0°\ C\ (273.15\ K)$ to $50°\ C\ (323.15\ K)$ in increments of $5\ °C$. Also, create a graph showing the value of density as a function of temperature.

9.3 Determine the specific gravity of the following materials: gold $(\rho = 1208\ lb/ft^3)$, platinum

$(\rho = 1340\ lb/ft^3)$, silver $(\rho = 654\ lb/ft^3)$, sand $(\rho = 94.6\ lb/ft^3)$, freshly fallen snow $(\rho = 31\ lb/ft^3)$, tar $(\rho = 75\ lb/ft^3)$, and hard rubber $(\rho = 74.4\ lb/ft^3)$.

9.4 Show that the specific weight and density are related according to specific weight = (density) (acceleration due to gravity).

9.5 Compute the values of momentum for the following situations:
   a. a 90-kg football player running at 6 m/s
   b. a 1500-kg car moving at a rate of 100 km/h
   c. a 200,000-kg Boeing 777 moving at a speed of 500 km/h
   d. a bullet with a mass of 15 g traveling at a speed of 500 m/s
   e. a 140-g baseball traveling at 120 km/h
   f. an 80-kg stunt performer falling off a ten-story building reaching a speed of 30 m/s

9.6 Investigate the mass flow rate of blood through your heart, and write a brief report to your instructor discussing your findings.

9.7 Investigate the mass flow rate of oil inside the Alaskan pipeline, and write a brief report to your instructor discussing your findings.

9.8 Investigate the mass flow rate of water through the Mississippi River during a normal year, and write a brief report discussing your findings. Explain flooding using the conservation of mass statement.

9.9 Determine the mass flow rate of fuel from the gasoline tank to the car's fuel injection system. Assume that the gasoline consumption of the car is 20 mpg when the car is moving at 60 mph. Use the specific gravity value of 0.72 for the gasoline.

9.10 Determine the mass moment of inertia for the following objects: a thin disk, circular cylinder, and a sphere. Refer to Equations (9.10) through (9.13) for appropriate relationships. If you were to place these objects alongside of each other on an inclined surface, which one of the objects would get to the bottom first, provided that they all have the same mass and diameter?

9.11 The use of ceiling fans to circulate air has become quite common. Suggest ways to correct the rotation of a wobbling fan using the concept of mass moment of inertia. As a starting point, you can use pocket change, such as dimes, nickels, and quarters, and chewing gum. How would you stop the fan from wobbling? Exercise caution!

9.12 Determine the mass moment of inertia of a steel shaft that is 1 m long and has a diameter of 5 cm. Determine the mass of the shaft using the density information provided in Table 9.1.

9.13 Determine the mass moment of inertia of steel balls used in ball bearings. Use a diameter of 2 cm.

9.14 Determine the mass moment of inertia of the earth about its axis of rotation, going through the poles. Assume the shape of the earth to be spherical. Look up information such as the mass of the earth and the radius of the earth at the equatorial plane.

9.15 Next time you put gasoline in your car, measure the mass flow rate (kg/s) of gasoline at the pumping station. Record the amount of gas in gallons (or liters if you are doing the experiment outside the United States) that you placed in your car's gas tank and the time that it took to do so. Make the appropriate conversion from volume flow rate to mass flow rate using the density of gasoline.

9.16 Measure the mass flow rate of water coming out of a drinking fountain by placing a cup under the running water and by measuring the time that it took to fill the cup. Measure the mass of the water by subtracting the total mass from the mass of the cup.

9.17 Obtain a graduated beaker and accurate scale from a chemistry lab and measure the density of the following liquids:

a. a cooking oil

b. SAE 10W-40 engine oil

c. water

d. milk

e. ethylene glycol (antifreeze)

Express your findings in kg/m^3, $slugs/ft^3$, and lbm/ft^3. Also determine the specific gravity of each liquid.

9.18 Obtain pieces of steel, wood, and concrete of known volume. Find pieces with simple shapes, so that you can measure the dimensions and calculate the volume quickly. Determine their mass by placing each on an accurate scale, and calculate their densities.

9.19 Take a 500-sheet ream of computer paper as it comes wrapped. Unwrap it and measure the height, width, and length of the stack. Determine the volume, measure the mass of the ream, and obtain the density. Determine how many reams come in a standard box. Estimate the total mass of the box. Discuss your assumptions and estimation procedure.

9.20 In this assignment you will investigate how much water may be wasted by a leaky faucet. Place a large cup under a leaky faucet. If you don't have a leaky faucet at home, open the faucet so that it just drips into the cup. Record the time that you started the experiment. Allow the water to drip into the cup for about an hour or two. Record the time when you remove the cup from under the faucet. Determine the mass flow rate. Estimate the water wasted by 100,000 people with leaky faucets during a period of one year.

9.21 Obtain a graduated beaker and accurate scale from a chemistry lab and measure the density of SAE 10W-30 engine oil. Repeat the experiment ten times. Determine the mean, variance, and standard deviation for your density measurement.

9.22 Referring to Figure 9.5, how much water is stored after 20 minutes in each of the tanks? How long will it take to fill the tanks completely, provided that the volume of tank is 24 m³ and tank (a) has a volume of 36 m³? Assume the density of water is 1000 kg/m³.

9.23 Investigate the size of the storage tank in a gas station. Apply the conservation of mass statement to the gasoline flow in the gas station. Draw a control volume showing appropriate components of the gasoline flow system. Estimate how much gasoline is removed from the storage tank per day. How often does the storage tank need to be filled? Is this a steady process?

9.24 Calculate the mass moment of inertia of the thin ring shown in the accompanying diagram. Express your answer in lbm · ft², lbm · in², and slugs · ft².

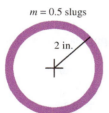

$m = 0.5$ slugs

2 in.

Problem 9.24

9.25 Determine the mass moment of inertia of a steel shaft that is 4 ft long and has a diameter of $2\frac{1}{2}$ in. Express your answer in lbm · ft², lbm · in², and slugs · ft².

9.26 Determine the mass moment of inertia of a steel ball with a diameter of 2 in. Express your answer in lbm·ft², lbm·in², and slugs·ft².

9.27 Determine the mass moment of inertia of a 4 in. square steel plate. Use Equation 9.13. Express your answer in lbm · ft², lbm · in², and slugs · ft².

9.28 Determine the specific gravity of the following gasses by comparing their densities to air at 1.23 kg/m³. Helium: 0.166 kg/m³, oxygen: 1.33 kg/m³, nitrogen: 1.16 kg/m³, natural gas: 0.667 kg/m³, and hydrogen: 0.0838 kg/m³.

9.29 Calculate the change in the momentum of a car whose speed changes from 100 km/h to 20 km/h. The car has a mass of 1000 kg and is moving along a straight line. Express your answer in SI, BG, and U.S. Customary units.

9.30 Calculate the change in the momentum of a 200,000 kg Boeing 777 whose speed changes from 450 mph to 180 mph. Express your answer in SI, BG, and U.S. Customary units.

9.31 Kinetic energy is another engineering quantity that is mass dependent. An object having a mass m and moving with a speed V has a kinetic energy equal to $1/2 \, mV^2$. In Chapter 13, we will explain the concept of kinetic energy in more detail after we introduce work. For now, compute the kinetic energy for the following situations: a 1500 kg car moving at a speed of 100 km/h and a 200,000 kg Boeing 777 moving at a speed of 700 km/h. Express your answers in SI, BG, and U.S. Customary units.

9.32 Rotational kinetic energy is yet another mass-dependent engineering quantity. An object having a mass moment of inertia I and rotating with angular speed of ω (rad/s) has a rotational kinetic energy which is equal to $\frac{1}{2} I \omega^2$. Determine the rotational kinetic energy of a steel shaft that is 4 ft long and has diameter of 2 in. and is rotating at an angular speed of 100 rpm. Express your answer in SI, BG, and U.S. Customary units.

9.33 Determine the change in mass moment of inertia of an object that could be modeled as a point mass m when its distance from a center of rotation is doubled.

9.34 Convert the density, specific gravity, and specific weight data in Table 9.1 from SI units to U.S. Customary units. Present your solution in a tabular form similar to Table 9.1. Do you need to change the values of specific gravity?

9.35 Convert the density, specific gravity, and specific weight data in Table 9.1 from SI units to British Gravitational units. Present your solution in a tabular form similar to Table 9.1. Do you need to change the values of specific gravity?

9.36 A plugged dishwasher sink with the dimensions of 14 in. × 16 in. × 6 in. is being filled with water from a faucet with an inner diameter of 1 in. If it takes 220 seconds to fill the sink to its rim, estimate the mass flow of water coming out of the faucet.

Problem 9.36
Saeed Moaveni

9.37 Imagine the plug of the sink described in Problem 9.36 leaks. If it now takes 250 seconds to fill the sink to its rim, estimate the mass-flow rate of the leak.

9.38 The tank shown in the accompanying figure is being filled by pipes 1 and 2. If the water level is to remain constant, what is the mass-flow rate of water leaving the tank at pipe 3? What is the average velocity of the water leaving the tank?

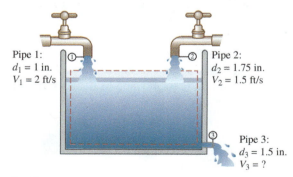

Pipe 1:
$d_1 = 1$ in.
$V_1 = 2$ ft/s

Pipe 2:
$d_2 = 1.75$ in.
$V_2 = 1.5$ ft/s

Pipe 3:
$d_3 = 1.5$ in.
$V_3 = ?$

Problem 9.38

9.39 Imagine the water level in the tank described in Problem 9.38 rises at a rate of 0.1 in/s. Knowing the diameter of the tank is 6 in., what is the average velocity of the water leaving the tank? Express your answer in in/s, ft/s, and m/s.

9.40 The density of air at a city (such as San Diego) located at sea level is 1.225 kg/m^3, while for a city (such as Denver) located at high altitude, it is 1.048 kg/m^3. Consider two auditoriums, each with dimensions of 100 ft × 150 ft × 20 ft, one located in San Diego and the other in Denver. Calculate the mass of air for each auditorium. Express your answer in kg, lbm, and slugs.

9.41 A rectangular duct with dimensions of 12 in. × 14 in. delivers conditioned air with a density of 1.2 kg/m^3 to a room at a rate of 1200 ft^3/min. What is the mass-flow rate of air in lbm/h? What is the average air speed in the duct? Express your answer in m/s and ft/s.

9.42 A rubber balloon has a mass of 10 grams. The balloon is filled with air and forms a shape that may be approximated as a sphere with a diameter of 1.5 ft. What is the total mass of the balloon? What would be the total mass of the balloon if it is filled with helium instead? Express your answers in grams.

9.43 A typical household of four people consumes approximately 80 gallons of water per day. Express the annual consumption rate of a city with a population of 100,000 people. Express your answer in slugs per year, lbm per year, and kg per year.

9.44 Lake Mead, near the Hoover Dam, which is the largest man-made lake in the United States, contains 28,537,000 acre-foot of water (an acre-foot is the amount of water required to cover 1 acre to a depth of 1 foot). Express the mass of the water volume in kg and lbm.

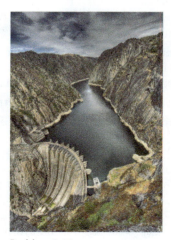

Problem 9.44
Lledo/Shutterstock.com

9.45 Air with a density of 0.45 kg/m³ enters an experimental jet engine at rate of 180 m/s. The inlet area of the engine is 5.0 ft². After fuel is burned, the air and combustion byproducts leave the engine with a velocity of 650 m/s through an outlet area of 4.0 ft². What is the average density of the gas exiting the engine? Express your answer in kg/m³, slugs/ft³, and lbm/ft³.

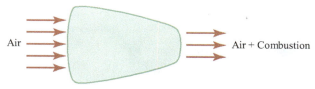

Problem 9.45

9.46 In a food processing (drying) application, air ($\rho = 1.10$ kg/m³, $V = 10$ ft/min) enters a chamber through a duct with dimensions of 20 in. × 18 in. and leaves ($\rho = 1.10$ kg/m³) through a ductwork with dimensions of 14 in. × 16 in. What is the average air speed in the exit duct? Express your answer in ft/min and m/s.

9.47 To measure the fuel consumption of an experimental lawn mower, fuel is drawn from a tank that is resting on a scale, as shown in the figure. The tank and the fuel have an initial total mass of 10 kg. If the mass scale shows a value of 9.45 kg after 30 minutes, what is the fuel consumption of the lawn mower in gallons per hour?

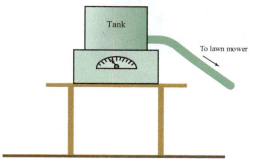

Problem 9.47

9.48 In a water bottling process, one-liter bottles are filled through a system that provides filtered water to each bottle at rate of 0.7 lbm/s. How long would it take to fill a bottle? How many bottles can be filled in one hour if the system can fill 20 bottles simultaneously with a required time of 3 seconds to move away the filled bottles and bring in the new batch of empty bottles?

9.49 Water is available at a rate of 400 gallons/min to a small apartment complex, consisting of four units. If at a given time, units 1, 2, and 3 are consuming 18 lbm/s, 14 lbm/s, 16 lbm/s, respectively, how much water is available to unit 4? Express your answer in gallons/min, ft³/s, m³/s, lbm/s, and kg/s.

9.50 *A hydrometer* is a device which uses the principle of buoyancy to measure the Specific Gravity (S.G.) of a liquid. Design a pocket-sized hydrometer that can be used to measure S.G. for liquids having 1.2 < S.G. < 1.5. Specify important dimensions and the mass of the device. Write a brief report to your instructor explaining how you arrived at the final design. Also include a sketch showing the calibrated scale from which one would read S.G. In addition, construct a model and attach the scale to the working model. To keep things simple, assume that the hydrometer has a cylindrical shape.

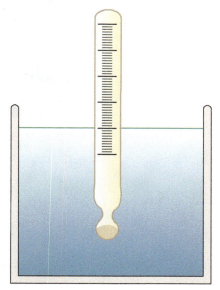

Problem 9.50

Fred Stein Archive/Contributor/Archive Photos/Getty Images

"Education is what remains after one has forgotten everything he learned in school." —ALBERT EINSTEIN (1879-1955)

Impromptu | Design IV

Objective: Given the following materials, design a vehicle that will transport a penny. The vehicle must remain in contact with the penny and the ground at all times. The design that travels a penny the farthest wins. Each design is allowed one practice run.

Provided Materials: Construction paper (one sheet—9″ × 12″ and one sheet—12″ × 18″), 1 balloon, 2 straws, and 10 inches of tape.

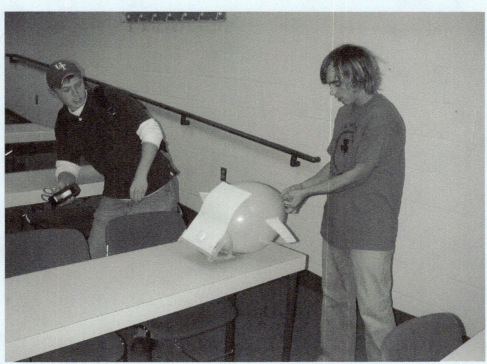

Saeed Moaveni

"If you study to remember, you will forget, but, if you study to understand, you will remember."—Unknown

Force and Force-Related Variables in Engineering

Stocktrek Images/Getty Images

To surface, air is blown into the ballast tank of a submarine to push water out of the tank to make the submarine lighter—the submarine floats to surface. To dive, the ballast tank is opened to let water in and to push out the air. The submarine sinks below the surface. The diving and surfacing of the submarine demonstrate the relationship between the weight of the submarine and the buoyancy force acting on it.

LEARNING OBJECTIVES

LO¹　**Force:** explain what is meant by force; give examples of different types of forces in engineering analysis and design

LO²　**Newton's Laws in Mechanics:** explain what is meant by mechanics and describe Newton's first, second, and third Laws

LO³　**Moment, Torque—Force Acting at a Distance:** describe the tendency of an unbalanced force which results in rotating, bending, or twisting an object; also explain how it is quantified

LO⁴　**Work—Force Acting Over a Distance:** describe the tendency of an unbalanced force which could result in moving an object through a distance; also explain how it is calculated

LO⁵　**Pressure and Stress—Force Acting Over an Area:** explain how pressure and stress provide measures of intensities of forces acting over areas

LO⁶　**Linear Impulse—Force Acting over Time:** describe the net effect of an unbalanced force acting over a period of time

SUBMARINES: HOW THEY WORK

Submarines are completely enclosed vessels with cylindrical shapes, narrowed ends, and two hulls: the inner hull and the outer hull. The inner hull protects the crew from the immense water pressure of the ocean depths and insulates the sub from the freezing temperatures. This hull is called the pressure hull. The outer hull shapes the submarine's body. The ballast tanks, which control the sub's buoyancy, are located between the inner and outer hulls.

To stay in control and stable, a submerged submarine must maintain a condition called trim. This means its weight must be perfectly balanced throughout the whole ship. It cannot be too light or too heavy aft or too light or too heavy forward. The submarine's crew must continually work to keep the submarine trim, because burning fuel and using supplies affect the sub's mass distribution. Tanks called trim tanks—one forward (front half of boat) and one aft (back half of boat)—help keep trim by allowing water to be added or expelled from them as needed. Once the submarine is underwater, it has two controls used for steering. The rudder controls side-to-side turning (or yaw), and diving planes control the sub's rise and descent (or pitch). There are two sets of diving planes: the sail planes, which are located on the sail, and the stern planes, which are located at the stern (back) of the boat with the rudder and propeller. Some submarines, including

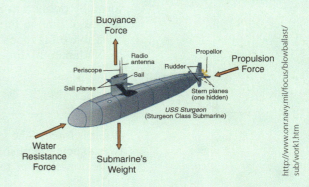

http://www.onr.navy.mil/focus/blowballast/sub/work1.htm

the new Virginia class, make use of bow planes (diving planes located at the bow—the front of the boat) rather than sail planes. Whether a submarine is floating or submerging depends on the ship's buoyancy. To surface, air is blown into the ballast tank of a submarine to push water out of the tank to make the submarine lighter—the submarine floats to surface. To dive, the ballast tank is opened to let water in and to push out the air. The submarine sinks below the surface. The diving and surfacing of the submarine demonstrate the relationship between the weight of the submarine and the buoyancy force acting on it. Also, if the submarine is moving at a constant speed, the propulsion forces must be equal to the resistance forces from the water.

http://www.onr.navy.mil/focus/blowballast/sub/work1.htm

To the students: Can you describe the net forces acting on a car? How about forces acting on an airplane? What are the forces acting on your body when you are sitting on a chair? What are the forces acting on a building?

The objectives of this chapter are to introduce the concept of force, its various types, and other force-related variables such as pressure and stress. We will discuss Newton's laws in this chapter. Newton's laws form the foundation of mechanics and analysis and design of many engineering problems, including structures, airplane airframes (fuselage and wings), car frames, medical implants for hips and other joint replacements, machine parts, and orbits of satellites. We will also explain the tendencies of unbalanced mechanical forces, which are to translate and rotate objects. Moreover, we will consider some of the mechanical properties of materials that show how stiff or flexible a material is when subjected to a force. We will then explain the effect of a force acting at a distance in terms of creating a moment about a point; the effect of a force acting over a distance, what is formally defined as mechanical work; and the effect of a force acting over a period of time in terms of what is generally referred to as **linear impulse**.

Table 6.7 is repeated here to show how the content of this chapter is related to fundamental dimensions discussed in previous chapters.

LO¹ 10.1   Force

What is force? The simplest form of a force that represents the interaction of two objects is a push or a pull. When you push or pull on a lawn mower or on a vacuum cleaner, that interaction between your hand and the lawn mower

TABLE 6.7   Fundamental Dimensions and How They Are Used in Defining Variables that Are Used in Engineering Analysis and Design

Chapter	Fundamental Dimension	Related Engineering Variables			
7	Length (L)	Radian (L/L), Strain (L/L)	Area (L^2)	Volume (L^3)	Area moment of inertia (L^4)
8	Time (t)	Angular speed ($1/t$), Angular acceleration ($1/t^2$), Linear speed (L/t), Linear acceleration (L/t^2)		Volume flow rate (L^3/t)	
9	Mass (M)	Mass flow rate (M/t), Momentum (ML/t), Kinetic energy (ML^2/t^2)		Density (M/L^3), Specific volume (L^3/M)	
10	Force (F)	Moment (LF), Work, energy (FL), Linear impulse (Ft), Power (FL/t)	Pressure (F/L^2), Stress (F/L^2), Modulus of elasticity (F/L^2), Modulus of rigidity (F/L^2)	Specific weight (F/L^3),	
11	Temperature (T)	Linear thermal expansion (L/LT), Specific heat (FL/MT)		Volume thermal expansion (L^3/L^3T)	
12	Electric Current (I)	Charge (It)	Current density (I/L^2)		

or the vacuum cleaner is called *force*. When an automobile pulls a U-Haul trailer, a force is exerted by the bumper hitch on the trailer (see Figure 10.1). The interaction representing the trailer being pulled by the bumper hitch is represented by a force. In these examples, the force is exerted by one body on another body by direct contact. Not all forces result from direct contact. For example, gravitational and magnetic forces are not exerted by direct contact. If you hold your book, say, 3 feet above the ground and let it go, what happens? It falls; that is due to gravitational force which is exerted by the earth on the book. The gravitational attractive forces act at a distance. A satellite orbiting the earth is continuously being pulled by the earth toward the center of the earth, and this allows the satellite to maintain its orbit. We will discuss the universal law of gravitational attraction in detail later. All forces, whether they represent the interaction of two bodies in direct contact or the interaction of two bodies at a distance (gravitational force), are defined by their magnitudes, their directions, and the points of application. The simple examples given in Figure 10.2 are used to show you the effects of a force's magnitude, direction, and point of application on the way an object behaves. Think about the different situations shown in Figure 10.2 and then in your own words explain the behavior of each bar and compare it to the other cases shown.

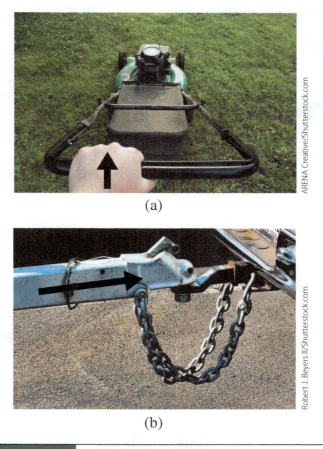

(a)

(b)

| FIGURE 10.1 | (a) Force exerted by your hand on a lawn mower, (b) Force exerted by bumper hitch on the trailer |

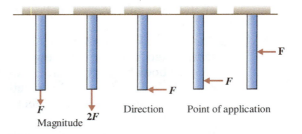

FIGURE 10.2	Simple examples to demonstrate the effect of magnitude, direction, and the point of application of a force on the same object. Explain the behavior of each case.

Tendencies of a Force

Now that you understand the concept of force, let us examine the tendencies of externally applied forces. The natural tendency of a force acting on an object, if unbalanced, will be to translate the object (move it along) and to rotate it. Moreover, forces acting on an object could squeeze or shorten, elongate, bend, or twist the object. The amount by which the object will translate, rotate, elongate, shorten, bend, or twist will depend on its support conditions, and material and geometric properties (length, area, and area moment of inertia). Machine components, tools, parts of the human body, and structural members are generally subjected to push—pull, bending, or twisting types of loading. In order to quantify these tendencies in engineering, we define terms such as *moment, work, impulse, pressure*, and *stress*. Those of you who are studying to become aerospace, civil, manufacturing, or mechanical engineers will take classes in basic mechanics and mechanics of materials in which you will explore forces, stresses, and materials behavior in more depth. The simple examples given in Figure 10.3 show the tendencies of a force. In your own words explain the behavior of each case.

> The simplest form of a force that represents the interaction of two objects is a push or a pull. All forces, whether they represent the interaction of two bodies in direct contact or the interaction of two bodies at a distance (e.g., gravitational force), are defined by their magnitudes, their directions, and their points of application.

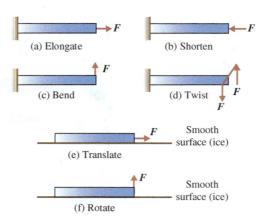

FIGURE 10.3	Simple examples to demonstrate the tendencies of a force.

Units of Force

The newton (N) is the unit of force in SI. One newton is defined as the force that will accelerate 1 kilogram of mass at a rate of 1 m/s². This relationship is based on Newton's law of motion.

$$1 \text{ newton} = (1 \text{ kg})(1 \text{ m/s}^2) \text{ or } 1 \text{ N} = 1 \text{ kg} \cdot \text{m/s}^2 \qquad \boxed{10.1}$$

In comparison, in the British Gravitational System of Units, 1 pound force will accelerate 1 slug at a rate of 1 ft/s², as given by

$$1 \text{ lbf} = (1 \text{ slug})(1 \text{ ft/s}^2) \qquad \boxed{10.2}$$

And 1 pound force is equal to 4.45 newtons (1 lbf = 4.448 N).

As engineering students, you will study different types of forces such as wind forces, drag forces, **viscous forces**, surface tension forces, contact forces, and normal forces during your course of studies. In the next sections, we will explain two forces (spring and friction) that every engineer regardless of his or her area of specialization should know.

Spring Forces and Hooke's Law

Most of you have seen springs that are used in cars, locomotives, clothespins, weighing scales, and clips for potato chip bags. Springs are also used in medical equipment, electronics equipment such as printers and copiers, and in many restoring mechanisms (a mechanism that returns a component to its original position) covering a wide range of applications. There are different types of springs, including extension (or compression) and torsional springs. Examples of different types of springs are shown in Figure 10.4.

Hooke's law (named after Robert Hooke, an English scientist who proposed the law) states that over the elastic range the deformation of a spring is directly proportional to the applied force, according to

$$F = kx \qquad \boxed{10.3}$$

where

F = applied force (N or lbf)

k = spring constant (N/mm or N/cm or lb/in.)

x = deformation of the spring
(mm or cm or in. — use units that are consistent with k)

> The natural tendency of force acting on an object will be to translate or rotate the object. Forces acting on an object could also squeeze or shorten, elongate, bend, or twist the object.

By *elastic range* we mean the range over which if the applied force is removed, the internal spring force will return the spring to its original unstretched shape and size and do so with no permanent deformation. Note the **spring force** is equal to the applied force. The value of the spring constant depends on the type of material used to make the spring. Moreover, the shape and winding of the spring will also affect its k value. The spring constant can be determined experimentally.

EXAMPLE 10.1

FIGURE 10.5

The spring setup.

For a given spring, in order to determine the value of the spring constant, we have attached dead weights to one end of the spring, as shown in Figure 10.5. We have measured and recorded the deflection caused by the corresponding weights as given in Table 10.1(a). What is the value of the spring constant? We have plotted the results of the experiment using Excel (shown in Figure 10.6).

The spring constant k is determined by calculating the slope of a force-deflection line (recall that the slope of a line is determined from slope = rise/run, and for this problem, the slope = change in force/change in deflection). This approach leads to a value of $k = 0.54$ N/mm.

TABLE 10.1(A)
The Results of the Experiment for Example 10.1

Weight (N)	The Deflection of the Spring (mm)
4.9	9
9.8	18
14.7	27
19.6	36

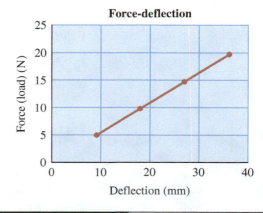

Force-deflection

FIGURE 10.6   The force-deflection diagram for the spring in Example 10.1.

Often, when connecting experimental force-deflection points, you may not obtain a straight line that goes through each experimental point. In this case, you will try to come up with the best fit to the data points. There are mathematical procedures (including least-squares techniques) that allow you to find the best fit to a set of data points. We will discuss curve fitting using Excel in Section 14.3, but for now just draw a line that you think best fits the data points. As an example, we have shown a set of data points in Table 10.1(b) and a good corresponding fit in Figure 10.7.

TABLE 10.1(B)
A Set of Force-deflection Data Points

Weight (N)	The Deflection of the Spring (mm)
5.0	9
10.0	17
15.0	29
20.0	35

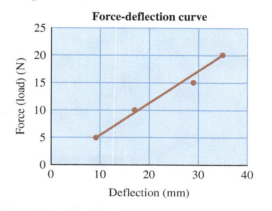

FIGURE 10.7   A good fit to a set of force-deflection data points.

Friction Forces—Dry Friction and Viscous Friction

There are basically two types of **frictional forces** that are important in engineering design: *dry frictional forces* and *viscous friction* (or fluid friction). Let us first take a closer look at dry friction, which allows us to walk or to drive our cars. Remember what happens when frictional forces get relatively small, as is the case when you try to walk on ice or drive your car on a sheet of ice. Stated in a simple way, dry friction exists because of irregularities between surfaces in contact. To better understand how frictional forces build up, imagine you were to perform the following experiment. Place a book on a table and begin by pushing the book gently. You will notice that the book will not move. The applied force is balanced by the friction force generated at the contact surface. Now push on the book harder; you will notice that the book will not move until the pushing force is greater than the frictional force that prevents the book from moving. The results of your experiment may be plotted in a diagram similar to the one shown in Figure 10.8. Note that the friction force is not constant and reaches a maximum value (see point *A* in Figure 10.8), which is given by

$$F_{max} = \mu N \qquad \boxed{10.4}$$

where N is the normal force—exerted by the table on the book to prevent it from falling—and, in the case of a book resting on the table, N is equal to the weight of the book, and μ is the coefficient of static friction for the two surfaces involved, the book and the desk surface. Figure 10.8 also shows that once the book is set in motion, the magnitude of the friction force drops to a value called the *dynamic* (or *kinetic*) *friction*.

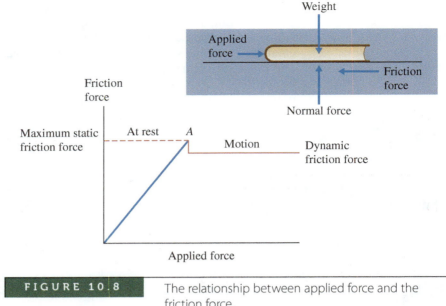

FIGURE 10.8	The relationship between applied force and the friction force.

The other form of friction which must be accounted for in engineering analysis is fluid friction, which is quantified by the property of a fluid called *viscosity*. The value of viscosity of a fluid represents a measure of how easily the given fluid can flow. The higher the viscosity value is, the more resistance the fluid offers to flow. For example, if you were to pour water and honey side by side on an inclined surface, which of the two liquids would flow easier? You know the answer: The water will flow down the inclined surface faster because it has less viscosity. Fluids with relatively smaller viscosity require less energy to be transported in pipelines. For example, it would require less energy to transport water in a pipe than it does to transport motor oil or glycerin. The viscosity of fluids is a function of temperature. In general, the viscosity of gases increases with increasing temperature, and the viscosity of liquids generally decreases with increasing temperature. Knowledge of the viscosity of a liquid and how it may change with temperature also helps when selecting lubricants to reduce wear and friction between moving parts.

EXAMPLE 10.2

The coefficient of static friction between a book and a desk surface is 0.6. The book weighs 20 N. If a horizontal force of 10 N is applied to the book, would the book move? And if not, what is the magnitude of the friction force? What should be the magnitude of the horizontal force to set the book in motion?

The maximum friction force is given by

$$F_{max} = \mu N = (0.6)(20) = 12 \text{ newtons}$$

Since the magnitude of the horizontal force (10 N) is less than the maximum friction force (12 N), the book will not move, and hence the friction force will equal the applied force. A horizontal force having a magnitude greater than 12 N will be required to set the book in motion.

LO² 10.2   Newton's Laws in Mechanics

As we mentioned in Chapter 6, physical laws are based on observations. In this section, we will briefly discuss **Newton's laws**, which form the foundation of mechanics. The design and analysis of many engineering problems, including structures, machine parts, and the orbit of satellites, begin with the application of Newton's laws. Most of you will have the opportunity to take a physics, statics, or dynamics class that will explore Newton's laws and their applications further.

What We Mean by Mechanics

Next, it is important that we define what we mean by the term **mechanics**. There are three concepts that you should fully understand: (1) Mechanics deals with the study of behavior of objects or structural members when subjected to forces. As mentioned previously, the tendencies of mechanical forces are to translate, rotate, squeeze or shorten, elongate, bend, or twist objects. (2) What do we mean by *behavior*? In mechanics, by the term *behavior* we mean obtaining information about the object's *linear* and *angular (rotational) displacements, velocity, acceleration, linear* and *angular deformations*, and *stresses*. As we mentioned in Chapter 6, engineers are book keepers. Whereas, in accounting practice, accountants keep track of dollars and cents, revenues, and expenditures, in mechanics, engineers keep track of forces, masses, energies, displacements, accelerations, deformations, rotations, and stresses. (3) Moreover, an object is defined by its *geometric characteristics* and *material properties*. Geometric characteristics provide information such as length, cross-sectional area, and first and second moment of area. On the other hand, the material properties provide information about material characteristics such as density, modulus of elasticity, and shear modulus of an object. The object's interaction with its surrounding is also defined by *boundary* and *initial conditions*, and the manner in which forces are applied. In the following sections, we will elaborate on each of these concepts and how they contribute to the way an object behaves when subjected to a force, and provide simple examples to demonstrate these concepts.

> Mechanics deals with the study of behavior of objects or structural members when subjected to forces. By behavior, we mean the object's linear and angular displacement, velocity, acceleration, deformation, and stresses.

Newton's First Law

If a given object is at rest, and if there are no unbalanced forces acting on it, the object will then remain at rest. If the object is moving with a constant speed in a certain direction, and if there are no unbalanced forces acting on it, the object will continue to move with its constant speed and in the same direction. Newton's first law is quite obvious and should be intuitive. For example, you know from your everyday experiences that if a book is resting on a table and you don't push, pull, or lift it, then the book will lie on that table in that position until it is disturbed with an unbalanced force.

Newton's Second Law

We briefly explained this law in Chapter 6, where we said that if you place a book on a smooth table and push it hard enough, it will move. Newton observed this and formulated his observation into what is called Newton's second law of motion. Newton observed that as he increased the mass of the object being pushed, while keeping the magnitude of the force constant, the object did not move as quickly and had a smaller rate of change of speed. Thus, Newton noticed that there was a direct relationship between the push (the force), the mass of the object being pushed, and the acceleration of the object. He also noticed that there was a direct relationship between the direction of the force and the direction of the acceleration. Newton's second law of motion simply states that the net effect of unbalanced forces is equal to mass times acceleration of the object, which is given by the expression

$$\sum \vec{F} = m\vec{a}$$

10.5

where the symbol Σ (sigma) means summation, F represents the forces in units of newton (or lbf), m is the mass of the object in kg (or slugs), and a is the resulting acceleration of the mass center of the object in m/s² (or ft/s²). In Equation (10.5), a summation of forces is used to allow for application of more than one force on an object.

Newton's Third Law

Returning to our example of a book resting on a desk, as shown in Figure 10.9, because the weight of the book is pushing down on the desk, simultaneously the desk also pushes up on the book. Otherwise, the desk won't be supporting the book. Newton's third law states that for every action there exists a reaction, and the forces of action and reaction have the same magnitude and act along the same line, but they have opposite directions.

Newton's Law of Gravitation

The weight of an object is the force that is exerted on the mass of the object by the earth's gravity. Newton discovered that any two masses, m_1 and m_2, attract each other with a force that is equal in magnitude and acts in the opposite direction, as shown in Figure 10.10, according to the following relationship:

$$F = \frac{Gm_1m_2}{r^2}$$

10.6

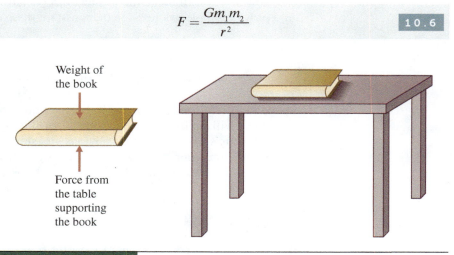

| FIGURE 10.9 | The forces of action and reaction; they have the same magnitude and the same line of action, but they act in opposite directions. |

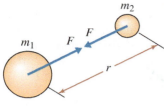

FIGURE 10.10

The gravitational attraction of two masses.

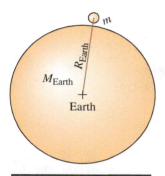

FIGURE 10.11

where

F = attractive force (N)
G = universal gravitational constant (6.673×10^{-11} m^3/kg·s^2)
m_1 = mass of particle 1 (kg, see Figure 10.10)
m_2 = mass of particle 2 (kg, see Figure 10.10)
r = distance between the center of each particle (m)

Using Equation (10.6), we can determine the weight of an object having a mass m, on the earth, by substituting m for the mass of particle 1 and substituting for the mass of particle 2 the mass of Earth ($M_{\text{Earth}} = 5.97 \times 10^{24}$ kg), and using the radius of Earth as the distance between the center of each particle as in Figure 10.11. Note that the radius of Earth is much larger than the physical dimension of any object on Earth, and, thus, for an object resting on or near the surface of Earth, r could be replaced with $R_{\text{Earth}} = 6378 \times 10^3$ m as a good approximation. Thus, the weight (W) of an object on the surface of Earth having a mass m is given by

$$W = \frac{GM_{\text{Earth}}\, m}{R^2_{\text{Earth}}}$$

And after letting $g = \dfrac{GM_{\text{Earth}}}{R^2_{\text{Earth}}}$, we can write

$$W = mg \qquad \boxed{10.7}$$

Equation (10.7) shows the relationship among the weight of an object, its mass, and the local acceleration due to gravity. Make sure you fully understand this simple relationship. Because the earth is not truly spherical in shape, the value of g varies with latitude and longitude. However, for most engineering applications, $g = 9.8$ m/s^2 or $g = 32.2$ ft/s^2 is used. Also note that the value of g decreases as you get further away from the surface of Earth. This fact is evident from examining Equation (10.6) and, the relationship for g leading to Equation (10.7).

EXAMPLE 10.3

Determine the weight of an exploration vehicle whose mass is 250 kg on the Earth. What is the mass of the vehicle on the Moon ($g_{\text{Moon}} = 1.6$ m/s^2) and the planet Mars ($g_{\text{Mars}} = 3.7$ m/s^2)? What is the weight of the vehicle on the Moon and Mars?

The mass of the vehicle is 250 kg on the Moon and on the planet Mars. The weight of the vehicle is determined from $W = mg$:

On Earth: $W = (250 \text{ kg})\left(9.8\dfrac{\text{m}}{\text{s}^2}\right) = 2450$ N

On Moon: $W = (250 \text{ kg})\left(1.6\dfrac{\text{m}}{\text{s}^2}\right) = 400$ N

On Mars: $W = (250 \text{ kg})\left(3.7\dfrac{\text{m}}{\text{s}^2}\right) = 925$ N

EXAMPLE 10.4

The space shuttle orbited the Earth at altitudes from as low as 250 km (155 miles) to as high as 965 km (600 miles), depending on its missions. Determine the value of g for an astronaut in a space shuttle. If an astronaut has a mass of 70 kg on the surface of Earth, what is her weight when in orbit around Earth?

When the space shuttle is orbiting at an altitude of 250 km above Earth's surface:

$$g = \frac{GM_{Earth}}{R^2} = \frac{\left(6.673 \times 10^{-11}\, \frac{m^3}{kg \cdot s^2}\right)(5.97 \times 10^{24}\, kg)}{\left[(6378 \times 10^3 + 250 \times 10^3)\, m\right]^2} = 9.07\ m/s^2$$

$$W = (70\ kg)\left(9.07\, \frac{m}{s^2}\right) = 635\ N$$

At an altitude of 965 km:

$$g = \frac{GM_{Earth}}{R^2} = \frac{\left(6.673 \times 10^{-11}\, \frac{m^3}{kg \cdot s^2}\right)(5.97 \times 10^{24}\, kg)}{\left[(6378 \times 10^3 + 965 \times 10^3)\, m\right]^2} = 7.38\ m/s^2$$

$$W = (70\ kg)\left(7.38\, \frac{m}{s^2}\right) = 517\ N$$

Note that at these altitudes an astronaut still has a significant weight. The near weightless conditions that you see on TV are created by the orbital speed of the shuttle. For example, when the space shuttle circles Earth at an altitude of 935 km at a speed of 7744 m/s, it creates a normal acceleration of 8.2 m/s². It is the difference between g and normal acceleration that creates the condition of weightlessness.

Before You Go On

Answer the following questions to test your understanding of the preceding sections.

1.   Explain what we mean by a force and give examples of different forces.

2.   What are the tendencies of a force?

3.   In your own words, explain what we mean by mechanics.

4.   Explain Newton's Laws.

Vocabulary—State the meaning of the following terms:

Hooke's Law _____

Spring Constant _____

Dry Friction _____

Viscous Friction _____

Universal Gravitational Constant _____

LO³ 10.3 Moment, Torque—Force Acting at a Distance

As we mentioned earlier, the two tendencies of an unbalanced force acting on an object are to translate the object (i.e., to move it in the direction of the unbalanced force) and to rotate or bend or twist the object. In this section, we will focus our attention on understanding the tendency of an unbalanced force to rotate, bend, or twist objects. Being able to calculate moments created by forces about various points and axes is important in many engineering analyses. For example, all of you have noticed that the street light poles or the traffic light poles are thicker near the ground than they are on the top. One of the main reasons for this is that wind loading creates a bending moment, which has a maximum value about the base of the pole. Thus, a bigger section is needed at the base to prevent failure. Nature understands the concept of bending moments well; that is why trees have big trunks near the ground to support the bending moment created by the weight of the branches and the wind loading. Understanding **moments**, or **torques**, is also important when designing objects that rotate, such as a shaft, gear, or wheel.

To better grasp the concept of moment, let us consider a simple example. When you open a door you apply a pulling or a pushing force on the doorknob (or handle). The application of this force will make the door rotate about its hinges. In mechanics, this tendency of force is measured in terms of a *moment of a force* about an axis or a point. Moment has both *direction* and *magnitude*.

In our example, the direction is defined by the sense of rotation of the door. As shown in Figure 10.12, looking at the door from the top, when you open the door by applying the force shown, the direction of moment is clockwise, and when you close the same door, the force creates a counterclockwise moment. The magnitude of the moment is obtained from the product of the moment arm times the force. Moment arm represents the perpendicular distance between the line of action of the force and the point about which the object could rotate.

In this book, we will only consider the moment of a force about a point. Most of you will take a mechanics class, where you will learn how to calculate the magnitude and the direction of the moment of a force about an arbitrary axis. For an object that

> Newton's law of gravitational attraction states that two masses attract each other with a force that is equal in magnitude and acts in opposite direction. Moreover, the force is directly proportional to the mass of each particle and inversely proportional to the distance squared between the two center of the masses.

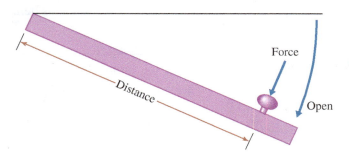

| FIGURE 10.12 | Moment of the force created by someone opening or closing a door. |

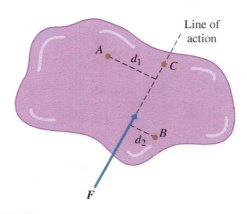

| FIGURE 10.13 | The moment of the force *F* about points *A*, *B*, and *C*. |

is subject to force **F,** the relationship between the line of action of the force and moment arm is shown in Figure 10.13. The magnitude of moment about arbitrary points A, B, and C is given by

$$M_A = d_1 F \circlearrowright \qquad \text{10.8}$$

$$M_B = d_2 F \circlearrowleft \qquad \text{10.9}$$

$$M_C = 0 \qquad \text{10.10}$$

Note that the moment of the force **F** about point C is zero because the line of action of the force goes through point C, and thus the value of the moment arm is zero. When calculating the sum of moments created by many forces about a certain point, make sure you use the correct moment arm for each force, and also make sure to ask yourself whether the tendency of a given force is to rotate the object clockwise or counterclockwise. As a bookkeeping procedure, you may assign a positive value to clockwise rotations and negative values to counterclockwise rotations or vice versa. The overall tendency of the forces could then be determined from the final sign of the sum of the moments. This approach is demonstrated in Example 10.5.

> Moment provides a measure of the tendency of a force to rotate, bend, or twist an object.

EXAMPLE 10.5

Determine the sum of the moment of the forces about point O shown in Figure 10.14.

$$\curvearrowright \sum M_O = (50 \text{ N})(0.05 \text{ m}) + (50 \text{ N})(0.07 \cos 35° \text{ m}) + (100 \text{ N})(0.1 \cos 35° \text{ m})$$
$$= 13.55 \text{ N} \cdot \text{m}$$

Note that in calculating the moment of each force, the perpendicular distance between the line of action of each force and the point O is used.

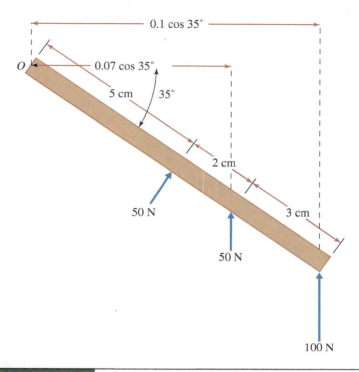

| **FIGURE 10.14** | The diagram for Example 10.5. |

Moreover, for bookkeeping, we assigned a positive sign to counterclockwise rotation.

EXAMPLE 10.6   Determine the moment of the two forces, shown in Figure 10.15, about points A, B, C, and D. As you can see, the forces have equal magnitudes and act in opposite directions to one another.

$$\curvearrowright \sum M_A = (100\ \text{N})(0) + (100\ \text{N})(0.1\ \text{m}) = 10\ \text{N} \cdot \text{m}$$

$$\curvearrowright \sum M_B = (100\ \text{N})(10\ \text{m}) + (100\ \text{N})(0) = 10\ \text{N} \cdot \text{m}$$

$$\curvearrowright \sum M_C = (100\ \text{N})(0.25\ \text{m}) - (100\ \text{N})(0.15\ \text{m}) = 10\ \text{N} \cdot \text{m}$$

$$\curvearrowright \sum M_D = (100\ \text{N})(0.35\ \text{m}) - (100\ \text{N})(0.25\ \text{m}) = 10\ \text{N} \cdot \text{m}$$

Referring to Example 10.6, note that two forces that are equal in magnitude and opposite in direction (not having the same line of action) constitute *a couple*. As you can see from the results of this example, the moment created by a couple is equal to the magnitude of either force times the perpendicular distance between the line of action of the forces involved. Also, note for this problem, we assigned a positive sign to clockwise rotation. Since the sum of the moments has a positive value, the overall tendency of forces is to turn the object clockwise.

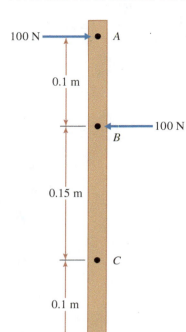

100 N ——→ ● A

0.1 m

100 N ←—— ● B

0.15 m

● C

0.1 m

● D

FIGURE 10.15   A schematic diagram for Example 10.6.

External Force, Internal Force, Reaction Force

We explained the concept of **external force** and moment previously. Let us now discuss the concept of *internal* and *reaction forces* and *moments*. When an object is subjected to an external force, **internal forces** are created inside the material to hold the material and the components together. Examples of internal forces are shown in Figure 10.16. Moreover, **reaction forces** are developed at the supporting boundaries to keep the object held in position as planned. Table 10.2 shows simple examples that demonstrate

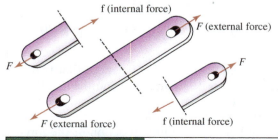

f (internal force)

F (external force)

F

F (external force)

f (internal force)

F

FIGURE 10.16   An example of Internal force. When you try to pull apart the bar, inside the bar material, internal forces develop that keep the bar together as one piece.

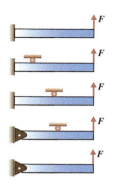

FIGURE 10.17

Simple examples to demonstrate the effect of boundary conditions. In your own words, explain the behavior of each case.

various support conditions and how they influence the behavior of an object. Try to explain the behavior of each case in your own words. First, look at the examples of a support given in the first column of Table 10.2, then look at the reaction forces. If you need help, read the explanation column.

Boundary and Initial Conditions

The manner in which an object is held in place also influences its behavior. Figure 10.17 demonstrates the effects of boundary conditions. How would you describe qualitatively in detail, by how much, and the manner by which, each beam bends? It is very important for you to clearly understand that boundary conditions refer to conditions at the boundaries of the object. These conditions provide information about how the object is supported at its boundaries. You also need to understand that for some problems, support conditions and loading could change with time, and as a result we need to specify the initial conditions before we could predict the behavior of the object.

TABLE 10.2   **Simple Examples to Demonstrate Various Support Conditions and How They Influence the Behavior of an Object**

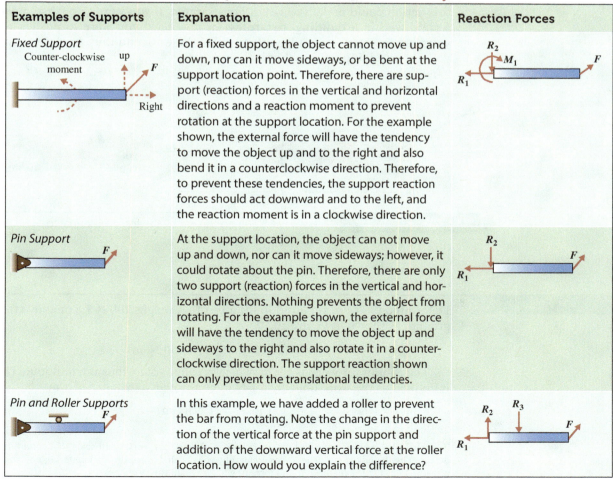

Examples of Supports	Explanation	Reaction Forces
Fixed Support	For a fixed support, the object cannot move up and down, nor can it move sideways, or be bent at the support location point. Therefore, there are support (reaction) forces in the vertical and horizontal directions and a reaction moment to prevent rotation at the support location. For the example shown, the external force will have the tendency to move the object up and to the right and also bend it in a counterclockwise direction. Therefore, to prevent these tendencies, the support reaction forces should act downward and to the left, and the reaction moment is in a clockwise direction.	
Pin Support	At the support location, the object can not move up and down, nor can it move sideways; however, it could rotate about the pin. Therefore, there are only two support (reaction) forces in the vertical and horizontal directions. Nothing prevents the object from rotating. For the example shown, the external force will have the tendency to move the object up and sideways to the right and also rotate it in a counterclockwise direction. The support reaction shown can only prevent the translational tendencies.	
Pin and Roller Supports	In this example, we have added a roller to prevent the bar from rotating. Note the change in the direction of the vertical force at the pin support and addition of the downward vertical force at the roller location. How would you explain the difference?	

LO⁴ 10.4   Work—Force Acting Over a Distance

When you push on a car that has run out of gas and move it through a distance, you perform mechanical work. When you push on a lawn mower and move it in a certain direction, you are doing work. Mechanical work is done when the applied force moves the object through a distance. Simply stated, mechanical **work** is defined as the component of the force that moves the object times the distance the object moves. Consider the car shown in Figure 10.18. The work done by the pushing force moving the car from position 1 to position 2 is given by

$$W_{1-2} = (F \cos \theta)(d)$$

10.11

> Mechanical work is defined as the component of the force that moves the object times the distance the object moves.

Note from examining Equation (10.11) that the normal component of the force does not perform mechanical work. That is because the car is not moving in a direction normal to the ground. So next time you are mowing the lawn, ask yourself if you should push on the lawn mower horizontally or at an angle? Another point you should remember is that, if you were to push hard against an object and were not able to move it, then by definition, you are not doing any mechanical work, even though this action could make you tired. For example, if you were to push against a rigid wall in a building, regardless of how hard you were to push with your hands, you wouldn't be able to move the wall, and thus, by definition, you are not doing any mechanical work.[*]

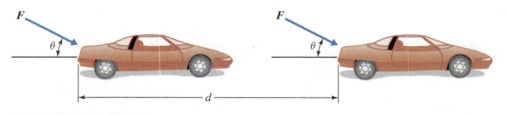

| FIGURE 10.18 | The work done on the car by the force *F* is equal to $W_{1-2} = (F \cos \theta)(d)$. |

EXAMPLE 10.7

Determine the work required to lift a box that weighs 100 N 1.5 m above the ground, as shown in Figure 10.19.

$$W = (100 \text{ N})(1.5 \text{ m}) = 150 \text{ N} \cdot \text{m}$$

The SI unit for work is N · m, which is called *joule*; thus the unit joule (J) is defined as the work done by a 1-N force through a distance of 1 m.

$$1 \text{ joule} = (1 \text{ N})(1 \text{ m})$$

---

[*]Of course, your effort goes into deforming the wall on a very, very small scale that cannot be detected by the naked eye. The forces created by your push deform the wall material, and your effort is stored in the material in the form of what is called the *strain energy*.

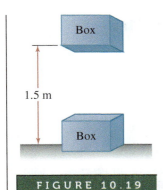

FIGURE 10.19

The box in Example 10.7.

Students often confuse the concept of work with power. Power represents how fast you want to do the work. It is the time rate of doing work, or said another way, it is the work done divided by the time that it took to perform it. For instance, in Example 10.7, if we want to lift the box in 3 seconds, then the power required is

$$\text{power} = \frac{\text{work}}{\text{time}} = \frac{150 \text{ J}}{3 \text{ s}} = 50 \text{ J/s} = 50 \text{ watts}$$

Note that 1 J/s is called 1 watt (W). If you wanted to lift the box in 1.5 seconds, then the required power to do this task is: 150 J/1.5 s or 100 W. Thus, twice as much power is required. It is important to note that the work done in each case is the same; however, the power requirement is different. Remember if you want to do work in a shorter time, then you are going to need more power. The shorter the time, the more power required to do that work. We will discuss power and its units in much more detail in Chapter 13.

Before You Go On

Answer the following questions to test your understanding of the preceding sections.

1. Explain what is meant by moment or torque and give examples from everyday life.

2. What is an internal force?

3. What is a reaction force?

4. Explain what is meant by boundary and initial conditions.

5. Explain what is meant by work and give examples from everyday life.

Vocabulary—State the meaning of the following terms:

Moment _____

External Force _____

Internal Force _____

Reaction Force _____

Work _____

Boundary Condition _____

Initial Condition _____

LO⁵ 10.5 Pressure and Stress—Force Acting Over an Area

Pressure provides a measure of intensity of a force acting over an area. It can be defined as the ratio of force over the contact surface area:

$$\text{pressure} = \frac{\text{force}}{\text{area}} \qquad \boxed{10.12}$$

To better understand what the magnitude of a **pressure** represents, consider the situations shown in Figure 10.20. Let us first look at the situation depicted in Figure 10.20(a), in which we lay a solid brick, in the form of a rectangular prism with dimensions of $21.6 \times 6.4 \times 10.2$ cm $(8\frac{1}{2} \times 2\frac{1}{2} \times 4$ in.) that weighs 28 N (6.4 lb), flat on its face. Using Equation (10.12) for this orientation, the pressure at the contact surface is

$$\text{pressure} = \frac{\text{force}}{\text{area}} = \frac{28 \text{ N}}{(0.216 \text{ m})(0.102 \text{ m})} = 1271 \frac{\text{N}}{\text{m}^2} = 1271 \text{ Pa}$$

Note that one newton per squared meter is called one pascal $(1 \text{ N}/1 \text{ m}^2 = 1 \text{ Pa})$. Now, if we were to lay the brick on its end as depicted in Figure 10.20(b), the pressure due to the weight of the brick becomes

$$\text{pressure} = \frac{\text{force}}{\text{area}} = \frac{28 \text{ N}}{(0.064 \text{ m})(0.102 \text{ m})} = 4289 \frac{\text{N}}{\text{m}^2} = 4289 \text{ Pa}$$

It is important to note here that the weight of the brick is 28 N, regardless of how it is laid. But the pressure that is created at the contact surface depends on the magnitude of the contact surface area. The smaller the contact area, the larger the pressure created by the same force. You already know this from your everyday experiences—which situation would create more pain, pushing on someone's arm with a finger or a thumbtack?

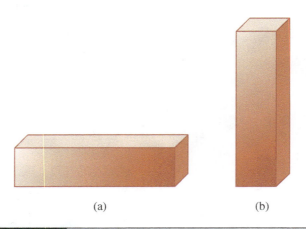

(a)   (b)

FIGURE 10.20   An experiment demonstrating the concept of pressure. (a) A solid brick resting on its face. (b) A solid brick resting on its end. In position (b), the brick creates a higher pressure on the surface.

In Chapter 7, when we were discussing the importance of area and its role in engineering applications, we briefly mentioned the importance of understanding pressure in situations such as foundations of buildings, hydraulic systems, and cutting tools. We also said that in order to design a sharp knife, one must design the cutting tool such that it creates large pressures along the cutting edge. This is achieved by reducing the contact surface area along the cutting edge. Understanding fluid pressure distributions is very important in engineering problems in hydrostatics, hydrodynamics, and aerodynamics. *Hydrostatic* refers to water at rest and its study; however, the understanding of other fluids is also considered in hydrostatics. A good example of a hydrostatic problem is calculating the water pressure acting on the surface of a dam and how it varies along the height of the dam. Understanding pressure is also important in hydrodynamics studies, which deal with understanding the motion of water and other fluids such as those encountered in the flow of oil or water in pipelines or the flow of water around the hull of a ship or a submarine. Air pressure distributions play important roles in the analysis of air resistance to the movement of vehicles or in creating lift forces over the wings of an airplane. Aerodynamics deals with understanding the motion of air around and over surfaces.

Common Units of Pressure

In the International System of units, pressure units are expressed in pascals (Pa). One pascal is the pressure created by one newton force acting over a surface area of 1 m²:

$$1 \ \text{Pa} = \frac{1 \ \text{N}}{\text{m}^2} \qquad \boxed{10.13}$$

In U.S. Customary units, pressure is commonly expressed in pounds per square inch or pounds per square foot; one lb/in² (1 psi) represents the pressure created by a 1-pound force over an area of 1 in², and 1 lb/ft² represents the pressure created by a 1-pound force over an area of 1 ft². To convert the pressure from lb/ft² to lb/in², we take the following step:

$$P\left(\frac{\text{lb}}{\text{in}^2}\right) = P\left(\frac{\text{lb}}{\text{ft}^2}\right)\left(\frac{1 \ \text{ft}^2}{144 \ \text{in}^2}\right) \qquad \boxed{10.14}$$

Note that 1 lb/in² is usually referred to as 1 **psi,** which reads one **p**ound per **s**quare **i**nch. Many other units are also commonly used for pressure. For example, the **atmospheric pressure** is generally given in inches of mercury (in · Hg) or millimeters of mercury (mm · Hg). In air-conditioning applications, often inches of water are used to express the magnitude of air pressure. These units and their relationships with Pa and psi will be explained further in Example 10.9, after we explain how the pressure of a fluid at rest changes with its depth.

> Pressure provides a measure of intensity of a force acting over an area. In a metric system, pressure units are expressed in pascal and in a U.S. Customary system, pressure is commonly expressed in pound per square inch.

For fluids at rest, there are two basic laws: (1) Pascal's law, which explains that the pressure of a fluid at a point is the same in all directions, and (2) the pressure of fluid increases with its depth. Even though these laws are simple, they are very powerful tools for analyzing various engineering problems.

FIGURE 10.21

In a static fluid, pressure at a point is the same in all directions.

Pascal's Law

Pascal's law states that for a fluid at rest, *pressure at a point* is the same in all directions. Note carefully that we are discussing pressure at a point. To demonstrate this law, consider the vessel shown in Figure 10.21; the pressure at point *A* is the same in all directions.

Another simple, yet important concept for a fluid at rest states that pressure increases with the depth of fluid. So the hull of a submarine is subjected to more water pressure when cruising at 300 m than at 100 m. Some of you have directly experienced the variation of pressure with depth when you have gone scuba diving or swimming in a lake. You recall from your experience that as you go deeper, you feel higher pressure on your body. Referring to Figure 10.22, the relationship between the gauge pressure (the pressure that is above atmospheric pressure) and the height of a fluid column above it is given by

$$P = \rho g h$$

10.15

where
P = is the fluid pressure at point B (see Figure 10.22) (in Pa or lb/ft^2)
ρ = is the density of the fluid (in kg/m^3 or slugs/ft^3)
g = is the acceleration due to gravity ($g = 9.8$ m/s^2 or $g = 32.2$ ft/s^2)
h = is the height of fluid column (in m or ft)

Also shown in Figure 10.22 is the force balance between the pressure acting on the bottom surface of a fluid column and its weight. Equation (10.15) is derived under the assumption of constant fluid density.

Another concept, which is closely related to fluid pressure distribution is *buoyancy*. As we discussed in Chapter 7, buoyancy is the force that a fluid exerts on a submerged object. The net upward buoyancy force arises from the fact that the fluid exerts a higher pressure at the bottom surfaces of the object than it does on the top surfaces of the object. Thus, the net effect of fluid pressure distribution acting over the submerged surface of an object is the buoyancy force. The magnitude of the buoyancy force is equal to the weight of the volume of the fluid displaced, which is given here again for convenience.

$$F_B = \rho V g$$

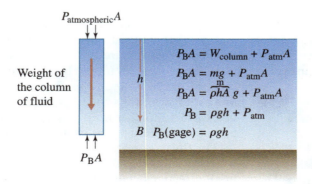

Weight of the column of fluid

$P_{atmospheric}A$

h

P_BA

B

$$P_BA = W_{column} + P_{atm}A$$
$$P_BA = mg + P_{atm}A$$
$$P_BA = \overline{\rho h A}\, g + P_{atm}A$$
$$P_B = \rho g h + P_{atm}$$
$$P_B(\text{gage}) = \rho g h$$

FIGURE 10.22   The variation of pressure with depth.

where F_B is the buoyancy force (N), ρ represents the density of the fluid (kg/m^3), g is acceleration due to gravity (9.8 m/s^2), and V is the volume of the object (m^3).

EXAMPLE 10.8

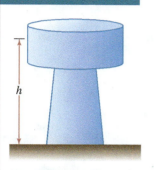

FIGURE 10.23

The water tower of Example 10.8.

Most of you may have seen water towers in and around small towns. The function of a water tower is to create a desirable municipal water pressure for household and other usage in a town. To achieve this purpose, water is stored in large quantities in elevated tanks. You may also have noticed that sometimes the water towers are placed on top of hills or at high points in a town. The municipal water pressure may vary from town to town, but it generally falls somewhere between 50 psi and 80 psi. We are interested in developing a table that shows the relationship between the height of water aboveground in the water tower and the water pressure in a pipeline located at the base of the water tower, as shown in Figure 10.23. We will calculate the water pressure, in psi, in a pipe located at the base of the water tower as we vary the height of water in increments of 10 ft. We will create a table showing our results up to 100 psi. And using the height–pressure table, we will answer the question, "What should be the water level in the water tower to create a 50-psi water pressure in a pipe at the base of the water tower?" The relationship between the height of water aboveground in the water tower and the water pressure in a pipeline located at the base of the water tower is given by Equation (10.15):

$$P = \rho g h$$

where

$P =$ is the water pressure at the base of the water tower (lb/ft^2)

$\rho =$ is the density of water (equal to 1.94 slugs/ft³)

$g =$ is the acceleration due to gravity $(g = 32.2 \text{ ft/s}^2)$

$h =$ is the height of water above ground (in ft)

Substituting the known values into Equation (10.15) leads to

$$P\left(\frac{\text{lb}}{\text{ft}^2}\right) = 1.94\left(\frac{\text{slugs}}{\text{ft}^3}\right)\left(32.2\frac{\text{ft}}{\text{s}^2}\right)[h(\text{ft})]$$

Recall that to convert the pressure from lb/ft² to lb/in², we take the following step:

$$P\left(\frac{\text{lb}}{\text{in}^2}\right) = P\left(\frac{\text{lb}}{\text{ft}^2}\right)\left(\frac{1 \text{ ft}^2}{144 \text{ in}^2}\right)$$

And performing the conversion,

$$P\left(\frac{\text{lb}}{\text{in}^2}\right) = 1.94\left(\frac{\text{slugs}}{\text{ft}^3}\right)\left(32.2\frac{\text{ft}}{\text{s}^2}\right)[h(\text{ft})]\left(\frac{1 \text{ ft}^2}{144 \text{ in}^2}\right) = (0.4338)[h(\text{ft})]$$

Note that the factor 0.4338 has the proper units that give the pressure in lb/in² when the value h is input in feet. Next, we generate Table 10.3 by substituting for h in increments of 10 ft in the preceding equation. From Table 10.3, we can see that in order to create a 50-psi water pressure in a pipeline at the base of the tower, the water level in the water tower should be approximately 120 ft.

TABLE 10.3	The Relationship Between the Height of Water in a Water Tower and the Pressure in a Pipeline at Its Base

Water Level in the Tower (ft)	Water Pressure (lb/in^2)	Water Level in the Tower (ft)	Water Pressure (lb/in^2)
10	4.3	130	56.4
20	8.7	140	60.7
30	13.0	150	65.0
40	17.3	160	69.4
50	21.7	170	73.7
60	26.0	180	78.1
70	30.4	190	82.4
80	34.7	200	86.8
90	39.0	210	91.1
100	43.4	220	95.4
110	47.7	230	99.8
120	52.0	240	104.1

Atmospheric Pressure

Earth's atmospheric pressure is due to the weight of the air in the atmosphere above the surface of the earth. It is the weight of the column of air (extending all the way to the outer edge of the atmosphere) divided by a unit area at the base of the air column. Standard atmosphere at sea level has a value of 101.325 kPa. From the definition of **atmospheric pressure** you should realize that it is a function of altitude. The variation of standard atmospheric pressure and air density with altitude is given in Table 10.4. For example, the atmospheric pressure and air density at the top of Mount Everest is approximately 31 kPa and 0.467 kg/m^3, respectively. The elevation of Mount Everest is approximately 9000 m (exactly 8848 m), and thus less air lies on top of Mount Everest than there is at sea level. Moreover, the air pressure at the top of Mount Everest is 30% of the value of atmospheric pressure at sea level, and the air density at its top is only 38% of air density at sea level. New commercial planes have a cruising altitude capacity of approximately 11,000 m. Referring to Table 10.4, you see the atmospheric pressure at that altitude is approximately one-fifth of the sea-level value, so there is a need for pressurizing the cabin. Also, because of the lower air density at that altitude, the power required to overcome air resistance (to move the plane through atmosphere) is not as much as it would be at lower altitudes. Atmospheric pressure is commonly expressed in one of the following units: pascals, pound per square inch, millimeters of mercury, and inches of mercury. Their values are

1 atm = 101325 Pa = 101.3 kPa

1 atm = 14.69 lb/in²

1 atm = 760 mm·Hg

1 atm = 29.92 in·Hg

Study Example 10.9 to see how these and other units of atmospheric pressure are related.

Absolute Pressure and Gauge Pressure

Most pressure gauges show the magnitude of the pressure of a gas or a liquid relative to the local atmospheric pressure. For example, when a tire pressure gauge shows 32 psi, this means the pressure of air inside the tire is 32 psi

TABLE 10.4	Variation of Standard Atmosphere with Altitude	
Altitude (m)	**Atmospheric Pressure (kPa)**	**Air Density (kg/m³)**
0 (sea level)	101.325	1.225
500	95.46	1.167
1000	89.87	1.112
1500	84.55	1.058
2000	79.50	1.006
2500	74.70	0.957
3000	70.11	0.909
3500	65.87	0.863
4000	61.66	0.819
4500	57.75	0.777
5000	54.05	0.736
6000	47.22	0.660
7000	41.11	0.590
8000	35.66	0.526
9000	30.80	0.467
10,000	26.50	0.413
11,000	22.70	0.365
12,000	19.40	0.312
13,000	16.58	0.266
14,000	14.17	0.228
15,000	12.11	0.195

Data from U.S. Standard Atmosphere (1962)

EXAMPLE 10.9

Starting with an atmospheric pressure of 101,325 Pa, express the magnitude of the pressure in the following units: (a) millimeters of mercury (mm Hg), (b) inches of mercury (in Hg), (c) meters of water, and (d) feet of water. The densities of water and mercury are $\rho_{H_2O} = 1,000$ kg/m^3 and $\rho_{Hg} = 13,550$ kg/m^3 respectively.

(a) We start with the relationship between the height of a fluid column and the pressure at the base of the column, which is

$$P = \rho g h = 101,325 \left(\frac{N}{m^2}\right) = 13,550 \left(\frac{kg}{m^3}\right)\left[9.81\left(\frac{m}{s^2}\right)\right]h\,(m)$$

And solving for h, we have

$$h = 0.76 \text{ m} = 760 \text{ mm}$$

Therefore, the pressure due to a standard atmosphere is equal to the pressure created at a base of a 760-mm-tall column of mercury. Stated another way, 1 atm = 760 mm Hg.

(b) Next, we will convert the magnitude of pressure from millimeters of mercury to units that express it in inches of mercury.

$$760\,(\text{mm Hg})\left(\frac{0.03937 \text{ in.}}{1 \text{ mm}}\right) = 30 \text{ in Hg}$$

(c) To express the magnitude of pressure in meters of water, as we did in part (a), we begin with the relationship between the height of a fluid column and the pressure at the base of the column, which is

$$P = \rho g h = 101,325 \left(\frac{N}{m^2}\right) = 1,000 \left(\frac{kg}{m^3}\right)\left[9.81\left(\frac{m}{s^2}\right)\right]h\,(m)$$

And then solve for h, which results in

$$h = 10.328 \text{ m}$$

Therefore, the pressure due to a standard atmosphere is equal to the pressure created at a base of a 10.328-m-tall column of water, that is, 1 **atm** = 10.328 m H$_2$O.

(d) Finally we convert the units of pressure from meters of water to feet of water.

$$10.328\,(\text{mm H}_2\text{O})\left(\frac{3.280 \text{ ft}}{1 \text{ m}}\right) = 33.87 \text{ ft H}_2\text{O}$$

above the local atmospheric pressure. In general, we can express the relationship between the **absolute** and the **gauge pressure** by

$$P_{absolute} = P_{gauge} + P_{atmospheric}$$ **10.16**

Vacuum refers to pressures below atmospheric level. Thus, negative gauge pressure readings indicate vacuum. Some of you, in your high school physics classes, may have seen demonstrations dealing with an enclosed container connected to a vacuum pump. As the vacuum pump draws air out of the container, thus creating a vacuum, the atmospheric pressure acting on the outside of the container made the container collapse. As more air is drawn, the container continued to collapse. An absolute zero pressure reading in a container indicates absolute vacuum, meaning there is no more air left in the container. In practice, achieving absolute zero pressure is not possible, meaning at least a little air will remain in the container.

EXAMPLE 10.10

We have used a tire gauge and measured the air pressure inside a car tire to be 35.0 psi. What is the absolute pressure of the air inside the tire, if the car is located in (a) a city located at sea level, (b) a city located in Colorado with an elevation of 1500 m? Express your results in both units of psi and pascals.

The absolute pressure is related to the gauge pressure according to

$$P_{\text{absolute}} = P_{\text{gauge}} + P_{\text{atmospheric}}$$

(a) For a city located at sea level, $P_{\text{atmospheric}} = 14.69 \text{ psi} \approx 14.7 \text{ psi}$,

$$P_{\text{absolute}} = 35.0 \text{ psi} + 14.7 \text{ psi} = 49.7 \text{ psi}$$

$$P_{\text{absolute}} = 49.7 (\text{psi}) \left(\frac{6895 \text{ Pa}}{1 \text{ psi}} \right)$$

$$= 342,680 \text{ Pa} = 342.68 \text{ kPa}$$

(b) We can use Table 10.4 to look up the standard atmospheric pressure for a city located in Colorado with an elevation of 1500 m. The atmospheric pressure at an elevation of 1500 m is $P_{\text{atmospheric}} = 84.55 \text{ kPa}$.

$$P_{\text{gauge}} = 35.0 (\text{psi}) \left(\frac{6895 \text{ Pa}}{1 \text{ psi}} \right) = 241,325 \text{ Pa}$$

$$P_{\text{absolute}} = 241,325 + 84,550 = 325,875 \text{ Pa}$$

$$= 325.875 \text{ kPa}$$

$$P_{\text{absolute}} = 325,875 (\text{Pa}) \left(\frac{1 \text{ psi}}{6895 \text{ Pa}} \right) = 47.3 \text{ psi}$$

EXAMPLE 10.11

A vacuum pressure gauge monitoring the pressure inside a container reads 200 mm Hg vacuum. What are the gauge and absolute pressures? Express results in pascals.

The stated vacuum pressure is the gauge pressure, and to convert its units from millimeters of mercury to pascal, we use the conversion factor 1 mm Hg = 133.3 Pa. This results in

$$P_{\text{gauge}} = -200 (\text{mm Hg}) \left(\frac{133.3 \text{ Pa}}{1 \text{ mm Hg}} \right) = -26,660 \text{ Pa}$$

The absolute pressure is related to the gauge pressure according to

$$P_{absolute} = P_{gauge} + P_{atmospheric}$$
$$P_{absolute} = -26,660 + 101,325 = 74,665 \text{ Pa}$$

Note that the negative gauge pressure indicates vacuum, or a pressure below atmospheric level.

Vapor Pressure of a Liquid

You may already know from your everyday experience that under the same surrounding conditions some fluids evaporate faster than others do. For example, if you were to leave a pan of water and a pan of alcohol side by side in a room, the alcohol would evaporate and leave the pan long before you would notice any changes in the level of water. That is because the alcohol has a higher **vapor pressure** than water. Thus, under the same conditions, fluids with low vapor pressure, such as glycerin, will not evaporate as quickly as those with high values of vapor pressure. Stated another way, at a given temperature the fluids with low vapor pressure require relatively smaller surrounding pressure at their free surface to prevent them from evaporating.

Blood Pressure We all have been to a doctor's office. One of the first things that a doctor or a nurse will measure is our blood pressure (Figure 10.24). Blood pressure readings are normally given by two numbers, for example 115/75 (mm Hg). The first, or the higher, number corresponds to **systolic pressure**, which is the maximum pressure exerted when our heart contracts. The second, or the lower number, measures the **diastolic** blood **pressure**, which is the pressure in the arteries when the heart is at rest. The pressure unit commonly used to represent blood pressure is millimeters of mercury (mm Hg). You may already know that blood pressure may change with level of activity, diet, temperature, emotional state, and so on.

FIGURE 10.24   Measuring blood pressure.
Andresr/Shutterstock.com

Hydraulic Systems

Hydraulic digging machines and bulldozers perform tasks in hours that used to take days or weeks. Today, hydraulic systems are readily used in many applications, including the braking and steering systems in cars, the aileron or flap control system of an airplane, and jacks to lift cars. To understand hydraulic systems, you must first fully understand the concept of pressure. In an enclosed fluid system, when a force is applied at one point, creating a pressure at that point, that force is transmitted through the fluid to other points, provided the fluid is incompressible. When you press your car's brake pedal, that action creates a force and subsequently a pressure in the brake master cylinder. The force is then transmitted through the hydraulic fluid from the master cylinder to the wheel cylinders where the pistons in the wheel cylinders, which push the brake shoes against the brake drums or rotors. The mechanical power is converted to hydraulic power and back to mechanical power. We will discuss the idea of what we mean by power later in this chapter and in Chapter 13. The typical hydraulic fluid pressure in a car's hydraulic brake lines is approximately 10 MPa (1500 psi).

Now let us formulate a general relationship among force, pressure, and area in a hydraulic system such as the simple hydraulic system shown in Figure 10.25. Because the pressure is nearly constant in the hydraulic system shown, the relationship between the forces F_1 and F_2 could be established in the following manner:

$$P_1 = \frac{F_1}{A_1} \tag{10.17}$$

$$P_2 = \frac{F_2}{A_2} \tag{10.18}$$

$$P_1 = P_2 = \frac{F_1}{A_1} = \frac{F_2}{A_2} \tag{10.19}$$

$$F_2 = \frac{A_2}{A_1} F_1 \tag{10.20}$$

It is important to realize that when the force F_1 is applied, the drive piston (piston 1) moves by a distance L_1, while the driven piston (piston 2) moves by a shorter distance, L_2. However, the volume of the fluid displaced is constant,

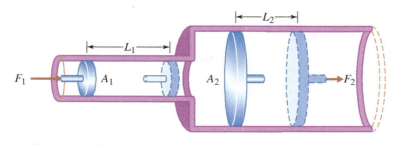

FIGURE 10.25 An example of a simple hydraulic system.

as shown in Figure 10.25. Knowing that in an enclosed system, the volume of the fluid displaced remains constant, we can then determine the distance L_2 by which piston 2 moves, provided the displacement of piston 1, L_1, is known.

$$\text{fluid volume displaced} = A_1 L_1 = A_2 L_2 \qquad \boxed{10.21}$$

$$L_2 = \frac{A_1}{A_2} L_1 \qquad \boxed{10.22}$$

Moreover, a relationship between the speed of the drive piston and the driven piston can be formulated by dividing both sides of Equation (10.22) by the time that it takes to move each piston. This leads to the following equation:

$$V_2 = \frac{A_1}{A_2} V_1 \qquad \boxed{10.23}$$

where V_1 represents the speed of the drive piston, and V_2 is the speed at which the driven piston moves.

Next, we will explain the operation and components of manually activated and pump-driven hydraulic systems. Examples of these types of hydraulic systems are depicted in Figures 10.26(a) and (b). Figure 10.26(a) shows a schematic diagram of a hand-driven hydraulic jack. The system consists of a reservoir, a hand pump, the load piston, a relief valve, and a high-pressure check valve. To raise the load, the arm of the hand pump is pushed downward; this action pushes the fluid into the load cylinder, which in turn creates a pressure that is transmitted to the load piston, and consequently the load is raised. To lower the load, the release valve is opened. The amount by which the release valve is opened will determine the speed at which the load will be lowered. Of course, the viscosity of the hydraulic fluid and the magnitude of the load will also affect the lowering speed. In the system shown, the fluid reservoir is necessary to supply the line with as much fluid as needed to extend the driven piston to any desired level.

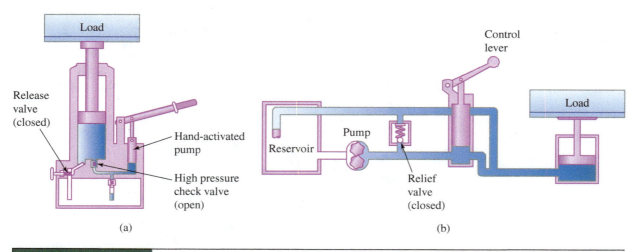

(a)   (b)

FIGURE 10.26   Two examples of hydraulic systems: (a) A hand-activated system. (b) A gear or rotary pump system.

The hydraulic system shown in Figure 10.26(b) replaces the hand-activated pump by a gear or a rotary pump that creates the necessary pressure in the line. As the control handle is moved up, it opens the passage that allows the hydraulic fluid to be pushed into the load cylinder. When the control handle is pushed down, the hydraulic fluid is routed back to the reservoir, as shown. The relief valves shown in Figure 10.26 could be set at desired pressure levels to control the fluid pressure in the lines by allowing the fluid to return to the reservoir as shown.

EXAMPLE 10.12

Determine the load that can be lifted by the hydraulic system shown in Figure 10.27. Use $g = 9.81$ m/s^2. We use Equation (10.20) to solve this problem.

$$F_1 = m_1 g = (100 \text{ kg})(9.81 \text{ m/s}^2) = 981 \text{ N}$$

$$F_2 = \frac{A_2}{A_1} F_1 = \frac{\pi(0.15 \text{ m})^2}{\pi(0.05 \text{ m})^2}(981 \text{ N}) = 8829 \text{ N}$$

$$F_2 = 8829 \text{ N} = (m_2 \text{kg})(9.81 \text{ m/s}^2) \Rightarrow m_2 = 900 \text{ kg}$$

Note that for this problem we could have started with the equation that relates F_2 to F_1 and then simplified the similar quantities such as π and g in the following manner:

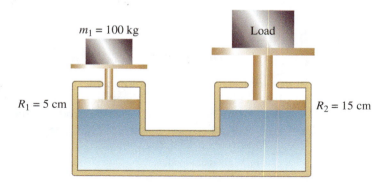

FIGURE 10.27   The hydraulic system of Example 10.12.

$$F_1 = \frac{A_2}{A_1} F_1 = m_2 g = \frac{\pi(R_2)^2}{\pi(R_1)^2}(m_1 g)$$

$$m_2 = \frac{(R_2)^2}{(R_1)^2} m_1 = \frac{(15 \text{ cm})^2}{(5 \text{ cm})^2}(100 \text{ kg}) = 900 \text{ kg}$$

This approach is preferred over the direct substitution of values into the equation right away because it allows us to change the value of a variable, such as m_1 or the dimensions of the pistons, and see what happens to the result. For example, using the second approach, we can see clearly that if m_1 is increased to a value of 200 kg, then m_2 changes to 1800 kg.

Stress

Geometric properties of an object such as its length, area, and first and second moment of area also play significant roles in the way the object responds to a force. As explained previously, when an object is subjected to an external force, internal forces are created inside the material to hold the material and the components together. **Stress** provides a measure of the intensity of internal forces acting over an area. Consider the situation shown in Figure 10.28. The plate shown is subjected to the compressive force. Because the force is applied at an angle, it has two components: a horizontal and a vertical component. The tendency of the horizontal component of the force is to shear the plate, and the tendency of the vertical component is to compress the plate.

The ratio of the normal (vertical) component of the force to the area is called the *normal stress*, and the ratio of the horizontal component of the force (the component of the force that is parallel to the plate surface) to the area is called **shear stress**. The normal stress component is often called **pressure**.

Another important geometric property of an object is its second moment of area. As we explained in Chapter 7, second moment of area provides information on the amount of resistance offered by a member to forces and moments that cause bending. The length also plays an important role when it comes to bending an object and the stresses that are associated with it. The force acting on each beam shown in Figure 10.29 will tend to stretch the bottom fibers and compress the top fibers of the beam. Which beam in Figure 10.29 is easier to bend and why? Explain.

Modulus of Elasticity, Modulus of Rigidity, and Bulk Modulus of Compressibility

In this section, we will discuss some important properties of materials. The way an object responds to an application of a force also depends on its material properties. Engineers, when designing products and structural members, need to know how a selected material behaves under applied forces or how well

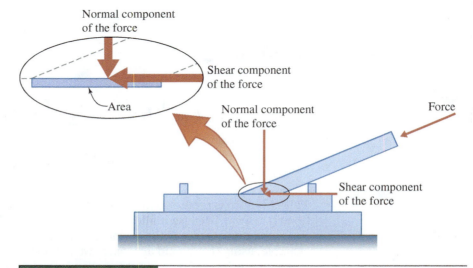

FIGURE 10.28   An anchoring plate subjected to a compressive and a shearing force.

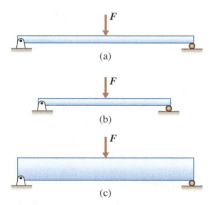

FIGURE 10.29 Simple examples to demonstrate the effect of length and cross-sectional area on bending.

the material conducts thermal energy or electricity. Here we will look at ways of quantifying (measuring) the response of solid materials to pulling or pushing forces or twisting torques. We will also investigate the behavior of fluids in response to applied pressures in terms of bulk modulus of compressibility.

Tensile tests are performed to measure the modulus of elasticity and strength of solid materials. A test specimen with a size in compliance with the American Society for Testing and Materials (ASTM) standards is made and placed in a tensile testing machine (Figure 10.30). When a material specimen

FIGURE 10.30 An example of a tensile test machine.
© age fotostock/Alamy

is tested for its strength, the applied tensile load is increased slowly. In the very beginning of the test, the material will deform elastically, meaning that if the load is removed, the material will return to its original size and shape without any permanent deformation. The point to which the material exhibits this elastic behavior is called *elastic point.* As the material stretches, the normal stress (the normal force divided by the cross-sectional area) is plotted versus the strain (deformation divided by the original length of the specimen). An example of such test results for a steel sample is shown in Figure 10.31.

The modulus of elasticity, or Young's modulus, is computed by calculating the slope of a stress-strain diagram over the elastic region. The **modulus of elasticity** is a measure of how easily a material will stretch when pulled (subject to a tensile force) or how well the material will shorten when pushed (subject to a compressive force). The larger the value of the modulus of elasticity, the larger the force required to stretch or shorten the material by a certain amount. In order to better understand what the value of the modulus of elasticity represents, consider the following example: Given a piece of rubber, a piece of aluminum, and a piece of steel, all having the same rectangular shape, cross- sectional area, and original length as shown in Figure 10.32, which piece will stretch more when subjected to the same force *F*?

As you already know from your experience, it will require much less effort to elongate the piece of rubber than it does to elongate the aluminum piece or the steel bar. In fact, the steel bar will require the greatest force to be elongated by the same amount when compared to the other samples. This is because steel has the highest modulus of elasticity value of the three samples. The results of our observation—in terms of the example cited—are expressed,

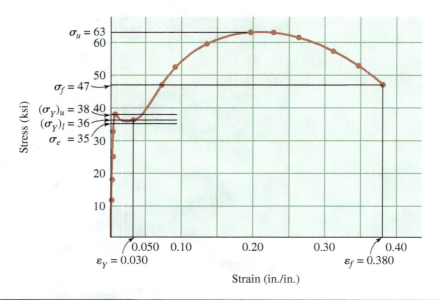

FIGURE 10.31   Stress-strain diagram for a mild steel sample
$(\text{ksi} = 1000 \text{ lb/in}^2, \sigma_e = \text{elastic stress},$
$(\sigma_Y)_u = \text{upper-yield stress}, (\sigma_Y)_l = \text{lower-yield stress},$
$\sigma_u = \text{ultimate stress}, \text{ and } \sigma_f = \text{fracture stress}).$
Based on R. C. Hibbler, *Mechanics of Materials.*

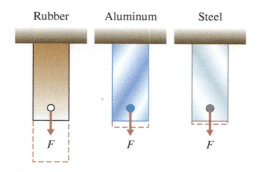

| FIGURE 10.32 | When subjected to the same force, which piece of material will stretch more? |

in a general form that applies to all solid materials, by Hooke's law. We discussed Hooke's law earlier when we were explaining springs. Hooke's law also applies to stretching or compressing a solid piece of material. However, for solid pieces of material, Hooke's law is expressed in terms of stress and strain according to

$$\sigma = E\varepsilon \qquad \boxed{10.24}$$

where

σ = normal stress (N/m^2 or lb/in^2)

E = modulus of elasticity (sometimes called Young's modulus), a property of the material (N/m^2 or lb/in^2)

ε = normal strain, the ratio of change in length to original length (dimensionless)

As we explained earlier, the slope of a stress—strain diagram is used to compute the value of the modulus of elasticity for a solid specimen. Equation (10.24) defines the equation of the line that relates the stress and strain on the stress—strain diagram.

As an alternative approach, we will next explain a simple procedure that can be used to measure E. Although the following procedure is not the formal way to obtain the modulus of elasticity value for a material, it is a simple procedure that provides additional insight into the modulus of elasticity. Consider a given piece of rectangular-shaped material with an original length L and cross-sectional area A, as shown in Figure 10.33. When subjected to a known force F, the bar will stretch; by measuring the amount the bar stretched (final length of the bar minus the original length), we can determine the modulus of elasticity of the material in the following manner. Starting with Hooke's law, $\sigma = E\varepsilon$, and substituting for σ and the strain ε, in terms of their elementary definitions, we have

$$\frac{F}{A} = E\frac{x}{L} \qquad \boxed{10.25}$$

Rearranging Equation (10.25) to solve for modulus of elasticity E, we have

$$E = \frac{FL}{Ax} \qquad \boxed{10.26}$$

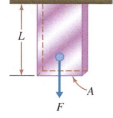

| FIGURE 10.33 |

A rectangular bar subjected to a tensile load.

where

    E = modulus of elasticity (N/m^2)
    F = the applied force (N)
    L = original length of the bar (m)
    A = cross-sectional area of the bar (m^2)
    x = elongation, final length minus original length (m)

Note that the physical variables involved include the fundamental base dimensions length and area from Chapter 7 and the physical variable force from this chapter. Note also from Equation (10.26) that the modulus of elasticity, E, is inversely proportional to elongation x; the more the material stretches, the smaller the value of E. Values of the modulus of elasticity E for selected materials are shown in Table 10.5.

> The modulus of elasticity is a measure of how easily a material will stretch when pulled or how easily the material will shorten when pushed. The larger the value of the modulus of elasticity, the larger the force required to stretch or shorten the material by a certain amount.

Another important mechanical property of material is the **modulus of rigidity** or *shear modulus*. The modulus of rigidity is a measure of how easily a material can be twisted or sheared. The value of the modulus of rigidity shows the resistance of a given material to shear deformation. Engineers consider the value of shear modulus when selecting materials for shafts or rods that are subjected to twisting torques. For example, the modulus of rigidity or shear modulus for aluminum alloys is in the range of 26 to 36 GPa, whereas the shear modulus for steel is in the range of 75 to 80 GPa. Therefore, steel is approximately three times more rigid in shear than aluminum. The shear modulus is measured using a torsional test machine. A cylindrical specimen of known dimensions is twisted with a known torque. The angle of twist is measured and is used to determine the value of the shear modulus (see Example 10.14). The values of shear modulus for various solid materials are given in Table 10.5.

Based on their mechanical behavior, solid materials are commonly classified as either *ductile* or *brittle*. A ductile material, when subjected to a tensile load, will go through significant permanent deformation before it breaks. Steel and aluminum are good examples of ductile materials. On the other hand, a brittle material shows little or no permanent deformation before it ruptures. Glass and concrete are examples of brittle materials.

The tensile and compressive strength are other important properties of materials. To predict failure, engineers perform stress calculations that are compared to the tensile and compressive strength of materials. The tensile strength of a piece of material is determined by measuring the maximum tensile load a material specimen in a shape of a rectangular bar or cylinder can carry without failure. The *tensile strength* or ultimate strength of a material is expressed as the maximum tensile force per unit of the original cross-sectional area of the specimen. As we mentioned earlier, when a material specimen is tested for its strength, the applied tensile load is increased slowly so that one can determine the modulus of elasticity and the elastic region. In the very beginning of the test, the material will deform elastically, meaning that if the load is removed, the material will return to its original size and shape without any permanent deformation. The point to which the material exhibits this elastic behavior is the elastic point. The elastic strength represents the maximum load that the material can carry without any permanent deformation. In many engineering design applications, the elastic strength or yield strength (the yield value is very close to the elastic strength value) is used as the tensile strength.

TABLE 10.5	Modulus of Elasticity and Shear Modulus of Selected Materials	
Material	**Modulus of Elasticity (GPa)**	**Shear Modulus (GPa)**
Aluminum alloys	70–79	26–30
Brass	96–110	36–41
Bronze	96–120	36–44
Cast iron	83–170	32–69
Concrete (compression)	17–31	
Copper alloys	110–120	40–47
Glass	48–83	19–35
Magnesium alloys	41–45	15–17
Nickel	210	80
Plastics		
Nylon	2.1–3.4	
Polyethylene	0.7–1.4	
Rock (compression)		
Granite, marble, quartz	40–100	
Limestone, sandstone	20–70	
Rubber	0.0007–0.004	0.0002–0.001
Steel	190–210	75–80
Titanium alloys	100–120	39–44
Tungsten	340–380	140–160
Wood (bending)		
Douglas fir	11–13	
Oak	11–12	
Southern pine	11–14	

Based on Gere, *Mechanics of Materials*, 5E. 2001, Cengage Learning.

Some materials are stronger in compression than they are in tension; concrete is a good example. The compression strength of a piece of material is determined by measuring the maximum compressive load a material specimen in a shape of a cube or cylinder can carry without failure. The ultimate *compressive strength* of a material is expressed as the maximum compressive force per unit of the cross-sectional area of the specimen. Concrete has a compressive strength in the range of 10 to 70 MPa. The strength of some selected materials is given in Table 10.6.

One of the goals of most structural analysis is to check for failure. The prediction of failure is quite complex in nature; consequently, many investigators have been studying this topic. In engineering design, to compensate for

TABLE 10.6   The Strength of Selected Materials

Material	Yield Strength (MPa)	Ultimate Strength (MPa)
Aluminum alloys	35–500	100–550
Brass	70–550	200–620
Bronze	82–690	200–830
Cast iron (tension)	120–290	69–480
Cast iron (compression)		340–1,400
Concrete (compression)		10–70
Copper alloys	55–760	230–830
Glass		30–1,000
Plate glass		70
Glass fibers		7,000–20,000
Magnesium alloys	80–280	140–340
Nickel	100–620	310–760
Plastics		
Nylon		40–80
Polyethylene		7–28
Rock (compression)		
Granite, marble, quartz		50–280
Limestone, sandstone		20–200
Rubber	1–7	7–20
Steel		
High-strength	340–1,000	550–1,200
Machine	340–700	550–860
Spring	400–1,600	700–1,900
Stainless	280–700	400–1,000
Tool	520	900
Steel wire	280–1,000	550–1,400
Structural steel	200–700	340–830
Titanium alloys	760–1,000	900–1,200
Tungsten		1,400–4,000
Wood (bending)		
Douglas fir	30–50	50–80
Oak	40–60	50–100
Southern Pine	40–60	50–100
Wood (compression parallel to grain)		
Douglas fir	30–50	40–70
Oak	30–40	30–50
Southern pine	30–50	40–70

Based on Gere, *Mechanics of Materials*, 5E. 2001, Cengage Learning.

what we do not know about the exact behavior of material and/or to account for future loading for which we may have not accounted but to which someone may subject the part or the structural member, we introduce a factor of safety (F.S.), which is defined as

$$\text{F.S.} = \frac{P_{max}}{P_{allowable}}$$

$\boxed{10.27}$

where P_{max} is the load that can cause failure. For certain situations, it is also customary to define the factor of safety in terms of the ratio of maximum stress that causes failure to the allowable stresses if the applied loads are linearly related to the stresses. The factor of safety for Example 10.13 using an alumnium alloy with a yield strength of 50 MPa is 3.1.

EXAMPLE 10.13

A structural member with a rectangular cross section, as shown in Figure 10.34, is used to support a load of 4000 N distributed uniformly over the cross-sectional area of the member. What type of material should be used to carry the load safely?

Material selection for structural members depends on a number of factors, including the density of the material, its strength, its toughness, its reaction to the surrounding environment, and its appearance. In this example, we will only consider the strength of the material as the design factor. The average normal stress in the member is given by

$$\sigma = \frac{4000 \text{ N}}{(0.05 \text{ m})(0.005 \text{ m})} = 16 \text{ MPa}$$

Aluminum alloy or structural steel material with the yield strength of 50 MPa and 200 MPa, respectively, could carry the load safely.

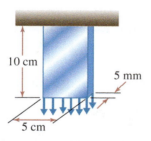

FIGURE 10.34   The structural member of Example 10.13.

Bulk Modulus of Compressibility

Most of you have pumped air into a bicycle tire at one time or another. From this and other experiences you know that gases are more easily compressed than liquids. In engineering, to see how compressible a fluid is, we look up the value of a **bulk modulus of compressibility** of the fluid. The value of fluid bulk modulus shows how easily the volume of the fluid can be reduced when the pressure acting on it is increased.

EXAMPLE 10.14

A setup similar to the one shown in Figure 10.35 is commonly used to measure the shear modulus, G. A specimen of known length L and diameter D is placed in the test machine. A known torque T is applied to the specimen and the angle of twist ϕ is measured. The shear modulus is then calculated from

$$G = \frac{32\,TL}{\pi\,D^4\,\phi} \qquad \text{10.28}$$

Using Equation (10.28), calculate the shear modulus for a given specimen and test results of $T = 3450$ N·m, $L = 20$ cm, $D = 5$ cm, $\phi = 0.015$ rad.

$$G = \frac{32\,TL}{\pi\,D^4\phi} = \frac{32\,(3450\ \text{N·m})(0.2\ \text{m})}{\pi\,(0.05\ \text{m})^4\,(0.015\ \text{rad})} = 75\ \text{GPa}$$

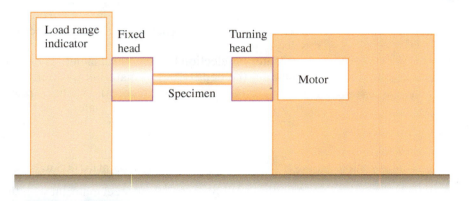

FIGURE 10.35   A setup to measure the shear modulus of a material.

The bulk modulus of compressibility is formally defined as

$$E_V = \frac{\text{increase in pressure}}{\dfrac{\text{decrease in volume}}{\text{original volume}}} = \frac{\text{increase in pressure}}{\dfrac{\text{increase in density}}{\text{original density}}} \qquad \text{10.29}$$

Equation (10.29) expressed mathematically is

$$E_V = \frac{dP}{-\dfrac{dV}{V}} = \frac{dP}{\dfrac{d\rho}{\rho}} \qquad \text{10.30}$$

where

E_V = modulus of compressibility (N/m^2)
dP = change in pressure (increase in pressure) (N/m^2)
dV = change in volume (decrease in volume, negative value) (m^3)

V = original volume (m^3)
$d\rho$ = change in density (increase in density) (kg/m^3)
ρ = original density (kg/m^3)

When the pressure is increased (a positive change in value), the volume of the fluid is decreased (a negative change in value), and thus the need for a minus sign in front of the dV term to make E_v a positive value. In Equation (10.30), also note that when the pressure is increased (a positive change in value), the density of the fluid is also increased (a positive change in value), and thus there is no need for a minus sign in front of the $d\rho$ term to make E_v a positive value. The compressibility values for some common fluids are given in Table 10.7.

TABLE 10.7   The Values of Compressibility Modulus for Some Common Fluids

Fluid	Compressibility Modulus (N/m^2)	Compressibility Modulus (lbf / in^2)
Ethyl alcohol	1.06×10^9	1.54×10^5
Gasoline	1.3×10^9	1.9×10^5
Glycerin	4.52×10^9	6.56×10^5
Mercury	2.85×10^{10}	4.14×10^6
SAE 30 oil	1.5×10^9	2.2×10^5
Water	2.24×10^9	3.25×10^5

EXAMPLE 10.15

When we were defining the bulk modulus of compressibility we said that as the pressure is increased, the volume of fluid is decreased, and consequently the density of the fluid is increased. Why does the density of a fluid increase when its volume is decreased?

Recall that density is defined as the ratio of mass to volume, as given by

$$\text{density} = \frac{\text{mass}}{\text{volume}}$$

From examining this equation, for a given mass, you can see that as the volume is reduced, the density is increased.

EXAMPLE 10.16

For water, starting with 1 m^3 of water in a container, what is the pressure required to decrease the volume of water by 1%?

From Table 10.7, the compressibility modulus for water is 2.24×10^9 N/m^2. It would require a pressure of 2.24×10^7 N/m^2 to decrease the original 1 m^3 volume of water to a final volume of 0.99 m^3, or, said another way, by 1%. It would require an equivalent pressure of 221 atm $(2.24 \times 10^7$ $N/m^2)$ to reduce the pressure of a unit volume of water by 1%.

LO⁶ 10.6   Linear Impulse—Force Acting Over Time

Up to this point, we have defined the effects of a force acting at a distance in terms of creating moment, and through a distance as doing work. Now we will consider force acting over a period of time. Understanding linear impulse and impulsive forces is important in the design of products such as air bags and sport helmets to prevent injuries. Understanding impulsive forces also helps with designing cushion materials to prevent damage to products when dropped or when subjected to impact. On TV or in the movies, you have seen a stuntman jumping off a roof of a multistory building onto an air mat on the ground and not getting hurt. Had he jumped onto concrete pavement, the same stuntman would most likely be killed. Why is that? Well, by using an inflated air mat, the stuntman is increasing his time of contact with the ground through staying in touch with the air mat for a long time before he is completely stopped. This statement will make more sense after we show the relationship between the linear impulse and linear momentum.

Linear impulse represents the net effect of a force acting over a period of time. There is a relationship between the linear impulse and the linear momentum. We explained what we mean by linear momentum in Section 9.5. A force acting on an object over a period of time creates a linear impulse that brings about a change in the linear momentum of the object, according to

$$\vec{F}_{\text{average}} \, \Delta t = m\vec{V}_f - m\vec{V}_i \qquad \boxed{10.31}$$

where

$\quad F_{\text{average}}$ = average magnitude of the force acting on the object (N)
$\quad \Delta t$ = time period over which the force acts on the object (s)
$\quad m$ = mass of the object (kg)
$\quad V_f$ = final velocity of the object (m/s)
$\quad V_i$ = initial velocity of the object (m/s)

It is important to note that Equation (10.31) is a vectoral relationship, meaning it has both magnitude and direction. Using Equation (10.31), let us now explain why the stuntman does not hurt himself when he jumps onto the air mat (Figure 10.36). For the sake of demonstration, let us assume that the stuntman is jumping off a ten-story building, where the average height of each floor is 15 ft. Moreover, let us assume that he jumps onto an inflated air mat that is 15 ft tall. Neglecting air resistance during his jump to make the calculation simpler, the stuntman's velocity right before he hits the air mat could be determined from

$$V_i = \sqrt{2gh} \qquad \boxed{10.32}$$

where V_i represents the velocity of the stuntman right before he hits the air mat, g is the acceleration due to gravity $(g = 32.2 \text{ ft/s}^2)$, and h is the height of the building minus the height of the air mat $(h = 135 \text{ ft})$. Substituting for g and h in Equation (10.32) leads to an initial velocity of $V_i = 93.2 \text{ ft/s}$. Also, we realize that the air mat reduces the velocity of the stuntman to a final velocity of zero $(V_f = 0)$. Now we can solve for F_{average} from Equation (10.31), assuming different values for Δt, and assuming a mass of $4.65 \text{ slugs} (150 \text{ lbm})$. The results of these calculations are summarized and given in Table 10.8. It is important to note that F_{average} represents the force that the stuntman exerts on the

> Linear impulse represents the net effect of a force acting over a period of time.

FIGURE 10.36   Two stuntmen practice a fall.

TABLE 10.8   **The Average Reaction Force Acting on a Stuntman**

Time of Contact (s)	Average Reaction Force (lbf)
0.1	4334
0.5	867
1.0	433
2.0	217
5.0	87
10.0	43

air mat and subsequently through the air to the ground. But you recall from Newton's third law that for every action there is a reaction equal in magnitude and opposite in direction. Therefore, the average force exerted by the ground on the stuntman is equal in magnitude to $F_{average}$. Also note that the reaction force will be distributed over the back surface of the stuntman's body and thus create a relatively small pressure distribution on his back. You can see from these calculations that the longer the time of contact, the lower the reaction force, and consequently the lower the pressure on the stuntman's back.

Before You Go On

Answer the following questions to test your understanding of the preceding sections:

1. Explain what is meant by pressure and stress and give examples from everyday life.

2. In your own words, explain Pascal's law.

3. What is atmospheric pressure?

4. What is the relationship between absolute pressure and gauge pressure?

5. Explain what is meant by modulus of elasticity and modulus of rigidity.

6. What is the net effect of force acting over a period of time?

Vocabulary—State the meaning of the following terms:

Pressure _____

Stress _____

Absolute Pressure _____

Gauge Pressure _____

Vapor Pressure _____

Modulus of Elasticity _____

Bulk Modulus of Rigidity _____

Linear Impulse _____

SUMMARY

LO¹   Force

The simplest form of a force that represents the interaction of two objects is a push or a pull. All forces, whether they represent the interaction of two bodies in direct contact or the interaction of two bodies at a distance (gravitational force), are defined by their magnitudes, directions, and their points of application. In SI units, one newton is defined as the force that will accelerate 1 kg of mass at a rate of 1 m/s^2. In comparison, one pound force will accelerate 1 slug at a rate of 1 ft/s^2. Engineering analyses could involve different types of forces, including external forces, internal forces, reaction forces, spring forces, friction forces, viscous forces, weight, and forces due to pressure.

LO²   Newton's Laws in Mechanics

Physical laws are based on observations. Newton's laws, which form the foundation of mechanics, are also based on observation. Mechanics is the study of behavior of objects when subjected to force. Mechanics deals with the study of behavior of objects or structural members when subjected to forces. By behavior, we mean obtaining information about the object's linear and angular displacement, velocity, acceleration, deformation, and stresses. Newton's first law states that, if a given object is at rest and if there are no unbalanced forces acting on it, the object will remain at rest. Or if the object is moving with a constant speed in a certain direction, and if there are no unbalanced forces acting on it, the object will continue to move with its constant speed and in the same direction. Newton's second law of motion states that the net effect of unbalanced forces is equal to mass times acceleration of the object. Moreover, the direction of the net forces and the direction of the acceleration will be the same. Newton's third law states that for every action there exists a reaction, and the forces of action and reaction have the same magnitude and act along the same line, but they have opposite directions.

LO³ Moment, Torque—Force Acting at a Distance

The two tendencies of an unbalanced force acting on an object are to translate the object (i.e., to move it in the direction of the unbalanced force) and to rotate or bend or twist the object. In mechanics, the tendency of force to rotate an object is measured in terms of a moment of a force about an axis or a point. Moment has both direction and magnitude. The magnitude of the moment is obtained from the product of the moment arm times the force. The moment arm represents the perpendicular distance between the line of action of the force and the point about which the object could rotate.

LO⁴ Work—Force Acting Over a Distance

Mechanical work is done when the applied force moves the object through a distance. Simply stated, mechanical work is defined as the component of the force that moves the object times the distance the object moves. Mechanical work is a scalar quantity and is equal to the change in the kinetic energy of the object.

LO⁵ Pressure and Stress—Force Acting Over an Area

Pressure and stress provide a measure of the intensity of a force acting over an area. The ratio of the normal (vertical) component of the force to the area is called the normal stress, and the ratio of the horizontal component of the force to the area is called shear stress. The normal stress component is often called pressure. In the International System of units, pressure and stress units are expressed in pascal. In U.S. Customary units, pressure or stress is commonly expressed in pounds per square inch; one lb/in^2 (1 psi) represents the pressure created by a 1-pound force over an area of 1 in^2.

Earth's atmospheric pressure is due to the weight of the air in the atmosphere above the surface of the earth. Moreover, pressure readings above atmospheric pressure are commonly referred to as gauge pressure, and vacuum refers to pressures below atmospheric level. The property of a fluid that shows how easily the volume of the fluid can be reduced when the pressure acting on it is increased is called the bulk modulus of compressibility.

The modulus of elasticity is a measure of how easily a material will stretch when pulled (subject to a tensile force) or how well the material will shorten when pushed (subject to a compressive force). The larger the value of the modulus of elasticity, the larger the force required to stretch or shorten the material by a certain amount. Moreover, the modulus of rigidity is a measure of how easily a material can be twisted or sheared.

LO⁶ Linear Impulse—Force Acting over Time

Linear impulse represents the net effect of a force acting over a period of time on an object. Linear impulse is a vector quantity and is related (equal) to the change in the linear momentum of the object.

KEY TERMS

Absolute Pressure 320
Atmospheric Pressure 315
Bulk Modulus of Compressibility 333
Diastolic Pressure 322
External Force 310
Friction Force 301
Gauge Pressure 320
Internal Force 310

Linear Impulse 296
Mechanics 303
Modulus of Elasticity 328
Modulus of Rigidity 330
Moment 307
Newton's Laws 303
Pressure 314
Reaction Force 310

Shear Stress 326
Spring Force 299
Stress 326
Systolic Pressure 322
Torque 307
Vapor Pressure 322
Viscous Force 299
Work 312

APPLY WHAT YOU HAVE LEARNED

In the past, scientists and engineers have used pendulums to measure the value of g at a location. The basic forces involved are the tension in the wire and the weight of the suspended mass. Design a pendulum that can be used to measure the value of g at your location. The formula to use to measure the acceleration due to gravity is

$$T = 2\pi \sqrt{\frac{L}{g}}$$

where T is the period of oscillation of the pendulum, which is the time that it takes the pendulum to complete one cycle. The distance between the pivot point and the center of the mass of the suspended deadweight is represented by L. Think carefully about ways to increase the accuracy of your system. Discuss your findings in a brief report.

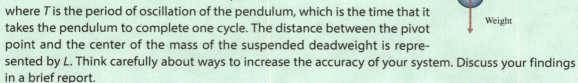

Tension

Weight

PROBLEMS

Problems that promote lifelong learning are denoted by ⚷—

10.1 Design a mass–spring system that can be taken to Mars to measure the acceleration due to gravity at the surface of Mars. Explain the basis of your design and how it should be calibrated and used.

10.2 Investigate what is meant by dead load, live load, impact load, wind load, and snow load in the design of structures.

10.3 An astronaut has a mass of 68 kg. What is the weight of the astronaut on Earth at sea level? What are the mass and the weight of the astronaut on the Moon, and on Mars? What is the ratio of the pressure exerted by the astronaut's shoe on Earth to Mars?

10.4 Calculate the attractive force between two students with masses of 70 kg and 80 kg. The students are standing about one meter apart. Compare their attractive force to their weights.

10.5 Former basketball player Shaquille O'Neal weighs approximately 335 lb and wears a size 23 shoe. Estimate the pressure he exerts on a floor. What pressure would he exert if his shoe size were 13? State all your assumptions, and show how you arrived at a reasonable solution.

10.6 Investigate the relationship between pressure for the following situations:

   a. when you are standing up barefooted

   b. when you are lying on your back

   c. when you are sitting on a tall chair with your feet off the ground and the chair

10.7 Investigate the pressure created on a surface by the following objects:

   a. a bulldozer

   b. a car

   c. a bicycle

   d. a pair of cross-country skis

10.8 Do you think there is relationship between a person's height, weight, and foot size? If so, verify your answer. Designers often learn from their natural surroundings; do the bigger animals have bigger feet?

10.9 Calculate the pressure exerted by water on the hull of a submarine that is cruising at a depth of 500 ft below ocean level. Assume the density of the ocean water is $\rho = 1025$ kg/m^3 .

10.10 Investigate the operation of pressure-measuring devices such as manometers, Bourdon tubes, and diaphragm pressure sensors. Write a brief report discussing your findings.

10.11 Convert the following pressure readings from inches of mercury units to psi and pascal units:

a. 28.5 in · Hg

b. 30.5 in · Hg

10.12 Convert the following air pressure readings from in · H$_2$O to psi and Pa.

a. 1.5 in · H$_2$O

b. 3.75 in · H$_2$O

10.13 A person with hypertension has a systolic/diastolic of 140/110 mm Hg. Express these pressures in Pa, psi, and in · H$_2$O.

10.14 If a pressure gauge on a compressed air tank reads 120 psi, assuming standard atmospheric pressure at sea level, what is the absolute pressure of air in

a. psi

b. pascal

c. feet of water

d. feet of mercury

10.15 An air compressor intakes atmospheric air at 14.7 psi and increases its pressure to 150 psi (gauge). The pressurized air is stored in a horizontal tank. What is the absolute pressure of air inside the tank? Express your answers in

a. psi

b. kPa

c. bar

d. standard atm (recall that 1 atm = 14.7 psi = 101.325 kPa)

10.16 Calculate the pressure exerted by water on a scuba diver who is swimming at a depth of 20 m below the water surface.

10.17 Investigate the typical operating pressure range in the hydraulic systems used in the following applications. Write a brief report to discuss your findings.

a. controlling wing flaps in airplanes

b. your car's steering system

c. jacks used to lift cars in a service station

10.18 Using the information given in Table 10.4, determine the ratio of local pressure and density to sea-level values. Estimate the value of air density at the cruising altitude of most commercial airliners.

Altitude (m)	$\dfrac{\rho}{\rho_{\text{sea level}}}$	$\dfrac{P}{P_{\text{sea level}}}$
0 (sea level)		
1000		
3000		
5000		
8000		
10,000		
12,000		
14,000		
15,000		

10.19 Bourdon-type pressure gauges are used in thousands of applications. A deadweight tester is a device that is used to calibrate pressure gauges. Investigate the operation of a deadweight pressure tester. Write a brief report to discuss your findings.

10.20 Investigate what the typical range of pressure is in the following applications: a bicycle tire, home water line, natural gas line, and refrigerant in your refrigerator's evaporator and condenser lines. Present your findings in a brief report.

10.21 Convert the 115/85 systolic/diastolic pressures given in millimeters of Hg to Pa, psi, and in · H_2O.

10.22 Determine the load that can be lifted by the hydraulic system shown in the accompanying figure. All of the necessary information is shown in the figure.

10.23 Determine the load that can be lifted by the system shown in Problem 10.22 if R_1 is reduced to 4 cm.

10.24 Determine the load that can be lifted by the system shown in Problem 10.22 if R_2 is increased to 25 cm.

10.25 Determine the pressure required to decrease the volume of the following fluids by 2%:

a. water

b. glycerin

c. SAE motor oil

10.26 SAE 30 oil is contained in a cylinder with inside diameter of 1 in. and length of 1 ft. What is the pressure required to decrease the volume of the oil by 2%?

10.27 Compute the deflection of a structural member made of aluminum if a load of 600 lb is applied to the bar.

The member has a uniform rectangular cross section with the dimensions of $\frac{1}{4}$ in. × 4 in., as shown in the accompanying figure.

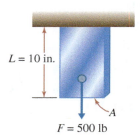

$L = 10$ in.

A

$F = 500$ lb

Problem 10.27

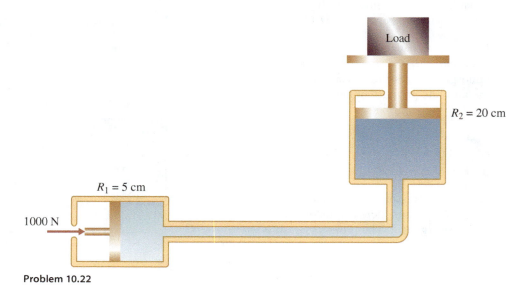

Load

$R_2 = 20$ cm

$R_1 = 5$ cm

1000 N

Problem 10.22

10.28 A 20-cm-diameter rod is subjected to a tensile load of 1000 N at one end and is fixed at the other end. Calculate the deflection of the rod if it is 50 cm long.

10.29 A structural member with a rectangular cross section as shown in the accompanying figure is used to support a load of 2500 N. What type of material do you recommend be used to carry the load safely? Base your calculations on the yield strength and a factor of safety of 2.0.

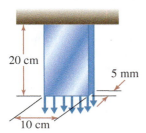

20 cm

5 mm

10 cm

Problem 10.29

10.30 The tire wrench shown in the accompanying figure is used to tighten the bolt on a wheel. Given the information on the diagram, determine the moment about point *O* for the two loading situations shown:

a. pushing perpendicular to the wrench arm

b. pushing at a 75° angle.

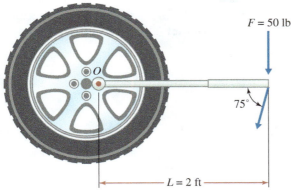

F = 50 lb

75°

L = 2 ft

Problem 10.30

10.31 Determine the moment created by the weight of the suspended sign about point *O*. Dimensions of the sign and the support are shown in the accompanying figure. The sign is 2-mm thick and is made of aluminum.

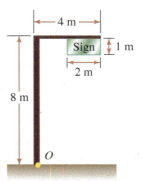

4 m

Sign   1 m

2 m

8 m

O

Problem 10.31

10.32 Determine the moment created by the weight of the lamp about point *O*. The dimensions of the street light post and arm are shown in the accompanying figure. The lamp weighs 20 lb.

5 ft

18 ft

O

Problem 10.32

10.33 Determine the moment created by the weight of the traffic light about point *O*. The dimensions of the traffic light post and arm are shown in the accompanying figure. The traffic light weighs 40 lb.

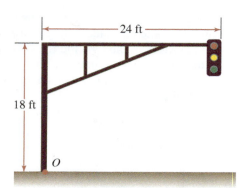

Problem 10.33

10.34 Determine the sum of the moments created by the forces shown in the accompanying figure about points A, B, C, and D.

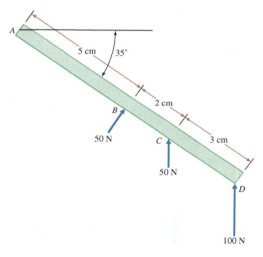

Problem 10.34

10.35 The coefficient of static friction between a concrete block and a surface is 0.8. The block weighs 20 lbf. If a horizontal force of 15 lbf is applied to the block, would the block move? And if not, what is the magnitude of friction force? What should be the magnitude of the horizontal force to set the block in motion?

10.36 Calculate the shear modulus for a given cylindrical metal speciman and test results of $T = 1500$ N $\cdot$ m, $L = 20$ cm, $D = 5$ cm, and $\varphi = 0.02$ rad. Can you tell what the material is?

10.37 Determine the amount of work done by you and a friend when each of you push on a car that has run out of gas. Assume you and your friend each push with a horizontal force with a magnitude of 200 N, a distance of 100 m. Also, suggest ways of actually measuring the magnitude and the direction of the pushing force.

10.38 Determine the work done by an electric motor lifting an elevator weighing 1000 lb through five floors; assume a distance of 15 ft between each floor. Compute the power requirements for the motor to go from the first to the fifth floor in

a. 5 s

b. 8 s

Express your results in horsepower $(1 \text{ hp} = 550 \text{ ft} \cdot \text{lb/s})$.

10.39 If a laptop computer weighing 22 N is dropped from a distance of 1 m onto a floor, determine the average reaction force from the floor if the time of contact is changed by using different cushion materials, as shown in the accompanying table.

Time of Contact (s)	Average Reaction Force (N)
0.01	
0.05	
0.1	
1.0	
2.0	

10.40 Obtain the values of vapor pressures of alcohol, water, and glycerin at a room temperature of 20° C.

10.41 When learning to play some sports, such as tennis, golf, or baseball, often you are told to follow through with your swing. Using Equation (10.31), explain why follow-through is important.

10.42 In many applications, calibrated springs are commonly used to measure the magnitude of a force. Investigate how typical force-measuring devices work. Write a brief report to discuss your findings.

10.43 Calculate the moment created by the forces shown in the accompanying figure about point O.

10.44 We have used an experimental setup similar to Example 10.1 to determine the value of a spring constant. The deflection caused by the corresponding weights are given in the accompanying table. What is the value of the spring constant?

Weight (lb)	The Deflection of the Spring (in.)
5.0	0.48
10.0	1.00
15.0	1.90
20.0	2.95

10.45 If an astronaut and her space suit weight 250 lb on Earth, what should be the volume of her suit if she is to practice for weightless conditions in an underwater neutral buoyant simulator similar to the one used by NASA, as shown in Chapter 7?

10.46 Calculate the work required to bench press a 200-lb weight, at a distance of 20 in.

10.47 Given that the three bars shown in the accompanying figure are made of the same material, comparing bar (a) to bar (b) which bar will stretch more, when subjected to the same force F? Bar (a) and (b) have the same cross-sectional area but different lengths. Comparing bar (a) to bar (c) having the same length but different cross-sectional areas, which bar will stretch more? Explain.

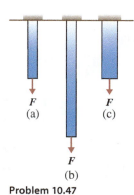

Problem 10.47

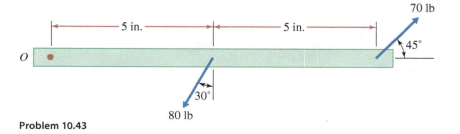

Problem 10.43

10.48 Create a table that shows the relative magnitudes of modulus of elasticity of steel to aluminum alloys, copper alloys, titanium alloys, rubber, and wood.

10.49 Consider the parallel springs shown in the accompanying figure. Realizing that the deflection of each spring in parallel is the same and the applied force must equal the sum of forces in individual springs, show that for springs in parallel the equivalent spring constant K_e is

$$K_e = K_1 + K_2 + K_3$$

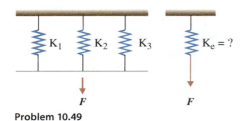

Problem 10.49

10.50 Consider the series springs shown in accompanying figure. Realizing that the total deflection of the springs are the sum of the deflections of the individual springs, and the force in each spring equals the applied force, show that for the springs in series, the equivalent spring constant K_e is

$$K_e = \cfrac{1}{\cfrac{1}{K_1} + \cfrac{1}{K_2} + \cfrac{1}{K_3}}$$

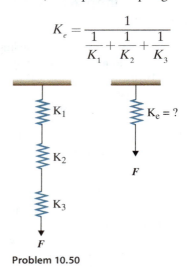

Problem 10.50

Impromptu | Design V

Objective: To design a structural (chain) link from the materials listed below that will support as much load as possible. The link must have at least a 2-in. diameter hole at its center to accommodate two pieces of rope. The pieces of ropes are used as connecting links to create a chain. Each piece of rope is passed through two adjacent designs (links) and its ends are tied. This procedure is repeated until you create a chain made of different link designs. The chain will then be pulled apart in a tug of war until a link fails. The surviving links will then be reconnected as described above and the tug of war is repeated until a winner is declared. Thirty minutes will be allowed for preparation.

Provided Materials: 4 rubber bands; 20 in. of adhesive tape; 40 in. of string; 2 sheets of paper ($8\frac{1}{2}" \times 11"$ each); 4 drinking straws.

An Engineering Marvel

Caterpillar 797 Mining Truck*

The Caterpillar® 797, which features many patented innovations, was developed using a clean- sheet approach. But the design also draws on the experience gained from hundreds of thousands of operating hours accumulated by more than 35,000 Caterpillar construction and mining trucks working worldwide. The 797 builds on such field-proven designs as the mechanical power train, electronics that manage and monitor all systems, and structures that provide durability and long life. Engineers from various fields, including mechanical, electrical, structural, hydraulic, and mining, collaborated to design the Caterpillar 797 mining truck.

Courtesy of Caterpillar

Caterpillar developed the 797 in response to mining companies that were seeking a means to reduce cost per ton in large-scale operations. Therefore, the 797 is sized to work efficiently with loading shovels used in large mining operations, and Caterpillar matched the body design to the material being hauled to optimize payloads.

The new 797 mining truck is the largest of the extensive Caterpillar off-highway truck line, and it is the largest mining truck ever constructed. The Caterpillar 797 mining truck has a payload capacity of 360 tons (326 metric tons) and design operating mass of 1,230,000 lb (557,820 kg). The Caterpillar 797 Mining Truck specifications include:

Engine model: 3524B high displacement
Gross power at 1750 rpm: 3400 hp (2537 kW)
Transmission: 7-speed planetary power shift
Top speed: 40 mph (64 km/h)
Operating mass: 610,082 lb (1,345,000 kg)
Nominal payload capacity: 360 tons (326 metric tons)
Body capacity (SAE 2:1): 290 yd³ (220 m³)
Tire size: 58/80R63

A summary of some information about various components of the 797 truck follows.

Comfortable and Efficient Operator Station
The 797 cab is designed to reduce operator fatigue,

enhance operator performance, and promote safe operation. The controls and layout provide greater operator comfort along with an automotive feel, while enhancing functionality and durability. The cab and frame design meets SAE standards for rollover and falling-object protection.

The truck and cab design provides exceptional all-around visibility. For example, the clean deck to the right of the cab improves sight lines. The spacious cab includes two full-sized air suspension seats, which allow a trainer to work with the operator.

The steering column tilts and telescopes so the operator can adjust it for comfort and optimum control. The Vital Information Management System (VIMS) display and keypad provide precise machine status information. The new hoist control is fingertip actuated, allowing the operator to easily and precisely control hoist functions—raise, hold, float, and lower. Electrically operated windows and standard air conditioning add to operator comfort. The cab is resiliently mounted to dampen noise and vibration, and sound-absorbing material in the doors and side panels cuts noise further.

Cast Frame Mild steel castings comprise the entire load-bearing frame for durability and resistance to impact loads. The nine major castings are machined for precise fit before being joined

Based on Caterpillar® documents.

using a robotic welding technology that ensures full-penetration welds. The frame design reduces the number of weld joints and ensures a durable foundation for the 797.

The suspension system, which uses oil-over-nitrogen struts similar to other Caterpillar mining trucks, is designed to dissipate haul road and loading impacts.

Electronically Controlled Engine   The new Cat 3524B high-displacement diesel engine produces 3400 gross hp (2537 kW). It is turbocharged and aftercooled, and features electronic unit injection (EUI) technology, which helps the engine meet year 2000 emissions regulations.

The engine uses two in-line blocks linked by an innovative coupler and precisely controlled by electronic controllers. These electronic controllers integrate engine information with mechanical power train information to optimize truck performance, extend component life, and improve operator comfort.

The engine uses a hydraulically driven fan for efficient cooling. The fan design and operation also reduce fuel consumption and noise levels. The engine and its components are designed to minimize service time, which helps keep availability of the truck high.

Efficient Mechanical Power Train   The 797 power train includes a new torque converter with lockup clutch that delivers high mechanical efficiency. The new automatic-shift transmission features seven speeds forward and one speed reverse. Electronic clutch pressure control (ECPC) technology smoothes shifts, reduces wear, and increases reliability. Large clutch discs give the transmission high torque capacity and extend transmission life. The transmission and torque converter enable the truck to maintain good speed up grades and to reach a top speed of 40 mph (64 km/h).

The differential is rear mounted, which improves access for maintenance. The differential is pressure lubricated, thus promoting greater efficiency and long life. Wide wheel-bearing spread reduces bearing loads and helps ensure durability.

A hydraulically driven lube and cooling system operates independently of ground speed and pumps a continuous supply of filtered oil to each final drive.

Oil-cooled, multiple disc brakes provide fade-resistant braking and retarding. The electronically managed automatic retarder control (ARC) is an integral part of the intelligent power train. ARC controls the brakes on grade to maintain optimum engine rpm and oil cooling. Automatic electronic traction aid (AETA) uses the rear brakes to optimize traction. A combination of constant-displacement and variable-displacement pumps delivers a regulated flow of brake-cooling oil for constant retarding capability and peak truck performance on downhill grades.

Electrohydraulic Hoist Control   New hoist hydraulics include an electronic hoist control, independent metering valve (IMV), and a large hoist pump. These features allow automatic body snubbing for reduced impact on the frame, hoist cylinders, and operator. This design also allows the operator to modulate flow and control overcentering when dumping.

Serviceability   Caterpillar designed the 797 to reduce service time thus ensuring maximum availability. Routine maintenance points, such as fluid fill and check points, are close to ground level. Easily accessed connectors allow technicians to download data and to calibrate machine functions, thus extending maintenance intervals which further increases availability. The 797 is also designed to make major components accessible, which effectively reduces removal, installation, and in-truck maintenance time for all drivetrain components.

To allow technicians to raise the truck body inside a standard-height maintenance building, the canopy portion of the body is hinged so that it can be folded back. In the folded and raised position, the 797 body height is no higher than the 793, Caterpillar's 240-ton-capacity (218-metric ton) mining truck.

Testing Several 797 mining trucks were tested at the Caterpillar proving grounds. Mine evaluation performances were also done during 1999–2000. The test results are used to correct unexpected problems and fine-tune various components of the truck. The 797 has been available commercially worldwide since January 2001.

PROBLEMS

1. Calculate the braking force required to bring the truck from its top speed to full stop in a distance of 250 ft.
2. Calculate the linear momentum of the truck when fully loaded and cruising near its top speed. Calculate the force required to bring the truck from its top speed to a full stop in 5 s.
3. Estimate the rotational speed of the tires when the truck is moving along near its top speed.
4. Estimate the net torque output of the engine from the following equation:

$$\text{power} = T\omega$$

where

power is in lbf · ft/s

T = torque (lbf · ft/s)

ω = angular speed (rad/s)

Temperature and Temperature-Related Variables in Engineering

iurii/Shutterstock.com
Yermolov/Shutterstock.com

Molten steel is being poured by a ladle into ingot molds for transport to another processing plant to make steel products. Depending on the carbon content of steel, the temperature of molten steel can exceed 1300° C.

LEARNING OBJECTIVES

LO¹ **Temperature as a Fundamental Dimension:** describe the role of temperature in engineering analysis and design and its units and measurement

LO² **Temperature Difference and Heat Transfer:** explain what causes heat transfer; the modes of heat transfer; and what is meant by thermal resistance and R-value

LO³ **Thermal Comfort:** describe factors that define thermal comfort

LO⁴ **Heating Values of Fuels:** explain what the heating values of fuels represent

LO⁵ **Degree-Days and Thermal Energy Estimation:** explain what is meant by degree-days; describe how engineers estimate monthly or annual energy consumptions to heat buildings

LO⁶ **Additional Temperature-Related Properties:** explain what is meant by coefficient of thermal expansion and specific heat

DISCUSSION STARTER

WIND CHILL TEMPERATURE INDEX

On November 1, 2001, the National Weather Service implemented a new Wind Chill Temperature (WCT) index designed to more accurately calculate how cold air feels on human skin. The former index used by the United States and Canada was based on 1945 research of Antarctic explorers Siple and Passel. They measured the cooling rate of water in a container hanging from a tall pole outside. A container of water will freeze faster than flesh. As a result, the previous wind chill index underestimated the time to freezing and overestimated the chilling effect of the wind. The current index is based on heat loss from exposed skin and was tested on human subjects. The wind chill chart shown includes a frostbite indicator showing the points where temperature, wind speed, and exposure time will produce frostbite on humans. The chart includes three shaded areas of frostbite danger. Each shaded area shows how long (30, 10, and 5 minutes) a person can be exposed before frostbite develops. For example, a temperature of 0° F and a wind speed of 15 mph will produce a wind chill temperature of −19° F. Under these conditions, exposed skin can freeze in 30 minutes.

Wind Chill Chart

Wind (mph) \ Temperature (°F): Calm	40	35	30	25	20	15	10	5	0	-5	-10	-15	-20	-25	-30	-35	-40	-45
5	36	31	25	19	13	7	1	-5	-11	-16	-22	-28	-34	-40	-46	-52	-57	-63
10	34	27	21	15	9	3	-4	-10	-16	-22	-28	-35	-41	-47	-53	-59	-66	-72
15	32	25	19	13	6	0	-7	-13	-19	-26	-32	-39	-45	-51	-58	-64	-71	-77
20	30	24	17	11	4	-2	-9	-15	-22	-29	-35	-42	-48	-55	-61	-68	-74	-81
25	29	23	16	9	3	-4	-11	-17	-24	-31	-37	-44	-51	-58	-64	-71	-78	-84
30	28	22	15	8	1	-5	-12	-19	-26	-33	-39	-46	-53	-60	-67	-73	-80	-87
35	28	21	14	7	0	-7	-14	-21	-27	-34	-41	-48	-55	-62	-69	-76	-82	-89
40	27	20	13	6	-1	-8	-15	-22	-29	-36	-43	-50	-57	-64	-71	-78	-84	-91
45	26	19	12	5	-2	-9	-16	-23	-30	-37	-44	-51	-58	-65	-72	-79	-86	-93
50	26	19	12	4	-3	-10	-17	-24	-31	-38	-45	-52	-60	-67	-74	-81	-88	-95
55	25	18	11	4	-3	-11	-18	-25	-32	-39	-46	-54	-61	-68	-75	-82	-89	-97
60	25	17	10	3	-4	-11	-19	-26	-33	-40	-48	-55	-62	-69	-76	-84	-91	-98

Frostbite Times: 30 minutes, 10 minutes, 5 minutes

$$\text{Wind Chill } (^\circ F) = 35.74 + 0.6215T - 35.75(V^{0.16}) + 0.4275T(V^{0.16})$$

Where, T= Air Temperature (°F) V= Wind Speed (mph) *Effective 11/01/01*

National Weather Service

Wind chill test subjects walking on a treadmill in a chilled wind tunnel Focial temperature readings were taken to help refine the new wind chill indes

In early summer of 2001, human trials were conducted at the Defence and Civil Institute of Environmental Medicine in Toronto, Canada. The trial results were used to improve the accuracy of the new formula and determine frostbite threshold values. During the human trials, six male and six female volunteers were placed in a chilled wind tunnel. Thermal transducers were stuck to their faces to measure heat flow from the cheeks, forehead, nose, and chin as they walked 3 mph on a treadmill. Each volunteer took part in four trials of 90 minutes each and was exposed to varying wind speeds and temperatures. The new wind chill does the following:

- Calculates wind speed at an average height of 5 feet (typical height of an adult human face) based on readings from the national standard height of 33 feet (height of an anemometer).
- Is based on a human face model.
- Incorporates modern heat transfer theory.
- Lowers the calm wind threshold from 4 mph to 3 mph.
- Uses a consistent standard for skin tissue resistance.
- Assumes no impact from the sun (i.e., clear night sky).

National Weather Service

To the students: What is temperature? Can you give other examples from everyday life where temperature and its values play important roles?

In this chapter, we will explain what temperature, thermal energy, and heat transfer mean. We will take a close look at the role of temperature and heat transfer in engineering design and examine how they play hidden roles in our everyday lives. We will discuss temperature and its various scales, including Celsius, Fahrenheit, Rankine, and Kelvin. We will also explain a number of temperature-related properties of materials, such as specific heat, thermal expansion, and thermal conductivity. After studying this chapter, you will have learned that thermal energy transfer occurs whenever there exists a temperature difference within an object or whenever there is a temperature difference between two bodies or a body and its surroundings. We will briefly discuss the various modes of heat transfer, what the R-value of insulation means, and what the term metabolic rate means. We will also explain the heating values of fuels.

In order to see how this chapter fits into what you have been studying so far, recall from our discussion in Chapter 6 that based on what we know about

---

TABLE 6.7   Fundamental Dimensions and How They Are Used in Defining Variables that Are Used in Engineering Analysis and Design

Chapter	Fundamental Dimension	Related Engineering Variables			
7	Length (L)	Radian (L/L), Strain (L/L)	Area (L^2)	Volume (L^3)	Area moment of inertia (L^4)
8	Time (t)	Angular speed ($1/t$), Angular acceleration ($1/t^2$) Linear speed (L/t), Linear acceleration (L/t^2)		Volume flow rate (L^3/t)	
9	Mass (M)	Mass flow rate (M/t), Momentum (ML/t) Kinetic energy (ML^2/t^2)		Density (M/L^3), Specific volume (L^3/M)	
10	Force (F)	Moment (LF), Work, energy (FL), Linear impulse (Ft), Power (FL/t)	Pressure (F/L^2), Stress (F/L^2), Modulus of elasticity (F/L^2), Modulus of rigidity (F/L^2)	Specific weight (F/L^3),	
11	Temperature (T)	Linear thermal expansion (L/LT), Specific heat (FL/MT)		Volume thermal expansion (L^3/L^3T)	
12	Electric Current (I)	Charge (It)	Current density (I/L^2)		

our physical world today, we need seven fundamental or base dimensions to correctly express our natural world. The seven fundamental dimensions are *length*, *mass*, *time*, *temperature*, *electric current*, *amount of substance*, and *luminous intensity*. Recall that with the help of these base dimensions, we can explain all other necessary physical quantities that describe how nature works. In the previous chapters, you studied the roles of length, mass, and time in engineering analysis and design and in your everyday lives. In this chapter, we will discuss the role of temperature, another fundamental, or base, dimension, in engineering analysis. Table 6.7 is repeated here again to remind you of the role of fundamental dimensions and how they are combined to define variables that are used in engineering analysis and design.

LO¹ 11.1   Temperature as a Fundamental Dimension

Understanding what temperature means and what its magnitude or value represents is very important in understanding our surroundings. Recall from our discussion in Chapter 6 that in order to describe how cold or hot something is, humans needed a physical quantity, or physical dimension, which we now refer to as temperature. Think about the important role of temperature in your everyday lives in describing various states of things. Do you know the answer to some of these questions: What is your body temperature? What is the room air temperature? What is the temperature of the water that you used this morning to take a shower? What is the temperature of the air inside your refrigerator that kept the milk cold overnight? What is the temperature inside the freezer section of your refrigerator? What is the temperature of the air coming out of your hair dryer? What is the surface temperature of your stove's heating element when set on high? What is the surface temperature of the iron used to press your shirt? What is the average operating temperature of the electronic chips inside your TV or your computer? What is the temperature of

I wonder what the surface temperature of this slice of pizza is!

combustion products coming out of your car's engine? Once you start thinking about the role of temperature in quantifying what goes on in our surroundings, you realize that you could ask hundreds of similar questions.

Regardless of which engineering discipline you are planning to pursue, you need to develop a good understanding of what is meant by temperature and how it is quantified. Figure 11.1 illustrates some of the systems for which this understanding is important. Electronic and computer engineers, when designing computers, televisions, or any electronic equipment, are concerned with keeping the temperature of various electronic components at a reasonable operating level so that the electronic components will function properly. In fact, they use heat sinks (fins) and fans to cool the electronic chips. Civil engineers need to have a good understanding of temperature when they design pavement, bridges, and other structures. They must design the structures in such a way as to allow for expansion and contraction of materials, such as concrete and steel, that occur due to changes in the surrounding temperatures. Mechanical engineers design heating, ventilating, and air-conditioning (HVAC) equipment to create the comfortable environment in which we rest, work, and play. They need to understand heat transfer processes and the properties of air, including its temperature and moisture content, when designing this equipment. Automotive engineers need to have a good understanding of temperature and heat transfer rates when designing the cooling system of an engine. Engineers working in a food processing plant need to monitor temperature of the drying and cooling processes. Materials engineers need to have a good grasp of temperature and heat transfer in order to create materials with desirable properties.

Material properties are a function of temperature. Physical and thermal properties of solids, liquids, and gases vary with temperature. For example, as you know, cold air is denser than warm air. The air resistance to your car's motion is greater in winter than it is in summer, provided the car is moving at the same speed. Most of you who live in a cold climate know it is harder to start your car in the morning in the winter. As you may know,

| FIGURE 11.1 | Examples of engineering systems for which understanding of temperature and heat transfer is important. |

starting difficulty is the result of the viscosity of oil increasing
as the temperature decreases. The variation in density of air
with temperature is shown in Figure 11.2. The temperature
dependence of SAE 10W, SAE 10W-30, and SAE 30 oil is
shown in Figure 11.3. These are but a few examples of why as
engineers you need to have a good understanding of temperature and its
role in design.

Let us now examine more closely what we mean by temperature.
Temperature provides a measure of molecular activity and the internal
energy of an object. Recall that all objects and living things are made of matter, and matter itself is made up of atoms or chemical elements. Moreover,
atoms are combined naturally, or in a laboratory setting, to create molecules.
For example, as you already know, water molecules are made of two atoms
of hydrogen and one atom of oxygen. Temperature represents the level of
molecular activity of a substance. The molecules of a substance at a high temperature are more active than at a lower temperature. Perhaps a simple way
to visualize this is to imagine the molecules of a gas as being the popcorn in a
popcorn popper; the molecules that are at a higher temperature move, rotate,
and bounce around faster than the colder ones in the popper. Therefore,
temperature quantifies or provides a measure of how active these molecules
are on a microscopic level. For example, air molecules are more active at,
say, 50° C than they are at 25° C. You may want to think of temperature this
way: We have bundled all the microscopic molecular movement into a single,
macroscopic, measurable value that we call **temperature**.

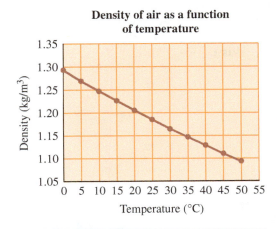

**Density of air as a function
of temperature**

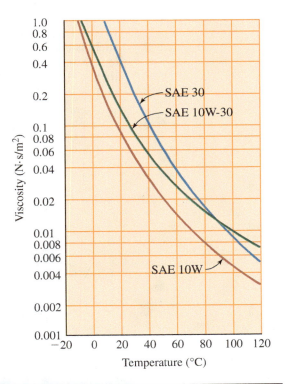

FIGURE 11.2   Density of air
at atmospheric pressure as a function of
temperature.

FIGURE 11.3   Viscosity of SAE 10W,
SAE 10W-30, and SAE 30 oil as a function of
temperature.

Measurement of Temperature and Its Units

Early humans relied on the sense of touch or vision to measure how cold or how warm something was. In fact, we still rely on touch today. When you are planning to take a bath, you first turn the hot and cold water on and let the bathtub fill with water. Before you enter the tub, however, you first touch the water to feel how warm it is. Basically, you are using your sense of touch to get an indication of the temperature. Of course, using touch alone, you can't quantify the temperature of water accurately. You cannot say, for example, that the water is at 21.5° C. So note the need for a more precise way of quantifying what the temperature of something is. Moreover, when we express the temperature of water, we need to use a number that is understood by all. In other words, we need to establish and use the same units and scales that are understood by everyone.

Another example of how people relied on their senses to quantify temperature is the way blacksmiths used to use their eyes to estimate how hot a fire was. They judged the temperature by the color of the burning fuel before they placed the horseshoe or an iron piece in the fire. In fact, this relationship between the color of heated iron and its actual temperature has been measured and established. Table 11.1 shows this relationship.

From these examples, you see that our senses are useful in judging how cold or how hot something is, but they are limited in accuracy and cannot quantify a value for a temperature. Thus, we need a measuring device that can provide information about the temperature of something more accurately and effectively.

This need led to the development of thermometers, which are based on thermal expansion or contraction of a fluid, such as alcohol, or a liquid metal, such as mercury. All of you know that almost everything will expand and its length increase when you increase its temperature, and it will contract and its length decrease when you decrease its temperature. We will discuss thermal expansion of material in more detail later in this chapter. But for now,

TABLE 11.1	The Relation of Color to Temperature of Iron
Color	**Temperature (°F)**
Dark blood red, black red	990
Dark red, blood red, low red	1050
Dark cherry red	1175
Medium cherry red	1250
Cherry, full red	1375
Light cherry, light red	1550
Orange	1650
Light orange	1725
Yellow	1825
Light yellow	1975
White	2200

Based on MARK'S STANDARD HANDBOOK FOR MECHANICAL ENGINEERS. 8TH EDITION by Baumeister et al.

remember that a mercury thermometer is a temperature sensor that works on the principle of expansion or contraction of mercury when its temperature is changed. Most of you have seen a thermometer, a graduated glass rod that is filled with mercury. On the Celsius scale, under standard atmospheric conditions, the value zero was *arbitrarily* assigned to the temperature at which water freezes, and the value of 100 was assigned to the temperature at which water boils. This procedure is called *calibration* of an instrument and is depicted in Figure 11.4. It is important for you to understand that the numbers were assigned arbitrarily; had someone decided to assign a value of 100 to the ice water temperature and a value of 1000 to boiling water, we would have had a very different type of temperature scale today! In fact, on a Fahrenheit temperature scale, under standard atmospheric conditions, the temperature at which water freezes is assigned a value of 32°, and the temperature at which the water boils is assigned a value of 212°. The relationship between the two temperature scales is given by

$$T(°C) = \frac{5}{9}\left[T(°F) - 32\right] \qquad \boxed{11.1}$$

$$T(°F) = \frac{9}{5}\left[T(°C)\right] + 32 \qquad \boxed{11.2}$$

As with other instruments, thermometers evolved over time into today's accurate instruments that can measure temperature to 1/100° C increments.

Today we also use other changes in properties of matter, such as electrical resistance or optical or emf (electromotive force) changes to

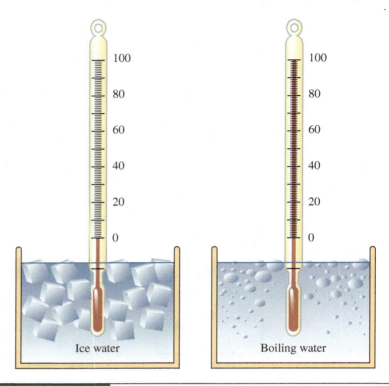

FIGURE 11.4   Calibration of a mercury thermometer.

measure temperature. These property changes occur within matter when we change its temperature. **Thermocouples** and **thermistors**, shown in Figure 11.5, are examples of temperature-measuring devices that use these properties. A thermocouple consists of two dissimilar metals. A relatively small voltage output is created when a difference in temperature between the two junctions of a thermocouple exists. The small voltage output is proportional to the difference in temperature between the two junctions. Two of the common combinations of dissimilar metals used in thermocouple wires include iron/constantan (J-type) and copper/constantan (T-type). (The J and the T are the American National Standards Institute (ANSI) symbols used to refer to these thermocouple wires.) A thermistor is a temperature-sensing device composed of a semiconductor material with such properties that a small temperature change creates large changes in the electrical resistance of the material. Therefore, the electrical resistance of a thermistor is correlated to a temperature value.

> Because both the Celsius and the Fahrenheit scales are arbitrarily defined, scientists recognized a need for a better temperature scale. This need led to the definition of absolute scales, the Kelvin and Rankine scales, that are based on the behavior of an ideal gas.

Absolute Zero Temperature

Because both the Celsius and the Fahrenheit scales are arbitrarily defined, as we have explained, scientists recognized a need for a better temperature scale. This need led to the definition of an absolute scale, the Kelvin and Rankine scales, which are based on the behavior of an ideal gas. You have observed what happens to the pressure inside your car's tires during a cold

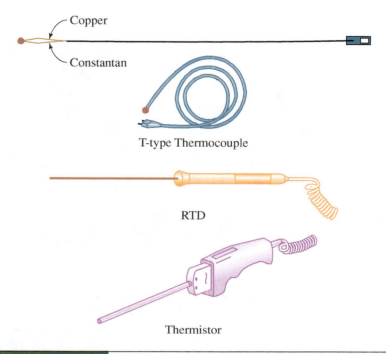

Copper

Constantan

T-type Thermocouple

RTD

Thermistor

| **FIGURE 11.5** | Examples of temperature-measuring devices. |

winter day, or what happens to the air pressure inside a basketball if it is left outside during a cold night. At a given pressure, as the temperature of an ideal gas is decreased, its volume will also decrease. Well, for gases under certain conditions, there is a relationship between the pressure of the gas, its volume, and its temperature as given by what is commonly called the *ideal gas law*. The ideal gas law is given by

$$PV = mRT \qquad \qquad \text{11.3}$$

where

P = absolute pressure of the gas (Pa or lb/ft^2)
V = volume of the gas (m^3 or ft^3)
m = mass (kg or lbm)
R = gas constant $\left(\dfrac{\text{J}}{\text{kg} \cdot \text{K}} \text{ or } \dfrac{\text{ft} \cdot \text{lb}}{\text{lbm} \cdot \text{R}} \right)$
T = absolute temperature (Kelvin or Rankine, which we will explain soon)

Consider the following experiment. Imagine that we have filled a rigid container (capsule) with a gas. The container is connected to a pressure gauge that reads the absolute pressure of the gas inside the container, as shown in Figure 11.6. Moreover, imagine that we immerse the capsule in a surrounding whose temperature we can lower. Of course, if we allow enough time, the temperature of the gas inside the container will reach the temperature of its surroundings. Also keep in mind that because the gas is contained inside a rigid container, it has a constant volume and a constant mass. Now, what happens to the pressure of the gas inside the container as indicated by the pressure gauge as we decrease

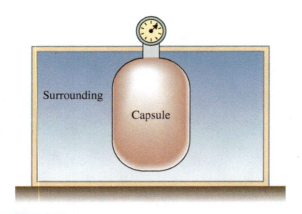

FIGURE 11.6   The capsule used in the absolute zero temperature example.

the surrounding temperature? The pressure will decrease as the temperature of the gas is decreased. We can determine the relationship between the pressure and the temperature using the ideal gas law, Equation (11.3).

Now, let us proceed to run a series of experiments. Starting with the surrounding temperature equal to some reference temperature—say, T_r—and allowing enough time for the container to reach equilibrium with the surroundings, we then record the corresponding pressure of the gas, P_r. The pressure and temperature of the gas are related according to the ideal gas law:

$$T_r = \frac{P_r V}{mR}$$ 11.4

Now imagine that we lower the surrounding temperature to T_1, once equilibrium is reached, recording the pressure of the gas and denoting the reading by P_1. Because the temperature of the gas was lowered, the pressure would be lowered as well.

$$T_1 = \frac{P_1 V}{mR}$$ 11.5

Dividing Equation (11.5) by Equation (11.4), we have

$$\frac{T_1}{T_r} = \frac{\dfrac{P_1 V}{mR}}{\dfrac{P_r V}{mR}}$$ 11.6

And after canceling out the m, the R, and the V, we have

$$\frac{T_1}{T_r} = \frac{P_1}{P_r}$$ 11.7

$$T_1 = T_r \left(\frac{P_1}{P_r}\right)$$ 11.8

Equation (11.8) establishes a relationship between the temperature of the gas, its pressure, and the reference pressure and temperature. If we were to proceed by lowering the surrounding temperature, a lower gas pressure would result, and if we were to extrapolate the results of our experiments, we would find that we eventually reach zero pressure at zero temperature. This temperature is called the *absolute thermodynamic temperature* and is related to the Celsius and Fahrenheit scales. The relationship between the Kelvin (K) and degree Celsius (°C) in SI units is

$$T(K) = T(°C) + 273.15 \qquad \boxed{11.9}$$

The relationship between degree Rankine (°R) and degree Fahrenheit (°F) in U.S. Customary units is

$$T(°R) = T(°F) + 459.67 \qquad \boxed{11.10}$$

Note that the experiment cannot be completely carried out, because as the temperature of the gas decreases, it reaches a point where it will liquify and thus the ideal gas law will not be valid. This is the reason for extrapolating the result to obtain a theoretical absolute zero temperature. It is also important to note that while there is a limit as to how cold something can be, there is no theoretical upper limit as to how hot something can be. Finally, we can also establish a relationship between the degree Rankine and the Kelvin by

$$T(K) = \frac{5}{9} T(°R) \qquad \boxed{11.11}$$

In Equations (11.9) and (11.10), unless you are dealing with very precise experiments, when converting from degree Celsius to Kelvin, you can round down the 273.15 to 273. The same is true when converting from degree Fahrenheit to Rankine; round up the 459.67 to 460.

EXAMPLE 11.1

What is the equivalent value of $T = 50°\,C$ in degrees Fahrenheit, Rankine, and Kelvin?

We can use Equations (11.2), (11.9), and (11.10):

$$T(°F) = \frac{9}{5}T(°C) + 32 = \left(\frac{9}{5}\right)(50) + 32 = 122°\,F$$

$$T(°R) = T(°F) + 460 = 122 + 460 = 582°\,R$$

$$T(K) = T(°C) + 273 = 50 + 273 = 323\,K$$

Note that we also could have converted the 582° R to Kelvin directly using Equation (11.11):

$$T(K) = \frac{5}{9}T(°R) = \left(\frac{5}{9}\right)(582) = 323\,K$$

EXAMPLE 11.2

On a summer day, in Phoenix, Arizona, the inside room temperature is maintained at 68° F while the outdoor air temperature is a sizzling 110° F. What is the outdoor-indoor temperature difference in (a) degree Fahrenheit, (b) degree Rankine, (c) degree Celsius, and (d) Kelvin? Is a 1° temperature difference in Celsius equal to a 1° temperature difference in Kelvin, and is a 1° temperature difference in Fahrenheit equal to a 1° temperature difference in Rankine? If so, why?

We will first answer these questions the long way, and then we will discuss the short way.

(a) $T_{outdoor} - T_{indoor} = 110°\,\text{F} - 68°\,\text{F} = 42°\,\text{F}$

(b) $T_{outdoor}(°\text{R}) = T_{outdoor}(°\text{F}) + 460 = 110 + 460 = 570°\,\text{R}$

$T_{indoor}(°\text{R}) = T_{indoor}(°\text{F}) + 460 = 68 + 460 = 528°\,\text{R}$

$T_{outdoor} - T_{indoor} = 570°\,\text{R} - 528°\,\text{R} = 42°\,\text{R}$

Note that the temperature difference expressed in degrees Fahrenheit is equal to the temperature difference expressed in degrees Rankine.

(c) $T_{outdoor}(°\text{C}) = \dfrac{5}{9}\left(T_{outdoor}(°\text{F}) - 32\right) = \dfrac{5}{9}(110 - 32) = 43.3°\,\text{C}$

$T_{indoor}(°\text{C}) = \dfrac{5}{9}\left(T_{indoor}(°\text{F}) - 32\right) = \dfrac{5}{9}(68 - 32) = 20°\,\text{C}$

$T_{outdoor} - T_{indoor} = 43.3°\,\text{C} - 20°\,\text{C} = 23.3°\,\text{C}$

(d) $T_{outdoor}(\text{K}) = T_{outdoor}(°\text{C}) + 273 = 43.3 + 273 = 316.3\,\text{K}$

$T_{indoor}(\text{K}) = T_{indoor}(°\text{C}) + 273 = 20 + 273 = 293\,\text{K}$

$T_{outdoor} - T_{indoor} = 316.3\,\text{K} - 293\,\text{K} = 23.3\,\text{K}$

Note that the temperature difference expressed in degrees Celsius is equal to the temperature difference expressed in Kelvin.

It should be clear by now that a 1° temperature difference in Celsius is equal to a 1° temperature difference in Kelvin, and a 1° temperature difference in Fahrenheit is equal to a 1° temperature difference in Rankine. Of course, this relationship is true because when you are computing the difference between two temperatures and converting to the **absolute temperature** scale, you are adding the same base value to each temperature. For example, to compute the temperature difference in degrees Rankine between two temperatures T_1 and T_2 given in degrees Fahrenheit, you first add 460 to each temperature to convert T_1 and T_2 from degrees Fahrenheit to degrees Rankine. This step is shown next.

$$T_1(°\text{R}) - T_2(°\text{R}) = \overbrace{[T_1(°\text{F}) + 460]}^{T_1\,(°\text{R})} - \overbrace{[T_2(°\text{F}) + 460]}^{T_2\,(°\text{R})}$$
$$= T_1(°\text{F}) + 460 - T_2(°\text{F}) - 460$$

And simplifying this relationship leads to

$$T_1(°\text{R}) - T_2(°\text{R}) = T_1(°\text{F}) - T_2(°\text{F})$$

Also note that you could have converted the temperature difference in degrees Rankine to Kelvin directly in the following manner:

$$\Delta T(\text{K}) = \frac{5}{9}\,\Delta T(^\circ\text{R}) = \left(\frac{5}{9}\right)(42) = 23.3 \text{ K}$$

LO² 11.2 Temperature Difference and Heat Transfer

Thermal energy transfer occurs whenever there exists a temperature difference within an object, or whenever there is a temperature difference between two bodies, or a temperature difference between a body and its surroundings. This form of energy transfer that occurs between bodies of different temperatures is called **heat transfer**. Additionally, heat always flows from a high- temperature region to a low-temperature region. This statement can be confirmed by observation of our surroundings. When hot coffee in a cup is left in a surrounding such as a room with a lower temperature, the coffee cools down. Thermal energy transfer takes place from the hot coffee through the cup and from its open surface to the surrounding room air. The thermal energy transfer occurs as long as there is a temperature difference between the coffee and its surroundings. At this point, make sure you understand the difference between *temperature* and *heat*. Heat is a form of energy that is transferred from one region to the next region as a result of a temperature difference between the regions, whereas temperature represents on a macroscopic level, by a single number, the level of microscopic molecular movement in a region.

There are three units that are commonly used to quantify thermal energy: (1) the British thermal unit (Btu), (2) the calorie, and (3) the joule. The **British thermal unit (Btu)** is defined as the amount of heat required to raise the temperature of 1 pound mass (1 lbm) of water by 1 degree Fahrenheit (1° F). The *calorie* is defined as the amount of heat required to raise the temperature of 1 gram of water by 1° C. Note, however, that the energy content of food is typically expressed in *Calories*, which is equal to 1000 calories. In SI units, no distinction is made between the units of thermal energy and mechanical energy and therefore energy is defined in terms of the fundamental dimensions of mass, length, and time. We will discuss this in more detail in Chapter 13. In the SI system of units, the *joule* is the unit of energy and is defined as

$$1 \text{ joule} = 1 \text{ N}\cdot\text{m} = 1 \text{ kg}\cdot\text{m}^2/\text{s}^2$$

The conversion factors among various units of heat are given in Table 11.2.

TABLE 11.2	Conversion Factors for Thermal Energy and Thermal Energy per Unit Time (Power)
Relationship Between the Units of Thermal Energy	**Relationship Between the Units of Thermal Energy per Unit Time (Power)**
1 Btu = 1055 J	1 W = 1 J/s
1 Btu = 252 cal	1 W = 3.4123 Btu/h
1 cal = 4.186 J	1 cal/s = 4.186 W

EXAMPLE 11.3

Using the units of energy and time, show that 1 watt (W) is equal to 3.4123 Btu/h, as shown in Table 11.2.

$$1 \text{ W} = 1\left(\frac{\text{J}}{\text{s}}\right)\left(\frac{1 \text{ Btu}}{1055 \text{ J}}\right)\left(\frac{3600 \text{ s}}{1 \text{ h}}\right) = 3.4123\frac{\text{Btu}}{\text{h}}$$

EXAMPLE 11.4

In many parts of the United States, in order to keep a house warm in the winter months, a gas furnace is used. If a gas furnace puts out 60,000 Btu/h to compensate for heat loss from a house, what is the equivalent value of the thermal power (energy per unit time) output of the furnace in watts?

From Table 11.2, you know that: 1 Btu = 1055 J; you also know that 1 h = 3600 s. Substituting for these values,

$$q = 60,000\left(\frac{\text{Btu}}{\text{h}}\right)\left(\frac{1,055 \text{ J}}{1 \text{ Btu}}\right)\left(\frac{1 \text{ h}}{3,600 \text{ s}}\right) = 17,583\left(\frac{\text{J}}{\text{s}}\right) = 17,583 \text{ W} = 17.583 \text{ kW}$$

Or you could have used the direct conversion factor between Btu/h and W, as shown:

$$q = 60,000\left(\frac{\text{Btu}}{\text{h}}\right)\left(\frac{1 \text{ W}}{3.4123\left(\frac{\text{Btu}}{\text{h}}\right)}\right) = 17,583 \text{ W} = 17.583 \text{ kW}$$

Now that you know heat transfer or thermal energy transfer occurs as a result of temperature difference in an object or between objects, let us look at different modes of heat transfer. There are three different mechanisms by which energy is transferred from a high-temperature region to a low-temperature region. These are referred to as the *modes* of heat transfer. The three modes of heat transfer are conduction, convection, and radiation.

Conduction

Conduction refers to that mode of heat transfer that occurs when a temperature difference (gradient) exists in a medium. The energy is transported within the medium from the region with more-energetic molecules to the region with less-energetic molecules. Of course, it is the interaction of the molecules with their neighbors that makes the transfer of energy possible. To better demonstrate the idea of molecular interactions, consider the following example of conduction heat transfer. All of you have experienced what happens when you heat up some soup in an aluminum container on a stove. Why do the handles or the lid of the soup container get hot, even though the handles and the lid are not in direct contact with the heating element? Well, let us examine what is happening.

Thermal energy transfer occurs whenever a temperature difference exists. This form of energy transfer that occurs between bodies of different temperatures or within a body with a temperature gradient is called heat transfer.

The rate of heat transfer by conduction is given by Fourier's law, which state that the rate of heat transfer through a material is proportional to the temperature gradient, thermal area *A*, and the type of material involved.

Because of the energy transfer from the heating element, the molecules of the container in the region near the heating element are more energetic than those molecules farther away. The more energetic molecules share or transfer some of their energy to the neighboring regions, and the neighboring regions do the same thing, until the energy transfer eventually reaches the handles and the lid of the container. The energy is transported from the high- temperature region to the low-temperature region by molecular activity. The rate of heat transfer by conduction is given by **Fourier's law**, which states that the rate of heat transfer through a material is proportional to the temperature difference, normal area *A*, through which heat transfer occurs, and the type of material involved. The law also states that the heat transfer rate is inversely proportional to the material thickness over which the temperature difference exists. For example, referring to Figure 11.7, we can write the Fourier's law for a single-pane glass window as

$$q = kA\frac{T_1 - T_2}{L}$$ 11.12

where

$$q = \text{heat transfer rate}\,(\text{W or Btu/h})$$

$$k = \text{thermal conductivity}\left(\frac{\text{W}}{\text{m}\cdot°\text{C}}\text{ or }\frac{\text{Btu}}{\text{h}\cdot\text{ft}\cdot°\text{F}}\right)$$

$$A = \text{cross-sectional area normal to heat flow (m}^2\text{ or ft}^2)$$

$$T_1 - T_2 = \text{temperature difference across the material of }L\text{ thickness}\,(°\text{C or }°\text{F})$$

$$L = \text{material thickness}\,(\text{m or ft})$$

The temperature difference $\frac{T_1 - T_2}{L}$ over material thickness is commonly referred to as the *temperature gradient*. Again, keep in mind that a temperature

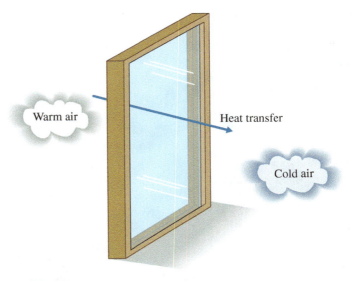

Warm air

Heat transfer

Cold air

FIGURE 11.7 Conduction heat transfer through a glass window.

gradient must exist in order for heat transfer to occur. **Thermal conductivity** is a property of materials that shows how good the material is in transferring thermal energy (heat) from a high-temperature region to a low-temperature region within the material. In general, solids have a higher thermal conductivity than liquids, and liquids have a higher thermal conductivity than gases. The thermal conductivity of some materials is given in Table 11.3.

TABLE 11.3	Thermal Conductivity of Some Materials at 300 K
Material	**Thermal Conductivity (W/m · k)**
Air (at atmospheric pressure)	0.0263
Aluminum (pure)	237
Aluminum alloy-2024-T6 (4.5% copper, 1.5% magnesium,0.6% manganese)	177
Asphalt	0.062
Bronze (90% copper, 10% aluminum)	52
Brass (70% copper, 30% zinc)	110
Brick (fire clay)	1.0
Concrete	1.4
Copper (pure)	401
Glass	1.4
Gold	317
Human fat layer	0.2
Human muscle	0.41
Human skin	0.37
Iron (pure)	80.2
Stainless steels (AISI 302, 304, 316, 347)	15.1, 14.9, 13.4, 14.2
Lead	35.3
Paper	0.18
Platinum (pure)	71.6
Sand	0.27
Silicon	148
Silver	429
Zinc	116
Water (liquid)	0.61

EXAMPLE 11.5

The single-pane window of Example 11.5.

Calculate the heat transfer rate from a single-pane glass window with an inside surface temperature of approximately $20°\,C$ and an outside surface temperature of $5°\,C$. The glass is 1 m tall, 1.8 m wide, and 8 mm thick, as shown in Figure 11.8. The thermal conductivity of the glass is approximately $k = 1.4$ W/m $\cdot$ K.

$$L = 8\,(\text{mm})\left(\frac{1\text{ m}}{1000\text{ mm}}\right) = 0.008\text{ m}$$

$$A = (1\text{ m})(1.8\text{ m}) = 1.8\text{ m}^2$$

$$T_1 - T_2 = 20°\,C - 5°\,C = 15°\,C = 15\text{ K}$$

Note, as we explained before, a $15°\,C$ temperature difference is equal to a 15 K temperature difference. Substituting for the values of k, A, $(T_1 - T_2)$, and L in Equation (11.12), we have

$$q = kA\frac{T_1 - T_2}{L} = (1.4)\left(\frac{W}{m \cdot K}\right)(1.8\text{ m}^2)\left(\frac{15\text{ K}}{0.008\text{ m}}\right) = 4725\text{ W}$$

(Be careful with lowercase k, denoting thermal conductivity, and the capital K, representing Kelvin, an absolute temperature scale.)

Thermal Resistance

In this section, we will explain what the **R-values** of insulating materials mean. Most of you understand the importance of having a well-insulated house because the better insulated a house is, the less the heating or cooling cost of the house. For example, you may have heard that in order to reduce heat loss through the attic, some people add enough insulation to their attic so that the R-value of insulation is 40. But what does the R-value of 40 mean, and what does the R-value of an insulating material mean in general? Let us start by rearranging Equation (11.12) in the following manner. Starting with Equation (11.12),

$$q = kA\frac{T_1 - T_2}{L} \qquad \boxed{11.12}$$

and rearranging it, we have

$$q = \frac{T_1 - T_2}{\dfrac{L}{kA}} = \frac{\text{temperature difference}}{\text{thermal resistance}} \qquad \boxed{11.13}$$

and **thermal resistance** $= L/kA$.

Figure 11.9 depicts the idea of thermal resistance and how it is related to the material's thickness, area, and thermal conductivity. When examining Equation (11.13), you should note the following: (1) The heat transfer (flow) rate is directly proportional to the temperature difference; (2) the heat flow rate is inversely proportional to the thermal resistance—the higher the value of thermal resistance, the lower the heat transfer rate will be.

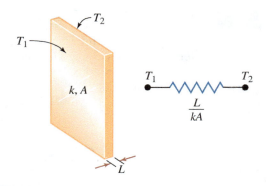

　A slab of material and its thermal resistance.

When expressing Fourier's law in the form of Equation (11.13), we are making an analogy between the flow of heat and the flow of electricity in a wire. Ohm's law, which relates the voltage V to current I and the electrical resistance R_e, is analogous to heat flow. Ohm's law is expressed as

$$V = R_e I \qquad \text{11.14}$$

or

$$I = \frac{V}{R_e} \qquad \text{11.15}$$

We will discuss Ohm's law in more detail in Chapter 12. Comparing Equation (11.13) to Equation (11.15), note that the heat flow is analogous to electric current, the temperature difference to voltage, and the thermal resistance to electrical resistance.

Now turning our attention back to thermal resistance and Equation (11.13), we realize that the thermal resistance for a unit area of a material is defined as

$$R' = \frac{L}{kA} \qquad \text{11.16}$$

R' has the units of °C/W(**K/W**) or °F·h/Btu(°R·h/Btu). When Equation (11.16) is expressed per unit area of the material, it is referred to as the R-value or the R-factor.

$$R = \frac{L}{k} \qquad \text{11.17}$$

where R has the units of

$$\frac{m^2 \cdot °C}{W}\left(\frac{m^2 \cdot K}{W}\right)$$

or

$$\frac{ft^2 \cdot °F}{\dfrac{Btu}{h}}\left(\frac{ft^2 \cdot °R}{\dfrac{Btu}{h}}\right)$$

The R-value of a material provides a measure of resistance to heat flow. The higher the value, the more resistance to heat flow the material offers.

Note that neither R' nor R is dimensionless and sometimes the R-values are expressed per unit thickness. The R-value or R-factor of a material provides a measure of resistance to heat flow: The higher the value, the more resistance to heat flow the material offers. Finally, when the materials used for insulation purposes consist of various components, the total R-value of the composite material is the sum of resistance offered by the various components.

EXAMPLE 11.6

Determine the thermal resistance R' and the R-value for the glass window of Example 11.5.

The thermal resistance R' and the R-value of the window can be determined from Equations (11.16) and (11.17), respectively.

$$R' = \frac{L}{kA} = \frac{0.008 \text{ m}}{1.4\left(\dfrac{\text{W}}{\text{m} \cdot \text{K}}\right)(1.8 \text{ m}^2)} = 0.00317 \frac{\text{K}}{\text{W}}$$

And the R-value or the R-factor for the given glass pane is

$$R = \frac{L}{k} = \frac{0.008 \text{ m}}{1.4\left(\dfrac{\text{W}}{\text{m} \cdot \text{K}}\right)} = 0.0057 \frac{\text{m}^2 \cdot \text{K}}{\text{W}}$$

EXAMPLE 11.7

For Example 11.6, convert the thermal resistance R' and R-factor results from SI units to U.S. Customary units.

$$R' = 0.00317\left(\frac{\text{K}}{\text{W}}\right)\left(\frac{1 \text{ W}}{3.4123\dfrac{\text{Btu}}{\text{h}}}\right)\left(\frac{\frac{5}{9}\,{}^\circ\text{R}}{\text{K}}\right) = 5.161 \times 10^{-4}\,\frac{{}^\circ\text{R}}{\dfrac{\text{Btu}}{\text{h}}}$$

$$R = \frac{L}{k} = 0.0057\left(\frac{\text{m}^2 \cdot \text{K}}{\text{W}}\right)\left(\frac{1 \text{ W}}{3.4123\dfrac{\text{Btu}}{\text{h}}}\right)\left(\frac{\frac{5}{9}\,{}^\circ\text{R}}{\text{K}}\right)\left(\frac{3.28 \text{ ft}}{1 \text{ m}}\right)^2 = 0.01\,\frac{{}^\circ\text{R} \cdot \text{ft}^2}{\dfrac{\text{Btu}}{\text{h}}}$$

Or, if the R-value is expressed in terms of in^2,

$$R = \frac{L}{k} = 0.0057\left(\frac{\text{m}^2 \cdot \text{K}}{\text{W}}\right)\left(\frac{1 \text{ W}}{3.4123\dfrac{\text{Btu}}{\text{h}}}\right)\left(\frac{\frac{5}{9}\,{}^\circ\text{R}}{\text{K}}\right)\left(\frac{39.37 \text{ in}}{1 \text{ m}}\right)^2 = 1.43\,\frac{{}^\circ\text{R} \cdot \text{in}^2}{\dfrac{\text{Btu}}{\text{h}}}$$

EXAMPLE 11.8

A double-pane glass window consists of two pieces of glass, each having a thickness of 8 mm, with a thermal conductivity of $k = 1.4$ W/m · K. The two glass panes are separated by an air gap of 10 mm, as shown in Figure 11.10. Assuming the thermal conductivity of air to be $k = 0.025$ W/m · K, determine the total R-value for this window.

The total thermal resistance of the window is obtained by adding the resistance offered by each pane of glass and the air gap in the following manner:

$$R_{total} = R_{glass} + R_{air} + R_{glass} = \frac{L_{glass}}{k_{glass}} + \frac{L_{air}}{k_{air}} + \frac{L_{glass}}{k_{glass}}$$

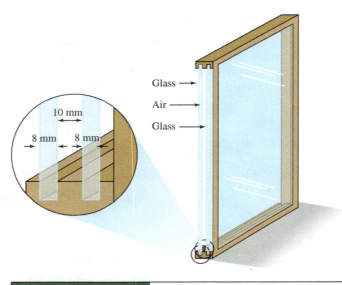

Glass →
Air →
Glass →

10 mm

8 mm   8 mm

FIGURE 11.10   The double-pane glass window of Example 11.8.

substituting for L_{glass}, k_{glass}, L_{air}, k_{air}, we have

$$R_{total} = \frac{0.008\,(\text{m})}{1.4\left(\dfrac{\text{W}}{\text{m} \cdot \text{K}}\right)} + \frac{0.01\,(\text{m})}{0.025\left(\dfrac{\text{W}}{\text{m} \cdot \text{K}}\right)} + \frac{0.008\,(\text{m})}{1.4\left(\dfrac{\text{W}}{\text{m} \cdot \text{K}}\right)} = 0.4\left(\frac{\text{m}^2 \cdot \text{K}}{\text{W}}\right)$$

Note the units of R-value. The R-value for the double-pane glass window in U.S. Customary units is

$$R_{total} = 0.4\left(\frac{\text{m}^2 \cdot \text{K}}{\text{W}}\right)\left(\frac{1\ \text{W}}{3.4123\,\dfrac{\text{Btu}}{\text{h}}}\right)\left(\frac{\frac{5}{9}\,{}^\circ\text{R}}{\text{K}}\right)\left(\frac{3.28\ \text{ft}}{1\ \text{m}}\right)^2 = 0.7\,\frac{{}^\circ\text{R} \cdot \text{ft}^2}{\dfrac{\text{Btu}}{\text{h}}}$$

As you can see from the results of Example 11.6 and 11.8, ordinary glass windows do not offer much resistance to heat flow. To increase the R-value of windows, some manufacturers make windows that use triple glass panes and fill the spacing between the glass panes with argon gas. See Example 11.11 for a sample calculation for total thermal resistance of a typical exterior frame wall of a house consisting of siding, sheathing, insulation material, and gypsum wall- board (drywall).

Convection

Convection heat transfer occurs when a fluid (a gas or a liquid) in motion comes into contact with a solid surface whose temperature differs from the moving fluid. For example, on a hot summer day when you sit in front of a fan to cool down, the heat transfer rate that occurs from your warm body to the cooler moving air is by convection. Or when you are cooling a hot food, such as freshly baked cookies, by blowing on it, you are using the principles of convection heat transfer. The cooling of computer chips by blowing air across them is another example of cooling something by convection heat transfer (Figure 11.11). There are two broad areas of convection heat transfer: *forced convection* and *free (natural) convection*. Forced convection refers to situations where the flow of fluid is caused or forced by a fan or a pump. Free convection, on the other hand, refers to situations where the flow of fluid occurs naturally due to density variation in the fluid. Of course, the density variation is caused by the temperature distribution within the fluid. When you leave a hot pie to cool on the kitchen counter, the heat transfer is by natural convection. The heat loss from the exterior surfaces of the hot oven is also by natural convection. To cite another example of natural convection, the large electrical transformers that sit outdoors at power substations are cooled by natural convection (on a calm day) as well.

For both the forced and the free convection situations, the overall heat transfer rate between the fluid and the surface is governed by Newton's law of cooling, which is given by

> Convection heat transfer occurs when a fluid in motion comes into contact with a solid surface whose temperature differs from the moving fluid. It is governed by Newton's law of cooling.

$$q = hA(T_s - T_f)$$
11.18

where h is the **heat transfer coefficient** in $W/m^2 \cdot K$ (or $Btu/h \cdot ft^2 \cdot {}^\circ R$), A is the area of the exposed surface in $m^2(ft^2)$, T_s is the surface temperature in $^\circ C$ ($^\circ F$), and T_f represents the temperature of moving fluid in $^\circ C$ ($^\circ F$). The value of the heat transfer coefficient for a particular situation is determined from experimental correlation; these values are available in many books about heat transfer. At this stage of your education, you need not be concerned about how to obtain the numerical values of heat transfer coefficients. However, it is important for you to know that the value of h is higher for forced convection than it is for free convection. Of course, you already know this! When you are trying to cool down rapidly, do you sit in front of a fan or do you sit in an area of the room where the air is still? Moreover, the heat transfer coefficient h is higher for liquids than it is for gases. Have you noticed that you can walk around comfortably in a T-shirt when the outdoor air temperature is $70^\circ F$, but if you went into a swimming pool whose water temperature was $70^\circ F$, you would feel cold? That is because the liquid water has a higher heat transfer coefficient than does air, and therefore, according to Newton's law of cooling, Equation (11.18), water removes more heat from your body. The typical range of heat transfer coefficient values is given in Table 11.4.

In the field of heat transfer, it is also common to define a resistance term for the convection process, similar to the R-value in conduction. The thermal convection resistance is defined as:

$$R' = \frac{1}{hA} \tag{11.19}$$

Again, R' has the units of $^\circ C/W(K/W)$ or $^\circ F \cdot h/Btu({}^\circ R \cdot h/Btu)$. Equation (11.19) is commonly expressed per unit area of solid surface exposure and is called *film resistance* or *film coefficient*.

$$R = \frac{1}{h} \tag{11.20}$$

TABLE 11.4	Typical Values of Heat Transfer Coefficients	
Convection Type	Heat Transfer Coefficient, $h(W/m^2 \cdot K)$	Heat Transfer Coefficient, $h(Btu/h \cdot ft^2 \cdot {}^\circ R)$
Free Convection		
Gases	2 to 25	0.35 to 4.4
Liquids	50 to 1000	8.8 to 175
Forced Convection		
Gases	25 to 250	4.4 to 44
Liquids	100 to 20,000	17.6 to 3500

where R has the units of

$$\frac{m^2 \cdot °C}{W} \left(\frac{m^2 \cdot K}{W} \right) \text{ or } \frac{ft^2 \cdot °F}{\dfrac{Btu}{h}} \left(\frac{ft^2 \cdot °R}{\dfrac{Btu}{h}} \right)$$

It is important to realize once again that neither R nor R is dimensionless, and they provide a measure of resistance to heat flow; the higher the values of R, the more resistance to heat flow to or from the surrounding fluid.

EXAMPLE 11.9

Determine the heat transfer rate by convection from an electronic chip whose surface temperature is 35° C and has an exposed surface area of 9 cm². The temperature of the surrounding air is 20° C. The heat transfer coefficient for this situation is $h = 40$ W/m² · K.

$$A = 9 \; (cm^2) \left| \frac{1 \; m^2}{10,000 \; cm^2} \right| = 0.0009 \; m^2$$

$$T_s - T_f = 35° \, C - 20° \, C = 15° \, C = 15 \, K$$

Note again that the 15° C temperature difference is equal to a 15 K temperature difference. We can determine the heat transfer rate from the chip by substituting for the values of h, A, and $(T_s - T_f)$ in Equation (11.18), which results in

$$q_{convecton} = hA\left(T_s - T_f\right) = 40 \left(\frac{W}{m^2 \cdot K} \right)(0.0009 \; m^2)(15 \; K) = 0.54 \; W$$

EXAMPLE 11.10

Calculate the R-factor (film resistance) for the following situations: (a) wind blowing over a wall, $h = 5.88$ Btu/h · °F · ft², and (b) still air inside a room near a wall, $h = 1.47$ Btu/h · °F · ft².

For the situation where wind is blowing over a wall:

$$R = \frac{1}{h} = \frac{1}{5.88 \dfrac{Btu}{h \cdot ft^2 \cdot °F}} = 0.17 \frac{h \cdot ft^2 \cdot °F}{Btu}$$

And for still air inside a room, near a wall:

$$R = \frac{1}{h} = \frac{1}{1.47 \dfrac{Btu}{h \cdot ft^2 \cdot °F}} = 0.68 \frac{h \cdot ft^2 \cdot °F}{Btu}$$

EXAMPLE 11.11

A typical exterior frame wall (made up of 2 × 4 studs) of a house contains the materials shown in Table 11.5 and in Figure 11.12. For most residential buildings, the inside room temperature is kept around 70° F. Assuming an outside temperature of 20° F and an exposed area of 150 ft², we are interested in determining the heat loss through the wall.

TABLE 11.5
Thermal Resistance of Wall Materials

Items	Thermal Resistance ($h \cdot ft^2 \cdot °F/Btu$)
1. Outside film resistance (winter, 15 mph wind)	0.17
2. Siding, wood (1/2 × 8 lapped)	0.81
3. Sheathing (1/2 in. regular)	1.32
4. Insulation batt ($3 - 3\frac{1}{2}$ in.)	11.0
5. Gypsum wallboard (1/2 in.)	0.45
6. Inside film resistance (winter)	0.68

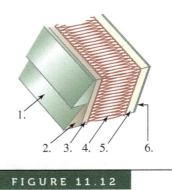

FIGURE 11.12
Wall layers to accompany table 11.5

In general, the heat loss through the walls, windows, doors, or roof of a building occurs due to conduction heat losses through the building materials—including siding, insulation material, gypsum wallboard (drywall), glass, and so on—and convection losses through the wall surfaces exposed to the indoor warm air and the outdoor cold air. The total resistance to heat flow is the sum of resistances offered by each component in the path of heat flow. For a plane wall we can write:

$$q = \frac{T_{inside} - T_{outside}}{\Sigma R'} = \frac{\left(T_{inside} - T_{outside}\right)A}{\Sigma R} \qquad \text{11.21}$$

The total resistance to heat flow is given by

$$\Sigma R = R_1 + R_2 + R_3 + R_4 + R_5 + R_6$$
$$= 0.17 + 0.81 + 1.32 + 11.0 + 0.45 + 0.68 = 14.43 \ h \cdot ft^2 \cdot °F/Btu$$
$$q = \frac{\left(T_{inside} - T_{outside}\right)A}{\Sigma R} = \frac{(70 - 20)(150)}{14.43} = 520 \frac{Btu}{h} \qquad \text{11.22}$$

The equivalent thermal resistance circuit for this problem is shown in Figure 11.13.

When performing heating load analysis to select a furnace to heat a building, it is common to calculate the heat loss through walls, roofs, windows, and doors of a building from $q = UA \, \Delta T$. In this equation, U represents the overall

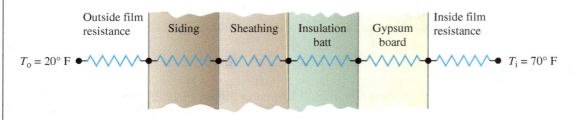

Outside film resistance | Siding | Sheathing | Insulation batt | Gypsum board | Inside film resistance

$T_o = 20°$ F $T_i = 70°$ F

FIGURE 11.13
The equivalent thermal resistance for Example 11.11.

heat transfer coefficient, or simply the U-factor for a wall, roof, window, or a door. The *U*-factor is the reciprocal of total thermal resistance and has the units of $\frac{\text{Btu}}{\text{h} \cdot \text{ft}^2 \cdot °\text{F}}$. For the above example problem, the *U*-factor for the wall is equal to:

$$U = \frac{1}{\sum R} = \frac{1}{14.43} = 0.0693 \frac{\text{Btu}}{\text{h} \cdot \text{ft}^2 \cdot °\text{F}}.$$

Using this *U* value, the heat loss through the wall is then calculated from

$$q = UA\,\Delta T = \left(0.0693 \frac{\text{Btu}}{\text{h} \cdot \text{ft}^2 \cdot °\text{F}}\right)(150\text{ft}^2)(70 - 20)°\,\text{F} = 520 \frac{\text{Btu}}{\text{h}}.$$

Radiation

All matter emits thermal **radiation**. This rule is true as long as the body in question is at a nonzero absolute temperature. The higher the temperature of the surface of the object, the more thermal energy is emitted by the object. A good example of thermal radiation is the heat you can literally feel radiated by a fire in a fireplace. The amount of radiant energy emitted by a surface is given by the equation

$$q = \varepsilon \sigma A T_s^4 \qquad \qquad \boxed{11.23}$$

where *q* represents the rate of thermal energy, per unit time, emitted by the surface; ε is the emissivity of the surface, $0 < \varepsilon < 1$, and σ is the **Stefan–Boltzmann constant** $(\sigma = 5.67 \times 10^{-8}\,\text{W/m}^2 \cdot \text{K}^4)$; *A* represents the area of the surface in m², and T_s is the surface temperature of the object expressed in Kelvin. **Emissivity**, ε, is a property of the surface of the object, and its value indicates how well the object emits thermal radiation compared to a black body (an ideal perfect emitter). It is important to note here that unlike the conduction and convection modes, heat transfer by radiation can occur in a vacuum. A daily example of this is the radiation of the sun reaching the earth's atmosphere as it travels through a vacuum in space. Because all objects emit thermal radiation, it is the net energy exchange among the bodies that is of interest to us. Because of this fact, thermal radiation calculations are generally complicated in nature and require an in-depth understanding of the underlying concepts and geometry of the problem.

> All matter omits thermal radiation. The amount of radiant energy omitted by a surface is governed by Stefan-Boltzmann law.

EXAMPLE 11.12

On a hot summer day, the flat roof of a tall building reaches 50° C in temperature. The area of the roof is 400 m². Estimate the heat radiated from this roof to the sky in the evening when the temperature of the surrounding air or sky is at 20° C. The temperature of the roof decreases as it cools down. Estimate the rate of energy radiated from the roof, assuming roof temperatures of 50, 40, 30, and 25° C. Assume $\varepsilon = 0.9$ for the roof.

We can determine the amount of thermal energy radiated by the surface from Equation (11.23). For roof temperature of 50° C, we get

$$q = \varepsilon \sigma A T_s^4 = (0.9)\left[5.67 \times 10^{-8}\left(\frac{\text{W}}{\text{m}^2 \cdot \text{K}^4}\right)\right](400 \text{ m}^2)(323 \text{ K})^4 = 222{,}000 \text{ W}$$

The rest of the solution is shown in Table 11.6.

TABLE 11.6	The Results of Example 11.12	
Surface Temperature (°C)	Surface Temperature (K) $T(K) = T(°C) + 273$	Energy Emitted by the Surface (W) $q = \varepsilon \sigma A T_s^4$
50	323	$(0.9)(5.67 \times 10^{-8})(400)(323)^4 = 222{,}000\ W$
40	313	$(0.9)(5.67 \times 10^{-8})(400)(313)^4 = 196{,}000\ W$
30	303	$(0.9)(5.67 \times 10^{-8})(400)(303)^4 = 172{,}000\ W$
25	298	$(0.9)(5.67 \times 10^{-8})(400)(298)^4 = 161{,}000\ W$

Most of you will take a heat transfer or a transport phenomenon class during your third year, where you will learn in more detail about various modes of heat transfer. You will also learn how to estimate heat transfer rates for various situations, including the cooling of electronic devices and the design of fins for transformers or motorcycle and lawn mower engine heads and other heat exchangers, like the radiator in your car or the heat exchangers in furnaces and boilers. The intent of this section was to briefly introduce you to the concept of heat transfer and its various modes.

Before You Go On

Answer the following questions to test your understanding of the preceding sections.

1. Explain what is meant by temperature and give examples of its important role in engineering analysis and design.

2. Explain why an absolute temperature is defined and give its SI and U.S. Customary units.

3. Give examples of how temperature is measured.

4. What does cause heat transfer?

5. What are the modes of heat transfer?

Vocabulary—State the meaning of the following terms:

Absolute Temperature _____

Thermocouple Wire_____

Conduction _____

Convection _____

Radiation _____

R-Value_____

LO³ 11.3   Thermal Comfort

Human thermal comfort is of special importance to bioengineers and mechanical engineers. For example, mechanical engineers design the heating, ventilating, and air-conditioning (HVAC) systems for homes, public buildings, hospitals, and manufacturing facilities. When sizing the HVAC systems, the engineer must design not only for the buildings' heat losses or gains but also for an environment that occupants feel comfortable within. What makes us thermally comfortable in an environment? As you know, the temperature of the environment and the humidity of the air are among the important factors that define thermal comfort. For example, most of us feel comfortable in a room that has a temperature of 70° F and a relative humidity of 40 to 50%. Of course, if you are exercising on a treadmill and watching TV, then perhaps you feel more comfortable in a room with a temperature lower than 70° F, say 50° to 60° F. Thus, the level of activity is also an important factor. In general, the amount of energy that a person generates depends on the person's age, gender, size, and activity level. A person's body temperature is controlled by (1) convective and radiative heat transfer to the surroundings, (2) sweating, (3) respiration by breathing surrounding air and exhaling it at near the body's temperature, (4) blood circulation near the surface of the skin, and (5) metabolic rate. **Metabolic rate** determines the rate of conversion of chemical to thermal energy within a person's body. The metabolic rate depends on the person's activity level. A unit commonly used to express the metabolic rate for an average person under sedentary conditions, per unit surface area, is called *met;* 1 met is equal to 58.2 W/m² or, in U.S. Customary units, 1 **met** = 18.4 Btu/h·ft². For an average person, a heat transfer surface area of 1.82 m² or 19.6 ft² was assumed when defining the unit of met. Table 11.7 shows the metabolic rate for various activities. As you might expect, clothing also affects thermal comfort. A unit that is generally used to express the insulating value of clothing is called *clo.* 1 clo is equal to 0.155 m² ·°C/W, or, in U.S. Customary units, 1 **clo** = 0.88° F · ft² · h/Btu. Table 11.8 on page 379 shows the insulating values of various types of clothing.

> The temperature of the environment and the humidity of the air are among the important factors that define thermal comfort. The level of a person's activity is also an important factor. Clothing is another factor that affects thermal comfort.

EXAMPLE 11.13

Use Table 11.7 to calculate the amount of energy dissipated by an average adult person doing the following things: (a) driving a car for 3 h, (b) sleeping for 8 h, (c) walking at the speed of 3 mph on a level surface for 2 h, (d) dancing for 2 h.

(a) Using average values, the amount of energy dissipated by an average adult person driving a car for 3 h is

$$\left(\frac{18 + 37}{2}\right)\left(\frac{\text{Btu}}{\text{h} \cdot \text{ft}^2}\right)(19.6 \text{ ft}^2)(3 \text{ h}) = 1617 \text{ Btu}$$

(b) Using average values, the amount of energy dissipated by an average adult person sleeping for 8 h is

$$13\left(\frac{\text{Btu}}{\text{h} \cdot \text{ft}^2}\right)(19.6 \text{ ft}^2)(8 \text{ h}) = 2038 \text{ Btu}$$

(c) Walking at the speed of 3 mph on a level surface for 2 h:

$$48\left(\frac{\text{Btu}}{\text{h} \cdot \text{ft}^2}\right)(19.6 \text{ ft}^2)(2 \text{ h}) = 1882 \text{ Btu}$$

(d) Dancing for 2 h:

$$\left(\frac{44 + 81}{2}\right)\left(\frac{\text{Btu}}{\text{h} \cdot \text{ft}^2}\right)(19.6 \text{ ft}^2)(2 \text{ h}) = 2450 \text{ Btu}$$

In Problem 11.26, you are asked to convert these results from Btu to calories.

TABLE 11.7	Typical Metabolic Heat Generation for Various Human Activities	
Activity	**Heat Generation (Btu/h · ft²)**	**Heat Generation (met)**
Resting		
Sleeping	13	0.7
Seated, quiet	18	1.0
Standing, relaxed	22	1.2
Walking on Level Surface		
2 miles/h	37	2.0
3 miles/h	48	2.6
4 miles/h	70	3.8
Office Work		
Reading, seated	18	1.0
Writing	18	1.0
Typing	20	1.1
Filing, seated	22	1.2
Filing, standing	26	1.4
Driving/Flying		
Car	18 to 37	1.0 to 2.0
Aircraft, routine	22	1.2
Aircraft, instrument landing	33	1.8
Aircraft, combat	44	2.4
Heavy vehicle	59	3.2
Miscellaneous Housework		
Cooking	29 to 37	1.6 to 2.0
House cleaning	37 to 63	2.0 to 3.4
Miscellaneous Leisure Activities		
Dancing, social	44 to 81	2.4 to 4.4
Calisthenics /exercise	55 to 74	3.0 to 4.0
Tennis, singles	66 to 74	3.6 to 4.0
Basketball	90 to 140	5.0 to 7.6
Wrestling, competitive	130 to 160	7.0 to 8.7

Based on American Society of Heating, Refrigerating, and Air-Conditioning Engineers.

TABLE 11.8	Typical Insulating Values for Clothing

Clothing	Insulation Values (clo) 1 clo = 0.155 m² •°C/W or 1 clo = 0.88° F • ft² • h/Btu
Walking shorts, short-sleeve shirt	0.36
Trousers, short-sleeve shirt	0.57
Trousers, long-sleeve shirt	0.61
Trousers, long-sleeve shirt, plus suit jacket	0.96
Same as above, plus vest and T-shirt	1.14
Sweatpants, sweatshirt	0.74
Knee-length skirt, short-sleeve shirt, panty hose, sandals	0.54
Knee-length skirt, short-sleeve shirt, full slip, panty hose	0.67
Knee-length skirt, long-sleeve shirt, half slip, panty hose, long-sleeve sweater	1.10
Same as above, replace sweater with suit jacket	1.04

Based on American Society of Heating, Refrigerating and Air-Conditioning Engineers, Inc. ASHRAE 1997, Handbook-Fundamentals.

LO⁴ 11.4 Heating Values of Fuels

As engineers you need to know what the heating value of a fuel means. Why? Where does the energy that drives your car come from? Where does the energy that makes your home warm and cozy during the cold winter months come from? How is electricity generated in a conventional power plant that supplies power to manufacturing companies, homes, and offices? The answer to all of these questions is that the initial energy comes from fuels. Most conventional fuels that we use today to generate power come from coal, natural gas, oil, or gasoline. All these fuels consist of carbon and hydrogen.

When a fuel is burned, whether it is gas, oil, etc., thermal energy is released. The heating value of a fuel quantifies the amount of energy that is released when a unit mass (kilogram or pound) or a unit volume (cubic meter or cubic foot) of a fuel is burned. Different fuels have different heating values. Moreover, based on the phase of water in the combustion product, whether it is in the liquid form or in vapor form, two different heating values are reported. The higher heating value of a fuel, as the name implies, is the higher end of energy released by the fuel when the combustion by-products include water in liquid form. The lower heating value refers to the amount of energy that is released during combustion when the combustion products include water vapor. The typical heating values of liquid fuels, coal, and natural gas are given in Tables 11.9 through 11.11. Also, a cord of wood (4 ft by 4 ft by 8 ft pile stacked neatly) has an average heating value of 20,000,000 Btu.

> When a fuel is burned, whether it is gas, oil, etc., thermal energy is released. The heating value of a fuel quantifies the amount of energy is released when a unit mass or a unit volume of a fuel is burned.

TABLE 11.9	Typical Heating Values of Standard Grades of Fuel Oil and Gasoline	
Grade No.	**Density (lb/gal)**	**Heating Value (Btu/gal)**
1	6.950 to 6.675	137,000 to 132, 900
2	7.296 to 6.960	141,800 to 137,000
4	7.787 to 7.396	148,100 to 143,100
5L	7.940 to 7.686	150,000 to 146,800
5H	8.080 to 7.890	152,000 to 149,400
6	8.448 to 8.053	155,900 to 151,300
Gasoline	6.0 to 6.2	108,500 to 117,000

Based on American Society of Heating, Refrigerating, and Air-Conditioning Engineers.

TABLE 11.10	
The Typical Heating Value of Some Coal	
Coal from County and State of	**Higher Heating Value (Btu/lbm)**
Musselshell, Montana	12,075
Emroy, Utah	13,560
Pike, Kentucky	15,040
Cambria, Pennsylvania	15,595
Williamson, Illinois	13,710
McDowell, West Virginia	15,600

Based on Babcock and Wilcox Company, *Streams: Its Generation and Use*.

TABLE 11.11		
The Typical Higher Heating Value of Natural Gas		
Source of Gas	**Heating Value (Btu/lbm)**	**Heating Value (Btu/ft³) @ 60° F and 30 in · Hg**
Pennsylvania	23,170	1129
Southern California	22,904	1116
Ohio	22,077	964
Louisiana	21,824	1002
Oklahoma	20,160	974

Based on Babcock and Wilcox Company, *Streams: Its Generation and Use*.

EXAMPLE 11.14

How much thermal energy is released when 5 lbm of a coal sample from Emroy, Utah is burned?

The total amount of thermal energy, $E_{thermal}$, released when some fuel is burned is determined by multiplying the mass of the fuel, m, by the heating value of fuel, H_V.

$$E_{thermal} = mH_V = (5 \text{ lbm})13,560\left(\frac{\text{Btu}}{\text{lbm}}\right) = 67,800 \text{ Btu}$$

EXAMPLE 11.15

Calculate the total amount of thermal energy released when 60 ft³ of natural gas from Louisiana is burned inside a gas furnace.

$$E_{thermal} = (60 \text{ ft}^3)1002\left(\frac{\text{Btu}}{\text{ft}^3}\right) = 60,120 \text{ Btu}$$

Answer the following questions to test your understanding of the preceding sections.

1.   Describe factors that affect thermal comfort.

2.   Explain what is meant by metabolic rate.

3.   What is a unit that is commonly used to express the insulation value of clothing?

4.   What does the heating value of a fuel represent?

Vocabulary—State the meaning of the following terms:

One Clo _____

Metabolic Rate_____

Heating Value _____

LO⁵ 11.5   Degree-Days and Energy Estimation

With the current energy and sustainability concerns, as future engineers, it is important for you to understand some of the simple **energy-estimation** procedures. For example, we use **degree-days** to estimate monthly and annual energy consumptions to heat a building. A degree-day (DD) is the difference between 65° F (typically) and the average temperature of the outside air during a 24-hour period. For example, for Mankato, MN, on an October day, the low temperature can be 28° F and the high temperature on that day can be 38° F. Then, the degree-day for that October day for Mankato, MN is: $DD = 65° \text{ F} - ((38° \text{ F} + 28° \text{ F})/2) = 32° \text{ F}$. Now, if we were to add up the degree-days for each day in a month, we would get the total degree-days for that month, and similarly, if we were to add up the degree-days for each month, we would then obtain the annual degree-days. In engineering practice, historical degree-day values (based on 30-year-averages) are used to estimate monthly and annual energy consumption rates to heat buildings from the following relationships:

> A degree-day is the difference between 65 (typically) and the average temperature of outdoor air during a 24-hour period.

$$Q_{DD} = \frac{\text{Heat Loss}\left(\dfrac{\text{Btu}}{\text{h}}\right) \times 24 \text{ hrs}}{\text{Design Temperature Difference}(°\text{F})} \qquad \boxed{11.24}$$

$$Q_{monthly} = (Q_{DD})(\text{Monthly Degree-Days}) \qquad \boxed{11.25}$$

$$Q_{yearly} = (Q_{DD})(\text{Yearly Degree-Days}) \qquad \boxed{11.26}$$

EXAMPLE 11.16

For a building located in Minnesota with annual heating degree-days of 8382 and a heating load (heat loss) of 62,000 Btu/h and a design temperature difference of 82° F (68° F indoor and −14° F outdoor), estimate the annual energy consumption. If the building is heated with a furnace with an efficiency of 94%, how much gas is burned to keep the home at 68° F?

We will solve this problem using Equations (11.24) and (11.26).

$$Q_{DD} = \frac{62,000 \left(\dfrac{Btu}{h}\right) \times 24 \text{ hrs}}{82 \, (°F)} = 18,146 \text{ Btu/DD}$$

$$Q_{yearly} = 18,146 \left(\frac{Btu}{DD}\right)(8382 \text{ DD/year}) = 152 \times 10^6 \text{ Btu/year}$$

Assuming the gas used in Minnesota has a heating value of 1,000 Btu/ft^3, then the amount of gas burned in the furnace can be estimated from:

$$\text{Volume of gas burned} = \left[\frac{(152 \times 10^6 \text{ Btu/year})}{0.94}\right]\left[\frac{1}{1000 \text{ Btu/ft}^3}\right]$$
$$= 161,700 \text{ ft}^3/\text{year}$$

LO⁶ 11.6 Additional Temperature-Related Material Properties

Thermal Expansion

As we mentioned earlier in this chapter, accounting for thermal expansion and contraction of materials due to temperature fluctuations is important in engineering problems, including the design of bridges, roads, piping systems (hot water or steam pipes), engine blocks, gas turbine blades, electronic devices and circuits, cookware, tires, and in many manufacturing processes. In general, as the temperature of a material is increased, the material will expand — increase in length — and if the temperature of the material is decreased, it will contract — decrease in length. The magnitude of this elongation or contraction due to temperature rise or temperature drop depends on the composition of the material. The **coefficient of** linear **thermal expansion** provides a measure of the change in length that occurs due to any temperature fluctuations. This effect is depicted in Figure 11.14.

The coefficient of linear expansion, α_L, is defined as the change in the length, ΔL, per original length L per degree rise in temperature, ΔT, and is given by the following relationship:

$$\alpha_L = \frac{\Delta L}{L \Delta T} \qquad \boxed{11.27}$$

Note that the coefficient of linear expansion is a property of a material and has the units of 1/°F or 1/°C. Because a 1° Fahrenheit

| FIGURE 11.14 | The expansion of a material caused by an increase in its temperature. |

temperature difference is equal to a 1° Rankine temperature difference, and a 1° Celsius temperature difference is equal to a 1 Kelvin temperature difference, the units of α_L can also be expressed using 1/°R and 1/K. The values of the coefficient of linear expansion itself depend on temperature; however, average values may be used for a specific temperature range. The values of the coefficient of thermal expansion for various solid materials for a temperature range of 0° C to 100° C are given in Table 11.12. Equation (11.27) could be expressed in such a way as to allow for direct calculation of the change in the length that occurs due to temperature change in the following manner:

$$\Delta L = \alpha_L L \Delta T$$

11.28

For liquids and gases, in place of the coefficient of linear expansion, it is customary to define the coefficient of volumetric thermal expansion, α_V. The

| TABLE 11.12 | The Coefficients of Linear Thermal Expansion for Various Solid Materials (mean value over 0° C to 100° C or 32° F to 212° F) |

Solid Material	Mean Value of α_L (1/°F)	Mean Value of α_L (1/°C)
Brick	5.3×10^{-6}	2.9×10^{-6}
Bronze	10.0×10^{-6}	5.5×10^{-6}
Cast iron	5.9×10^{-6}	3.3×10^{-6}
Concrete	8.0×10^{-6}	4.4×10^{-6}
Glass (plate)	5.0×10^{-6}	2.8×10^{-6}
Glass (Pyrex)	1.8×10^{-6}	1.0×10^{-6}
Glass (thermometer)	4.5×10^{-6}	2.5×10^{-6}
Masonry	$2.5 \times 10^{-6} - 5.0 \times 10^{-6}$	$1.4 \times 10^{-6} - 2.8 \times 10^{-6}$
Solder	13.4×10^{-6}	7.4×10^{-6}
Stainless steel (AISI 316)	2.9×10^{-6}	1.6×10^{-6}
Steel (hard-rolled)	5.6×10^{-6}	3.1×10^{-6}
Steel (soft-rolled)	6.3×10^{-6}	3.5×10^{-6}
Wood (oak) normal to fiber	3.0×10^{-6}	1.7×10^{-6}
Wood (oak) parallel to fiber	2.7×10^{-6}	1.5×10^{-6}
Wood (pine) parallel to fiber	3.0×10^{-6}	1.7×10^{-6}

Based on T. Baumeister, et al., *Mark's Handbook*.

coefficient of volumetric expansion is defined as the change in the volume, ΔV, per original volume V per degree rise in temperature, ΔT, and is given by the following relationship:

$$\alpha_v = \frac{\Delta V}{V \Delta T}$$

11.29

Note that the coefficient of volumetric expansion also has the units of $1/°F$ or $1/°C$. Equation (11.29) could also be used for solids. Moreover, for homogeneous solid materials, the relationship between the coefficient of linear expansion and the coefficient of volumetric expansion is given by

$$\alpha_v = 3\alpha_L$$

11.30

EXAMPLE 11.17

Calculate the change in length of a 1000-ft-long stainless steel cable when its temperature changes by 100° F.

We will use Equation (11.28) and Table 11.12 to solve this problem. From Table 11.12 the coefficient of thermal expansion for stainless steel is $\alpha = 2.9 \times 10^{-6}\ 1/°F$. Using Equation (11.28), we have

$$\Delta L = \alpha_L L \Delta T = (2.9 \times 10^{-6}\ 1/°F)(1000\ \text{ft})(100°\ F) = 0.29\ \text{ft} = 3.48\ \text{in.}$$

Specific Heat

Have you noticed that some materials get hotter than others when exposed to the same amount of thermal energy? For example, if we were to expose 1 kg of water and 1 kg of concrete to a heat source that puts out 100 J every second, you would see that the concrete would experience a higher temperature rise. The reason for this material behavior is that when compared to water, concrete has a lower heat capacity. More explanation regarding our observation will be given in Example 11.16.

Specific heat provides a quantitative way to show how much thermal energy is required to raise the temperature of a 1 kg mass of a material by 1° Celsius. Or, using U.S. Customary units, the specific heat is defined as the amount of thermal energy required to raise the temperature of a 1-lb mass of a material by 1° Fahrenheit. The values of specific heat for various materials at constant pressure are given in Table 11.13. For solids and liquids in the absence of any phase change, the relationship among the required thermal energy (E_{thermal}), mass of the given material (m), its specific heat (c), and the temperature rise ($T_{\text{final}} - T_{\text{initial}}$) that will occur is given by

$$E_{\text{thermal}} = mc(T_{\text{final}} - T_{\text{initial}})$$

11.31

where

$$E_{\text{thermal}} = \text{thermal energy (J or Btu)}$$
$$m = \text{mass (kg or lbm)}$$
$$c = \text{specific heat (J/kg} \cdot \text{K or J/kg} \cdot °C \text{ or Btu/lbm} \cdot °R \text{ or Btu/lbm} \cdot °F)$$
$$T_{\text{final}} - T_{\text{initial}} = \text{temperature rise (}°C \text{ or K, } °F \text{ or } °R)$$

Next we will look at two examples that demonstrate the use of Equation (11.31).

TABLE 11.13	Specific Heat (at Constant Pressure) of Some Materials at 300 K
Material	**Specific Heat (J/kg · K)**
Air (at atmospheric pressure)	1007
Aluminum (pure)	903
Aluminum alloy-2024-T6 (4.5% copper, 1.5 % Mg, 0.6% Mn)	875
Asphalt	920
Bronze (90% copper, 10% aluminum)	420
Brass (70% copper, 30% zinc)	380
Brick (fire clay)	960
Concrete	880
Copper (pure)	385
Glass	750
Gold	129
Iron (pure)	447
Stainless steels (AISI 302, 304, 316, 347)	480, 477, 468, 480
Lead	129
Paper	1340
Platinum (pure)	133
Sand	800
Silicon	712
Silver	235
Zinc	389
Water (liquid)	4180

EXAMPLE 11.18

An aluminum circular disk with a diameter, d, of 15 cm and a thickness of 4 mm is exposed to a heat source that puts out 200 J every second. The density of the aluminum is 2700 kg/m³. Assuming no heat loss to the surrounding air, estimate the temperature rise of the disk after 15 s.

We will make use of Equation (11.31) and Table 11.13 to solve this problem, but first we need to calculate the mass of the disk using the information given.

$$\text{mass} = m = (\text{density})\,(\text{volume})$$

$$\text{Volume} = V = \frac{\pi}{4}d^2\,(\text{thickness}) = \frac{\pi}{4}(0.15\text{ m})^2\,(0.004\text{ m}) = 7.06858 \times 10^{-5}\text{ m}^3$$

$$m = (7.06858 \times 10^{-5}\text{ m}^3)(2700\text{ kg/m}^3) = 0.191\text{ kg}$$

$$E_{\text{thermal}} = mc\left(T_{\text{final}} - T_{\text{initial}}\right)$$

$$200 \text{ J} = (0.191 \text{ kg})(875 \text{ J/kg} \cdot \text{K})\left(T_{\text{final}} - T_{\text{initial}}\right)$$

$$\left(T_{\text{final}} - T_{\text{initial}}\right) = 1.2 \text{ K (every second)}$$

And after 15 s the temperature rise will be $\left(15 \text{ seconds} \times \dfrac{1.2\text{k}}{\text{s}} = 18\text{k}\right)$ $18°\text{C}$ or 18 K.

EXAMPLE 11.19

We have exposed 1 kg of water, 1 kg of brick, and 1 kg of concrete each to a heat source that puts out 100 J every second. Assuming that all of the supplied energy goes to each material and they were all initially at the same temperature, which one of these materials will have a greater temperature rise after 10 s?

We can answer this question using Equation (11.31) and Table 11.13. We will first look up the values of the specific heat for water, brick, and concrete, which are $c_{\text{water}} = 4180 \text{ J/kg} \cdot \text{K}$, $c_{\text{brick}} = 960 \text{ J/kg} \cdot \text{K}$ and $c_{\text{concrete}} = 880 \text{ J/kg} \cdot \text{K}$. Now applying Equation (11.31), $E_{\text{thermal}} = mc\left(T_{\text{final}} - T_{\text{initial}}\right)$ to each situation, it should be clear that although each material has the same amount of mass and is exposed to the same amount of thermal energy, the concrete will experience a higher temperature rise because it has the lowest heat capacity value among the three given materials.

Before You Go On

Answer the following questions to test your understanding of the preceding sections.

1. Explain what is meant by a degree-day.

2. Describe the method that engineers use to estimate monthly and annual energy consumption to heat buildings.

3. Define coefficient of volumetric expansion.

4. Explain what is meant by specific heat.

Vocabulary—State the meaning of the following terms:

Degree-Day _____

Coefficient of Linear Expansion_____

Coefficient of Volumetric Expansion _____

Specific Heat _____

SUMMARY

LO¹ Temperature as a Fundamental Dimension

Temperature provides a measure of both molecular activity and the internal energy of an object. We have bundled all of the microscopic molecular movement into a single, macroscopic, and measurable value that we call temperature. As explained in detail in this chapter, temperature plays an important role in engineering analysis and design. On the Celsius scale, under standard atmospheric conditions, the value zero is arbitrarily assigned to the temperature at which water freezes, and the value of 100 is assigned to the temperature at which water boils. On a Fahrenheit temperature scale, under standard atmospheric conditions, the temperature at which water freezes is assigned a value of 32, and the temperature at which the water boils is assigned a value of 212. Because both the Celsius and the Fahrenheit scales are arbitrarily defined, scientists recognized a need for a better temperature scale. This need led to the definition of an absolute scale, which is represented by the Kelvin and Rankine scales, that are based on the behavior of an ideal gas. We use changes in properties of matter, such as electrical resistance, optical changes, or emf (electromotive force) changes to measure temperature.

LO² Temperature Difference and Heat Transfer

Thermal energy transfer occurs whenever a temperature difference exists. This form of energy transfer that occurs between bodies of different temperatures is called heat transfer. There are three different mechanisms by which energy is transferred from a high-temperature region to a low-temperature region. These are referred to as the modes of heat transfer. Conduction refers to that mode of heat transfer that occurs when a temperature difference exists in a medium and is governed by Fourier's Law. The R-value of a material provides a measure of resistance to heat flow: the higher the value, the more resistance to heat flow the material offers. Convection heat transfer occurs when a gas or a liquid in motion comes into contact with a solid surface whose temperature differs from the moving fluid. This mode of heat transfer is governed by Newton's law of cooling. All matter emits thermal radiation. The Stefan-Boltzmann law determines the amount of radiant energy emitted by a surface.

LO³ Thermal Comfort

The temperature of the environment and the humidity of the air are among the important factors that define thermal comfort. The level of a person's activity is also an important factor. A unit commonly used to express the metabolic rate for an average person under sedentary conditions, per unit surface area, is the *met*; 1 met is equal to 58.2 W/m^2. Metabolic rate determines the rate of conversion of chemical to thermal energy within a person's body. Clothing is another factor that affects thermal comfort. A unit that is generally used to express the insulating value of clothing is the *clo*.

LO⁴ Heating Values of Fuels

When a fuel is burned (gas, oil, etc.), thermal energy is released. The heating value of a fuel quantifies the amount of energy that is released when a unit mass (kilogram or pound) or a unit volume (cubic meter or cubic foot) of a fuel is burned. Different fuels have different heating values.

LO⁵ Degree-Days and Thermal Energy Estimation

A degree-day is difference between 65° F (typically) and the average temperature of outside air during a 24-hour period. In engineering practice, historical degree-day values based on 30-year-average are used to estimate monthly and annual energy consumption to heat a building.

LO⁶ Additional Temperature-Related Properties

The coefficient of expansion and specific heat are examples of material properties that are temperature related. The coefficient of linear expansion is defined as the change in the length per original length per degree rise in temperature. For liquids and gases, in place of the coefficient of linear expansion, it is customary to define the coefficient of volumetric thermal expansion. The coefficient of volumetric expansion is defined as the change in the volume per original volume per degree rise in temperature. Specific heat provides a quantitative way to show how much thermal energy is required to raise the temperature of a 1 kg mass of a material by 1°Celsius, or using U.S. Customary units, to show the amount of thermal energy required to raise the temperature of a 1-lb mass of a material by 1° F.

KEY TERMS

Absolute Temperature 362	Energy Estimation 381	Specific Heat 384
British Thermal Unit (btu) 363	Fourier's Law 365	Stefan-Boltzmann Constant 375
Coefficient of Thermal Expansion 382	Heat Transfer 363	Temperature 355
Conduction 364	Heat Transfer Coefficient 372	Thermal Conductivity 366
Convection 371	Heating Value 379	Thermal Resistance 367
Degree-Days 381	Metabolic Rate 377	Thermistor 358
Emissivity 375	Radiation 375	Thermocouple Wire 358
	R-Value 367	

APPLY WHAT YOU HAVE LEARNED

In your area, identify a home that is heated with natural gas and look up its furnace size and efficiency. Also, look up the monthly and annual heating degree-days for the area for last year or the year before. Using the energy estimation discussed in this chapter, estimate the monthly and annual natural gas consumption to heat the building if the home is to be kept 68°F. Next, obtain the natural gas bill for last year or the year before, and compare your analysis with the actual amount of natural gas consumed. State your assumptions and discuss your findings in a brief report.

Dmitry Bruskov/Shutterstock.com

John Mann

Professional Profile

When I graduated as a Bachelor of Applied Science — Chemical Engineering from the University of Toronto in 1979, the job situation for young engineers in my field was very similar to the opportunities today. The oil industry in Canada was booming, with significant refinery expansion and the development of upgraders to extract oil from Alberta's tar sands via the Syncrude project. That boom subsided, but as world oil prices have increased again, the oil sands are the subject of renewed attention as a reliable energy resource for the North American market. It promises to be an exciting time for engineers in the oil industry.

My fascination with science was sparked by the achievements of the U.S. space program in the 1960s and 1970s. An insightful high school guidance counselor noted my interest in math and science and steered me toward chemical engineering as a career choice. After four years at the University of Toronto, I joined Imperial Oil (Esso Canada). Much of my work focused on debottlenecking and increasing capacity at the company's Vancouver refinery. This was a great opportunity to go into and

Courtesy of John Mann

study the whole refinery from end to end, learning about every process at a detailed level. From there I spent five years at the Sarnia, Ontario refinery as part of a task force analyzing ways to improve safety, reliability, and efficiency. Then it was back to head office in Toronto, working on process simulation. My section was reorganized into corporate IT, so there were new opportunities for training in areas such as database design, data flow diagrams, and modeling design, including simulating models in plant design.

Esso's career development and mentoring programs gave me a solid basis for evaluating economics and working to schedule in the oil industry. With these skills I moved to a Canadian consulting company focusing on designing and developing process plant information systems in the oil industry. The work often involved travelling to meet with clients in places such as the United States, Italy, the Netherlands, France, and the Middle East. Working for a smaller company meant that I experienced much more client interaction—making proposals, designing and quoting on projects, and negotiating contract terms. I was working closely with other members of the team, but at the same time the job demanded that I be capable in many different areas. You could almost call it trial by fire—I was learning on the job, but enjoyed the challenge of going outside of my particular "box."

I am now working for Honeywell Canada as part of the software development group that adds value to the process control systems Honeywell is renowned for. We bring customer requirements into the development process, soliciting customer input and extrapolating to the larger market to genericize processes so that they can accommodate the different site requirements and work practices of our clients. I work with software developers in India and have traveled to support sales efforts and consult on projects around the world . From China to Chile. From a beginning in the oil industry, my travels have taken me to iron mining in Australia coal liquification in South Africa, aluminum smelting in Argentina, and phosphate mining in Northern Ontario.

To engineering students today, I would say that we are entering an era in which resources are becoming more and more valuable and the human factor—the way that we use resources—must be the focus of new technological innovations. Harnessing resources more effectively, with less pollution, will be *the* challenge for the foreseeable future.

Courtesy of John Mann

PROBLEMS

Problems that promote lifelong learning are denoted by 🔑

11.1 Investigate the value of temperature for the following items. Write a brief report discussing how these values are used in their respective areas.

    a. What is a normal body temperature?

    b. What is the temperature range that clinically is referred to as fever?

    c. What is a normal surface temperature of your body? Is this value constant?

    d. What is a comfortable room temperature range? What is its significance in terms of human thermal comfort? What is the role of humidity?

    e. What is the operating temperature range of the freezer in a household refrigerator?

11.2 Using Excel or a spreadsheet of your choice, create a degrees Fahrenheit to degrees Celsius conversion table for the following temperature range: from $-40°$ F to $130°$ F in increments of $5°$ F.

11.3 Alcohol thermometers can measure temperatures in the range of $-100°$ F to $200°$ F Determine the temperature at which an alcohol thermometer with a Fahrenheit scale will read the same number as a thermometer with a Celsius scale.

11.4 What is the equivalent value of $T = 60°$ C in degrees Fahrenheit, Rankine, and Kelvin?

11.5 What is the equivalent value of $T = 120°$ F in degrees Rankine, Celsius, and Kelvin?

11.6 The inside temperature of an oven is maintained at 450° F while the kitchen air temperature is 76° F. What is the oven/kitchen air temperature difference in (a) degree Fahrenheit, (b) degree Rankine, (c) degree Celsius, and (d) Kelvin?

11.7 Obtain information about *K-*, *E-*, and *R*-type thermocouple wires. Write a brief report discussing their accuracy, temperature range of application, and in what application they are commonly employed.

11.8 A manufacturer of loose-fill cellulose insulating material provides a table showing the relationship between the thickness of the material and its R-value. The manufacturer's data is shown in the accompanying table.

R-value (units?)	Thickness (in.)
R-40	11
R-32	9
R-24	6.5
R-19	5.25
R-13	3.5

Calculate the thermal conductivity of the insulating material. Also, determine how thick the insulation should be to provide R-values of

a. R-30

b. R-20

11.9 Calculate the *R*-value for the following materials:

a. 4 in. thick brick

b. 10 cm thick brick

c. 12 in. thick concrete slab

d. 20 cm thick concrete slab

e. 1 cm thick human fat layer

11.10 Calculate the thermal resistance due to convection for the following situations:

a. warm water with $h = 200$ W/m² · K

b. warm air with $h = 10$ W/m² · K

c. warm moving air (windy situation) $h = 30$ W/m² · K

11.11 A typical exterior masonry wall of a house, shown in the accompanying figure, consists of the items in the accompanying table. Assume an inside room temperature of 68° F and an outside air temperature of 10° F, with an exposed area of 150 ft². Calculate the heat loss through the wall.

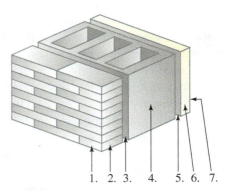

1. 2. 3.   4.   5. 6. 7.

Problem 11.11

Items		Resistance (h·ft²·°F/Btu)
1.	Outside film resistance (winter, 15 mph wind)	0.17
2.	Face brick (4 in.)	0.44
3.	Cement mortar (1/2 in.)	0.1
4.	Cinder block (8 in.)	1.72
5.	Air space (3/4 in.)	1.28
6.	Gypsum wallboard (1/2 in.)	0.45
7.	Inside film resistance (winter)	0.68

11.12 In order to increase the thermal resistance of a typical exterior frame wall, such as the one shown in Example 11.11, it is customary to use 2 × 6 studs instead of 2 × 4 studs to allow for placement of more insulation within the wall cavity. A typical exterior (2 × 6) frame wall of a house consists of the materials shown in the accompanying figure. Assume an inside room temperature of 68° F and an outside air temperature of 20° F with an exposed area of 150 ft². Determine the heat loss through this wall.

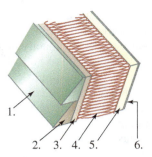

Problem 11.12

Items	Resistance (h · ft² · °F/Btu)
1. Outside film resistance (winter, 15 mph wind)	0.17
2. Siding, wood (1/2 × 8 lapped)	0.81
3. Sheathing (1/2 in. regular)	1.32
4. Insulation batt (5 1/2 in.)	19.0
5. Gypsum wallboard (1/2 in.)	0.45
6. Inside film resistance (winter)	0.68

11.13 A typical ceiling of a house consists of items shown in the accompanying table. Assume an inside room temperature of 70° F and an attic air temperature of 15° F, with an exposed area of 1000 ft². Calculate the heat loss through the ceiling.

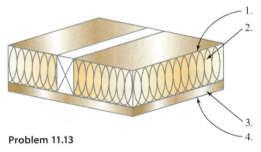

Problem 11.13

Items	Resistance (h · ft² · °F/Btu)
1. Inside attic film resistance	0.68
2. Insulation batt (6 in.)	19.0
3. Gypsum wallboard (1/2 in.)	0.45
4. Inside film resistance (winter)	0.68

11.14 Estimate the change in the length of a power transmission line in your state when the temperature changes by 50° F. Write a brief memo to your instructor discussing your findings.

11.15 Calculate the change in 5 m long copper wire when its temperature changes by 120° F.

11.16 Determine the temperature rise that would occur when 2 kg of the following materials are exposed to a heating element putting out 500 J. Discuss your assumptions.

a. The material is copper.

b. The material is aluminum.

c. The material is concrete.

11.17 The thermal conductivity of a solid material can be determined using a setup similar to the one shown in the accompanying figure. The thermocouples are placed at 2.5 cm intervals in the known material $\left(\text{copper alloy, } k = 52\frac{w}{m \cdot k}\right)$ and the unknown sample, as shown. The known material is heated on the top by a heating element, and the bottom surface of the sample is cooled by running water through the heat sink shown. Determine the thermal conductivity of the unknown sample for the set of data given in the accompanying table. Assume no heat loss to the surroundings and perfect thermal contact at the common interface of the sample and the copper.

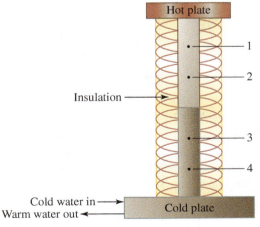

Problem 11.17

Thermocouple Location	Temperature (°C)
1	120 ⎤
2	100 ⎦ Copper
3	85
4	72

11.18 Use the basic idea of Problem 11.17 to design an apparatus that could be constructed to measure the thermal conductivity of solid samples in the range of 50 to 300 W/m · K. Write a brief report discussing in detail the overall size of the apparatus; components of the apparatus, including the heating source, the cooling source, insulation materials; and the measurement elements. In the report, include drawings and a sample experiment. Estimate the cost of such a setup and briefly discuss it in your report. Also, write a brief experimental procedure that could be used by someone who is unfamiliar with the apparatus.

11.19 A copper plate, with dimensions of 3 cm × 3 cm × 5 cm (length, width, and thickness, respectively), is exposed to a thermal energy source that puts out 150 J every second, as shown in the accompanying figure. The density of copper is 8900 kg/m³. Assuming no heat loss to the surrounding block, determine the temperature rise in the plate after 10 seconds.

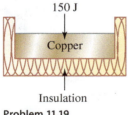

150 J

Copper

Insulation

Problem 11.19

11.20 An aluminum plate, with dimensions of 3 cm × 3 cm × 5 cm (length, width, and thickness, respectively), is exposed to a thermal energy source that puts out 150 J every second. The density of aluminum is 2700 kg/m³. Assuming no heat loss to the surrounding block, determine the temperature rise in the plate after 10 seconds.

11.21 Use the basic idea of Problem 11.19 to design an apparatus that could be constructed to measure the heat capacity of solid samples in the range of 500 to 800 J/kg · K. Write a brief report discussing in detail the overall size of the apparatus; components of the apparatus, including the heating source, the supporting block, insulation materials; and the measurement elements. In the report include drawings and a sample experiment. Estimate the cost of such a setup, and briefly discuss it in your report. Also, write a brief experimental procedure that could be used by someone who is unfamiliar with the apparatus.

11.22 How would you use the principle given in Problem 11.19 to measure the heat output of something? For example, use the basic idea behind Problem 11.19 to design a setup that can be used to measure the heat output of an ironing press.

11.23 Calorimeters are devices that are commonly used to measure the heating value of fuels. For example, a Junkers flow calorimeter is used to measure the heating value of gaseous fuels. A bomb calorimeter, on the other hand, is used to measure the heating value of liquid or solid fuels, such as kerosene, heating oil, or coal. Perform a search to obtain information about these two types of calorimeters. Write a brief report discussing the principles behind the operation of these calorimeters.

11.24 Refer to Tables 11.9 and 11.10 to answer this question. What is the maximum amount of energy released when a 10 lbm sample of coal from McDowell, West Virginia is burned? Also calculate the amount of energy released when 15 ft³ of natural gas from Oklahoma is burned.

11.25 Contact the natural gas provider in your city and find out how much you are being charged for each ft³ of natural gas. Also, contact your electric company and determine how much they are charging on average per kWh usage of electricity. If a hot-air gas furnace has an efficiency of 94% and an electric heater has an efficiency of 100%, what is the more economical way of heating your home: a gas furnace or an electric heater?

11.26 Convert the results of Example 11.13 from Btu to calories.

11.27 Calculate the heat transfer rate from a 1000 ft², 6-in- thick concrete wall with inside and outside surface temperatures of 20° C and 0° C.

11.28 For Problem 11.27, calculate the reduction in the heat transfer rate if a 2-in.-insulation batt with thermal conductivity of $k = 0.03 \frac{Btu}{h \cdot ft \cdot °F}$ is added to the wall.

11.29 The side walls of a refrigerator are made from two thin layers of sheet metal, each 1/8 in. thick, and with thermal conductivity of $k = 42 \frac{Btu}{h \cdot ft \cdot °F}$, and a 2-in. foam insulation with $k = 0.027 \frac{Btu}{h \cdot ft \cdot °F}$. If the inside and outside surface temperatures of the refrigerator wall are 35° F and 70° F, respectively, calculate the heat loss through an area of 8 ft².

11.30 Calculate the amount of thermal energy required to raise the temperature of 20 gallon of water from 60° F to 120° F. Express your answer in Btu, J, and cal.

11.31 In a commercial water heater, 20 gallons/min of water is heated from 60° F to 140° F. Calculate the amount of energy required per hour $\left(\frac{Btu}{h} \right)$.

11.32 A gas furnace puts out 60,000 $\frac{Btu}{h}$ to compensate for heat loss from a house located in Minnesota. What is the equivalent value of thermal power (thermal energy per unit time) output of the furnace in watts?

11.33 Calculate the heat loss from a double-pane glass window consisting of two pieces of glass, each having a thickness of 10 mm with a thermal conductivity of $k = 1.3 \frac{W}{m \cdot K}$. The two glass panes are separated by an air gap of 7 mm. Assume thermal conductivity of air to be $k = 0.022 \frac{W}{m \cdot K}$. Also, express the total R and U values in both SI and U.S. Customary units.

11.34 Determine the heat-transfer rate from an electronic chip whose surface temperature is 30° C and has an exposed surface area of 4 cm². The temperature of surrounding air is 25° C. The heat-transfer coefficient for this situation is $h = 25 \frac{W}{m² \cdot K}$. Express your answer in both SI and U.S. Customary units. What is the R-factor (film resistance) for this situation?

11.35 Calculate the amount of radiation emitted for a unit surface (1 m²) for the following situations: (a) a hot pavement in Arizona at 50° C (122° F) and $\varepsilon \approx 0.8$, (b) a hood of a car at 40° C (104° F) $\varepsilon \approx$ and 0.9, and (c) a sunbather at 38° C (100° F) $\varepsilon \approx$ and 0.9. Express your answers in both SI and U.S. Customary units.

11.36 For Problems 11.11, 11.12, and 11.13, calculate the U-factors.

11.37 For Problem 11.12, calculate the heat loss through the frame wall if the R-19 insulation batt is replaced by foam insulation having a R-value of 22.

11.38 For Problem 11.13, calculate the heat loss through the ceiling, if the R-19 insulation batt is replaced by R-40 fiberglass insulation.

11.39 Nine old 12 ft² windows with $U = 1.2 \frac{Btu}{h \cdot ft² \cdot °F}$ were replaced with new windows having $U = 0.3 \frac{Btu}{h \cdot ft² \cdot °F}$. Calculate the energy savings on a day during a 5 hour period, when $T_{in} = 68°$ F, $T_{outside} = 10°$ F.

11.40 For Problem 11.39, calculate the savings in ft³ of natural gas (from Louisiana). Assume the furnace has an efficiency of 92%.

11.41 A family uses 80 gallons of hot water per day. The water is heated from a line temperature of 55° F to 140° F. Calculate the amount of natural gas (Oklahoma) that is required to heat the water in a heater with an efficiency of 80%.

11.42 For Problem 11.17, determine the thermal conductivity of the unknown sample for the set of data given in accompanying table.

Thermocouple Location	Temperature (°C)
1	125
2	105
3	80
4	75

11.43 Compare heat-transfer rates through layers of human skin, human muscle, and human fat tissue. Assume a 5 mm thickness and a temperature difference of 2° C across each layer.

11.44 Calculate the heat-transfer rate through a 24 ft² door with $U = 0.75\frac{Btu}{h \cdot ft^2 \cdot \,^\circ F}$. The indoor and outdoor temperatures are 68° F and 10° F.

11.45 Look up the low and high daily temperature values for the month of October 2014 for your town and calculate the degree days for the given month.

11.46 For your town calculate the degree-days for the months of December, January, and February 2014 and compare it to the 2014 annual degree-days value.

11.47 For a building located in Baltimore, Maryland with annual heating degree-days (dd) of 4654, a heating load (heat loss) of 30,000 Btu/h, and a design temperature difference of 52° F (68° F indoor and 16° F outdoor), estimate the annual energy consumption. If the building is heated with a furnace with an efficiency of 92%, how much gas is burned to keep the home at 68° F? State your assumptions.

11.48 For a building located in Boston, Massachusetts with annual heating degree-days (dd) of 5634, a heating load (heat loss) of 42,000 Btu/h, and a design temperature difference of 62° F (68° F indoor and 6° F outdoor), estimate the annual energy consumption. If the building is heated with a furnace with an efficiency of 98%, how much gas is burned to keep the home at 68° F? State your assumptions.

11.49 For a building located in Detroit, Michigan with annual heating degree-days (dd) of 6232, a heating load (heat loss) of 52,000 Btu/h, and a design temperature difference of 73° F (68° F indoor and −5° F outdoor), estimate the annual energy consumption. If the building is heated with a furnace with an

efficiency of 92%, how much gas is burned to keep the home at 68° F. State your assumptions.

11.50 Visit the National Fenestration Rating Council (NFRC) website at www.nfrc.org and lookup the definition of energy performance ratings given by **A** through **E**. Write a brief report discussing your findings.

   a. U-Factor

   b. Solar Heat Gain Coefficient (SHGC)

   c. Visible Transmittance (VT)

   d. Air Leakage

   e. Condensation Resistance

Problem 11.50
Courtesy of The National Fenestration Rating Council (NFRC).

Electric Current and Related Variables in Engineering

gwycech/Shutterstock.com

WebButtonsCO/Shutterstock.com

Dmitry Melnikov/Shutterstock.com

Ljupco Smokovski/Shutterstock.com

Alex Mit/Shutterstock.com

Daboost/Shutterstock.com

Engineers understand the importance of electricity and electrical power and the role they play in our everyday lives. As future engineers, you should know what is meant by electric current, voltage, and know the difference between direct and alternating current. You should also know the various sources of electricity and understand how electricity is generated.

LEARNING OBJECTIVES

LO¹ **Electric Current, Voltage, and Electric Power:** describe the role of current as a fundamental dimension in engineering analysis; explain basic principles of electricity

LO² **Electrical Circuits and Components:** explain what is meant by electrical circuits and give examples of its components

LO³ **Electric Motors:** describe the role of motors in our everyday life and give examples of types of motors

LO⁴ **Lighting Systems:** explain lighting terminology and give examples of different lighting systems and their power consumption rates

DISCUSSION STARTER

THE PROCESS OF TRANSPORTING ELECTRICITY

Getting electricity from power-generating stations to our homes and workplaces is quite a challenging process. Electricity must be produced at the same time as it is used because large quantities of electricity cannot be stored effectively. High-voltage transmission lines (those lines between tall metal towers that you often see along the highway) are used to carry electricity from power-generating stations to the places where it is needed. However, when electricity flows over these lines, some of it is lost. One of the properties of high voltage lines is that the higher the voltage, the more efficient they are at transmitting electricity, that is, the lower the losses are. Using transformers, high-voltage electricity is "stepped down" several times to a lower voltage before arriving over the distribution system of utility poles and wires to your home and workplace so it can be used safely.

There is no "national" power grid. There are actually three power grids operating in the 48 contiguous states: (1) the Eastern Interconnected System (for states east of the Rocky Mountains), (2) the Western Interconnected System (from the Pacific Ocean to the Rocky Mountain states), and (3) the Texas Interconnected System. These systems generally operate independently of each other, although there are limited links between them. Major areas in Canada are totally interconnected with our Western and Eastern power grids, while parts of Mexico have limited connection to the Texas and the Western power grids.

The "Smart Grid"

The "Smart Grid" consists of devices connected to transmission and distribution lines that allow utilities and customers to receive digital information from and communicate with the grid. These devices allow a utility to find out where an outage or other problem is on the line and sometimes even fix the problem by sending digital instructions. Smart devices in the home, office, or factory inform consumers of times when an appliance is using relatively high-cost energy and allow consumers to remotely adjust its settings. Smart devices make a Smart Grid as they help utilities reduce line losses, detect and fix problems faster, and help consumers conserve energy, especially at times when demand reaches significantly high levels or an energy demand reduction is needed to support system reliability.

U.S. Energy Information Administration

Shutter_M/Shutterstock.com

Western Interconnection

Eastern Interconnection

Texas Interconnection

- 230,000 volts
- 345,000 volts
- 500,000 volts
- 765,000 volts
- High-voltage direct current

"The National Power Grid," Microsoft® Encarta® Encyclopedia. http://encarta.msn.com
©1993-2004 Microsoft Corporation. All rights reserved.

Anita Potter/Shutterstock.com, data from OffGridWorld,
What is the Electric Power Grid, http://www.offgridworld
.com/what-is-the-electric-power-grid-u-s-grid-map/, EIA
U.S. Energy Information Administration

To the students: How are you consuming electricity? What do you think is your daily, monthly, annual electricity consumption footprint? Do you have any suggestions as to how you might reduce your electricity consumption footprint?

very engineer needs to understand the fundamentals of electricity and magnetism. Look around, and you will see all the devices, appliances, and machines that are driven by electrical power. The objective of this chapter is to introduce the fundamentals of electricity. We will briefly discuss what is meant by electric charge, electric current (both alternating current, ac, and direct current, dc), electrical resistance, and voltage. We will also define what is meant by an electrical circuit and its components. We will then look at the role of electric motors in our everyday lives and identify the factors that engineers consider when selecting a motor for a specific application.

As explained in Chapter 6, based on our understanding of our physical world today, we need seven fundamental or base dimensions to correctly express the physical laws that govern our world. They are length, mass, time, temperature, electric current, amount of substance, and luminous intensity. Moreover, with the help of these base dimensions we can derive all other necessary physical quantities that describe how nature works.

In the previous chapters, we discussed the role of the fundamental dimension length and length-related variables, the fundamental dimension mass and mass-related variables, the fundamental dimension time and time-related variables, and the fundamental dimension temperature and

| TABLE 6.7 | Fundamental Dimensions and How They Are Used in Defining Variables that Are Used in Engineering Analysis and Design |

Chapter	Fundamental Dimension	Related Engineering Variables			
7	Length (L)	Radian (L/L), Strain (L/L)	Area (L^2)	Volume (L^3)	Area moment of inertia (L^4)
8	Time (t)	Angular speed ($1/t$), Angular acceleration ($1/t^2$), Linear speed (L/t), Linear acceleration (L/t^2)		Volume flow rate (L^3/t)	
9	Mass (M)	Mass flow rate (M/t), Momentum (ML/t), Kinetic energy (ML^2/t^2)		Density (M/L^3), Specific volume (L^3/M)	
10	Force (F)	Moment (LF), Work, energy (FL), Linear impulse (Ft), Power (FL/t)	Pressure (F/L^2), Stress (F/L^2), Modulus of elasticity (F/L^2), Modulus of rigidity (F/L^2)	Specific weight (F/L^3),	
11	Temperature (T)	Linear thermal expansion (L/LT), Specific heat (FL/MT)		Volume thermal expansion (L^3/L^3T)	
12	Electric Current (I)	Charge (It)	Current density (I/L^2)		

temperature-related variables in engineering analysis and design. We now turn our attention to the fundamental dimension current.

Table 6.7 is repeated again to remind you of the role of fundamental dimensions and how they are combined to define engineering quantities used in analysis and design.

LO¹ 12.1   Electric Current, Voltage, and Electric Power

Electric Current as a Fundamental Dimension

As explained in Chapter 6, it was not until 1946 that the proposal for ampere as a base unit for electric current was approved by the General Conference on Weights and Measures (CGPM). In 1954, CGPM included ampere among the base units. The **ampere** is defined formally as that constant current which, if maintained in two straight parallel conductors of infinite length, of negligible circular cross section, and placed 1 meter apart in a vacuum, would produce between these conductors a force equal to 2×10^{-7} newton per meter of length.

To better understand what the ampere represents, we need to take a closer look at the behavior of material at the subatomic level. In Chapter 9 we explained what is meant by atoms and molecules. An atom has three major subatomic particles, namely, electrons, protons, and neutrons. Neutrons and protons form the nucleus of an atom. How a material conducts electricity is influenced by the number and the arrangement of electrons. Electrons have negative charge, whereas protons have a positive charge, and neutrons have no charge.

Simply stated, the basic law of electric **charges** states that *unlike charges attract each other while like charges repel.* In SI units, the unit of charge is the coulomb (C). One coulomb is defined as the amount of charge that passes a point in a wire in 1 second when a current of 1 ampere is flowing through the wire. In Chapter 10, we explained the universal law of gravitational attraction between two masses. Similarly, there exists a law that describes the attractive electric force between two opposite-charge particles. The electric force exerted by one point charge on another is proportional to the magnitude of each charge and is inversely proportional to the square of the distance between the point charges. Moreover, the electric force is attractive if the charges have opposite signs, and it is repulsive if the charges have the same sign. The electric force between two point charges is given by Coulomb's law:

$$F_{12} = \frac{kq_1q_2}{r^2}$$

12.1

where $k = 8.99 \times 10^9 \ \text{N} \cdot \text{m}^2/\text{C}^2$, q_1 and q_2 (C) are the point charges, and r is the distance (m) between them. Another important fact that one must keep in mind is that the electric charge is conserved, meaning the electric charge is not created nor destroyed; it can only be transferred from one object to another.

You may already know that in order for water to flow through a pipe, a pressure difference must exist. Moreover, the water flows from the high-pressure

> The flow of electric charge is called current and its unit is denoted as ampere in both SI and U.S. Customary Systems.

region to the lower-pressure region. In Chapter 11, we also explained that whenever there is a temperature difference in a medium or between bodies, thermal energy flows from the high-temperature region to the low-temperature region. In a similar way, whenever there exists a difference in electric potential between two bodies, electric charge will flow from the higher electric potential to the lower potential region. The flow of charge will occur when the two bodies are connected by an electrical conductor such as a copper wire. The flow of electric charge is called **electric current** or simply, **current**. The electric current, or the flow of charge, is measured in amperes. One ampere or "amp" (A) is defined as the flow of 1 unit of charge per second. For example, a toaster that draws 6 amps has 6 units of charge flowing through the heating element each second. The amount of current that flows through an electrical element depends on the electrical potential, or voltage, available across the element and the resistance the element offers to the flow of charge.

Voltage

Voltage represents the amount of work required to move charge between two points, and the amount of charge that is moving between the two points per unit time is called current. **Electromotive force (emf)** represents the electric potential difference between an area with an excess of free electrons (negative charge) and an area with an electron deficit (positive charge). The voltage, or the electromotive force, induces current to flow in a circuit. The most common sources of electricity are chemical reaction, light, and magnetism.

> Voltage represents the amount of work required to move charge between two points, and the amount of charge that is moving between the two points per unit time is called current.

Batteries All of you have used batteries for different purposes at one time or another (Figure 12.1). In all batteries, electricity is produced by the chemical reaction that takes place within the battery. When a device that uses batteries is on, its circuits create paths for the electrons to flow through. When the device is turned off, there is no path for the electrons to flow, thus the chemical reaction stops.

A battery cell consists of chemical compounds, internal conductors, positive and negative connections, and the casing. Examples of cells include sizes N, AA, AAA, C, and D. A cell that cannot be recharged is called a *primary cell.* An alkaline battery is an example of a primary cell. On the other hand, a *secondary cell* is a cell that can be recharged. The recharging is accomplished by reversing the current flow from the positive to the negative areas. Lead acid

FIGURE 12.1

cells in your car battery and nickel-cadmium (NiCd), and nickel-metal hydride (NiMH) cells are examples of secondary cells. The NiCd batteries are some of the most common rechargeable batteries used in cordless phones, toys, and some cellular phones. The NiMH batteries, which are smaller, are used in many smaller cellular phones because of their size and capacity.

To increase the voltage output, batteries are often placed in a series arrangement. If we connect batteries in a series arrangement, the batteries will produce a net voltage that will be the sum of the individual batteries placed in series. For example, if we were to connect four 1.5-volt batteries in series, the resulting potential will be 6 volts, as shown in Figure 12.2. Batteries connected in a parallel arrangement, as shown in Figure 12.3, produce the same voltage but more current.

Photoemission **Photoemisssion** is another principle used to generate electricity. When light strikes a surface that has certain properties, electrons can be freed; thus electric power is generated. You may have seen examples of photovoltaic devices such as light meters used in photography, photovoltaic cells in hand-held calculators, and solar cells used in remote areas to generate electricity.

Photovoltaic devices are becoming increasingly common in many applications because they do not pollute the environment. In general, there are two ways in which the sun's radiation is converted to electricity: photothermal and photovoltaic. In photothermal plants, solar radiation (the sun's radiation) is used to make steam by heating water flowing through a pipe. The pipe runs through a parabolic solar collector, and then the steam is used to drive a generator. In a photovoltaic cell, the light is converted directly to electricity. Photovoltaic cells,

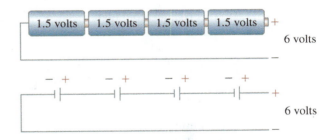

FIGURE 12.2   Batteries connected in a series arrangement.

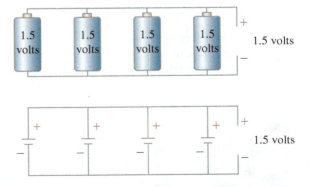

FIGURE 12.3   Batteries connected in a parallel arrangement.

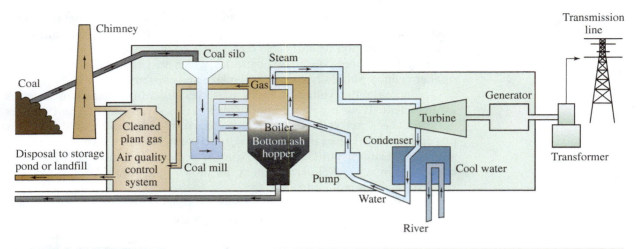

which are made of gallium arsenide, have conversion efficiencies of approximately 20%. The cells are grouped together in many satellites to create an array that is used to generate electric power. Photovoltaic solar farms are also becoming common. A solar farm consists of a vast area where a great number of solar arrays are put together to convert the sun's radiation into electricity.

Power Plants Electricity that is consumed at homes, schools, malls, and by various industries is generated in a power plant. Water is used in all steam power-generating plants to produce electricity. A simple schematic of a power plant is shown in Figure 12.4. Fuel is burned in a boiler to generate heat, which in turn is added to liquid water to change its phase to steam; steam passes through turbine blades, turning the blades, which in effect runs the generator connected to the turbine, creating electricity. The electricity is generated by turning a coil of wire inside a magnetic field. A conductor placed in a changing magnetic field will have a current induced in it. Magnetism is the most common method for generating electricity. The low-pressure steam leaving the turbine liquefies in a condenser and is pumped through the boiler again, closing a cycle, as shown in Figure 12.4. U.S. electricity generation by fuel type for the year 2013 is shown in Figure 12.5.

Direct Current and Alternating Current

Direct current (dc) is the flow of electric charge that occurs in one direction, as shown in Figure 12.6(a). Direct current is typically produced by batteries and direct current generators. In the late 19th century, given the limited understanding of fundamentals and technology and for economic reasons, direct current could not be transmitted over long distances. Therefore, it was succeeded by alternating current (ac). Direct current was not economically feasible to transform because of the high voltages needed for long-distance transmission. However, developments in the 1960s have led to techniques that now allow the transmission of direct current over long distances.

> Direct current is the flow of charge that occurs in one direction, whereas alternating current is the flow of electric charge that periodically reverses.

Alternating current (ac) is the flow of electric charge that periodically reverses. As shown in Figure 12.6(b), the magnitude

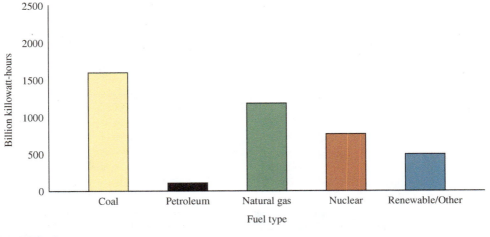

U.S. electricity generation by fuel type (2013).

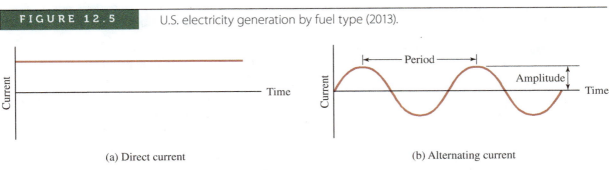

The direct and alternating currents.

of the current starts from zero, increases to a maximum value, and then decreases to zero; the flow of electric charge reverses direction, reaches a maximum value, and returns to zero again. This flow pattern is repeated in a cyclic manner. The time interval between the peak value of the current on two successive cycles is called the *period,* and the number of cycles per second is called the *frequency.* The peak (maximum) value of the alternating current in either direction is called the *amplitude.* Alternating current is created by generators at power plants. The current drawn by various electrical devices at your home is alternating current. The alternating current in domestic and commercial power use is 60 cycles per second (hertz) in the United States.

Kirchhoff's Current Law

One of the basic laws in electricity that allows for the analysis of currents in electrical circuits is **Kirchhoff's current law**. The law states that at any given time, the sum of the currents entering a node must be equal to the sum of the current leaving the node. This statement is demonstrated in Figure 12.7. As explained earlier, physical laws are based on observations, and the Kirchhoff's current law is no exception. This law represents the physical fact that charge is always conserved in an electrical circuit. The charge cannot accumulate or deplete at an electrical node; consequently, the sum of the currents entering a node must equal the sum of the currents leaving the node. For example, Kirchhoff's current law applied to the circuit shown in Figure 12.7 leads to: $i_1 + i_2 = i_3 + i_4$.

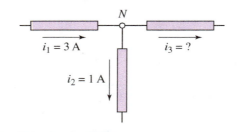

i_4

i_1 i_3

i_2

| FIGURE 12.7 | The sum of currents entering a node must equal the sum of the current leaving the node: $i_1 + i_2 = i_3 + i_4$. |

EXAMPLE 12.1

Determine the value of current i_3 in the circuit shown in Figure 12.8.

N

$i_1 = 3\ A$ $i_3 = ?$

$i_2 = 1\ A$

| FIGURE 12.8 | The circuit for Example 12.1. |

We can solve this simple problem by applying Kirchhoff's current law to node N. This leads to:

$$i_1 = i_2 + i_3$$
$$i_3 = 2A$$

Residential Power Distribution

An example of a typical residential power distribution system is shown in Figure 12.9, which gives examples of amperage requirements for outlets, lights, kitchen appliances, and central air conditioning. To wire a building, an electrical plan for the building must first be developed. In the plan, the locations and the types of switches and outlets, including outlets for the range and dryer, must be specified. We will discuss engineering symbols in more detail in Chapter 16. For now, examples of electrical symbols used in a house plan are shown in Figure 12.10. Moreover, an example of an electrical plan for a residential building is shown in Figure 12.11.

> The most common means by which we produce electricity are magnetism (e.g., in power plants), chemical reaction (e.g., batteries), light (e.g., photovoltaic cells), and converting wind energy.

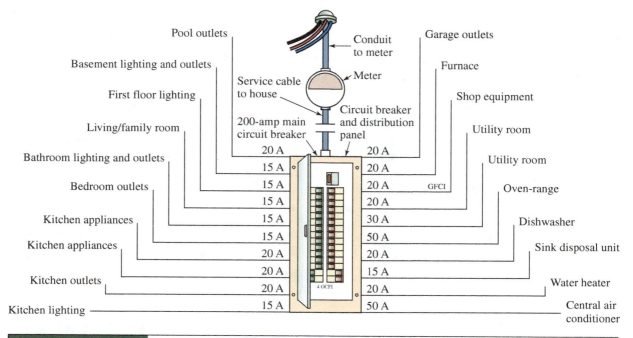

Pool outlets
Basement lighting and outlets
First floor lighting
Living/family room
Bathroom lighting and outlets
Bedroom outlets
Kitchen appliances
Kitchen appliances
Kitchen outlets
Kitchen lighting

Conduit to meter
Service cable to house
Meter
Circuit breaker and distribution panel
200-amp main circuit breaker

Garage outlets
Furnace
Shop equipment
Utility room
Utility room
Oven-range
Dishwasher
Sink disposal unit
Water heater
Central air conditioner

20 A	20 A
15 A	20 A
15 A	20 A GFCI
15 A	20 A
15 A	30 A
15 A	50 A
20 A	20 A
20 A	15 A
20 A	20 A
15 A	50 A

4 GCFI

FIGURE 12.9 An example of an electrical distribution system for a residential building.
From Electrical Wiring, Second Edition, 1981, p18, Copyright © 1981 AAVIM. Reprinted by permission.

Common Electrical Symbols			
○	Ceiling outlet	▨	Service entrance panel
⊸○	Wall outlet	S	Single-pole switch
○PS	Ceiling outlet with pull switch	S2	Double-pole switch
⊸○PS	Wall outlet with pull switch	S3	3-way switch
⊜	Duplex convenience outlet	S4	4-way switch
⊜WP	Weatherproof convenience outlet	SP	Switch with pilot light
⊜1,3	Convenience outlet 1 = single 3 = triple	◉	Push button
⊜R	Range outlet	CH⊃	Bell or chimes
⊜S	Convenience outlet with switch	◀	Telephone
⊜D	Dryer outlet	TV	Television outlet
⊖	Split-wired duplex outlet	S⌐	Switch wiring
◉	Special-purpose outlet	⊐⊏	Fluorescent ceiling fixture
D	Electric door opener	⊙⌣⊙	Fluorescent wall fixture

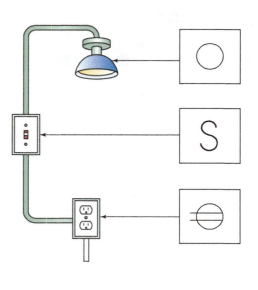

FIGURE 12.10 Examples of electrical symbols in a house plan.
From Electrical Wiring, Second Edition, 1981, p10, Copyright © 1981 AAVIM. Reprinted by permission.

General-Purpose Circuits

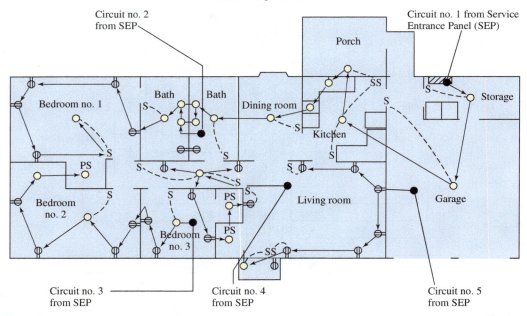

FIGURE 12.11   An example of an electrical plan for a house.

From Electrical Wiring, Second Edition, 1981, p28, Copyright © 1981 AAVIM. Reprinted by permission.

Before You Go On

Answer the following questions to test your understanding of the preceding sections.

1. In what unit is electric current measured?

2. Explain what is meant by voltage.

3. Explain what is meant by direct current.

4. Explain what is meant by alternating current.

5. In your own words explain how electricity is produced in a conventional power plant.

6. Explain the Kirchhoff's current law.

Vocabulary—State the meaning of the following terms:

Ampere_____

Direct current _____

Alternating current_____

Voltage_____

LO² 12.2 Electrical Circuits and Components

An **electrical circuit** refers to the combination of various electrical components that are connected together. Examples of electrical components include wires (conductors), switches, outlets, resistors, and capacitors. First, let us take a closer look at electrical wires. In a wire the **resistance** to electrical current depends on the material from which the wire is made and its length, diameter, and temperature. Materials show varying amount of resistance to the flow of electric current. **Resistivity** is a measure of resistance of a piece of material to electric current.

The resistivity values are usually measured using samples made of a 1 cm cube or a cylinder having a diameter of 1 mil and length of 1 ft. The resistance of the sample is then given by

$$R = \frac{\rho \ell}{a}$$

12.2

where ρ is the resistivity, ℓ is the length of the sample, and a is the cross-sectional area of the sample. The resistance for 1 ft long wire having a diameter of 1 mil made of various materials is given in Table 12.1. The electrical resistance of a material varies with temperature. In general, the resistance of conductors (except for carbon) increases with increasing temperature. Some materials exhibit near-zero resistance at very low temperatures (temperatures approaching absolute zero). This behavior is commonly referred to as *superconductivity*.

TABLE 12.1	Electrical Resistance for 1 ft long Wire Made of Various Metals Having a Diameter of 1 mil at 20° C

Metal	Resistance (ohms, Ω)
Aluminum	17.01
Brass	66.80
Copper	10.37
Gold	14.7
Iron	59.9
Lead	132.0
Nickel	50.8
Platinum	63.8
Silver	9.8
Tin	70.0
Tungsten	33.2
Zinc	35.58

Ohm's Law

Ohm's law describes the relationship among voltage, V, resistance, R, and current, I, according to

$$V = RI$$

12.3

Note from Ohm's law, Equation (12.3), that current is directly proportional to voltage and inversely proportional to resistance. As electric potential is increased, so is the current; and if the resistance is increased, the current will decrease. The electric resistance is measured in units of **ohms** (Ω). An element with 1 ohm resistance allows a current flow of 1 amp when there exists a potential of 1 volt across the element. Stated another way, when there exists an electrical potential of 1 volt across a conductor with a resistance of 1 ohm, then 1 ampere of electric current will flow through the conductor.

EXAMPLE 12.2

The electric resistance of a light bulb is $145 \, \Omega$. Determine the value of current flowing through the lamp when it is connected to a 120-volt source.

Using Ohm's law, Equation (12.3), we have

$$V = RI$$

$$I = \frac{V}{R} = \frac{120}{145} = 0.83 \text{ A}$$

The American Wire Gage (AWG)

Electrical wires are typically made of copper or aluminum. The actual size of wires is commonly expressed in terms of gage number. The American Wire Gage (**AWG**) is based on successive gage numbers having a constant ratio of approximately 1.12 between their diameters. For example, the ratio of the diameter of No. 1 AWG wire to No. 2 is 1.12 (289 mils/258 mils = 1.12), or, as another example, the ratio of diameter of No. 0000 to No. 000 is 1.12 (460 mils/410 mils = 1.12). Moreover, the ratio of cross sections of successive gage numbers is approximately $(1.12)^2 = 1.25$. The ratio of the diameters of wires differing by 6 gage numbers is approximately 2.0. For example, the ratio of diameters of wire No. 1 to No. 7 is 2 (289 mils/144 mils $\approx$ 2.0), or the ratio of diameters of No. 30 to No. 36 is 2 (10.0 mils/5.0 mils = 2.0). Table 12.2 shows the gage number, the diameter, and the resistance for copper wires. When examining Table 12.2, note that the smaller the gage number, the bigger the wire diameter. Also note from data given in Table 12.2 that the electric resistance of a wire is increased as its diameter is decreased.

The National Electrical Code, published by the Fire Protection Association, contains specific information on the type of wires used for general wiring. The code describes the wire types, maximum operating temperatures, insulating materials, outer cover sheaths, the type usage, and the specific location where a wire should be used.

Today, a typical home has a total 200 amperage rating. You also will find that various types of wires are used for general wiring ranging from American Wire Gauge Numbers of 00 to 14.

TABLE 12.2	Examples of American Wire Gage (AWG) for Solid Copper Wire			
American Wire Gauge (AWG) Number	**Diameter (mils)**	**Resistance per 1000 ft (ohms) @ 77° F**	**Current**	**Common Use**
0000	460.0	0.0500		
000	410.0	0.0630		
00	365.0	0.0795	200 A	Service Entrance
1	289.0	0.126		
2	258.0	0.159	100 A	Service Panels
5	182.0	0.319		
6	162.0	0.403	60 A	Electric Furnaces
7	144.0	0.508	40 A	Kitchen Appliances, Receptacles light
10	91.0	1.28	30 A	Fixtures
12	81.0	1.62	20 A	Residential Wiring
14	64.0	2.58	15 A	Lamps; Light Fixtures
16	51.0	4.09		
18	40.0	6.51		
20	32.0	10.4		
22	25.3	16.5		
24	20.1	26.2		
26	15.9	41.6		
28	12.6	66.2		
30	10.0	133		
32	8.0	167		
34	6.3	266		
36	5.0	423		
38	4.0	673		
40	3.1	1070		

Electric Power

The **electric power** consumption of various electrical components can be determined using the following power formula:

$$P = VI \qquad \text{12.4}$$

where P is power in watts, V is the voltage, and I is the current in amps. The kilowatt hour units are used in measuring the rate of consumption of electricity by homes and business. One kilowatt hour represents the amount of power consumed during 1 hour by a device that uses one kilowatt (kW), or 1000 joules per second.

Appliance and Home Electronics Power Consumption Let us now turn our attention to the power consumption of typical home appliances and electronic devices. The range of power consumption for common appliances and electronics such as clock radio, coffee maker, clothes washer, clothes dryer, fan, hair dryer, television, toaster (and toaster oven), and vacuum cleaner are shown in Table 12.3. As you can see, hot water heaters, dishwashers, and clothes dryers are among the energy hogs. For example, clothes dryers, depending on their sizes, could consume between 2000 to 5000 watts; so if you run a clothes dryer with a power rating of 5000 watts for 2 h, it will consume: $(5000 \text{ W}) \times (2 \text{ h}) = 10,000 \text{ W-h} = 10 \text{ kWh of energy}$. An example of home electronics, a 46 inch LCD TV, could consume about 250 watts, so if you watch this TV for 4 hours, then the TV will consume: $(250 \text{ W}) \times (4 \text{ h}) = 1000 \text{ Wh} = 1 \text{kWh of energy}$.

EXAMPLE 12.3

Assuming that your electric power company is charging you 10 cents for each kWh usage, estimate the cost of leaving five 100-W light bulbs on from 6 P.M. until 11 P.M. every night for 30 nights.

$$(5 \text{ light bulbs})\left(100 \frac{\text{W}}{\text{light bulb}}\right)\left(\frac{1 \text{ kW}}{1000 \text{ W}}\right)\left(5 \frac{\text{h}}{\text{nights}}\right)(30 \text{ nights})\left(10 \frac{\text{cents}}{\text{kWh}}\right)$$

$$= 750 \text{ cents}$$
$$= \$7.50$$

TABLE 12.3	Examples of home appliances and electronic devices and their range of power consumption.

Item	Range of power consumption (watts)
Aquarium	50–1210
Clock radio	10
Coffee maker	900–1200
Clothes washer	350–500
Clothes dryer	1800–5000
Dishwasher	1200–2400*
Dehumidifier	785
Electric blanket-Single/Double	60/100
Fans	
Ceiling	65–175
Window	55–250
Furnace	750
Whole house	240–750
Hair dryer	1200–1875
Heater (portable)	750–1500
Clothes iron	1000–1800
Microwave oven	750–1100
Personal computer	
CPU - awake/asleep	120/30 or less
Monitor - awake/asleep	150/30 or less
Laptop	50
Radio (stereo)	70–400
Refrigerator (frost-free, 16 cubic feet)	725
Televisions	65–250
Toaster	800–1400
Toaster oven	1225
VCR/DVD	17–21/20–25
Vacuum cleaner	1000–1440
Water heater (40 gallon)	4500–5500
Water pump (deep well)	250–1100
Water bed (with heater, no cover)	120–380

*Using the drying feature greatly increases energy consumption.

U.S. Department of Energy

Series Circuit

Electrical components can be connected in either a series or a parallel arrangement. As we mentioned earlier in this chapter, there are many different types of electrical components. Here in this section we will only look at resistors and capacitors.

A **resistor** is an electrical component that resists the flow of either direct or alternating current. Resistors are commonly used to protect sensitive components or to control the flow of current in a circuit. Moreover, a resistor can be used to divide or control voltages in a circuit. There are two broad groups of resistors: fixed-value resistors and variable resistors. As the name implies, the fixed-value resistor has a fixed value. On the other hand, the variable resistors, sometimes called *potentiometers,* can be adjusted to a desired value. The variable resistors are used in various circuits to adjust the current in the circuit. For example, variable resistors are used in light switches that allow you to dim the light. They are also used to adjust the volume in a radio. A resistor dissipates heat as electric current flows through it. The amount of the heat dissipated depends on the magnitude of the resistor and the amount of the current that is passing through the resistor.

Similar to a constant flow of water in a series of pipes of varying size, the electric current flowing through a series of elements in an electric circuit is the same (constant). Moreover, for a circuit that has elements in a series arrangement, the following is true:

- The voltage drop across each element can be determined using Ohm's law.
- The sum of the voltage drop across each element is equal to the total voltage supplied to the circuit.
- The total resistance is the sum of resistance in the circuit.

For resistors in series, the total equivalent resistance is equal to the sum of the individual resistors, as shown in Figure 12.12.

$$R_{total} = R_1 + R_2 + R_3 + R_4 + R_5 \qquad \boxed{12.5}$$

One of the problems with a circuit that has elements in a series arrangement is that if one of the elements fails, that failure prevents the current from flowing through other elements in the circuit; thus the entire circuit fails. You may have experienced this problem with a series of light bulbs in a string of Christmas tree lights. In a series arrangement, when one light fails, the current to other lights stops, resulting in all of the lights being off.

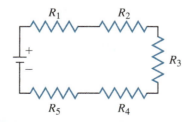

FIGURE 12.12   Resistors in series, $R_{total} = R_1 + R_2 + R_3 + R_4 + R_5$

EXAMPLE 12.4

Determine the total resistance and the current flowing in the circuit shown in Figure 12.13.

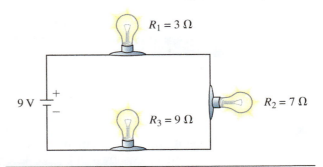

FIGURE 12.13 The circuit for Example 12.4.

The light bulbs are connected in a series arrangement; therefore, the total resistance is given by

$$R_{total} = R_1 + R_2 + R_3 = 3\,\Omega + 7\,\Omega + 9\,\Omega = 19\,\Omega$$

We can now use Ohm's law to determine the current flowing through the circuit.

$$I = \frac{V}{R_{total}} = \frac{9}{19} = 0.47\,\text{A} \approx 0.5\,\text{A}$$

We can also obtain the voltage drop across each lamp using Ohm's law:

$$V_{1-2} = R_1 I = (3)(0.47) = 1.41\,\text{V}$$
$$V_{2-3} = R_2 I = (7)(0.47) = 3.29\,\text{V}$$
$$V_{3-4} = R_3 I = (9)(0.47) = 4.23\,\text{V}$$

Note that, neglecting rounding-off errors, the sum of the voltage drops across each light bulb should be 9 volts.

Parallel Circuit

Consider the circuit shown in Figure 12.14. The resistive elements in the given circuit are connected in a parallel arrangement. For this situation, the electric current is divided among each branch. For the parallel arrangement shown in Figure 12.14, the electric potential, or the voltage, across each branch is the same. Moreover, the sum of the current in each branch is equal to the total current flowing in the circuit. The current flow in each branch can be determined using Ohm's law. Note that unlike a series circuit, if one branch fails in a parallel circuit, depending on the failure mode, the other branches could still remain operational. This is the reason parallel circuit arrangement is used when wiring different zones of a building. However, it is worth noting that if one branch fails in a parallel circuit, it could result in an increased current flow in other branches, which could be undesirable.

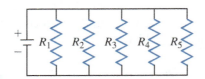

| FIGURE 12.14 | Resistors in parallel, $\dfrac{1}{R_{total}} = \dfrac{1}{R_1} + \dfrac{1}{R_2} + \dfrac{1}{R_3} + \dfrac{1}{R_4} + \dfrac{1}{R_5}$ |

EXAMPLE 12.5

The light bulbs in the circuit shown in Example 12.4 are placed in a parallel arrangement, as shown in Figure 12.15. Determine the current flow through each branch. Also compute the total resistance offered by all light bulbs to current flow.

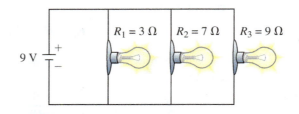

| FIGURE 12.15 | The circuit for Example 12.5. |

Because the light bulbs are connected in a parallel arrangement, the voltage drop across each light bulb is equal to 9 volts. We use Ohm's law to determine the current in each branch in the following manner:

$$V = R_1 I_1 \quad \Rightarrow \quad 9 = 3I_1 \quad \Rightarrow \quad I_1 = 3.0 \text{ A}$$
$$V = R_2 I_2 \quad \Rightarrow \quad 9 = 7I_1 \quad \Rightarrow \quad I_2 = 1.3 \text{ A}$$
$$V = R_3 I_3 \quad \Rightarrow \quad 9 = 3I_3 \quad \Rightarrow \quad I_3 = 1.0 \text{ A}$$

The total current drawn by the circuit is

$$I_{total} = I_1 + I_2 + I_3 = 3.0 + 1.3 + 1.0 = 5.3 \text{ A}$$

The total resistance is given by

$$\frac{1}{R_{total}} = \frac{1}{R_1} + \frac{1}{R_2} + \frac{1}{R_3} = \frac{1}{3} + \frac{1}{7} + \frac{1}{9} \quad \Rightarrow \quad R_{total} = 1.7\,\Omega$$

Note that we could have obtained the total current drawn by the circuit using the total resistance and the Ohm's law in the following manner:

$$V = R_{total} I_{total} \quad \Rightarrow \quad 9 \text{ V} = (1.7\ \Omega)(I_{total}) \quad \Rightarrow \quad I_{total} = 5.3 \text{ A}$$

EXAMPLE 12.6

Determine the total current drawn by the circuit shown in Figure 12.16.

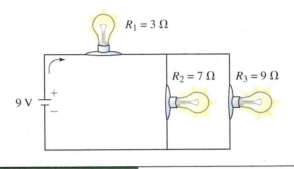

FIGURE 12.16   The circuit for Example 12.6.

The circuit given in this example has components that are in both series and parallel arrangements. We will first combine the light bulbs in the parallel branches into one equivalent resistance.

$$\frac{1}{R_{\text{equivalent}}} = \frac{1}{R_2} + \frac{1}{R_3} = \frac{1}{7} + \frac{1}{9} \Rightarrow R_{\text{equivalent}} = 3.9 \, \Omega$$

Next, we add the equivalent resistance to R_1, noting that the two resistors are in series now.

$$R_{\text{total}} = R_1 + R_{\text{equivalent}} = 3 + 3.9 = 6.9 \, \Omega$$
$$V = R_{\text{total}} I_{\text{total}} \Rightarrow 9 \, \text{V} = (6.9 \, \Omega)(I_{\text{total}}) \Rightarrow I_{\text{total}} = 1.3$$

Capacitors

Capacitors are electrical components that store electrical energy. A capacitor has two oppositely charged electrodes with a dielectric material inserted between the electrodes. Dielectric material is a poor conductor of electricity. Capacitors are used in many applications to serve as filters to protect sensitive components in electrical circuits against power surges. They are also used in large computer memories to store information during a temporary loss of electric power to the computer. You will also find capacitors in tuned circuits for radio receivers, audio filtering applications (e.g., bass and treble controls), timing elements for strobe lights, and intermittent wipers in automobiles.

The *farad* (F) is the basic unit used to designate the size of a capacitor. One farad is equal to 1 coulomb per volt. Because the farad is a relatively large unit, many capacitor sizes are expressed in microfarad (μF $= 10^{-6}$ F) or picofarad (pF $= 10^{-12}$ F).

Before You Go On

Answer the following questions to test your understanding of the preceding sections.

1. Explain what is meant by electric circuits.

2. What is a typical amperage rating for a residential building?

3. Explain what electric wires are typically made of and how wire sizes are expressed.

4. Give examples of power ratings for home appliances and electronics.

5. Describe the difference between a parallel and a series circuit.

6. Describe the Ohm's Law.

Vocabulary—State the meaning of the following terms:

AWG_____

Direct Current _____

Alternating Current_____

Electric Power_____

LO³ 12.3   Electric Motors

As engineering students, most of you will take at least one class in basic electrical circuits wherein you will learn about circuit theory and various elements, such as resistors, capacitors, inductors, and transformers. You will also be introduced to different types of motors and their operations. Even if you are studying to be a civil engineer, you should pay special attention to this class in electrical circuits, especially the part about motors, because motors drive many devices and equipment that make our lives comfortable and less laborious. To understand the significant role motors play in our everyday lives, look around you. You will find motors running all types of equipment in homes, commercial buildings, hospitals, recreational equipment, automobiles, computers, printers, copiers, and so on. For example, in a typical home you can identify a large number of motors operating quietly all around you that you normally don't think about. Here, we have identified a few household appliances with motors:

*   Refrigerator: compressor motor, fan motor
*   Garbage disposer

- Microwave with a turning tray
- Stove hood with a fan
- Exhaust fan in the bathroom
- Room ceiling fan
- DVD player
- Hand-held power screwdriver or hand-held drill
- Heating, ventilating, or cooling system fan
- Vacuum cleaner
- Hair dryer
- Electric shaver
- Computer: cooling fan, hard drive

Some of the factors that engineers consider when selecting a motor for an application are: (a) motor type, (b) motor speed (rpm), (c) motor performance in terms of torque output, (d) efficiency, (e) duty cycles, (f) cost, (g) life expectancy, (h) noise level, and (i) maintenance and service requirements. Most of these are self-explanatory. We discussed speed and torque in previous chapters; next we will discuss motor types and duty cycles.

Motor Type

Selecting a motor for an application depends on a number of factors, including the speed of the motor (in rpm), the power requirement, and the type of load. Some applications deal with difficult starting loads, such as conveyors, while others deal with easy starting loads, such as a fan. For this reason, there are many different types of motors. Here we will not discuss the principles of how various motors work. Instead, we will provide examples of various motors and their applications, as given in Table 12.4. Most of you will take some classes later that will be devoted to the theory and operation of motors.

Duty Cycle

Manufacturers of motors classify them according to the amount of time the motor needs to be operated. The motors are generally classified as *continuous duty* or *intermittent duty*. The continuous duty motors are used in applications where the motor is expected to operate over a period of an hour or longer. In some applications, continuous operation of the motor may be required. In applications where the motor is expected to operate for short periods of time and then rest, the intermittent duty motors are found. The continuous duty motors are more expensive than the intermittent duty motors.

In this chapter, we introduced you to electricity and some basic electrical components. Most of you will take a basic electrical circuit class where you will learn in more detail about electricity, electrical components, and motors, so you now see the importance of paying attention and studying carefully.

| TABLE 12.4 | Examples of Motors Used in Different Applications |

Schematic Diagrams for AC Motors

Split-phase	Repulsion-start induction	Capacitor-start	Repulsion-induction	Capacitor-motor

| Load Type | Motor Type (1) | Starting Ability (Torque) (2) | Starting Current (3) | Size hp (4) | Electrical Power Requirements | |
					Phase (5)	Voltage (6)
Easy Starting Loads	Shaded-pole induction	Very low, 1/2 to 1 times running torque	Low	.035–.2 kW (1/20–1/4 hp)	Single	Usually 120
	Split-phase	Low, 1 to 1½ times running torque	High 6 to 8 times running current	.035–.56 kW (1/20–3/4 hp)	Single	Usually 120
	Permanent-split, capacitor-induction	Very low, 1/2 to 1 times running torque	Low	.035–.8 kW (1/20–1 hp)	Single	Single voltage 120 or 240
	Soft-start	Very low, 1/2 to 1 times running torque	Low 2 to 2½ times running current	5.6–25 kW (7½–50 hp)	Single	240
	Capacitor-start, induction-run	High, 3 to 4 times running torque	Medium 3 to 6 times running current	.14–8 kW (1/6–10 hp)	Single	120–240
	Repulsion-start, induction-run	High, 4 times running torque	Low 2½ to 3 times running current	.14–16 kW (1/6–20 hp)	Single	120–240
Difficult Starting Loads	Capacitor-start, capacitor-run	High, 3½ to 4½ times running torque	Medium 3 to 5 times running current	.4–20 kW (1/2–25 hp)	Single	120–240
	Repulsion-start, capacitor-run	High, 4 times running torque	Low 2½ to 3 times running current	.8–12 kW (1–15 hp)	Single	Usually 240
	Three-phase, general-purpose	Medium, 2 to 3 times running torque	High 3 to 6 times running current	.4–300 kW (1/2–400 hp)	Three	120–240; 240–480 or higher

From Electric Motors, Fifth Edition, 1982. Copyright © 1982 AAVIM. Reprinted by permission.

TABLE 12.4   Examples of Motors Used in Different Applications (continued)

Schematic Diagrams for AC Motors				
Universal motor	Two-value capacitor	Shaded pole	Repulsion motor	3-phase squirrel cage

Load Type	Speed Range (7)	Reversible (8)	Relative Cost (9)	Other Characteristics (10)	Typical Uses (11)
Easy Starting Loads	900, 1200, 1800, 3600	No	Very Low	Light duty, low in efficiency	Small fans, freezer blowers, arc welder blower, hair dryers
	900, 1200, 1800, 3600	Yes	Low	Simple construction	Fans, furnace blowers, lathes, small shop tools, jet pumps
	Variable 900–1800	Yes	Low	Usually custom-designed for special application	Small compressors, fans
	1800, 3600	Yes	High	Used in motor sizes normally served by 3-phase power when 3-phase power not available	Centrifugal pumps, crop dryer fans, feed grinder
	900, 1200, 1800, 3600	Yes	Moderate	Long service, low maintenance, very popular	Water systems, air compressors, ventillating fans, grinders, blowers
	1200, 1800, 3600	Yes	Moderate to high	Handles large load variations with little variation in current demand	Grinders, deep-well pumps, silo unloaders, grains conveyors, barn cleaners
Difficult Starting Loads	900, 1200, 1800, 3600	Yes	Moderate	Good starting ability and full-load efficiency	Pumps, air compressors, drying fans, large conveyors, feed mills
	1200, 1800, 3600	Yes	Moderate to high	High efficiency, requires more service than most motors	Conveyors, deep-well pumps, feed mills, silo unloaders
	900, 1200, 1800, 3600	Yes	Very Low	Very simple construction, dependable, service-free	Conveyors, dryers, elevators, hoists, irrigation pumps

Source: AAVIM, *Electric Motors.*

LO⁴ 12.4   Lighting Systems

In this section, we will provide a brief introduction to lighting systems. Lighting systems (Figure 12.17) account for a major portion of electricity use in buildings, and have received much attention recently due to the energy and sustainability concerns. Energy could be saved by reducing illumination levels, increasing lighting efficiency, or by taking advantage of daylighting. Daylighting refers to using windows and skylights to bring light into a building to reduce the need for artificial lighting. As is the case with any new areas you explore, *lighting* has its own terminology. Therefore, make sure you spend a little time to familiarize yourself with the terminology, so you can follow the example problems later.

Let us begin by defining illumination. *Illumination* refers to distribution of light on a horizontal surface, and the amount of light emitted by a lamp is expressed in **lumens**. As a reference, a 100-watt incandescent lamp may emit 1700 lumens. Another important lighting characteristic is the intensity of illumination. The *intensity of illumination* is a measure of how light is distributed over an area. A common unit of illumination intensity is called **footcandle** and is equal to one lumen distributed over an area of 1-square-foot. To give you an idea of what a footcandle represents, to find your way around at night you would need between 5 to 20 footcandles. As another example, 30 to 50 footcandles will be needed for office work. If you have to do detailed work such as fixing electronic equipment or a spring-driven watch, you would then need around 200 footcandles of illumination intensity.

Efficacy is another term used by lighting engineers. **Efficacy** is the ratio of how much light is produced by a lamp (in lumens) to how much energy is consumed by the lamp (in watts).

> Lighting systems account for a major portion of electricity use in buildings. The lighting has its own terminology; therefore, make sure you spend a little time to familiarize yourself with the terminology.

$$\text{efficacy} = \frac{\text{light produced (lumens)}}{\text{energy consumed by the lamp (watts)}}$$

Efficacy is used by lighting engineers when designing an optimal lighting system for a building or by an engineer who is doing an energy audit of a building to determine if the lighting system is energy efficient. When engineers design a lighting system for a building, they consider many factors such as activity,

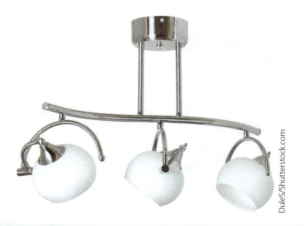

FIGURE 12.17   A lighting system.

DuleSy/Shutterstock.com

safety, and task. Sometimes the lighting system is designed to draw attention to a feature or something special in the building, then the engineer designs for so-called accent lighting.

As you know, there are many types of light bulbs and fixtures. According to the U.S. Department of Energy, incandescent lights accounted for 85% of lights used in homes in 2009. Unfortunately, the incandescent lights have very low efficacy values (10–17 lumens/watt). They also have short service life (750–2500 hours). Another important factor in choosing a lighting system for an application is its source color. As shown in Figure 12.18, in a incandescent lamp, the electric current runs through the lead wires and heats up the filament (a tiny coil of tungsten wire), which in turn makes the tungsten to glow or produces light. The light produced in this manner is a yellowish color. In general, the colors of light sources are classified into warm or cool categories. The yellow-red colors are considered warm, whereas the blue-green colors are considered cool. For a light source it is common to define a color temperature in Kelvin. The higher Kelvin temperatures (3600–5500 K) are considered cool, while lower color temperatures (2700–3000 K) are considered warm. Warm light sources are preferred for general indoor tasks. Be careful with the counter-intuitive way the warm and cool light sources are defined (high temperatures are cool, whereas low temperatures are warm!). How true the colors of an object appear when illuminated by a light source is more important than color temperature of light source. For this reason, a variable called **color rendition index (CRI)** is defined. The CRI provides a measure of how well a light source renders true colors of an object as compared with direct sunlight. The color rendition index has a scale of 1 to 100 with a 100-W incandescent light bulb having a CRI value of approximately 100. For most indoor applications, light sources with CRI of 80 or higher are preferred.

There are different types of incandescent light bulbs. The standard incandescent light is referred to as a screw-in A-type. There are also tungsten halogen and type R incandescent light bulbs. The tungsten halogen lamps have higher efficiencies than A-types, because they have inner coatings that reflect heat, consequently require less energy to keep the filament hot at a certain temperature. The type R incandescent lights also spread and direct light over a specific

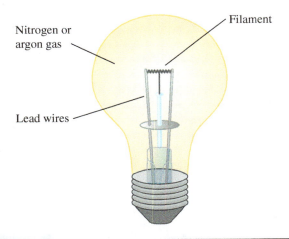

Nitrogen or argon gas

Filament

Lead wires

FIGURE 12.18	A schematic of incandescent lamp.

DOE's Office of Energy Efficiency and Renewable Energy.

area. They are commonly used as floodlights or spotlights. The comparisons of performance of incandescent lights are shown in Table 12.5.

The second most common type of lighting system is fluorescent lamps, which use 25% to 35% of energy consumption of incandescent lamps and produce the same amount of illumination. The efficacy of fluorescent lamps is somewhere between 30 to 110 lumens/watts. When compared to incandescent lamps, they also have longer service life, in the range of 7,000 to 24,000 hours. In a fluorescent tube, electric current is conducted through mercury and inert gases to produce light. The fluorescent lights used to have poor color rendition, but because of improvements in technology, they now have high CRI values. The 40 W, 4-foot (1.2-meter) lamps, and 75 W, 8-foot (2.4-meter) lamps are the two most common fluorescent lamps. These lamps require special fixtures, but the new generation of compact fluorescent lamps (CFLs) fit into the incandescent fixtures (Figure 12.19). Although CFLs are more expensive than incandescent light bulbs (3 to 10 times), because of their long service lives (16,000 to 15,000 hours) and high efficacy values, their use results in net savings. The comparison among different types of fluorescent lights is shown in Table 12.6.

TABLE 12.5	Comparison of Incandescent Lights			
Incandescent Lighting Type	Efficacy (lumens/W)	Life (hours)	Color Rendition Index (CRI)	Color Temperature (K)
Standard A	10–17	750–2500	98–100	2700–2800 (warm)
Tungsten halogen	12–22	2000–4000	98–100	2900–3200 (warm to neutral)
Reflector	12–19	2000–3000	98–100	2800 (warm)

U.S. Department of Energy.

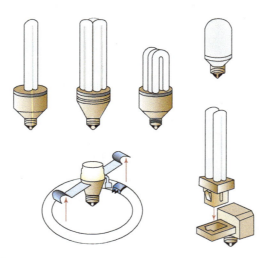

| FIGURE 12.19 | Examples of compact fluorescent lights that fit into screw-in-A-type. |

DOE's Office of Energy Efficiency and Renewable Energy.

TABLE 12.6	Comparison of Fluorescent Lights			
Fluorescent Lighting Type	Efficacy (lumens/watt)	Lifetime (hours)	Color Rendition Index (CRI)	Color Temperature (K)
Straight tube	30–110	7,000–24,000	50–90 (fair to good)	2700–6500 (warm to cold)
Compact fluorescent lamp (CFL)	50–70	10,000	65–88 (good)	2700–6500 (warm to cold)

Another common type of lighting system is the high-intensity discharge (HID) lamps (Figure 12.20). They have the highest efficacy values and the longest service life of any lighting systems. The HID lamps are commonly used in indoor arenas and outdoor stadiums. As you know from your experience, they have low color rendition index, and when you turn them on, it would take a few minutes before they produce light. The comparison among different types of HID lights is shown in Table 12.7.

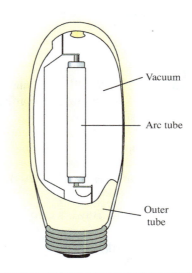

Vacuum

Arc tube

Outer tube

FIGURE 12.20 A schematic of high intensity discharge lamp.
DOE's Office of Energy Efficiency and Renewable Energy.

TABLE 12.7	Comparison Between High Intensity Discharging Lights			
High Intensity Discharging Lighting Type	Efficacy (lumens/W)	Life (hours)	Color Rendition Index (CRI)	Color Temp. (K)
Mercury Vapor	25–60	16,000–24,000	50 (poor to fair)	3200–7000 (warm to cold)
Metal Halide	70–115	5,000–20,000	70 (fair)	3700 (cold)
High Pressure Sodium	50–140	16,000–24,000	25 (poor)	2100 (warm)

The newest type of lighting systems are those which use LED (light emitting diode) lights. They have become a popular alternative to incandescent Christmas lights. They last longer than conventional incandescent lights with service life of approximately 20,000 hours. They also use much less power and operate at cooler temperatures, so they reduce fire hazard during the holiday season. Increasingly, they are becoming popular alternatives in other applications such as traffic lights, street lights, indoor lights, large display screens, and TV screens. The U.S. Department of Energy estimates that the widespread use of LED lights by 2027 could result in energy savings of 350×10^9 kWh.

EXAMPLE 12.7

According to Sylvania, a light bulb manufacturer, its 75 W CFL floodlight consumes 23 W and produces 1250 lumens. What is the efficacy of the floodlight?

$$\text{efficacy} = \frac{\text{Light Produced (lumens)}}{\text{Energy Consumed (watts)}} = \frac{1250}{23} = 54$$

EXAMPLE 12.8

A 100 W CFL light manufactured by Buyer's Choice consumes 23 watts and has an illumination rating of 1600 lumens and service life of 8000 hours and costs $1.81. As an alternative, a generic 100 W incandescent light bulb costs $0.38 and produces 1500 lumens and has a service life of 750 hours. Let us compare the performance of each light bulb by calculating the efficacy for each light, and also estimating how much it would cost to run each light for 8 hours per day for 220 days in a year. We will assume electricity costs 9 cents per kWh.

For the Buyer's Choice 100 W CFL light:

$$\text{efficacy} = \frac{1600}{23} = 70$$

$$\text{cost} = \left(\frac{8 \text{ hours}}{\text{day}}\right)(220 \text{ days})(23 \text{ W})\left(\frac{1 \text{ kW}}{1000 \text{ W}}\right)\left(\frac{\$0.09}{\text{kWh}}\right) = \$3.64$$

For the generic incandescent 100 W light bulb:

$$\text{efficacy} = \frac{1500}{100} = 15$$

$$\text{cost} = \left(\frac{8 \text{ hours}}{\text{day}}\right)(220 \text{ days})(100 \text{ W})\left(\frac{1 \text{ kW}}{1000 \text{ W}}\right)\left(\frac{\$0.09}{\text{kWh}}\right) = \$15.84$$

It should be obvious that the CFL light is more efficient and economical to operate than the generic incandescent.

Lighting System Audit

This is a good place to say a few words about lighting energy audits. As we said at the beginning of this section, lighting systems account for a major portion of electricity use in buildings and have received much attention recently due to the energy and sustainability concerns. A lighting energy audit starts with space classification. That is, what is the space used for? Is it used as an office, warehouse, manufacturing plant, and so on? Next, an energy auditor determines the space characteristics (length, width, height), the light fixtures (lamp types, their number, and lamp wattage), and their controls. The auditor then talks to the users about the lighting level, their tasks, occupancy profile, and using a light meter measures the light level in the space. Then, the comparison between the measurements and the Illuminating Engineering Society (IES) recommendation values for a given task is made. The auditor also calculates power consumption of the lighting system per unit area $(watts/ft^2)$ and compares it to design guidelines. Finally, the energy auditor prepares a report discussing his findings including estimate of annual lighting energy cost and ways by which the energy consumption due to lighting system may be reduced (for example, by reducing illumination levels, taking advantage of daylighting, or increasing the efficiency of lighting systems).

Before You Go On

Answer the following questions to test your understanding of the preceding sections.

1.   Give examples of the role of motors in our daily life.

2.   Explain what is meant by illumination and how it is expressed.

3.   What are the common types of lighting systems?

4.   Which type of lighting system has the highest CRI?

5.   Which type of lighting system has the highest life span?

6.   Which type of lighting system has the highest efficacy?

Vocabulary—State the meaning of the following terms:

Duty Cycle_____

Footcandle_____

Efficacy_____

CRI_____

CFL_____

LED_____

J. Duncan Glover, Ph.D., P.E.

My career in electrical engineering began with a keen interest in mathematics and the challenges of solving technical problems. But I also wanted to apply my technical abilities towards practical applications and technologies that would be useful to society, which led me to earning a B.S., a M.S., and a Ph.D. in Electrical Engineering.

Over the years I've worked for various companies, including a two-year assignment in Rio de Janeiro, Brazil as a consulting engineer on a large hydroelectric project. I also taught and performed research in electrical engineering for fifteen years at Northeastern University, first as an Assistant Professor and later as a tenured Associate Professor. But in 2004 I founded my own company, Failure Electrical, LLC.

I started the company after spending many years investigating a wide and diverse array of electrical and electronic equipment failures, including explosions, fires, and injuries. I now team up with cause and origin investigators, mechanical engineers, thermal experts, and other engineering specialties to provide a multi-disciplinary approach to solving complex technical problems.

Failure Electrical is rather like the CSI of electrical engineering, investigating what went wrong, why, and what can be done to prevent a reoccurrence.

By specializing in issues pertaining to electrical engineering, as they relate to failure analysis of electrical systems, subsystems, and components, (including causes of electrical fires) I cover investigations of the following: electric utility service interruptions and blackouts; heavy equipment failures including generators, transformers, circuit breakers, and motors; electrocutions; consumer appliance failures; and failures of semiconductors including printed circuit boards.

There are so many avenues to explore in my field; I've been fortunate in being a part of so many of them.

Courtesy of John Duncan Glover

SUMMARY

LO¹ Electric Current, Voltage, and Electric Power

By now, you should be familiar with basic principles of electricity and know what we mean by current, voltage, an electric circuit, and what a typical amperage rating for a residential building is. The flow of electric charge is called electric current or simply, current. The electric current, or the flow of charge, is measured in amperes. One ampere or "amp" (A) is defined as the flow of 1 unit of charge per second. Voltage represents the amount of work required to move charge between two points, and the amount of charge that is moving between the two points per unit time is called current. Moreover, direct current (dc) is the flow of electric charge that occurs in one direction. Batteries

and photovoltaic systems create direct current. Alternating current (ac) is the flow of electric charge that periodically reverses. Alternating current is created by generators at power plants. The current drawn by various electrical devices at your home is alternating current. The alternating current in domestic and commercial power use is 60 cycles per second (hertz) in the United States.

LO² Electrical Circuits and Components

An electrical circuit refers to the combination of various electrical components that are connected together. Examples of electrical components include wires (conductors), switches, outlets, resistors, and capacitors. Electrical components can be connected in either a series or a parallel arrangement.

Electrical wires are typically made of copper or aluminum. The actual size of wires is commonly expressed in terms of gage number. The American Wire Gage (AWG) is based on successive gage numbers having a constant ratio of approximately 1.12 between their diameters.

A resistor is an electrical component that resists the flow of either direct or alternating current. Resistors are commonly used to protect sensitive components or to control the flow of current in a circuit. Resistivity is a measure of resistance of a piece of material to electric current. Ohm's law describes the relationship among voltage, V, resistance, R, and current, I, according to

$$voltage = (resistance)(current)$$

The electric resistance is measured in units of ohms (Ω). An element with 1 ohm resistance allows a current flow of 1 amp when there exists a potential of 1 volt across the element. The electric power consumption of various electrical components can be determined using the following power formula:

$$P = (voltage)(current)$$

Capacitors are electrical components that store electrical energy.

LO³ Electric Motors

Motors are found running all types of equipment in homes, commercial buildings, hospitals, recreational equipment, automobiles, computers, printers, copiers, and so on. Some of the factors that engineers consider when selecting a motor for an application are (a) motor type, (b) motor speed (rpm), (c) motor performance in terms of torque output, (d) efficiency, (e) duty cycles, (f) cost, (g) life expectancy, (h) noise level, and (i) maintenance and service requirements.

LO⁴ Lighting Systems

You also should be familiar with basic lighting terminology and know how to calculate power consumption rates for lighting systems. Illumination refers to the distribution of light on a horizontal surface, and the amount of light emitted by a lamp is expressed in lumens. As a reference, a 100-watt incandescent lamp may emit 1700 lumens. A common unit of illumination intensity is called a footcandle and is equal to one lumen distributed over an area of 1 square foot. For example, to find your way around at night, you would need between 5 to 20 footcandles. Efficacy is another term used by lighting engineer. Efficacy is the ratio of how much light is produced by a lamp (in lumens) to how much energy is consumed by the lamp (in watts). How true the colors of an object appear when illuminated by a light source is represented by the color rendition index (CRI). The color rendition index has a scale of 1 to 100 with a 100-W incandescent light bulb having a CRI value of approximately 100. There are different types of lighting systems, including incandescent light bulbs, fluorescent lamps, compact fluorescent lamps high-intensity discharge (HID) lamps, and LED (light emitting diode) lights.

KEY TERMS

Alternating Current 402
Ampere 399
AWG 408
Capacitor 415
Charges 399
Color Rendition Index 421
Current 400
Direct Current 402
Duty Cycle 417

Electric Charge 399
Electric Circuit 407
Electric Current 400
Electric Power 409
Electromotive Force (emf) 400
Efficacy 420
Footcandle 420
Kirchhoff's Law 403
Lumens 420

Ohm 408
Ohm's Law 408
Photoemission 401
Resistance 407
Resistivity 407
Resistor 412
Voltage 400

APPLY WHAT YOU HAVE LEARNED

This is a class project. You are to perform a lighting energy audit for your indoor sports arena. Collect information about the arena size, occupancy profile—that is, the number of people who use the facility every 15 or 30 minutes. Also, obtain information about the lighting systems (fluorescent, incandescent, mercury vapor, sodium, metal halide, other) and controlling devices used in the space. Calculate the watts/ft² for the space as a function of time. Write a brief report to your instructor discussing your findings. Suggest ways the lighting energy consumption may be improved.

PROBLEMS

Problems that promote lifelong learning are denoted by 🔑

12.1 How are batteries connected in the following products: a hand-held calculator such as a TI 85, a flashlight, and a portable radio? Are the batteries connected in a series arrangement or in a parallel arrangement?

12.2 Identify the types of batteries used in the following products:

 a. a laptop computer

 b. an electric shaver

 c. a cordless drill

 d. a camcorder

 e. a cellular phone

 f. a flashlight

 g. a wristwatch

 h. a camera

12.3 Investigate the size and the material used for heating elements in the following products: a toaster, hair dryer, pressing iron, coffeemaker, and an electric stovetop.

12.4 As explained in this chapter, potentiometers (rheostats) are used in applications to adjust the electric current in a circuit. Investigate the different types of resistance element used for various applications. The resistance element of the rheostat is typically made from metal wires or ribbons, carbon, or a conducting liquid. For example, in applications where the current in the circuit is relatively small, the carbon type is used. Write a brief report discussing your findings.

12.5 Electrical furnaces are used in the production of steel that is consumed in the structural, automotive, tool, and aircraft industries in the United States. Electrical furnaces are typically classified into resistance furnaces, arc furnaces, and induction furnaces. Investigate the operation of these three types of furnaces. Write a brief report describing your findings.

12.6 Identify examples of motors used in a new automobile. For example, a motor is used to run the fan that delivers warm or cold air into the car.

12.7 Obtain information about the electric current and voltage ratings of your own residential building or a building belonging to someone you know. If possible draw a diagram showing the power distribution, similar to Figure 12.9.

12.8 What is the current that flows through each of the following light bulbs: 40 W, 60 W, 75 W, 100 W? Each light is connected to a 120 V line.

12.9 If a 1500 W hair dryer is connected to a 120 V line, what is the maximum current drawn?

12.10 Refering to Table 12.1, create a table that shows the relative resistance of 1 ft long wires having diameters of 1 mil made of the metals given in Table 12.1 to 1 ft long copper with a diameter of 1 mil. For example, using the data given in Table 12.1, the relative resistance of 1 foot long, 1 mil diameter aluminum wire to 1 ft long, 1 mil copper wire is 1.64 ($17.01\ \Omega/10.37\Omega = 1.64$).

12.11 Investigate how an alkaline battery and an automobile maintenance-free battery (lead acid, gel cell) work. Write a brief report discussing your findings.

12.12 When subjected to pressure, certain materials create a relatively small voltage. Materials that behave in this manner are called piezoelectrics. Investigate the applications in which piezoelectrics are used. Write a brief report discussing your findings.

12.13 Prepare an electrical circuit plan similar to the one shown in Figure 12.11 for your apartment, your home, or a section of your dormitory.

12.14 The National Electrical Code (NEC) covers the safe and proper installation of wiring, electrical devices, and equipment in private and public buildings. NEC is published by the National Fire Protection Association (NFPA) every three years. As an example of an NEC provision, the receptacle outlets in a room in a dwelling should be placed such that no point on the wall space is more than 6 ft away from the outlet in order to minimize the use of extension cords. After performing a Web search or obtaining a copy of the NEC handbook, give at least three other examples of National Electric Codes for a family dwelling.

12.15 Obtain a multimeter (voltmeter, ohm, and current meter) and measure the resistance and the voltage of the heating element in the back window of a car. Determine the power output of the heater.

12.16 Visit a hardware store and obtain information on the sizes of heating elements used in home hot water heaters. If a hot water heater is connected to a line with 240 V, determine the current drawn by the hot water heater and its power consumption.

12.17 As explained in this chapter, the National Electrical Code gives specific information on the type of wires used for general wiring. Perform a search and obtain information on wire types, maximum operating temperatures, insulating materials, outer cover sheaths, and usage types. Prepare a table that shows examples of these codes; show wire size, temperature rating, application provisions, insulation, and outer covering.

12.18 Determine the total resistance and the current flow for the circuit shown in the accompanying figure.

12.19 Determine the total resistance and the current flow in each branch for the circuit shown in the accompanying figure.

12.20 Use Kirchhoff's current law to determine the missing current in the circuit shown in the accompanying figure.

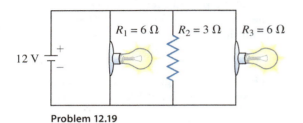

Problem 12.18

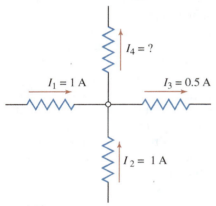

Problem 12.19

Problem 12.20

12.21 Obtain a flashlight and a voltmeter. Measure the resistance of the light bulb used in the flashlight. Draw the electrical circuit for the flashlight. Estimate the current drawn by the light when the flashlight is on.

12.22 Contact your electric utility company and obtain information on what the company charges for each kilowatt hour usage. Estimate the cost of your electric power consumption for a typical day. Make a list of your activities for the day and estimate the power consumption for the devices that you used. Write a brief report discussing your findings.

12.23 For Example 12.6, we computed the total current drawn by the circuit. Determine the current in each branch of the circuit given in Example 12.6.

12.24 For an automobile battery, investigate what is meant by the following terms: ampere-hour rating, cold-cranking rating, and reserve capacity rating. Write a brief report discussing your findings.

12.25 For a primary cell such as an alkaline battery, what does the term amp-hour represent? Gather information on the amp-hour rating for some alkaline batteries. Write a one-page summary of your findings.

12.26 As you know, a fuse is a safety device that is commonly placed in an electrical circuit to protect the circuit against excessive current. Investigate the various types of fuses, their shapes, materials, and sizes. Write a brief report discussing your findings.

12.27 There are many different types of capacitors, including ceramic, air, mica, paper, and electrolytic. Investigate their usage in electrical and electronic applications. Write a brief report discussing your findings.

12.28 In class, you are given three items: a battery, a light bulb, and a piece of wire. Make a flashlight using these items.

12.29 According to Sylvania, a light-bulb manufacturer, its 40 W CFL light consumes 9 watts and produces 495 lumens. What is the efficacy for this light?

12.30 Calculate the efficacy for the following lights: (a) Sylvania accent LED light, which uses 2 watts and produces 60 lumens, and (b) Sylvania 40 W CFL, which uses 9 watts and produces 495 lumens.

12.31 The Sylvania Super Saver 75 W light uses 20 watts, produces 1280 lumens, and costs $4.49. As an alternative, a generic 75 W incandescent light bulb costs $0.40, produces 1200 lumens, and has a service life of 750 hours. Compare the performance of each light by calculating the efficacy for each light and estimate how much it would cost to run each light for 4 hours per day for 300 days in a year. Assume electricity costs 9 cents per kWh.

12.32 There are a number of ways that you can reduce the energy waste associated with lighting systems. One way is to make use of smart controllers and use devices such as lighting-dimmer controls, lighting-motion-sensor controls, lighting-occupancy controls, lighting-photosensor controls, and lighting-timer controls. Write a brief report describing how these various types of lighting controllers work and give examples of their usage.

12.33 Visit the lighting section of a hardware store and look up the following information for comparable incandescent bulb, CFL, and LED lights. Read the manufacturer's ratings on the packaging and record lumens, light source color temperature, and power consumption in terms of watts. Write a brief report discussing your findings.

12.34 Perform a lighting energy audit for your classroom. First, define the space characteristics by recording the length, width, and the height of your classroom. Next, obtain information about the type of lighting system, the number of light sources, and their controls. Also, obtain information about the occupancy profile—that is, the number of people every hour using the classroom. Calculate the watts/ft^2 for the classroom as a function of time. Write a brief report to your instructor discussing your findings. Suggest ways the lighting energy consumption for your classroom may be improved.

Nika Zolfaghari

My name is Nika Zolfaghari, and I completed my undergraduate degree in biomedical engineering at Ryerson University. The truth is that I didn't always want to be an engineer. In high school, I was always good at math and science, but I never thought I would enjoy doing them for the rest of my life, I just knew that I loved helping people. When I first heard about engineering, I thought about all the stereotypes and thought that is was about "fixing" things. Luckily, my dad is an engineer, and he cleared those stereotypes up for me rather quickly. When I heard about biomedical engineering, I was in awe. It was the perfect blend of innovative design and science to help solve real world problems related to the human body.

When I entered engineering, I was fully aware that it was traditionally a male-dominated field, with the national average of females in engineering at 17%. I wasn't going to let that stop me though from speaking up in class or succeeding, because I knew that academically I could do anything a male can. Plus, females bring something different to the table, especially in engineering design projects. Although it may have been intimidating being the only girl in some of my classes and labs, I got used to it very quickly and didn't think much of it. I realized that it didn't faze the guys either, we were all equals, and all had the same engineering mindset and goals.

As part of our Capstone design project, my partners and I designed and built from scratch a prototype of a prosthetic arm, which we entered into a design

Courtesy of Nika Zolfaghari

competition. At the competition, there was a gentleman passing by who started asking us about our project. He was in a wheelchair and had an amputated arm. As we talked, he told us that he had tried many prosthetic arms in the past, but none of them worked for him because they never took into account what the user wanted. He then told us that he liked our design, and would actually wear it. Before he left, he turned to us and thanked us from the bottom of his heart for the work that we are doing in our field to help improve the lives of others such as himself. It was at that instance that I truly felt the most gratification. All of my hard work in engineering had paid off because I was able to give someone hope. To me, that's what engineering is all about.

It was through such experiences that I saw the impact that engineering can have on the quality of life of others. Therefore, I decided to pursue my master's degree in electrical and computer engineering at Ryerson University with specialization in biomedical engineering. I am currently doing research and conducting experiments in the area of spinal cord injury and testing how muscle contractions are affected. Upon graduation, I hope to obtain a job either for a medical device company or a hospital working in the area of design.

Courtesy of Nika Zolfaghari

Energy and Power

Vitaliiy/Shutterstock.com

julius fekete/Shutterstock.com

bikeriderlondon/Shutterstock.com

the808/Shutterstock.com

Sofiaworld/Shutterstock.com

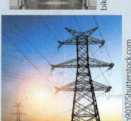

gyn9037/Shutterstock.com

jenifoto1/Shutterstock.com

Maksim Toome/
Shutterstock.com

Dmitrijs Mihejevs/Shutterstock.com

Glovatskiy/Shutterstock.com

abutyrin/Shutterstock.com

iStock/Thinkstock

We need energy to build shelter, to cultivate and process food, to make goods, and to maintain our living places at comfortable settings. To quantify the requirements to move objects, to lift objects, or to heat or cool something, energy is defined and classified into different categories. Power is the time rate of doing work. The value of power required to perform a task represents how fast you want the task done. If you want a task done in a shorter period of time, more power is required.

LEARNING OBJECTIVES

LO¹ **Work, Mechanical Energy, and Thermal Energy:** describe how we quantify what it takes to move things (kinetic energy), to lift things (potential energy), and to heat or cool things (thermal energy)

LO² **Conservation of Energy:** describe the conservation of energy principle

LO³ **Power:** describe what is meant by power; explain the difference between work, energy, and power

LO⁴ **Efficiency:** explain what is meant by efficiency and how efficiency is defined for a power plant, an automobile engine, an electric motor, a pump, and heating and cooling systems

LO⁵ **Energy Sources, Generation, and Consumption:** describe U.S. energy production and its consumption by source and sector

ENERGY CONSUMPTION IN THE UNITED STATES

The U.S. primary energy consumption in 2013 by source and sector (transportation, industrial, residential, and commercial) is shown in accompanying figures. In 2013, the U.S. consumed 99.3 quadrillion Btu. That is 93,000,000,000,000,000 Btu. As you already know, one Btu represents the amount of thermal energy needed to raise the temperature of 1 pound-mass of water by 1°F. Most of energy consumed came from petroleum (37%), natural gas (24%), and coal (23%). Nuclear and renewable energy accounted only for 9% and 7%, respectively. The 99.3 quadrillion Btu is equivalent to 29,000,000,000,000 kW-hr. One kilowatt-hour represents the amount of energy consumed during 1 hour by a device that uses 1000 watts or 1 kilowatt (kW). Moreover, in 2013, most of the carbon dioxide emissions (78%) came from burning coal and petroleum fuels to generate electricity and to transport goods and people.

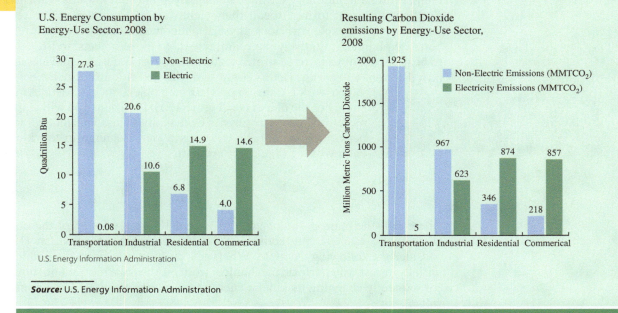

U.S. Energy Consumption by Energy-Use Sector, 2008

U.S. Energy Information Administration

Resulting Carbon Dioxide emissions by Energy-Use Sector, 2008

Source: U.S. Energy Information Administration

To the students: What do you think is your annual energy footprint?

The objectives of this chapter are to introduce the concept of energy, the various types of energy, and what is meant by the term power. We will explain various mechanical forms of energy, including kinetic energy, potential energy, and elastic energy. We will also revisit the definition of thermal energy forms from Chapter 11, including heat and internal energy. Next, we will present conservation of energy and its applications. We will define power as the rate of doing work and explain in detail the difference between work, energy, and power. After our discussion, these differences should be clear to you. The common units of power, watts and horsepower, are also explained. Once you have a good grasp of the concepts of work, energy, and power, then you can better understand the manufacturer's power ratings of engines and motors. Moreover, in this chapter we will also explain what is meant by the term efficiency and look at the efficiencies of power plants, internal combustion engines (car engines), electric motors, pumps, and heating, air-conditioning, and refrigeration systems.

LO¹ 13.1   Work, Mechanical Energy, and Thermal Energy

As we explained in Chapter 10, mechanical **work** is performed when a force moves an object through a distance. But what is **energy**? Energy is one of those abstract terms that you already have a good feel for. For instance, you already know that we need energy to create goods, to build shelter, to cultivate and process food, and to maintain our living places at comfortable temperature and humidity settings. But what you may not know is that energy can have different forms. Recall that scientists and engineers define terms and concepts to explain various physical phenomena that govern nature. To better explain quantitatively the requirements to move objects, to lift things, to heat or cool objects, or to stretch materials, energy is defined and classified into different categories. Let us begin with the definition of **kinetic energy**.

> Kinetic energy quantifies the amount of energy required to move something.

When work is done on or against an object, it changes the kinetic energy of the object (see Figure 13.1). In fact, as you will learn in more detail in your physics or dynamics class, mechanical work performed on an object brings about a change in the kinetic energy of the object according to

$$\text{work}_{1-2} = \frac{1}{2}mV_2^2 - \frac{1}{2}mV_1^2 \qquad \boxed{13.1}$$

where m is the mass of the object and V_1 and V_2 are the speed of the object at positions 1 and 2, respectively. To better demonstrate Equation (13.1), consider the following example. When you push on a lawn mower, which is initially at rest, you perform mechanical work on the lawn mower and move it, consequently changing its kinetic energy from a zero value to some nonzero value.

Kinetic Energy

An object having a mass m and moving with a speed V has a kinetic energy, which is equal to

$$\text{kinetic energy} = \frac{1}{2}mV^2 \qquad \boxed{13.2}$$

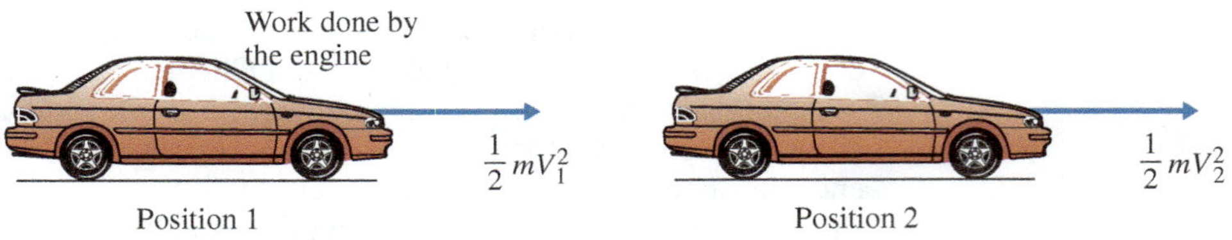

Work done by
the engine

$\frac{1}{2}mV_1^2$

Position 1

$\frac{1}{2}mV_2^2$

Position 2

FIGURE 13.1   The relationship between work and change in kinetic energy.

The SI unit for kinetic energy is the **joule**, which is a derived unit. The unit of joule is obtained by substituting kg for the units of mass and m/s for the units of speed, as shown here.

$$\text{kinetic energy} = \frac{1}{2}mV^2 = (\text{kg})\left(\frac{\text{m}}{\text{s}}\right)^2 = \overbrace{(\text{kg})\left(\frac{\text{m}}{\text{s}^2}\right)}^{\text{N}}(\text{m}) = \text{N}\cdot\text{m} = \text{joule} = \text{J}$$

Note that the $\frac{1}{2}$ factor in the kinetic energy equation is unitless. In the British Gravitational system of units, the unit for kinetic energy is lbf · ft. This unit is obtained by substituting slug for the unit of mass and ft/s for the unit of speed, recognizing that lbf = (slug)(ft/s²). Moreover, when dealing with problems in which mass is given in pound mass (U.S. Customary unit), convert the lbm to slug first. Then you can express your kinetic energy results in lbf · ft. As we discussed in the previous section, it is the change in kinetic energy (ΔKE) that is used in engineering analysis as given by Equation (13.1). The change in the kinetic energy is given by

$$\Delta\text{KE} = \frac{1}{2}mV_2^2 - \frac{1}{2}mV_1^2 \qquad \boxed{13.3}$$

EXAMPLE 13.1

Determine the net force needed to bring a car that is traveling at 90 km/h to a full stop in a distance of 100 m. The mass of the car is 1400 kg.

We begin the analysis by first changing the units of speed to m/s and then use Equation (13.1) to analyze the problem.

$$V_1 = V_{\text{initial}} = 90\left(\frac{\text{km}}{\text{h}}\right)\left(\frac{1\text{ h}}{3600\text{ s}}\right)\left(\frac{1000\text{ m}}{1\text{ km}}\right) = 25\,\frac{\text{m}}{\text{s}}$$

$$V_2 = V_{\text{final}} = 0$$

$$\text{work}_{1-2} = (\text{force})(\text{distance}) = \frac{1}{2}mV_2^2 - \frac{1}{2}mV_1^2$$

$$(\text{force})(100\text{ m}) = 0 - \frac{1}{2}(1400\text{ kg})\left(25\,\frac{\text{m}}{\text{s}}\right)^2$$

$$\text{force} = -4375\text{ N}$$

Note that in Example 13.1, change in the kinetic energy was used in the analysis. Also note, the negative value of force indicates that the force must be applied in opposite direction of motion, as expected.

Potential Energy

The work required to lift an object with a mass m by a vertical distance Δh is called **gravitational potential energy**. It is the mechanical work that must be performed to overcome the gravitational pull of the earth on the object (see Figure 13.2). The change in the potential energy of the object when its elevation is changed is given by

$$\text{change in potential energy} = \Delta\text{PE} = mg\,\Delta h \qquad \boxed{13.4}$$

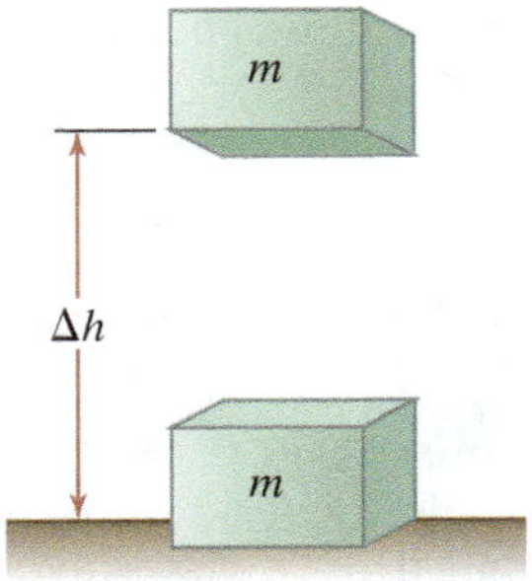

FIGURE 13.2

Change in potential energy of an object.

where

m = mass of the object (kg)
g = acceleration due to gravity (9.81 m/s^2)
Δh = change in the elevation (m)

The SI unit for potential energy is also the joule, a derived unit, and is obtained by substituting kg for the units of mass, m/s^2 for the units of acceleration due to gravity, and m for the elevation change:

$$\text{potential energy} = \Delta mg\Delta h = \overbrace{(\text{kg})\left(\frac{\text{m}}{\text{s}^2}\right)}^{\text{N}}(\text{m}) = \text{N}\cdot\text{m} = \text{J}$$

In the British Gravitational and U.S. Customary systems, the unit of potential energy is expressed in lbf · ft. As in the case with kinetic energy, keep in mind that it is the change in the potential energy that is of significance in engineering calculations. For example, the energy required to lift an elevator from the first floor to the second floor is the same as lifting the elevator from the third floor to the fourth floor, provided that the distance between each floor is the same.

Potential energy quantifies the amount of energy required to lift something.

EXAMPLE 13.2

Calculate the energy required to lift an elevator and its occupants with a mass of 2000 kg for the following situations: (a) between the first and the second floor, (b) between the third and the fourth floor, (c) between the first and the fourth floor. (See Figure 13.3.) The vertical distance between each floor is 4.5 m.

We can use Equation (13.4) to analyze this problem; the energy required to lift the elevator is equal to the change in its potential energy, starting with

(a) change in potential energy $= mg\,\Delta h$

$$= (200\ \text{kg})\left(9.81\frac{\text{m}}{\text{s}^2}\right)(4.5\ \text{m}) = 88{,}290\ \text{J}$$

(b) change in potential energy $= mg\,\Delta h$

$$= (200\ \text{kg})\left(9.81\frac{\text{m}}{\text{s}^2}\right)(4.5\ \text{m}) = 88{,}290\ \text{J}$$

(c) change in potential energy $= mg\,\Delta h$

$$= (200\ \text{kg})\left(9.81\frac{\text{m}}{\text{s}^2}\right)(13.5\ \text{m}) = 264{,}870\ \text{J}$$

Note that the amount of energy required to lift the elevator from the first to the second floor and from the third to the fourth floor is the same. Also realize that we have neglected any frictional effect in our analysis. The actual energy requirement would be greater in the presence of frictional effect.

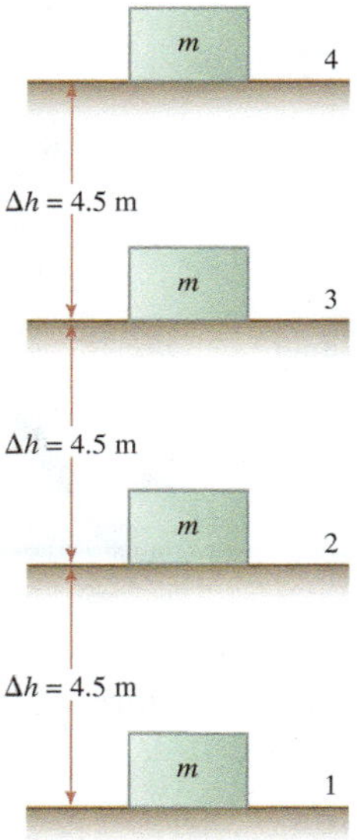

FIGURE 13.3

A schematic diagram for Example 13.2.

Elastic Energy

As we explained in Chapter 10, springs are used in a variety of products such as cars, weighting scales, clothespins, and printers. When a spring is stretched or compressed from its unstretched position, **elastic energy** is stored in the spring, energy that will be released when the spring is allowed to return to its unstretched position (see Figure 13.4). The elastic energy stored in a spring when stretched by a distance x or compressed is given by

$$\text{elastic energy} = \frac{1}{2}kx^2 \qquad \boxed{13.5}$$

where

$k = \text{spring constant}\,(\mathbf{N/m})$

$x = \text{deflection of spring from its unstretched position}\,(\mathbf{m})$

The SI unit for elastic energy is also the joule. It is obtained by substituting N/m for the units of spring constant and m for the units of deflection, as shown:

$$\text{elastic energy} = \frac{1}{2}kx^2 = \left(\frac{\mathbf{N}}{\mathbf{m}}\right)(\mathbf{m})^2 = \mathbf{N}\cdot\mathbf{m} = \mathbf{J}$$

Note once again that the $\frac{1}{2}$ factor in the elastic energy equation is unitless. In the U.S. Customary and British Gravitational systems of units, the unit of elastic energy is expressed in $\text{lbf}\cdot\text{ft}$. Let us now consider the spring shown in Figure 13.5; the spring is stretched by x_1 to position 1 and then stretched by x_2 to position 2. The elastic energy stored in the spring in position 1 is given by

$$\text{elastic energy} = \frac{1}{2}kx_1^2$$

$$\text{elastic energy} = \frac{1}{2}kx_2^2 \qquad \boxed{13.6}$$

$$\text{change in elastic energy} = \Delta\text{EE} = \frac{1}{2}kx_2^2 - \frac{1}{2}kx_1^2$$

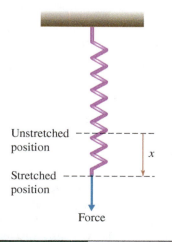

FIGURE 13.4   The elastic energy of a spring.

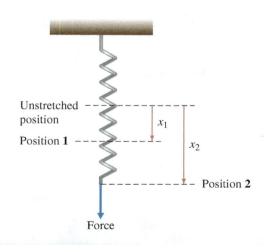

FIGURE 13.5   The change in the elastic energy of a spring.

EXAMPLE 13.3

Determine the change in the elastic energy of the spring shown in Figure 13.6 when it is stretched from: (a) position 1 to position 2, (b) position 2 to position 3, and position 1 to position 3. The spring constant is $k = 100$ N/cm. Additional information is shown in Figure 13.6.

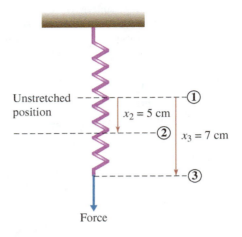

FIGURE 13.6   The spring in Example 13.3.

We begin by converting the units of the spring constant from N/cm to N/m in the following manner:

$$(100 \text{ N/cm})(100 \text{ cm/m}) = 10,000 \text{ N/m}$$

Using Equation (13.6), we can now answer the questions:

$$\text{change in elastic energy} = \Delta EE = \frac{1}{2}kx_2^2 - \frac{1}{2}kx_1^2$$

(a) $\Delta EE = \frac{1}{2}kx_2^2 - \frac{1}{2}kx_1^2 = \frac{1}{2}(10,000 \text{ N/m})(0.05)^2 - 0 = 12.5 \text{ J}$

(b) $\Delta EE = \frac{1}{2}kx_3^2 - \frac{1}{2}kx_2^2 = \frac{1}{2}(10,000 \text{ N/m})(0.07)^2 - \frac{1}{2}(10,000 \text{ N/m})(0.05)^2 = 12 \text{ J}$

(c) $\Delta EE = \frac{1}{2}kx_3^2 - \frac{1}{2}kx_1^2 = \frac{1}{2}(10,000 \text{ N/m})(0.07)^2 - 0 = 24.5 \text{ J}$

Thermal Energy Units

In Chapter 11, we explained that **thermal energy** transfer occurs whenever there exists a temperature difference within an object, or whenever there is a temperature difference between two bodies, or a temperature difference between a body and its surroundings. This form of energy transfer is called *heat*. Remember the fact that heat always flows from a high-temperature region to the low-temperature region. Moreover, we discussed the three different modes of heat transfer: conduction, convection, and radiation. We also discussed three units that are commonly used to quantify thermal energy (1) the British thermal unit, (2) the calorie, and (3) the joule.

Thermal energy quantifies the amount of energy required to heat or cool something.

As we explained the Btu (British thermal unit) in Chapter 11, one Btu is formally defined as the amount of thermal energy needed to raise the temperature of 1 lbm of water by 1° F. The **calorie** is defined as the amount of heat required to raise the temperature of 1 g of water by 1° C. And as you may also recall from our discussion in Chapter 11, in SI units no distinction is made between the units of thermal energy and mechanical energy, and therefore the units of thermal energy are defined in terms of fundamental dimensions of mass, length, and time. In the SI system of units, the joule is the unit of energy and is defined as

$$1 \text{ joule} = 1 \text{ N} \cdot \text{m} = 1 \text{ kg} \cdot \text{m}^2/\text{s}^2$$

The U.S. Customary unit of thermal energy is related to mechanical energy through

$$1 \text{ Btu} = 778 \text{ lb} \cdot \text{ft}$$
$$1 \text{ Btu} = 1055 \text{ J}$$

Finally, internal energy is a measure of the molecular activity of a substance and related to the temperature of a substance. As we explained in Chapter 11, the higher the temperature of an object, the higher its molecular activity and thus its internal energy.

LO² 13.2   Conservation of Energy

Conservation of Mechanical Energy

In the absence of heat transfer, and assuming negligible losses and no work, the *conservation of mechanical energy* states that the total mechanical energy of a system is constant. Stated another way, the change in the kinetic energy of the object, plus the change in the elastic energy, plus the change in the potential energy of the object is zero. This statement is represented mathematically as follows.

$$\Delta KE + \Delta PE + \Delta EE = 0 \qquad \boxed{13.7}$$

Equation (13.7) states that the energy content of a system can change form, but the total energy content of the system is constant.

EXAMPLE 13.4

In a manufacturing process, carts are rolling down an inclined surface, as shown in Figure 13.7. Estimate the height from which the cart must be

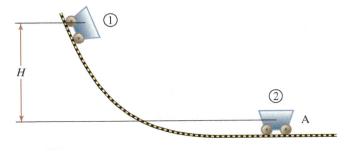

FIGURE 13.7   A schematic diagram for Example 13.4.

released so that, as it reaches point A, the cart has a velocity of 2.5 m/s. Neglect rolling friction.

We can solve this problem using Equation (13.7)

$$\Delta KE + \Delta PE + \Delta EE = 0$$

$$\Delta KE = \frac{1}{2}mV_2^2 - \frac{1}{2}mV_1^2 = \frac{1}{2}m(2.5 \text{ m/s})^2 - 0$$

$$\Delta EE = 0$$

$$\Delta PE = mg\,\Delta h = -m\left(9.81\frac{\text{m}}{\text{s}^2}\right)H$$

$$\frac{1}{2}m(2.5 \text{ m/s})^2 + -m\left(9.81\frac{\text{m}}{\text{s}^2}\right)H = 0$$

And solving for H we have

$$H = 0.318 \text{ m}$$

Note, the minus sign associated with change in potential energy indicates (shows) that potential energy of the cart is decreasing.

First Law of Thermodynamics

Previously, we discussed the conservation of mechanical energy. We stated that in the absence of heat transfer, and assuming negligible losses and no work, the conservation of mechanical energy states that the total mechanical energy of the system is constant. In this section, we will discuss the effects of heat and work in conservation of energy. There are a number of different ways that we can describe the general form of the conservation of energy, or the first law of thermodynamics. Expressed simply, the *first law of thermodynamics* states that energy is conserved. It cannot be created or destroyed; energy can only change forms. Another more elaborate statement of the first law says that for a system having a fixed mass, the net heat transfer to the system minus the work done by the system is equal to the change in total energy of the system (see Figure 13.8) according to

$$Q - W = \Delta E \qquad \boxed{13.8}$$

where

$Q = $ net heat transfer into the system $\left(\Sigma Q_{\text{in}} - \Sigma Q_{\text{out}}\right)$ in joules (J)

$W = $ net work done by the system $\left(\Sigma W_{\text{out}} - \Sigma W_{\text{in}}\right)$ in joules (J)

$\Delta E = $ net change in total energy of the system in joules (J), where E represents the sum of the internal energy, kinetic energy, potential energy, elastic energy, and other forms of energy of the system.

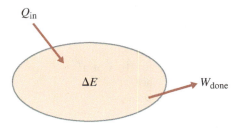

| FIGURE 13.8 | The first law of thermodynamics for a system having a fixed mass. |

There is a sign convention associated with Equation (13.8) that must be followed carefully. Heat transfer into the system, or the work done by the system, is considered a positive quantity, whereas the heat transfer out of the system, or the work done on the system, is a negative quantity. The reason for this type of sign convention is to show that work done on the system increases the total energy of the system, while work done by the system decreases the total energy of the system. Also note that using this sign convention, heat transfer into the system increases the total energy of the system, while heat transfer out of the system decreases its total energy.

You may also think of the first law of thermodynamics in the following manner: When it comes to energy, the best you can do is to break even. You cannot get more energy out of a system than the amount you put into it. For example, if you put 100 J into a system as work, you can get 100 J out of the system in the form of change in internal, kinetic, or potential energy of the system. As you learn more about energy you will also realize that according to the second law of thermodynamics, unfortunately you cannot even break even, because there are always losses associated with processes. We will discuss the effect of losses in terms of performance and efficiency of various systems later in this chapter.

> The first law of thermodynamics states that, for a system having a fixed mass, the net heat transfer to the system minus the work done by the system is equal to the change in total energy of the system.

EXAMPLE 13.5

Determine the change in the total energy of the system shown in Figure 13.9. The heater puts 150 W (J/s) into the water pot. The heat loss from the water pot to the atmosphere is 60 W. Calculate the change in total energy of the water in the pot after 5 minutes.

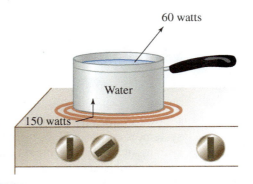

60 watts

Water

150 watts

FIGURE 13.9   A schematic diagram for Example 13.5.

We can use Equation (13.8) to solve this problem.

$$Q - W = \Delta E$$
$$W = 0$$
$$(150 \text{ J/s})(300 \text{ s}) - (60 \text{ J/s})(300 \text{ s}) = \Delta E$$
$$\Delta E = 27 \text{ kJ}$$

Note, for this problem, there is no change in the kinetic energy and potential energy of the system (water inside the pot). Consequently, the change in total energy is equal to the change in internal energy of water. Moreover, the rise in the internal energy manifests itself in the rise of the water temperature.

Before You Go On

Answer the following questions to test your understanding of the preceding sections.

1. In your own words, explain what is meant by work.

2. What is the difference between work and energy?

3. Give examples of forms of energy and explain what they quantify.

4. What are the SI and U.S. Customary units for energy?

5. In your own words, explain the conservation of energy.

Vocabulary—State the meaning of the following terms:

Work _____

Kinetic Energy _____

Potential Energy _____

Elastic Energy _____

Thermal Energy _____

First Law of Thermodynamics _____

LO³ 13.3   Power

In Section 13.1, we reviewed the concept of work as presented in Chapter 10 and explained the different forms of energy. We now consider what is meant by the term power. **Power** is formally defined as the time rate of doing work, or stated simply, the required work, or energy, divided by the time required to perform the task (work).

$$\text{power} = \frac{\text{work}}{\text{time}} = \frac{(\text{force})(\text{distance})}{\text{time}} \qquad \boxed{13.9}$$

or

$$\text{power} = \frac{\text{energy}}{\text{time}}$$

From Equation (13.9), the definition of power, it should be clear that the value of power required to perform a task represents how fast you want the task done. If you want a task done in a shorter time, then more power is required. For the sake of demonstrating this point better, imagine that in order to perform a task 3600 J are required. The next question then becomes, how fast do we want this task done? If we want the task done in 1 second, 3600 J/s power is required; if we want the task done in 1 minute, then 60 J/s power is needed; and if we want the task done in 1 hour, then the required power is 1 J/s. From this simple example, you should see clearly that to perform the same task in a shorter period of time, more power is required. More power means more energy expenditure per second. Another example that you have a direct experience with is the following situation: Which requires more power, to walk up a flight of stairs or to run up the stairs? Of course, as you already know, it requires more power to run up the stairs, because as compared to walking, when running up the stairs, you perform the same amount of work in a shorter time period. Many engineering managers understand the concept of power well, for they understand the benefit of teamwork. In order to finish a project in a shorter period of time, instead of assigning a task to an individual, the task is divided among several team members. More useful energy expenditure per day is expected from a team than from a single person, thus the project or the task can be done in less time.

> **Power represents the amount of work done or energy expended per unit time.**

Watts and Horsepower

As we explained in the previous section, power is defined as the time rate of doing work, or stated another way, work or energy divided by time. The units of power in SI units are defined in the following manner:

$$\text{power} = \frac{\text{work}}{\text{time}} = \frac{(\text{force})(\text{distance})}{\text{time}} = \frac{N \cdot m}{s} = \frac{J}{s} = W \qquad \boxed{13.10}$$

Note the following: 1 N·m is called 1 joule (J), and 1 J/s is called 1 **watt** (W). In U.S. Customary units, the units of power are expressed in lbf · ft/s and **horsepower** (hp), in the following manner:

$$\text{power} = \frac{\text{work}}{\text{time}} = \frac{(\text{force})(\text{distance})}{\text{time}} = \frac{\text{lbf} \cdot \text{ft}}{s} \qquad \boxed{13.11}$$

and

$$1 \text{ hp} = 550 \frac{\text{lbf} \cdot \text{ft}}{s} \qquad \boxed{13.12}$$

The U.S. Customary units of power are related to the SI unit of power, watt (W), in the manner:

$$1 \frac{\text{lbf} \cdot \text{ft}}{s} = 1.3558 \text{ W} \cong 1.36 \text{ W} \qquad \boxed{13.13}$$

$$1 \text{ hp} = 745.69 \text{ W} \cong 746 \text{ W} \qquad \boxed{13.14}$$

Remember that $1(\text{lbf} \cdot \text{ft/s})$ is slightly greater in magnitude than 1 W. Also keep in mind that 1 hp is slightly smaller than 1 kW. Another unit that is sometimes confused for the unit of power is kilowatt hour, used in measuring the consumption of electricity by homes and the manufacturing sector. First, kilowatt hour (kWh) is a unit of energy—not power. One **kilowatt hour** represents the amount of energy consumed during 1 hour by a device that uses one **kilowatt** (kW) or 1000 joules per second (J/s). Therefore,

$$1 \text{ kW} = 1000 \text{ W} = 1000 \text{ J/s}$$
$$1 \text{ kWh} = (1000 \text{ J/s})(3600 \text{ s}) = 3,600,000 \text{ J} = 3.6 \text{ MJ}$$
$$1 \text{ kWh} = 3.6 \text{ MJ}$$

In heating, ventilating, and air-conditioning (HVAC) applications, Btu per hour (Btu/h) is used to represent the heat loss from a building during cold months and the heat gained by the building during summer months. The units of Btu/h are related to the unit of watt in the manner:

$$1 \text{ Btu/s} = 1055 \text{ W} = 1.055 \text{ kW}$$

Another common unit used in the United States in air-conditioning and refrigeration systems is **ton of refrigeration or cooling**. One ton of refrigeration represents the capacity of a refrigeration system to freeze 2000 lbm or 1 ton of liquid water at 32° F into 32° F ice in 24 hours. It is

$$1 \text{ ton of refrigeration} = 12,000 \text{ Btu/h}$$

In the case of an air-conditioning unit, one ton of cooling represents the capacity of the air-conditioning system to remove 12,000 Btu of thermal energy from a building in 1 hour. Clearly, the capacity of a residential air-conditioning system depends on the size of the building, its construction, shading, the orientation of its windows, and its climatic location. Residential air-conditioning units generally have a 1- to 5-ton capacity. The sizes of home gas furnaces in the United States are also expressed in units of Btu/h. The size of a typical single-family-home gas furnace used in moderate winter conditions is 60,000 Btu/h.

To get a feel for the relative magnitudes that watt and horsepower physically represent, consider the following examples.

We will revisit this problem, after we discuss efficiency, to determine the amount of fuel needed in a power plant to provide the amount of energy that we just calculated in Example 13.6.

EXAMPLE 13.6

Determine the power required to move 30 people, with an average mass of 61 kg (135 lbm) per person, between two floors of a building, a vertical distance of 5 m (16 ft) in 2 s.

The required power is determined by

$$\text{power} = \frac{\text{work}}{\text{time}} = \frac{(30 \text{ persons})\left(61 \dfrac{\text{kg}}{\text{person}}\right)\left(9.81 \dfrac{\text{m}}{\text{s}^2}\right)(5 \text{ m})}{2 \text{ s}} \cong 45{,}000 \text{ W}$$

The minimum energy requirement for this task is equivalent to providing electricity to fifteen 100-W lightbulbs for 1 minute (90,000 J). Next time you feel lazy and are thinking about taking the elevator to go up one floor, reconsider and think about the total amount of energy that could be saved if people would take the stairs instead of taking the elevator to go up one floor. As an example, if 1 million people decided to take the stairs on a daily basis, the minimum amount of energy saved during a year, based on an estimate of 220 working days in a year, would be

$$\text{energy savings} = \left(\frac{90{,}000 \text{ J}}{30 \text{ persons}}\right)\left(\frac{1}{\text{day}}\right)(1{,}000{,}000 \text{ persons})(220 \text{ days})$$

$$= 660 \times 10^9 \text{ J} = 660 \text{ GJ}$$

EXAMPLE 13.7

Determine the power required to move a person who weighs 220 lbf a vertical distance of 2.5 ft in 1 s.

$$\text{power lbf} = \frac{\text{work}}{\text{time}} = \frac{(220 \text{ lbf})(2.5 \text{ ft})}{1 \text{ s}} = 550\frac{\text{lbf} \cdot \text{ft}}{\text{s}} = 1 \text{ hp}$$

Therefore, 1 horsepower represents the power required to lift a person weighing 220 lbf, a distance of 2.5 ft in 1 second. There are a number of other ways to think about what 1 horsepower physically represents. It could also be interpreted as the power required to lift an object weighting 100 lbf a distance of 5.5 ft in 1 second. How powerful are you?

| EXAMPLE 13.8 | Determine the power required to move an object that weighs 800 N (179.85 lbf) a vertical distance of 4 m (13.12 ft) in 2 s.
The power requirement expressed in SI units is given by |

$$\text{power} = \frac{\text{work}}{\text{time}} = \frac{(800 \text{ N})(4 \text{ m})}{2s} = 1600 \text{ W}$$

And using U.S. Customary units, the power requirement is

$$\text{power} = \frac{\text{work}}{\text{time}} = \frac{(179.85 \text{ lbf})(13.12 \text{ ft})}{2s} = 1180 \frac{\text{lbf} \cdot \text{ft}}{s}$$

The power expressed in horsepower is

$$\text{power} = 1180 \left(\frac{\text{lbf} \cdot \text{ft}}{s}\right)\left(\frac{1 \text{ hp}}{550 \dfrac{\text{lbf} \cdot \text{ft}}{s}}\right) = 2.14 \text{ hp}$$

In this example, think about the relationships among various units of power needed to perform the same amount of work in the same amount of time.

| EXAMPLE 13.9 | Most of you have seen car advertisements where the car manufacturer brags about how fast one of its cars can go from 0 to 60 mph. According to the manufacturer, the BMW 750iL model can go from 0 to 60 mph in 6.7 seconds. This performance is usually measured on a test track. The engine in the car is rated at 326 hp at 5000 rpm. The car has a reported weight of 4597 lbf. Is the claim by the manufacturer justifiable?
Well, to answer this question you may agree it would be more fun to drive to a BMW dealership and take the car for a test run on a racing track. |

But let us answer this question with the knowledge you have gained so far in this course. We need to make some assumptions first; we can assume that the driver weighs 180 lbf and the reported weight of the car includes enough gasoline for this test. Next, we convert the speed and the mass values into appropriate units.

$$V_1 = V_{initial} = 0\,\frac{ft}{s}$$

$$V_2 = V_{final} = \left(60\,\frac{mi}{h}\right)\left(\frac{1\,h}{3600\,s}\right)\left(\frac{5280\,ft}{1\,mi}\right) = 88\,\frac{ft}{s}$$

$$m = \frac{weight}{g} = \frac{(4597 + 180)\,lbf}{32.2\,\frac{ft}{s^2}} = 148\,slugs$$

Using Equation (13.1), we can determine the required work to go from 0 to 60 mph.

$$work_{1-2} = \frac{1}{2}mV_2^2 - \frac{1}{2}mV_1^2$$

$$work_{1-2} = \frac{1}{2}(148\,slugs)\left(88\,\frac{ft}{s}\right)^2 - 0 = 573,056\,lbf \cdot ft$$

The power requirement to perform this work in 6.7 seconds is

$$power = \frac{work}{time} = \frac{573,056\,lbf \cdot ft}{6.7\,s} = 85,530\,\frac{lbf \cdot ft}{s}$$

The power expressed in horsepower is

$$power = 85,530\left(\frac{lbf \cdot ft}{s}\right)\left(\frac{1\,hp}{550\,\frac{lbf \cdot ft}{s}}\right) = 155.5\,hp$$

Keeping in mind that power is needed to overcome air resistance and that there are always additional mechanical losses in the car, it is still safe to say the claim is good.

LO⁴ 13.4 Efficiency

As we mentioned earlier, there is always some loss associated with a dynamic system. In engineering, when we wish to show how well a machine or a system is functioning, we express its **efficiency**. In general, the overall efficiency of a system is defined as

$$efficiency = \frac{actual\ output}{required\ input} \qquad \boxed{13.15}$$

All machines and engineering systems require more input than what they put out. In the next few sections, we will look at the efficiencies of common engineering components and systems.

Power Plant Efficiency

Water is used in all steam power-generating plants to produce electricity. A simple schematic of a power plant is shown in Figure 13.10. Fuel is burned in a boiler to generate heat, which in turn is added to liquid water to change its phase to steam; steam passes through turbine blades, turning the blades, which in effect runs the generator connected to the turbine, creating electricity. The low-pressure steam liquefies in a condenser and is pumped through the boiler again, completing a cycle, as shown in Figure 13.10. The overall efficiency of a steam power plant is defined as

$$\text{power plant efficiency} = \frac{\text{energy generated}}{\text{energy input from fuel}}$$

13.16

The efficiency of today's power plants where a fossil fuel (oil, gas, coal) is burned in the boiler is near 40%, and for nuclear power plants the overall efficiency is nearly 34%.

> Efficiency is a measure of how much input is required to have a desired output.

Electricity is also generated by liquid water stored behind dams. The water is guided into water turbines located in hydroelectric power plants housed within the dam to generate electricity. The potential energy of the water stored behind the dam is converted to kinetic energy as the water flows through the turbine and consequently spins the turbine, which turns the generator.

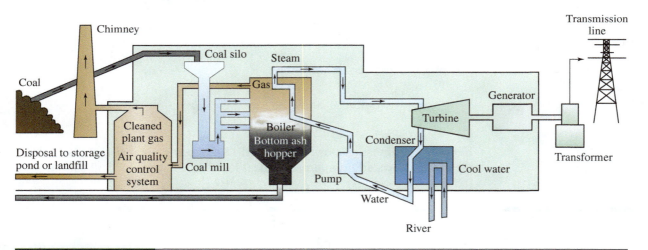

FIGURE 13.10   A schematic diagram of a steam water plant.
Courtesy Xcel Energy.

EXAMPLE 13.10   In Example 13.6, we determined the power required to move 30 people, with an average mass of 61 kg (135 lbm) per person, between two floors of a building: a vertical distance of 5 m (16 ft) in 2 s. The energy and the power requirements were 90,000 J and 45,000 W, respectively. Moreover, we estimated savings in energy if 1 million people decided to walk up a floor instead of taking the elevator on a daily basis. The minimum amount of energy saved during a year, based on an estimate of 220 working days in a year, would be 660 GJ. Let us now estimate the amount of fuel, such as coal, that can be saved in a power plant, assuming a 38% overall efficiency for the power plant and a heating value of approximately 7.5 MJ/kg for coal.

$$\text{power plant efficiency} = \frac{\text{energy generated}}{\text{energy input from fuel}}$$

$$0.38 = \frac{660 \text{ GJ}}{\text{energy input from fuel}} \Rightarrow \text{energy input from fuel}$$

$$= 1.74 \times 10^{12} \text{ J} = 1.74 \text{ TJ}$$

$$\text{amount of coal required} = \frac{1.74 \times 10^{12} \text{ J}}{7.5 \times 10^6 \dfrac{\text{J}}{\text{kg}}} = 232,000 \text{ kg } (511,472 \text{ lbm})$$

As you can see, the amount of coal that could be saved is quite large! Before you get on an elevator next time, think about the amount of fuel—not to mention the pollution—that can be saved if people just walk up a floor!

Internal Combustion Engine Efficiency

The thermal efficiency of a typical gasoline engine is approximately 25 to 30% and for a diesel engine is 35 to 40%. The thermal efficiency of an internal combustion engine is defined as

$$\text{thermal efficiency} = \frac{\text{power output}}{\text{heat power input as fuel is burned}} \qquad \boxed{13.17}$$

Keep in mind that when expressing the overall efficiency of a car, one must account for the mechanical losses as well.

Motor and Pump Efficiency

As we explained in Chapter 12, motors run many devices and equipment that make our lives comfortable and less laborious (Figure 13.11). As an example, we identified a large number of motors in various devices at home, including motors that run the compressor of your refrigerator, garbage disposer, exhaust fans, tape player in a VCR, vacuum cleaner, turntable of a microwave, hair dryer, electric shaver, computer fan, and computer hard drive. When selecting motors for these products, engineers consider the efficiency of the motor as one of the design criteria. The efficiency of an electric motor can be simply defined as

$$\text{efficiency} = \frac{\text{power input to the device being driven by the motor}}{\text{electric power input to the motor}} \qquad \boxed{13.18}$$

The efficiency of motors is a function of load and speed. The electric motor manufacturers provide performance curves and tables for their products that show, among other information, the efficiency of the motor.

You find pumps in hydraulic systems (Figure 13.12), the fuel system of your car, and systems that deliver water to a city piping network. Pumps are also used in food processing and in petrochemical plants. The function of a pump is to increase the pressure of a liquid entering the pump. The pressure rise in the fluid is used to overcome pipe friction and losses in fittings and

FIGURE 13.11   A power saw.
Eimantas Buzas/Shutterstock.com

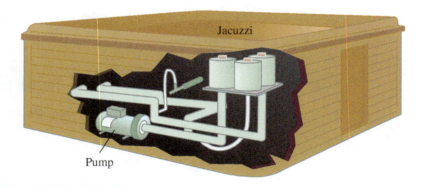

FIGURE 13.12   A Jacuzzi®.

valves and to transport the liquid to a higher elevation. Pumps themselves are driven by motors and engines. The efficiency of a pump is defined by

$$\text{efficiency} = \frac{\text{power input to the fluid by the pump}}{\text{power input to the pump by the motor}}$$

13.19

The efficiency of a pump, at a given operating speed, is a function of flow rate and the pressure rise (head) of the pump. Manufacturers of pumps provide performance curves or tables that show, among other performance data, the efficiency of the pump.

Efficiency of Heating, Cooling, and Refrigeration Systems

Before we discuss the efficiency of heating, cooling, and refrigeration systems, let us briefly explain the main components of these systems and how they operate. We begin with the cooling and refrigeration systems because

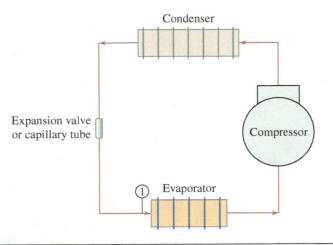

Condenser

Expansion valve
or capillary tube

Compressor

① Evaporator

FIGURE 13.13   A schematic diagram of a vapor–compression cycle.

---

their designs and operations are similar. Most of today's air-conditioning and refrigeration systems are designed according to a vapor–compression cycle. A schematic diagram of a simple vapor–compression cycle is shown in Figure 13.13. The main components of the vapor–compression cycle include a condenser, an evaporator, a compressor, and a throttling device, such as an expansion valve or a capillary tube, as shown in Figure 13.13.

Refrigerant is the fluid that transports thermal energy from the evaporator, where thermal energy or heat is absorbed, to the condenser, where the thermal energy is ejected to the surroundings. Referring to Figure 13.13, at state 1, the refrigerant exists as a mixture of liquid and gas. As the refrigerant flows through the evaporator, its phase completely changes to vapor. The phase change occurs because of the heat transfer from the surroundings to the evaporator and consequently to the refrigerant. The refrigerant enters the evaporator tube in a liquid/vapor mixture phase at very low temperature and pressure. The temperature of the air surrounding the evaporator is higher than the temperature of the evaporator, and thus heat transfer occurs from the surrounding air to the evaporator, changing the refrigerant's phase into vapor.

The evaporator in a refrigerator is a coil that is made from series of tubes that are located inside the freezer section of the refrigerator (see Figure 13.14). In an air-conditioning unit (Figure 13.15), the evaporator coil is located in the ductwork, near the fan–furnace unit inside the house. After leaving the evaporator, the refrigerant enters the compressor, where the temperature and the pressure of the refrigerant is raised. The discharge side of the compressor is connected to the inlet side of the condenser, where the refrigerant enters the condenser in gaseous phase at a high temperature and pressure. Because the refrigerant in the condenser has a higher temperature than the surrounding air, heat transfer to the surrounding air occurs, and consequently thermal energy is ejected to the surroundings. Like an evaporator, a condenser is also made of a series of tubes with good thermal conductivity. As the refrigerant flows through the condenser, more and more heat is removed (or transferred to the surroundings); consequently, the refrigerant changes phase from gas to liquid and it leaves the condenser coil in liquid phase. In the older version of a household refrigerator, the condenser is the series of black tubes located on the back of the refrigerator. In an air-conditioning unit, the condenser is

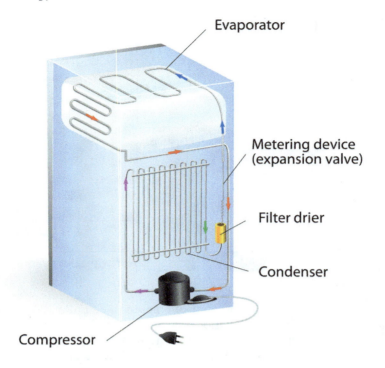

high temperature   low temperature   low temperature   low temperature
high pressure   high pressure   low pressure   low pressure
vapor state   liquid state   vapor state   liquid state

FIGURE 13.14   The typical locations of the evaporator and the condenser in a household refrigerator.
Designua/Shutterstock.com

AIR CONDITIONING

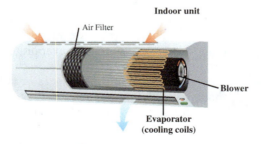

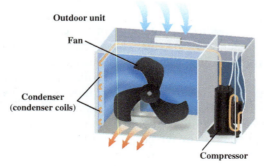

FIGURE 13.15   Air-conditioning unit.
Shutterstock.com

located outside the building in a housing unit that also contains the compressor and a fan that forces air over the condenser. After leaving the condenser, the liquid refrigerant flows through an expansion valve or a long capillary tube, which makes the refrigerant expand. The expansion is followed by a drop in the refrigerant's temperature and pressure. The refrigerant leaves the expansion valve or the capillary tube and flows into the evaporator to complete the cycle shown in Figure 13.13.

Another point worth mentioning is that as the warm air flows over the evaporator section of an air-conditioning unit and as the air cools, the moisture (the water vapor) in the air condenses on the outside of the evaporator coil. The condensation forming on the outside of the evaporator is eventually drained. Thus, the evaporator acts as a dehumidification device as well. This process is similar to what happens when hot, humid air comes into contact with a glass of ice water. You have all seen the condensation that forms on the outside surface of a glass of ice water as the neighboring hot, humid air cools down. We will discuss what we mean by absolute and relative humidity in Chapter 17.

The efficiency of a refrigeration system or an air-conditioning unit is given by the coefficient of performance (COP) which is defined as

$$COP = \frac{\text{heat removal from the evaporator}}{\text{energy input to the compressor}} \qquad \boxed{13.20}$$

You should use consistent units to calculate the coefficient of performance. The COP of most vapor compression units is 2.9 to 4.9. In the United States, it is customary to express the coefficient of performance of a refrigeration or an air-conditioning system using mixed SI and U.S. units. Quite often, the coefficient of performance is called energy efficiency ratio (EER) or the **seasonal energy efficiency ratio (SEER)**. In such cases, the heat removal is expressed in Btu, and the energy input to the compressor is expressed in watt-hours, and because 1 Wh = 3.412 Btu, EER or SEER values of greater than 10 are obtained for the coefficient of performance. Therefore, keep in mind that in the United States, the EER is defined in the following manner:

$$EER = \frac{\text{heat removal from the evaporator (Btu)}}{\text{energy input to the compressor (Wh)}} \qquad \boxed{13.21}$$

The reason for using the units of Wh for energy input to the compressor is that compressors are powered by electricity, and electricity consumption is measured (even in the United States) in kWh. Today's air-conditioning units have SEER values that range from approximately 10 to 17. In fact, all new air-conditioning units sold in the United States must have a SEER value of at least 10. In 1992, the United States government established the minimum standard efficiencies for various appliances, including air-conditioning units and gas furnaces.

In a gas furnace, natural gas is burned, and as a result the hot combustion products go through the inside of a heat exchanger where thermal energy is transported to cold indoor air that is passing over the heat exchanger. The warm air is distributed to various parts of the house through conduits. As mentioned, in 1992 the United States government established a minimum **annual fuel utilization efficiency (AFUE)** rating of 78% for furnaces installed in new homes, so manufacturers must design their gas furnaces to adhere to this standard. Today, most high-efficiency furnaces offer AFUE ratings in the range of 80 to 96%.

EXAMPLE 13.11

An air-conditioning unit has a cooling capacity of 24,000 Btu/h. If the unit has a rated energy efficiency ratio (EER) of 10, how much electrical energy is consumed by the unit in 1 h? If a power company charges 12 cents per kWh usage, how much would it cost to run the air conditioning unit for a month (30 days), assuming the unit runs 10 hours a day? What is the coefficient of performance (COP) for the given air-conditioning unit?

We can compute the energy consumption of the given air-conditioning unit using Equation (13.21).

$$\text{EER} = \frac{\text{heat removal from the evaporator (Btu)}}{\text{energy input to the compressor (Wh)}}$$

$$10 = \frac{24,000\,(\text{Btu})}{\text{energy input to the compressor (Wh)}}$$

Energy input to the unit for 1 h of operation = 2400 Wh = 2.4 kWh.

The cost to run the unit for a month for a period of 10 hours a day is calculated in the following manner:

$$\text{cost to operate the unit} = \left(\frac{2.4\ \text{kWh}}{1\ \text{h}}\right)\left(\frac{10\ \text{h}}{\text{day}}\right)\left(\frac{\$0.12}{\text{kWh}}\right)(30\ \text{days}) = \$86.40$$

The coefficient of performance (COP) is calculated from Equation (13.21):

$$\text{COP} = \frac{\text{heat removal from the evaporator}}{\text{energy input to the compressor}} = \frac{24,000\ \text{Btu}}{(2400\ \text{Wh})\left(\dfrac{3.412\ \text{Btu}}{1\ \text{Wh}}\right)} = 2.9$$

Note the relationship between the EER and COP:

$$\text{COP} = \frac{\text{EER}}{3.412} = \frac{10}{3.412} = 2.9$$

Before You Go On

Answer the following questions to test your understanding of the preceding sections.

1. What is the difference between energy and power?

2. What are the SI and U.S. Customary units for power?

3. Which represents more power, kilowatt, or horsepower?

4. What do we mean by efficiency, and why is it important to know the efficiency of products that we use in our daily life?

Vocabulary—State the meaning of the following terms:

Kilowatt-hour _____

Power _____

Kilowatt _____

Horsepower _____

Efficiency _____

LOˢ 13.5   Energy Sources, Generation, and Consumption

As we have been emphasizing throughout this book, there are certain concepts that every engineer regardless of his or her area of specialization should know. In this chapter, we discussed the importance of energy and power in engineering analysis and in our everyday life. We stressed the fact that we need energy to build structures, make goods, move or lift things, cultivate and process food, and heat or cool buildings. So whether you are planning to be a civil engineer, mechanical engineer, or electrical engineer, you need to have a firm grasp of how we quantify the amount of energy we need to address our needs. It is also equally important for every engineer to know about energy sources, generation, and consumptions rates. This is especially true during this period in our history where the world's growing demand for energy is among one of the most difficult challenges that we face. As future engineers, you are faced with two problems, energy sources and emissions; the solutions to these problems require innovative approaches. The energy use per capita in the world has been increasing steadily as the economies of the world grow. Add to these concerns, the expected rise in the population of the world from the current 6.5 billion to about 9 billion people by mid-century. We are counting on you to address these concerns while allowing for the standard of living to increase in the underdeveloped countries. To shed light on the energy sources, generation, and consumption we will focus on U.S. data (from last decade). However, realize this is a global issue that requires global solutions by all engineers in the world. We are merely using the data from last decade here as a means to convey important information to you. Moreover, you are encouraged to look up the data for the current year and compare them to the data given here.

The U.S. primary energy consumption by source and sector is shown in Figure 13.16, and its breakdown is shown in Figure 13.17. Next, we will briefly explain some of these major sources.

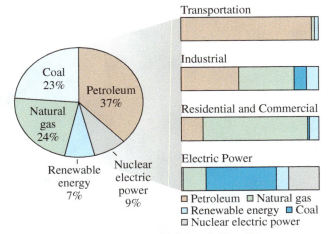

Total U.S. Energy = 99.3 Quadrillion Btu

FIGURE 13.16   The U.S. energy consumption by source and sector in 2008.
Energy Information Administration, Annual Energy Review (2013).

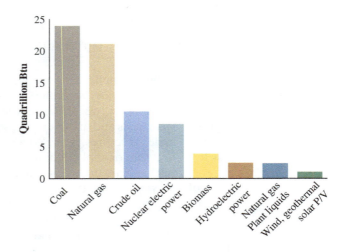

FIGURE 13.17   The U.S. energy production by major source (2008).
Energy Information Administration, Annual Energy Review (2013).

Coal

In 2007, the U.S. electric power industry generated nearly 4157 billion kilowatthours. Coal, natural gas, petroleum, nuclear, and renewable sources were used to generate electricity. From examining Figure 13.18, you can see that almost half (48.5%) of all electricity generated in the United States was created from coal. Coal-fired power plants burn coal in boilers or steam

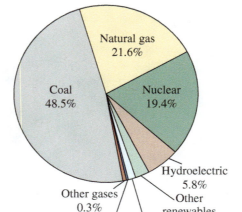

FIGURE 13.18   U.S electric power industry net generation (2007).
EIA, Form EIA-923, "Power Plant Operations Report" and predecessor form(s) including Form EIA-906, Power Plant report" and Form EIA-920, "Combined Heat and Power Plant Report".

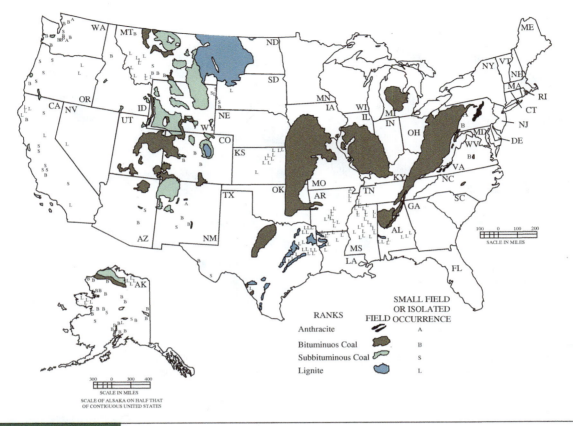

FIGURE 13.19   Regions where coal is mined in the United States.
Energy Information Administration.

generators to make steam. The steam then turns turbines that are connected to generators to create electricity. Figure 13.19 shows the major regions where coal is mined in the United States. According to the U.S. Department of Energy, nearly 93% of the coal mined in the U.S. is used for generating electricity. The rest of the coal is used in other industries including steel, cement, and paper, to process materials.

Natural Gas

The U.S. natural gas transportation network consists of 1.5 million miles of mainline and secondary pipelines. These pipelines connect production areas and markets, and in 2008 delivered more than 23 trillion cubic feet of natural gas to about 70 million customers (Figure 13.20). Salt caverns, depleted oil reservoirs, or aquifer reservoirs serve as underground storage facilities to store natural gas as a seasonal backup supply. Aboveground liquefied natural gas storage facilities are also used to store natural gas. In 2007, there were approximately 400 active storage fields. The major gas transportation pipelines in U.S. are shown in Figure 13.20, and the percentage of natural gas transmission pipeline mileage in each state is shown in Figure 13.21.

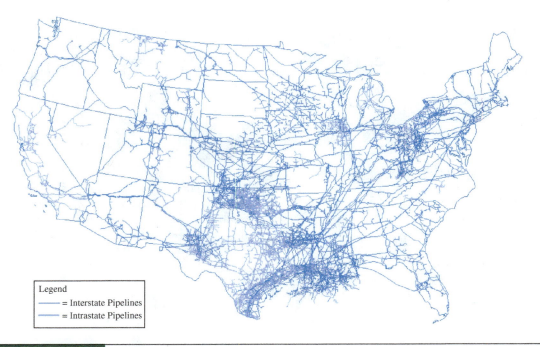

Legend
——— = Interstate Pipelines
——— = Intrastate Pipelines

| FIGURE 13.20 | U.S. natural gas transportation network. 1.5 million miles of mainline and other pipelines, which link production areas and markets.
Energy Information Administration. |

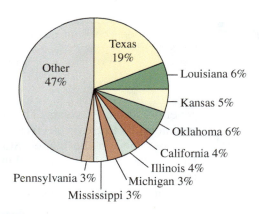

| FIGURE 13.21 | Percent of U.S. natural gas transmission pipeline mileage in each state (2008).
Energy Information Administration. |

Heating Oil

Heating oil is a petroleum product used to heat homes in America, especially in the Northeast. At refineries, crude oil is refined into lubricating oil and different types of fuels, including gasoline, diesel, jet fuel/kerosene, and heating oil. Heating oil and diesel fuel, are similar in composition; the main difference between the two fuels is sulfur content. Heating oil has more sulfur than diesel fuel. In addition, because heating oil is tax-exempt and cannot be legally used to fuel cars and trucks on highways, the U.S. Internal Revenue Service requires heating oil to be dyed red. The red color makes it clear that the product is tax-exempt and cannot legally be used as highway diesel.

Nuclear Energy

There are two processes by which **nuclear energy** is harnessed, nuclear fission and nuclear fusion. Nuclear power plants (Figure 13.22) use nuclear fission to produce electricity. In nuclear fission, in order to release energy, atoms of uranium are bombarded by a small particle called a neutron. This process splits the atoms of uranium and releases more neutrons and energy in the form of heat and radiation. The additional neutrons go on to bombard other uranium atoms, and the process keeps repeating itself, leading to a chain reaction. This process is shown in Figure 13.23. The fuel most widely used by nuclear power plants is uranium 235 or simply U-235. The U-235 is relatively rare and must be processed from the uranium that is mined. According to the U.S. Department of Energy, the owners and operators of U.S. civilian nuclear power reactors purchased the equivalent of 53 million pounds of uranium during 2008, of which 14% came from the United States, and the remaining 86% was of foreign origin (42% from Australia and Canada, 33% from Kazakhstan, Russia and Uzbekistan, and 11% from Brazil, the Czech Republic, Namibia, Niger, South Africa, and the United Kingdom).

The energy in the nucleus or core of atoms can also be released by nuclear fusion. In nuclear fusion, energy is released when atoms are combined or fused together to form a larger atom. This process is called nuclear fusion and is the process by which the sun's energy is produced. The percentage of electricity generated by nuclear fuel during 1973–2008 is shown in Figure 13.24. Next, we will briefly discuss renewable energy sources, such as hydropower, solar, wind, ethanol, and biodiesel.

| FIGURE 13.22 | A nuclear power plant. |

jorisvo/Shutterstock.com

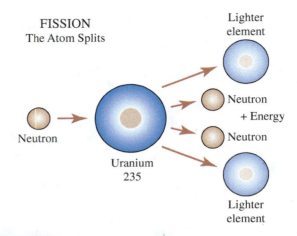

FISSION
The Atom Splits

Neutron

Uranium
235

Lighter
element

Neutron
+ Energy

Neutron

Lighter
element

FIGURE 13.23	The nuclear fission process.

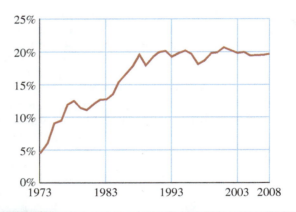

FIGURE 13.24	Percentage of electricity generated by nuclear fuel during 1973–2008.

Energy Information Administration, Monthly Energy Review

Hydropower

Hydropower accounts for 6% of total U.S. electricity generation. In 2008, it accounted for 67% of energy generation of all the renewable energy sources. In a hydroelectric power plant, to generate electricity, water stored behind dams is guided through water turbines that are connected to generators located within the plant. Approximately 31% of the total U.S. hydropower is generated by the water stored in the state of Washington behind the Grand Coulee Dam, the largest hydroelectric facility in the country.

Solar Energy

Because of the current energy and sustainability concerns, there has been a renewal of interest in solar energy. Solar energy starts with the sun at an average distance of 93 million miles from earth. The sun is a nuclear fusion reactor, with its surface temperature at approximately 10,000° F (5500° C).

There are two basic types of active solar heating systems: liquid and air.

Solar energy that reaches the earth is in the form of electromagnetic radiation consisting of a wide spectrum of wavelengths and energy intensities. Almost half of the solar energy received on earth is in the band of visible light. The solar radiation could be divided into three bands: the ultraviolet band, the visible band, and the infrared band. Many of you have a firsthand experience with the ultraviolet band that causes sunburn. The visible band comprises about 48% of useful radiation for heating, and the near infrared makes up the rest. As you know, the earth's orbit around the sun is elliptical. When the sun is nearer the earth, the earth's surface receives a little more solar energy. The earth is closer to the sun when it's summer in the southern hemisphere and winter in the northern hemisphere. Because the distance from the earth to the sun changes during the year, the energy reaching the outer atmosphere of the earth varies from 410 to 440 Btu/ft² · h. At the average earth–sun distance, out in the space at the edge of earth's atmosphere, the intensity of solar energy is 428 Btu/ft² · h or 1350 W/m². The amount of radiation available at a location on the surface of the earth, depends on many factors including geographical position, season, local landscape and weather, and time of day.

As solar energy passes through the earth's atmosphere, some of it is absorbed, some of it is scattered, and some of it is reflected by clouds, dust, pollutants, forest fires, volcanoes, or water vapor in the atmosphere. The solar radiation that reaches the earth's surface without being diffused is called direct beam solar radiation. Atmospheric conditions can reduce direct beam radiation by 10% on clear, dry days, and by 100% during thick, cloudy days. The direct and diffuse radiation beams are depicted in Figure 13.25.

Solar Systems

Using various technologies, solar radiation can be converted into useful forms of energy, such as heating water or air, or generating electricity. The economic feasibility of solar systems depends on the amount of solar radiation available at a location. In many countries, radiation data for solar-water heating and space-heating systems is expressed in watts per square meter per day (kW/m²/day). In the U.S., some of the radiation data is given in British thermal units per square foot per day (Btu/ft²/day). The radiation data for solar-electric (photovoltaic) systems are represented as kilowatt-hours per square meter (kWh/m²).

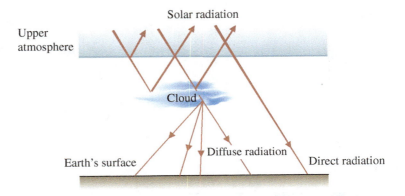

FIGURE 13.25   The direct and diffuse radiation.

Flat-Plate Collector

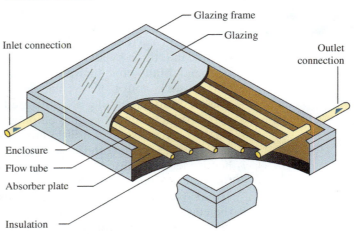

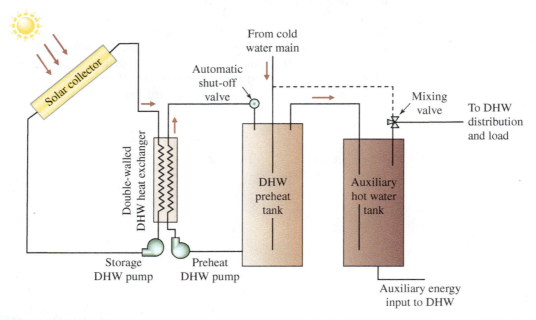

FIGURE 13.26	A schematic of a solar collector.
	U.S. Department of Energy

There are two basic types of **active solar** heating **systems**, liquid and air. The liquid systems make use of water or water-antifreeze mixture (in cold climates) to collect solar energy. In such systems, the liquid is heated in a solar collector (Figure 13.26) and then transported via a pump to a storage system. In contrast, in air systems, the air is heated in air collectors and is transported to storage or space using blowers. Most solar systems cannot provide adequate space or hot-water heating. Consequently, an auxiliary or back-up heating system is needed. The main components of an active liquid hot water solar system are shown in Figure 13.27. A photograph of a home in Golden, Colorado using a liquid solar system is shown in Figure 13.28.

FIGURE 13.27	A schematic of a solar hot water system. DHW is the abbreviation for domestic hot water.

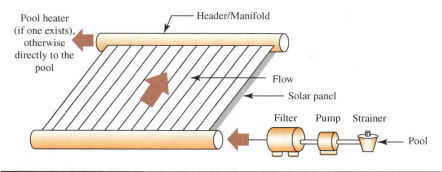

FIGURE 13.29   A swimming pool solar heater.

In moderate climates, solar hot water systems are also used to heat swimming pools. The goal of this type of system is to extend the swimming season. Swimming pool solar heaters operate at slightly warmer temperatures than the surrounding air temperature. These types of systems typically use inexpensive, unglazed collectors made from plastic materials. Because these systems are not insulated, they require large collector areas, approximately 50 to 100% of the pool area (Figure 13.29).

Passive Solar Systems

The **passive solar systems** do not make use of any mechanical components such as collectors, pumps, blowers, or fans to collect, transport or distribute solar heat to various parts of a building. Instead, a *direct passive solar*

A direct passive solar system uses large glass areas on the south wall of a building and a thermal mass to collect the solar energy.

system uses large glass areas on the south wall of a building and a thermal mass to collect the solar energy. The solar energy is stored in interior thick masonry walls and floors during the day and is released at night. In cold climates, the passive systems also use insulated curtains at night to cover the glass areas to reduce the heat loss. Another feature of a passive solar system is an overhang to shade the windows during summer as shown in Figure 13.30. *Indirect gain passive* designs utilize a storage mass placed between the glass wall and the heated space (Figure 13.31). As the air between the glass and masonry wall is heated, it rises and enters the room through a vent at the top of the wall. The room air enters the lower vent and is heated as it rises between the window and the masonry wall. Not all of the solar heat is transferred to the air; some is stored.

Another common type of solar system is a sunspace. The space may be used as a greenhouse, atrium, sun porch, or sun room. Masonry or concrete floors and walls, water containers, or covered pools of water may serve as thermal storage. A photograph of an interior section of a passive house is shown in Figure 13.32.

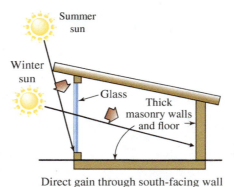

Direct gain through south-facing wall

| FIGURE 13.30 | A schematic of a building with direct solar passive gain. |

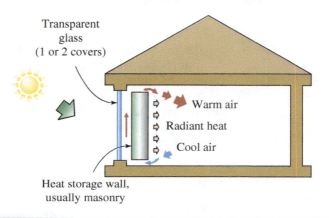

| FIGURE 13.31 | A schematic of a building with indirect passive gain. |

Photovoltaic Systems

A photovoltaic system converts light energy directly into electricity. It consists of a photovoltaic array, batteries, charge controller, and an inverter (a device that converts direct current into alternating current). Examples of photovoltaic systems are shown in Figure 13.33. The photovoltaic systems come in all sizes and shapes and are generally classified into stand-alone systems, hybrid systems, or grid-tied systems. The systems that are not connected to a utility grid are called stand alone. Hybrid systems are those which use a combination of photovoltaic arrays and some other form of energy, such as diesel generation or wind. As the name implies, the grid-tied systems are connected to a utility grid. One of the largest grid-tied photovoltaic power plants in the United States is the Alamosa photovoltaic plant (Figure 13.34), which is located in an area of 82 acres in south central Colorado. It went on-line in 2007 and generates about 8.2 megawatts.

> A photovoltaic system converts light energy directly into electricity.

The backbone of any photovoltaic system is the cells. The **photovoltaic cells** are combined to form a **module**, and modules are combined to form an **array**. The photovoltaic cells are classified as crystalline, polycrystalline silicon, and amorphous silicon. Examples of photovoltaic cells are shown in Figure 13.35.

Wind Energy

Wind energy is a form of solar energy. As you all know, because of the earth's tilt and orbit, sun heats the earth and its atmosphere at different rates. You also know that hot air rises and cold air sinks to replace it. As the air moves, it has kinetic energy. Part of this kinetic energy can then be

FIGURE 13.33   Examples of photovoltaic systems: (a) parking rooftop, (b) solar bike, (c) space station, (d) a building rooftop, (e) photovoltaic roof shingles, (f) a remote communication facility.

FIGURE 13.34   The Alamosa photovoltaic plant in Colorado.
Courtesy of DOE/NREL

converted into mechanical energy and into electricity. A U.S. wind resource map is shown in Figure 13.36. As shown in this figure, the potential for generating electricity from wind is categorized as marginal to superb based on wind speeds. Two types of wind turbines are used to extract the energy from the wind, a vertical axis turbine (Figure 13.37) and a horizontal axis turbine. Schematic diagrams of a vertical axis and a horizontal axis turbine

FIGURE 13.35   Examples of photovoltaic materials.
Courtesy of DOE/NREL

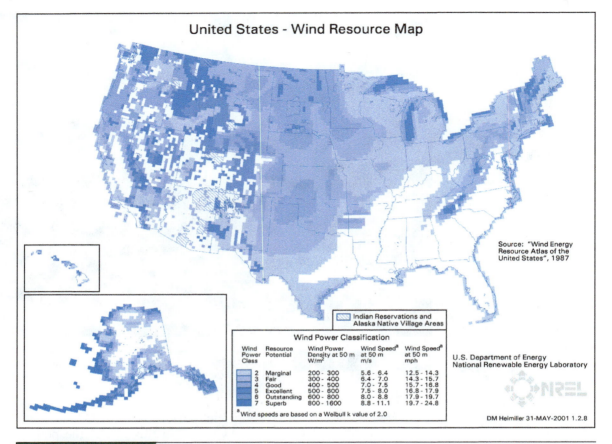

United States - Wind Resource Map

Source: "Wind Energy Resource Atlas of the United States", 1987

Indian Reservations and Alaska Native Village Areas

Wind Power Classification

Wind Power Class	Resource Potential	Wind Power Density at 50 m W/m²	Wind Speed[a] at 50 m m/s	Wind Speed[a] at 50 m mph
2	Marginal	200 - 300	5.6 - 6.4	12.5 - 14.3
3	Fair	300 - 400	6.4 - 7.0	14.3 - 15.7
4	Good	400 - 500	7.0 - 7.5	15.7 - 16.8
5	Excellent	500 - 600	7.5 - 8.0	16.8 - 17.9
6	Outstanding	600 - 800	8.0 - 8.8	17.9 - 19.7
7	Superb	800 - 1600	8.8 - 11.1	19.7 - 24.8

[a] Wind speeds are based on a Weibull k value of 2.0

U.S. Department of Energy
National Renewable Energy Laboratory

DM Heimiller 31-MAY-2001 1.2.8

FIGURE 13.36   U.S. wind resource map.
Courtesy of DOE/NREL

are shown in Figure 13.38. The vertical axis turbine can accept wind from any angle, requires light-weight towers, and is easy to service. The main disadvantage of this type of turbine is that because the rotors are near the ground, where the wind speeds are relatively low, it has poor performance. Most of wind turbines in use in the U.S. are of horizontal axis type. The wind turbines are typically classified as small (< 100 kW), intermediate (< 250 kW), and large (250 kW to 2 MW).

Here is some terminology that you would find useful while studying the major components of a wind turbine:

- The *blades* and *hub* are called *rotors*. Most horizontal axis turbines have either two or three blades.
- A *gear box* connects the low speed shaft attached to the rotor and to the high speed shaft of the generator.
- A *yaw motor* runs the yaw drive to keep the blades facing into the wind as the wind direction changes.
- A *controller* starts the wind turbine at speeds of 8 to 16 mph and stops the turbine at about 65 mph to prevent damage to the blades and components. An anemometer measures the wind speed and transmits the data to the controller.

FIGURE 13.37   Example of a vertical axis wind turbine.
Bart Everett/Shutterstock.com

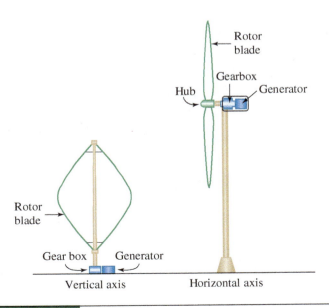

FIGURE 13.38   Schematics of a vertical axis and a horizontal axis wind turbine.

• A *brake* stops the rotor in emergencies or high wind speeds. The brake is applied mechanically, electrically, or hydraulically.

Ethanol and Biodiesel

Ethanol is an alcohol-based fuel that is made from the sugars found in corn and barley. Other sources, such as rice, sugar cane, and potato skins, are also used to produce ethanol. Most of the ethanol used in the U.S. today is distilled from corn. You have seen E10 signs at gas stations. This E10 designation refers to a fuel that is a mixture of 10% ethanol and 90% gasoline. Another renewable fuel that you may have heard about is biodiesel. Biodiesel is a fuel that is commonly made from vegetable oils or recycled restaurant grease and can be in diesel engines. The B20 designation refers to a blend of 20% biodiesel with 80% petroleum diesel.

As we mentioned at the beginning of this section, as future engineers you are faced with two problems, energy sources and emissions. The solutions to these problems require innovative approaches, so study hard as you take your engineering classes, prepare well, and along the way think about ways that you might be able to solve these problems.

Before You Go On

Answer the following questions to test your understanding of the preceding sections.

1. What is the major source of energy production in the United States?

2. What percentage of the U.S. electric power is generated using nuclear fission?

3. What percentage of the U.S. electric power is generated using renewable sources?

4. Describe the active solar heating systems. How do they function?

5. How is electricity generated in a photovoltaic system?

Vocabulary—State the meaning of the following terms:

Nuclear Energy _____

Active Solar System _____

Passive Solar System _____

Photovoltaic Cell _____

Photovoltaic Module _____

Photovoltaic Array _____

Ming Dong

My name is Ming Dong. I am currently a sophomore at Carnegie Mellon University. I am double majoring in mechanical engineering and biomedical engineering. I came to the Unites States when I was 12, and I have grown up in a traditional Chinese family. My parents were hoping that I would major in chemistry or biology. They felt that engineering is too "boyish." But I decided to major in engineering because for me it is the combination of math, science, and art—three of my favorite subjects.

Now I like engineering even more; it is an exciting field. I learn something new every day. Engineering is creativity; it is the combination of math and science, and how to apply them to real life. It is about creating something that's real and helps solve real problems. The biggest challenge I faced was being comfortable in class with males. There are about 15% of girls in

Courtesy of Ming Dong

mechanical engineering class. It was hard at first, because I have to prove that "I am not just a girl. I am as good as your guys." But once I got used to having classes with boys, it seemed less obvious to me the difference between girls and guys; we are all engineering students.

After graduation, I am hoping to find a job and work for a couple of years, and then return to graduate school to enter the joint PhD/MD program, and conduct research that directs me toward neuroengineering.

Courtesy of Ming Dong

Dominique Green

Engineering has always been a way of thinking for me. I remember growing up completely fascinated by science, math, and with TV programs like Mr. Wizard and the many educational programs on PBS (Public Broadcasting Service). It's funny because I was given an opportunity to work with PBS on one of my many projects with Accenture (my current employer). One major up-close and personal chance to pursue my passion as an engineer came when I entered high school and was able to do more hands-on experiments in programs like BEAMS (Becoming Enthusiastic About Math and Science), sponsored by the Thomas Jefferson National Laboratory (a Department of Energy-sponsored research facility). While in that internship, I was able to actually meet Secretary O'Leary, who served one term under President Clinton. We took several pictures, which still hang proudly at my mother's home in Newport News, VA. My excitement later turned into a series of internships with Thomas Jefferson National Laboratory, NASA, and the Newport

Courtesy of Dominique Green

News Shipbuilding, which all collectively helped to solidify my commitment to math, science, and engineering.

My biggest asset as a youth and until now has been my excitement for math and science, which translated into my academic acceleration. My decision to pursue engineering was rooted in how well I did in science fair and technology competitions, my ability to take college-level courses in math and science while in high school, and the real hands-on (life-changing) internships I pursued and won, all before and during college.

The icing on the cake was that I was always employed by engineering-technology companies, always had internships in the field of engineering, and received tons of scholarship money.

The biggest challenge I faced as a student was trying to manage my time with studies and extra-curricular activities. There really wasn't enough time for me to excel in any outside activities the way I dreamed. As an engineer, you really have to commit 100% of your academic regimen to learning a ton of information in a short period of time. Looking back on my four-year spurt in engineering, I wish that I could have done more of the "stuff" that turned me on to engineering, such as joining two mechanical auto teams as opposed to one (I was a member of the Hybrid Electric Vehicle Team). I wish I had the time to create another technology and have it patented before graduating college. I had so many ideas on paper that never came to fruition. Thankfully, I took good notes when those ideas surfaced and now all I need is the time once again to bring the ideas on paper to life. It's not a total wash because there is always graduate school, which does provide the resources needed for research to make ideas happen. The undergraduate curriculum is such that most professors aren't allowed to offer students exciting experiments and projects because they have to meet requirements for accreditation.

The thing I like most about engineering is that it's wide open. I can remember judging a technology future-city competition some months ago, which included middle and elementary students. Some of these ideas were just amazing. Today, engineering is the backbone of any technology or innovation. Engineering is the track and the wheels needed to make ideas move. I like its diversity because I've found myself speaking to people with all sorts of accents, religion, and gender in our attempt to solve complex problems by sharing our knowledge base. In the engineering world, you're judged on your understanding, abilities, and knowledge. It's all about what you know and not who you know that makes your career move. Engineering is something you can't fake when someone asks you to explain the underlying principles of electromagnetic fields and its role in the universe.

I still don't think my most interesting project has come yet. But it's funny because all of the projects I've enjoyed have been simple hands-on projects. I haven't been fortunate enough to do much hands-on engineering for the past couple of years. Most of my work has been along the path of business process design oversight and software implementation. My work with the Hybrid Electric Vehicle (where automobiles use fuel cells and have a by-product of water) project was probably the most rewarding because it still has real-life application. Of course, we can all look around today and see how gas prices have affected our choice to look for alternative energy sources. My excitement really is for the future of the technology and in knowing that my ideas may one day make their way to the show-room floor of some car dealership or household.

Courtesy of Dominique Green

SUMMARY

LO¹ Work, Mechanical Energy, and Thermal Energy

We need energy to create goods, build shelter, cultivate and process food, and maintain our living places at comfortable temperature and humidity settings. Energy can have different forms, and to better explain quantitatively the requirements to move objects such as our cars, to lift things like an elevator, or to heat or cool our homes. Energy is defined and classified into different categories, such as kinetic energy, potential energy, and thermal energy. Work is performed when a force moves an object through a distance. Moreover, when we do work on an object, we change its energy; in other words, we need energy to do the work.

Kinetic energy is a way by which we quantify how much energy or work is required to move something. The SI and U.S. Customary units for kinetic energy are the joule and lbf · ft, respectively. Moreover, when we do work on or against an object, we change its kinetic energy.

The work required to lift an object over a vertical distance is called potential energy. The SI and U.S. Customary units for potential energy are also joule and lbf · ft, respectively.

When a spring is stretched or compressed from its unstretched position, elastic energy is stored in the spring. This energy will be released when the spring is allowed to return to its original position.

Thermal energy or heat transfer occurs whenever a temperature difference exists within an object or between a body and its surroundings. Also recall that heat always flows from a high-temperature region to the low-temperature region. The three units that are commonly used to quantify thermal energy are the British thermal unit (Btu), the calorie, and the joule. The Btu (British thermal unit) represents the amount of thermal energy needed to raise the temperature of one pound mass (1 lbm) of water by one degree Fahrenheit $(1° F)$. The calorie represents the amount of heat required to raise the temperature of one gram (1 g) of water by one degree Celsius $(1° C)$.

LO² Conservation of Energy

In the absence of heat transfer, and assuming negligible losses and no work, the conservation of mechanical energy states that the total mechanical energy of a system is constant. That is, the change in the kinetic energy of the object, plus the change in the elastic energy, plus the change in the potential energy of the object must be zero.

The effects of heat and work in the conservation of energy are represented in the first law of thermodynamics. The first law states that, for a system having a fixed mass, the net heat transfer to the system minus the work done by the system is equal to the change in total energy of the system.

LO³ Power

By now, you should have a clear understanding of power, its common units, and how it is related to work and energy. Power is the time rate of doing work or how fast you are expending energy. The value of power required to do the work (perform a task) represents how fast you want the work (task) done. If you want the work done in a shorter period of time, you need to spend more power.

$$\text{power} = \frac{\text{work}}{\text{time}} = \frac{\text{energy}}{\text{time}}$$

The SI and U.S. Customary units of power are watts and lbf · ft/s, respectively, where

$$1 \text{horsepower} = 1 \text{hp} = 550 \frac{\text{lbf} \cdot \text{ft}}{\text{s}}$$

LO⁴ Efficiency

You should know the basic definition of efficiency and be familiar with its various forms, including the definitions for thermal efficiency, SEER, and AFUE that are commonly used to express the efficiencies of systems including heating, ventilating, and air-conditioning (HVAC) equipment and household appliances. In general, the overall efficiency of a system is defined as

$$\text{efficiency} = \frac{\text{actual output}}{\text{required input}}$$

All machines and systems require more input than what they put out. For example, the thermal efficiency of a typical gasoline engine is approximately 25 to 30%. The thermal efficiency of an internal combustion engine is defined as

$$\text{efficiency} = \frac{\text{power output of the car}}{\text{heat power input as fuel is burned}}$$

LO⁵ Energy Sources, Generation, and Consumption

You should be familiar with the U.S. primary energy consumption by source (e.g., coal, natural gas, crude oil, nuclear, etc.) and sector (e.g., transportation, industrial, residential, and commercial). You should know about active and passive solar systems and photovoltaic systems. You also should be familiar with the different types of turbines and their major components (i.e., blades and hub, gear box, yaw motor, controller, and the brake).

KEY TERMS

Active Solar System 462

AFUE 453

calories 439

Efficiency 447

Elastic Energy 437

Energy 434

Horsepower 444

Joule 435

Kilowatt 444

Kilowatt-hour 444

Kinetic Energy 434

Nuclear Energy 459

Passive Solar System 463

Photovoltaic Array 465

Photovoltaic Cell 465

Photovoltaic Module 465
Potential Energy 435
Power 443

SEER 453
Thermal Energy 438
Ton of Cooling 444

Watt 444
Work 434

APPLY WHAT YOU HAVE LEARNED

Identify ways that you can save energy: for example, walking up a floor instead of taking the elevator or walking or riding your bike an hour a day instead of taking the car. Estimate the amount of energy that you could save every year with your proposal. Also, estimate the amount of fuel that can be saved in the same manner. State your assumptions, and present your analysis in a report.

PROBLEMS

Problems that promote liflong learning are denoted by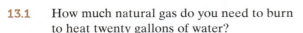

13.1 How much natural gas do you need to burn to heat twenty gallons of water?

13.2 Look up the manufacturer's data for the most recent year of the following cars:

a. Toyota Camry

b. Honda Accord

c. BMW 750 Li

You can visit the cars.com Website to gather information. For each car, perform calculations similar to Example 13.9 to determine the power required to accelerate the car from 0 to 60 mph. State all of your assumptions.

13.3 An elevator has a rated capacity of 2200 lb. It can transport people at the rated capacity between the first and the fifth floors, with a vertical distance of 15 ft between each floor, in 7 s. Estimate the power requirement for such an elevator.

13.4 Determine the gross force needed to bring a car that is traveling at 120 km/h to a full stop in a distance of 100 m. The mass of the car is 2000 kg. What happens to the initial kinetic energy? Where does it go or to what form of energy does the kinetic energy convert?

13.5 A centrifugal pump is driven by a motor. The performance of the pump reveals the following information:

Power input to the pump by the motor (kW):

0.5, 0.7, 0.9, 1.0, 1.2

Power input to the fluid by the pump (kW):

0.3, 0.55, 0.7, 0.9, 1.0

Plot the efficiency curve. The efficiency of a pump is a function of the flow rate. Assume that the flow-rate readings corresponding to power data points are equally spaced.

13.6 A power plant has an overall efficiency of 30%. The plant generates 30 MW of electricity, and uses coal from Montana (see Table 11.9) as fuel. Determine how much coal must be burned to sustain the generation of 30 MW of electricity.

13.7 Estimate the amount of gasoline that could be saved if all of the passenger cars in the United States were driven 1000 miles less each year. State your assumptions and write a brief report discussing your findings.

13.8 Investigate the typical power consumption range of the following products:

   a. home refrigerator

   b. 25-inch television set

   c. clothes washer

   d. electric clothes dryer

   e. vacuum cleaner

   f. hair dryer

   Discuss your findings in a brief report.

13.9 Investigate the typical power consumption range of the following products:

   a. personal computer with a 19-inch monitor

   b. laser printer

   c. cellular phone

   d. palm calculator

   Discuss your findings in a brief report.

13.10 Look up the furnace size and the size of the air-conditioning unit in your own home or apartment. Investigate the SEER and the AFUE of the units.

13.11 Investigate the size of a gas furnace used in a typical single-family dwelling in upstate New York, and compare that size to the furnaces used in Minnesota and in Kansas.

13.12 An air-conditioning unit has a cooling capacity of 24,000 Btu/h. If the unit has a rated energy efficiency ratio (EER) of 11, how much electrical energy is consumed by the unit in 1 h? If a power company charges 14 cents per kWh usage, how much would it cost to run the air-conditioning unit for a month (31 days), assuming the unit runs 8 h a day? What is the coefficient of performance (COP) for the given air-conditioning unit?

13.13 Visit a store that sells window-mount air-conditioning units. Obtain information on their rated cooling capacities and EER values. Contact your local power company and determine the cost of electricity in your area. Estimate how much it will cost to run the air-conditioning unit during the summer. Write a brief report to your instructor discussing your findings and assumptions.

For Problems 13.14 through 13.20 use the data from the accompanying table shown below.

13.14 Calculate and plot the percentage of each fuel used in generating electricity for each year shown in the table.

13.15 Calculate and plot the percentage of increase in coal consumption for the data shown in the table.

13.16 Convert the data given in the table from kilowatt-hours to Btu.

13.17 Assuming an average 35% efficiency for power plants and a heating value of approximately 7.5 MJ/kg, calculate the amount of coal (in kg) required for generating electricity for each year shown in the table.

Electricity Generation by Fuel, 1980–2030 (billion kilowatt-hours)—Data from U.S. Department of Energy

Year	Coal	Petroleum	Natural Gas	Nuclear	Renewable/Other
1980	1161.562	245.9942	346.2399	251.1156	284.6883
1990	1594.011	126.6211	372.7652	576.8617	357.2381
2000	1966.265	111.221	601.0382	753.8929	356.4786
2005	2040.913	115.4264	751.8189	774.0726	375.8663
2010	2217.555	104.8182	773.8234	808.6948	475.7432
2020	2504.786	106.6799	1102.762	870.698	515.1523 projected values
2030	3380.674	114.6741	992.7706	870.5909	559.1335 projected values

Data from U.S. Department of Energy

13.18 How many kilograms of coal could be saved if we were to increase the average efficiency of power plants by 1% to 36%?

13.19 Assuming an average 35% efficiency for power plants and a heating value of approximately 1000 Btu/ft³ (22,000 Btu/lbm), calculate the amount of natural gas required in ft³ and lbm for generating electricity for each year shown in the table.

13.20 How many ft³ and pounds of natural gas could be saved if we were to increase the average efficiency of power plants by 1 to 36%?

13.21 Visit a flat panel solar collector manufacturer and obtain data sheets including cost and efficiencies for single-pane and double-pane collectors.

13.22 The masonry wall shown in Figure 13.27 is commonly referred to as a Trombe wall in honor of Felix Trombe, who developed this concept. Investigate the factors that must be considered when sizing (thickness) the Trombe wall. Write a brief report discussing your findings.

13.23 The photovoltaic systems are designed based on "Peak Sun Hours." What is peak sun hour?

13.24 You may have noticed that solar collectors are sloped at a certain angle. Investigate the tilt of solar collectors. What is a collector tilt based on? Write a brief report discussing your findings.

13.25 For a wind turbine, look up the definition for the following terms: rotor solidity, tip speed ratio, capacity factor, and Betz limit. Write a brief report discussing your findings.

13.26 Calculate the amount of natural gas that you need to burn to heat twenty gallons of water from room temperature at 70° F to 120° F to take shower if the water heater has an efficiency of (a) 78 percent, (b) 85 percent, and (c) 90 percent.

Impromptu | Design VI

Objective: To design a catapult system from the materials listed that throws a ping-pong ball a maximum distance in a specified direction. During the launch process, the catapult system must be handled by one team member alone. Each team is allowed one practice launch. Thirty minutes will be allowed for preparation.

Provided Materials: 8 rubber bands; 20 inches of adhesive tape; 20 inches of string; 2 Dixie cups; 2 sheets of paper ($8\frac{1}{2}'' \times 11''$ each); 4 plastic drinking straws; 2 plastic spoons; 8 popsicle sticks ; 4 thumb tacks; a ping-pong ball.

The system that throws the ping-pong ball the maximum distance along the specified direction wins.

An Engineering Marvel

Hoover Dam[*]

Hoover Dam is one of the Bureau of Reclamations' multipurpose projects on the Colorado River. These projects control floods; they store water for irrigation, municipal, and industrial use; and they provide hydroelectric power, recreation, and fish and wildlife habitat.

The Hoover Dam is a concrete arch–gravity type of dam, in which the water load is carried by both gravity action and horizontal arch action. The first concrete for the dam was placed on June 6, 1933, and the last concrete was placed in the dam on May 29, 1935. The following is a summary of some facts about the Hoover Dam.

Andrew Zarivny/Shutterstock.com

Dam dimensions: Height: 726.4 ft; length at crest: 1244 ft; width at top: 45 ft; width at base: 660 ft

Weight: 6.6 million tons

Reservoir statistics: Capacity: 28,537,000 acre-feet; length: 110 mi; shoreline: 550 mi; max depth: 500 ft; surface area: 157,000 acres

Quantities of materials used in project: Concrete: $4,440,000 \text{ yd}^3$; explosives: 6,500,000 lb; plate steel and outlet pipes: 88,000,000 lb; pipe and fittings: 6,700,000 lb (840 mi); reinforcement steel: 45,000,000 lb; concrete mix proportions:

[*] Materials were adapted from U.S. Bureau of Reclamation.

cement: 1.00 part, sand: 2.45 parts, fine gravel: 1.75 parts, intermediate gravel: 1.46 parts, coarse gravel: 1.66 parts, cobbles (3 to 9 in.): 2.18 parts; water: 0.54 parts

The dam was built in blocks, or vertical columns, varying in size from approximately 60 ft square at the upstream face of the dam to about 25 ft square at the downstream face. Adjacent columns were locked together by a system of vertical keys on the radial joints and horizontal keys on the circumferential joints. After the concrete was cooled, a cement and water mixture called grout was forced into the spaces created between the columns by the contraction of the cooled concrete to form a monolithic (one-piece) structure.

Hoover Dam itself contains 3.25 million yd^3 of concrete. Altogether, there are 4,360,000 yd^3 of concrete in the dam, power plant, and appurtenant works. This much concrete would build a monument 100 ft square and 2-1/2 mi high; would rise higher than the Empire State Building (which is 1250 ft) if placed on an ordinary city block; or would pave a standard highway, 16 ft wide, from San Francisco to New York City.

The Reservoir   At elevation 1221.4, Lake Mead, the largest man-made lake in the United States, contains 28,537,000 acre-feet (an acre-foot is the amount of water required to cover 1 acre to a depth of 1 foot). This reservoir will store the entire average flow of the river for two years. That much water would cover the entire state of Pennsylvania to a depth of 1 ft.

Lake Mead extends approximately 110 mi upstream toward the Grand Canyon and approximately 35 mi up the Virgin River. The width of Lake Mead varies from several hundred feet in the canyons to a maximum of 8 mi. The reservoir covers about 157,900 acres, or 247 square miles.

Recreation, although a by-product of this project, constitutes a major use of the lakes and controlled flows created by Hoover and other dams on the lower Colorado River today. Lake Mead is one of America's most popular recreation areas, with a 12-month season that attracts more than 9 million visitors each year for swimming, boating, skiing, and fishing. The lake and surrounding area are administered by the National Park Service as part of the Lake Mead National Recreation Area, which also includes Lake Mohave downstream from Hoover Dam.

The Power Plant   There are 17 main turbines in the Hoover Power plant. The original turbines were all replaced through an upgrading program between 1986 and 1993. With a rated capacity of 2,991,000 hp, and two station-service units rated at 3500 hp each, for a plant total of 2,998,000 hp, the plant has a nameplate capacity of 2,074,000 kW. This includes the two station-service units, which are rated at 2400 kW each.

Hoover Dam generates low-cost hydroelectric power for use in Nevada, Arizona, and California. Hoover Dam alone generates more than 4 billion kWh a year—enough to serve 1.3 million people. From 1939 to 1949, the Hoover Power plant was the world's largest hydroelectric installation; with an installed capacity of 2.08 million kW, it is still one of this country's largest.

The $165 million cost of Hoover Dam has been repaid, with interest, to the federal treasury through the sale of its power. Hoover Dam energy is marketed by the Western Area Power Administration to 15 entities in Arizona, California, and Nevada under contracts that expire in 2017. More than half, 56%, goes to southern California users; Arizona contractors receive 19%, and Nevada users get 25%. The revenues from the sale of this power now pay for the dam's operation and maintenance. The power contractors also paid for the uprating of the power plant's nameplate capacity from 1.3 million to over 2.0 million kW.

PROBLEMS

1. Calculate the water pressure at the bottom of the dam when the water level is two-thirds of the height of the dam. Express your result in

lb/in² and Pa. Also, calculate the magnitude of the force due to water pressure acting on a narrow strip (1 ft by 100 ft wide) located at the base of the dam.

2. As mentioned in the article describing Hoover Dam, Lake Mead contains 28,537,000 acre-feet of water (an acre-foot is the amount of water required to cover 1 acre to a depth of 1 ft). Express this water volume in gallons and m³.

3. Hoover Dam generates more than 4 billion kWh a year. How many 100-watt lightbulbs could be powered every hour by the Hoover Dam's power plant?

4. How much coal must be burned in a steam power plant with a thermal efficiency of 34% to generate enough power to equal the 4 billion KWh a year generated by Hoover Dam? Assume the coal comes from Montana (see Chapter 11 for heating value data).

PART 3

Computational Engineering Tools

Using Available Software to Solve Engineering Problems

In Part Three of this book, we will introduce Microsoft Excel and MATLAB®, two computational tools that are commonly used by engineers to solve engineering problems. These computational tools are used to record, organize, analyze data using formulas, and present the results of an analysis in chart forms. MATLAB is also versatile enough that you can use it to write your own program to solve complex problems.

CHAPTER 14   ELECTRONIC SPREADSHEETS

CHAPTER 15   MATLAB

Computational Engineering Tools
Electronic Spreadsheets

In recent years, the use of spreadsheets as an analysis and a design tool has grown rapidly. Easy-to-use spreadsheets such as Excel are used by engineers to record, organize, and analyze data, and present the results of an analysis in a graph or a bar chart form.

LEARNING OBJECTIVES

LO¹ **Microsoft Excel Basics:** explain the basics of Excel workbook environment including cells and their addresses (absolute, relative, and mixed), a range, and how to create formulas

LO² **Excel Functions:** know how to use Excel's built-in mathematical, trigonometric, statistical, engineering, logical, and financial functions

LO³ **Plotting with Excel:** know how to plot two sets of data with different ranges on the same chart

LO⁴ **Matrix Computation with Excel:** know how to use Excel to perform matrix operations and solve a set of linear equations

LO⁵ **An Introduction to Excel's Visual Basic for Applications:** know how to use Excel's VBA; a programming language that allows you to use Excel more effectively

WORKBOOK AND VBA

An electronic spreadsheet is a tool that can be used to solve engineering problems. Spreadsheets are commonly used to record, organize, and analyze data using formulas. Spreadsheets are also used to present the results of an analysis in a graph form as shown here. Although engineers still write computer programs to solve complex engineering problems, simpler problems can be solved with the help of a spreadsheet.

The Excel spreadsheet software consists of two parts: the workbook and the Visual Basic Editor. In the Excel workbook environment, you can solve many simple engineering problems. The Excel VBA is a programming language that allows you to use Excel even more effectively and solve more complicated problems.

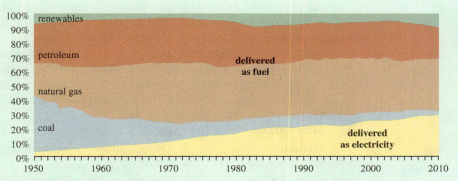

Electricity's share of U.S. delivered energy has risen significantly since 1950

Source: U.S. Energy Information Administration (2012)

To the students: How proficient are you in the use of Excel? Do you know the difference between absolute cell reference, relative cell reference, and mixed cell reference? Do you know what is meant by a range? Do you know how to use Excel to solve a set of linear equations simultaneously? Can you use Excel to find a function that best fits a set of data? What is VBA?

In this chapter, we will discuss the use of spreadsheets in solving engineering problems. Before the introduction of electronic spreadsheets, engineers wrote their own computer programs. Computer programs were typically written for problems where more than a few hand calculations were required. FORTRAN was a common programming language that was used by many engineers to perform numerical computations. Although engineers still write computer programs to solve complex engineering problems, simpler problems can be solved with the help of a spreadsheet. Compared to writing a computer program and debugging it, spreadsheets are much easier to use to record, organize, and analyze data using formulas that are input by the user. Spreadsheets are also used to show the results of an analysis in the form of charts. Because of their ease of use, spreadsheets are common in many other disciplines, including business, marketing, and accounting.

This chapter begins by discussing the basic makeup of Microsoft Excel, a common spreadsheet program. We will explain how a spreadsheet is divided into rows and columns and how to input data or a formula into an active cell. We will also explain the use of other tools such as Excel's mathematical, statistical, and logical functions. Plotting the results of an engineering analysis using Excel is also presented. If you are already familiar with Excel, you can skip Section 14.1 and 14.2 without the loss of continuity.

LO¹ 14.1   Microsoft Excel Basics

We will begin by explaining the basic components of Excel; then once you have a good understanding of these concepts, we will use Excel to solve some engineering problems. As is the case with any new areas you explore, the spreadsheet has its own terminology. Therefore, make sure you spend a little time at the beginning to familiarize yourself with the terminology, so you can follow the examples later. A typical Excel window is shown in Figure 14.1. The main components of the Excel window, which are marked by arrows and numbered as shown in Figure 14.1 are

1. **Title bar:** Contains the name of the current active workbook.

2. **Menu bar (tab):** Contains the commands used by Excel to perform certain tasks.

3. **Toolbar buttons:** Contains push buttons (icons) that execute commands used by Excel.

4. **Active cell:** A worksheet is divided into rows and columns. A cell is the box that you see as the result of the intersection of a column and a cell. *Active cell* refers to a specific selected cell.

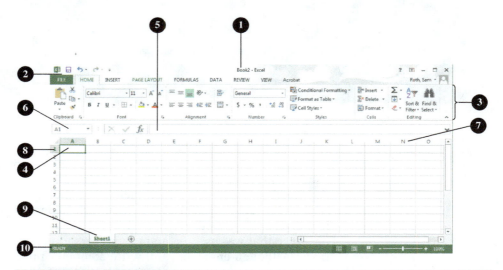

FIGURE 14.1   The components of the Excel window.

5. **Formula bar:** Shows the data or the formula used in the active cell.

6. **Name box:** Contains the address of the active cell.

7. **Column header:** A worksheet is divided into rows and columns. The columns are marked by A, B, C, D, and so on.

8. **Row header:** The rows are identified by numbers 1, 2, 3, 4, and so on.

9. **Worksheet tabs:** Allow you to move from one sheet to another sheet. As you will learn later, you can name these worksheets.

10. **Status bar:** Gives information about the command mode. For example, "Ready" indicates the program is ready to accept input for a cell, or "Edit" indicates Excel is in an edit mode.

A *workbook* is the spreadsheet file that you create and save. A workbook could consist of many worksheets and charts. A *worksheet* represents the rows and columns where you input information such as data, formulas, and the result of various calculations. As you will see soon, you may also include charts as a part of a given worksheet as well.

Naming Worksheets

To name a worksheet, double-click the sheet tab to be named, type the desired name, and press the Enter key. You can move (or change the position of) a worksheet in the workbook by selecting the sheet tab and while holding down the left button on the mouse move the tab to the desired position among other sheets.

Cells and Their Addresses

As shown in Figure 14.1, a worksheet is divided into rows and columns. The columns are marked by A, B, C, D, and so on, while the rows are identified by numbers 1, 2, 3, 4, and so on. A **cell** represents the box that you see as the result of the intersection of a **row** and a **column**. You can input (enter) various entities in a cell. For example, you can type in words or enter numbers or a formula. To enter words or a number in a cell, simply choose the cell where you want to enter the information, type the information, and then press the Enter key on your keyboard. Perhaps the simplest and the easiest way to move around in a worksheet is to use a mouse. For example, if you want to move from cell A5 to cell C8, move the mouse such that the mouse pointer is in the desired cell and then click the left mouse button. To edit the content of a cell, choose the cell, double-click the left mouse button, and then similar to editing a word processing document, use any combination of delete, backspace, or arrow keys to edit the content of the cell. As an alternative to double clicking, you can use the F2 key to select the edit mode.

Keep in mind that as you become more proficient in using Excel, you will learn that for certain tasks there is more than one way to do something. In this chapter, we will explain one of the ways, which can be followed easily.

> The Excel software consists of two parts: the workbook and the Visual Basic Editor. A workbook — the spreadsheet file — is divided into worksheets, and each worksheet is divided into columns, rows, and cells. Visual Basic for Applications (VBA) is a programming language that allows you to use Excel more effectively.

A Range

As you will soon see when formatting, analyzing, or plotting data, it is often convenient to select a number of cells simultaneously. The cells that are selected simultaneously are called

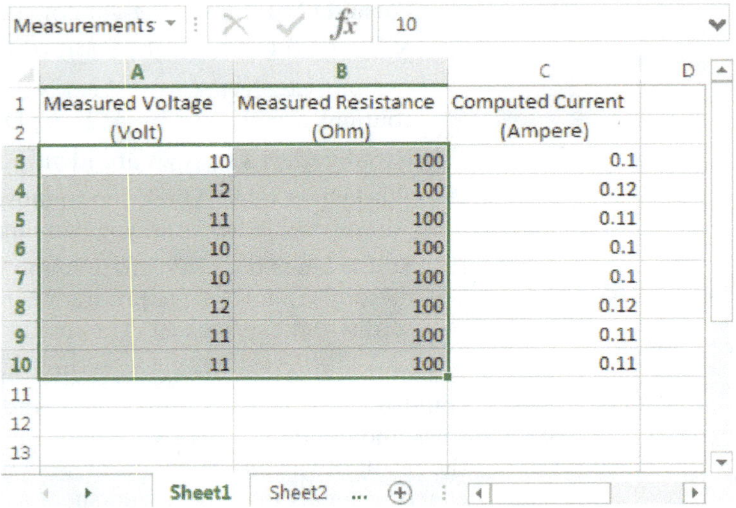

	A	B	C	D
1	Measured Voltage	Measured Resistance	Computed Current	
2	(Volt)	(Ohm)	(Ampere)	
3	10	100	0.1	
4	12	100	0.12	
5	11	100	0.11	
6	10	100	0.1	
7	10	100	0.1	
8	12	100	0.12	
9	11	100	0.11	
10	11	100	0.11	
11				
12				
13				

Measurements | fx | 10

Sheet1   Sheet2   ...

FIGURE 14.2   An example showing the selection of a range of cells.

a **range**. To define a range, begin with the first cell that you want included in the range and then drag the mouse (while pressing down the left button) to the last cell that should be included in the range. An example of selecting a range is shown in Figure 14.2. Note that in spreadsheet language, a range is defined by the cell address of the top-left selected cell in the range followed by a colon, :, and ends with the address of the bottom-right cell in the range. For example, to select cells A3 through B10, we first select A3 and then drag the mouse diagonally to B10. In spreadsheet language, this range is specified in the following manner—A3:B10. There are situations where you may want to select a number of cells that are not side by side. In such cases, you must first select the contiguous cells, and then while holding (pressing) the Ctrl key select the other noncontiguous cells by dragging the mouse button.

Excel allows the user to assign names to a range (selected cells). To name a range, first select the range as just described, and then click on the Name box in the Formula bar and type in the name you want to assign to the range. You can use upper- or lowercase letters along with numbers, but no spaces are allowed between the characters or the numbers. For example, as shown in Figure 14.2, we have grouped the measured voltages and the resistance into one range, which we have called *Measurements*. You can then use the name in formulas or in plotting data.

Inserting Cells, Columns, and Rows

After entering data into a spreadsheet, you may realize that you should have entered some additional data in between two cells, columns, or rows that you have just created. In such a case, you can always insert new cells, column(s), or row(s) among already existing data cells, columns, and rows in a worksheet. To insert new cell(s) between other existing cells, you must first select the cell(s) where the new cell(s) are to be inserted. Next, from the **Insert** menu (click the right button on the mouse) choose the **Cells** option. Indicate whether you want the selected cells to be shifted to the right or down. For example, let's say you want to insert three new cells in the location E8 through E11 (E8:E11) and shift the existing content of E8:E11 down. First select cells E8:E11; then from the **Insert** menu choose then choose the **Shift cell** down option. To insert a column, click on the column indicator button to the right of where you would like to have the new column inserted. Then click on the right button of the mouse and choose **Insert**. The procedure is similar for inserting a new row among already existing rows. For example, if you would like to insert a new column between columns D and E, you must first select column E, and then click on the right button of the mouse and choose **Insert**; the new column will be inserted to the left of column E. To insert more than one column or row simultaneously, you should select as many column indicator buttons as necessary to the right of where you would like to have the columns inserted. For example, if you would like to insert three new columns between columns D and E, then you must first select columns E, F, G; then click on the right button of the mouse and choose **Insert**, and three new columns will be inserted to the left of column E.

Creating Formulas in Excel

By now, you know that engineers use formulas that represent physical and chemical laws governing our surroundings to analyze various problems. You can use Excel to input engineering formulas and compute the results. In Excel, a formula always begins with an equal sign, =. To enter a formula, select the cell where you want the result of the formula to be displayed. In the Formula bar, then type the equal sign and the formula. Remember when typing your formula to use parentheses to dictate the order of operations. For example, if you were to type $= 100 + 5^*2$, Excel will perform the multiplication first, which results in a value of 10, and then this result is added to 100, which yields an overall value of 110 for the formula. If however, you wanted Excel to add the 100 to 5 first and then multiply the resulting 105 by 2, you should have placed parentheses around the 100 and 5 in the following manner: $= (100 + 5)^* 2$, which results in a value of 210. The basic Excel arithmetic operations are shown in Table 14.1.

TABLE 14.1		The Basic Excel Arithmetic Operations	
Operation	Symbol	Example: Cells A5 and A6 contain the values 10 and 2, respectively	Cell A7 contains the result of the formula given in the example
Addition	+	$= A5 + A6 + 20$	32
Subtraction	−	$= A5 - A6$	8
Multiplication	*	$= (A5 * A6) + 9$	29
Division	/	$= (A5/2.5) + A6$	6
Raised to a power	^	$= (A5 \wedge A6) \wedge 0.5$	10

EXAMPLE 14.1

As we explained in the previous chapters, thermophysical properties of a substance, including density, viscosity, thermal conductivity, and heat capacity, play a key role in engineering calculations. As discussed, the thermophysical property values represent information such as how compact the material is for a given volume (density), or how easily a fluid flows (viscosity), or how good a material is in conducting heat (thermal conductivity), or how good the material is in storing thermal energy (heat capacity). The values of thermophysical properties are commonly measured in laboratories at given conditions. Moreover, the values of thermophysical properties of a substance generally change with temperature. The following example will show how the density of standard air changes with temperature. The density of standard air is a function of temperature and may be approximated using the ideal gas law according to

$$\rho = \frac{P}{RT}$$

where

P = standard atmospheric pressure (101.3 kPa)

R = gas constant and its value for air is $286.9 \left(\dfrac{J}{kg \cdot K} \right)$

T = air temperature in kelvin (K)

Using Excel, we want to create a table that shows the density of air as a function of temperature in the range of $0°\,C\,(273.15\,K)$ to $50°\,C\,(323.15\,K)$ in increments of $5°\,C$.

Refer to the Excel sheets shown in the accompanying figures when following the steps.

1. In cell A1, type **Density of air as a function of temperature**.
2. In cells A3 and B3, type **Temperature (C), Density (kg/m^3)**, respectively.
3. In cells A5 and A6, type **0** and **5,** respectively (Figure 14.3).

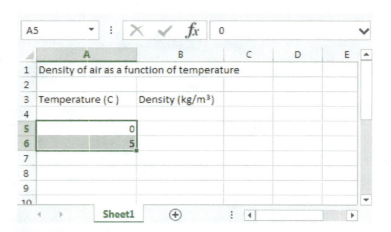

FIGURE 14.3   Steps 1, 2, and 3.

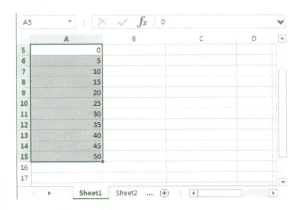

FIGURE 14.4

4. Pick cells A5 and A6 and use the **Fill** command with the + handle to copy the pattern into cells A7 to A15 (Figure 14.4).

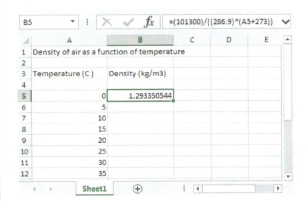

FIGURE 14.5

5. In cell B5, type the formula $= (101300)/((286.9)*(A5 + 273))$, as shown in Figure 14.5.

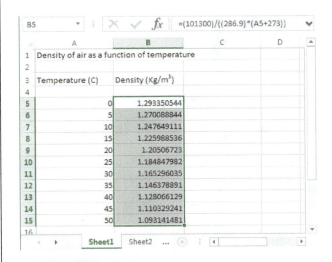

FIGURE 14.6

6. Use the **Home** menu (tab) and the **Fill** command to copy the formula into cells B6 to B15 (Figure 14.6). You could also use the **Fill** command with the + handle to copy the formula into cells B6 to B15.

7. Pick cells B5: B15, right-click and pick **Format Cells.** Change the number of decimal places to 2, as shown in Figure 14.7.

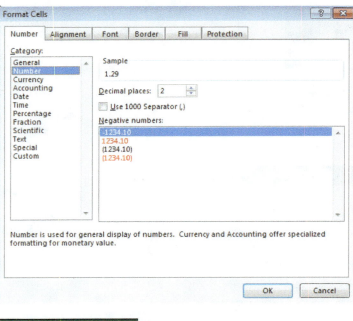

FIGURE 14.7

The final results for Example 14.1 are shown in Figure 14.8. The cell contents were centered using the center button (icon) from the Toolbar.

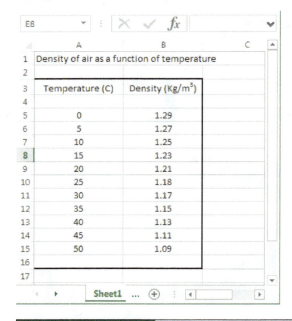

A	B
1 Density of air as a function of temperature	
3 Temperature (C)	Density (Kg/m^3)
5 0	1.29
6 5	1.27
7 10	1.25
8 15	1.23
9 20	1.21
10 25	1.18
11 30	1.17
12 35	1.15
13 40	1.13
14 45	1.11
15 50	1.09

FIGURE 14.8   The final result for Example 14.1.

Absolute Cell Reference, Relative Cell Reference, and Mixed Cell Reference

When creating formulas you have to be careful how you refer to the address of a cell, especially if you are planning to use the **Fill** command to copy the pattern of formulas in the other cells. There are three ways that you can refer to a cell address in a formula: *absolute*, *relative*, and *mixed reference*.

> There are three ways that you can refer to a cell address in a formula: absolute, relative, and mixed reference.

To better understand the differences among the absolute, relative, and mixed reference, consider the examples shown in Figure 14.9. As the name implies, **absolute reference** is absolute, meaning it does not change when the **Fill** command is used to copy the formula into other cells. Absolute reference to a cell is made by $column-letter$row-number. For example, A3 will always refer to the content of cell A3, regardless of how the formula is copied. In the example shown, cell A3 contains the value 1000, and if we were to input the formula $= 0.06^*$ \$A\$3 in cell B3, the result would be 60. Now if we were to use the **Fill** command and copy the formula down in cells B4 through B11, this would result in a value of 60 appearing in cells B4 through B11, as shown in Figure 14.9(a).

On the other hand, if we were to make a **relative reference** to A3, that would change the formula when the **Fill** command is used to copy the formula into other cells. To make a relative reference to a cell, a special character, such as \$, is not needed. You simply refer to the cell address. For example, if we were to input the formula $= 0.06^*$A3 in cell B3, the result would be 60; and if we use the **Fill** command to copy the formula into cell B4, the A3 in the formula will automatically be substituted by A4, resulting in a value 75. Note that the formula in cell B4 now becomes $= 0.06^*$A4. The result of applying the **Fill** command to cells B4 through B11 is shown in Figure 14.9(b).

The **mixed cell reference** could be done in one of two ways: (1) You can keep the column as absolute (unchanged) and have a relative row, or (2) you can keep the row as absolute and have a relative column. For example, if you were to use $A3 in a formula, it would mean that column A remains absolute and unchanged, but row 3 is a reference row and changes as the formula is copied into other cells. On the other hand, A$3 means row 3 remains absolute while column A changes as the formula is copied into other cells. The use of mixed cell reference is demonstrated in the following example.

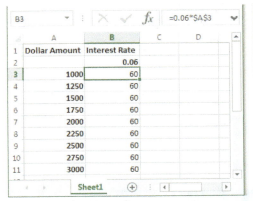

FIGURE 14.9   Examples showing the difference between the results of a formula when absolute and relative cell references are made in the formula.

EXAMPLE 14.2

Using Excel, create a table that shows the relationship between the interest earned and the amount deposited, as shown in Table 14.2.

In order to create the table for Example 14.2 using Excel, we will first create the dollar amount column and the interest row, as shown in Figure 14.10. Next we will type into cell B3 the formula $= \$A3*B\$2$. We can now use the **Fill** command to copy the formula in other cells, resulting in the table shown in Figure 14.10. Note that the dollar sign before A3 means column A is to remain unchanged in the calculations when the formula is copied into other cells. Also note that the dollar sign before 2 means that row 2 is to remain unchanged in calculations when the **Fill** command is used.

TABLE 14.2 **The Relationship between the Interest Earned and the Amount Deposited**

	Interest Rate			
Dollar Amount	**0.06**	**0.07**	**0.075**	**0.08**
1000	60	70	75	80
1250	75	87.5	93.75	100
1500	90	105	112.5	120
1750	105	122.5	131.25	140
2000	120	140	150	160
2250	135	157.5	168.75	180
2500	150	175	187.5	200
2750	165	192.5	206.25	220
3000	180	210	225	240

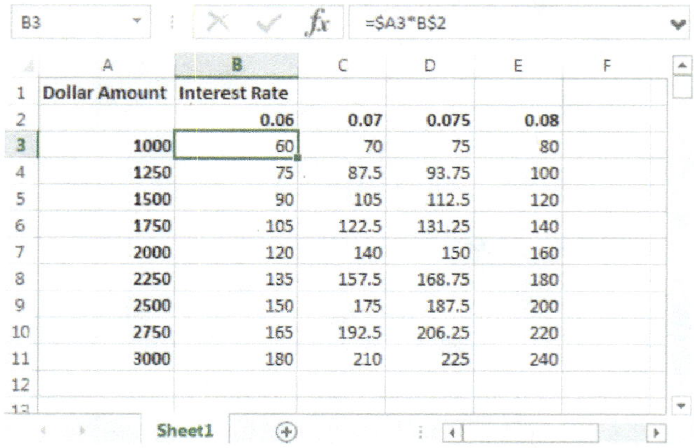

B3 fx =$A3*B$2

	A	B	C	D	E	F
1	Dollar Amount	Interest Rate				
2		0.06	0.07	0.075	0.08	
3	1000	60	70	75	80	
4	1250	75	87.5	93.75	100	
5	1500	90	105	112.5	120	
6	1750	105	122.5	131.25	140	
7	2000	120	140	150	160	
8	2250	135	157.5	168.75	180	
9	2500	150	175	187.5	200	
10	2750	165	192.5	206.25	220	
11	3000	180	210	225	240	
12						
13						

Sheet1

FIGURE 14.10   Excel spreadsheet for Example 14.2.

LO² 14.2   Excel Functions

Excel offers a large selection of built-in functions that you can use to ana-lyze data. By built-in functions, we mean standard functions such as the sine or cosine of an angle as well as formulas that calculate the total value, the average value, or standard deviation of a set of data points. The Excel functions are grouped into various categories, including mathematical and trigonometric, statistical, financial, and logical functions. In this chapter, we will discuss some of the common functions that you may use during your engineering education or later as a practicing engineer. You can enter a function in any cell by simply typing the name of the function if you already know it. If you do not know the name of the function, then you can press the **Insert Function** $\left(f_x\right)$ button, and then from the menu select the Function category and the Function name. There is also a Help button, on the lower left corner of the **Insert Function** menu, which once activated and followed leads to information about what the function computes and how the function is to be used.

> The Excel functions are grouped into various categories, including mathematical and trigonometric, engineering, statistical, financial, and logical functions. For example, logical functions allow you to test various conditions when programming formulas to analyze data.

Some examples of commonly used Excel functions, along with their proper use and descriptions, are shown in Table 14.3. Refer to Example 14.3 and Figure 14.11 when studying Table 14.3.

More examples of Excel's functions are shown in Table 14.4.

EXAMPLE 14.3

As set of values is given in the worksheet shown in Figure 14.11. Familiarize yourself with some of Excel's built-in functions, as described in Table 14.3. When studying Table 14.3, note that columns A and B contain the data range, which we have named *values*; cell D1 contains the angle 180. Also note that the functions were typed in the cells E1 through E14; consequently, the results of the executed Excel functions are shown in those cells.

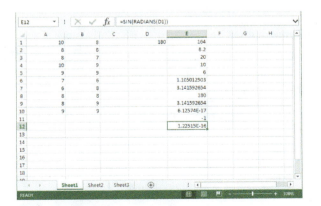

FIGURE 14.11   The Excel worksheet for Example 14.3.

TABLE 14.3	Some Excel Functions that You May Use in Engineering Analyses		
Function	**Description of the Function**	**Example**	**Result of the Example**
SUM(range)	It sums the values in the given range.	=SUM(A1:B10) or =SUM(values)	164
AVERAGE(range)	It calculates the average value of the data in the given range.	=AVERAGE(A1:B10) or =AVERAGE(values)	8.2
COUNT	It counts the number of values in the given range.	=COUNT (A1:B10) or =COUNT(values)	20
MAX	It determines the largest value in the given range.	=MAX(A1:B10) or =MAX(values)	10
MIN	It determines the smallest value in the given range.	=MIN(A1:B10) or =MIN(values)	6
STDEV	It calculates the standard deviation for the values in the given range.	=STDEV(A1:B10) or =STDEV(values)	1.105
PI	It returns the value of π, 3.141519265358979, accurate to 15 digits.	=PI()	3.141519265358979
DEGREES	It converts the value in the cell from radians to degrees.	=DEGREES(PI())	180
RADIANS	It converts the value from degrees to radians.	RADIANS(90) or =RADIANS(D1)	1.57079 / 3.14159
COS	It returns the cosine value of the argument. The argument must be in radians.	=COS(PI()/2) or =COS(RADIANS(D1))	0 / −1
SIN	It returns the sine value of the argument. The argument must be in radians.	=SIN(PI()/2) or =SIN(RADIANS(D1))	1 / 0

TABLE 14.4	More Examples of Additional Excel Functions
Function	**Description of the Function**
SQRT(x)	Returns the square root of value x.
FACT(x)	Returns the value of the factorial of x. For example, FACT (5) will return: (5)(4)(3)(2)(1) = 120.
Trigonometric Functions	
TAN(x)	Returns the value for the tangent of x. The argument must be in radians.
DEGREES (x)	Converts the value of x from radians to degrees. It returns the value of x in degrees.

TABLE 14.4	More Examples of Additional Excel Functions (continued)
ACOS(x)	This is the inverse cosine function of *x*. It is used to determine the value of an angle when its cosine value is known. It returns the angle value in radians, when the value of cosine between –1 and 1 is used for argument *x*.
ASIN(x)	This is the inverse sine function of *x*. It is used to determine the value of an angle when its sine value is known. It returns the angle value in radians when the value of sine falls between –1 and 1.
ATAN(x)	This is the inverse tangent function of *x*. It is used to determine the value of an angle when its tangent value is known.
Exponential and Logarithmic Functions	
EXP(x)	Returns the value of e^x.
LN(x)	Returns the value of the natural logarithm of *x*. Note that *x* must be greater than 0.
LOG(x)	Returns the value of the common logarithm of *x*.

The Now and Today Functions

When you work on an important Excel document, it is a good idea to indicate when the document was last modified. In one of the top cells, you may want to type **"Last Modified:"**, and in the adjacent cell, you can use the **=now()** or **=today()** function. Then, each time you access the Excel document, the **now()** function will automatically update the date and the time the file was last used. So, when you print the sheet, it will show the date and the time. If you use the **today()** function, it will only update the date.

EXAMPLE 14.4

Using Excel, compute the average (arithmetic mean) and the standard deviation of the density of water data given in Table 14.5. Refer to Chapter 19, Section 19.5, to learn about what the value of standard deviation for a set of data points represents.

Refer to Figure 14.12 when following the steps.

1. In cell B1, type **Group A findings**, and in cell C1 type **Group B findings**.
2. In each of cells B3 and C3, type **Density (kg/m³)**. Highlight the 3 in the kg/m³, and use the following command to make 3 a superscript. Right click on the mouse right button and choose **Format Cells**..." Next click on the **Font** tab, and turn on the superscript toggle switch. In cells B5 to C14, type density values for Group A and Group B.
3. Next, we want to compute the arithmetic means for the Group A and Group B data, but first we need to create a title for this computation. Because we are calculating the average, we might as well just use the word AVERAGE for the title of our calculations, thus in cell B15 type **AVERAGE:**.
4. In order to have Excel compute the average, we use the AVERAGE function in the following manner. In cell B16, we type **=AVERAGE(B5:B14),** and similarly in cell C16, we type **=AVERAGE(C5:C14)**.
5. Next, we will make a title for the standard deviation calculation by simply typing in cell B18 **STAND. DEV**.

TABLE 14.5	Data for Example 14.4
Group A Findings	**Group B Findings**
$\rho \ (kg/m^3)$	$\rho \ (kg/m^3)$
1020	950
1015	940
990	890
1060	1080
1030	1120
950	900
975	1040
1020	1150
980	910
960	1020

6. To compute the standard deviation for the Group A findings, in cell B19 type **= STDEV(B5:B14)**, and similarly to calculate the standard deviation for the Group B findings, in cell C19 type **= STDEV(C5:C14)**. Note that we used the function STDEV and the appropriate data range.

The final results for Example 14.4 are shown in Figure 14.12.

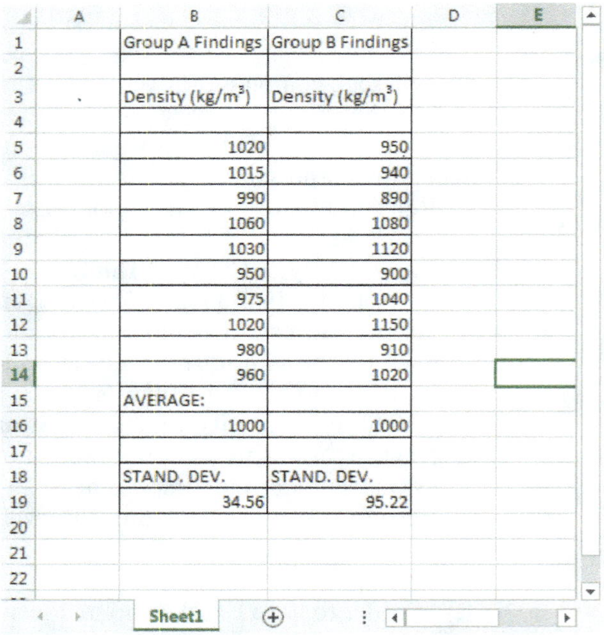

	A	B	C	D	E
1		Group A Findings	Group B Findings		
2					
3	.	Density (kg/m³)	Density (kg/m³)		
4					
5		1020	950		
6		1015	940		
7		990	890		
8		1060	1080		
9		1030	1120		
10		950	900		
11		975	1040		
12		1020	1150		
13		980	910		
14		960	1020		
15		AVERAGE:			
16		1000	1000		
17					
18		STAND. DEV.	STAND. DEV.		
19		34.56	95.22		
20					
21					
22					

Sheet1

FIGURE 14.12   The Excel worksheet for Example 14.4.

Using Excel Logical Functions

In this section, we will look at some of Excel's logical functions. These are functions that allow you to test various conditions when programming formulas to analyze data. Excel's logical functions and their descriptions are shown in Table 14.6.

Excel also offers relational or comparison operators that allow for testing the relative magnitude of various arguments. These relational operators are shown in Table 14.7. We will use Example 14.5 to demonstrate the use of Excel's logical functions and relational operators.

TABLE 14.6	Excel's Logical Functions
Logical Functions	**Description of the Function**
AND(logic1, logic2, logic3, …)	Returns true if all arguments are true and returns false if any of the arguments are false.
False()	Returns the logical value false.
IF(logical test, value_if_true, value_if_false)	It first evaluates the logical test; if true, then it returns the value_if_true; if the evaluation of the logical test deems false, then it returns the value_if_false value.
NOT(logical)	Reverses the logic of its argument; returns true for a false argument and false for a true argument.
OR(logical1, logical2, …)	Returns TRUE if any argument is true and returns FALSE if all arguments are false.
TRUE()	Returns the logical value TRUE.

TABLE 14.7	Excel's Relational Operators and Their Descriptions
Relational Operator	**Description**
<	Less than
< =	Less than or equal to
=	Equal to
>	Greater than
> =	Greater than or equal to
<>	Not equal to

EXAMPLE 14.5

The pipeline shown in Figure 14.13 is connected to a control (check) valve that opens when the pressure in the line reaches 20 psi. Various readings were taken at different times and recorded.

FIGURE 14.13   A schematic diagram for Example 14.5.

Using Excel's logical functions, create a list that shows the corresponding open and closed position of the check valve (see Figure 14.14).

The solution to Example 14.5 is shown in Figure 14.14. The pressure readings were entered in column A. In cell B3, we type the formula = **IF (A3 >= 20,"OPEN","CLOSED"**) and use the **Fill** command to copy the formula in cells B4 through B10. Note that we made use of the relational operator >= and relative reference in the **IF** function.

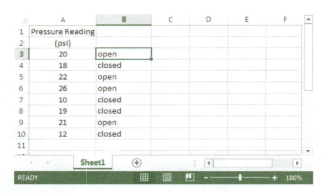

FIGURE 14.14   The solution to Example 14.5.

Answer the following questions to test your understanding of the preceding sections.

Before You Go On

1. What is the difference between a workbook and a worksheet?

2. What is a range?

3. Explain what is meant by absolute cell reference, relative cell reference, and mixed cell reference.

4. Give examples of Excel's mathematical and statistical functions.

5. Give examples of Excel's logical functions and relational operators.

Vocabulary—State the meaning of the following terms:

A Range _____

Absolute Cell Reference _____

Mixed Cell Reference _____

Logical Function _____

LO³ 14.3 Plotting with Excel

Today's spreadsheets offer many choices when it comes to creating charts. You can create column charts (or histograms), pie charts, line charts, or *xy* charts. As an engineering student, and later as a practicing engineer, most of the charts that you will create will be of *xy*-type charts. Therefore, next we will explain in detail how to create an *xy* chart.

Excel offers Chart Wizard, which is a series of **dialog box** that walks you through the necessary steps to create a chart. To create a chart using the Excel Chart Wizard, follow the procedure explained here.

- Select the data range as was explained earlier in this chapter.
- Click the Insert tab.
- Select the **XY (Scatter)** Chart type. The XY Chart type offers five Chart sub-type options. (It is important to note here that the **Line** chart is often mistakenly used instead of **XY (Scatter)).**
- From the four Chart sub-type options, select the **"data points connected by smooth lines"** Chart option.
- Next you can use the Chart Tools (Design, Layout, Format) to modify the chart.
- For example, you can use the **Layout** tools to enter Chart Title, Axis Titles, Gridlines.

You can create column charts (or histograms), pie charts, line charts, or *xy* charts. As an engineer, most of the charts that you will create will be of *xy*-type charts.

When creating an engineering chart, whether you are using Excel or using freehand methods, you must include proper labels with proper units for each axis. The chart must also contain a figure number with a title explaining what the chart represents. If more than one set of data is plotted on the same chart, the chart must also contain a legend or list showing symbols used for different data sets.

EXAMPLE 14.6

Using the results of Example 14.1, create a graph showing the value of air density as a function of temperature.

1. First you will select the data range as shown in Figure 14.15.

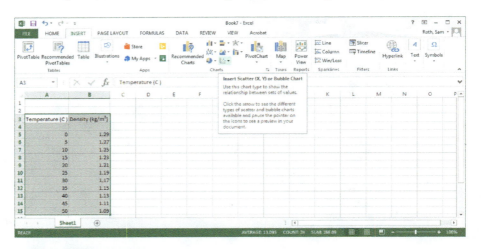

FIGURE 14.15

2. Next, pick the **Insert** tab and then select Scatter (Figure 14.16) with the Smooth Lines and Markers button.

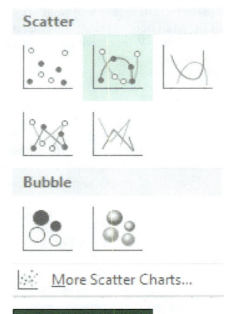

FIGURE 14.16

3. You will now see the chart (Figure 14.17). Next, add X-axis and Y-axis Titles and modify the chart title and gridlines as desired. To do so, choose the Layout tab and click on the Axis Titles button, Chart Title button, or Gridlines.

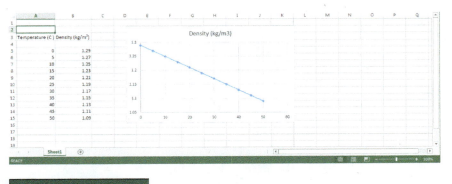

FIGURE 14.17

4. Finally, you can place the chart in an appropriate location as shown in Figure 14.18. If for some reason you need to make changes, pick the item you want to change, right click, and a menu will appear.

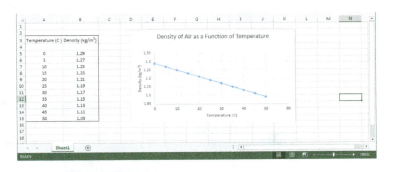

FIGURE 14.18

It is worth noting that you can plot more than one set of data on the same chart. To do so, first pick the chart by clicking anywhere on the chart area, and then from the **Chart** menu use the **Select Data ...** and follow the steps to plot the other data set to the chart.

Plotting Two Sets of Data with Different Ranges on the Same Chart

At times, it is convenient to show the plot of two variables versus the same variable on a single chart. For example, in Figure 14.19, we have shown how air temperature and wind speed change with the same variable time. Using Example 14.7, we will show how you can plot two sets of data series with different ranges on the same chart.

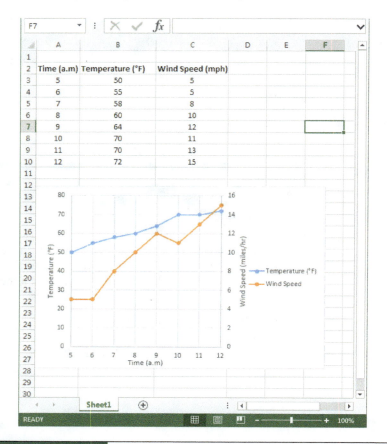

FIGURE 14.19   Using Excel to plot two sets of data with different ranges.

EXAMPLE 14.7

Use the following empirical relationship to plot the fuel consumption in both miles per gallon and gallons per mile for a car for which the following relationship applies. Note: V is the speed of the car in miles per hour and the given relationship is valid for $20 \leq V \leq 75$.

$$\text{Fuel Consumption (Miles per Gallon)} = \frac{1000 \times V}{900 + V^{1.85}}$$

Refer to the Excel sheets shown in the accompanying figures when following the steps.

1. First, in Figure 14.20 using Excel and the given formula, we compute the fuel consumption in miles per gallon and gallons per mile for the given range of speeds. Note that the values in cells C3 through C14 are the inverse of cell values in B3 through B14.
2. We plot the fuel consumption in miles per gallon versus speed, as shown in Figure 14.21.
3. With the mouse pointer in the chart area, click the right mouse button and choose **Select Data** …, also shown in Figure 14.21.
4. In the **Select Data Source,** under Legend Entries, click on the Add button (Figure 14.22), and then type in the series name, and choose the Series X values, and Series Y values as shown in Figure 14.23.

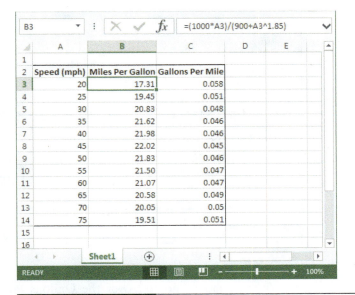

FIGURE 14.20 Fuel consumption computation.

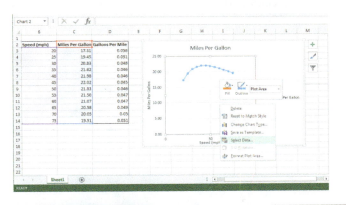

FIGURE 14.21 Plotting fuel consumption.

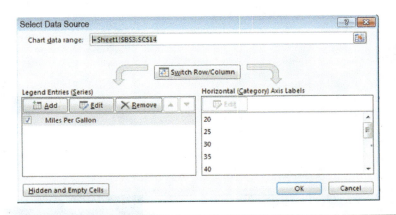

FIGURE 14.22 Select Data Source.

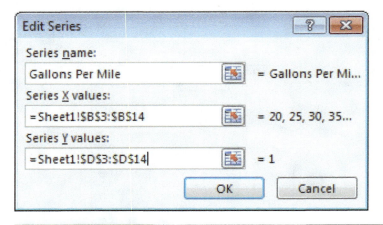

FIGURE 14.23   Choosing Series X and Y values.

5. With the mouse pointer over the Gallons Per Mile curve, double-click the left mouse button. Choose **Format Data Series...,** then under Series Options, turn on the Secondary Axis (Figure 14.24). You may also want to change the line style (Figure 14.25) to dash lines, so when you print your chart, it will be easier to compare the two curves.

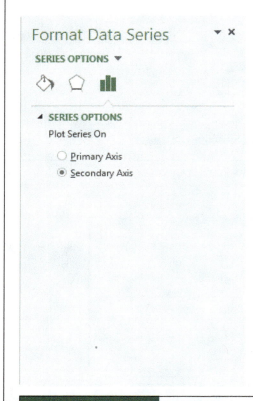

FIGURE 14.24   Format Data Series Options.

FIGURE 14.25   Choosing line style.

The final results for Example 14.7 are shown in Figure 14.26.

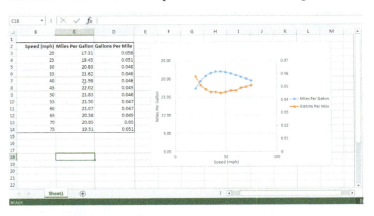

FIGURE 14.26   The final results for Example 14.7.

Curve Fitting with Excel

Curve fitting deals with finding an equation that best fits a set of data. There are a number of techniques that you can use to determine these functions. You will learn about them in your numerical methods and other future engineering classes. The purpose of this section is to demonstrate how to use Excel to find an equation that best fits a set of data which you have plotted. We will demonstrate the curve-fitting capabilities of Excel using the following examples.

> You can use Excel to find an equation that best fits a set of data.

EXAMPLE 14.8

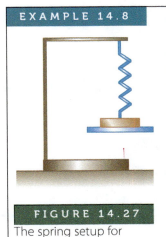

FIGURE 14.27

The spring setup for Example 14.8.

In Chapter 10, we discussed linear springs. We will revisit Example 10.1 to show how you can use Excel to obtain an equation that best fits a set of force–deflection data for a linear spring. For a given spring, in order to determine the value of the spring constant, we attached dead weights to one end of the spring, as shown in Figure 14.27. We have measured and recorded the deflection caused by the corresponding weights, as given in Table 14.8. What is the value of the spring constant?

Recall, the spring constant k is determined by calculating the slope of a force–deflection line (i.e., slope = change in force/change in deflection). We have used Excel to plot the deflection–load results of the experiment using the **XY (Scatter)** without the data points connected, as shown in Figure 14.28. If you were to connect the experimental force–deflection points, you would not obtain a straight line that goes through each experimental point. In this case, you will try to come up with the best fit to the data points. There are mathematical procedures (including least squares techniques) that allow you to find the best fit to a set of data points; Excel makes use of such techniques.

To add the trendline or the best fit, with the mouse pointer over a data point, click the right button and choose **Add Trendline …,** as shown in Figure 14.29. Next, from the **Format Trendline** dialog box, under **Trend/Regression** type, select **Linear,** and toggle on the **Set intercept=** and the **Display equation on chart,** as shown in Figure 14.30.

TABLE 14.8
The Results of the Experiment for Example 14.8

Weight (N)	The Deflection of the Spring (mm)
5.0	9.0
10.0	17.0
15.0	29.0
20.0	35.0

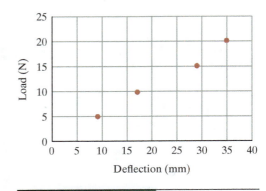

FIGURE 14.28   The XY (Scatter) plot of experienced data points connected.

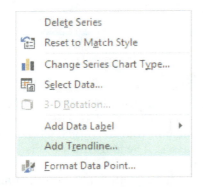

FIGURE 14.29

Add Trendline . . . .

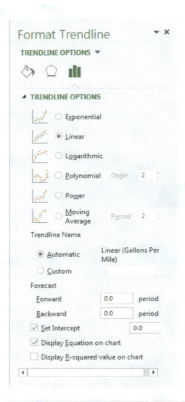

FIGURE 14.30   The Format Trendline dialog box-Type.

After you press close, you should see the equation $y = 0.5542x$ on the chart, as shown in Figure 14.31. We have edited the variables of the equation to reflect the experimental variables as shown in Figure 14.32.

To edit the equation, left-click on the equation ($y = 0.5542x$) and change it to read $F = 0.5542\ x$, where $F = $ load (N), and x = deflection (mm), as shown in Figure 14.32.

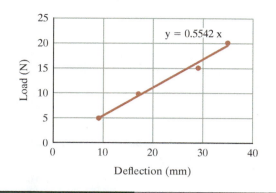

FIGURE 14.31 The linear fit to data of Example 14.8.

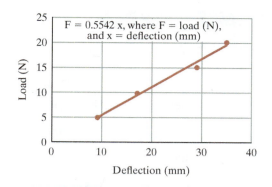

FIGURE 14.32 The edited linear fit to data of Example 14.8.

TABLE 14.9 **The Comparison between the Measured and Predicted Spring Force**

The Measured Deflection of the Spring, x (mm)	Measured Spring Force (N)	The Predicted Force (N) using: $F = 0.5542 x$
9.0	5.0	5.0
17.0	10.0	9.4
29.0	15.0	16.1
35.0	20.0	19.4

In order to examine how well the linear equation $F = 0.5542x$ fits the data, we compare the force results obtained from the equation to the actual data points as shown in Table 14.9. As you can see the equation fits the data reasonably well.

EXAMPLE 14.9 Find the equation that best fits the following set of data points in Table 14.10.

We first plot the data points using the XY (Scatter) without the data points connected, as shown in Figure 14.33.

Right-click on any of the data points to add a trendline. From the plot of the data points, it should be obvious that an equation describing the relationship between x and y is nonlinear. Select a polynomial of second order (Order: 2), toggle on the **Display equation on chart** and the **Display R-squared value on chart,** as shown in Figure 14.34. After you press Close, you should see the equation $y = x^2 - 3x + 2$ and $R^2 = 1$ on the chart, as shown in Figure 14.35. The R^2 is called the coefficient of determination, and its value provides an indication of how good the fit is. $R^2 = 1$ indicates perfect fit, and

TABLE 14.10
A Set of Data Points

X	Y
0.00	2.00
0.50	0.75
1.00	0.00
1.50	− 0.25
2.00	0.00
2.50	0.75
3.00	2.00

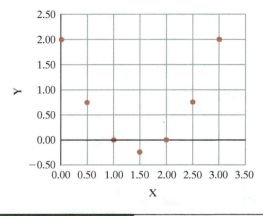

FIGURE 14.33   The XY (Scatter) plot of data points for Example 14.9.

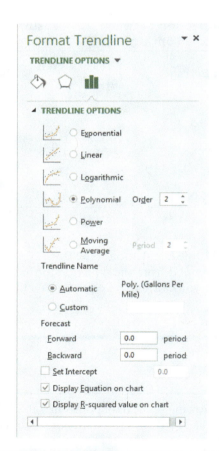

FIGURE 14.34   The Add Trendline dialog box-Type.

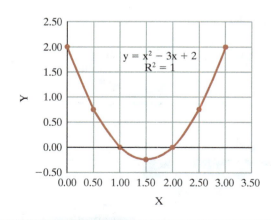

FIGURE 14.35   The results for Example 14.9

R^2 values that are near zero indicate extremely poor fits. The comparison between the actual and predicted *y* values (using equation $y = x^2 - 3x + 2$) is shown in Table 14.11.

TABLE 14.11	The Comparison between Actual and Predicted y Values	
X	**Actual y**	**Predicted value of y using** $y = X^2 - 3x + 2$
0.00	2.00	2.00
0.50	0.75	0.75
1.00	0.00	0.00
1.50	−0.25	−0.25
2.00	0.00	0.00
2.50	0.75	0.75
3.00	2.00	2.00

LO⁴ 14.4 Matrix Computation with Excel

During your engineering education, you will learn about different types of physical variables. There are those that are identifiable by either a single value or the magnitude. For example, time can be described by a single value, such as two hours. These types of physical variables that are identifiable by a single value are called *scalars*. Temperature is another example of a scalar variable. On the other hand, if you were to describe the velocity of a vehicle, you not only have to specify how fast it is moving (speed) but also its direction. The physical variables that possess both magnitude and direction are called *vectors*. There are also other quantities that in order to describe them accurately we need to specify more than two pieces of information. For example, if you were to describe the location (with respect to the entrance of a garage) of a car parked in a multi-story garage, you need to specify the floor (the z coordinate) and then the location of the car on that floor, specifying the section and the row (x and y coordinates). A matrix is often used to describe situations that require many values. A **matrix** is an array of numbers, variables, or mathematical terms. The numbers or the variables that make up the matrix are called the *elements of a matrix*. The *size* of a matrix is defined by its number of rows and columns. A matrix may consists of m rows and n columns. For example,

$$[N] = \begin{bmatrix} 6 & 5 & 9 \\ 1 & 26 & 14 \\ -5 & 8 & 0 \end{bmatrix} \quad \{L\} = \begin{Bmatrix} x \\ y \\ z \end{Bmatrix}$$

Here, matrix $[N]$ is a three by three (or 3×3) matrix whose elements are numbers, and $\{L\}$ is a three by one matrix with its elements representing variables x, y, and z. The $[N]$ is called a square matrix. A *square* matrix has the same number of rows and columns. The element of a matrix is denoted by its location. For example, the element in the first row and the third column of matrix $[N]$ is denoted by n_{13}, which has a value of nine. In this book, we denote a matrix by a **boldface letter** in brackets [] or {}, for example: $[N]$, $[T]$, $\{F\}$, and the elements of matrices are represented by regular lower case letters. The {} brackets are used to distinguish a column matrix. A column matrix is defined as a matrix that has one column but could have many rows. On the other hand, a row matrix is a matrix that has one row but could have many columns.

$$\{A\} = \begin{Bmatrix} 1 \\ 5 \\ -2 \\ 3 \end{Bmatrix} \text{ and } \{X\} = \begin{Bmatrix} x_1 \\ x_2 \\ x_3 \end{Bmatrix}$$

are examples of column matrices, whereas

$$[C] = [5 \quad 0 \quad 2 \quad -3] \text{ and } [Y] = [y_1 \ y_2 \ y_3]$$

are examples of row matrices.

In Section 18.5, we will discuss matrix algebra in more detail. If you do not have an adequate background in matrix algebra, you may want to read Section 18.5 before studying the following examples.

Matrix Algebra

Using Example 14.10, we will show how to use Excel to perform certain matrix operations.

EXAMPLE 14.10

Given matrices: $[A] = \begin{bmatrix} 0 & 5 & 0 \\ 8 & 3 & 7 \\ 9 & -2 & 9 \end{bmatrix}$, $[B] = \begin{bmatrix} 4 & 6 & -2 \\ 7 & 2 & 3 \\ 1 & 3 & -4 \end{bmatrix}$, and $\{C\} = \begin{Bmatrix} -1 \\ 2 \\ 5 \end{Bmatrix}$, use

Excel to perform the following operations.

(a) $[A] + [B] = ?$
(b) $[A] - [B] = ?$
(c) $[A][B] = ?$
(d) $[A]\{C\} = ?$

If you do not have any background in matrix algebra, you may want to study Section 18.5 to learn about matrix operation rules. The manual (hand) calculations for this example problem are also shown in that section.

Refer to the Excel sheets shown in the accompanying figures when following the steps.

1. In the cells shown, type the appropriate characters and values. Use the **Format Cells** and **Font** option to create the boldface variables, as shown in Figure 14.36.
2. In cell A10, type $[A] + [B] =$, and using the left mouse button, pick cells B9 through D11, as shown in Figure 14.37.

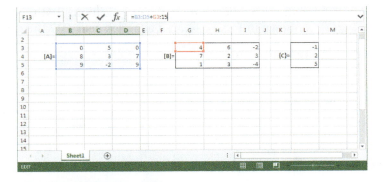

FIGURE 14.36

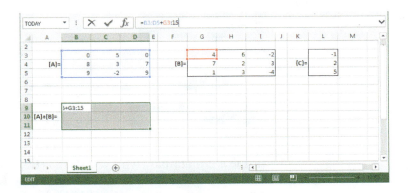

FIGURE 14.37

3. Next, in the Formula bar, type = **B3:D5** + **G3:I5** and, while holding down the **Ctrl** and the **Shift** keys, press the **Enter** key. Note that using the mouse, you could also pick the ranges of B3:D5 or G3:I5 instead of typing them. This sequence of operations will create the result shown in Figure 14.38. You follow a procedure similar to the one outlined in step 2 to perform [*A*] − [*B*], except in the Formula bar type = **B3:D5** − **G3:I5** .

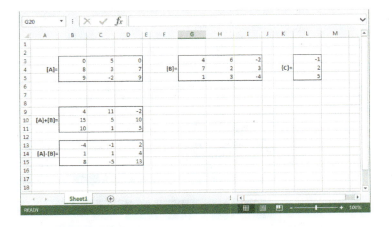

FIGURE 14.38

4. To carry out the matrix multiplication, first type [*A*][*B*] = in cell A18, as shown in Figure 14.39. Then pick cells B17 through D19.

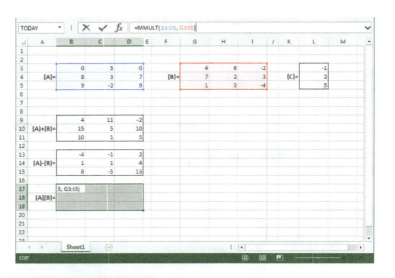

FIGURE 14.39

5. In the Formula bar, type **=MMULT(B3:D5,G3:I5),** and, while holding down the **Ctrl** and the **Shift** keys, press the **Enter** key. Similarly, you can perform the matrix operation [*A*]{*C*}. First you pick cells B21 through B23, in the Formula bar, type **=MMULT(B3:D5, L3:L5),** and, while holding down the **Ctrl** and the **Shift** keys, press the **Enter** key. This sequence of operations will create the result shown in Figure 14.40.

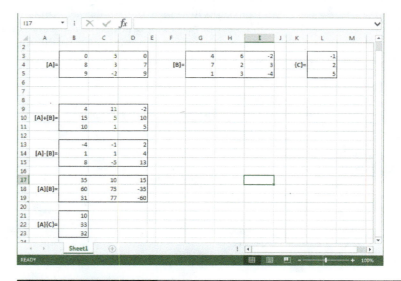

FIGURE 14.40   Results for Example 14.10.

The formulation of many engineering problems leads to a system of algebraic equations. As you will learn later in your math and engineering classes, there are a number of ways that you can use Excel to solve a set of linear equations. Using Example 14.11, we will show how you can use Excel to obtain a solution to a simultaneous set of linear equations.

EXAMPLE 14.11

Consider the following three linear equations with three unknowns: x_1, x_2, and x_3. Our intent here is to show you how to use Excel to solve a set of linear equations.

$$2x_1 + x_2 + x_3 = 13$$
$$3x_1 + 2x_2 + 4x_3 = 32$$
$$5x_1 - x_2 + 3x_3 = 17$$

The solution to this problem is discussed in detail in Chapter 18 and is given by: $x_1 = 2$, $x_2 = 5$, and $x_3 = 4$. The solution easily can be verified by substituting the results (the values of x_1, x_2, and x_3) back into the three linear equations.

$$2(2) + 5 + 4 = 13$$
$$3(2) + 2(5) + 4(4) = 32$$
$$5(2) - 5 + 3(4) = 17 \qquad \textbf{Q.E.D.}$$

Now the Excel solution. Refer to the Excel sheets shown in the accompanying figures when following the steps.

1. In the cells shown in Figure 14.41, type the appropriate characters and values. Use the **Format Cells** and **Font** option to create the bold and subscript variables.

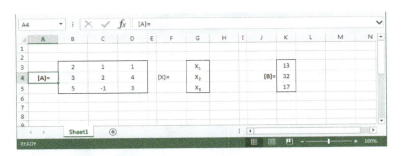

FIGURE 14.41

2. In cell A10, type $[A]^{-1} =$, and using the left mouse button, pick cells B9 through D11 as shown in Figure 14.42.
3. Next, in the formula bar, type **=MINVERSE(B3:D5)** and, while holding down the **Ctrl** and the **Shift** keys, press the **Enter** key. This sequence of operations will create the result shown in Figure 14.43. The inverse of matrix **[A]** is computed.
4. Type the information shown in cells A14, B13 through B15, C14, D14, and E14, as shown in Figure 14.44.

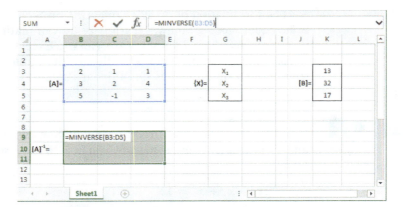

FIGURE 14.42

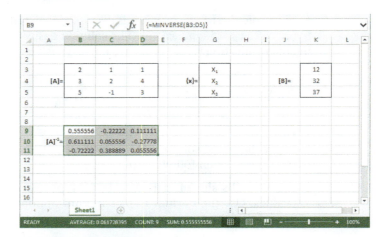

FIGURE 14.43

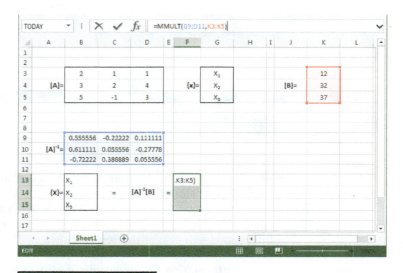

FIGURE 14.44

5. Then, pick cells F13 through F15; in the formula bar, type **=MMULT (B9:D11, K3:K5);** and, while holding down the **Ctrl** and the **Shift** keys, press the **Enter** key. This sequence of operations will create the result shown in Figure 14.45. The values of x_1, x_2, and x_3 are now calculated.

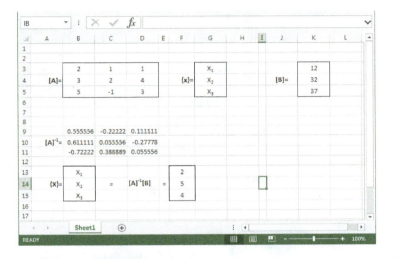

FIGURE 14.45   Results for Example 14.11.

Before You Go On

Answer the following questions to test your understanding of the preceding sections.

1. What types of charts can be created with Excel?

2. What is the most common type of chart in engineering?

3. What is a matrix?

4. Give examples of matrix operations that can be carried out by Excel.

5. How do you solve a set of linear equations with Excel?

Vocabulary—State the meaning of the following terms:

Matrix _____

Element of a Matrix _____

Matrix Algebra _____

LO⁵ 14.5   An Introduction to Excel's Visual Basic for Applications

As mentioned previously, the Excel software consists of two parts: the workbook and the Visual Basic Editor. In previous sections, you learned how to use the Excel workbook environment to solve engineering problems. In this section, we will discuss the **Visual Basic for Applications (VBA)**. Excel's VBA is a programming language that allows you to use Excel more effectively. There are many good textbooks that discuss the capabilities of VBA to solve a full range of problems. Here, our intent is to introduce only some basic ideas so that you can perform some essential operations.

In the following sections, we will explain how to input and retrieve data, display results, create a subroutine, and how to use Excel's built-in functions in your VBA program. We will also explain how to create a loop and use arrays. Creating a custom dialog box is also presented in this section. Moreover, we will use previous examples to emphasize that VBA is just another tool that you can use to solve a variety of engineering problems.

> Visual Basic for Applications (VBA) is a programming language that allows you to use Excel more effectively. With VBA, you can write a program to input and retrieve data using Excel's built-in functions to perform analysis and display results.

Before proceeding with a VBA session, you need to display the **Developer** in your Excel Ribbon. Use the Excel's option to add the **Developer** to your Excel Ribbon. There is an additional item that you need to understand. It relates to the security feature of Microsoft Office. When you create a VBA program (a macro in VBA), Excel appends a special extension of *.xlsm*. To avoid undesirable security problems, in the **Trust Center**, you need to set the Macro settings to **Disable all macros with notification**. It is also important to note that each time you re-open the *.xlsm* file, you need to enable the macros; otherwise, you will not be able to run a module.

A Simple VBA Code

In this example, we write a simple program to convert a temperature value given in a worksheet from degree Celsius to degree Fahrenheit.

1. Open a new Excel workbook, and in Sheet1, type in the content shown in Figure 14.46, and save it as VBA_Lesson_1.xlsm. Make sure to save it with the extension .xlsm.

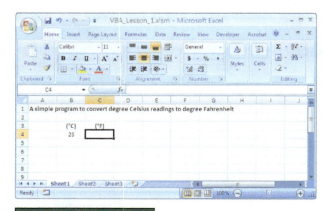

FIGURE 14.46

2. Choose the **Developer** tab, and then launch the **Visual Basic** editor by clicking the **Visual Basic** icon. Next, in the **Visual Basic** editor, click on **Insert** and then choose **Module** and type in the program (code) shown in Figure 14.47. Notice that, after you type **Sub temperature_conversion ()** and when you hit the Enter key, the VBA Editor will automatically place an End Sub at the end of the page. As you type the program, notice the different colors that are used by the editor. Each color conveys a different meaning. For example, the blue color is used for keywords, red for errors, and green is used for documentation and comments. The single quotation mark at the beginning of a line indicates a comment, and as a result, that line will not be treated as a command and will not be executed.

```
Sub temperature_conversion()
'This statement starts the subroutine and names it
temperature_conversion
Sheets("sheet1").Select
'This statement selects sheet1
Range("B4").Select
'This statement selects cell B4
deg_C = ActiveCell.Value
'This statement assigns the value in cell B4 to a
variable that we named deg_C
deg_F = (9/5) * deg_C + 32
'This statement makes the conversion from degree
Celsius to degree Fahrenheit
'It converts the deg_C value to degree Fahrenheit
and assigns it to a variable we named deg_F
Range("C4").Select
'This statement selects cell C4
ActiveCell.Value = deg_F
'This statement assigns the value of the variable
deg_F to the active cell, which is cell C4
End Sub
'The last statement ends the subroutine
```

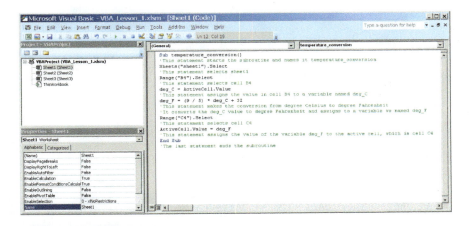

FIGURE 14.47

3. To run the code, click on **Run** and then **Run Sub/UserForm**. You will then see the result in Sheet1, as shown in Figure 14.48.

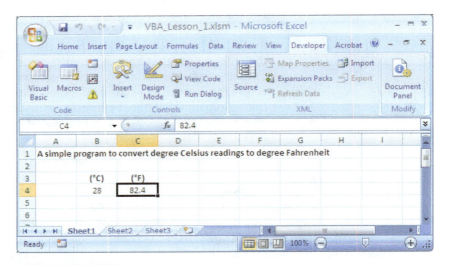

FIGURE 14.48

4. Delete the B4 and C4 values of 28 and 82.4, type in a different value in B4, and run the code again.

Using Excel Functions in VBA

In this example, we write a simple program to make use of Excel's built-in function. We will use the Average function to calculate the mean air density over a given temperature range, using the data from Example 14.1.

1. Open a new Excel Workbook, and in Sheet1 type in the content shown in Figure 14.49. Save it as VBA_Lesson_2.xlsm.

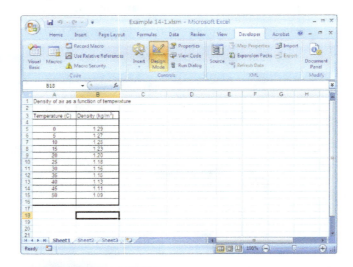

FIGURE 14.49

2. Launch the Visual Basic editor, click on **Insert**, and then choose **Module** and type in the code shown here (see also Figure 14.50).

```vba
Sub Using_Excel_Builtin_Function_Average()
Sheets("sheet1").Select
Average_Density = Application.WorksheetFunction.Average(Range("B5:B15"))
'This statement selects the "Average", a built-in function of Excel, to calculate the
'average of density values in cells B5 through B15 and assigns the result to the variable Average_Density
MsgBox "The average air density is = " & Average_Density & " (kg/m^3)"
'This statement uses the Message Box (MsgBox) to display both text and the average value.
'Notice the text that is to be displayed must be enclosed in double quotation marks.
'Notice the ampersand (&) is used to include additional information such as numeric values or other texts
'Notice that in order to display numeric values, you need not to enclose the variable in double quotation marks.
End Sub
```

FIGURE 14.50

3. To run the code, click on **Run** and then **Run Sub/UserForm.** You will then see the result in Figure 14.51.

FIGURE 14.51

4. If you would like to display the result in Sheet1, say, cell B18, then you need to remove the command

```
MsgBox "The average air density is = "& Average _
Density & " (kg/m^3)"
```

and in its place issue the command

```
Range("B18").Select
ActiveCell.value=Average _ Density
```

as shown in Figure 14.52. The final results are given in Figure 14.53.

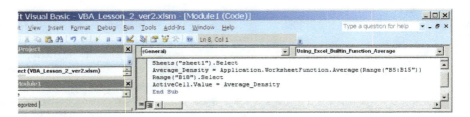

FIGURE 14.52

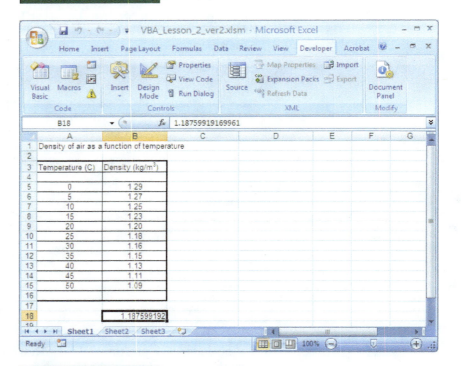

FIGURE 14.53

Learning by Recording Macros

A simple way to learn about the VBA's object-oriented commands and syntax is by using the **Record Macro** option. Let's use Worksheet1 from Example 14.1 to demonstrate how you may use the **Record Macro** to learn about VBA's commands and syntax.

1. Open the previous file, VBA_Lesson_2.xlsm, and save it as VBA_Lesson_3.xlsm.

2. Click on the **Developer** tab, and then in the Code section, click on **Record Macro**. A dialog box will appear next. In the Record Macro dialog box, for the Macro name type in *ChangeFormat*; in the Shortcut key block (after Ctrl+) type, *F* and for the Description type, *Change Text Format to Bold and Size 14*. The information that you need to type in the Record Macro fields is shown in Figure 14.54.

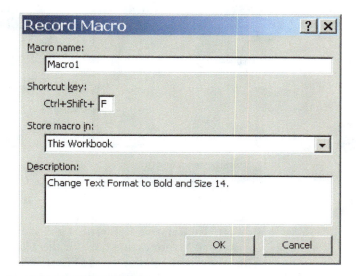

FIGURE 14.54

3. Next, choose cell A1; then right mouse-click Format Cells … and Font, change the font style to Bold and size to 14, and then click OK.

4. Click on **Stop Recording** in the Code section.

5. Now you can view the code that was created by going to the VBA editor and double-clicking on Module2. You should see the code shown in Figure 14.55.

Note that, in order to make simple format changes to cell A1, you first need to select the cell A1 (the object) by issuing the command **Range("A1").Select**. Then you can change the properties of the object by issuing commands such as

```
With Selection.Font
  .Name = "Arial"
  .FontStyle = "Bold"
  .Size = 14
End With
```

Also, it is important to note that, in this example, we have made use of an absolute reference. This means that the macro that we have created only applies to cell A1. For example, if you were to select cell A3 and use the shortcut key **Ctrl+Shift+F**, no change in formatting will occur. If you would like to use the macro to change the format of other cells, you need to make use of a relative reference using the following step.

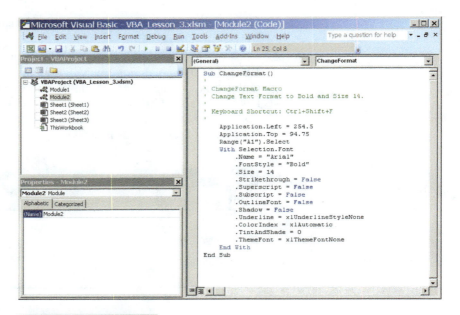

6. Click on the **Developer** tab and then in the Code section click first on **Use Relative References** and then on **Record Macro**. In the **Record Macro** dialog box, for the *Macro name*, type in **ChangeFormatRelative**; in the Shortcut key block (after Ctrl+) type, R; and for the Description, type **Change Text Format Using Relative Reference**.

The results are shown in Figure 14.56.

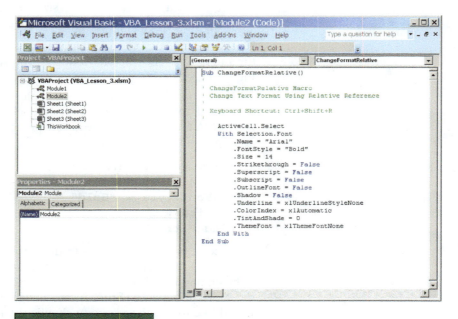

Note that compared to the previous example, instead of **Range("A1"). Select**, the active cell is now selected and the command **ActiveCell.Select** is issued. Also note that the rest of the code looks identical to the previous example. To try the new macro, select, say, cell A3 and apply the shortcut key **Ctrl+Shift+R**. The font size of the text Temperature will change to 14 and the font style will become Bold now.

VBA'S Object-Oriented Syntax

VBA's code makes use of **object-oriented** syntax. To shed light on what we mean by object-oriented syntax and object-oriented programming, consider the following example. A multi-story apartment building—we will call it "EngineersHouse"—could consist of 50 individual apartments (Apartment #1, Apartment #2,..., Apartment #49, and Apartment #50). The entire building (EngineersHouse) is considered an object. Moreover, each apartment also may be considered an object containing additional objects, such as lighting systems, heating and cooling units, appliances, etc., with different properties that we can change. For example, at a certain instant, Apartment#27 could have the following settings.

```
EngineersHourse.Apartment#27.LivingRoom.Lighting = on
```

or

```
EngineersHourse.Apartment#27.Kitechen.Refrigerator.
Temperatature.Value = 5
```

Similarly, using VBA's object-oriented syntax, you can change the properties of various objects. For example, in our *VBA_Lesson_1.xlsm* when you issued the command **Sheets("sheet1").Select**, you selected the object called Sheet1, or when you issued the commands **Range("B4").Select** and **deg_C = ActiveCell.Value**, you selected cell B4 and assigned its value (content) to a variable named **deg_C**. As another example, regarding macros, we selected cell A1 (the object) and then changed the font style and the font size by issuing the following commands.

```
Range("A1").Select
With Selection.Font
        .Name = "Arial"
        .FontStyle = "Bold"
        .Size = 14
End With
```

As we mentioned previously, the intent of this section is to provide some of the basic ideas about VBA. For detailed coverage, you need to see textbooks that are devoted entirely to VBA. Now that you have some understanding of object-oriented syntax, we will discuss an example of a repetitive loop in VBA next.

An Example of a For Loop

As we mentioned previously, as engineers, it often becomes necessary when you write a computer program to execute a line or a block of your computer code many times. VBA provides the **for** command for such situations.

Using the **for loop**, you can execute a line or a block of code a specified (defined) number of times. The syntax of a for loop is

```
for index = start-value to end-value
a line or a block of your computer code
next
```

For example, suppose you want to evaluate the function $y = x^2 + 10$ for x values of 22.00, 22.50, 23.00, 23.50, and 24.00. This operation will result in corresponding y values of 494.00, 516.25, 539.00, 562.25, and 586.00. The VBA code for this example then would have the form shown in Figure 14.57.

```vba
Sub for_loop_example()
Worksheets("Sheet1").Activate
'This statement activates Sheet1
Range("A1,A6").Activate
'This Statement activates cells A1 through A6
Range("B1,B6").Activate
'This Statement activates cells B1 through B6
Cells(1, 1) = "x values"
'This statement writes the text x values in the
cell located in row 1, column 1 (column A) of
Sheet1
Cells(1, 2) = "y values"
'This statement writes the text y values in the
cell located in row 1, column 2 (column B) of
Sheet1
x = 22
For i = 1 To 5
y = x ^ 2 + 10
Cells(i + 1, "A") = x
'This statement assigns (writes) the x value to the
cell located in row i+1 and column A;
'As the i values changes so does the row number
Cells(i + 1, "B") = y
'This statement assigns (writes) the y value to the
cell located in row i+1 and column B;
'As the i values changes so does the row number
x = x + 0.5
Next i
End Sub
```

Now, if you run the code, you should see the results in Sheet1 in Figure 14.58.

Note that in this example, the index is the integer i and its start value is 1. It is incremented by a value of 1, and its end value is 5.

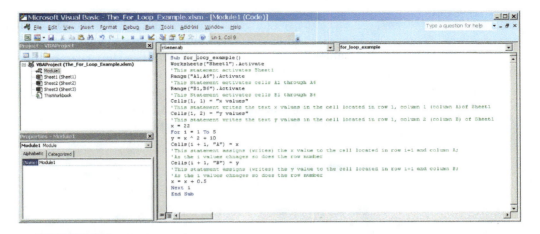

FIGURE 14.57

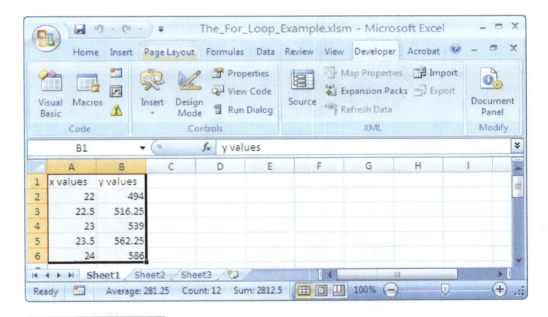

FIGURE 14.58

A Custom-Made Dialog Box

Next, we will use Example 20.8 from Chapter 20 to demonstrate how to create a custom-made dialog box in Example 14.11.

EXAMPLE 14.12

Determine the monthly payments for a five-year, $10,000 loan at an interest rate of 8% compounding monthly. (To calculate the monthly payments, see Chapter 20. Please review Equation (20.5) to see how the monthly payments are related to loan, interest rate, and time.)

1. Start a new Excel Workbook and name it Example 14.11.xlsm.
2. Click on **Developer** and then **Visual Basic**.
3. Click on **Insert** and then **UserForm** (Figure 14.59).

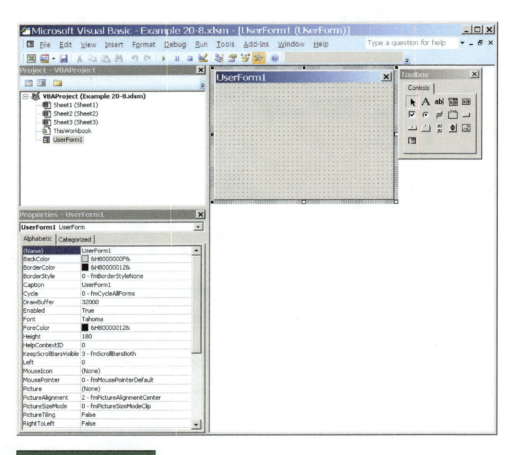

FIGURE 14.59

4. In the Properties window, under **Alphabetic** and (Name), assign a name such as Loan_Payments. This is the name of the UserForm that we are creating. Notice no space is allowed in the Name field.
5. In the Properties window, under **Alphabetic** and Caption, assign a name such as Monthly Loan Payment Calculator (Figure 14.60). This is the name of the dialog box. Notice spaces are allowed in the dialog box name.
6. Next, create the labels and corresponding input boxes for Loan Amount ($), Interest Rate (%), and Years. In the Toolbox, click on the Label tool (the big **A** button in Figure 14.61(a)). You should see a cross-hair now; move it over to the Monthly Loan Payment Calculator dialog box and then press the left mouse button and draw a box (Figure 14.61(b)) and resize as desired.

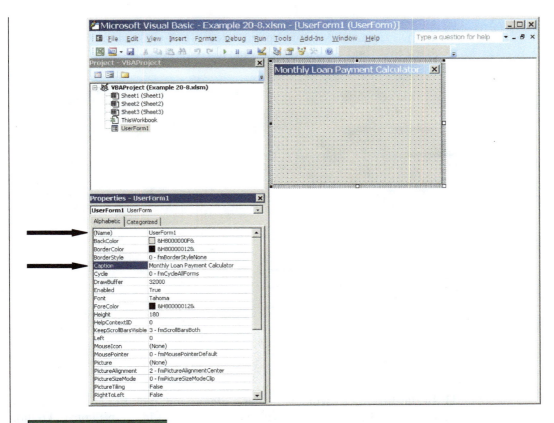

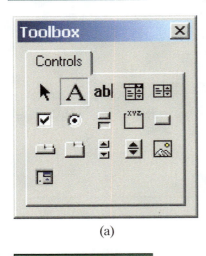

(a)

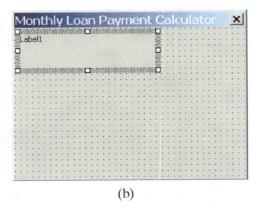

(b)

7. In the Properties window shown in Figure 14.62(a), under **Alphabetic** and (Name), assign a name such as Loan. This is the name for the label. Notice no space is allowed in the Name field.

8. In the Properties window, under **Alphabetic**, and Caption, assign a name such as Loan Amount ($). This is the name of the label box (Figure 14.62(b)). Notice space is allowed in the box name.

9. Change the BackColor property as shown.
10. In the Properties window, under **Categorized**, click on the BackColor and Back Style and change their properties as shown.

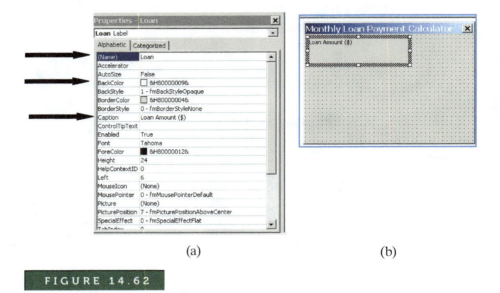

(a)   (b)

FIGURE 14.62

11. In the Properties window, under **Categorized**, pick **Font** and choose the desired style and size. We will choose Bold and 14, as in Figure 14.63(a).

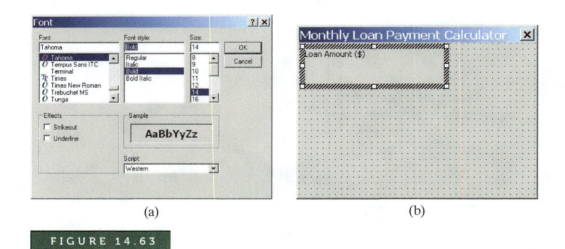

(a)   (b)

FIGURE 14.63

12. Next, create an input box(a) next to the Loan Amount ($) label box. In the Toolbox, click on the TextBox (the **ab** button)(Figure 14.61(a)). You should see a cross-hair now; move it to the right of the Loan Amount ($) label box, and then press the left mouse button, draw a box, and resize as desired. Name this textbox Loan_Amount (Figure 14.64(a)). Also, change the BackColor property as shown in Figure 14.64(b).

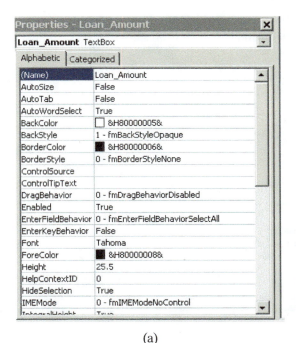

(a)

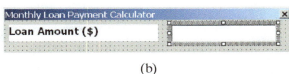

(b)

FIGURE 14.64

13. In a similar way, create a label and text input boxes for Interest Rate (%). For the **Name,** use Interest_Rate, and for the **Caption**, use Interest Rate (%). Name the corresponding textbox Nominal_Interest_Rate.
14. Similarly, create a label and text input boxes for Years. Use Years for the **Name.** For the **Caption,** use Years, and name the corresponding textbox Number_of_Years.
15. Create one more label and text box. For the **Name,** use Monthly_Payments, and for the **Caption,** use Monthly Loan Payments. Name the corresponding text box Payments.

The results are shown in Figure 14.65.

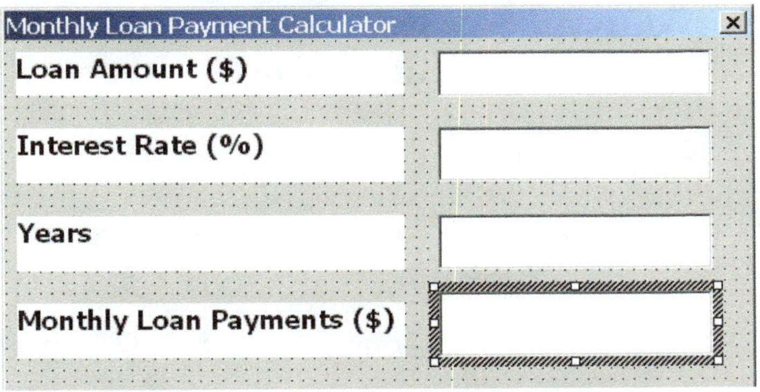

Monthly Loan Payment Calculator

Loan Amount ($)

Interest Rate (%)

Years

Monthly Loan Payments ($)

FIGURE 14.65

16. Next, using the Toolbox Command button shown in Figure 14.66, create a Calculate and a Done button.

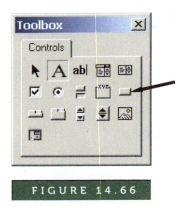

FIGURE 14.66

17. Use the Visual Basic Text Editor and type the code given in Figure 14.67.

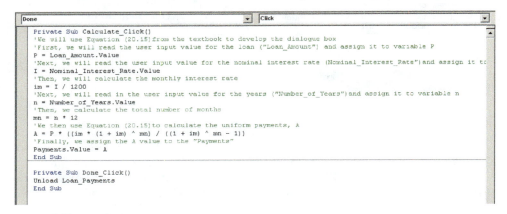

FIGURE 14.67

18. Run the code. Input the following data, and then click on the Calculate button shown in Figure 14.68.

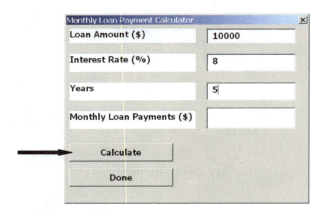

FIGURE 14.68

19. The monthly loan payment is now calculated and displayed as in Figure 14.69. Click on the Done button to exit the code.

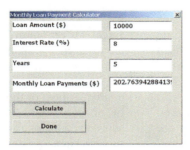

FIGURE 14.69

20. Now, see if you can change the properties of the dialogue, label, and text boxes to create a Monthly Loan Payment Calculator that looks like the one shown below.

FIGURE 14.70

Before You Go On

Answer the following questions to test your understanding of the preceding sections.

1. What is VBA?

2. What is a subroutine?

3. What is a macro?

4. Explain what is meant by object-oriented syntax.

Vocabulary—State the meaning of the following terms:

Object-Oriented Programming_____

A Macro_____

A For Loop_____

An Object Property_____

SUMMARY

LO¹ Microsoft Excel Basics

Spreadsheets are used to record, organize, and analyze data using formulas. They are also used to present the results of an analysis in chart forms. A workbook—the spreadsheet file—is divided into worksheets, and each worksheet is divided into columns, rows, and cells. The columns are marked by A, B, C, D, and so on, while the rows are identified by numbers 1, 2, 3, 4, and so on. A cell represents the box that one sees as the result of the intersection of a row and a column. The cells that are selected simultaneously are called a range. When creating formulas, you have to be careful how you refer to the address of a cell, especially if you are planning to use the **Fill** command to copy the pattern of formulas in the other cells. There are three ways that you can refer to a cell address in a formula: absolute, relative, and mixed references. By now, you also know that you can use Excel to input engineering formulas and compute the results.

LO² Excel Functions

Excel offers a large selection of built-in functions that you can use to analyze data. By built-in functions, we mean standard functions such as the sine or cosine of an angle as well as formulas that calculate the total value, the average value, or standard deviation of a set of data points. The Excel functions are grouped into various categories, including mathematical and trigonometric, engineering, statistical, financial, and logical functions. For example, logical functions allow you to test various conditions when programming formulas to analyze data.

LO³ Plotting with Excel

Excel offers many choices when it comes to creating charts. You can create column charts (or histograms), pie charts, line charts, or *xy* charts. As an engineering student, and later as a practicing engineer, most of the charts that you will create will be of *xy*-type charts. At times, it is convenient to show the plot of two or more variables versus the same variable on a single chart. By now, you should know how to create such plots. You should also know how to use Excel to find an equation that best fits a set of data.

LO⁴ Matrix Computation with Excel

You should also know how to use Excel to perform matrix operations such as addition, subtraction, multiplication, and solving a set of linear equations. When performing matrix computation with Excel, you should have noticed that you need to follow a sequence of operations that include holding the Ctrl and Shift keys before pressing the Enter key.

LO⁵ An Introduction to Excel's Visual Basic for Applications

Excel software consists of two parts: the workbook and the Visual Basic Editor. Visual Basic for Applications (VBA) is a programming language that allows you to use Excel more effectively. By now, you should know how to create a subroutine, input and retrieve data, use Excel's built-in functions in your program, and display the results. You should also feel comfortable with creating loops, using arrays, and making a custom dialog box to solve a common problem.

KEY TERMS

Absolute Cell Reference 491
Cell 485
Column 485
Curve Fitting 505
Developer 516
Dialog Box 499

For Loop 524
Matrix 509
Matrix Algebra 510
Mixed Cell Reference 491
Object-Oriented Programming 523
Range 486

Record Macro 520
Relative Cell Reference 491
Row 485
Stop Recording 521
VBA 516

APPLY WHAT YOU HAVE LEARNED

Atmospheric pressure is commonly expressed in one of the following units: pascals (Pa), pound per square inch (lb/in²), millimeters of mercury (mm.Hg), and inches of mercury (in.Hg). Use VBA and create a custom-made dialog box to convert values of atmospheric pressure from pound per square inch to pascals, millimeters of mercury, and inches of mercury.

PROBLEMS

Problems that promote lifelong learning are denoted by 🔑

14.1 Using the Excel **Help** menu, discuss how the following functions are used. Create a simple example and demonstrate the proper use of the function.

    a. TRUNC(number, num_digits)

    b. ROUND(number, num_digits)

    c. COMBIN(number, number_chosen)

    d. DEGREES(angle)

    e. SLOPE(known_y's, known_x's)

    f. CEILING(number, significance)

14.2 In Chapter 20, we will cover engineering economics. For now, using the Excel **Help** menu, familiarize yourself with the following functions. Create a simple example and demonstrate the proper use of the function.

    a. FV(rate, nper, pmt, pv, type)

    b. IPMT(rate, per, nper, pv, fv, type)

    c. NPER(rate, pmt, pv, fv, type)

    d. PV(rate, nper, pmt, fv, type)

14.3 In Chapter 10, we discussed fluid pressure and the role of water towers in small towns. Recall that the function of a water tower is to create a desirable municipal water pressure for household and other usage in a town. To achieve this purpose, water is stored in large quantities in elevated tanks. Also recall that the municipal water pressure may vary from town to town, but it generally falls somewhere between 50 and 80 lb/in² (psi).

In this assignment, use Excel to create a table that shows the relationship between the height of water above ground in the water tower and the water pressure in a pipeline located at the base of the water tower. The relationship is given by

$$P = \rho g h$$

where

P = the water pressure at the base of the water tower in pounds per square foot (lb/ft²)

ρ = the density of water in slugs per cubic foot, $\rho = 1.94$ slugs/ft³

g = the acceleration due to gravity, $g = 32.2$ ft/s²

h = the height of water above ground in feet (ft)

Create a table that shows the water pressure in lb/in² in a pipe located at the base of the water tower as you vary the height of water in increments of 10 ft. Also plot water pressure (lb/in²) vs. the height of water in feet. What should be the water level in the water tower to create 80 psi water pressure in a pipe at the base of the water tower?

14.4 As we explained in Chapter 10, viscosity is a measure of how easily a fluid flows. For example, honey has a higher value of viscosity than does water because if you were to pour water and honey side by side on an inclined surface, the water will flow faster. The viscosity of a fluid plays a significant role in the analysis of many fluid dynamics problems. The viscosity of water can be determined from the following correlation:

$$\mu = c_1 10^{c_2/(T-c_3)}$$

where

$$\mu = \text{viscosity } (\text{N/s} \cdot \text{m}^2)$$
$$T = \text{temperature } (\text{K})$$
$$c_1 = 2.414 \times 10^{-5} \, (\text{N/s} \cdot \text{m}^2)$$
$$c_2 = 247.8 \, \text{K}$$
$$c_3 = 140 \, \text{K}$$

Using Excel, create a table that shows the viscosity of water as a function of temperature in the range of $0°$ C (273.15 K) to $100°$ C (373.15 K) in increments of $5°$ C. Also create a graph showing the value of viscosity as a function of temperature.

14.5 Using Excel, create a table that shows the relationship between the units of temperature in degrees Celsius and Fahrenheit in the range of -50 to $150°$ C. Use increments of $10°$ C.

14.6 Using Excel, create a table that shows the relationship among the units of height of people in centimeters, inches, and feet in the range of 150 cm to 2 m. Use increments of 5 cm.

14.7 Using Excel, create a table that shows the relationship among the units of mass to describe people's mass in kilogram, slugs, and pound mass in the range of 20 kg to 120 kg. Use increments of 5 kg.

14.8 Using Excel, create a table that shows the relationship among the units of pressure in Pa, psi, and inches of water in the range of 1000 to 10,000 Pa. Use increments of 500 Pa.

14.9 Using Excel, create a table that shows the relationship between the units of pressure in Pa and psi in the range of 10 kPa to 100 kPa. Use increments of 5 kPa.

14.10 Using Excel, create a table that shows the relationship between the units of power in watts and horsepower in the range of 100 W to 10,000 W. Use smaller increments of 100 W up to 1000 W, and then use increments of 1000 W all the way up to 10,000 W.

14.11 As we explained in Chapter 7, the air resistance to the motion of a vehicle is something important that engineers investigate. As you may also know, the drag force acting on a car is determined experimentally by placing the car in a wind tunnel. The air speed inside the tunnel is changed, and the drag force acting on the car is measured. For a given car, the experimental data is generally represented by a single coefficient that is called the *drag coefficient*. It is defined by the following relationship:

$$C_d = \frac{F_d}{\frac{1}{2}\rho V^2 A}$$

where

$$C_d = \text{drag coefficient (unitless)}$$
$$F_d = \text{measured drag force (N or lb)}$$
$$\rho = \text{air density } (\text{kg/m}^3 \text{ or slugs/ft}^3)$$

	0	5	10	15	20	25	30	35	40	45

Table-1 Power requirement (kW)

Car speed (m/s) / Ambient Temperature (C)

Car speed (m/s)	0	5	10	15	20	25	30	35	40	45
15	2.0	2.0	2.0	1.9	1.9	1.9	1.8	1.8	1.8	1.7
20	4.8	4.7	4.7	4.6	4.5	4.4	4.3	4.3	4.2	4.1
25	9.4	9.3	9.1	8.9	8.8	8.6	8.5	8.4	8.2	8.1
30	16.3	16.0	15.7	15.4	15.2	14.9	14.7	14.4	14.2	14.0
35	25.9	25.4	24.9	24.5	24.1	23.7	23.3	22.9	22.6	22.2

Table-2 Power requirement (hp)

	0	5	10	15	20	25	30	35	40	45
15	2.7	2.7	2.6	2.6	2.5	2.5	2.5	2.4	2.4	2.3
20	6.5	6.4	6.2	6.1	6.0	5.9	5.8	5.7	5.6	5.6
25	12.6	12.4	12.2	12.0	11.8	11.6	11.4	11.2	11.0	10.8
30	21.8	21.4	21.1	20.7	20.3	20.0	19.7	19.4	19.0	18.7
35	34.7	34.0	33.4	32.9	32.3	31.8	31.2	30.7	30.2	29.8

Problem 14.11

V = air speed inside the wind tunnel (m/s or ft/s)

A = frontal area of the car (m² or ft²)

The frontal area A represents the frontal projection of the car's area and could be approximated simply by multiplying 0.85 times the width and the height of a rectangle that outlines the front of a car. This is the area that you see when you view the car from a direction normal to the front grills. The 0.85 factor is used to adjust for rounded corners, open space below the bumper, and so on. To give you some idea, typical drag coefficient values for sports cars are between 0.27 to 0.38, and for sedans are between 0.34 to 0.5.

The power requirement to overcome air resistance is computed by

$$P = F_d V$$

where

P = power (watts or ft · lbf/s)

1 horsepower (hp) = 550 ft · lbf/s

and

1 horsepower (hp) = 746 W

The purpose of this exercise is to see how the power requirement changes with the car speed and the air temperature. Determine the power requirement to overcome air resistance for a car that has a listed drag coefficient of 0.4 and width of 74.4 in. and height of 57.4 in. Vary the air speed in the range of 15 m/s < V < 35 m/s, and change the air density range of 1.11 kg/m³ < ρ < 1.29 kg/m³. The given air density range corresponds to 0° to 45° C. You may use the ideal gas law to relate the density of the air to its temperature. Present your findings in both kilowatts and horsepower as shown in the accompanying spreadsheet. Discuss your findings in terms of power consumption as a function of speed and air temperature.

14.12 The cantilevered beam shown in the accompanying figure is used to support a load acting on a balcony. The deflection of the centerline of the beam is given by the following equation:

$$y = \frac{-wx^2}{24EI}(x^2 - 4Lx + 6L^2)$$

where

y = deflection at a given x location (m)

w = distributed load (N/m)

E = modulus of elasticity (N/m²)

I = second moment of area (m⁴)

x = distance from the support as shown (x)

L = length of the beam (m)

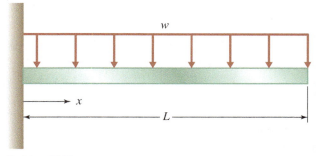

Problem 14.12

Using Excel, plot the deflection of a beam whose length is 5 m with the modulus of elasticity of E = 200 GPa and I = 99.1 × 10⁶ mm⁴. The beam is designed to carry a load of 10,000 N/m. What is the maximum deflection of the beam?

14.13 Fins, or extended surfaces, are commonly used in a variety of engineering applications to enhance cooling. Common examples include a motorcycle or lawn mower engine head, extended surfaces used in electronic equipment, and finned tube heat exchangers in room heating and cooling applications. Consider a rectangular profile of aluminum fins shown in the accompanying figure, which are used to remove heat from a surface whose temperature is 100° C (T_{base} = 100° C). The temperature of the ambient air is 20° C. We are interested in determining how the temperature of the fin varies along its length and plotting this temperature variation. For long fins, the temperature distribution along the fin is given by

$$T - T_{ambient} = (T_{base} - T_{ambient})e^{-mx}$$

where

$$m = \sqrt{\frac{hp}{kA}}$$

h = the heat transfer coefficient (W/m² · K)

p = perimeter of the fin $2 \cdot (a + b)$, (m)

A = cross-sectional area of the fin $(a \cdot b)$, (m^2)

k = thermal conductivity of the fin material $(W/m \cdot K)$

Plot the temperature distribution along the fin using the following data: $k = 168$ W/m $\cdot$ K, $h = 12$ W/m^2 $\cdot$ K, $a = 0.05$ m, $b = 0.01$ m. Vary x from 0 to 0.1 m in increments of 0.01 m.

Problem 14.13

14.14 A person by the name of Huebscher developed a relationship between the equivalent size of round ducts and rectangular ducts according to

$$D = 1.3 \frac{(ab)^{0.625}}{(a + b)^{0.25}}$$

where

D = diameter of equivalent circular duct (mm)

a = dimension of one side of the rectangular duct (mm)

b = the other dimension of the rectangular duct (mm)

Using Excel, create a table that shows the relationship between the circular and the rectangular duct dimensions, similar to the one shown in the accompanying table.

	Length of One Side of Rectangular Duct (length a), mm				
Length b	**100**	**125**	**150**	**175**	**200**
400	207				
450					
500					
550					
600					

14.15 A Pitot tube is a device commonly used in a wind tunnel to measure the speed of the air flowing over a model. The air speed is measured from the following equation:

$$V = \sqrt{\frac{2 P_d}{\rho}}$$

where

V = air speed (m/s)

P_d = dynamic pressure (Pa)

ρ = density of air (1.23 kg/m^3)

Using Excel, create a table that shows the air speed for the range of dynamic pressure of 500 to 800 Pa. Use increments of 50 Pa.

14.16 Use Excel to solve Example 7.1. Recall we applied the trapezoidal rule to determine the area of the shape given.

14.17 We will discuss engineering economics in Chapter 20. Using Excel, create a table that can be used to look up monthly payments on a car loan for a period of five years. The monthly payments are calculated from

$$A = P \left[\frac{\left(\frac{i}{1200}\right)\left(1 + \frac{i}{1200}\right)^{60}}{\left(1 + \frac{i}{1200}\right)^{60} - 1} \right]$$

where

A = monthly payments in dollars

P = the loan in dollars

i = interest rate, e.g., 7, 7.5, ..., 9

	Interest Rate				
Loan	**7**	**7.5**	**8**	**8.5**	**9**
10,000					
15,000					
20,000					
25,000					

14.18 A person by the name of Sutterland has developed a correlation that can be used to evaluate the viscosity of air as a function of temperature. It is given by

$$\mu = \frac{c_1 T^{0.5}}{1 + \dfrac{c_2}{T}}$$

where

$$\mu = \text{viscosity (N/s} \cdot \text{m}^2)$$

$$T = \text{temperature (K)}$$

$$c_1 = 1.458 \times 10^{-6} \left(\frac{\text{kg}}{\text{m} \cdot \text{s} \cdot \text{K}^{1/2}} \right)$$

$$c_2 = 110.4 \text{ K}$$

Create a table that shows the viscosity of air as a function of temperature in the range of $0°\,C\,(273.15\,K)$ to $100°\,C\,(373.15\,K)$ in increments of $5°\,C$. Also create a graph showing the value of viscosity as a function of temperature as shown in the accompanying spreadsheet.

14.19 In Chapter 11, we explained the concept of windchill factors. We said that the heat transfer rates from your body to the surroundings increase on a cold, windy day. Simply stated, you lose more body heat on the cold, windy day than you do on a calm day. The windchill index accounts for the combined effect of wind speed and the air temperature. It accounts for the additional body heat loss that occurs on a cold, windy day. The old windchill values were determined empirically, and a common correlation used to determine the windchill index was

$$WCI = (10.45 - V + 10\sqrt{V})(33 - T_a)$$

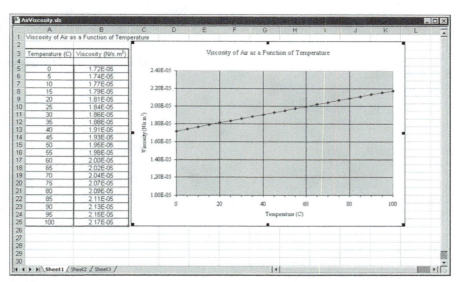

Problem 14.18

Problem 14.19

where

WCI = windchill index $(\text{kcal/m}^2 \cdot \text{h})$

V = wind speed (m/s)

T_a = ambient air temperature $(^\circ\text{C})$

and the value 33 is the body surface temperature in degrees Celsius.

The more common equivalent windchill temperature $T_{equivalent}$ $(^\circ\text{C})$ was given by

$$T_{equivalent} = 0.045\left(5.27\,V^{0.5} + 10.45 - 0.28\,V\right) \times \left(T_a - 33\right) + 33$$

Note that V is expressed in km/h.

Create a table that shows the windchill temperatures for the range of ambient air temperature $-30^\circ\,\text{C} < T_a < 10^\circ\,\text{C}$ and wind speed of $20\,\text{km/h} < V < 80\,\text{km/h}$, as shown in the accompanying spreadsheet.

14.20 Use the data given in Figure 14.10 and duplicate the chart shown there.

14.21 Use Excel to plot the following data. Use two different y axes. Use a scale of zero to 100° C for temperature, and zero to 12 mph for wind speed.

Time (P.M.)	Temperature (°F)	Wind Speed (mph)
1	75	4
2	80	5
3	82	8
4	82	5
5	78	5
6	75	4
7	70	3
8	68	3

14.22 Use Excel to plot the following data for a pump. Use two different y axes. Use a scale of zero to 140 ft for the head and zero to 100 for efficiency.

Flow Rate (GPM)	Head of Pump (ft)	Efficiency (%)
0	120	1
2	119.2	10
4	116.8	30
6	112.8	50
8	107.2	70
10	100	80
12	91.2	79
14	80.8	72
16	68.8	50
18	55.2	30

14.23 Use the following empirical relationship to plot the fuel consumption in both miles per gallon and gallons per mile for a car for which the following relationship applies. *Note:* V is the speed of the car in miles per hour and the given relationship is valid for $30 \le V \le 70$.

Fuel Consumption (Miles per Gallon)

$$= \frac{1050 \times V}{910 + V^{1.88}}$$

14.24 Starting with a $10\,\text{cm} \times 10\,\text{cm}$ sheet of paper, what is the largest volume you can create by cutting out $x\,\text{cm} \times x\,\text{cm}$ from each corner of the sheet and then folding up the sides. Use Excel to obtain the solution. *Hint:* The volume created by cutting out $x\,\text{cm} \times x\,\text{cm}$ from each corner of the $10\,\text{cm} \times 10\,\text{cm}$ sheet of paper is given by $V = (10 - 2x)(10 - 2x)x$

14.25 Given matrices:

$$[A] = \begin{bmatrix} 4 & 2 & 1 \\ 7 & 0 & -7 \\ 1 & -5 & 3 \end{bmatrix}, \quad [B] = \begin{bmatrix} 1 & 2 & -1 \\ 5 & 3 & 3 \\ 4 & 5 & -7 \end{bmatrix}, \text{ and}$$

$$[C] = \begin{Bmatrix} 1 \\ -2 \\ 4 \end{Bmatrix}, \text{ perform the following}$$

operations using Excel.

a. $[A] + [B] = ?$

b. $[A] - [B] =$

c. $3[A] = ?$

d. $[A][B] = ?$

e. $[A]\{C\} = ?$

14.26 Solve the following set of equations using Excel.

$$\begin{bmatrix} 1 & 1 & 1 \\ 2 & 5 & 1 \\ -3 & 1 & 5 \end{bmatrix} \begin{Bmatrix} x_1 \\ x_2 \\ x_3 \end{Bmatrix} = \begin{Bmatrix} 6 \\ 15 \\ 14 \end{Bmatrix}$$

14.27 Solve the following set of equations using Excel.

$$\begin{bmatrix} 7.11 & -1.23 & 0 & 0 & 0 \\ -1.23 & 1.99 & -0.76 & 0 & 0 \\ 0 & -0.76 & 0.851 & -0.091 & 0 \\ 0 & 0 & -0.091 & 2.31 & -2.22 \\ 0 & 0 & 0 & -2.22 & 3.69 \end{bmatrix} \begin{Bmatrix} T_1 \\ T_2 \\ T_3 \\ T_4 \\ T_5 \end{Bmatrix}$$

$$= \begin{Bmatrix} (5.88)(20) \\ 0 \\ 0 \\ 0 \\ (1.47)(70) \end{Bmatrix}$$

14.28 Find the equation that best fits the following set of data points. Compare the actual and predicted *y* values.

x	0	1	2	3	4	5	6	7	8	9	10
y	10	12	15	19	23	25	27	32	34	36	41

14.29 Find the equation that best fits the following set of data points. Compare the actual and predicted *y* values.

x	0	1	2	3	4	5	6	7	8
y	5	8	15	32	65	120	203	320	477

14.30 Find the equation that best fits the following set of data points. Compare the actual and predicted *y* values.

x	0	5	10	15	20	25	30	35	40	45	50
y	100	101.25	105	111.25	120	131.25	145	161.25	180	201.25	225

14.31 Use VBA to create a custom-made dialog box for Problem 14.10.

14.32 Use VBA to create a custom-made dialog box for Problem 14.11.

14.33 Use VBA to create a custom-made dialog box for Problem 14.14.

14.34 Use VBA to create a custom-made dialog box for Problem 14.15.

14.35 Use VBA to create a custom-made dialog box to convert data (for length, mass, temperature, force, and power) from SI units to U.S. Customary units.

Computational Engineering Tools MATLAB

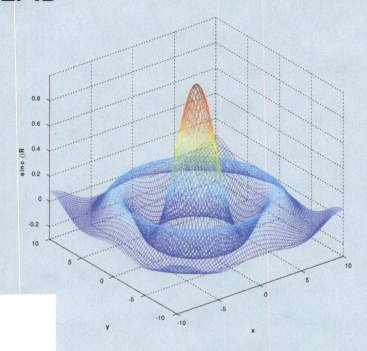

MATLAB® is mathematical software that can be used to solve a wide range of engineering problems. Before the introduction of MATLAB, engineers wrote their own computer programs to solve engineering problems. Even though engineers still write computer code to solve complex problems, they take advantage of built-in functions of the software. MATLAB is also versatile enough that you can use it to write your own programs.

LEARNING OBJECTIVES

LO¹   **MATLAB Basics:** explain the basics of MATLAB environment

LO²   **MATLAB Functions, Loop Control, and Conditional Statements:** describe how to use MATLAB's built-in functions; explain how to execute a block of code many times, based on whether a condition is met

LO³   **Plotting with MATLAB:** know how to plot in MATLAB environment

LO⁴   **Matrix Computation with MATLAB:** know how to use MATLAB to perform matrix operations and solve a set of liner equations

LO⁵   **Symbolic Mathematics with MATLAB:** describe how to use MATLAB's symbolic capabilities to present the problem and its solution using symbols

BUILDING THE FUTURE SPACE SUIT BY DAVA NEWMAN

For the past dozen years, I have been working with colleagues and students here at the Massachusetts Institute of Technology (MIT) and with collaborators in various disciplines from around the world to develop a new kind of spacesuit. My hope is that the astronauts who someday walk on the surface of Mars will be protected by a future version of what we are calling the "BioSuit™."

The suits that kept NASA astronauts alive on the moon and those worn by Space Shuttle and International Space Station crewmembers for extravehicular activities (EVAs), including the Hubble repair missions, are technological marvels; in effect, they are miniature spacecraft that provide the pressure, oxygen, and thermal control that humans need to survive in the vacuum of space. The greatest problem with these suits is their rigidity. The air that supplies the necessary pressure to the bodies of wearers turns them into stiff balloons that make movement difficult and tiring. These suits are officially known as EMUs—extravehicular mobility units—but they allow only limited mobility. Astronauts who perform repair work in space find the stiffness of spacesuit gloves especially challenging: imagine manipulating tools and small parts for hours wearing gas-filled gloves that fight against the flexing of your fingers.

Future space exploration will be expensive. If we send humans to Mars, we will want to maximize the work effort and science return. One contributor to that efficiency will need to be a new kind of spacesuit that allows our explorer-astronauts to move freely and quickly on the Martian surface. That could be the BioSuit. The BioSuit is based on the idea that there is another way to apply the necessary pressure to an astronaut's body. In theory at least, a form-fitting suit that presses directly on the skin can accomplish the job. What is needed is an elastic fabric and a structure that can provide about one-third of sea-level atmospheric pressure, or 4.3 psi (approximately the pressure at the top of Mt. Everest). The skintight suit would allow for a

degree of mobility impossible in a gas-filled suit. It also would be potentially safer. While an abrasion or micrometeor puncture in a traditional suit would threaten sudden decompression puncturing the balloon and causing a major emergency and immediate termination of the EVA—a small breach in the BioSuit could be readily repaired with a kind of high-tech Ace™ bandage to cover a small tear.

Collaborators outside the MIT community include Trotti and Associates, Inc., an architectural and industrial design firm in Cambridge, Mass.; engineers from Draper Laboratories; and Dainese, an Italian manufacturer of motorcycle racing "leathers"—leather and carbon-fiber suits designed to protect racers traveling at up to 200 mph. Bringing together designers from Trotti and Associates, students from the Rhode Island School of Design, and my MIT engineering students has greatly influenced the way our groups work. In our early sessions together to realize a second-skin spacesuit, my engineering students spent much of their time hunched over their laptops, calculating and analyzing the governing equations, while the designers—visual thinkers— took out sketchbooks and immediately started drawing to attack the problem. After working together for weeks, the engineers got more comfortable with the idea of sketching solutions and some of the designers added MATLAB and its more analytical approach to their repertoires. We all ended up better off.

To see the entire article visit: http://www.nasa.gov

To the students: Do you see the importance of computational tools such as MATLAB in solving complicated problems? Can you give other reasons as to why you should learn MATLAB?

In this chapter, we will discuss the use of MATLAB in solving engineering problems. MATLAB is mathematical software available in most university computational labs today. MATLAB is a very powerful tool, especially for manipulating matrices; in fact, it originally was designed for that purpose. There are many good textbooks that discuss the capabilities of MATLAB to solve a full range of problems. Here, our intent is to introduce only some basic ideas so that you can perform some essential operations. As you continue your engineering education in other classes, you will learn more about how to use MATLAB effectively to solve a wide range of engineering problems. As we discussed in Chapter 14, before the introduction of electronic spreadsheets and mathematical software such as MATLAB, engineers wrote their own computer programs to solve engineering problems. Even though engineers still write computer code to solve complex problems, they take advantage of built-in functions of software that are readily available with computational tools, such as MATLAB. MATLAB is also versatile enough that you can use it to write your own programs.

This chapter begins by discussing the basic makeup of MATLAB. We will explain how to input data or a formula in MATLAB and how to carry out some typical engineering computations. We will also explain the use of MATLAB's mathematical, statistical, and logical functions. Plotting the results of an engineering analysis using MATLAB is also presented in this chapter. Finally, we will discuss briefly the curve fitting and symbolic capabilities of MATLAB. Symbolic mathematics refers to the use of symbols rather than numbers to set up problems. Moreover, we will use examples from Chapter 14 (Excel) to emphasize that MATLAB is just another tool that you can use to solve a variety of engineering problems.

LO¹ 15.1   MATLAB Basics

We begin by explaining some basic ideas; then once you have a good understanding of these concepts, we will use MATLAB to solve some engineering problems. As is the case with any new software you explore,

MATLAB has its own syntax and terminology. A typical MATLAB window is shown in Figure 15.1. The main components of the MATLAB window in the default mode are marked by arrows and numbered, as shown in Figure 15.1.

1. **Menu Tabs/Bar:** Contains the commands you can use to perform certain tasks, for example, to save your workspace or import data.
2. **Current Folder:** Shows the active directory, but you can also use it to change the directory.
3. **Current Folder Window:** Shows all files, their types, sizes, and descriptions in the current directory.
4. **Command Window:** This is where you enter variables and issue MATLAB commands.
5. **Command History Window:** Shows the time and the date at which commands were issued during the previous MATLAB sessions. It also shows the history of commands in the current (active) session.
6. **Workspace:** Shows variables you created during your MATLAB session.

As shown in Figure 15.1, MATLAB's desktop layout, in default mode, is divided into four windows: the Current Folder, the **Command Window**, workspace, and the **Command History**. You type (enter) commands in the Command Window. For example, you can assign values to variables or plot a set of variables. The Command History window shows the time and the date of the commands you issued during previous MATLAB sessions. It also shows the history of commands in the current (active) session. You can transfer old commands, which you issued during previous sessions, from the Command History Window to the Command Window. To do

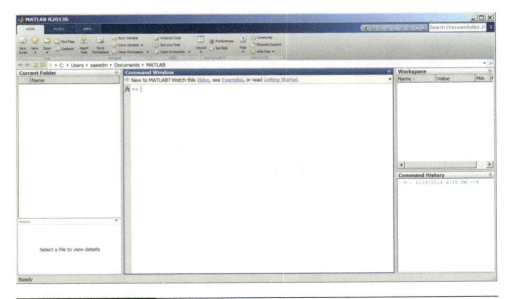

FIGURE 15.1   The desktop layout for MATLAB.

In the MATLAB environment, you can assign values to a variable or define the elements of a matrix. You can also use MATLAB to input engineering formulas and compute the results. When typing your formula, use parentheses to dictate the order of operations.

this, move your mouse pointer over the command you want to move, then click the left button, and, while holding down the button, drag the old command line into the Command Window. Alternatively, when you press the up-arrow key, the previously executed command will appear again. You can also copy and paste commands from the current Command Window, edit them, and use them again. To clear the contents of the Command Window, type **clc**.

In the MATLAB environment, you can assign values to a variable or define the elements of a matrix. For example, as shown in Figure 15.2, to assign a value of 5 to the variable x, in the Command Window after the prompt sign >> you simply type **x = 5**. The basic MATLAB scalar (arithmetic) operations are shown in Table 15.1.

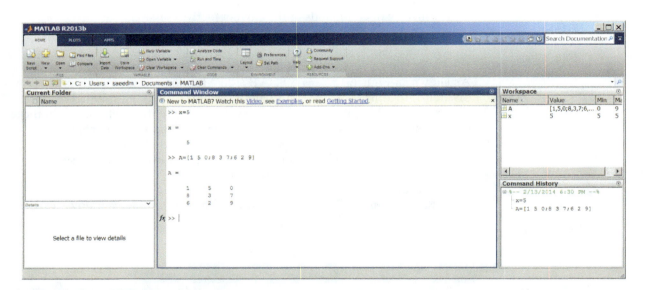

FIGURE 15.2   Examples of assigning values or defining elements of a matrix in MATLAB.

TABLE 15.1	MATLAB's Basic Scalar (Arithmetic) Operators		
Operation	Symbol	Example: $x = 5$ and $y = 3$	Result
Addition	+	$x+y$	8
Subtraction	−	$x-y$	2
Multiplication	*	$x*y$	15
Division	/	$(x+y)/2$	4
Raised to a power	^	x^2	25

As we explained in Section 14.4, during your engineering education you will learn about different types of physical variables. Some variables are identifiable by a single value or by magnitude. For example, time can be described by a single value such as two hours. These types of physical variables, which are identifiable by a single value are called *scalars*. Temperature is another example of a scalar variable. On the other hand, if you were to describe the velocity of a vehicle, you not only have to specify how fast it is moving (speed), but also its direction. The physical variables that possess both magnitude and direction are called *vectors*.

There are also other quantities that require specifying more than two pieces of information. For example, if you were to describe the location of a car parked in a multi-story garage (with respect to the garage entrance), you would need to specify the floor (the *z* coordinate), and then the location of the car on that floor (x and y coordinates). A **matrix** is often used to describe situations that require many values. In general, a matrix is an array of numbers, variables, or mathematical terms. To define the elements of a matrix, for example

$$[A] = \begin{bmatrix} 1 & 5 & 0 \\ 8 & 3 & 7 \\ 6 & 2 & 9 \end{bmatrix},$$

you type

```
A = [1 5 0;8 3 7;6 2 9]
```

Note that in MATLAB, the elements of the matrix are enclosed in brackets [ ] and are separated by blank space, and the elements of each row are separated by a semicolon (;) as shown in Figure 15.2.

In Section 18.5, we will discuss matrix algebra in more detail. If you do not have an adequate background in matrix algebra, you may want to read Section 18.5 before studying the MATLAB examples that deal with matrices.

The `format`, `disp`, and `fprintf` Commands

MATLAB offers a number of commands that you can use to display the results of your calculation. The MATLAB `format` command allows you to display values in certain ways. For example, if you define x = 2/3, then MATLAB will display x = 0.6667. By default, MATLAB will display four decimal digits. If you want more decimal digits displayed, then you type **format long**. Now, if you type x again, the value of *x* is displayed with 14 decimal digits, that is x = 0.66666666666667. You can control the way the value of *x* is displayed in a number of other ways, as shown in Table 15.2. It is important to note that the `format` command does not affect the number of digits maintained when calculations are carried out by MATLAB. It only affects the way the values are displayed.

The `disp` command is used to display text or values. For example, given **x = [1 2 3 4 5],** then the command **disp(x)** will display 1 2 3 4 5. Or the command disp('Result =') will display Result =. Note that the text that you want displayed must be enclosed within a set of single quotation marks.

The `fprintf` is a print (display) command that offers a great deal of flexibility. You can use it to print text and/or values with a desired number of digits. You can also use special formatting characters, such as **\n** and **\t,** to produce linefeed and tabs. The following example will demonstrate the use of the fprintf command.

TABLE 15.2	The Format Command Options	
MATLAB Command	**How the Result of x = 2/3 is Displayed**	**Explanation**
format short	0.6667	Shows four decimal digits-default format.
format long	0.66666666666667	Shows 14 decimal digits.
format rat	2/3	Shows as fractions of whole numbers.
format bank	0.67	Shows two decimal digits.
format short e	6.6667e-001	Shows scientific notation with four decimal digits.
format long e	6.666666666666666e-001	Shows scientific notation with 14 decimal digits.
format hex	3fe5555555555555	Shows hexadecimal output.
format +	+	Shows +, −, or blank based on whether the number is positive, negative, or zero.
format compact		Suppresses blank lines in the output.

EXAMPLE 15.1

In the MATLAB Command Window, type the following commands as shown.

```
x = 10
fprintf ('The value of x is %g \n', x)
```

MATLAB will display

```
The value of x is 10
```

A screen capture of the Command Window for Example 15.1 is shown in Figure 15.3.

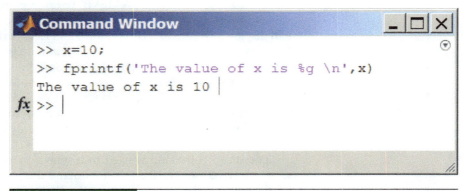

FIGURE 15.3   The screen capture of the Command Window for Example 15.1.

Note that the text and formatting code are enclosed within a set of single quotation marks. Also note that the %g is the number format character and is replaced by the value of x, 10. It is also important to mention that MATLAB will not produce an output until it encounters the \n.

Additional capabilities of `disp` and `fprintf` commands will be demonstrated using other examples later.

Saving Your MATLAB Workspace

You can save the workspace to a file by issuing the command: **save your _ filename**. The your_filename is the name that you would like to assign to the **workspace**. Later, you can load the file from the disk to memory by issuing the command: **load your _ filename**. Over time you may create many files; then you can use the **dir** command to list the content of the directory. For simple operations you can use the Command Window to enter variables and issue MATLAB commands. However, when you write a program that is longer than a few lines, you use an M-file. Later in this chapter, we will explain how to create, edit, run, and debug an M-file.

Generating a Range of Values

When creating, analyzing, or plotting data, it is often convenient to create a range of numbers. To create a *range* of data or a row matrix, you only need to specify the starting number, the increment, and the end number. For example, to generate a set of *x* values in the range of zero to 100 in increments of 25 (i.e., 0 25 50 75 100), in the Command Window, type

```
x = 0:25:100
```

Note that in MATLAB language, the range is defined by a starting value, followed by a colon (:), the increment followed by another colon, and the end value. As another example, if you were to type

```
Countdown =5:-1:0
```

then the countdown row matrix will consist of values 5 4 3 2 1 0.

Creating Formulas in MATLAB

As we have said before, engineers use formulas that represent physical and chemical laws governing our surroundings to analyze various problems. You can use MATLAB to input engineering formulas and compute the results. When typing your formula, use parentheses to dictate the order of operation. For example, in MATLAB's Command Window, if you were to type **count = 100+5*2,** MATLAB will perform the multiplication first, which results in a value of 10, and then this result is added to 100, which yields an overall value of 110 for the variable count. If, however, you wanted MATLAB to add the 100 to 5 first and then multiply the resulting 105 by 2, you should have placed parentheses around the 100 and 5 in the following manner, **count = (100+5)*2,** which results in a value of 210. The basic MATLAB arithmetic operations are shown in Table 15.3.

TABLE 15.3	The Basic MATLAB Arithmetic Operations		
Operation	**Symbol**	**Example:** **x = 10 and y = 2**	**z, the Result of the Formula** **Given in the Example**
Addition	+	$z = x + y + 20$	32
Subtraction	–	$z = x - y$	8
Multiplication	*	$z = (x * y) + 9$	29
Division	/	$z = (x / 2.5) + y$	6
Raised to a power	^	$z = (x ^ y) ^ 0.5$	10

Element-by-Element Operation

In addition to basic scalar (arithmetic) operations, MATLAB provides **element-by-element operations** and **matrix operations**. MATLAB's symbols for element-by-element operations are shown in Table 15.4. To better understand their use, suppose you have measured and recorded the mass (kg) and speed (m/s) of five runners who are running along a straight path: **m = [60 55 70 68 72]** and **s = [4 4.5 3.8 3.6 3.1]**. Note that the mass **m** array and speed **s** array each have five elements. Now suppose you are interested in determining the magnitude of each runner's momentum. We can calculate the momentum for each runner using MATLAB's element-by-element multiplication operation in the following manner: **momentum = m.*s** resulting in **momentum = [240 247.5 266 244.8 223.2]**. Note the multiplication symbol (.*) for element-by-element operation.

To improve your understanding of the element-by-element operations, in MATLAB's Command Window type **a = [7 4 3 –1]** and **b = [1, 3,5, 7]** and then try the following operations.

```
>>a+b
ans =  8  7  8  6
>>a-b
ans =  6  1  -2  -8
>>3*a
ans =  21  12  9  -3
>>3.*a
ans =  21  12  9  -3
```

TABLE 15.4	MATLAB's Element-by-Element Operations	
Operation	**Arithmatic** **Operations**	**Equivalent Element-by-Element** **Symbol for the Operation**
Addition	+	+
Subtraction	–	–
Multiplication	*	.*
Division	/	./
Raised to a power	^	.^

```
>>a.*b
ans = 7  12  15  -7
>>b.*a
ans = 7  12  15  -7
>>3.^a
ans = 1.0e+003*
  2.1870  0.0810  0.0270  0.0003
>>a.^b
ans = 7  64  243  -1
>>b.^a
ans = 1.0000  81.0000  125.0000  0.1429
```

Try the following example on your own.

EXAMPLE 14.1 (REVISITED)

This example will show how the density of standard air changes with temperature. It also makes use of MATLAB's element-by-element operation. The density of standard air is a function of temperature and may be approximated using the ideal gas law according to

$$\rho = \frac{P}{RT}$$

where

$P =$ standard atmospheric pressure (101.3 kPa)

$R =$ gas constant and its value for air is $286.9 \left(\dfrac{J}{kg \cdot K} \right)$

$T =$ air temperature in kelvin (K)

Using MATLAB, we want to create a table that shows the density of air as a function of temperature in the range of $0°\,C$ (273.15 K) to $50°\,C$ (323.15 K) in increments of $5°\,C$.

In MATLAB's Command Window, we type the following commands.

```
>> Temperature = 0:5:50;
>> Density = 101300./((286.9)*(Temperature+273));
>> fprintf('\n\n');disp(' Temperature(C)
    Density(kg/m^3)');disp([Temperature',Density'])
```

In the commands just shown, the semicolon **;** suppresses MATLAB's automatic display action. If you type **Temperature = 0:5:50** without the semicolon at the end, MATLAB will display the values of temperature in a row. It will show

```
Temperature =
0   5   10   15   20   25   30   35   40   45   50
```

The **./** is a special element-by-element division operation that tells MATLAB to carry the division operation for each temperature value.

In the **disp** command, the prime or the single quotation mark over the variables **Temperature'** and **Density'** will change the values of the temperature and the density, which are stored in rows, to column format before they are displayed. In matrix operations, the process of changing the rows into columns is called the transpose of the matrix. The final results for

Example 14.1 (revisited) are shown in Figure 15.4. Note that the values of the temperature and the density are shown in columns. Also note the use of the `fprintf` and `disp` commands.

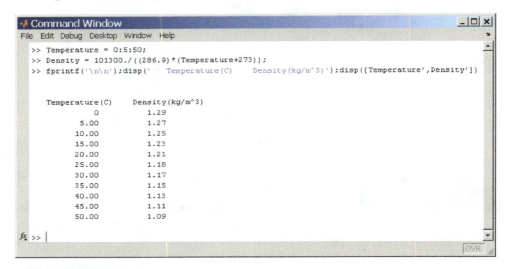

FIGURE 15.4   The result of Example 14.1 (revisited).

Matrix Operations

MATLAB offers many tools for **matrix operations** and manipulations. Table 15.5 shows examples of these capabilities. We will explain MATLAB's matrix operations in more detail in Section 15.5.

TABLE 15.5	Examples of the MATLAB's Matrix Operations	
Operation	**Symbols or Commands**	**Example: A and B are Matrices Which You Have Defined**
Addition	+	A + B
Subtraction	−	A − B
Multiplication	*	A * B
Transpose	*matrix name'*	A'
Inverse	`inv`*(matrix name)*	inv(A)
Determinant	`det`*(matrix name)*	det(A)
Eigenvalues	`eig`*(matrix name)*	eig(A)
Matrix left division (uses Gauss elimination to solve a set of linear equations)	\	See Example 15.5

EXAMPLE 14.2 (REVISITED)

Using MATLAB, create a table that shows the relationship between the interest earned and the amount deposited, as shown in Table 15.6.

To create a table that is similar to Table 15.6, we type the following commands.

```
>> format bank
>> Amount = 1000:250:3000;
>> Interest_Rate = 0.06:0.01:0.08;
>> Interest_Earned = (Amount')*(Interest _ Rate);
>> fprintf('\n\n\t\t\t\t\t\t\t Interest
   Rate');fprintf('\n\t Amount\t\t');...
fprintf('\t\t %g',Interest_Rate);fprintf('\n');
   disp ([Amount',Interest_Earned])
```

TABLE 15.6 The Relationship between the Interest Earned and the Amount Deposited

	Interest Rate		
Dollar Amount	**0.06**	**0.07**	**0.08**
1000	60	70	80
1250	75	87.5	100
1500	90	105	120
1750	105	122.5	140
2000	120	140	160
2250	135	157.5	180
2500	150	175	200
2750	165	192.5	220
3000	180	210	240

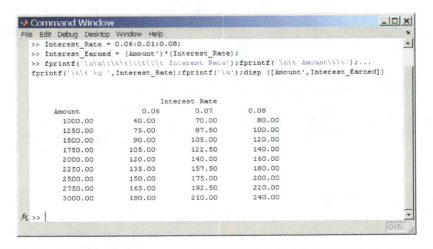

FIGURE 15.5   The commands and result for Example 14.2 (revisited).

> Note that the three periods … (an ellipsis) represents a continuation marker in MATLAB. The ellipsis means there is more to follow on this command line.

On the last command line, note that the three periods . . . (an ellipsis) represents a continuation marker in MATLAB. The ellipsis means there is more to follow on this command line. Note the use of `fprintf` and `disp` commands. The final result for Example 14.2 (revisited) is shown in Figure 15.5.

Before You Go On

Answer the following questions to test your understanding of the preceding section.

1. How do you generate a range of values in MATLAB?

2. Explain the difference between the element-by-element and matrix operations.

3. How do you display text and values in MATLAB?

4. How do you control the number of decimal digits displayed in MATLAB?

5. How do you input elements of a matrix in MATLAB?

LO² 15.2   MATLAB Functions, Loop Control, and Conditional Statements

MATLAB offers a large selection of built-in functions that you can use to analyze data. As we discussed in the previous chapter, by built-in functions we mean standard functions such as the sine or cosine of an angle, as well as formulas that calculate the total value, the average value, or the standard deviation of a set of data points. The MATLAB functions are available in various categories, including mathematical, trigonometric, statistical, and logical functions. In this chapter, we will discuss some of the common functions. MATLAB offers a Help menu that you can use to obtain information on various commands and functions.

> MATLAB offers a large selection of built-in functions that you can use to analyze data. They include mathematical, statistical, trigonometric, and engineering functions.

The Help button is marked by a question mark ? located on the main tab bar. You can also type **help** followed by a command name to learn how to use the command.

Some examples of commonly used MATLAB functions, along with their proper use and descriptions, are shown in Table 15.7. Refer to Example 15.2 when studying Table 15.7.

EXAMPLE 15.2

The following set of values will be used to introduce some of MATLAB's built-in functions. Mass = [102 115 99 106 103 95 97 102 98 96]. When studying Table 15.7, the results of the executed functions are shown under the "Result of the Example" column.

More examples of MATLAB's Functions are shown in Table 15.8.

TABLE 15.7	Some MATLAB Functions that You May Use in Engineering Analyses		
Function	**Description of the Function**	**Example**	**Result of the Example**
sum	Sums the values in a given array.	sum(Mass)	1013
mean	Calculates the average value of the data in a given array.	mean(Mass)	101.3
max	Determines the largest value in the given array.	max(Mass)	115
min	Determines the smallest value in the given array.	min(Mass)	95
std	Calculates the standard deviation for the values in the given array.	std(Mass)	5.93
sort	Sorts the values in the given array in ascending order.	sort(Mass)	95 96 97 98 99 102 102 103 106 115
pi	Returns the value of π, 3.14151926535897 …	pi	3.14151926535897…
tan	Returns tangent value of the argument. The argument must be in radians.	tan(pi/4)	1
cos	Returns cosine value of the argument. The argument must be in radians.	cos(pi/2)	0
sin	Returns sine value of the argument. The argument must be in radians.	sin(pi/2)	1

TABLE 15.8	More Examples of MATLAB Functions
Function	**Description of the Function**
sqrt(x)	Returns the square root of value *x*.
Factorial(x)	Returns the value of factorial of *x*. For example, factorial(5) will return: (5)(4)(3)(2)(1) = 120.
Trigonometric Functions	
acos(x)	The inverse cosine function of *x*. Determines the value of an angle when its cosine value is known.
asin(x)	The inverse sine function of *x*. Determines the value of an angle when its sine value is known.
atan(x)	The inverse tangent function of *x*. Determines the value of an angle when its tangent value is known.
Exponential and Logarithmic Functions	
exp(x)	Returns the value of e^x.
log(x)	Returns the value of the natural logarithm of *x*.
log10(x)	Returns the value of the common (base 10) logarithm of *x*.
log2(x)	Returns the value of the base-2 logarithm of *x*.

EXAMPLE 14.4 (REVISITED)

Using MATLAB, compute the average (arithmetic mean) and the standard deviation of the density of water data given in Table 15.9. Refer to Chapter 19, Section 19.4, to learn about what the value of the standard deviation for a set of data points represents.

TABLE 15.9	Data for Example 14.4 (revisited)
Group-A Findings	**Group-B Findings**
ρ (kg/m³)	ρ (kg/m³)
1020	950
1015	940
990	890
1060	1080
1030	1120
950	900
975	1040
1020	1150
980	910
960	1020

The final results for Example 14.4 (revisited) are shown in Figure 15.6. The MATLAB commands leading to these results are:

```
>> Density_A = [1020 1015 990 1060 1030 950 975 1020 980
               960];
>> Density_B = [950 940 890 1080 1120 900 1040 1150 910
               1020];
>> Density_A_Average = mean(Density_A)
Density_A_Average =
   1000.00
>> Density_B_Average = mean(Density_B)
Density_B_Average =
   1000.00
>> Standard_Deviation_For_Group_A = std(Density_A)
Standard_Deviation_For_Group_A =
   34.56
>> Standard_Deviation_For_Group_B = std(Density_B)
Standard_Deviation_For_Group_B =
   95.22
>>
```

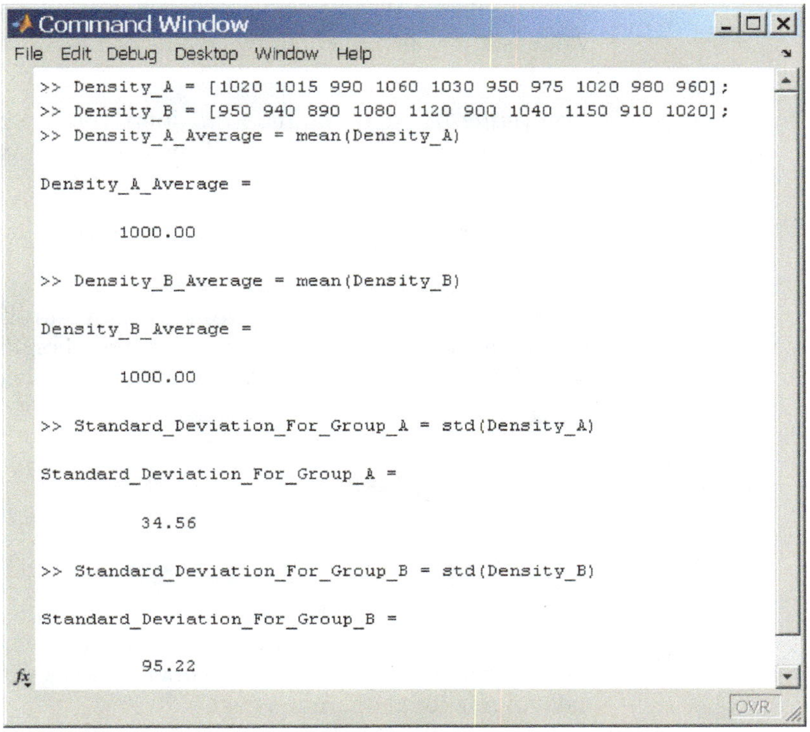

| **FIGURE 15.6** | MATLAB's Command Window for Example 14.4 (revisited). |

The Loop Control – *for* and *while* Commands

When writing a computer program, it often becomes necessary to execute a line or a block of your computer code many times. MATLAB provides *for* and *while* commands for such situations.

The *for* Loop   Using the *for* loop, you can execute a line or a block of code a specified (defined) number of times. The syntax of a *for* loop is

```
for index = start-value : increment : end-value
    a line or a block of your computer code
end
```

For example, suppose you want to evaluate the function $y = x^2 + 10$ for x values of 22.00 22.50, 23.00, 23.50, and 24.00. This operation will result in corresponding y values of 494.00, 516.25, 539.00, 562.25, and 586.00. The MATLAB code for this example then could have the form:

> When required by the logic of your program, MATLAB provides *for* and *while* commands to execute a block of your computer code many times.

```
x = 22.0;
for i = 1:1:5
    y=x^2+10;
    disp([x', y'])
    x = x + 0.5;
end
```

Note that in the preceding example, the index is the integer i and its start-value is 1, it is incremented by a value of 1, and its end-value is 5.

The *while* Loop Using the *while* loop, you can execute a line or a block of code until a specified condition is met. The syntax of a *while* loop is

```
while controlling-expression
   a line or a block of your computer code
end
```

With the *while* command, as long as the controlling-expression is true, the line or a block of code will be executed. For the preceding example, the MATLAB code using the *while* command becomes:

```
x = 22.0;
while x <= 24.00
      y=x^2+10;
      disp([x',y'])
      x = x + 0.5;
end
```

In the preceding example the <= symbol denotes less than or equal to and is called a relational or comparison operator. We will explain MATLAB's logical and relational operators next.

Using MATLAB Logical and Relational Operators

In this section, we will look at some of MATLAB's logical operators. These are operators that allow you to test various conditions. MATLAB's logical operators and their descriptions are shown in Table 15.10.

MATLAB also offers relational or comparison operators that allow for the testing of relative magnitude of various arguments. These relational operators are shown in Table 15.11.

TABLE 15.10		MATLAB's Logical Operators
Logical Operator	**Symbol**	**Description**
and	&	TRUE if both conditions are true
or	|	TRUE if either or both conditions are true
not	~	FALSE if the result of a condition is TRUE, and TRUE if the result of a condition is FALSE
exclusive or	xor	FALSE if both conditions are TRUE or both conditions are FALSE, and TRUE if exactly one of the two conditions is true
any		TRUE if any element of vector is nonzero
all		TRUE if all elements of vector is nonzero

TABLE 15.11	MATLAB's Relational Operators and their Descriptions
Relational Operator	**Its Meaning**
<	Less than
< =	Less than or equal to
= =	Equal to
>	Greater than
> =	Greater than or equal to
~ =	Not equal to

MATLAB's logical operators allow the programmer to test various conditions. MATLAB provides *if* and *else* commands to execute a block of code based on whether a condition or set of conditions is met.

The Conditional Statements – *if, else*

When writing a computer program, sometimes it becomes necessary to execute a line or a block of code based on whether a condition or set of conditions is met (true). MATLAB provides *if* and **else** commands for such situations.

The *if* Statement The *if* statement is the simplest form of a conditional control. Using the *if* statement, you can execute a line or a block of your program as long as the expression following the *if* statement is true. The syntax for the *if* statement is

```
if expression
  a line or a block of your computer code
end
```

For example, suppose we have a set of scores for an exam: 85, 92, 50, 77, 80, 59, 65, 97, 72, 40. We are interested in writing a code that shows that scores below 60 indicate failing. The MATLAB code for this example could have the following form:

```
scores = [85 92 50 77 80 59 65 97 72 40];
for i=1:1:10
  if scores (i) <60
     fprintf('\t %g \t\t\t\t\t FAILING\n', scores (i))
  end
end
```

The *if, else* Statement The **else** statement allows us to execute other line(s) of computer code(s) if the expression following the *if* statement is not true. For example, suppose we are interested in showing not only the scores that indicate failing but also the scores that show passing. We can then modify our code in the following manner:

```
scores = [85 92 50 77 80 59 65 97 72 40];
for i=1:1:10
    if scores (i) >=60
      fprintf('\t %g \t\t\t\t\t PASSING\n', scores (i));
    else
      fprintf('\t %g \t\t\t\t\t FAILING\n', scores (i))
    end
end
```

In the MATLAB Command Window, try the preceding examples on your own.

MATLAB also provides the *elseif* command that could be used with the *if* and *else* statements. It is left for you to learn about it. Try the MATLAB help menu: type *help elseif* to learn about the elseif statement.

EXAMPLE 14.5 (REVISITED)

The pipeline shown in Figure 15.7 is connected to a control (check) valve that opens when the pressure in the line reaches 20 psi. Various readings were taken at different times and recorded. Using MATLAB's logical functions, create a list that shows the corresponding open and closed positions of the check valve.

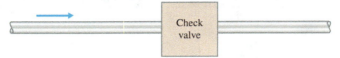

FIGURE 15.7 A schematic diagram for Example 14.5 (revisited).

The solution to Example 14.5 (revisited) is shown in Figure 15.8. The commands leading to the solution are:

```
>> pressure=[20 18 22 26 19 19 21 12];
>> fprintf('\t Line Pressure (psi) \t Valve Position\n\n');
for i=1:8
if pressure(i) >=20
fprintf('\t %g \t\t\t\t\t OPEN\n',pressure(i))
else
fprintf('\t %g \t\t\t\t\t CLOSED\n',pressure(i))
end
end
```

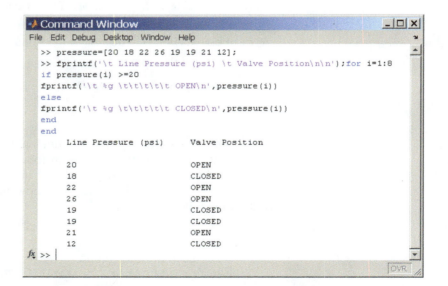

FIGURE 15.8 The solution of Example 14.5 (revisited).

The M-File

As explained previously, for simple operations you can use MATLAB's Command Window to enter variables and issue commands. However, when you write a program that is more than a few lines long, you use an **M-file**. It is called an M-file because of its *.m* extension. You can create an M-file using any text editor or using MATLAB's Editor/Debugger. To create an M-file, open the M-file Editor by going to **HOME→New Script**, and MATLAB opens a new window in which you can type your program. As you type your program, you will notice that MATLAB assigns line numbers in the left column of the window. The line numbers are quite useful for debugging your program. To save the file, simply click **EDITOR→Save As** … and type in the file-name. The name of your file must begin with a letter and may include other characters such as underscore and digits. Be careful not to name your file the same as a MATLAB command. To see if a file name is used by a MATLAB command, type **exist ('file name')** in the MATLAB's Command Window. To run your program, in the EDITOR tab click on **Run**. Don't be discouraged to find mistakes in your program the first time you attempt to run it. This is quite normal! You can use the Debugger to find your mistakes. To learn more about debugging options, type *help debug* in the MATLAB Command Window.

EXAMPLE 15.3

It has been said that, when Pascal was 7 years old, he came up with the formula $\frac{n(n+1)}{2}$ to determine the sum of 1, 2, 3, ..., through *n*. The story suggests that one day he was asked by his teacher to add up numbers 1 through 100, and Pascal came up with the answer in few minutes. It is believed that Pascal solved the problem in the following manner:

First, on one line he wrote the numbers 1 through 100, similar to

$$1 \ 2 \ 3 \ 4 \dots\dots\dots 99 \ 100$$

Then, on the second line he wrote the numbers backward

$$100 \ 99 \ 98 \ 97 \dots\dots\dots 2 \ 1$$

Then, he added up the numbers in the two lines, resulting in one hundred identical values of 101

$$101 \ 101 \ 101 \ 101 \dots\dots\dots 101 \ 101$$

Pascal also realized that the result should be divided by 2—since he wrote down the numbers 1 through 100 twice—leading to the answer: $\frac{100(101)}{2} = 5050$. Later, he generalized his approach and came up with the formula $\frac{n(n+1)}{2}$.

Next, we will write a computer program using an M-file that asks a user to input a value for *n* and computes the sum of 1 through *n*. To make the program interesting, we will not make use of Pascal's formula; instead, we

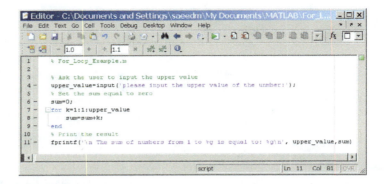

FIGURE 15.9   The M-file for Example 15.3.

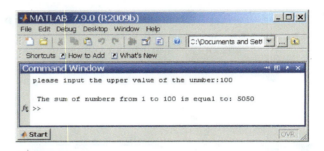

FIGURE 15.10   The results of Example 15.3.

will use a *for* loop to solve the problem. We have used MATLAB's Editor to create the program and have named it **For_Loop_Example.m**, as shown in Figure 15.9. In the shown program, the % symbol denotes comments, and any text following the % symbol will be treated as comments by MATLAB. Also, note that you can find the Line (Ln) and Column (Col) numbers corresponding to a specific location in your program by moving the cursor. The line and the column numbers are shown in the right side, bottom corner of the Editor window. As you will see, the knowledge of line and column numbers are useful for debugging your program. We run the program by clicking on **Debug → Save File and Run**, and the result is shown in Figure 15.10.

Before You Go On

Answer the following questions to test your understanding of the preceding section.

1. Give examples of MATLAB built-in functions.

2. Give examples of MATLAB loop control.

3. Give examples of MATLAB conditional statements

4. What is an M-file?

LO³ 15.3 Plotting with MATLAB

MATLAB offers many choices when it comes to creating charts. For example, you can create *x*–*y* charts, column charts (or histograms), contour, or surface plots. As we mentioned in Chapter 14, as an engineering student, and later as a practicing engineer, most of the charts that you will create will be *x*–*y* type charts. Therefore, we will explain in detail how to create an *x*–*y* chart.

EXAMPLE 15.4

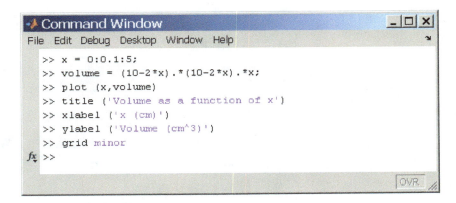

FIGURE 15.11

The 10 cm × 10 cm sheet in Example 15.4.

Starting with a 10 cm × 10 cm sheet of paper, what is the largest volume you can create by cutting out *x* cm × *x* cm from each corner of the sheet and then folding up the sides? See Figure 15.11. Use MATLAB to obtain the solution.

The volume created by cutting out *x* cm × *x* cm from each corner of the 10 cm × 10 cm sheet of paper is given by $V = (10-2x)(10-2x)x$. Moreover, we know that, for $x = 0$ and $x = 5$, the volume will be zero. Therefore, we need to create a range of *x* values from 0 to 5 using some small increments, such as 0.1. We then plot the volume versus *x* and look for the maximum value of volume. The MATLAB commands that lead to the solution are:

```
>> x = 0:0.1:5;
>> volume = (10-2*x).*(10-2*x).*x;
>> plot (x,volume)
>> title ('Volume as a function of x')
>> xlabel ('x (cm)')
>> ylabel ('Volume (cm^3)')
>> grid minor
>>
```

```
Command Window
File  Edit  Debug  Desktop  Window  Help
>> x = 0:0.1:5;
>> volume = (10-2*x).*(10-2*x).*x;
>> plot (x,volume)
>> title ('Volume as a function of x')
>> xlabel ('x (cm)')
>> ylabel ('Volume (cm^3)')
>> grid minor
fx >>
                                    OVR
```

FIGURE 15.12   The MATLAB Command Window for Example 15.4.

The MATLAB Command Window for Example 15.4 is shown in Figure 15.12. The plot of volume versus *x* is shown in Figure 15.13.

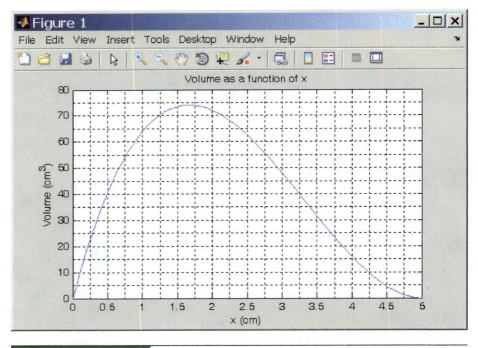

FIGURE 15.13   The plot of volume versus *x* for Example 15.4.

Let us now discuss the MATLAB commands that commonly are used when plotting data. The `plot(x,y)` command plots *y* values versus *x* values. You can use various line types, plot symbols, or colors with the command `plot(x, y, s)`, where **s** is a character string that defines a particular line type, plot symbol, or line color. The **s** can take on one of the properties shown in Table 15.12.

TABLE 15.12	MATLAB Line and Symbol Properties				
s	**Color**	**s**	**Data Symbol**	**s**	**Line Type**
b	Blue	.	Point	-	Solid
g	Green	O	Circle	:	Dotted
r	Red	x	x-mark	-.	Dash-dot
c	Cyan	+	Plus	--	Dashed
m	Magenta	*	Star		
y	Yellow	s	Square		
k	Black	d	Diamond		
		v	Triangle (down)		
		^	Triangle (up)		
		<	Triangle (left)		
		>	Triangle (right)		

For example, if you issue the command **plot (*x,y,* 'k*-')**, MATLAB will plot the curve using a black solid line with an * marker shown at each data point. If you do not specify a line color, MATLAB automatically assigns a color to the plot.

Using the **title('text')** command, you can add text on top of the plot. The **xlabel('text')** command creates the title for the *x*-axis. The text that you enclose between single quotation marks will be shown below the *x*-axis. Similarly, the **ylabel ('text')** command creates the title for the *y*-axis. To turn on the grid lines, type the command **grid on** (or just **grid**). The command **grid off** removes the grid lines. To turn on the minor grid lines, as shown in Figure 15.13, type the command **grid minor**.

Generally, it is easier to use the Graph Property Editor. For example, to make the curve line thicker, change the line color, and to add markers to the data points (with the mouse pointer on the curve) double-click the left mouse button. Make sure you are in the picking mode first. You may need to click on the arrow next to the print icon to activate the picking mode. After double-clicking on the line, you should see the line and the Marker Editor window. As shown in Figure 15.14, we increased the line thickness from 0.5 to 2, changed the line color to black, and set the data-point marker style to Diamond. These new settings are reflected in Figure 15.15.

Next, we will add an arrow pointing to the maximum value of the volume by selecting the **Text Arrow** under the **Insert** option (see Figure 15.16), and add the text "Maximum volume occurs at x = 1.7 cm." These additions are reflected in Figure 15.17.

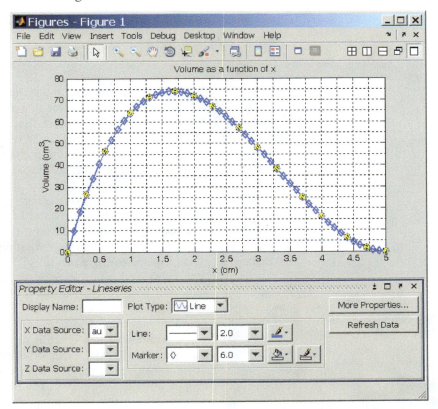

FIGURE 15.14 MATLAB's Line Property Editor.

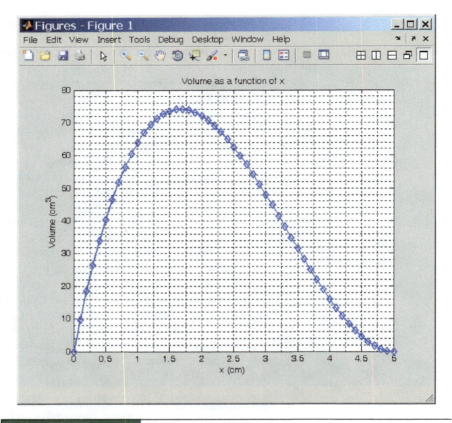

FIGURE 15.15   The plot of Example 15.4 with modified properties.

We can also change the font size and style, and make the title or the axes labels boldface. To do so from the Menu bar, select **Edit** and then **Figure Properties…**. Then, we pick the object that we want to modify and using the Property Editor shown in Figure 15.18 on page 485, we can modify the properties of the selected object. We have changed the font size and the font weight of the title and the labels for Example 15.4 and shown the changes in Figure 15.19 on page 566.

With MATLAB, you can generate other types of plots, including contour and surface plots. You can also control the *x*- and *y*-axis scales. For example, MATLAB's **loglog(x,y)** command uses the base-10 logarithmic scales for *x*- and *y*-axes. Note *x* and *y* are the variables that you want to plot. The command **loglog(x,y)** is identical to the **plot(x,y)**, except it uses logarithmic axes. The command **semilogx(x,y)** or **semilogy(x,y)** creates a plot with base-10 logarithmic scales for either only the *x*-axis or *y*-axis. Finally, it is worth noting that you can use the **hold** command to plot more than one set of data on the same chart.

A reminder, when creating an engineering chart, whether you are using MATLAB, Excel, other drawing software, or a freehand drawing: an engineering chart must contain proper labels with proper units for each axis. The chart must also contain a figure number with a title explaining what the chart represents. If more than one set of data is plotted on the same chart, the chart must also contain a legend or list showing symbols used for different data sets.

MATLAB's **plot(x,y,s)** command plots *y* values versus *x* values. The s is a character string that defines a particular line type, plot symbol, or line color. For example, if you issue the command **plot(x,y,'k*-')**, MATLAB will plot the curve using a black solid line with an * marker shown at each data point.

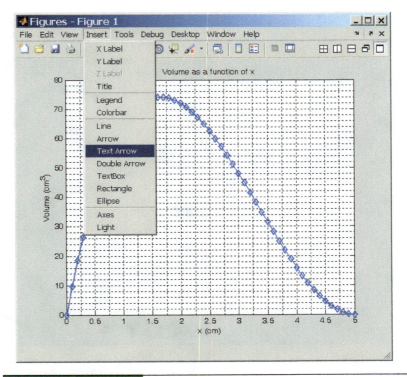

FIGURE 15.16   Using the Insert Text Arrow options, you can add arrows or text to the plot.

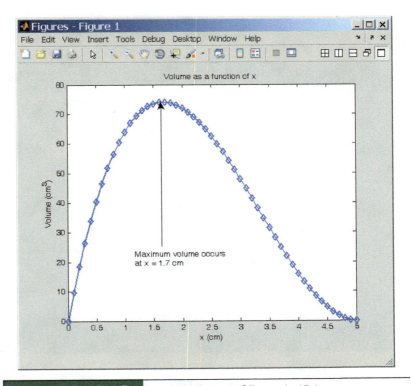

FIGURE 15.17   The solution of Example 15.4.

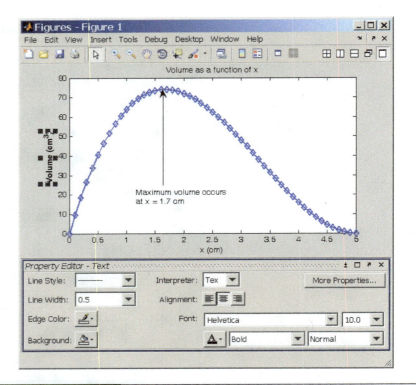

FIGURE 15.18   MATLAB's Text Property Editor.

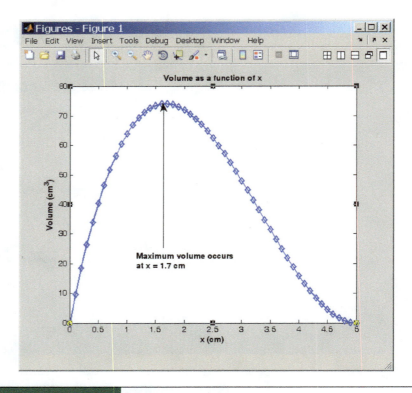

FIGURE 15.19   The result of Example 15.4.

**EXAMPLE 14.6
(REVISITED)**

Using the results of Example 14.1, create a graph showing the value of the air density as a function of temperature.

The Command Window and the plot of the density of air as a function of temperature are shown in Figures 15.20 and 15.21, respectively.

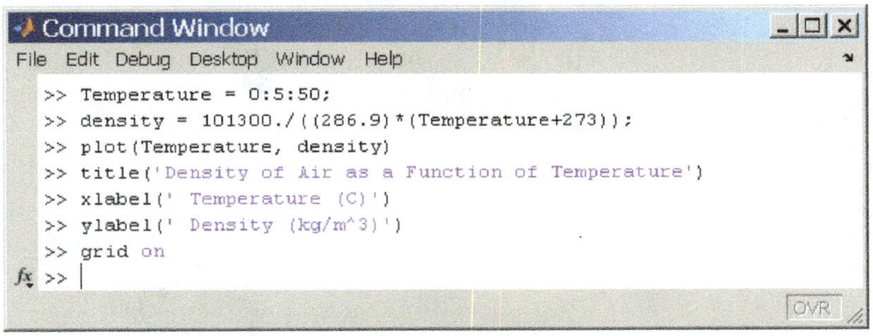

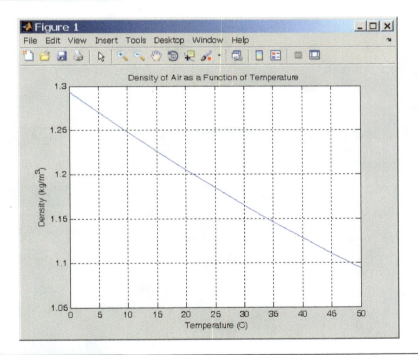

```
>> Temperature = 0:5:50;
>> density = 101300./((286.9)*(Temperature+273));
>> plot(Temperature, density)
>> title('Density of Air as a Function of Temperature')
>> xlabel(' Temperature (C)')
>> ylabel(' Density (kg/m^3)')
>> grid on
```

FIGURE 15.20   Command Window for Example 14.6 (revisited).

FIGURE 15.21   Plot of density of air for Example 14.6 (revisited).

Importing Excel and Other Data Files into MATLAB

At times, it might be convenient to import data files that were generated by other programs, such as Excel, into MATLAB for additional analysis. To demonstrate how we go about importing a data file into MATLAB, consider the Excel file shown in Figure 15.22. The Excel file was created for Example 15.4 with two columns: the *x* values and the corresponding volumes. To import this file into MATLAB, from the HOME tab, we select **Import Data**, then go to the appropriate directory, and open the file we want. MATLAB will import the data and will save them as **x** and **volume** variables (Figure 15.23).

Now let's say that we want to plot the volume as a function of *x*. We then simply type the MATLAB commands that are shown in Figure 15.24. The resulting plot is shown in Figure 15.25 on page 569.

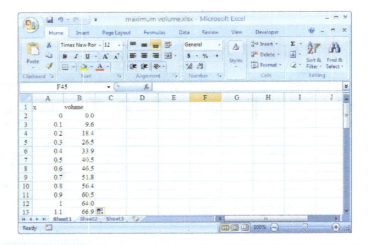

FIGURE 15.22   The Excel data file used in Example 15.4.

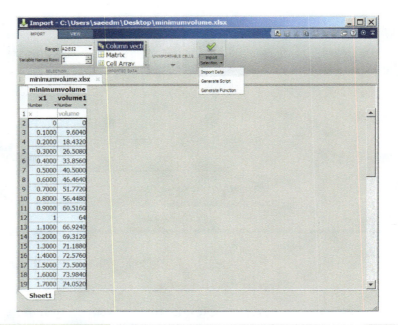

FIGURE 15.23   MATLAB's Import Wizard.

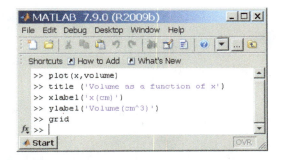

FIGURE 15.24   The commands leading to the plot shown in Figure 15.25.

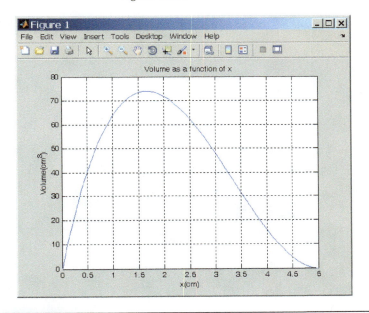

FIGURE 15.25   Plot of volume versus *x* using data imported from an Excel file.

LO⁴ 15.4   Matrix Computations with MATLAB

As explained earlier, MATLAB offers many tools for matrix operations and manipulations. Table 15.4 shows examples of these capabilities. We will demonstrate a few of MATLAB's matrix commands with the aid of the following examples.

EXAMPLE 15.5

Given the following matrices:

$$[A] = \begin{bmatrix} 0 & 5 & 0 \\ 8 & 3 & 7 \\ 9 & -2 & 9 \end{bmatrix}, [B] = \begin{bmatrix} 4 & 6 & -2 \\ 7 & 2 & 3 \\ 1 & 3 & -4 \end{bmatrix}, \{C\} = \begin{Bmatrix} -1 \\ 2 \\ 5 \end{Bmatrix} \text{ using MATLAB, perform}$$

the following operations. (a) $[A] + [B] = ?$, (b) $[A] - [B] = ?$, (c) $3[A] = ?$, (d) $[A][B] = ?$, (e) $[A]\{C\} = ?$ (f) determinate of $[A]$.

The solution is shown in Figure 15.26. When studying these examples, note that the response given by MATLAB is shown in regular typeface. Information that the user needs to type is shown in **boldface.**

```
>> A=[0 5 0;8 3 7;9 -2 9]
A =
   0  5  0
   8  3  7
   9 -2  9
>> B=[4 6 -2;7 2 3;1 3 -4]
B =
   4  6  -2
   7  2   3
   1  3  -4
>> C=[-1; 2; 5]
C =
  -1
   2
   5
>> A+B
ans =
   4 11 -2
  15  5 10
  10  1  5
>> A-B
ans =
  -4 -1  2
   1  1  4
   8 -5 13
>>3*A
ans =
   0 15  0
  24  9 21
  27 -6 27
>>A*B
ans =
  35 10  15
  60 75 -35
  31 77 -60
>>A*C
ans =
  10
  33
  32
>>det(A)
ans =
 -45
>>
```

| **FIGURE 15.26** | The solution to Example 15.5. |

EXAMPLE 15.6

The formulation of many engineering problems leads to a system of algebraic equations. As you will learn later in your math and engineering classes, there are a number of ways to solve a set of linear equations. Solve the following set of equations using the Gauss elimination, by inverting the [*A*] matrix (the coefficients of unknowns), and multiplying it by the {*b*} matrix (the values on the right-hand side of equations). The Gauss elimination method is discussed in detail in Section 18.5. Here, our intent is to show how to use MATLAB to solve a set of linear equations.

$$2x_1 + x_2 + x_3 = 13$$
$$3x_1 + 2x_2 + 4x_3 = 32$$
$$5x_1 - x_2 + 3x_3 = 17$$

For this problem, the coefficient matrix [*A*] and the right-hand side matrix {*b*} are

$$[A] = \begin{bmatrix} 2 & 1 & 1 \\ 3 & 2 & 4 \\ 5 & -1 & 3 \end{bmatrix} \text{ and } \{b\} = \begin{Bmatrix} 13 \\ 32 \\ 17 \end{Bmatrix}$$

We will first use the MATLAB matrix left division operator \ to solve this problem. The \ operator solves the problem using the Gauss elimination. We then solve the problem using the **inv** command.

```
>> A = [2 1 1;3 2 4;5 -1 3]
A =
   2   1 1
   3   2 4
   5  -1 3
>> b = [13;32;17]
b =
   13
   32
   17
>> x = A\b
x =
   2.0000
   5.0000
   4.0000
>> x = inv(A)*b
x =
   2.0000
   5.0000
   4.0000
```

Note that if you substitute the solution $x_1 = 2$, $x_2 = 5$, and $x_3 = 4$ into each equation, you find that they satisfy them. That is: $2(2) + 5 + 4 = 13$, $3(2) + 2(5) + 4(4) = 32$, and $5(2) - 5 + 3(4) = 17$.

Curve Fitting with MATLAB

In Section 14.3, we discussed the concept of **curve fitting**. MATLAB offers a variety of curve-fitting options. We will use Example 14.9 to show how you can also use MATLAB to obtain an equation that closely fits a set of data points. For Example 14.9 (revisited), we will use the command **POLYFIT(x, y, n)**, which determines the coefficients $(c_0, c_1, c_2, \ldots, c_n)$ of a polynomial of order n that best fits the data according to:

$$y = c_0 x^n + c_1 x^{n-1} + c_2 x^{n-2} + c_3 x^{n-3} + \cdots + c_n$$

EXAMPLE 14.9 (REVISITED)

Find the equation that best fits the following set of data points in Table 15.13.

In Section 14.3, plots of data points revealed that the relationship between y and x is quadratic (second order polynomial). To obtain the coefficients of the second order polynomial that best fits the given data, we will type the following sequence of commands. The MATLAB Command Window for Example 14.9 (revisited) is shown in Figure 15.27.

TABLE 15.13

A Set of Data Points for Example 14.9 (revisited)

X	Y
0.00	2.00
0.50	0.75
1.00	0.00
1.50	−0.25
2.00	0.00
2.50	0.75
3.00	2.00

```
>> format compact
>> x=0:0.5:3
>> y = [2 0.75 0 -0.25 0 0.75 2]
>> Coefficients = polyfit(x,y,2)
```

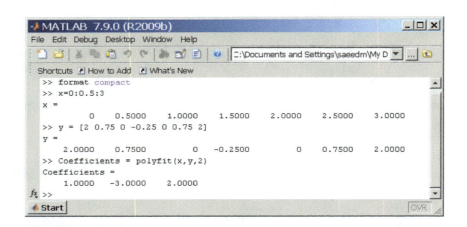

FIGURE 15.27   The Command Window for Example 14.9 (revisited).

Upon execution of the `polyfit` command, MATLAB will return the following coefficients, $c_0 = 1$, $c_1 = -3$, and $c_2 = 2$, which leads to the equation $y = x^2 - 3x + 2$.

LO⁵ 15.5   Symbolic Mathematics with MATLAB

In the previous sections, we discussed how to use MATLAB to solve engineering problems with numerical values. In this section, we briefly explain the symbolic capabilities of MATLAB. In symbolic mathematics, as the name implies, the problem and the solution are presented using symbols, such as *x*, instead of numerical values. We will demonstrate MATLAB's symbolic capabilities using Examples 15.7 and 15.8.

EXAMPLE 15.7

We will use the following functions to perform MATLAB's symbolic operations, as shown in Table 15.14.

$$f_1(x) = x^2 - 5x + 6$$
$$f_2(x) = x - 3$$
$$f_3(x) = (x + 5)^2$$
$$f_4(x) = 5x - y + 2x - y$$

TABLE 15.14   Examples of MATLAB's Symbolic Operations

Function	Description of the Function	Example	Result of the Example
sym	Creates a symbolic function.	F1x = sym('x^2 – 5*x + 6') F2x = sym ('x – 3') F3x = sym('(x + 5)^2') F4x = sym('5*x – y + 2*x – y')	F1x = x^2 – 5*x + 6 F2x = x – 3 F3x = (x + 5)^2 F4x = 5*x – y + 2*x – y
factor	When possible, factorizes the function into simpler terms.	factor(Fx1)	(x – 2)*(x – 3)
simplify	Simplifies the function.	simplify(F1x /F2x)	x–2
expand	Expands the function.	expand(F3x)	x^2 + 10*x+25
collect	Simplifies a symbolic expression by collecting like coefficients.	collect(F4x)	7*x–2*y
solve	Solves the expression for its roots.	solve(F1x)	x = 2 and x = 3
ezplot(f, min, max)	Plots the function f for values of its argument between min and max	ezplot(F1x,0,5)	See Figure 15.28

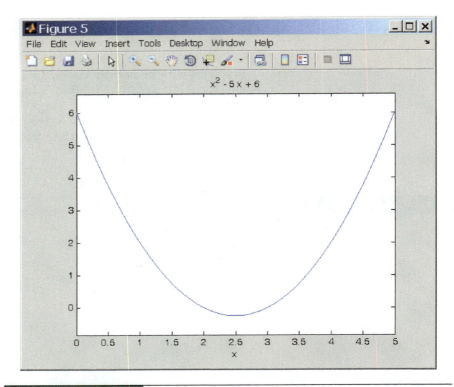

FIGURE 15.28   The `ezplot` for Example 15.7; see the last row in Table 15.14

Solutions of Simultaneous Linear Equations

In this section, we will show how you can use MATLAB's symbolic solvers to obtain solutions to a set of linear equations.

EXAMPLE 15.8

Consider the following three linear equations with three unknowns: x, y, and z.

$$2x + y + z = 13$$
$$3x + 2y + 4z = 32$$
$$5x - y + 3z = 17$$

In MATLAB, the solve command is used to obtain solutions to symbolic algebraic equations. The basic form of the **solve** command is `solve('eqn1','eqn2', . . . ,'eqn')`. As shown below, we define each equation first and then use the **solve** command to obtain the solution.

```
>>equation_1 = '2*x+y+z=13';
>>equation_2 = '3*x+2*y+4*z=32';
>>equation_3 = '5*x-y+3*z=17';
>>[x,y,z] = solve(equation_1,equation_2,equation_3)
```

The solution is given by $x = 2$, $y = 5$, and $z = 4$. The MATLAB Command Window for Example 15.8 is shown in Figure 15.29.

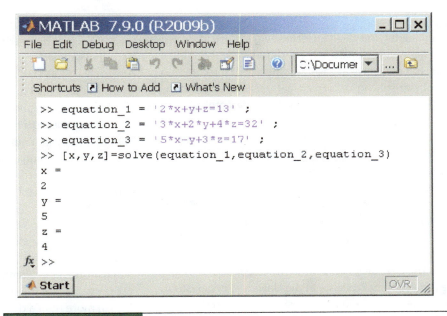

FIGURE 15.29   The solution of the set of linear equations discussed in Example 15.8.

MATLAB's symbolic capabilities allow you to solve engineering problems and present their solution using symbols, such as *x*, instead of numerical values.

As we said at the beginning of this chapter, there are many good textbooks that discuss the capabilities of MATLAB to solve a full range of problems. Here, our intent was to introduce only some basic ideas so that you can perform some essential operations. As you continue your engineering education in other classes, you will learn more about how to use MATLAB effectively to solve a wide range of engineering problems.

Before You Go On

Answer the following questions to test your understanding of the preceding section.

1. Explain the steps that you would follow to plot in MATLAB.

2. Explain how you would import an Excel file into MATLAB.

3. Give examples of MATLAB's matrix operations.

4. Explain how you would fit a curve to a set of data in MATLAB.

5. Give examples of MATLAB's symbolic operations.

Steve Chapman

Engineering has been an interesting and exciting career for me. I have worked in many different areas during my life, but they have all taken advantage of the common problem-solving skills that I learned as an engineering student and have applied to solving life's problem since.

I graduated in Electrical Engineering from Louisiana State University in 1975, and then served as an officer in the U.S. Navy for 4 years. The Navy got me out of Louisiana and gave me a chance to see the country for the first time, since I was posted to Mare Island, California and Orlando, Florida. I used one aspect of my engineering skill in this job, because I served primarily as an instructor in Electrical Engineering at the U.S. Navy Nuclear Power Schools.

The next major episode in my career was serving as an Assistant Professor at the University of Houston College of Technology, while simultaneously studying digital signal processing at Rice University. The academic life was very different, but it also utilized the basic engineering skills that I learned as an undergraduate.

Next, I moved to the Massachusetts Institute of Technology's Lincoln Laboratory in Lexington, Massachusetts. There I became a radar researcher, applying the signal-processing skills picked up at Rice University to the development of new radar systems and algorithms. For nine years of this time, my family and I got to live on Kwajalein Atoll in the Republic of the Marshall Islands, working with first-rate radars used to track ICBM tests, new satellite launches, and so forth. It was a great life in a tropical paradise; we biked down to the airport every morning and flew to work!

I also spent three years doing seismic signal processing research at Shell Oil Company in Houston, TX. Here I turned the same engineering and signal-processing skills to a totally different domain–finding oil. This job was exciting in a very different way than the work for MIT, but equally satisfying.

In 1995, my family and I immigrated to Australia, and we now live in Melbourne. I work for BAE SYSTEMS Australia, designing software programs that model the defense of naval ships or taskgroups against attacking aircraft and missiles. This work takes me around the world to navies and research labs in more than a dozen countries. It builds on all the disparate components of my earlier career: naval experience, radar, signal processing, missiles, and so forth. It is very exciting and challenging, and yet I am building on exactly the same skills I began learning at LSU so long ago.

Add in a few odds and ends along the way (such as textbook writing on electrical machinery, MATLAB, Fortran, Java, etc.), and I have had as fun, diverse, and exciting a career as anyone could ask. We have seen the United States and the world along the way. My employers have paid me a good salary to go to work each day and have fun. What more could you ask for in a career?

Source: Steve Chapman

SUMMARY

LO¹   MATLAB Basics

You can use MATLAB to solve a wide range of engineering problems. Once in the MATLAB environment, you can assign values to a variable or define the elements of a matrix. By now, you should know that the elements of the matrix are enclosed in brackets [ ] and are separated by blank space, and the elements of each row are separated by a semicolon. MATLAB also offers a number of commands that you can use to display the results of your calculation.

LO²   MATLAB Functions, Loop Control, and Conditional Statements

MATLAB offers a large selection of mathematical, trigonometric, statistical, engineering, and logical functions that you can use to analyze data. Also, when writing a computer program to solve an engineering problem, it often becomes necessary to execute a line or a block of your computer code many times. MATLAB provides *for* and *while* commands for such situations. MATLAB also offers logical operators that allow you to test various conditions. Moreover, sometimes when solving engineering problems it becomes necessary to execute a line or a block of code based on whether a condition or set of conditions is met. MATLAB provides *if* and *else* commands for such situations.

LO³ Plotting with MATLAB

MATLAB also offers many choices when it comes to creating charts. For example, you can create *x–y* charts, column charts (or histograms), contour, or surface plots. The MATLAB command that most commonly is used to plot *y* values versus *x* values is the plot(*x*,*y*) command. After you have created a chart, you can use MATLAB's graph property editor to change line type, thickness, color, etc.

LO⁴ Matrix Computation with MATLAB

The formulation of many engineering problems leads to a system of algebraic equations that must be solved simultaneously. MATLAB is a very powerful tool, especially for manipulating matrices; in fact, it originally was designed for that purpose. Common matrix operations that can be carried out by MATLAB include addition, subtraction, multiplication, transpose, inverse, and determinant. With MATLAB, you also can use the Gauss elimination method to solve a set of linear equations.

LO⁵ Symbolic Mathematics with MATLAB

At times, it becomes necessary to present a problem and its solution using symbols such as *x* instead of numerical values. MATLAB also offers a wide range of capabilities to deal with such situations.

KEY TERMS

Command History 543
Command Window 543
Curve Fitting 572
disp 545
Element-by-Element Operation 548
for Command 555

format 545
fprintf 545
if Statement 557
if, else Statement 557
Importing Excel Files 567
Matrix 545

Matrix Operation 548
M-file 559
Plotting 561
Symbolic Mathematics 573
while Command 555
Workspace 547

APPLY WHAT YOU HAVE LEARNED

As we discussed in Chapter 11, on November 1, 2001, the National Weather Service implemented a new windchill temperature (WCT) index designed to more accurately calculate how cold air feels on human skin. The current index is based on heat loss from exposed skin and was tested on human subjects. The windchill chart shown includes a frostbite indicator, showing the points where temperature, wind speed, and exposure time will produce frostbite on humans. The given chart includes three shaded areas of frostbite danger. Each shaded area shows how long (30, 10, and 5 minutes) a person can be exposed before frostbite develops. For example, a temperature of 0° F and a wind speed of -15 mph will produce a windchill temperature of -19° F. Under these conditions, exposed skin can freeze in -30 minutes. Write a MATLAB program based on a user input for temperature and wind speed that calculates the windchill temperature and provides an appropriate warning.

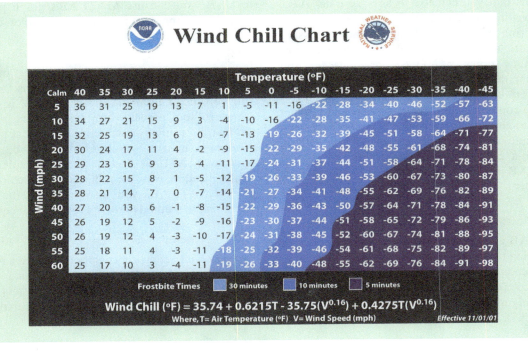

Wind Chill Chart

Temperature (°F)

Calm	40	35	30	25	20	15	10	5	0	-5	-10	-15	-20	-25	-30	-35	-40	-45
5	36	31	25	19	13	7	1	-5	-11	-16	-22	-28	-34	-40	-46	-52	-57	-63
10	34	27	21	15	9	3	-4	-10	-16	-22	-28	-35	-41	-47	-53	-59	-66	-72
15	32	25	19	13	6	0	-7	-13	-19	-26	-32	-39	-45	-51	-58	-64	-71	-77
20	30	24	17	11	4	-2	-9	-15	-22	-29	-35	-42	-48	-55	-61	-68	-74	-81
25	29	23	16	9	3	-4	-11	-17	-24	-31	-37	-44	-51	-58	-64	-71	-78	-84
30	28	22	15	8	1	-5	-12	-19	-26	-33	-39	-46	-53	-60	-67	-73	-80	-87
35	28	21	14	7	0	-7	-14	-21	-27	-34	-41	-48	-55	-62	-69	-76	-82	-89
40	27	20	13	6	-1	-8	-15	-22	-29	-36	-43	-50	-57	-64	-71	-78	-84	-91
45	26	19	12	5	-2	-9	-16	-23	-30	-37	-44	-51	-58	-65	-72	-79	-86	-93
50	26	19	12	4	-3	-10	-17	-24	-31	-38	-45	-52	-60	-67	-74	-81	-88	-95
55	25	18	11	4	-3	-11	-18	-25	-32	-39	-46	-54	-61	-68	-75	-82	-89	-97
60	25	17	10	3	-4	-11	-19	-26	-33	-40	-48	-55	-62	-69	-76	-84	-91	-98

Wind (mph)

Frostbite Times   ☐ 30 minutes   ☐ 10 minutes   ☐ 5 minutes

Wind Chill (°F) = $35.74 + 0.6215T - 35.75(V^{0.16}) + 0.4275T(V^{0.16})$

Where, T= Air Temperature (°F) V= Wind Speed (mph)   *Effective 11/01/01*

PROBLEMS

Problems that promote lifelong learning are denoted by 🔑

15.1 Using the MATLAB Help menu, discuss how the following functions are used. Create a simple example, and demonstrate the proper use of the function.

a. ABS (X)

b. TIC, TOC

c. SIZE (x)

d. FIX (x)

e. FLOOR (x)

f. CEIL (x)

g. CALENDAR

15.2 In Chapter 10, we discussed fluid pressure and the role of water towers in small towns. Use MATLAB to create a table that shows the relationship between the height of water above ground in the water tower and the water pressure in a pipeline located at the base of the water tower. The relationship is given by

$$P = \rho g h$$

where

P = water pressure at the base of the water tower in pounds per square foot (lb/ft²)

ρ = density of water in slugs per cubic foot ($\rho = 1.94$ slugs/ft³)

g = acceleration due to gravity ($g = 32.2$ ft/s²)

h = height of water above ground in feet (ft)

Create a table that shows the water pressure in lb/in² in a pipe located at the base of the water tower as you vary the height of the water in increments of 10 ft. Also, plot the water pressure (lb/in²) versus the height of water in feet. What should the water level in the water tower be to create 80 psi of water pressure in a pipe at the base of the water tower?

15.3 As we explained in Chapter 10, viscosity is a measure of how easily a fluid flows. The viscosity of water can be determined from the following correlation:

$$\mu = c_1 10^{\left(\frac{c_2}{T - c_3}\right)}$$

where

$$\mu \equiv \text{viscosity (N/s} \cdot \text{m}^2)$$
$$T \equiv \text{temperature (K)}$$
$$c_1 \equiv 2.414 \times 10^{-5} \text{ N/s} \cdot \text{m}^2$$
$$c_2 \equiv 247.8 \text{ K}$$
$$c_3 \equiv 140 \text{ K}$$

Using MATLAB, create a table that shows the viscosity of water as a function of temperature in the range of 0° C (273.15 K) to 100° C (373.15 K) in increments of 5° C. Also, create a graph showing the value of viscosity as a function of temperature.

15.4 Using MATLAB, create a table that shows the relationship between the units of temperature in degrees Celsius and Fahrenheit in the range of −50° C to 150° C. Use increments of 10° C.

15.5 Using MATLAB, create a table that shows the relationship among units of the height of people in centimeters, inches, and feet in the range of 150 cm to 2 m. Use increments of 5 cm.

15.6 Using MATLAB, create a table that shows the relationship among the units of mass to describe people's mass in kilograms, slugs, and pound mass in the range of 20 kg to 120 kg. Use increments of 5 kg.

15.7 Using MATLAB, create a table that shows the relationship among the units of pressure in Pa, psi, and in. of water in the range of 1000 Pa to 10000 Pa. Use increments of 500 Pa.

15.8 Using MATLAB, create a table that shows the relationship between the units of pressure in Pa and psi in the range of 10 kPa to 100 kPa. Use increments of 0.5 kPa.

15.9 Using MATLAB, create a table that shows the relationship between the units of power in Watts and horsepower in the range of 100 W to 10000 W. Use smaller increments of 100 W up to 1000 W, and then use increments of 1000 W all the way up to 10000 W.

15.10 As we explained in earlier chapters, the air resistance to the motion of a vehicle is something important that engineers investigate. The drag force acting on a car is determined experimentally by placing the car in a wind tunnel. The air speed inside the tunnel is changed, and the drag force acting on the car is measured. For a given car, the experimental data generally is represented by a single coefficient that is called drag coefficient. It is defined by the following relationship:

$$C_d = \frac{F_d}{\frac{1}{2}\rho V^2 A}$$

where

$$C_d = \text{drag coefficient (unitless)}$$
$$F_d = \text{measured drag force (N or lb)}$$
$$\rho = \text{air density (kg/m}^3 \text{ or slugs/ft}^3)$$
$$V = \text{air speed inside the wind tunnel}$$
$$\text{(m/s or ft/s)}$$
$$A = \text{frontal area of the car (m}^2 \text{ or ft}^2)$$

The frontal area A represents the frontal projection of the car's area and could be approximated simply by multiplying 0.85 times the width and the height of a rectangle that outlines the front of the car. This is the area that you see when you view the car from a direction normal to the front grill. The 0.85 factor is used to adjust for rounded corners, open space below the bumper, and so on. To give you some idea, typical drag coefficient values for sports cars are between 0.27 to 0.38 and for sedans are between 0.34 to 0.5.

The power requirement to overcome air resistance is computed by

$$P = F_d V$$

where

$$P = \text{power (W or ft} \cdot \text{lb/s)}$$
$$1 \text{ horsepower (hp)} = 550 \text{ ft} \cdot \text{lb/s}$$

and

$$1 \text{ horsepower (hp)} = 746 \text{ W}$$

The purpose of this problem is to see how the power requirement changes with the car speed and the air temperature. Determine the power requirement to overcome the air resistance for a car that has a listed drag coefficient of 0.4 and width of 74.4 inches and height of 57.4 inches. Vary the air speed in the range of $15 \text{ m/s} < V < 35 \text{ m/s}$, and change the air density range of $1.11 \text{ kg/m}^3 < \rho < 1.29 \text{ kg/m}^3$. The given air density range corresponds to 0° C to 45° C. You may use the ideal gas law to relate the density of the air to its temperature. Present your findings in both kilowatts and horsepower. Discuss your findings in terms of power consumption as a function of speed and air temperature.

15.11 The cantilevered beam shown in the accompanying figure is used to support a load acting on a balcony. The deflection of the centerline of the beam is given by the following equation

$$y = \frac{-wx^2}{24EI}(x^2 - 4Lx + 6L^2)$$

where

$y =$ deflection at a given x location (m)

$w =$ distributed load (N/m)

$E =$ modulus of elasticity (N/m^2)

$I =$ second moment of area (m^4)

$x =$ distance from the support as shown (m)

$L =$ length of the beam (m)

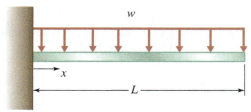

Problem 15.11

Using MATLAB, plot the deflection of a beam whose length is 5 m with the modulus of elasticity of $E = 200$ GPa and $I = 99.1 \times 10^6 \text{ mm}^4$. The beam is designed to carry a load of 10000 N/m. What is the maximum deflection of the beam?

15.12 Fins, or extended surfaces, commonly are used in a variety of engineering applications to enhance cooling. Common examples include a motorcycle engine head, a lawn mower engine head, extended surfaces used in electronic equipment, and finned tube heat exchangers in room heating and cooling applications. Consider aluminum fins of a rectangular profile shown in Problem 14.13, which are used to remove heat from a surface whose temperature is 100° C. The temperature of the ambient air is 20° C. We are interested in determining how the temperature of the fin varies along its length and plotting this temperature variation. For long fins, the temperature distribution along the fin is given by

$$T - T_{\text{ambient}} = (T_{\text{base}} - T_{\text{ambient}})e^{-mx}$$

where

$$m = \sqrt{\frac{hp}{kA}}$$

and

$h =$ the heat transfer coefficient (W/m$^2 \cdot$ K)

$p =$ perimeter $2 * (a + b)$ of the fin (m)

$A =$ cross-sectional area of the fin $(a * b)$ (m^2)

$k =$ thermal conductivity of the fin material (W/m $\cdot$ K)

Plot the temperature distribution along the fin using the following data: $k = 168$ W/m $\cdot$ K, $h = 12$ W/m$^2 \cdot$ K, $a = 0.05$ m, and $b = 0.01$ m. Vary x from 0 to 0.1 m in increments of 0.01 m.

15.13 A person by the name of Huebscher developed a relationship between the equivalent size of round ducts and rectangular ducts according to

$$D = 1.3\frac{(ab)^{0.625}}{(a + b)^{0.25}}$$

where

$D =$ diameter of equivalent circular duct (mm)

$a =$ dimension of one side of the rectangular duct (mm)

$b =$ dimension of the other side of the rectangular duct (mm)

Using MATLAB, create a table that shows the relationship between the circular and the rectangular duct, similar to the one shown in the accompanying table.

	Length of One Side of Rectangular Duct (length a), mm				
Length b	100	125	150	175	200
400					
450					
500					
550					
600					

15.14 A Pitot tube is a device commonly used in a wind tunnel to measure the speed of the air flowing over a model. The air speed is measured from the following equation:

$$V = \sqrt{\frac{2P_d}{\rho}}$$

where

V = air speed (m/s)
P_d = dynamic pressure (Pa)
ρ = density of air (1.23 kg/m^3)

Using MATLAB, create a table that shows the air speed for the range of dynamic pressure of 500 Pa to 800 Pa. Use increments of 50 Pa.

15.15 Use MATLAB to solve Example 7.1. Recall that we applied the trapezoidal rule to determine the area of the shape given. Create an Excel file with the given data, and then import the file into MATLAB.

15.16 We will discuss engineering economics in Chapter 20. Using MATLAB, create a table that can be used to look up monthly payments on a car loan for a period of five years. The monthly payments are calculated from

$$A = P\left[\frac{\left(\frac{i}{1200}\right)\left(1+\frac{i}{1200}\right)^{60}}{\left(1+\frac{i}{1200}\right)^{60}-1}\right]$$

where

A = monthly payments in dollars
P = loan in dollars
i = interest rate; e.g., 7, 7.5, ..., 9

	Interest Rate				
Loan	7	7.5	8	8.5	9
10,000					
15,000					
20,000					
25,000					

15.17 A person by the name of Sutterland has developed a correlation that can be used to evaluate the viscosity of air as a function of temperature. It is given by

$$\mu = \frac{c_1 T^{0.5}}{1 + \frac{c_2}{T}}$$

where

μ = viscosity $(N/s \cdot m^2)$
T = temperature (K)
$c_1 = 1.458 \times 10^{-6} \left(\frac{kg}{m \cdot s \cdot K^{1/2}}\right)$
$c_2 = 110.4 \text{ K}$

Create a table that shows the viscosity of air as a function of temperature in the range of $0° C$ (273.15 K) to $100° C$ (373.15 K) in increments of $5° C$. Also, create a graph showing the value of viscosity as a function of temperature.

15.18 In Chapter 11, we explained the concept of windchill factors. The old windchill values were determined empirically, and the common equivalent windchill temperature $T_{equivalent}$ (°C) was given by

$$T_{equivalent} = 0.045(5.27 \, V^{0.5} + 10.45 - 0.28 \, V) \cdot (T_a - 33) + 33$$

Create a table that shows the windchill temperatures for the range of ambient air temperature of $-30° \, C < T_a < 10° \, C$ and wind speeds of $5 \, m/s < V < 20 \, m/s$.

15.19 Given the matrices:

$$[A] = \begin{bmatrix} 4 & 2 & 1 \\ 7 & 0 & -7 \\ 1 & -5 & 3 \end{bmatrix}, \, [B] = \begin{bmatrix} 1 & 2 & -1 \\ 5 & 3 & 3 \\ 4 & 5 & -7 \end{bmatrix}, \text{ and}$$

$$\{C\} = \begin{Bmatrix} 1 \\ -2 \\ 4 \end{Bmatrix}, \text{ perform the following operations}$$

using MATLAB.

a. $[A] + [B] = ?$

b. $[A] - [B] = ?$

c. $3[A] = ?$

d. $[A][B] = ?$

e. $[A]\{C\} = ?$

15.20 Given the following matrices:

$$[A] = \begin{bmatrix} 2 & 10 & 0 \\ 16 & 6 & 14 \\ 12 & -4 & 18 \end{bmatrix}, \text{ and } [B] = \begin{bmatrix} 2 & 10 & 0 \\ 4 & 20 & 0 \\ 12 & -4 & 18 \end{bmatrix},$$

calculate the determinant of [A] and [B] using MATLAB.

15.21 Solve the following set of equations using MATLAB.

$$\begin{bmatrix} 10875000 & -1812500 & 0 \\ -1812500 & 6343750 & -4531250 \\ 0 & -4531250 & 4531250 \end{bmatrix} \begin{Bmatrix} u_2 \\ u_3 \\ u_4 \end{Bmatrix}$$

$$= \begin{Bmatrix} 0 \\ 0 \\ 800 \end{Bmatrix}$$

15.22 Solve the following set of equations using MATLAB.

$$\begin{bmatrix} 1 & 1 & 1 \\ 2 & 5 & 1 \\ -3 & 1 & 5 \end{bmatrix} \begin{Bmatrix} x_1 \\ x_2 \\ x_3 \end{Bmatrix} = \begin{Bmatrix} 6 \\ 15 \\ 14 \end{Bmatrix}$$

15.23 Solve the following set of equations using MATLAB.

$$\begin{bmatrix} 7.11 & -1.23 & 0 & 0 & 0 \\ -1.23 & 1.99 & -0.76 & 0 & 0 \\ 0 & -0.76 & 0.851 & -0.091 & 0 \\ 0 & 0 & -0.091 & 2.31 & -2.22 \\ 0 & 0 & 0 & -2.22 & 3.69 \end{bmatrix} \begin{Bmatrix} T_1 \\ T_2 \\ T_3 \\ T_4 \\ T_5 \end{Bmatrix} = \begin{Bmatrix} (5.88)(20) \\ 0 \\ 0 \\ 0 \\ (1.47)(70) \end{Bmatrix}$$

15.24 Solve the following set of equations using MATLAB.

$$
10^5
\begin{bmatrix}
7.2 & 0 & 0 & 0 & -1.49 & -1.49 \\
0 & 7.2 & 0 & -4.22 & -1.49 & -1.49 \\
0 & 0 & 8.44 & 0 & -4.22 & 0 \\
0 & -4.22 & 0 & 4.22 & 0 & 0 \\
-1.49 & -1.49 & -4.22 & 0 & 5.71 & 1.49 \\
-1.49 & -1.49 & 0 & 0 & 1.49 & 1.49
\end{bmatrix}
\begin{Bmatrix}
x_1 \\ x_2 \\ x_3 \\ x_4 \\ x_5 \\ x_6
\end{Bmatrix}
=
\begin{Bmatrix}
0 \\ 0 \\ 0 \\ -500 \\ 0 \\ -500
\end{Bmatrix}
$$

15.25 Find the equation that best fits the following set of data points. Compare the actual and predicted y values. Plot the data first.

x	0	1	2	3	4	5	6	7	8	9	10
y	10	12	15	19	23	25	27	32	34	36	41

15.26 Find the equation that best fits the following set of data points. Compare the actual and predicted y values. Plot the data first.

x	0	1	2	3	4	5	6	7	8
y	5	8	15	32	65	120	203	320	477

15.27 Find the equation that best fits the following set of data points. Compare the actual and predicted y values. Plot the data first.

x	0	5	10	15	20	25	30	35	40	45	50
y	100	101.25	105	111.25	120	131.25	145	161.25	180	201.25	225

15.28 Find the equation that best fits the force-deflection data points given in Table 10.1(a). Compare the actual and predicted force values. Plot the data first.

Force (N)	Spring Deflection (mm)
5	9
10	17
15	29
20	35

15.29 The data given in the accompanying table represents the velocity distribution inside a pipe. Find the equation that best fits the fluid velocity–radial distance data given. Compare the actual and predicted velocity values. Plot the data first.

Radial Distance (m)	Fluid Velocity (m/s)
−0.1	0
−0.08	0.17
−0.06	0.33
−0.04	0.42
−0.02	0.49
0 center of pipe	0.5
0.02	0.48
0.04	0.43
0.06	0.32
0.08	0.18
0.1	0

15.30 The data given in the accompanying table represents the cooling temperature of a plate as a function of time during a material processing stage. Find the equation that best fits the temperature-time data given. Compare the actual and predicted temperature values. Plot the data first.

Temperature (°C)	Time (hr)
900	0
722	0.2
580	0.4
468	0.6
379	0.8
308	1.0
252	1.2
207	1.4
172	1.6
143	1.8
121	2.0
103	2.2
89	2.4
78	2.6
69	2.8
62	3.0

15.31 Write a function named **Weight** that when called in the command window will compute the weight in pound force based on an input of mass in kilograms.

15.32 The Body Mass Index (BMI) is a way of determining obesity and whether someone is overweight. It is computed from

$$BMI = \frac{mass \text{ (in kg)}}{[height \text{ (in meter)}]^2}.$$

Write a program that will create the table shown in accompanying figure. The BMI values in the range of 18.5–24.9, 25.0–29.9, and > 30.0 are considered healthy, overweight, and obese, respectively.

Height (m)	Mass (kg)						
	50	**55**	**60**	**65**	**70**	**75**	**80**
1.5	22.2	24.4	26.7	28.9	31.1	33.3	35.6
1.6	19.5	21.5	23.4	25.4	27.3	29.3	31.3
1.7	17.3	19.0	20.8	22.5	24.2	26.0	27.7
1.8	15.4	17.0	18.5	20.1	21.6	23.1	24.7
1.9	13.9	15.2	16.6	18.0	19.4	20.8	22.2

For problems 15.33 through 15.40, go through each line of MATLAB code and show the result, or indicate if an error will occur by executing the commands.

15.33
```
A = [3 1 0;4 0 1];
B = [−3 1 4;5 6 −1];
C = [A,B]
```

15.34
```
B = [1;0;−5;4];
D = B′
```

15.35
```
X = [4 5 − 2 8];
Y = [2 0 1 10];
Z = (X < Y)
Z = Y(X ~= Y)
```

15.36
```
i = 5
for k = 1:3:7
i = i + k;
end
value = i
```

15.37
```
A = [5 0 2];
B = [1 5 4];
C = B.^A
```

15.38
```
i = 10;
k = 2;
while k < 4
i = i + k;
k = k + 1;
end
value = i
```

15.39
```
A = [3 1 0;4 0 1];
B = [−3; 1; 4];
C = A*B
```

15.40
```
X = [0:1:5];
Y = X.^2 + 4;
plot(X,Y)
xlabel('X')
ylabel('Y')
```

Engineering Graphical Communication

Conveying Information to other Engineers, Machinists, Technicians, and Managers

Super3D/Shutterstock.com

Engineers use technical drawings to convey useful information to others in a standard manner. An engineering drawing provides information, such as the shape of a product, its dimensions, materials from which to fabricate the product, and assembly steps. Some engineering drawings are specific to a particular discipline. For example, civil engineers deal with land or boundary, topographic, construction, and route survey drawings. Electrical and electronic engineers, on the other hand, could deal with printed circuit-board assembly drawings, printed circuit-board drill plans, and wiring diagrams. Engineers also use special symbols and signs to convey their ideas, analyses, and solutions to problems. In Part Four of this book, we will introduce you to the principles and rules of engineering graphical communication and engineering symbols. A good grasp of these principles will enable you to convey and understand information effectively.

CHAPTER 16   ENGINEERING DRAWINGS AND SYMBOLS

Engineering Drawings and Symbols

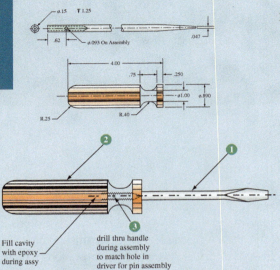

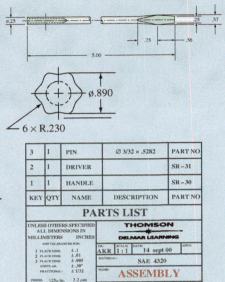

Fill cavity
with epoxy
during assy

drill thru handle
during assembly
to match hole in
driver for pin assembly

3	1	PIN	Ø 3/32 × .5282	PART NO
2	1	DRIVER		SR – 31
1	1	HANDLE		SR – 30
KEY	QTY	NAME	DESCRIPTION	PART NO

PARTS LIST

UNLESS OTHERS SPECIFIED ALL DIMENSIONS IN MILLIMETERS INCHES			THOMSON DELMAR LEARNING	
AND TOLERANCES FOR:		DR. AKR	SCALE: 1 : 1	DATE: 14 sept 00	APPD:
1 PLACE DIMS ±.1 2 PLACE DIMS ±.01 3 PLACE DIMS ±.005 ANGULAR: ±.30° FRACTIONAL: ±1/32		MATERIAL: SAE 4320		
FINISH: 125μ in. 3.2 μm		NAME: ASSEMBLY		
START USED ON:	SIMILAR TO:	SIZE B	PART NO SR – ASSY	REV 0

Based on Madsen, Engineering Drawing and Design, 4e. Delmar Learning, a part of Cengage Learning, Inc., 2007.

Engineering drawings, such as the schematics shown here, are important in conveying useful information to other engineers or machinists in a standard manner that allows for visualization of the proposed product. Important information, such as the shape of the product, its size, type of material used, and the assembly steps required, are provided by these drawings.

LEARNING OBJECTIVES

LO[1] **Mechanical Drawings:** explain the basic rules that are followed by engineers, draftspersons, and machinists to draw or read mechanical drawings

LO[2] **Civil, Electrical, and Electronic Drawings:** explain why we need discipline-specific drawings and give some examples

LO[3] **Solid Modeling:** explain the basic ideas that solid-modeling software uses to create models of objects with surfaces and volumes that look almost indistinguishable from the actual objects

LO[4] **Engineering Symbols:** explain why engineering symbols are needed and give some examples of common symbols in civil, electrical, and mechanical engineering

ENGINEERING GRAPHICAL COMMUNICATION

Conveying Information to other Engineers, Machinists, Technicians, and Managers

Engineers use technical drawings to convey useful information to others in a standard manner. An engineering drawing provides information, such as the shape of a product, its dimensions, materials to be used to fabricate the product, and assembly steps. Today, with solid-modeling software, we can create models of objects with surfaces and volumes that look almost indistinguishable from the actual objects. The solid-modeling software allows for experimenting on a computer screen with the assembly of parts to examine any unforeseen problems before the parts are actually made and assembled. Some engineering drawings are specific to a particular discipline. For example, civil engineers deal with land or boundary, topographic, construction, and route survey drawings. Electrical and electronic engineers, on the other hand, could deal with printed circuit-board assembly drawings, printed circuit-board drill plans, and wiring diagrams. Engineers also use special symbols and signs to convey their ideas, analyses, and solutions to problems. A good grasp of these principles will enable you to convey and understand information effectively.

To The Students: Have you ever had an idea for a product or a service? How did you go about conveying your idea to others?

Traffic signs are designed and developed based on some acceptable national and international standards to convey information not only effectively but also quickly. For example, a stop sign, which has an octagon shape with a red background, tells you to bring your car to a complete stop. When autonomous (self-driving) cars become reality, will we still need road signs?

Engineers use a special kind of drawings, called engineering drawings, to convey their ideas and design information about products. These drawings portray vital information, such as the shape of the product, its size, type of materials used, and assembly steps. Moreover, machinists use the information provided by engineers or draftspersons on the engineering drawings to make the parts. For complicated systems made of various parts, the drawings also serve as a how-to-assemble guide, showing how the various parts fit together. Most of you will eventually take a semester-long class in engineering drawing where you will learn in much more detail how to create such drawings. For now, the following sections provide a brief introduction to engineering graphical communication principles. We will discuss why engineering drawings are important, how they are drawn, and what rules must be followed to create such drawings. Engineering symbols and signs also provide valuable information. These symbols are a "language" used by engineers to convey their ideas, solutions to problems, or analyses of certain situations. In this chapter, we will also discuss the need for conventional engineering symbols and will show some common symbols used in civil, electrical, and mechanical engineering.

LO¹ 16.1 Mechanical Drawings

Have you ever had an idea about a new product that could make a certain task easier? How did you get your idea across to other people? What were the first things you did to make your idea clearly known to your audience? Imagine you are having a cup of coffee with a friend, and you decide to share your idea about a product with her. After talking about the idea for a while, to clarify your idea, you will naturally draw a picture or a diagram to show what the product would look like. You have heard the saying "a picture is worth a thousand words"; well, in engineering, a good drawing is worth even more words! Technical drawings or engineering drawings are important in conveying useful information to other engineers or machinists in a standard acceptable manner to allow the readers of these drawings to visualize what the proposed product would look like. More significantly, information such as the dimensions of the proposed product or what it would look like when viewed from the top or from the side or the front is provided. The drawings will also specify what type of material is to be used to make this product. In order to draw or read engineering drawings, you must first learn a set of standard rules that are followed by all engineers, draftspersons, and machinists. In the next sections, we will briefly discuss these rules.

Orthographic Views

Orthographic views (diagrams) show what an object's projection looks like when seen from the top, the front, or the side. To better understand what we mean by orthographic views, imagine that you have placed the object shown in Figure 16.1 in the center of a glass box.

> Orthographic views show what an object's projection looks like when seen from the top, the front, or the side.

Now if you were to draw perpendicular lines from the corners of the object into the faces of the glass box, you would see the outlines shown in Figure 16.1. The outlines are called the orthographic projection of the object into the *horizontal*, *vertical*, and the *profile planes*.

Now imagine that you open up or unfold the faces of the glass box that have the projections of the object. The unfolding of the glass faces will result in the layout shown in Figure 16.2. Note the relative locations of the top view, the bottom view, the front view, the back view, the right-side view, and the left-side view.

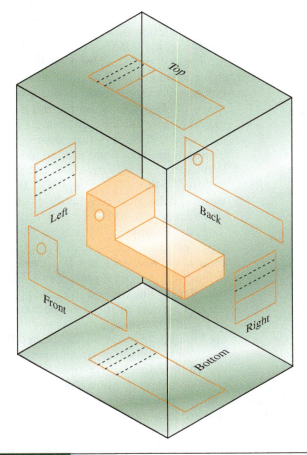

FIGURE 16.1   The orthographic projection of an object into the horizontal, vertical, and profile planes.

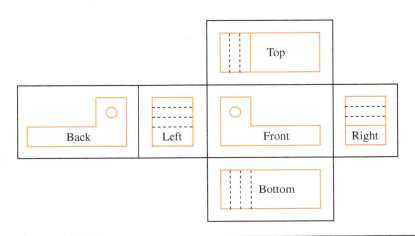

FIGURE 16.2   The relative locations of the top, bottom, front, back, right-side, and left-side views.

At this point, you may realize that the top view is similar to the bottom view, the front view is similar to the back view, and the right-side view is similar to the left-side view. Therefore, you notice some redundancy in the information provided by these six views (diagrams). Therefore, you conclude that you do not need to draw all six views to describe this object. In fact, the number of views needed to describe an object depends on how complex the shape of an object is. So the question is, then, how many views are needed to completely describe the object? For the object shown in Figure 16.1, three views are sufficient to fully describe the object, because only three principal planes of projection are needed to show the object. For the example shown in Figure 16.1, we may decide to use the top, the front, and the right-side views to describe the object completely. In fact, the top, the front, and the right-side views are the most commonly created views to describe most objects. These views are shown in Figure 16.3.

From examining Figure 16.3, we should also note that three different types of lines are used in the orthographic views to describe the object: *solid lines, hidden* or *dashed lines,* and *centerlines.* The solid lines on the orthographic views represent the visible edges of planes or the intersection of two planes. The dashed lines (hidden lines), on the other hand, represent an edge of a plane or the extreme limits of a cylindrical hole inside the object, or the intersection of two planes that are not visible from the direction you are looking. In other words, dashed lines are used when some material exists between the observer (from where he or she is looking) and the actual location of the edge. Referring to Figure 16.3, when you view the object from the right side, its projection contains the limits of the hole within the object. The right-side projection of the object also contains the intersection of two planes that are located on the object. Therefore, the solid and dashed lines are used to show these edges and limits. The third type of line that is employed in orthographic projections is the centerline, or the line of symmetry, which shows where the centers of holes or the centers of cylinders are. Pay close attention to the difference in the line patterns between a dashed line and a centerline. Examples of solid lines, dashed lines, and lines of symmetry are shown in Figure 16.3.

As we said earlier, the number of views that you should draw to represent an object will depend on how complex the object is. For example, if you want to show a bolt washer or a gasket, you need to draw only a single top view and specify the thickness of the washer or the gasket. For other objects, such as bolts, we may draw only two views. Examples of objects requiring one or two views are shown in Figure 16.4.

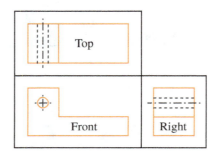

FIGURE 16.3   The top, front, and the right-side views of an object.

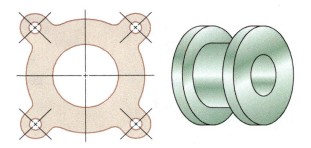

FIGURE 16.4   Examples of objects requiring one or two views.

EXAMPLE 16.1   Draw the orthographic views of the object shown in Figure 16.5(a).

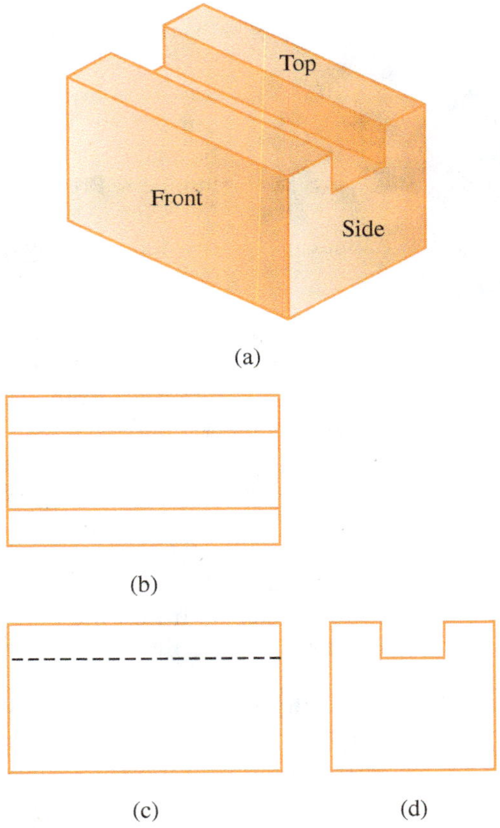

(a)

(b)

(c)   (d)

FIGURE 16.5   An object and its orthographic views: (a) the object, (b) the top view, (c) the front view, and (d) the side view.

Dimensioning and Tolerancing

Engineering drawings provide information about the shape, size, and material of a product. In the previous section, we discussed how to draw the orthographic views. We did not say anything about how to show the actual size of the object on

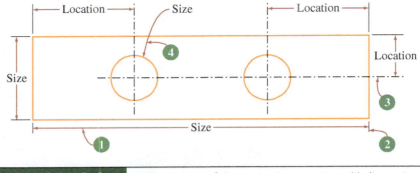

FIGURE 16.6   The basics of dimensioning practices: (1) dimension line, (2) extension line, (3) centerline, and (4) leader.

> ANSI sets the standards for the dimensioning and tolerancing practices for engineering drawings. Every engineering drawing must include dimensions, tolerances, the materials from which the product will be made, and the finished surfaces marked.

the drawings. The **American National Standards Institute (ANSI)** sets the standards for the **dimensioning** and **tolerancing** practices for engineering drawings. Every engineering drawing must include dimensions, tolerances, the materials from which the product will be made, the finished surfaces marked, and other notes such as part numbers. Providing this information on the diagrams is important for many reasons. A machinist must be able to make the part from the detailed drawings without needing to go back to the engineer or the draftsperson who drew the drawings to ask questions regarding the size, or the tolerances, or what type of material the part should be made from. There are basically two concepts that you need to keep in mind when specifying dimensions in an engineering drawing: *size* and *location*. As shown in Figure 16.6, not only do you need to specify how wide or how long an object is but you must also specify the location of the center of a hole or center of a fillet in the part. Moreover, a drawing is dimensioned with the aid of *dimension lines, extension lines, centerlines,* and *leaders.*

Dimension lines provide information on the size of the object; for example, how wide it is and how long it is. You need to show the overall dimensions of the object because the machinist can then determine the overall size of the stock material from which to make the piece. As the name implies, *extension lines* are those lines that extend from the points to which the dimension or location is to be specified. Extension lines are drawn parallel to each other, and the dimension lines are placed between them, as shown in Figure 16.6. The *leaders* are the arrows that point to a circle or a fillet for the purpose of specifying their sizes. Often the drawings are shown *Not To Scale* (NTS), and therefore a scaling factor for the drawing must also be specified. In addition to dimensions, all engineering drawings must also contain an information box with the following items: name of the person who prepared the drawing, title of the drawing, date, scale, sheet number, and drawing number. This information is normally shown on the upper- or lower-right corner of a drawing. An example of an information box is shown in Figure 16.7.

Let us now say a few words about fillets, which are often overlooked in engineering drawings, a shortcoming that could lead to problems. *Fillet* refers to the rounded edges of an object; their sizes, the radius of roundness, must be specified in all drawings. If the size of fillets is not specified in a drawing, the machinist may not round the edges; consequently, the absence of fillets could create problems or failure in parts. As some of you will learn

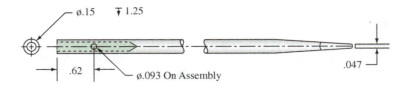

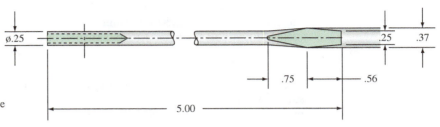

(a) The DRIVER detail drawing based on the engineer's sketch before an engineering change request (ECR) is issued. Notice the "o" in the lower right corner. This indicates an original unchanged drawing. Dimension values in this figure are in inches.

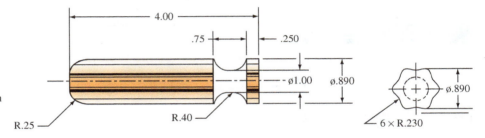

(b) Detail drawing of the HANDLE created from the engineer's sketch. Dimension values in this figure are in inches.

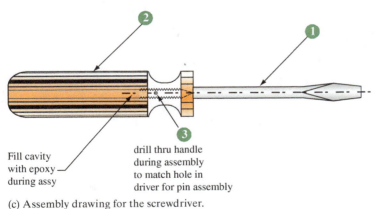

Fill cavity with epoxy during assy

drill thru handle during assembly to match hole in driver for pin assembly

(c) Assembly drawing for the screwdriver.

3	1	PIN	Ø 3/32 × .5282	PART NO
2	1	DRIVER		SR – 31
1	1	HANDLE		SR – 30
KEY	QTY	NAME	DESCRIPTION	PART NO

PARTS LIST

UNLESS OTHERS SPECIFIED
ALL DIMENSIONS IN
MILLIMETERS   INCHES
AND TOLERANCES FOR:

1 PLACE DIMS: ± .1
2 PLACE DIMS: ± .01
3 PLACE DIMS: ± .005
ANGULAR: ± .30°
FRACTIONAL: ± 1/32

FINISH: 125μ in.   3.2 μm

FIRST USED ON: SIMILAR TO:

THOMSON

DELMAR LEARNING

DR: AKR   SCALE: 1 : 1   DATE: 14 sept 00   APPD:

MATERIAL: SAE 4320

NAME: **ASSEMBLY**

SIZE: **B**   PART NO: **SR – ASSY**   REV: **0**

FIGURE 16.7   Examples of engineering drawings.
Based on Madsen, Engineering Drawing and Design, 4e. Delmar Learning, a part of Cengage Learning, Inc., 2007.

later in your mechanics of materials class, mechanical parts with sharp edges or a sudden reduction in their cross-sectional areas could fail when subjected to loads because of high stress concentrations near the sharp regions. As you will learn later, a simple way of reducing the stress in these regions is by rounding the edges and creating a gradual reduction in cross-sectional areas.

Engineered products generally consist of many parts. In today's globally driven economy, some of the parts made for a product in one place must be easily assembled with parts made elsewhere. When you specify a dimension on a drawing—say, 2.50 centimeter—how close does the actual dimension of the machined part need to be to the specified 2.50 cm for the part to fit properly with other parts in the product? Would everything fit correctly if the actual dimension of the machine part were 2.49 cm or 2.51 cm? If so, then you must specify a tolerance of ±0.01 cm on your drawing regarding this dimension. Tolerancing is a broad subject with its own rules and symbols, and as we mentioned earlier, the American National Standard Institute sets the tolerancing standards that must be followed by those creating or reading engineering drawings. Here, we have briefly introduced these ideas; you must consult the standards if you are planning to prepare an actual engineering drawing.

EXAMPLE 16.2   Show the dimensions of the object in Figure 16.8 on its orthographic views.

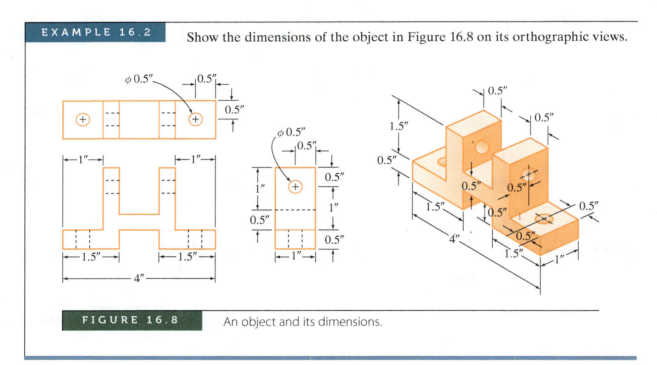

FIGURE 16.8   An object and its dimensions.

Isometric View

When it is difficult to visualize an object using only its orthographic views, an isometric sketch is also drawn. The **isometric drawing** shows the three dimensions of an object in a single view. The isometric drawings are sometimes referred to as technical illustrations and are used to show what parts or products look like in parts manuals, repair manuals, and product catalogs. Examples of isometric drawings are shown in Figure 16.9.

> The isometric drawing shows the three dimensions of an object in a single view.

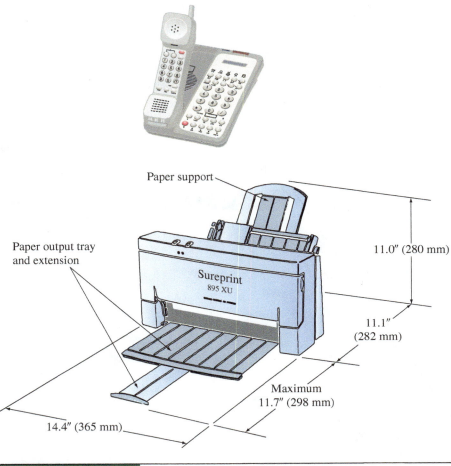

FIGURE 16.9   Examples of isometric drawings.

As was the case with orthographic views, there are specific rules that one must follow to draw the isometric view of an object. We will use the object shown in Figure 16.10 to demonstrate the steps that you need to follow to draw the isometric view of an object.

Step 1: Draw the width, height, and the depth axes, as shown in Figure 16.11(a). Note that the isometric grid consists of the width and the depth axes; they form a 30° angle with a horizontal line. Also note that the height axis makes a 90° angle with a horizontal line and a 60° angle with each of the depth and width axes.

Step 2: Measure and draw the total width, height, and depth of the object. Hence, draw lines 1–2, 1–3, and 1–4 as shown in Figure 16.11(b).

Step 3: Create the front, the top, and the side work faces. Draw line 2–5 parallel to line 1–3; draw line 4–6 parallel to line 1–3; draw line 3–5 parallel to line 1–2; draw line 3–6 parallel to line 1–4; draw line 5–7 parallel to line 3–6; and draw line 6–7 parallel to line 3–5 as shown in Figure 16.11(c).

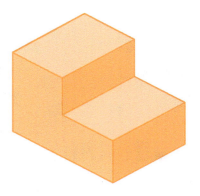

FIGURE 16.10   The object used in demonstrating the steps in creating an isometric view.

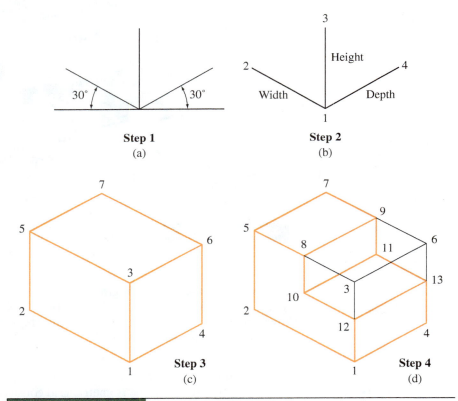

FIGURE 16.11   Steps in creating an isometric view: (1) Create the isometric axes and grid; (2) mark the height, width, and depth of the object on the isometric grid; (3) create the front, the top, and the side work faces; and (4) complete the drawing.

Step 4: Complete the drawing as marked by the remaining line numbers, and remove the unwanted lines 3–6, 3–8, 3–12, 6–13, and 6–9 as shown in Figure 16.11(d).

Next, we will use an example to demonstrate the steps in creating an isometric view of an object.

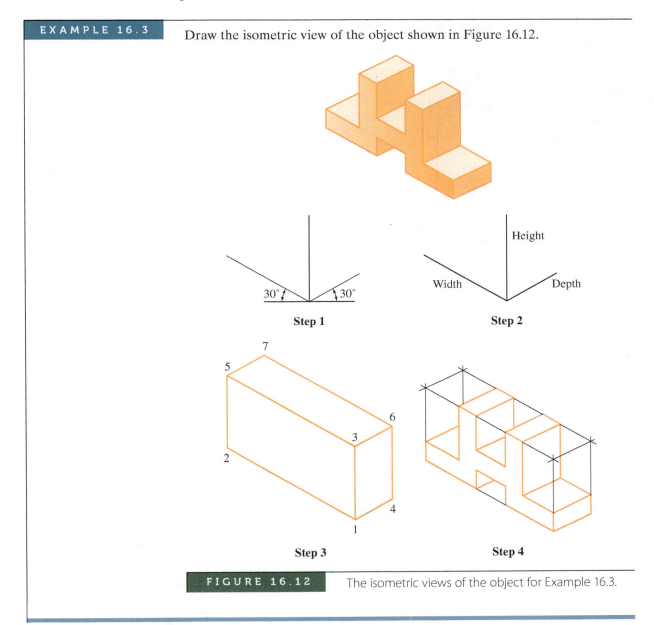

EXAMPLE 16.3

Draw the isometric view of the object shown in Figure 16.12.

Step 1

Step 2

Step 3

Step 4

FIGURE 16.12   The isometric views of the object for Example 16.3.

Sectional Views

You recall from the section on orthographic views that you use the dashed (hidden) lines to represent the edge of a plane or intersection of two planes or limits of a hole that are not visible from the direction you are looking. Of course, as mentioned earlier, this happens when some material exists between you and the edge. For objects with a complex interior—for example, with many interior holes or edges—the use of dashed lines on orthographic views could result in a confusing drawing. The dashed lines could make the drawing difficult to read and thus difficult for the reader to form a visual image of

For objects with complex interiors, sectional views are used. Sectional views reveal the inside of the object. A sectional view is created by making an imaginary cut through the object, in a certain direction, to reveal its interior.

what the inside of the object looks like. An example of an object with a complex interior is shown in Figure 16.13. For objects with complex interiors, **sectional views** are used. Sectional views reveal the inside of the object. A sectional view is created by making an imaginary cut through the object, in a certain direction, to reveal its interior. The sectional views are drawn to show clearly the solid portions and the voids within the object.

Let us now look at the procedure that you need to follow to create a sectional view. The first step in creating a sectional view involves defining the cutting plane and the direction of

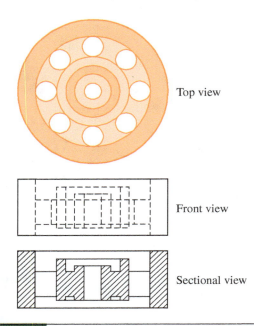

Top view

Front view

Sectional view

FIGURE 16.13   An object with a complex interior.

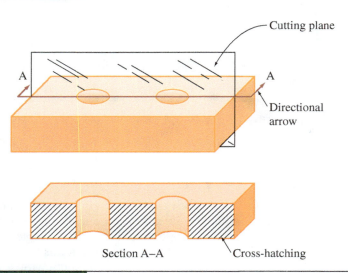

Cutting plane

A   A

Directional arrow

Section A–A   Cross-hatching

FIGURE 16.14   A sectional view of an object. Crosshatch patterns also indicate the type of material that parts are made from.

the sight. The direction of the sight is marked using directional arrows, as shown in Figure 16.14. Moreover, identifying letters are used with the directional arrows to name the section. The next step involves identifying and showing on the sectional view which portion of the object is made of solid material and which portion has the voids. The solid section of the view is then marked by parallel inclined lines. This method of marking the solid portion of the view is called *cross-hatching.* An example of a cutting plane, its directional arrow, its identifying letter, and cross-hatching is shown in Figure 16.14.

Based on how complex the inside of an object is, different methods are used to show sectional views. Some of the common section types include *full section, half section, broken-out section, rotated section,* and *removed section.*

- Full-section views are created when the cutting plane passes through the object completely, as shown in Figure 16.14.
- Half-sectional views are used for symmetrical objects. For such objects, it is customary to draw half of the object in sectional view and the other half as exterior view. The main advantage of half-sectional views is that they show the interior and exterior of the object using one view. An example of a half-section view is shown in Figure 16.15.
- Rotated section views may be used when the object has a uniform cross section with a shape that is difficult to visualize. In such cases, the cross section is rotated by 90° and is shown in the plane of view. An example of a rotated section is shown in Figure 16.16.

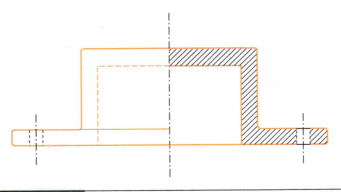

FIGURE 16.15   An example of a half-sectional view.

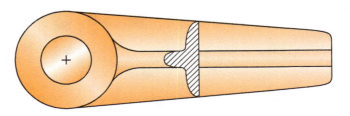

FIGURE 16.16   An example of a rotated section view.

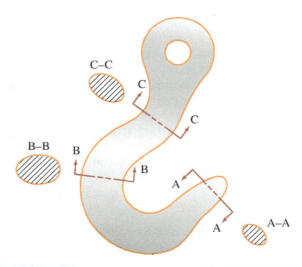

FIGURE 16.17   An example of an object with removed sections.

- Removed sections are similar to rotated sections, except instead of drawing the rotated view on the view itself, removed sections are shown adjacent to the view. They may be used for objects with a variable cross section, and generally many cuts through the section are shown. It is important to note that the cutting planes must be properly marked, as shown in Figure 16.17.

EXAMPLE 16.4   Draw the sectional view of the object shown in Figure 16.18, as marked by the cutting plane.

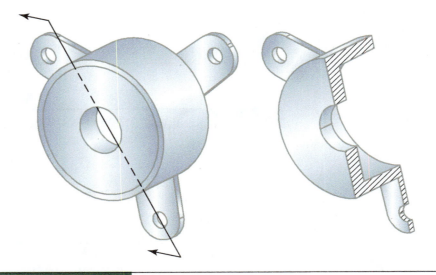

FIGURE 16.18   The object used in Example 16.4.

LO² 16.2   Civil, Electrical, and Electronic Drawings

In addition to the type of drawings that we have discussed so far, there are also discipline-specific drawings. For example, civil engineers typically deal with land or boundary, topographic, construction, connection and reinforcement details, and route survey drawings. Examples of **drawings** used in **civil engineering** are shown in Figure 16.19. To produce these types of drawings, a survey is first performed. A survey is a process by which something (such as land) is measured. During a survey, information such as distance, direction, and elevation are measured and recorded. Other examples of discipline-specific drawings include printed circuit-board assembly drawings, printed circuit-board drill plans, and wiring diagrams; which are commonly used by electrical and electronic engineers. Examples of the **electrical** and **electronic drawings** are shown in Figure 16.20.

LO³ 16.3   Solid Modeling

In recent years, the use of solid modeling software as a design tool has grown dramatically. Easy-to-use packages, such as AutoCAD, SolidWorks, and Creo, have become common tools in the hands of engineers. With these software tools you can create models of objects with surfaces and volumes that look almost indistinguishable from the actual objects. These solid models provide great visual aids for what the parts that make up a product look like before they are manufactured. The solid modeling software also allows for experimenting on a computer screen with the assembly of parts to examine any unforeseen problems before the parts are actually made and

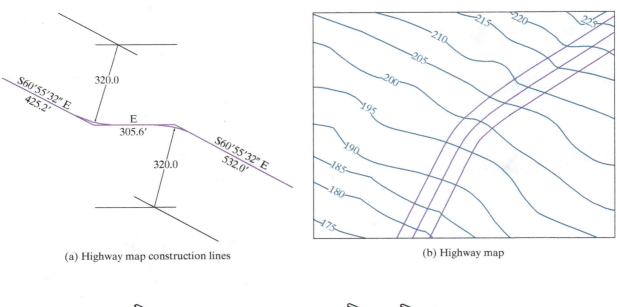

(a) Highway map construction lines

(b) Highway map

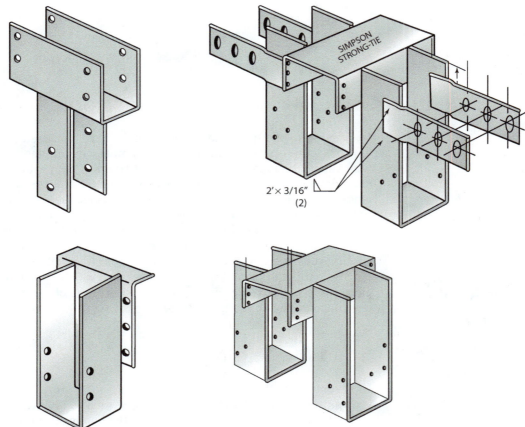

(c) Common manufactured metal beam connectors.

FIGURE 16.19 Examples of drawings used in civil engineering.
Based on Simpson Strong-Tie Company, Inc.

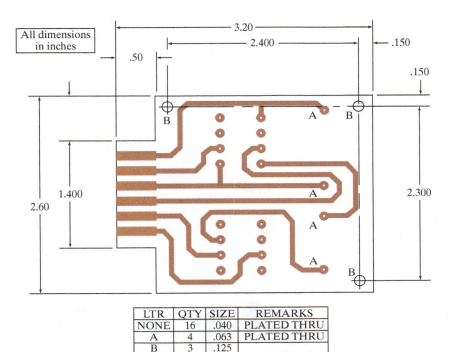

(a) A printed circuit board drill plan

LTR	QTY	SIZE	REMARKS
NONE	16	.040	PLATED THRU
A	4	.063	PLATED THRU
B	3	.125	

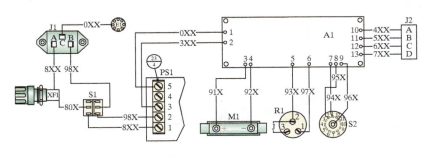

(b) A wiring diagram

FIGURE 16.20   Examples of drawings used in electrical and electronic engineering.

assembled. Moreover, changes to the shape and the size of a part can be made quickly with such software. Once the final design is agreed upon, the computer-generated drawings can be sent directly to computer numerically controlled (CNC) machines to make the parts.

Solid modeling software is also used by architects and engineers to present concepts. For example, an architect uses such software to show a client a model of what the exterior or interior of a proposed building would look like. Design engineers employ the solid modeling software to show concepts for shapes of cars, boats, computers, and so on. The computer-generated models save time and money. Moreover, there is additional software that makes use of these solid models to

Examples of discipline-specific drawings include printed circuit-board assembly drawings and wiring diagrams that are commonly used by electrical and electronic engineers.

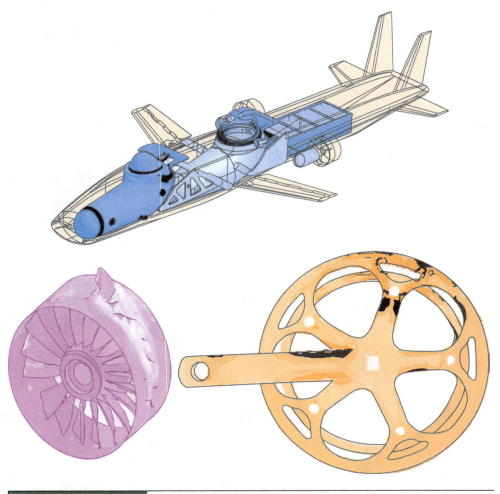

FIGURE 16.21   Examples of computer-generated solid models.

perform additional engineering analysis, such as stress calculations or temperature distribution calculations for products subjected to loads and/or heat transfer. Examples of solid models generated by commonly used software are shown in Figure 16.21.

Let us now briefly look at how solid modeling software generates solid models. There are two ways to create a solid model of an object: *bottom-up modeling* and *top-down modeling*. With bottom-up modeling you start by defining keypoints first, then lines, areas, and volumes in terms of the defined keypoints. Keypoints are used to define the vertices of an object. Lines, next in the hierarchy of bottom-up modeling, are used to represent the edges of an object. You can then use the created lines to generate a surface. For example, to create a rectangle, you first define the corner points by four keypoints, next you connect the keypoints to define four lines, and then you define the area of

> Solid models provide great visual aids showing what the parts that make up a product look like before they are manufactured. The solid-modeling software also allows for experimenting on a computer screen with the assembly of parts to examine any unforeseen problems before the parts are actually made and assembled.

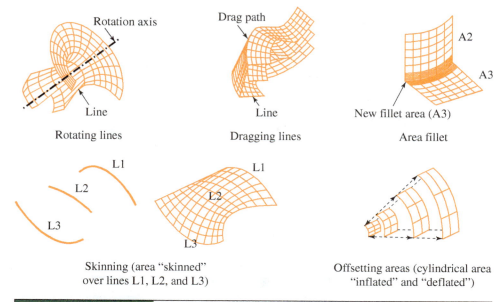

Rotation axis

Line

Rotating lines

Drag path

Line

Dragging lines

A2

A3

New fillet area (A3)

Area fillet

L1

L2

L3

Skinning (area "skinned"
over lines L1, L2, and L3)

L1

L2

L3

Offsetting areas (cylindrical area
"inflated" and "deflated")

FIGURE 16.22   Additional area-generation methods.

the rectangle by the four lines that enclose the area. There are additional ways to create areas: (1) dragging a line along a path, (2) rotating a line about an axis, (3) creating an area fillet, (4) skinning a set of lines, and (5) offsetting areas. With the area-fillet operation, you can create a constant-radius fillet tangent to two other areas. You can generate a smooth surface over a set of lines by using the skinning operation. Using the area offset command, you can generate an area by offsetting an existing area. These operations are all shown in Figure 16.22.

The created areas then may be put together to enclose and create a volume. As with areas, you can also generate volumes by dragging or extruding an area along a line (path) or by rotating an area about a line (axis of rotation). Examples of these volume-generating operations are shown in Figure 16.23.

With top-down modeling, you can create surfaces or three-dimensional solid objects using area and volume *primitives*. Primitives are simple geometric shapes. Two-dimensional primitives include rectangles, circles, and polygons, and three-dimensional volume primitives include blocks, prisms, cylinders, cones, and spheres, as shown in Figure 16.24.

Regardless of how you generate areas or volumes, you can use Boolean operations to add (union) or subtract entities to create a solid model. Examples of Boolean operations are shown in Figure 16.25.

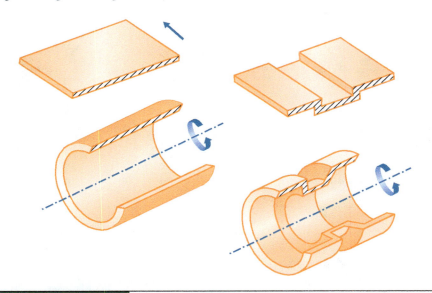

FIGURE 16.23   Examples of volume-generating operations.

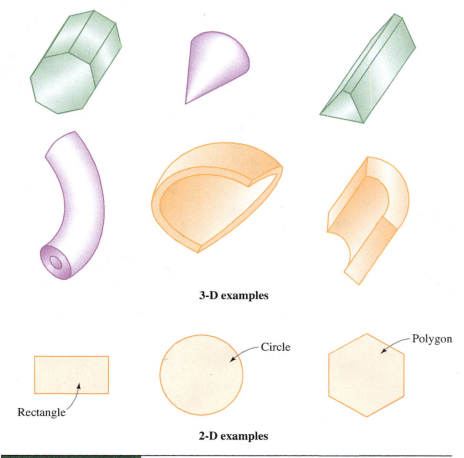

3-D examples

Rectangle

Circle

Polygon

2-D examples

FIGURE 16.24   Examples of two- and three-dimensional primitives.

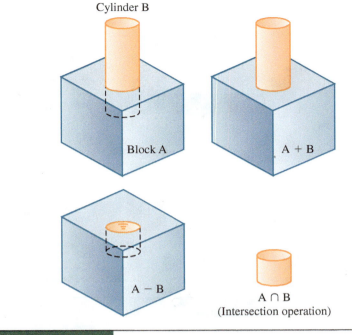

Cylinder B

Block A

A + B

A − B

A ∩ B
(Intersection operation)

FIGURE 16.25   Examples of Boolean union (add), subtract, and intersection operations.

EXAMPLE 16.5   Discuss how to create the solid model of the objects shown in Figure 16.26, using the operations discussed in this section.

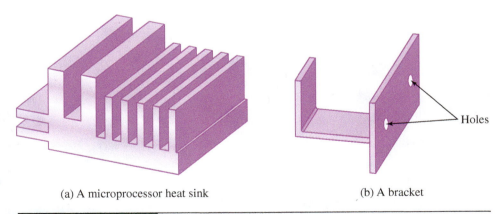

(a) A microprocessor heat sink

(b) A bracket

Holes

FIGURE 16.26   The objects for Example 16.5.

(a) In order to create the solid model of the heat sink shown in Figure 16.26(a), we draw its front view first, as shown in Figure 16.27.
We then extrude this profile in the normal direction, which leads to the solid model of the heat sink as depicted in Figure 16.26(a).

(b) Similarly for the bracket given in Figure 16.26(b), we begin by creating the profile shown in Figure 16.28.

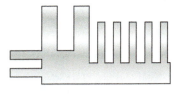

FIGURE 16.27 Front view.

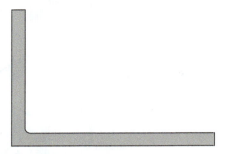

FIGURE 16.28 Partial profile of the bracket.

Next, we will extrude the profile in the normal direction, as shown in Figure 16.29.

We then create the block and the holes. In order to create the holes, first we create two solid cylinders and then use the Boolean operation to subtract the cylinders from the block. Finally, we add the new block volume to the volume, which we created by the extrusion method, as in Figure 16.30.

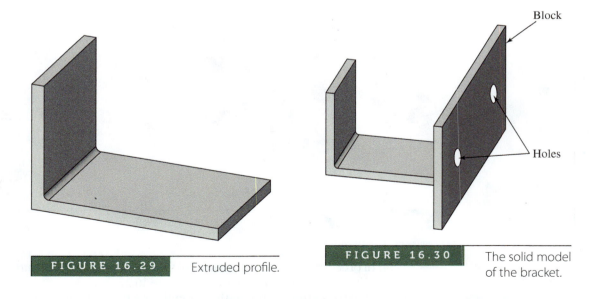

FIGURE 16.29 Extruded profile.

FIGURE 16.30 The solid model of the bracket.

Answer the following questions to test your understanding of the preceding sections.

1.   Explain why there is a need for discipline-specific drawings.

2.   Give examples of civil engineering drawings.

3.   Give examples of electrical and electronic drawings.

4.   Explain what is meant by solid modeling.

5.   There are two ways to create a solid model of an object; explain them.

Vocabulary—State the meaning of the following terms:

Bottom-Up Approach _____

Primitive _____

Solid Model _____

Top-Down Approach _____

LO⁴ 16.4   Engineering Symbols

The symbols in Figure 16.31 are a "language" used by engineers to convey their ideas, their solutions to problems, or analyses of certain situations. As engineering students you will learn about the graphical ways by which engineers communicate among themselves. In this section, we will discuss the need for conventional **engineering symbols** as a means to convey information and to effectively communicate to other engineers. We will begin by explaining why there exists a need for engineering signs and symbols. We will then discuss some common symbols used in civil, electrical, and mechanical engineering.

You may have a driver's license or you have probably been in a car on a highway. If this is the case, then you are already familiar with conventional traffic signs that provide valuable information to drivers. Examples of traffic signs are shown in Figure 16.32. These signs are designed and developed, based on some acceptable national and international standards, to convey information not only effectively but also quickly. For example, a stop sign, which has an octagon shape with a red background, tells you to bring your car to a complete stop.

A sign that signals a possibly icy road ahead is another example that warns you to slow down because the road conditions may be such that you may end up in a hazardous situation. That same information could have been conveyed to you in other ways. In place of the sign indicating a slippery road, the highway department could have posted the following words on a board: "Hey you, be careful. The road is slippery, and you could end up in a ditch!" Or they could have installed a loudspeaker warning drivers: "Hey, be careful,

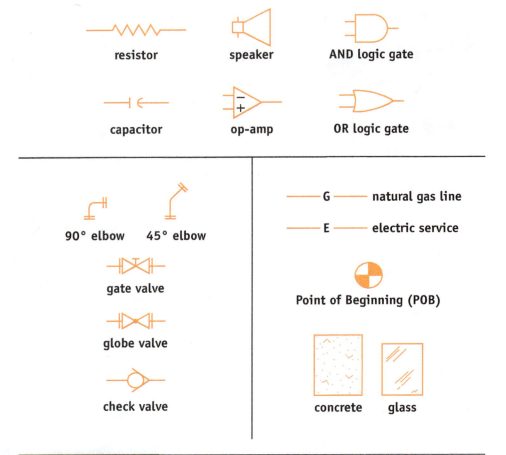

FIGURE 16.31   Engineering symbols and signs provide valuable information.

FIGURE 16.32   Examples of traffic signs.

slippery road ahead. You could end up in a ditch." Which is the most effective, efficient, and least expensive way of conveying the information? You understand the point of these examples and the question. Road and traffic information can be conveyed inexpensively, quickly, and effectively using signs and symbols. Of course, to understand the meaning of the signs, you had to study them and learn what they mean before you could take your driving test. Perhaps in the future the new advancements in technology will reach a stage that road and traffic information can be conveyed directly in a wireless digital format to a computer in your car to allow your car to respond accordingly to the given information! In that case, would you still need to know the meaning of the signs, and would the highway departments need to post them?

Examples of Common Symbols in Civil, Electrical, and Mechanical Engineering

As the examples in the previous section demonstrated, valuable information can be provided in a number of ways: through a long written sentence, orally, graphically or symbolically, or any logical combination of these. But which is the more effective way? As you study various engineering topics, not only will you learn many new concepts but you will also learn about the graphical way that engineers communicate among themselves. You will learn about engineering signs and symbols that provide valuable information and save time, money, and space. These symbols and signs are like a language that engineers use to convey their ideas, their solutions to a problem, or their analyses of certain situations. For example, electrical engineers use various symbols to represent the components that make up an electrical or an electronic system, such as a television set, a cellular phone, or a computer. Examples of engineering symbols are shown in Table 16.1.

Mechanical engineers use diagrams to show the layout of piping networks in buildings or to show the placement of air supply ducts, air return ducts, and fans in a heating or cooling system.

> Engineering symbols are a "language" used by engineers to convey their ideas, their solutions to problems, or analyses of certain situations.

TABLE 16.1	Examples of Engineering Symbols

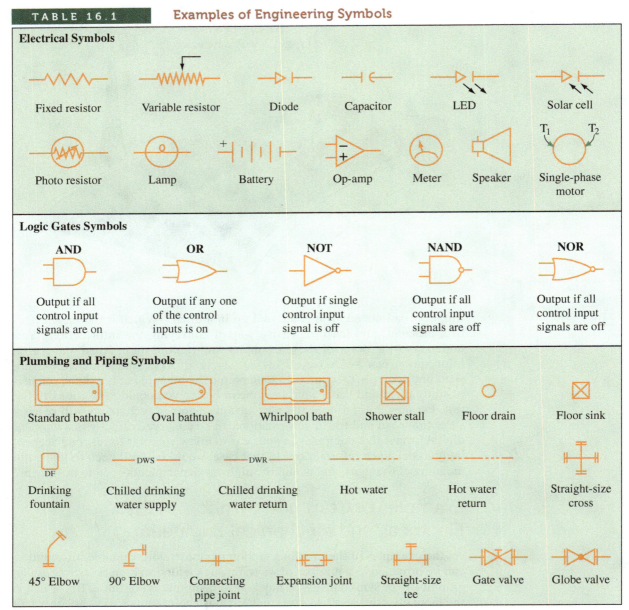

Electrical Symbols

Fixed resistor   Variable resistor   Diode   Capacitor   LED   Solar cell

Photo resistor   Lamp   Battery   Op-amp   Meter   Speaker   Single-phase motor

Logic Gates Symbols

AND — Output if all control input signals are on

OR — Output if any one of the control inputs is on

NOT — Output if single control input signal is off

NAND — Output if all control input signals are off

NOR — Output if all control input signals are off

Plumbing and Piping Symbols

Standard bathtub   Oval bathtub   Whirlpool bath   Shower stall   Floor drain   Floor sink

Drinking fountain   Chilled drinking water supply   Chilled drinking water return   Hot water   Hot water return   Straight-size cross

45° Elbow   90° Elbow   Connecting pipe joint   Expansion joint   Straight-size tee   Gate valve   Globe valve

Based on American Technical Publishers, Ltd.

An example of a HVAC (heating, ventilation, and air conditioning) system drawing is shown in Figure 16.33.

For more detailed information on symbols, see the following documents:

Graphic Electrical Symbols for Air-Conditioning and Refrigeration Equipment by ARI (ARI 130-88).

Graphic Symbols for Electrical and Electronic Diagrams by IEEE (ANSI/IEEE 315-1975).

Graphic Symbols for Pipe Fittings, Valves, and Piping by ASME (ANSI/ASME ASME Y32.2.2.3-1949 (R 1988)).

Symbols for Mechanical and Acoustical Elements as Used in Schematic Diagrams by ASME (ANSI/ASME Y32.18-1972 (R 1985)).

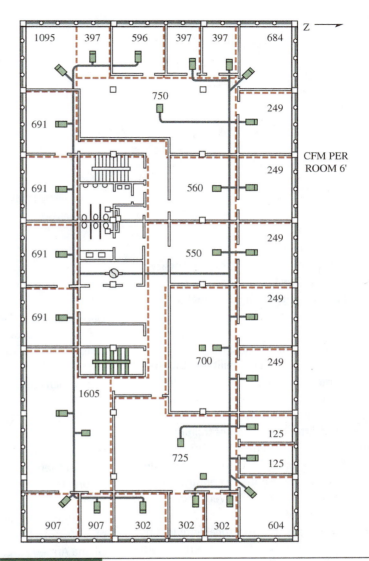

FIGURE 16.33	Single-line ducted HVAC system showing a layout of the proposed trunk and runout ductwork.

Based on The Trane Company, La Crosse, WI

Before You Go On

Answer the following questions to test your understanding of the preceding sections.

1. Explain why there is a need for engineering symbols.

2. Give examples of common civil engineering symbols.

3. Give examples of common electrical engineering symbols.

4. Give examples of common mechanical engineering symbols.

Vocabulary—State the meaning of the following terms:

Diode _____

LED _____

HVAC _____

Jerome Antonio

As a young teenager about to complete secondary school in Ghana I was torn between a military career and a career in engineering. Fortunately for me, by the time I had to make a decision the military had announced a new scheme under which young academically promising secondary school graduates would be trained in the military academy for commission into the armed forces while pursuing university education leading to degrees in engineering, medicine, and other professions. I took advantage of this scheme and, after three years of training, was commissioned as an officer in the Army Corps of Electrical and Mechanical Engineers. I was subsequently sponsored by the military to enter the engineering program in the Kwame Nkrumah University of Science and Technology in Kumasi, Ghana.

My personal interest was in electrical engineering, but I had to follow the mechanical engineering program because that was the discipline in which my sponsors had the greatest need for engineers. Fortunately, after the common first-year program during which all students took courses in all the engineering disciplines, I realized that I liked the mechanical engineering courses better than the others. I graduated four years after enrolling on the program with a First Class degree in mechanical engineering and returned to the Army. I was assigned to duty in a large workshop dealing with the maintenance, repair and modification of a wide range of equipment from military vehicles and communication equipment to guns.

Two years after leaving the university in Ghana I was sponsored to go to the United Kingdom for graduate studies. I was admitted into Imperial College of Science and Technology, which was then one of the constituent

Courtesy of Jerry Antonio

colleges of the University of London. At Imperial College I studied for the Masters Degree in Advanced Applied Mechanics. I followed up with research work leading to the PhD in mechanical engineering after which I returned to Ghana.

The early 1980s was a time when an unprecedented downturn in the country's economy had resulted in a massive exodus of professionals leaving to seek better lives for themselves abroad. The universities were among the most severely affected by this problem. Many academic departments in the universities faced imminent collapse as a result of the shortage of professors. It was under these conditions that the Ghana Armed Forces agreed to second me to the School of Engineering in my alma mater to assist with teaching. I was eventually released from military service so that I could take up a permanent teaching position in the university.

Although a large part of my working life has been spent in university teaching and administration, I have also had many opportunities to engage in professional engineering activities. One of the assignments that I found most interesting was a consultancy assignment to carry out an analysis of the problems of a thermal power plant. It is very fulfilling to see action being taken on the basis of one's professional advice. As I progressed through my engineering career I found myself increasingly being called upon to serve my community through membership in various bodies. I have served on several bodies responsible for advising government on a wide range of issues. For example, for several years I served on the National Board for Small-Scale Industries and the National Board for Professional and Technician Examinations. I have found that the basic problem-solving skills learned in engineering schools have always come in handy when performing both engineering and non-engineering assignments.

In my professional work I have identified several things that contribute to a successful engineering career. Among them is knowledge of and sensitivity to the wide variety of contexts within which engineering is practiced. As an engineer, I have often found that the success of my work depended on my ability to appreciate how my work impacts and is impacted upon by seemingly non-engineering issues such as history, social and cultural practices, legal constraints, and environmental concerns. Students can start preparing themselves to face this challenge by being smart with their choice of non-technical electives, by participating in multicultural activities on campus, and by attending seminars and public lectures in areas outside their own disciplines. Students can also prepare themselves for practicing engineering in a global context by availing themselves of any opportunities for interacting with students and professors from universities in other parts of the world. An increasing number of universities have student exchange programs which can be very beneficial in broadening the outlook of participants.

One of the experiences that had a big positive impact on my own training as an engineer was the long vacation that I spent in Zurich, Switzerland, working with other students in a factory which manufactured turbomachinery. This experience was made possible by the International Association for the Exchange of Students for Technical Experience (IAESTE).

After spending all these years teaching engineering in Ghana I decided to seek new challenges elsewhere prior to my retirement. I worked as a visiting professor in North Carolina Agricultural and Technical State University in the United States of America. Teaching engineering in America was quite a different experience because of the differences in the teaching and learning resources and the different backgrounds of the students. However, my experiences confirm to me the fact that no matter where they study, all engineering students require the same thing to prepare them for a successful professional life: development of critical thinking and good problem-solving skills. As a student you should realize that when your professors always insist that you present your assignments in a clear and orderly manner, or when they insist on meeting deadlines for turning in assignments, they are helping you to acquire the skills necessary for you to succeed not only in the practice of engineering but also in the non-professional aspects of your life.

SUMMARY

LO¹ Mechanical Drawings

Mechanical drawings are important in conveying useful information to other engineers or machinists in a standard acceptable manner to allow the readers of these drawings to visualize what the proposed product would look like. Orthographic views show what an object's projection looks like when seen from the top, the front, or the side. On the other hand, the isometric drawing shows the three dimensions of an object in a single view. For objects with complex interiors, sectional views are used. Sectional views reveal the inside of the object. A sectional view is created by making an imaginary cut through the object, in a certain direction, to reveal its interior. Moreover, the American National Standards Institute sets the standards for the dimensioning and tolerancing practices for engineering drawings. Every engineering drawing must include dimensions, tolerances, the materials from which the product will be made, and the finished surfaces marked.

LO² Civil, Electrical, and Electronic Drawings

In addition to mechanical engineering drawings, there are also discipline-specific drawings. For example, civil engineers deal with land or boundary, topographic, construction, connection and reinforcement details and route survey drawings. Other examples of discipline-specific drawings include printed circuit-board assembly drawings and wiring diagrams that are used by electrical and electronic engineers.

LO³   Solid Modeling

With solid-modeling software, we can create models of objects with surfaces and volumes that look almost indistinguishable from the actual objects. Solid models provide great visual aids for what the parts that make up a product look like before they are manufactured. The solid-modeling software also allows for experimenting on a computer screen with the assembly of parts to examine any unforeseen problems before the parts are actually made and assembled. There are two ways to create a solid model of an object: bottom-up modeling and top-down modeling. With bottom-up modeling, you start by defining keypoints first, then creating lines, areas, and volumes. With top-down modeling, you can create surfaces or three-dimensional solid objects using area and volume primitives. Primitives are simple geometric shapes such as rectangles, circles, polygons, blocks, prisms, cylinders, cones, and spheres.

LO⁴   Engineering Symbols

Engineering symbols are a "language" used by engineers to convey their ideas, solutions to problems, or analyses of certain situations. For example, mechanical engineers use symbols and diagrams to show the layout of piping networks in buildings or to show the placement of air-supply ducts, air-return ducts, and fans in a heating or cooling system. Electrical engineers use various symbols to represent the components that make up an electrical or an electronic system such as a television set, a cellular phone, or a computer.

KEY TERMS

ANSI 594
Civil Engineering Drawings 603
Dimensioning and
 Tolerancing 594

Electrical and Electronic
 Engineering Drawings 603
Engineering Symbols 611
Isometric View 596

Orthographic Views 590
Sectional Views 600
Solid Modeling 603

APPLY WHAT YOU HAVE LEARNED

Select a tool or a tool box and provide all necessary drawings for the item. Your final report must include at least:

1. orthographic views,
2. isometric drawing,
3. dimensions and tolerances,
4. materials from which item was made from, and
5. assembly steps.

maxim ibragimov/Shutterstock.com

Yanas/Shutterstock.com

PROBLEMS

Problems that promote lifelong learning are denoted by 🔑

For Problems 16.1 through 16.19, draw the top, the front, and the right-side orthographic views of the objects shown. Indicate when an object needs only one or two views to be fully described.

16.1

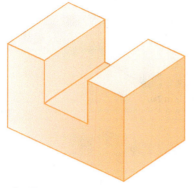

Problem 16.1

16.2

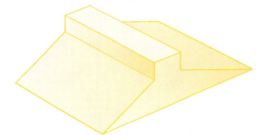

Problem 16.2

16.3

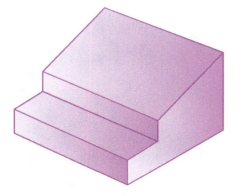

Problem 16.3

16.4

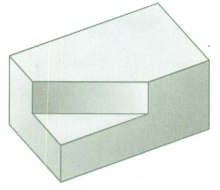

Problem 16.4

16.5

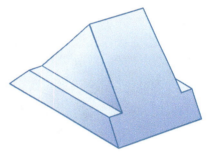

Problem 16.5

16.6

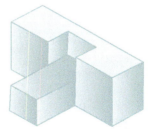

Problem 16.6
Based on Madsen, Engineering Drawing and Design, 4e. Delmar Learning, a part of Cengage Learning, Inc., 2007.

16.7

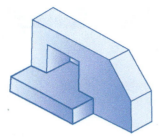

Problem 16.7
Based on Madsen, Engineering Drawing and Design, 4e. Delmar
Learning, a part of Cengage Learning, Inc., 2007.

16.8

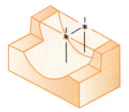

Problem 16.8
Based on Madsen, Engineering Drawing and Design, 4e. Delmar
Learning, a part of Cengage Learning, Inc., 2007.

16.9

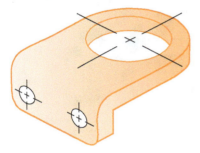

Problem 16.9
Based on Madsen, Engineering Drawing and Design, 4e. Delmar
Learning, a part of Cengage Learning, Inc., 2007.

16.10

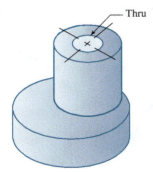

Problem 16.10
Based on Madsen, Engineering Drawing and Design, 4e. Delmar
Learning, a part of Cengage Learning, Inc., 2007.

16.11

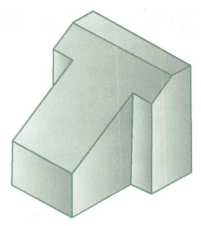

Problem 16.11
Based on Madsen, Engineering Drawing and Design, 4e. Delmar
Learning, a part of Cengage Learning, Inc., 2007.

16.12

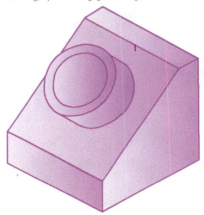

Problem 16.12
Based on Madsen, Engineering Drawing and Design, 4e. Delmar
Learning, a part of Cengage Learning, Inc., 2007.

16.13

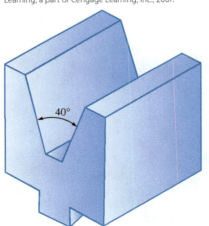

Problem 16.13
Based on Madsen, Engineering Drawing and Design, 4e. Delmar
Learning, a part of Cengage Learning, Inc., 2007.

16.14

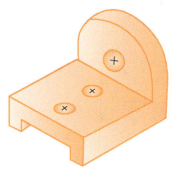

Problem 16.14
Based on Madsen, Engineering Drawing and Design, 4e. Delmar
Learning, a part of Cengage Learning, Inc., 2007.

16.15

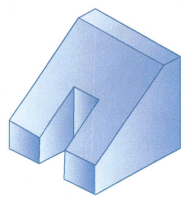

Problem 16.15
Based on Madsen, Engineering Drawing and Design, 4e. Delmar
Learning, a part of Cengage Learning, Inc., 2007.

16.16

Problem 16.16

16.17

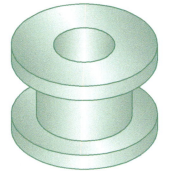

Problem 16.17

16.18

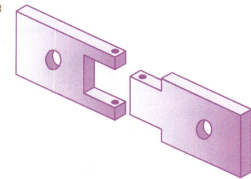

Problem 16.18

16.19

Problem 16.19

For Problems 16.20 through 16.23, use the cutting
planes shown to draw the sectional views.

16.20

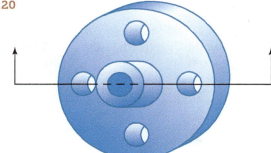

Problem 16.20

16.21

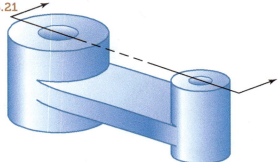

Problem 16.21

16.22

Problem 16.22

16.23

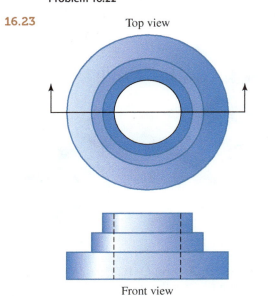

Top view

Front view

Problem 16.23

For Problems 16.24 through 16.28, using the rules discussed in this chapter, show the dimensions of the views shown.

16.24   0   1   2   3   4   5 cm

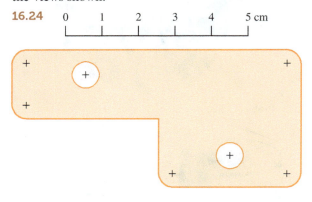

Problem 16.24

16.25   0   1   2 in.

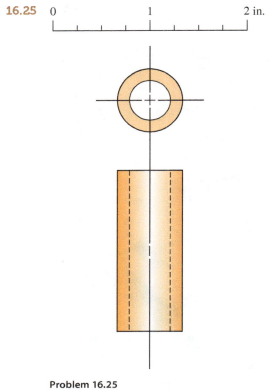

Problem 16.25

16.26   0   1   2   3 in.

Problem 16.26

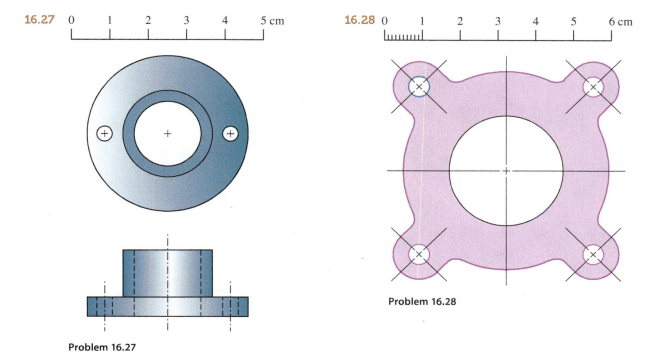

16.27 0   1   2   3   4   5 cm

16.28 0   1   2   3   4   5   6 cm

Problem 16.27

Problem 16.28

For Problems 16.29 through 16.32, show the dimensions of the object on its orthographic views.

16.29

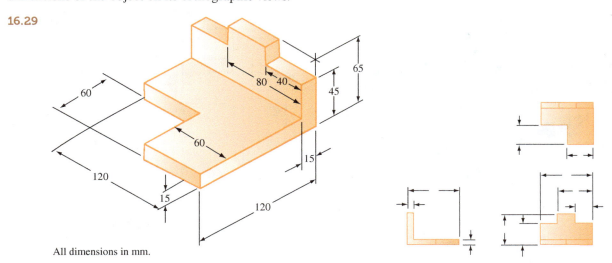

60   80   40   65   45   60   15   120   15   120

All dimensions in mm.

Problem 16.29
Based on Madsen, Engineering Drawing and Design, 4e. Delmar Learning, a part of Cengage Learning, Inc., 2007.

16.30

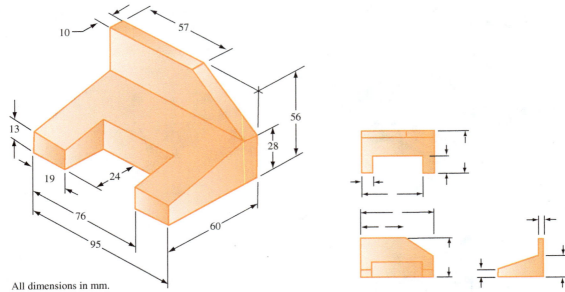

All dimensions in mm.

Problem 16.30
Based on Madsen, Engineering Drawing and Design, 4e. Delmar Learning, a part of Cengage Learning, Inc., 2007.

16.31

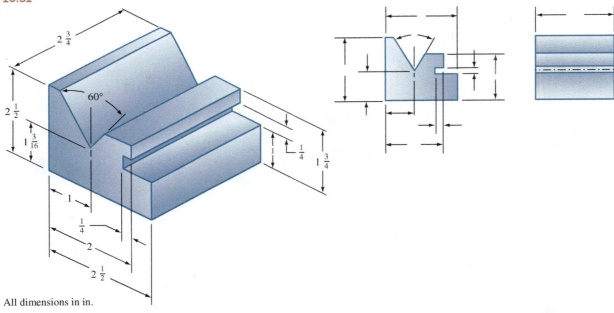

All dimensions in in.

Problem 16.31
Based on Madsen, Engineering Drawing and Design, 4e. Delmar Learning, a part of Cengage Learning, Inc., 2007.

16.32

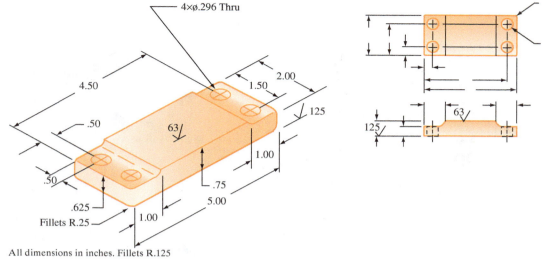

4×ø.296 Thru

All dimensions in inches. Fillets R.125

Problem 16.32
Based on Madsen, Engineering Drawing and Design, 4e. Delmar Learning, a part of Cengage Learning, Inc., 2007.

For Problems 16.33 through 16.38, draw the isometric view of the following objects. Make the necessary measurements or estimations of dimensions.

16.33	A television set.	**16.34**	A computer.
16.35	A telephone.	**16.36**	A razor.
16.37	A chair.	**16.38**	A car.

16.39 Follow the steps discussed in Section 16.4 and draw the isometric view for Problem 16.4.

16.40 Follow the steps discussed in Section 16.4 and draw the isometric view for Problem 16.6.

16.41 Follow the steps discussed in Section 16.4 and draw the isometric view for Problem 16.7.

16.42 Follow the steps discussed in Section 16.4 and draw the isometric view for Problem 16.11.

16.43 Follow the steps discussed in Section 16.4 and draw the isometric view for Problem 16.13.

For Problems 16.44 through 16.48, discuss how you would create the solid model of the given objects. See Example 16.5 to better understand what you are being asked to do.

16.44 A bracket.

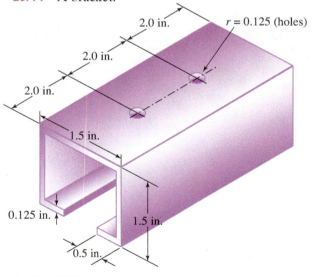

Problem 16.44

16.45 A wheel.

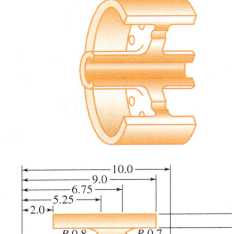

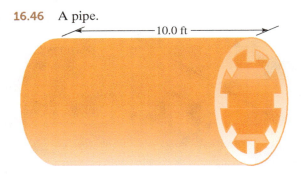

Dimensions are in inches.

Problem 16.45

16.46 A pipe.

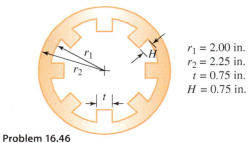

$r_1 = 2.00$ in.
$r_2 = 2.25$ in.
$t = 0.75$ in.
$H = 0.75$ in.

Problem 16.46

16.47 A socket.

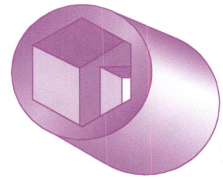

Problem 16.47

16.48 A heat exchanger. The pipes pass through all of the fins.

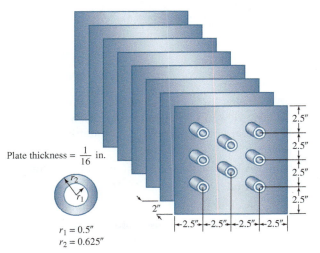

Plate thickness = $\frac{1}{16}$ in.

$r_1 = 0.5''$
$r_2 = 0.625''$

Problem 16.48

16.49 Using Table 16.1, identify the engineering symbols shown in the accompanying figure.

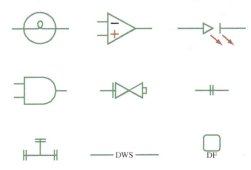

Problem 16.49

16.50 Using Table 16.1, identify the components of the logic system shown in the figure.

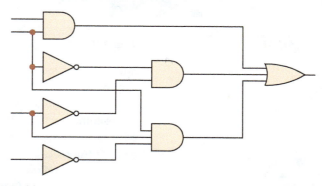

Problem 16.50

An Engineering Marvel

Boeing 777* Commercial Airplane

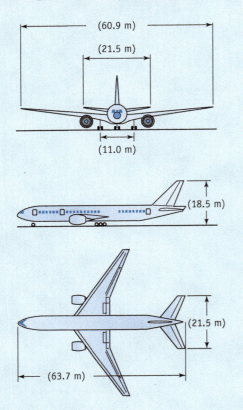

Overview

The Boeing 777 was the first commercial airplane that was fully designed using three-dimensional digital solid modeling technology. The core of the design group consisted of 238 teams that included engineers of various backgrounds. A number of international aerospace companies, from Europe, Canada, and Asia/Pacific, contributed to the design and production of the 777. The Japanese aerospace industry was among the largest of the overseas participants. Representatives from airline customers such as Nippon Airways and British Airways also provided input to the design of 777. Throughout the design process, the different components of the airplane were designed, tested, assembled (to ensure proper fitting),

and disassembled on a network of computers. Approximately 1700 individual workstations and 4 IBM mainframe computers were used. The use of the computers and the engineering software eliminated the need for the development of a costly, full-scale prototype. The digital solid modeling technology allowed the engineers to improve the quality of work, experiment with various design concepts, and reduce changes and errors, all of which resulted in lower costs and increased efficiency in building and installing various parts and components.

The engineers used, among other software, CATIA (Computer-Aided Three-Dimensional Interactive Application) and ELFINI (Finite Element Analysis System), both developed by Dassault Systems of France and licensed in the United States through IBM. Designers also used EPIC (Electronic Preassembly Integration on CATIA) and other digital preassembly applications developed by Boeing.

The 777 series, the world's largest twinjet (at the time it was designed), is available in three models: the initial model, 777-200; the 777-200ER (Extended Range) model; and the larger 777-300 model. The 777-300 is stretched (10 m) from the initial 777-200 model to a total of 73.9 m. In an all-economy layout, the 777-300 can accommodate as many as 550 passengers. However, it may be configured for 368 to 386 passengers in three classes to provide more comfort.

In terms of range capability, the 777-300 can serve routes up to 10,370 km. The 777-300 has nearly the same passenger capacity and range capability as the 747-100/-200 models but burns one-third less fuel and has 40% lower maintenance costs. Of course, this results in a lower operating cost.

Baseline maximum takeoff mass for the 777-300 is 263,080 kg; the highest maximum takeoff weight being offered is 299,370 kg. Maximum fuel capacity is 171,160 L. The 777-300 has a total available cargo volume of 200.5 m³.

*Materials were adapted with permission from Boeing documents.

Satellite communication and global positioning systems are basic to the airplane. The 777 wing uses the most aerodynamically efficient airfoil ever developed for subsonic commercial aviation. The wing has a span of 60.9 m. The advanced wing design enhances the airplane's ability to climb quickly and cruise at higher altitudes than its predecessor airplanes. The wing design also allows the airplane to carry full passenger payloads out of many high-elevation, high-temperature airfields. Fuel is stored entirely within the wing and its structural center section. The longer-range model and the 777-300 model can carry up to 171,155 L.

The Boeing Company, upon request from airplane buyers, can install engines from three leading engine manufacturers, namely, Pratt & Whitney, General Electric, and Rolls-Royce. These engines are rated in the 333,000- to 346,500-N thrust class. For the longer-range model and the 777-300, these engines will be capable of thrust ratings in the 378,000- to 441,000-N category.

New, lightweight, cost-effective structural materials are used in several 777 applications. For example, an improved aluminum alloy, 7055, is used in the upper wing skin and stringers. This alloy offers greater compression strength than previous alloys, enabling designers to save weight and also improve resistance to corrosion and fatigue. Lightweight composites are found in the vertical and horizontal tails. The floor beams of the passenger cabin also are made of advanced composite materials.

The principal flight, navigation, and engine information is all presented on six large, liquid-crystal, flat-panel displays. In addition to saving space, the new displays weigh less and require less power, and because they generate less heat, less cooling is required compared to the older conventional cathode-ray-tube screens. The flat-panel displays remain clearly visible in all conditions, even direct sunlight.

The Boeing 777 uses an Integrated Airplane Information Management System that provides flight and maintenance crews all pertinent information concerning the overall condition of the airplane, its maintenance requirements, and its key operating functions, including flight, thrust, and communication management.

The flight crew transmits control and maneuvering commands through electrical wires, augmented by computers, directly to hydraulic actuators for the elevators, rudder, ailerons, and other controls surfaces. The three-axis, "fly-by-wire" flight-control system saves weight, simplifies factory assembly compared to conventional mechanical systems relying on steel cables, and requires fewer spares and less maintenance in airline service.

A key part of the 777 system is a Boeing-patented, two-way digital data bus, which has been adopted as a new industry standard: ARINC 629. It permits airplane systems and their computers to communicate with one another through a common wire path (a twisted pair of wires) instead of through separate one-way wire connections. This further simplifies assembly and saves weight, while increasing reliability through a reduction in the amount of wires and connectors. There are 11 of these ARINC 629 pathways in the 777.

The interior of the Boeing 777 is one of the most spacious passenger cabins ever developed; the 777 interior offers configuration flexibility. Flexibility zones have been designed into the cabin areas specified by the airlines, primarily at the airplane's doors. In 25 mm. increments, galleys and lavatories can be positioned anywhere within these zones, which are preengineered to accommodate wiring, plumbing, and attachment fixtures. Passenger service units overhead storage compartments are designed for quick removal without disturbing ceiling panels, air- conditioning ducts, or support structure. A typical 777 configuration change is expected to take as little as 72 hours, while such a change might take two to three weeks on other aircraft. For improved, more efficient, in-flight service, the 777 is equipped with an advanced cabin management system. Linked to a computerized control console, the cabin management system assists cabin crews with many tasks and allows airlines to provide new services of passengers, including a digital sound system comparable to the most state-of-the-art home stereo or compact disc players.

The main landing gear for the 777 is in a standard two-post arrangement but features six-wheel trucks, instead of the conventional four-wheel units. This provides the main landing gear with a total of 12 wheels for better weight distribution on runways and taxi areas and avoids the need for a supplemental two-wheel gear under the center of the fuselage. Another advantage is that the six-wheel trucks allow for a more economical brake design. The 777 landing gear is the largest ever incorporated into a commercial airplane.

The Boeing-United Airlines 1000-cycle flight tests for the Pratt & Whitney engine were completed on May 22, 1995. In addition, engine makers and the many suppliers of parts for the airplane intensified their own development and testing efforts to ensure that their products met airline requirements.

This thorough test program demonstrated the design features needed to obtain approval for extended-range twin-engine operations (ETOPS). All 777s are ETOPS-capable, as part of the basic design. To ensure reliability, the 777 with Pratt & Whitney engines was tested and flown under all appropriate conditions to prove it is capable of flying ETOPS missions. A summary of Boeing 777 specifications is shown in Table 1.

PROBLEMS

1. Using the data given for the Boeing 777, estimate the flight time from New York City to London.

2. Estimate the mass of passengers, fuel, and cargo for a full flight.

3. Using the maximum range and fuel capacity data, estimate the fuel consumption of the Boeing 777 on a per hour and per km basis.

TABLE 1	**Boeing 777-200/300 Specifications**	
Design Variable	**777-200**	**777-300**
Seating	305 to 320 passengers in three classes	368 to 386 passengers in three classes
Length	63.7 m	73.9 m
Wingspan	60.9 m	60.9 m
Tail height	18.5 m	18.5 m
Engines	Pratt & Whitney 4000	Pratt & Whitney 4000
	General Electric GE90	General Electric GE90
	Rolls-Royce Trent 800	Rolls-Royce Trent 800
Maximum takeoff mass	229,520 kg	263,080 kg
Fuel capacity	117,335 L	171,160 L
Altitude capability	11,975 m	11,095 m
Cruise speed	893 km/h	893 km/h
	Mach 0.84	Mach 0.84
Cargo capacity	160 m³	214 m³
Maximum range	9525 km	10,370 km

4. Calculate the linear momentum of the Boeing 777 at cruise speed and at two-thirds maximum takeoff mass.

5. Calculate the Mach number of the Boeing 777 at cruise speed using

$$\text{Mach number} = \frac{\text{cruise speed}}{\sqrt{kRT}}$$

where

k = specific heat ratio = 1.4
R = air gas constant = 287 J/kg · K
T = air temperature at cruising altitude (K)

Compare your Mach number to the one given in the Boeing data table.

6. As mentioned previously, the flight crew transmits control and maneuvering commands through electrical wires, augmented by computers, directly to hydraulic actuators for the elevators, rudder, ailerons, and other controls surfaces. The elevators, rudder, and ailerons for a small plane are shown in the accompanying figure. In a small plane the ailerons are moved by turning the control wheel in the cockpit. When the wheel is turned left, the left aileron moves up and the right aileron moves down. This is how the pilots starts a turn to the left. In a small plane the rudder is operated by the pilot's feet. When the pilot presses the left rudder pedal, the nose of the plane moves left. When the right rudder pedal is pressed the nose moves right. The elevators make the nose of the plane move up and down. When the pilot pulls back on the control wheel in the cockpit, the nose of the plane moves up. When the control wheel is pushed forward, the nose moves down.

Investigate the aerodynamics of maneuvering flight in more detail. Explain what happens to air pressure distribution over these surfaces as their orientations are changed. What are the directions of the resulting force due to the pressure distributions over these surfaces? Write a brief report explaining your findings.

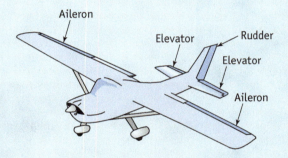

Engineering Material Selection

An Important Design Decision

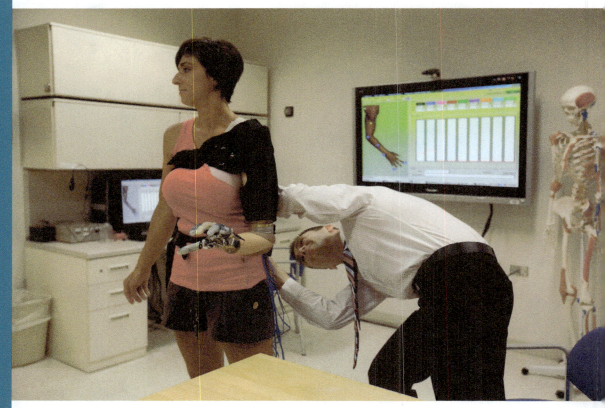

As an engineer, whether you are designing a machine part, a toy, a frame of a car, a structure, or artificial limbs the selection of materials is an important design decision. In this part of the book, we will look more closely at materials such as metals and their alloys, plastics, glass, wood, composites, and concrete that commonly are used in various engineering applications. We will also discuss some of the basic characteristics of the materials that are considered in design.

CHAPTER 17   ENGINEERING MATERIALS

Engineering Materials

dibrova/Shutterstock.com

Vladimir Gjorgiev/Shutterstock.com

nito/Shutterstock.com

Antonio Abrignani/Shutterstock.com

leonello calvetti/Shutterstock.com

Alena Brozova/Shutterstock.com

Tatiana53/Shutterstock.com

Tatiana53/Shutterstock.com

Engineers, when selecting materials for a product, consider many factors including cost, weight, corrosiveness, and load-bearing capacity.

LEARNING OBJECTIVES

LO¹ **Material Selection and Origin:** explain factors that are considered when selecting a material for a product and where materials come from

LO² **The Properties of Materials:** describe important properties of materials

LO³ **Metals:** describe different metals and their compositions and applications

LO⁴ **Concrete:** describe its basic ingredients and use

LO⁵ **Wood, Plastics, Silicon, Glass, and Composites:** describe their compositions and applications

LO⁶ **Fluid Materials: Air and Water:** explain their role in our daily life and properties and applications

WHAT DO YOU THINK?

DISCUSSION STARTER

For much of the last century, the straightforward solution to making a car perform better has been to install a bigger engine. In the hybrids and electric cars of coming years, however, the answer might be installing motors with more powerful magnets. Until the 1980s, the most powerful magnets available were those made from an alloy containing samarium and cobalt. But mining and processing those metals presented challenges: samarium is one of 17 rare-earth elements and was costly to refine, while most cobalt came from mines in unstable regions of Africa.

In 1982, when researchers at General Motors developed a magnet based on neodymium, it seemed that an ideal alternative had arrived. While neodymium is also one of the rare-earth metals—a misleading name, as they are actually fairly common, just widely dispersed—it is more abundant than samarium, and at the time, it was cheaper. When combined with iron and boron,—both readily available elements—it produced powerful magnets.

In the electric drive motor of a hybrid car, for instance, just a kilogram (2.2 pounds) of neodymium-based magnets can deliver 80 horsepower, which is enough to move a 3,000-pound vehicle like the Toyota Prius. Neodymium is an ideal magnetic material, because it helps to retain a magnetic charge during all driving conditions, and when dysprosium is added to the alloy, performance at high temperatures is preserved.

In recent decades, the demand for neodymium has increased sharply, which is a result of its usefulness in producing the compact, lightweight magnets used in devices like computer hard drives and audio system speakers. Today, China controls more than 90 percent of the world's production of rare-earth metals and tightly regulates

Rare-earth oxides, clockwise from top center: praseodymium, cerium, lanthanum, neodymium, samarium, and gadolinium.

Image by Peggy Greb, USDA

their export. In 2010, China suspended exports of rare earths to Japan over a territorial dispute. The ruckus caused neodymium prices to soar to nearly $500 a kilogram by the summer of 2011, from less than $50 a kilogram at the start of 2010.

Although shipments have resumed, and prices have since come down, the uncertainties of supply have prompted a search for alternatives. Companies like Molycorp, which is reopening and expanding its rare-earth mine in Mountain Pass, California (about 55 miles south of Las Vegas), are seeking new sources for these metals. New supplies are not the only searches under way: last week Honda announced that it would start recycling rare earth metals from used car parts, like the nickel-metal-hydride batteries used in hybrid cars, that contain small amounts of neodymium along with lanthanum and cerium. In 2011, Toyota said it was developing induction motors that do not require rare-earth magnets.

Source: Jim Witkin, "A Push to Make Motors With Fewer Rare Earths," The New York Times, April 20, 2012, available at http://www.nytimes.com/2012/04/22/automobiles/a-push-to-make-motors-with-fewer-rare-earths.html?pagewanted=1&_r=0

To the students: What are your thoughts: recycle more, design with alternative materials, or expand mining? How much of other materials such as metals, plastics, glass, and wood do you think you consume in your lifetime?

As we discussed in Chapter 1, engineers design millions of products and services that we use in our everyday lives: cars, computers, aircraft, clothing, toys, home appliances, surgical equipment, heating and cooling equipment, health care devices, tools, and machines that make various products. Engineers also design and supervise the construction of buildings, dams, highways, power plants, and mass transit systems.

As design engineers, whether you are designing a machine part, a toy, a frame for a car, or a structure, the selection of material is an important design decision. There are a number of factors that engineers consider when selecting a material for a specific application. For example, they consider properties of material such as density, ultimate strength, flexibility, machinability, durability, thermal expansion, electrical and thermal conductivity, and resistance to corrosion. They also consider the cost of the material and how easily it can be repaired. Engineers are always searching for ways to use advanced materials to make products lighter and stronger for different applications.

In this chapter, we will look more closely at materials that commonly are used in various engineering applications. We will also discuss some of the basic physical characteristics of materials that are considered in design. We will examine solid materials like metals and their alloys, plastics, glass, and wood, and those that solidify over time (such as concrete). We will also investigate in more detail basic fluids like air and water that not only are needed to sustain life but also play important roles in engineering. Did you ever stop to think about the important role that air plays in food processing, driving power tools, or in your car's tire to provide a cushiony ride? You may not think of water as an engineering material either, but we not only need water to live, we also need water to generate electricity in steam and hydroelectric power plants, and we use high-pressurized water, which functions like a saw, to cut materials.

LO¹ 17.1   Material Selection and Origin

Design engineers, when faced with selecting materials for their products, often ask questions such as: How strong will the material be when subjected to an expected load? Would it fail, and if not, how safely would the material carry the load? How would the material behave if its temperature were changed? Would the material remain as strong as it would under normal conditions if its temperature is increased? How much would it expand when its temperature is increased? How heavy and flexible is the material? What are its energy-absorbing properties? Would the material corrode? How would it react in the presence of some chemicals? How expensive is the material? Would it dissipate heat effectively? Would the material act as a conductor or as an insulator to the flow of electricity?

It is important to note here that we have only posed a few generic questions; we could have asked additional questions had we considered the specifics of the application. For example, when selecting materials for implants in bioengineering applications, one must consider many additional factors, including: Is the material toxic to the body? Can the material be sterilized? When the material comes into contact with bodily fluids, will it corrode or deteriorate? Because the human body is a dynamic system, we should also ask: How would the material react to mechanical shock and fatigue? Are the mechanical properties of the implant material

Engineers, when selecting materials for products, consider many factors including cost, weight, corrosiveness, and load-bearing capacity.

compatible with those of bone to ensure appropriate stress distributions at contact surfaces? These are examples of additional specific questions that one could ask to find a suitable material for a specific application.

By now, it should be clear that material properties and material cost are important design factors. However, in order to better understand material properties, we must first understand the phases of a substance. We discussed the phases of matter in Chapter 9; as a review and for the sake of continuity and convenience, we will briefly present the phases of matter again here.

As we discussed in Chapter 9, when you look around, you will find that matter exists in various forms and shapes. You will also notice that matter can change shape when its condition or its surroundings are changed. We also explained that all solid objects, liquids, gases, and living things are made of matter, and matter itself is made up of atoms or chemical elements. There are 106 known chemical elements to date. Atoms of similar characteristics are grouped together and shown in a table called the periodic table of chemical elements. Atoms are made up of even smaller particles we call *electrons*, *protons*, and *neutrons*. In your first chemistry class, you will study these ideas in more detail (if you have not yet done so). Some of you may decide to study chemical engineering, in which case you will spend a great deal of time studying chemistry. But for now, remember that atoms are the basic building blocks of all matter. Atoms are combined naturally or in a laboratory setting to create molecules. For example, as you already know, water molecules are made of two atoms of hydrogen and one atom of oxygen. A glass of water is made of billions and billions of homogeneous water molecules. A molecule is the smallest portion of a given matter that still possesses its microscopic characteristic properties.

Matter can exist in four states, depending on its own and the surrounding conditions: solid, liquid, gaseous, or plasma. Let us consider the water that

> Matter can exists in four states: solid, liquid, gaseous (vapor), or plasma.

FIGURE 17.1   Ice cubes.

we drink every day. As you already know, under certain conditions, water exists in a solid form that we call *ice* (Figure 17.1). At a standard atmospheric pressure, water exists in a solid form as long its temperature is kept under 0 °C. Under standard atmospheric pressure, if you were to heat the ice and consequently change its temperature, the ice would melt and change into a liquid form. Under standard pressure at sea level, the water remains liquid up to a temperature of 100° C as you continue heating the water. If you were to carry out this experiment further by adding more heat to the water, eventually the phase changes from a liquid into a gas. This phase of water we commonly refer to as *steam* (Figure 17.2). If you had the means to heat the water to even higher temperatures—temperatures exceeding 2000° C—you would find that you can break up the water molecules into their atoms, and eventually, the atoms break up into free electrons and nuclei that we call *plasma*.

In general, the mechanical and thermophysical properties of a material depend on its phase. For example, as you know from your everyday experience, the density of ice is different from liquid water (ice cubes float in liquid water), and the density of liquid water is different from that of steam. Moreover, the properties of a material in a single phase could depend on its temperature and the surrounding pressure. For example, if you were to look up the density of liquid water in the temperature range of, say, 4 to 100° C under standard atmospheric pressure, you would find that its density decreases with increasing temperature in that range. Therefore, properties of materials depend not only on their phase but also on their temperature and pressure. This is another important fact to keep in mind when selecting materials.

Next, to better understand where materials come from, we need to take a closer look at our home, Earth. As you learned in high school, Earth is the third planet from the Sun. It has a spherical shape with an average diameter diameter of 7926.4 miles (12,756.3 km) and an approximate mass of

FIGURE 17.2   Steam.

13.17×10^{24} pounds (5.98×10^{24} **kg**). Moreover, to better represent the Earth's structure, it is divided into major layers that are located above and below its surface (Figure 17.3). For example, the **atmosphere** represents the air that covers the surface of the Earth. The air extends approximately 90 miles (140 km) from the surface of the earth to a point called the edge of space.

> The structure below the earth's surface is commonly grouped into the crust, mantle, outer core, and inner core.

Our knowledge of what is inside the Earth and its composition continues to improve. Each day, we learn from studies that deal with the earth's surface and near-surface about rocks, interior heat-transfer rates, gravity and magnetic fields, and earthquakes. The results of these studies suggest that the Earth is made up of different layers with different characteristics and that its mass is composed mostly of iron, oxygen, and silicon (approximately 32% of iron, 30% of oxygen, and 15% of silicon). It also contains other elements such as sulfur, nickel, magnesium, and aluminum. The structure below the earth's surface (Figure 17.4) is generally grouped into four layers: the *crust, mantle, outer core,* and *inner core* (see Table 17.1). This classification is based on the properties of materials and the manner by which the materials move or flow.

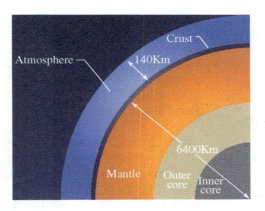

FIGURE 17.3   Earth's layers.

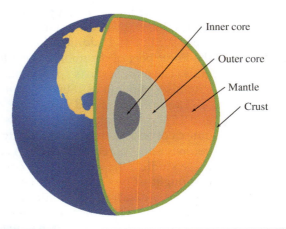

FIGURE 17.4   Earth's core.

| TABLE 17.1 | The Approximate Mass for Each Layer of the Earth |

	Approximate Mass (pounds)	Approximate Mass (kg)	Percentage of the Earth's Total Mass
Atmosphere	1.12×10^{19}	5.1×10^{18}	0.000086
Oceans	3.08×10^{21}	1.4×10^{21}	0.024
Crust	5.73×10^{22}	2.6×10^{22}	0.44
Mantle	8.9×10^{24}	4.04×10^{24}	68.47
Core			
Outer Core	4.03×10^{24}	1.83×10^{24}	31.01
Inner Core	2.12×10^{23}	9.65×10^{22}	1.63

The *crust* makes up about 0.5% of the earth's total mass and 1% of its volume; and because of the ease of access to materials near the surface, its composition and structure has been studied extensively. It has a maximum thickness of 25 miles (40 km). Scientists have been able to collect samples of the crust up to the depth of 12 kilometers; however, because the drilling expenses increase with depth, the progress to deeper locations has slowed down. The Earth's crust—the oceanic floors and the continents—is made up of about twelve plates that continuously move at slow rates (a few centimeters per year). Moreover, the boundaries of these plates—where they come together—mark regions of earthquake and volcanic activities. Over time, the collisions of these plates have created mountain ranges around the world. It is also important to note that the crust is thicker under the continents and thinner under the oceanic floors.

As shown in Table 17.1, most of the mass of the earth comes from the mantle. The *mantle* is made up of molten rock that lies underneath the crust and makes up nearly 84% of the Earth's volume. Unlike the crust, what we know of the mantle's composition is based on our studies of sound propagation, heat flow, earthquakes, and magnetic and gravity fields. Rooted in these studies and additional laboratory investigations is the suggestion that the lower part of the mantle is made up of iron and magnesium silicate minerals. The mantle starts approximately 25 miles below the earth's surface and extends to a depth of 1800 miles (40 to 2900 km).

The *inner core* and *outer core* make up about 33% of the earth's mass and 15% of its volume. Our knowledge of the structures of the inner and outer cores comes from the study of earthquakes and, in particular, the behavior and speed of shear and compression waves in the core. Based on these studies, the inner core is considered to be solid, while the outer core is thought to be fluid and to be composed mainly of iron. The outer core starts at a depth of 1800 miles and extends to a depth of 3200 miles (2900 to 5200 km). The inner core is located between 3200 and 4000 miles (5200 to 6400 km) below the earth's surface.

Before You Go On

Answer the following questions to test your understanding of the preceding section.

1. Give examples of properties that engineers consider when selecting a material for a product or an application.

2. What are the phases of matter?

3. Name the different layers that make up the Earth.

4. Which layer of the Earth contains the most amount of mass?

5. What are the major chemical components of the Earth?

Vocabulary—State the meaning of the following terms:

Plasma _____

Continental crust _____

Oceanic crust _____

Mantle _____

Inner Core _____

Outer Core _____

LO² 17.2   The Properties of Materials

As we have been explaining up to this point, when selecting a material for an application and as an engineer, you need to consider a number of material properties. In general, the properties of a material may be divided into three groups: *electrical, mechanical,* and *thermal.* In electrical and electronic applications, for example, the electrical resistivity of materials is important. How much resistance to the flow of electricity does the material offer? In many mechanical, civil, and aerospace engineering applications, the mechanical properties of materials are important. These properties include modulus of elasticity, modulus of rigidity, tensile strength, compression strength, the strength-to-weight ratio, modulus of resilience, and modulus of toughness. In applications dealing with fluids (liquids and gases), thermophysical properties such as thermal conductivity, heat capacity, viscosity, vapor pressure, and compressibility are important properties. Thermal expansion of a material, whether solid or fluid, is also an important design factor. Resistance to corrosion is another important factor that must be considered when selecting materials.

Material properties depend on many factors, including how the material was processed, its age, its exact chemical composition, and any nonhomogeneity or defect within the material.

The properties of a material are commonly divided into three groups: electrical, mechanical, and thermal.

Material properties also change with temperature and time as the material ages. Most companies that sell materials will provide upon request information on the important properties of their manufactured materials. Keep in mind that, when practicing as an engineer, you should use the manufacturer's material property values in your design calculations. The property values given in this and other textbooks should be used as typical values—not as exact values.

In the previous chapters, we have explained what some properties of materials mean. The meaning of those and other properties that we have not explained already are summarized next.

Electrical Resistivity The value of **electrical resistivity** is a measure of the resistance of material to the flow of electricity. For example, plastics and ceramics typically have high resistivity, whereas metals typically have low resistivity, and among the best conductors of electricity are silver and copper.

Density **Density** is defined as mass per unit volume; it is a measure of how compact the material is for a given volume. For example, the average density of aluminum alloys is 2700 kg/m³, and when compared to the steel density of 7850 kg/m³, we find aluminum has a density that is approximately one-third the density of steel.

Modulus of Elasticity (Young's Modulus) The **modulus of elasticity** is a measure of how easily a material will stretch when pulled (subject to a tensile force) or how well the material will shorten when pushed (subject to a compressive force). The larger the value of the modulus of elasticity is, the larger the required force would be to stretch or shorten the material. For example, the modulus of elasticity of aluminum alloy is in the range of 70 to 79 GPa, whereas steel has a modulus of elasticity in the range of 190 to 210 GPa; therefore, steel is approximately three times stiffer than aluminum alloys.

Modulus of Rigidity (Shear Modulus) The **modulus of rigidity** is a measure of how easily a material can be twisted or sheared. The value of the modulus of rigidity is also called the *shear modulus* and shows the resistance of a given material to shear deformation. Engineers consider the value of the shear modulus when selecting materials for shafts and rods that are subjected to twisting torques. For example, the modulus of rigidity or shear modulus for aluminum alloys is in the range of 26 to 36 GPa, whereas the shear modulus for steel is in the range of 75 to 80 GPa. Therefore, steel is approximately three times more rigid in shear than aluminum.

Tensile Strength The **tensile strength** of a piece of material is determined by measuring the maximum tensile load a material specimen in the shape of a rectangular bar or cylinder can carry without failure. The tensile strength or ultimate strength of a material is expressed as the maximum tensile force per unit cross-sectional area of the specimen. When a material specimen is tested for its strength, the applied tensile load is increased slowly. In the very beginning of the test, the material will deform elastically, meaning that if the load is removed the material will return to its original size and shape without any permanent deformation. The point to which the material exhibits this elastic behavior is called the *yield point*. The yield strength represents the maximum load that the material can carry without any permanent deformation. In certain engineering design applications (especially involving brittle materials), the yield strength is used as the tensile strength.

Compression Strength Some materials are stronger in compression than they are in tension; concrete is a good example. The **compression strength** of a piece of material is determined by measuring the maximum compressive load a material specimen in the shape of cylinder or cube can carry without failure. The ultimate compressive strength of a material is expressed as the maximum compressive force per unit cross-sectional area of the specimen. Concrete has a compressive strength in the range of 10 to 70 MPa.

Modulus of Resilience The **modulus of resilience** is a mechanical property of a material that shows how effective the material is in absorbing mechanical energy without sustaining any permanent damage.

Modulus of Toughness The **modulus of toughness** is a mechanical property of a material that indicates the ability of the material to handle overloading before it fractures.

Strength-to-Weight Ratio As the term implies, this is the ratio of the strength of the material to its specific weight (weight of the material per unit volume). Based on the application, engineers use either the yield or the ultimate strength of the material when determining the **strength-to-weight ratio** of a material.

Thermal Expansion The coefficient of linear expansion can be used to determine the change in the length (per original length) of a material that would occur if the temperature of the material were changed. The **thermal expansion** is an important material property to consider when designing products and structures that are expected to experience a relatively large temperature swing during their service lives.

Thermal Conductivity **Thermal conductivity** is a property of materials that shows how good the material is in transferring thermal energy (heat) from a high-temperature region to a low-temperature region within the material.

Heat Capacity Some materials are better than others in storing thermal energy. The value of **heat capacity** represents the amount of thermal energy required to raise the temperature of 1 kilogram mass of a material by $1°$ C or, using U.S. Customary Units, the amount of thermal energy required to raise one pound mass of a material by $1°$ F. Materials with large heat capacity values are good at storing thermal energy.

Viscosity, vapor pressure, and bulk modulus of compressibility are additional fluid properties that engineers consider in design.

Viscosity The value of **viscosity** of a fluid represents a measure of how easily the given fluid can flow. The higher the viscosity value is, the more resistance the fluid offers to flow. For example, it would require less energy to transport water in a pipe than it would to transport motor oil or glycerin.

Vapor Pressure Under the same conditions, fluids with low **vapor pressure** values will not evaporate as quickly as those with high values of vapor pressure. For example, if you were to leave a pan of water and a pan of glycerin side by side in a room, the water will evaporate and leave the pan long before you would notice any changes in the level of glycerin.

Bulk Modulus of Compressibility A fluid's bulk modulus represents how compressible the fluid is. How easily can one reduce the volume of the fluid when the fluid pressure is increased? As we discussed in Chapter 10, it would take a pressure of 2.24×10^7 N/m^2 to reduce 1 m^3 volume of water by 1% or, said another way, to a final volume of 0.99 m^3.

In this section, we explained the meaning and significance of some of the physical properties of materials. Tables 17.2 through 17.5 show some properties of solid materials. In the following sections, we will examine the application and chemical composition of some common engineering materials.

TABLE 17.2	Modulus of Elasticity and Shear Modulus of Selected Materials	
Material	**Modulus of Elasticity (GPa)**	**Shear Modulus (GPa)**
Aluminum alloys	70–79	26–30
Brass	96–110	36–41
Bronze	96–120	36–44
Cast iron	83–170	32–69
Concrete (compression)	17–31	
Copper alloys	110–120	40–47
Glass	48–83	19–35
Magnesium alloys	41–45	15–17
Nickel	210	80
Plastics		
Nylon	2.1–3.4	
Polyethylene	0.7–1.4	
Rock (compression)		
Granite, marble, quartz	40–100	
Limestone, sandstone	20–70	
Rubber	0.0007–0.004	0.0002–0.001
Steel	190–210	75–80
Titanium alloys	100–120	39–44
Tungsten	340–380	140–160
Wood (bending)		
Douglas fir	11–13	
Oak	11–12	
Southern pine	11–14	

Based on Gere, Mechanics of Materials, 5E. 2001, Cengage Learning.

TABLE 17.3	Densities of Selected Materials	
Material	**Mass Density (kg/m³)**	**Specific Weight (kN/m³)**
Aluminum alloys	2600–2800	25.5–27.5
Brass	8400–8600	82.4–84.4
Bronze	8200–8800	80.4–86.3
Cast iron	7000–7400	68.7–72.5
Concrete		
Plain	2300	22.5
Reinforced	2400	23.5
Lightweight	1100–1800	10.8–17.7
Copper	8900	87.3
Glass	2400–2800	23.5–27.5
Magnesium alloys	1760–1830	17.3–18.0
Nickel	8800	86.3
Plastics		
Nylon	880–1100	8.6–10.8
Polyethylene	960–1400	9.4–13.7
Rock		
Granite, marble, quartz	2600–2900	25.5–28.4
Limestone, sandstone	2000–2900	19.6–28.4
Rubber	960–1300	9.4–12.7
Steel	7850	77.0
Titanium alloys	4500	44.1
Tungsten	1900	18.6
Wood (air dry)		
Douglas fir	480–560	4.7–5.5
Oak	640–720	6.3–7.1
Southern pine	560–640	5.5–6.3

Based on Gere, Mechanics of Materials, 5E. 2001, Cengage Learning.

TABLE 17.4   The Strength of Selected Materials

Material	Yield Strength (MPa)	Ultimate Strength (MPa)
Aluminum alloys	35–500	100–550
Brass	70–550	200–620
Bronze	82–690	200–830
Cast iron (tension)	120–290	69–480
Cast iron (compression)		340–1400
Concrete (compression)		10–70
Copper alloys	55–760	230–830
Glass		30–1000
Plate glass		70
Glass fibers		7000–20,000
Magnesium alloys	80–280	140–340
Nickel	100–620	310–760
Plastics		
Nylon		40–80
Polyethylene		7–28
Rock (compression)		
Granite, marble, quartz		50–280
Limestone, sandstone		20–200
Rubber	1–7	7–20
Steel		
High-strength	340–1000	550–1200
Machine	340–700	550–860
Spring	400–1600	700–1900
Stainless	280–700	400–1000
Tool	520	900
Steel wire	280–1000	550–1400
Structural steel	200–700	340–830
Titanium alloys	760–1000	900–1200
Tungsten		1400–4000
Wood (bending)		
Douglas fir	30–50	50–80

TABLE 17.4	The Strength of Selected Materials *(Continued)*	
Oak	40–60	50–100
Southern pine	40–60	50–100
Wood (compression parallel to grain)		
Douglas fir	30–50	40–70
Oak	30–40	30–50
Southern pine	30–50	40–70

Based on Gere, Mechanics of Materials, 5E. 2001, Cengage Learning.

TABLE 17.5	Coefficients of Thermal Expansion for Selected Materials*	
Material	**Coefficient of Thermal Expansion $(1/°C) \times 10^6$**	**Coefficient of Thermal Expansion $(1/°F) \times 10^6$**
Aluminum alloys	23	13
Brass	19.1–21.2	10.6–11.8
Bronze	18–21	9.9–11.6
Cast iron	9.9–12	5.5–6.6
Concrete	7–14	4–8
Copper alloys	16.6–17.6	9.2–9.8
Glass	5–11	3–6
Magnesium alloys	26.1–28.8	14.5–16.0
Nickel	13	7.2
Plastics		
Nylon	70–140	40–80
Polyethylene	140–290	80–160
Rock	5–9	3–5
Rubber	130–200	70–110
Steel	10–18	5.5–9.9
High-strength	14	8.0
Stainless	17	9.6
Structural	12	6.5
Titanium alloys	8.1–11	4.5–6.0
Tungsten	4.3	2.4

*Note that you must multiply the coefficients given in this table by 10^{-6} to obtain the actual values of coefficients of thermal expansion.
Based on Gere, Mechanics of Materials, 5E. 2001, Cengage Learning.

LO³ 17.3 Metals

In this section, we briefly examine the chemical composition and common application of metals. We will discuss **light metals**, copper and its alloys, iron and steel.

Lightweight Metals

Aluminum, titanium, and magnesium, because of their small densities (relative to steel), are commonly referred to as *lightweight metals*. Because of their relatively high strength-to-weight ratios, lightweight metals are used in many structural and aerospace applications.

Aluminum and its alloys have densities that are approximately one-third the density of steel. Pure aluminum is very soft; thus, it is generally used in electronics applications and in making reflectors and foils. Because pure aluminum is soft and has a relatively small tensile strength, it is alloyed with other metals to make it stronger, easier to weld, and to increase its resistance to corrosive environments. Aluminum is commonly alloyed with copper (Cu), zinc (Zn), magnesium (Mg), manganese (Mn), silicon (Si), and lithium

> Aluminum, titanium, and magnesium are called lightweight metals because of their (relative to steel) small densities.

(Li). The American National Standards Institute (ANSI) assigns designation numbers to specify aluminum alloys. Generally speaking, aluminum and its alloys resist corrosion; they are easy to mill and cut and can be brazed or welded. Aluminum parts also can be joined using adhesives. They are good conductors of electricity and heat and thus have relatively high thermal conductivity and low electrical resistance values. Aluminum is fabricated in sheets, plates, foil, rods, and wire and is extruded to make window frames or automotive parts. You are already familiar with everyday examples of common aluminum products (Figure 17.5), including beverage cans, household aluminum foil, staples in tea bags (that don't rust), aluminum engine block, and so on. The use of aluminum in variuos sectors of our economy is shown in Figure 17.6, indicating that most of the aluminum produced is consumed by the packing, transportation, building, and electrical sectors.

Titanium has an excellent strength-to-weight-ratio. Titanium is used in applications where relatively high temperatures, exceeding $400°\,C$ up to $600°\,C$, are expected. Titanium alloys are used in the fan blades and the compressor blades of the gas turbine engines of commercial and military airplanes. In

(a)   (b)   (c)

FIGURE 17.5   Uses of aluminium.

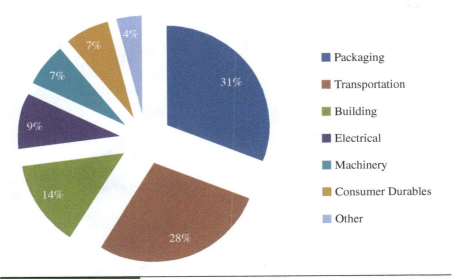

FIGURE 17.6   Aluminum use by various sectors.

fact, without the use of titanium alloys, the engines on commercial airplanes would not have been possible. Like aluminum, titanium is alloyed with other metals to improve its properties. Titanium alloys show excellent resistance to corrosion. Titanium is quite expensive compared to aluminum. It also is heavier than aluminum, having a density which is roughly one-half that of steel. Because of their relatively high strength-to-weight ratios, titanium alloys are used in both commercial and military airplane airframes (fuselage and wings) and landing gear components. Titanium alloys are becoming a metal of choice in many products (Figure 17.7); you can find them in golf clubs, bicycle frames, tennis racquets, and spectacle frames. Because of their excellent corrosion resistance, titanium alloys have been used in the tubing in desalination plants as well. Replacement hips and other joints are examples of medical applications where titanium is currently being used.

As shown in Figure 17.8, about 94% of titanium mineral concentrate was consumed as titanium dioxide (TiO_2), which is commonly used as pigments in paint, plastics, and paper. The remaining 6% is used in the manufacturing of welding rods, chemicals, and metals.

With its silvery white appearance, *magnesium* is another lightweight metal that looks like aluminum, but it is lighter, having a density of approximately 1700 kg/m³. Pure magnesium does not provide good strength for structural applications; because of this fact, it is alloyed with other elements such as aluminum, manganese, and zinc to improve its mechanical characteristics.

lasha/Shutterstock.com   nikkytok/Shutterstock.com   PeJo/Shutterstock.com

FIGURE 17.7   Uses of titanium.

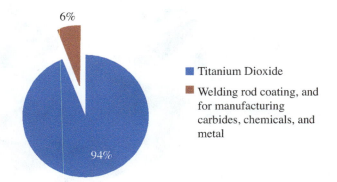

6%

■ Titanium Dioxide
■ Welding rod coating, and for manufacturing carbides, chemicals, and metal

94%

FIGURE 17.8   The percentage of titanium mineral concentrate use by sector.

Magnesium and its alloys are used in nuclear applications, in drycell batteries, in aerospace applications, and in some automobile parts as sacrificial anodes to protect other metals from corrosion. The mechanical properties of the lightweight metals are shown in Tables 17.2 through 17.5.

Copper and Its Alloys

Copper is a good conductor of electricity, and because of this, it is commonly used in many electrical applications, including home wiring (Figure 17.9(a)). Copper and many of its alloys are also good conductors of heat, and this thermal property makes copper a good choice for heat exchanger applications in air conditioning and refrigeration systems. Copper alloys are also used as tubes, pipes, and fittings in plumbing (Figure 17.9(c)) and heating applications.

Copper Alloys Copper is alloyed with zinc, tin, aluminum, nickel, and other elements to modify its properties. When copper is alloyed with zinc, it is commonly called **brass**. The mechanical properties of brass depend on the exact composition of percent copper and percent zinc. **Bronze** is an alloy of copper and tin. Copper is also alloyed with aluminum and is referred to as *aluminum bronze*. Copper and its alloys are also used in hydraulic brake lines, pumps, and screws.

The percentage of copper consumption in various sectors of our economy is shown in Figure 17.10. The majority of the copper extracted is consumed in building construction, electrical, and electronic products.

As mentioned previously, copper is alloyed with zinc, aluminum, nickel, and other elements to modify its properties. Zinc also is alloyed with other

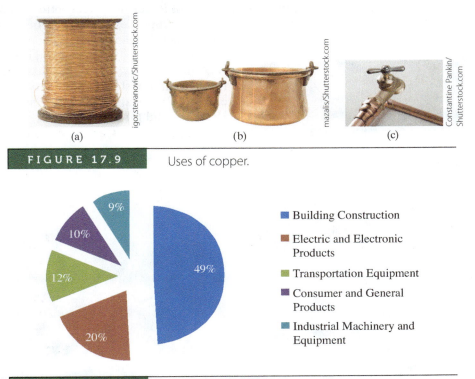

(a)   (b)   (c)

igor.stevanovic/Shutterstock.com

mazalis/Shutterstock.com

Constantine Pankin/Shutterstock.com

FIGURE 17.9 Uses of copper.

- Building Construction
- Electric and Electronic Products
- Transportation Equipment
- Consumer and General Products
- Industrial Machinery and Equipment

49%
20%
12%
10%
9%

FIGURE 17.10 Copper use by various sectors.

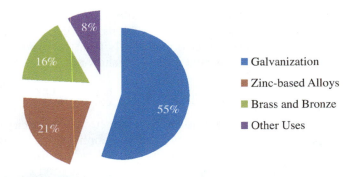

FIGURE 17.11   The percentage of zinc consumption by end-use.

materials to increase the resistance of a material to corrosion. As shown in Figure 17.11, 55% of the zinc consumed was for galvanizing, and 16% for making brass and bronze. Zinc is also consumed by the rubber, chemical, and paint industries.

Iron and Steel

Steel is a common material (Figure 17.12) that is used in the framework of buildings, bridges, and the body of appliances such as refrigerators, ovens, dishwashers, washers and dryers, and cooking utensils. *Steel* is an alloy of iron with approximately 2% or less carbon. Pure iron is soft, and thus, it is not good for structural applications. But the addition of even a small amount of carbon to iron hardens it and gives steel better mechanical properties, such as greater strength. The properties of steel can be modified by adding elements such as chromium, nickel, manganese, silicon, and tungsten. For example, chromium is used to increase the resistance of steel to corrosion. In general, steel can be classified into three broad groups: (1) the carbon steels containing approximately 0.015 to 2% carbon, (2) low-alloy steels having a maximum of 8% alloying elements, and (3) high-alloy steels containing more than 8% of alloying elements. Carbon steels constitute most of the world's

> Steel is an alloy of iron with approximately 2% or less carbon. The addition of carbon to iron gives steel greater strength.

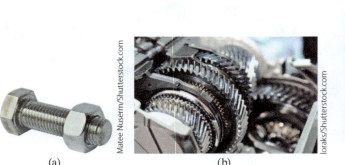

(a)   (b)   (c)

FIGURE 17.12   Uses for steel.

steel consumption; thus, you will commonly find them in the bodies of appliances and cars. Low-alloy steels have good strength and are commonly used as machine or tool parts and as structural members. High-alloy steels, such as **stainless steels**, could contain approximately 10 to 30% chromium and could contain up to 35% nickel. The 18/8 stainless steels, which contain 18% chromium and 8% nickel, are commonly used for tableware, kitchen utensils, and other household products. Finally, **cast iron** is also an alloy of iron that has 2 to 4% carbon. Note that the addition of extra carbon to the iron changes its properties completely. In fact, cast iron is a brittle material, whereas most iron alloys containing less than 2% carbon are ductile.

As shown in Figure 17.13, most of the steel consumption went to steel service centers, construction, and automotive.

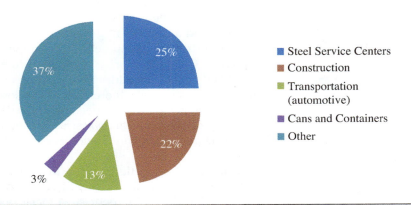

■ Steel Service Centers
■ Construction
■ Transportation (automotive)
■ Cans and Containers
■ Other

FIGURE 17.13   The percentage of steel consumption by sector.

Before You Go On

Answer the following questions to test your understanding of the preceding section.

1. What is a lightweight metal?

2. What is the difference between steel and iron?

3. Give examples of applications in which titanium is used.

4. Give examples of aluminum use.

5. Give examples of copper use.

Vocabulary—State the meaning of the following terms:

Steel _____

Bronze _____

Brass _____

18/8 Stainless Steel _____

LO⁴ 17.4 Concrete

Today, concrete is commonly used in construction (Figure 17.14) of roads, bridges, buildings, tunnels, and dams. What is normally called **concrete** consists of three main ingredients: aggregate, cement, and water. Aggregate refers to materials such as gravel and sand, while cement refers to the bonding material that holds the aggregate together. The types and size (fine to coarse) of aggregate used in making concrete varies depending on the application. The amount of water used in making concrete (water-to-cement ratio) could also influence its strength. Of course, the mixture must have enough water so that the concrete can be poured and have a consistent cement paste that completely wraps around all aggregates. The ratio of cement-to-aggregate used in making concrete also affects the strength and durability of the concrete. Another factor that could influence the cured strength of concrete is the temperature of its surroundings when it is poured. Calcium chloride is added to cement when the concrete is poured in cold climates. The addition of calcium chloride will accelerate the curing process to counteract the effect of the low temperature of the surroundings. You may have also noticed as you walk by newly poured concrete for a driveway or sidewalk that water is sprayed onto the concrete for some time after it is poured. This is to control the rate of contraction of the concrete as it sets.

> Concrete is a mixture of cement, aggregate (such as sand and gravel) and water.

Concrete is a brittle material that can support compressive loads much better than it does tensile loads. Because of this fact, concrete is commonly **reinforced** with steel bars or steel mesh that consists of thin metal rods to increase its load-bearing capacity, especially in the sections where tensile stress is expected. The concrete is poured into forms that contain the metal mesh or steel bars. Reinforced concrete is used in foundations, floors, walls, and columns. Another common construction practice is the use of **precast concrete**. Precast concrete slabs, blocks, and structural members are fabricated in less time and as noted earlier with less cost in factory settings where the surrounding conditions are controlled. The precast concrete parts are then moved to the construction site where they are erected. As we mentioned, concrete has a higher compressive strength than tensile strength. Because of

(a)

(b)

(c)

MarchCattle/Shutterstock.com

SARIN KUNTHONG/Shutterstock.com

Pavel L Photo and Video/Shutterstock.com

FIGURE 17.14   Uses for concrete.

this fact, concrete is also **prestressed** in the following manner. Before concrete is poured into forms that have the steel rods or wires, the steel rods or wires are stretched; after the concrete has been poured and after enough time has elapsed, the tension in the rods or wires is released. This process, in turn, compresses the concrete. The prestressed concrete then acts as a compressed spring, which will become un-compressed under the action of tensile loading. Therefore, the prestressed concrete section will not experience any tensile stress until the section has been completely uncompressed. It is important to note once again the reason for this practice is that concrete is weak under tension.

The percentage of cement use by various sectors is shown in Figure 17.15. As you would expect, the economy and construction projects in particular define most of the cement production amount.

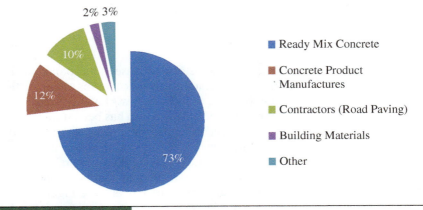

2% 3%

10%

12%

73%

- Ready Mix Concrete
- Concrete Product Manufactures
- Contractors (Road Paving)
- Building Materials
- Other

FIGURE 17.15   The percentage of cement use by sectors.

Before You Go On

Answer the following questions to test your understanding of the preceding section.

1. What are the main ingredients of concrete?

2. Why is water sprayed on newly poured concrete for some time after it is poured?

3. Is concrete better in supporting compressive loads or tensile loads?

Vocabulary—State the meaning of the following terms:

Aggregate _____

Precast Concrete _____

Prestressed Concrete _____

(a)   (b)   (c)   (d)   (e)

FIGURE 17.16   Uses for wood.

LO⁵ 17.5   Wood, Plastics, Silicon, Glass, and Composites

Wood

Throughout history, because of its abundance in many parts of the world, wood has been a material of choice for many applications. Wood is a renewable source, and because of its ease of workability and its strength, it has been used to make many products. Wood also has been used as fuel in stoves and fireplaces. Today, wood is used in a variety of products ranging from telephone poles to toothpicks. Common examples of wood products include hardwood flooring, roof trusses, furniture frames, wall supports, doors, decorative items, window frames, trimming in luxury cars, tongue depressors, clothespins, baseball bats, bowling pins, fishing rods, and wine barrels (see Figure 17.16). Wood is also the main ingredient that is used to make various paper products. Whereas a steel structural member is susceptible to rust, wood (on the other hand) is prone to fire, termites, and rotting. Wood is an *anisotropic* material, meaning that its properties are direction-dependent. For example, as you may already know, under axial loading (when pulled), wood is stronger in a direction parallel to a grain than it is in a direction across the grain. However, wood is stronger in a direction normal to the grain when it is bent. The properties of wood also depend on its moisture content; the lower the moisture content, the stronger the wood is. Density of wood is generally a good indication of its strength. As a rule of thumb, the higher the density of wood, the higher its strength. Moreover, any defects, such as knots, would affect the load-carrying capacity of wood. Of course, the location of the knot and the extent of the defect will directly affect its strength.

> Wood is anisotropic material, meaning that its properties are direction-dependent.

Timber is commonly classified as softwood and hardwood. **Softwood** timber is made from trees that have cones (coniferous), such as pine, spruce, and Douglas fir. On the other hand, **hardwood** timber is made from trees that have broad leaves or have flowers. Examples of hard-woods include walnut, maple, oak, and beech. This classification of wood into softwood and hardwood should be used with caution, because there are some hardwood timbers that are softer than softwoods.

Plastics

In the latter part of the 20th century, plastics increasingly became the material of choice for many applications. They are very lightweight, strong, inexpensive, and easily made into various shapes (Figure 17.17). Over

(a)

(b)

(c)

FIGURE 17.17   Uses for polymer-based products.

100 million metric tons of plastic are produced annually worldwide. Of course, this number increases as the demand for inexpensive, durable, and disposable material grows. Most of you are already familiar with examples of plastic products, including grocery and trash bags, plastic soft drink containers, home cleaning containers, vinyl siding, polyvinyl chloride (PVC) piping, valves, and fittings that are readily available in home improvement centers. Styrofoam™ plates and cups, plastic forks, knives, spoons, and sandwich bags are other examples of plastic products that are consumed every day.

Polymers are the backbone of what we call plastics. They are chemical compounds that have very large and chainlike molecular structures. Plastics are often classified into two categories: **thermoplastics** and **thermosets**. When heated to certain temperatures, thermoplastics can be molded and remolded. For example, when you recycle styrofoam dishes, they can be heated and reshaped into cups or bowls or other shapes. By contrast, thermosets cannot be remolded into other shapes by heating. The application of heat to thermosets does not soften the material for remolding; instead, the material will simply break down. There are many other ways of classifying plastics; for instance, they may be classified on the basis of their chemical composition, molecular structure (the way molecules are arranged), or their densities. For example, based on their chemical composition, polyethylene, polypropylene, polyvinyl chloride, and polystyrene are the most commonly produced plastics. A grocery bag is an example of a product made from high-density polyethylene (HDPE). However, note that in a broader sense polyethylene and polystyrene, for example, are thermoplastics. In general, the way the molecules of a plastic are arranged will influence its mechanical and thermal properties.

Plastics have relatively small thermal and electrical conductivity values. Some plastic materials, such as styrofoam cups, are designed to have air trapped in them to reduce the heat conduction even more. Plastics are easily colored by using various metal oxides. For example, titanium oxide and zinc

Oleksiy Mark/Shutterstock.com

FIGURE 17.18   Uses for silicon.

oxide are used to make plastic appear white. Carbon is used to give plastic sheets a black color, as is the case in black trash bags. Depending on the application, other additives to the polymers are used to obtain specific characteristics such as rigidity, flexibility, enhanced strength, or a longer life span that excludes any change in the appearance or mechanical properties of the plastic over time. As with other materials, research is being performed every day to make them plastics stronger and more durable, to control the aging process, to make them less susceptible to sun damage, and to control water and gas diffusion through them. The latter is especially important when the goal is to add shelf life to food that is wrapped in plastics. Those of you who are planning to study chemical engineering will take semester-long classes that will explore polymers in much more detail.

Silicon

Silicon is a nonmetallic chemical element that is used quite extensively in the manufacturing of transistors and various electronic and computer chips (Figure 17.18). Pure silicon is not found in nature; it is found in the form of silicon dioxide in sands and rocks or combined with other elements, such as aluminum, calcium, sodium, or magnesium in the form that is commonly referred to as *silicates*. Silicon, because of its atomic structure, is an excellent semiconductor, which is a material whose electrical conductivity properties can be changed to act either as a conductor of electricity or as an insulator (preventor of electricity flow). Silicon is also used as an alloying element with other elements such as iron and copper to give steel and brass certain desired characteristics.

> Silicone is a synthetic compound that consists of silicon, oxygen, carbon, and hydrogen. Be sure not to confuse it with silicon, which is a nonmetallic chemical element.

Be sure not to confuse silicon with **silicones**, which are synthetic compounds consisting of silicon, oxygen, carbon, and hydrogen. You find silicones in lubricants, varnishes, and water-proofing products.

Glass

Glass is commonly used in products such as windows, light bulbs, housewares (such as drinking glasses), chemical containers, beverage containers, and decorative items (see Figure 17.19). The composition of

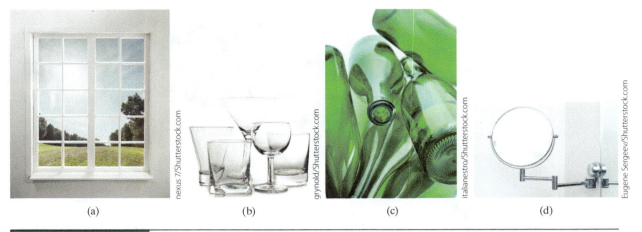

(a) (b) (c) (d)

nexus 7/Shutterstock.com
grynold/Shutterstock.com
italianestro/Shutterstock.com
Eugene Sergeev/Shutterstock.com

FIGURE 17.19   Uses for glass.

the glass depends on its application. The most widely used form of glass is soda-lime-silica glass. The materials used in making soda-lime-silica glass include sand (silicon dioxide), limestone (calcium carbonate), and soda ash (sodium carbonate). Other materials are added to create desired characteristics for specific applications. For example, bottle glass contains approximately 2% aluminum oxide, and glass sheets contain about 4% magnesium oxide. Metallic oxides are also added to give glass various colors. For example, silver oxide gives glass a yellowish stain, and copper oxide gives glass its blueish or greenish color—the degree of which depends on the amount added to the composition of the glass. Optical glasses have very specific chemical compositions and are quite expensive. The composition of optical glass will influence its refractive index and its light-dispersion properties. Glass that is made completely from silica (silicon dioxide) has properties that are sought after by many industries (such as fiber optics), but it is quite expensive to manufacture, because the sand has to be heated to temperatures exceeding 1700° C. Silica glass has a low coefficient of thermal expansion, high electrical resistivity, and high transparency to ultraviolet light. Because silica glass has a low coefficient of thermal expansion, it can be used in high-temperature applications. Ordinary glass has a relatively high coefficient of thermal expansion; therefore, when its temperature is changed suddenly, it could break easily due to thermal stresses developed by the temperature change. Cookware glass contains boric oxide and aluminum oxide to reduce its coefficient of thermal expansion.

Glass Fiber Silica **glass fibers** are commonly used today in fiber optics, which is that branch of science that deals with transmitting data, voice, and images through thin glass or plastic fibers. Every day, copper wires are being replaced by transparent glass fibers in telecommunication to connect computers together in networks. The glass fibers (Figure 17.20) typically have an outer diameter of 0.0125 mm (12 micron) with an inner transmitting core diameter of 0.01 mm (10 micron). Infrared light signals in the wavelength ranges of 0.8 to 0.9 m or 1.3 to 1.6 m are generated by light-emitting diodes or semiconductor lasers and travel through the inner core of glass fiber.

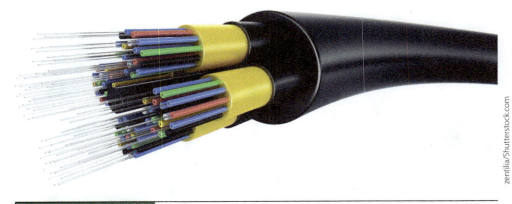

zentilia/Shutterstock.com

FIGURE 17.20   Fiber glass uses.

The optical signals generated in this manner can travel to distances as far as 100 km without any need to amplify them again. Plastic fibers made of polymethylmethacrylate, polystyrene, or polycarbonate are also used in fiber optics. These plastic fibers are, in general, cheaper and more flexible than glass fibers. But when compared to glass fibers, plastic fibers require more amplification of signals due to their greater optical losses. They are generally used in networking computers in a building.

Composites

Because of their light weight and good strength, composite materials are increasingly the materials of choice for a number of products and aerospace applications. Today, you will find composite materials in military planes, helicopters, satellites, commercial planes, fast-food restaurant tables and chairs, and many sporting goods. They also commonly are used to repair bodies of automobiles. In comparison to conventional materials (such as metals), composite materials can be lighter and stronger. For this reason, composite materials are used extensively in aerospace applications.

> Composites are created by combining two or more solid materials to make a new material that has properties that are superior to those of the individual components.

Composites are created by combining two or more solid materials to make a new material that has properties that are superior to those of the individual components. **Composite** materials consist of two main ingredients: matrix material and fibers. Fibers are embedded in matrix materials, such as aluminum or other metals, plastics, or ceramics. Glass, graphite, and silicon carbide fibers are examples of fibers used in the construction of composite materials. The strength of the fibers is increased when embedded in the matrix material, and the composite material created in this manner is lighter and stronger. Moreover, once a crack starts in a single material due to either excessive loading or imperfections in the material, the crack will propagate to the point of failure. In a composite material, on the other hand, if one or a few fibers fail, it does not necessarily lead to failure of other fibers or the material as a whole. Furthermore, the fibers in a composite material can be oriented either in a certain direction or many directions to offer more strength in the direction of expected loads. Therefore, composite materials are designed for specific load applications. For instance, if the expected load is uni-axial, meaning that it is applied in a single direction, then all of the fibers are aligned in the direction of the expected load. For applications expecting multidirection loads, the fibers are aligned in different directions to make the material equally strong in various directions.

Depending upon what type of host matrix material is used in creating the composite material, the composites may be classified into three classes: (1) polymer–matrix composites, (2) metal–matrix composites, and (3) ceramic–matrix composites. We discussed the characteristics of matrix materials earlier when we covered metals and plastics.

Before You Go On

Answer the following questions to test your understanding of the preceding section.

1. Explain the difference between softwood and hardwood.

2. In your own words, explain what is meant by the word polymer.

3. What is the difference between thermoplastics and thermosets?

4. What is the difference between silicon and silicone?

5. What are the materials that are used in making sod-lime-silica glass?

6. What are the main constituents of composite materials?

Vocabulary—State the meaning of the following terms:

Fiber Glass _____

Thermoplastics _____

Polymer _____

Silicon _____

Silicates _____

Anisotropic Material _____

Composite Material _____

LO⁶ 17.6 Fluid Materials: Air and Water

Fluid refers to both liquids and gases. Air and water are among the most abundant fluids on earth. They are important in sustaining life and are used in many engineering applications. We will briefly discuss them next.

Air

We all need air and water to sustain life. Because air is readily available to us, it is also used in engineering as a cooling and heating medium in food processing, in controlling thermal comfort in buildings, as a controlling medium to turn equipment on and off, and to drive power tools. Compressed air in the tires of a car provides a cushioned medium to transfer the weight of the car to the road. Understanding the properties of air and how it behaves is important in many engineering applications, including understanding the lift and drag forces. Better understanding of how air behaves under certain conditions leads to the design of better planes and automobiles. The earth's atmosphere, which we refer to as air, is a mixture of approximately 78% nitrogen, 21% oxygen, and less than 1% argon. Small amounts of other gases are present in earth's atmosphere, as shown in Table 17.6.

> Air is a mixture of mostly nitrogen, oxygen, and small amounts of other gases such as argon, carbon dioxide, sulfur dioxide, and nitrogen oxide.
>
> The air surrounding the earth, depending on its temperature, can be divided into four regions: troposphere, stratosphere, mesosphere, and thermosphere.

There are other gases present in the atmosphere, including carbon dioxide, sulfur dioxide, and nitrogen oxide. The atmosphere also contains water vapor. The concentration level of these gases depends on the altitude and geographical location. At higher altitudes (10 to 50 km), the earth's atmosphere also contains ozone. Even though these gases make up a small percentage of earth's atmosphere, they play a significant role in maintaining a thermally comfortable environment for us and other living species. For example, the ozone absorbs most of the ultraviolet radiation arriving from the sun that could harm us. The carbon dioxide plays an important role in sustaining

TABLE 17.6	The Composition of Dry Air	
Gases		**Volume by Percent**
Nitrogen (N_2)		78.084
Oxygen (O_2)		20.946
Argon (Ar)		0.934
Small amounts of other gases are present in atmosphere including:		
Neon (Ne)		0.0018
Helium (He)		0.000524
Methane (CH_4)		0.0002
Krypton (Kr)		0.000114
Hydrogen (H_2)		0.00005
Nitrous oxide (N_2O)		0.00005
Xenon (Xe)		0.0000087

plant life; however, if the atmosphere contains too much carbon dioxide, it will not allow the earth to cool down effectively by radiation.

Atmosphere Let us now look at the Earth's **atmosphere** in greater detail. The air surrounding the earth can be divided into four distinct regions: **troposphere**, **stratosphere**, **mesosphere**, and **thermosphere** (see Figure 17.21). The layer of air closest to the earth's surface is called the **troposphere** and plays an important role in shaping our weather. The radiation from the sun heats the earth's surface and in turn heats the air near the surface. As the air heats up, it moves away from the earth's surface and cools down. Water vapor in the atmosphere in the form of clouds allows for transport of water from the ocean to land in the form of rain and snow. As shown in Figure 17.21, the temperature of troposphere decreases with altitude. The **stratosphere** starts at an altitude of about 20 kilometers (12 miles), and the air temperature in this region

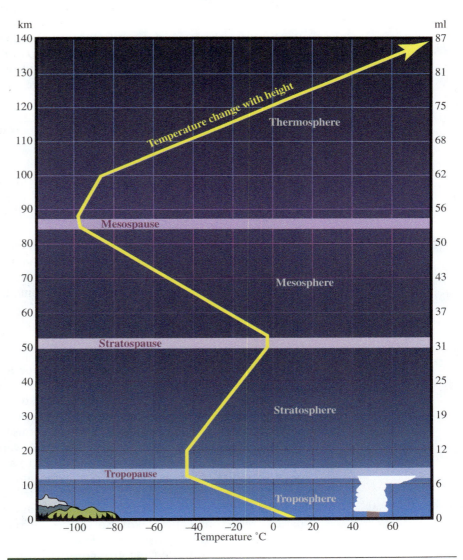

FIGURE 17.21   The different layers of Earth's atmosphere.

increases with altitude as shown. The reason for the increase in temperature in the stratosphere is that the ozone in this layer absorbs ultraviolet (UV) radiation, and as the result, it warms it up. The region above the stratosphere is called the **mesosphere** and contains relatively small amounts of ozone, and consequently, the air temperature decreases again as shown in Figure 17.21. The last layer of air surrounding the Earth is called the **thermosphere**, and the temperature in this layer increases again with altitude because of the absorption of solar radiation by oxygen molecules in the thermosphere.

Humidity There are two common ways of expressing the amount of water vapor in air: absolute humidity (or humidity ratio) and relative humidity. The **absolute humidity** is defined as the ratio of mass of water vapor in a unit mass of dry air, according to

$$\text{absolute humidity} = \frac{\text{mass of water vapor (kg)}}{\text{mass of dry air (kg)}} \qquad \boxed{17.1}$$

For humans, the level of a comfortable environment is better expressed by **relative humidity**, which is defined as the ratio of the amount of water vapor or moisture in the air to the maximum amount of moisture that the air can hold at a given temperature. Therefore, relative humidity is defined as

$$\text{relative humidity} = \frac{\text{the amount of moisture in the air (kg)}}{\text{the maximum amount of moisture that air can hold (kg)}}$$

$$\boxed{17.2}$$

Most people feel comfortable when the relative humidity is around 30 to 50%. The higher the temperature of air, the more water vapor the air can hold before it is fully saturated. Because of its abundance, air is commonly used in food processing, especially in food dehydating processes to make dried fruits, spaghetti, cereals, and soup mixes. Hot air is transported over the food to absorb water vapors and thus remove them from the source.

Understanding how air behaves at given pressures and temperatures is also important when designing cars to overcome air resistance or designing buildings to withstand wind loading.

Water

You already know that every living thing needs water to sustain life. In addition to drinking water, we also need water for washing, laundry, grooming, cooking, and fire protection. You may also know that two-thirds of the earth's surface is covered with water, but most of this water cannot be consumed directly; it contains salt and other minerals that must be removed first. Radiation from the sun evaporates water; water vapors form into clouds, and eventually—under favorable conditions—water vapors turn into liquid water or snow and fall back on the land and the ocean. On land, depending on the amount of precipitation, part of the water infiltrates the soil, part of it may be absorbed by vegetation, and part runs as streams or rivers and collects into natural reservoirs called lakes. **Surface water** refers to water in reservoirs, lakes, rivers, and streams. **Groundwater**, on the other hand, refers to the water that has infiltrated the ground; surface and groundwaters eventually return to the ocean, and the water cycle is completed (see Figure 17.22).

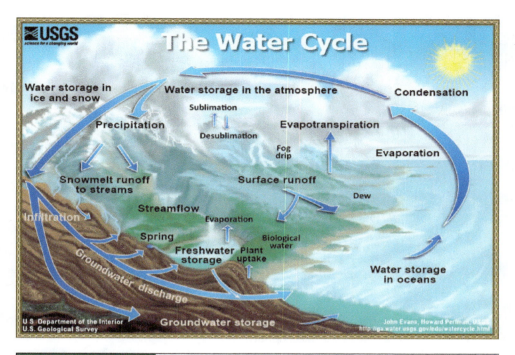

FIGURE 17.22   The water cycle.
Source: USGS

As we said earlier, everyone knows that we need water to sustain life, but what you may not realize is that water can be thought of as a common engineering material! Water is used in all steam power-generating plants to produce electricity. As we explained in Chapter 13, fuel is burned in a boiler to generate heat, which in turn is added to liquid water to change its phase to steam. Steam passes through turbine blades, turning the blades, which in effect runs the generator connected to the turbine, creating electricity. The low-pressure steam liquefies in a condenser and is pumped through the boiler again, closing a cycle. Liquid water stored behind dams is also guided through water turbines located in hydroelectric power plants to generate electricity. Mechanical engineers need to understand the thermophysical properties of liquid water and steam when designing power plants.

We also need water to grow fruits, vegetables, nuts, cotton, trees, and so on. Irrigation channels are designed by civil engineers to provide water to farms and agricultural fields. Water is also used as a cutting tool. High-pressure water containing abrasive particles is used to cut marble or metals. Water is commonly used as a cooling or cleaning agent in a number of food processing plants and industrial applications. Thus, water is not only transported to our homes for our domestic use, but it is also used in many engineering applications. So you see, understanding the properties of water and how it can be used to transport thermal energy, or what it takes to transport water from one location to the next, is important to mechanical engineers, civil engineers, manufacturing engineers, agricultural engineers, and so on. We discussed the Environmental Protection Agency (EPA) standards for drinking water in Chapter 3.

Before You Go On

Answer the following questions to test your understanding of the preceding section.

1. What are the major gases that make up the atmosphere?

2. What is the role of ozone in the atmosphere?

3. In your own words, explain the water cycle.

4. In your own words, explain the difference between absolute humidity and relative humidity.

Vocabulary—State the meaning of the following terms:

Relative Humidity _____

Troposphere _____

Stratosphere _____

Mesosphere _____

Thermosphere _____

Groundwater _____

SUMMARY

LO¹ Material Selection and Origin

Engineers, when selecting materials for a product, consider many factors including cost, weight, corrosiveness, and load-bearing capacity. You should know what is meant by the phase of a material. A matter may exist as solid, liquid, gas, or plasma and can change phase when its condition or its surroundings are changed. Water is a good example. It may exist in a solid form (ice), liquid form, or gaseous from (commonly refer to as steam). You should also have a good understanding of the structure of the Earth: its size, layers, and chief chemical composition. You also should be familiar with the basic elements such as aluminum, zinc, iron, zinc, copper, nickel, and magnesium that we extract from earth to make products. The Earth has a spherical shape with a diameter of 7926.4 miles (12,756.3 km) and an approximate mass of 13.17×10^{24} pounds $(5.98 \times 10^{24}$ **kg**). It is divided into major layers that are located above and below its surface. The Earth's mass is composed mostly of iron, oxygen, and silicon (approximately 32% iron, 30% oxygen, and 15% silicon). It also contains other elements such as sulfur, nickel, magnesium, and aluminum. The structure below the earth's surface is generally grouped into four layers: the crust, mantle, outer core, and inner core. The crust makes up about 0.5% of the earth's total mass and 1% of its volume. The mantle is made up of molten rock that lies underneath the crust, and makes up nearly 84% of the Earth's volume. The inner core and outer core make up about 33% of the earth's mass and 15% of its volume. The inner core is considered to be solid, while the outer core is thought to be fluid and is composed mainly of iron.

LO² The Properties of Materials

You should understand the basic properties of materials, such as density, modulus of elasticity, thermal conductivity, and viscosity. Material properties depend on many factors, including its exact chemical composition and how the material was processed. Material properties also change with temperature and time as the material ages. In your own words, you should be able to explain some of the basic material properties. As examples, the value of electrical resistivity is a measure of the resistance of material to the flow of electricity, the density is a measure of how compact the material is for a given volume, a modulus of elasticity is a measure of how easily a material will stretch when pulled (subject to a tensile force) or how well the material will shorten when pushed (subject to a compressive force), and thermal conductivity is a property of material that shows how good the material is in transferring thermal energy (heat) from a high-temperature region to a low-temperature region within the material.

LO³   Metals

You should be familiar with common applications of basic materials, such as light metals, steel and their alloys. Aluminum, titanium, and magnesium, because of their small densities (relative to steel), are commonly referred to as *lightweight metals* and are used in many structural and aerospace applications. Aluminum and its alloys have densities that are approximately one-third the density of steel. Aluminum is commonly alloyed with other metals such as copper, zinc, and magnesium (Mg). Everyday examples of common aluminum products include beverage cans, household aluminum foil, staples in tea bags, building insulation, and so on.

Titanium has an excellent strength-to-weight-ratio. Titanium is used in applications where relatively high temperatures—exceeding 400 and up to 600° C—are expected. Titanium alloys are used in the fan blades and the compressor blades of engines of commercial and military airplanes. Titanium alloys also are used in both commercial and military airplane airframes (fuselage and wings) and landing gear components.

Magnesium is another lightweight metal that looks like aluminum, but it is lighter. It is commonly alloyed with other elements such as aluminum, manganese, and zinc to improve its properties. Magnesium and its alloys are used in nuclear applications, in drycell batteries, and in aerospace applications.

Copper is a good conductor of electricity and heat, and because of these properties, it is commonly used in many electrical, heating, and cooling applications. Copper alloys are also used as tubes, pipes, and fittings in plumbing. When copper is alloyed with zinc, it is commonly called *brass*. *Bronze* is an alloy of copper and tin.

Steel is a common material that is used in the framework of buildings, bridges, in the body of appliances such as refrigerators, ovens, dishwashers, washers and in dryers, and in cooking utensils. Steel is an alloy of iron with approximately 2% or less carbon. The properties of steel can be modified by adding other elements, such as chromium, nickel, manganese, silicon, and tungsten. The 18/8 stainless steels, which contain 18% chromium and 8% nickel, are commonly used for tableware, kitchen utensils and other household products. Cast iron is also an alloy of iron that has 2 to 4% carbon.

LO⁴   Concrete

Concrete is used in the construction of roads, bridges, buildings, tunnels, and dams. It consists of three main ingredients: aggregate, cement, and water. Aggregate refers to materials such as gravel and sand, and cement refers to the bonding material that holds the aggregate together. Concrete is usually *reinforced* with steel bars or steel mesh that consists of thin metal rods to increase its load-bearing capacity. Another common construction practice is the use of *precast concrete*. Precast concrete slabs, blocks, and structural members are fabricated in less time with less cost in factory settings where the surrounding conditions are controlled. Because concrete has a higher compressive strength than tensile strength, it is *prestressed* by pouring it into forms that have steel rods or wires that are stretched. The prestressed concrete then acts as a compressed spring, which will become un-compressed under the action of tensile loading.

LO⁵   Wood, Plastics, Silicon, Glass, and Composites

Plastic products include grocery and trash bags, soft drink containers, home-cleaning containers, vinyl siding, polyvinyl chloride (PVC) piping, valves, and fittings. Styrofoam™ plates and cups, plastic forks, knives, spoons, and sandwich bags are other examples of plastic products that are consumed every day. *Polymers* are the backbone of what we call plastics. They are chemical compounds that have large and chain-like molecular structures. Plastics are often classified into two categories: *thermoplastics* and *thermosets*. When heated to certain temperatures, the thermoplastics can be molded and remolded. By contrast, thermosets cannot be remolded into other shapes by heating.

Silicon is a nonmetallic chemical element that is used quite extensively in the manufacturing of transistors and various electronic and computer chips. It is found in the form of silicon dioxide in sands and rocks or found combined with other elements such as aluminum or calcium or sodium or magnesium in the form that is commonly referred to as *silicates*. Silicon, because of its atomic structure, is an excellent semiconductor, which is a material whose electrical conductivity properties can be changed to act either as a conductor of electricity or as an insulator (preventer of electricity flow).

Glass is commonly used in products such as windows, light bulbs, housewares (such as drinking glasses), chemical containers, beverage containers, and decorative items. The composition of the glass depends on its application. The most widely used form of glass is soda-lime-silica glass. The materials used in making soda-lime-silica glass include sand (silicon dioxide), limestone (calcium carbonate), and soda ash (sodium carbonate). Other materials are added to create desired characteristics for specific applications. Silica glass fibers are commonly used today in fiber optics, which is that branch of science that deals with transmitting data, voice, and images through thin glass or plastic fibers. Every day, copper wires are being replaced by transparent glass fibers in telecommunication applications to connect computers together in networks.

Composite materials are found in military planes, helicopters, satellites, commercial planes, fast-food restaurant tables and chairs, and many sporting goods. In comparison to conventional materials such as metals, composite materials can be lighter and stronger. Composite materials consist of two main ingredients: matrix material and fibers. Fibers are embedded in matrix materials, such as aluminum or other metals, plastics, or ceramics. Glass, graphite, and silicon carbide fibers are examples of fibers used in the construction of composite materials. The strength of the fibers is increased when embedded in the matrix material, and the composite material created in this manner is lighter and stronger.

Common examples of wood products include hardwood flooring, roof trusses, furniture frames, wall supports, doors, decorative items, window frames, trimming in luxury cars, tongue depressors, clothespins, baseball bats, bowling pins, fishing rods, and wine barrels. Timber is commonly classified as *softwood* and *hardwood*. Softwood timber is made from trees that have cones (coniferous), such as pine, spruce, and Douglas fir. On the other hand, hardwood timber is made from trees that have broad leaves or flowers.

LO⁶ Fluid Materials: Air and Water

You should recall the characteristics of the atmosphere. Air is a mixture of mostly nitrogen, oxygen, and small amounts of other gases such as argon, carbon dioxide, sulfur dioxide, and nitrogen oxide. The air surrounding the earth (depending on its temperature) can be divided into four regions: troposphere, stratosphere, mesosphere, and thermosphere. You should also know that the carbon dioxide plays an important role in sustaining plant life; however, if the atmosphere contains too much carbon dioxide, it will not allow the earth to cool down effectively by radiation and results in the greenhouse effect. News reports often contain words such as weather conditions or climate change. What is the difference between weather and climate? Weather represents atmospheric conditions such as temperature, pressure, wind speed, and moisture level that could occur during a period of hours or days. For example, a weather report could state an approaching snowstorm or a thunderstorm with details about temperature, wind speed, and the amount of rain or snow. Climate, on the other hand, represents the average weather conditions over a long period of time. By a long period of time, we mean many decades. For example, when we say Chicago is cold and windy in winter, or Houston is hot and humid in summer, then we are talking about the climate of these cities. Note that, even though Chicago may experience a mild winter one year, we know from past historical data averaged over many years that the city is cold and windy in winter. Now, if Chicago were to experience many consecutive winters that were mild and calm, we could say that perhaps the climate of Chicago is changing. Why is the distinction between weather and climate important to know? When scientists warn us about global warming, they are talking about a warming trend such as average temperature of the earth that is on the rise. The trend is based on data averaged over many decades. Warmer earth temperature, which also indicates warmer oceans, mean stronger storms and weather anomalies.

You should know about the water cycle and realize that the total amount of water available on earth remains constant. Even though water can change phase from liquid to solid (ice) or from liquid to vapor, we don't lose or gain water on earth. You also should be familiar with water resources terminology, for example, what we mean by surface water or groundwater.

KEY TERMS

Absolute Humidity 664

Atmosphere 663

Brass 651

Bronze 651

Cast Iron 653

Composites 660

Compression Strength 643

Concrete 654

Density 642

Electrical Resistivity 642

Glass Fiber 659

Groundwater 664

Hardwood 656

Heat Capacity 643

Light Metals 648

Mesosphere 664

Modulus of Elasticity 642

Modulus of Resilience 643

Modulus of Rigidity 642

Modulus of Toughness 643

Precast Concrete 654

Prestressed Concrete 655

Reinforced Concrete 654

Relative Humidity 664

Silicon 658

Silicone 658

Softwood 656

Stainless Steel 653

Stratosphere 663

Strength-to-Weight Ratio 643

Surface Water 664

Tensile Strength 642

Thermal Conductivity 643

Thermal Expansion 643

Thermoplastics 657

Thermosets 657

Thermosphere 664

Troposphere 663

Vapor Pressure 643

Viscosity 643

APPLY WHAT YOU HAVE LEARNED

Every day, we use a wide range of paper products at home or school. These paper products are made from different paper grades. Wood pulp is the main ingredient used in making a paper product. It is common practice to grind the wood first then cook it with some chemicals. Investigate the composition, processing methods, and the annual consumption rate of the following grades of paper products in the United States, and write a brief report discussing your findings. The paper products to investigate should include printing papers, sanitary papers, glassine and waxing papers, bag paper, boxboard, and paper towels.

PROBLEMS

Problems that promote lifelong learning are denoted by

17.1   Identify and list at least ten different materials that are used in a car.

17.2   Name at least five different materials that are used in a refrigerator.

17.3   Identify and list at least five different materials that are used in a TV set or computer.

17.4   List at least ten different materials that are used in a building envelope (walls, floors, roofs, windows, doors).

17.5   List at least five different materials used to fabricate window and door frames.

17.6   List the materials used in the fabrication of compact fluorescent lights.

17.7   Identify at least ten products around your home that make use of plastics.

17.8   In a brief report, discuss the advantages and disadvantages of using styrofoam, paper, glass, stainless steel, and ceramic materials for coffee or tea cups.

17.9   As you already know, roofing materials keep water from penetrating into the roof structure. There is a wide range of roofing products available on the market today.

For example, asphalt shingles, which are made by impregnating a dry felt with hot asphalt, are used in some houses. Other houses use wood shingles, such as red cedar or redwood. A large number of houses in California use interlocking clay tiles as roofing materials. Investigate the properties and characteristics of various roofing materials. Write a brief report discussing your findings.

17.10   Visit a home improvement center (hardware/lumber store) in your town, and try to gather information about various types of insulating materials that can be used in a house. Write a brief report discussing the advantages, disadvantages, and characteristics of various insulating materials, including their thermal characteristics in terms of R-value.

17.11   Investigate the characteristics of titanium alloys used in sporting equipment, such as bicycle frames, tennis racquets, and golf shafts. Write a brief summary report discussing your findings.

17.12   Investigate the characteristics of titanium alloys used in medical implants for hips and other joint replacements. Write a brief summary report discussing your findings.

17.13   Cobalt-chromium alloys, stainless steel, and titanium alloys are three common biomaterials that have been used as surgical implants. Investigate the use of these biomaterials, and write a brief report discussing the advantages and disadvantages of each.

17.14   According to the Aluminum Association, every year over 100 billion aluminum cans are produced, and of these, approximately 60% were recycled. Measure the mass of ten aluminum cans, and use an average mass for an aluminum can to estimate the total mass of aluminum cans that was recycled.

17.15   Endoscopy refers to medical examination of the inside of a human body by means of inserting a lighted optical instrument through a body opening. Fiberscopes operate in the visible wavelengths and consist of two major components. One component consists of a bundle of fibers that illuminates the examined area,

and the other component transmits the images of the examined area to the eye of the physician or to some display device. Investigate the design of fiberscopes or the fiber-optic endoscope, and discuss your findings in a brief report.

17.16   Crystal tableware glass that sparkles is sought after by many people as a sign of affluence. This crystal commonly contains lead monoxide. Investigate the properties of crystal glass in detail, and write a brief report discussing your findings.

17.17   You all have seen grocery bags that have labels and printed information on them. Investigate how information is printed on plastic bags. For example, a common practice includes using a wet-inking process; another process makes use of lasers and heat-transfer decals. Discuss your findings in a brief report.

17.18   Teflon and Nylon are trade names of plastics that are used in many products. Look up the actual chemical name of these products, and give at least five examples of where they are used.

17.19   Investigate how the following basic wood products are made: plywood, particle board, veneer, and fiberboard. Discuss your findings in a brief report. Also investigate common methods of wood preservation, and discuss your findings in your report. What is the environmental impact of both the production and use of treated wood products in this question?

17.20   Investigate the common uses of cotton and its typical properties. Discuss your findings in a brief report.

17.21   Look around your home and estimate how many meters or feet of visible copper wire that are in use. Consider extension and power cords for common items such as hairdryer, TV, cell phone charger, computer charger, lamps, printer cable, refrigerator, microwave oven, and so on. Write a brief report, and discuss your findings.

17.22   How many cans or glasses of soda do you drink every day? Estimate your annual aluminum and or glass consumption. Express your results in kilograms or pounds per year.

17.23 Investigate how much steel was used in making the following appliances: clothes washer, dryer, dishwasher, refrigerator, and oven. Discuss your findings in a brief report.

17.24 This is a group assignment. Investigate how much concrete is used to make a sidewalk or walkway. Estimate the amount of concrete used to make walkways on your campus. Discuss your findings and assumptions in a brief report.

17.25 How much paper do you use every year? Estimate the amount of paper that you consume every year. Consider your printing habits and needs, use of loose and bonded paper, and magazine and newspaper consumption. How much wood would it take to meet your demand? Discuss your assumptions and findings in a brief report.

17.26 By some estimates, we consume twice as many goods and services as we did fifty years ago. As a result, our appetite for raw materials from wood to steel keeps increasing. The rise in world population and standard of living will only exacerbate this problem. What can you do to reverse this trend? Discuss your suggestions as backed up by data in a brief report.

17.27 As we discussed in this chapter, when selecting materials for mechanical applications, the value of modulus of resilience for a material shows how good the material is at absorbing mechanical energy without sustaining any permanent damage. Another important characteristic of a material is its ability to handle overloading before it fractures. The value of modulus of toughness provides such information. Look up the values of modulus of resilience and modulus of toughness for (a) titanium and (b) steel

17.28 Investigate and discuss some of the characteristics of the materials that are used in bridge construction.

17.29 As we discussed in this chapter, the strength-to-weight ratio of material is an important criterion when selecting material for aerospace applications. Calculate the average strength-to-weight ratio for the following materials: aluminum alloy, titanium alloy, and steel. Use Tables 17.3 and 17.4 to look up appropriate values.

17.30 How much heavier, on average, will an aluminum-alloy tennis racquet be if it is made from titanium alloy? Obtain a tennis racquet, and take appropriate measurements to perform your analysis.

17.31 Tensile test machines are used to measure the mechanical properties of materials, such as the modulus of elasticity and tensile strength. Visit the Web site of the MTS Systems Corporation to obtain information on test machines used to test the strength of materials. Write a brief report discussing your findings.

17.32 As most of you know, commercial transport planes cruise at an altitude of approximately 10,000 m (~30,000 ft). The power required to maintain level flight depends on air drag or resistance at that altitude, which may be estimated by the relationship

$$power = \frac{1}{2}\rho_{air}C_D A U^3$$

where ρ_{air} is density of air at the given altitude, C_D represents the drag coefficient of the plane, A is the planform area, and U represents the cruising speed of the plane. Assume that a plane is moving at constant speed and, C_D remaining constant, determine the ratio of power that would be required when the plane is cruising at 8000 m and when the plane is cruising at 11,000 m.

17.33 Investigate the average daily water consumption per capita in the United States. Discuss the personal and public needs in a brief report. Also discuss factors such as geographical location, time of the year, time of the day, and cost of consumption patterns. For example, more water is consumed during the early morning hours. Civil engineers need to consider all of these factors when designing water systems for cities. Assuming that the life expectancy of people has increased by five years over the past decades, how much additional water is needed to sustain the lives of 50 million people?

17.34   When a ring gets stuck on a finger, most people resort to water and soap as a lubricant to get the ring off. In earlier times, animal fat was a common lubricant used in wheel axles. Moving parts in machinery, the piston inside your car's engine, and bearings are examples of mechanical components that require lubrication. A lubricant is a substance that is introduced between the parts that have relative motion to reduce wear and friction. The lubricants must have characteristics that are suitable for a given situation. For instance, for liquid lubricants, viscosity is one of the important properties. The flash and fire and the cloud and pour points are examples of other characteristics that are examined when selecting lubricants. Investigate the use of petroleum-based lubricants in reducing wear and friction in today's mechanical components. Write a brief report discussing the application and characteristics of liquid-petroleum-based and solid lubricants that are commonly used, such as SAE 10W-40 oil and graphite.

Impromptu   Design VII

Objective: To design a vehicle from the materials listed below that will transport an egg safely. The vehicle is to be dropped from a height of 4 m. Thirty minutes will be allowed for preparation

Provided Materials: 4 rubber bands; 900 mm of adhesive tape; 6 m of string; 2 paper plates; 2 Styrofoam cups; 2 sheets of paper (255 mm × 330 mm); 4 drinking straws; 4 small paper clips; 4 large paper clips; a raw egg. The team with the unbroken egg and the slowest drop time wins.

"Anyone who has never made a mistake has never tried anything new."

—ALBERT EINSTEIN (1879–1955)

Celeste Baine

Courtesy of Celeste Baine

I decided to become an engineer because I knew it would be challenging, and the idealist in me wanted to make the world a better place. Engineers were problem solvers, and I always had a creative solution for whatever was happening in my life. In high school, I wasn't the best student in math and science, but I really enjoyed the classes and knew that I could do anything that I set my mind to. I visited the biomedical engineering department of a local hospital and knew I was hooked before the tour was over.

After a particularly hard calculus test my sophomore year, I had a moment that set my course. I was struggling with my classes, and it seemed that everyone was doing better than me with less effort. I went to my advisor's office and told him that I didn't think I would make a good engineer because I wasn't like everyone else. I was worried that all employers only wanted to know your GPA. Was a 3.4 good enough? He said, "Celeste, the world needs all kinds of engineers. You don't need to be like everyone else." I felt myself lift on his words, and suddenly I realized that I could communicate better (speaking and writing) than a large majority of the class. I realized that I had skills that were almost impossible to learn from a book. I could be one of the most valuable types of engineers because I knew how to work with people and how to manage my strengths and weaknesses.

Currently, I am the director of the Engineering Education Service Center. I work on numerous projects that help promote engineering to the K-12 market. I develop multimedia presentations and books that show engineering as a fun, satisfying, and lucrative career. I've authored over 20 books on engineering careers and education and I've been named one of the Nifty-Fifty individuals who have made a major impact on the field of engineering by the USA Science and Engineering Festival.

Courtesy of Celeste Baine

An Engineering Marvel

The Jet Engine*

Introduction

You think you've got problems? How would you like to be a molecule of air minding your own business at 9000 m when all of a sudden you get mugged by five tons of a Pratt & Whitney jet engine?

Over the course of the next 40 thousandths of a second you, Mr. or Ms. Molecule, will be beaten through 18 stages of compression, singed in a furnace heated to nearly 1650° C, expanded through

Based on Pratt & Whitney, a United Techonologies Company. Text by Matthew Broder.

*Materials were adapted with permission from Pratt & Whitney, a United Techonologies Company. Text by Matthew Broder.

a turbine and pushed out the back with a wicked headache and the bitter knowledge as the aircraft screams on that your ugly ordeal has been merely for the sake of getting Aunt Marge to Phoenix for a much needed rest.

That's basically what happens every day in the skies above as thousands of Pratt engines power aircraft from place to place while the passengers and crew inside those aircraft experience only the steady and reassuring thrum that results from Pratt's precision manufacturing.

The principle of jet propulsion has been demonstrated by anyone who has blown up a balloon for a child and accidentally let go before tying it closed. The air stored up inside the balloon accelerates as it rushes to escape through the narrowest part of the balloon's neck. This acceleration, or change in the speed of air, combines with the weight of the air itself to produce thrust. It is the thrust that sends the renegade balloon zipping madly about the room.

The original turbojet engine, which debuted in scheduled commercial service in 1952, accomplished this same trick on a breathtaking scale and with pinpoint control. The progress of jet engine technology in the past four decades has been to increase the amount of air going into the engine, and change the speed of that air with ever greater efficiency. All the examples in this article are drawn from the mightiest family of engines ever to fly, the Pratt & Whitney 4000s.

The Jet Engine

Fan   The propulsion process begins with the huge, 2.7 m diameter fan at the front of the engine, spinning 2,800 times a minute at takeoff speed. That fan sucks in air at the rate of 1180 kg per second, or enough to vacuum out the air from a 4-bedroom house in less than half a second.

Compression   As the air leaves the fan it is now separated into two streams. The smaller stream, about 15 percent of the total volume of air, is called primary or core air and enters the first of two compressors that are spinning in the same direction as the fan itself. As the primary air passes through each stage of the two compressors, both its temperature and pressure rise.

Combustion   When compression is complete, the air, now 30 times higher in pressure and 610° C hotter, is forced through a furnace or combustor. In the combustion chamber, fuel is added and burnt. The air's temperature soars even higher, and the air is finally ready to do the two jobs for which it has been so hastily prepared.

Turbine   The first job is to blast through the blades of two turbines, sending them whirling just like the wind spinning the arms of a windmill. The whirling turbines turn the shafts that drive both compressors and the fan at the front of the engine. This process, in which the engine extracts energy from the air it has just captured, is what allows modern jets to operate with such high fuel efficiency.

Exhaust   The second job is to push the airplane. After passing through the turbines, the hot air is forced through the exhaust opening at the back of the engine. The narrowing walls of the exhaust force the air to accelerate and, just as with the balloon, the mass of the air combined with its acceleration drives the engine, and the airplane attached to it, forward.

Fan Air or Bypass Air   The larger air stream exiting the fan, representing 85 percent of the total, is called fan air or bypass air, because it bypasses this entire process.

The engine itself is shrouded in a metal casing called the nacelle, shaped roughly like a sideways ice cream cone with the bottom cut off. Bypass air is forced through the ever narrower space between the nacelle wall and the engine, picking up speed along the way.

Because of its huge volume, bypass air needs only to accelerate a small amount to produce an enormous kick of thrust. In the PW4084 engine, bypass air accounts for 90 percent of the thrust, and has the added benefits of keeping the engine cooler, quieter and more fuel efficient.

PART 6

Mathematics, Statistics, and Engineering Economics

Why Are They Important?

Christopher Halloran/Shutterstock.com

Engineering problems are mathematical models of physical situations with different forms that rely on a wide range of mathematical concepts. Therefore, good understanding of mathematical concepts is essential in the formulation and solution of many engineering problems. Moreover, statistical models are becoming common tools in the hands of practicing engineers to solve quality control and reliability issues, and to perform failure analyses. Civil engineers use statistical models to study the reliability of construction materials and structures and to design for flood control. Electrical engineers use statistical models for signal processing and for developing voice-recognition software. Manufacturing engineers use statistics for quality control assurance of the products they produce. Mechanical engineers use statistics to study the failure of materials and machine parts. Economic factors also play important roles in engineering design decision making. If you design a product that is too expensive to manufacture, then it cannot be sold at a price that consumers can afford and still be profitable to your company. In the last part of this book, we will introduce you to important mathematical, statistical, and economic concepts.

CHAPTER 18   MATHEMATICS IN ENGINEERING

CHAPTER 19   PROBABILITY AND STATISTICS IN ENGINEERING

CHAPTER 20   ENGINEERING ECONOMICS

CHAPTER

18

Mathematics in Engineering

Speed (mph)	Speed (ft/s)	Stopping Sight Distance (ft)
0	0.0	0
5	7.3	21
10	14.7	47
15	22.0	78
20	29.3	114
25	36.7	155
30	44.0	201
35	51.3	252
40	58.7	309
45	66.0	370
50	73.3	436
55	80.7	508
60	88.0	584
65	95.3	666
70	102.7	753
75	110.0	844
80	117.3	941

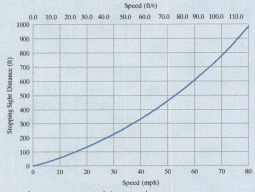

"I THINK YOU SHOULD BE MORE EXPLICIT HERE IN STEP TWO."

Initial speed = V   Final speed = 0

S

$G = 0$, $f = 0.33$, and $T = 2.5$ seconds were used to generate this graph.

In general, engineering problems are mathematical models of physical situations. For example, a model known as *stopping sight distance* is used by civil engineers to design roadways. This simple model estimates the distance a driver needs in order to stop his or her car traveling at a certain speed after detecting a hazard. The model proposed by the American Association of State Highway and Transportation Officials (AASHTO) is given by

$$S = \frac{V^2}{2g(f \pm G)} + TV$$

where

S = stopping sight distance (ft)
V = initial speed (ft/s)
g = acceleration due to gravity, 32.2 ft/s²
f = coefficient of friction between tires and roadway
G = grade of road $\left(\dfrac{\%}{100}\right)$
T = driver reaction time (s)

LEARNING OBJECTIVES

LO¹   **Mathematical Symbols and Greek Alphabet:** know important mathematical symbols and Greek alphabetic characters

LO²   **Linear Models:** explain linear equations, their characteristics, and how they are used to describe engineering problems

LO³   **Nonlinear Models:** explain nonlinear equations, their characteristics, and how they are used to represent engineering problems

LO⁴   **Exponential and Logarithmic Models:** know about exponential and logarithmic functions, their important characteristics, and how they are used to model engineering problems

LO⁵   **Matrix Algebra:** state basic definitions and operations and why matrix algebra plays an important in solving engineering problems

LO⁶   **Calculus:** explain key concepts related to differential and integral calculus

LO⁷   **Differential Equations:** explain what we mean by differential equations and give examples of boundary and initial conditions

DISCUSSION STARTER

WHAT DO YOU THINK?

U.S. Performance in Numeracy U.S. adults scored below the international average in numeracy, with 18 countries ranking above the United States. We performed on the same level as Ireland and France, but higher than Italy and Spain.

Proficiency Levels Fifteen countries had higher percentages of adults reaching the top proficiency level (4/5) in numeracy than the United States, and the U.S. percentage (9 percent) was lower than the international average (12 percent). The percentage of adults at the top numeracy proficiency level ranged from 19 percent (Finland, Japan, and Sweden) to 4 percent (Spain). The percentage of the population at the lowest proficiency levels was higher in the United States than it was on average internationally.

Performance of Subgroups in the United States Among young adults (ages 16–24), the U.S. numeracy score was below the international average, and only one country (Italy) had a numeracy score that was not significantly different from ours, putting the United States at the bottom of the ranking. In fact, unlike literacy, in numeracy U.S. adults across all age groups performed lower than the international average.

U.S. adults also performed lower than the international average across all educational levels, labor force participation statuses, income levels, and levels of health. Looking at racial and ethnic groups within the United States, whites scored higher in numeracy than either blacks or Hispanics.

The gap in average numeracy scores across parental educational levels was larger in the United States than it was on average internationally, while the gap in the United States between those born in the country and outside the country was not significantly different from the international average.

The gap in numeracy scores between genders in the United States was also similar to the international average. Our youngest and oldest adults performed more similarly to each other than did the youngest and oldest adults internationally. Finally, the gap between those reporting poor or fair health and those reporting excellent or very good health was larger in the United States than it was internationally.

Numeracy is defined as "the ability to access, use, interpret, and communicate mathematical information and ideas, to engage in and manage mathematical demands of a range of situations in adult life."

Jack Buckley, Commissioner, National Center for Education Statistics, NCES Statement on PIAAC 2012, U.S. Department of Education

To the students: Why do you think U.S. adults scored below the international average in numeracy?

In general, engineering problems are mathematical models of physical situations. Mathematical models of engineering problems may have many different forms. Some engineering problems lead to linear models, whereas others result in nonlinear models. Some engineering problems are formulated in the form of differential equations and some in the form of integrals. Formulation of many engineering problems results in a set of linear algebraic equations that are solved simultaneously. A good understanding of matrix algebra is essential in the formulation and solution of these problems. Therefore, in this chapter, we will discuss various mathematical models that commonly are used to solve engineering problems. We begin our discussion with an explanation of the need for conventional math symbols as a means to convey information and to effectively communicate to other engineers. Examples of math symbols will also be given. Next, we discuss the importance of knowing the Greek alphabet and its use in engineering formulas and drawings. This section will be followed with a discussion of simple linear and nonlinear models. Next, we introduce matrix algebra, with its own terminology and rules, which we will define and explain. We will then briefly discuss calculus and its importance in solving engineering problems. Calculus commonly is divided into two areas: differential and integral calculus. Finally, the role of differential equations in formulating engineering problems and their solutions is presented.

It is important to keep in mind that the purpose of this chapter is to focus on important mathematical concepts and to point out why mathematics is so important in your engineering education. The focus of this chapter is not to explain mathematical concepts in detail; that will take place later in your math classes.

LO¹ 18.1   Mathematical Symbols and Greek Alphabet

As you already know, mathematics is a language that has its own symbols and terminology. In elementary school, you learned about the arithmetic operational symbols, such as plus, minus, division, and multiplication. Later, you learned about degree symbols, trigonometry symbols, and so on. In the next four years, you will learn additional mathematical symbols and their meanings. Make sure that you understand what they mean and use them properly when communicating with other students or with your instructor. Examples of some **math symbols** are shown in Table 18.1.

TABLE 18.1		Some Math Symbols		
$+$	Plus or positive	$\geq$	Equal to or greater than
$-$	Minus or negative	$|x|$	Absolute value of x
$\pm$	Plus or minus	α	Proportional to
$\times$ or $\cdot$	Multiplication	$\therefore$	Therefore
$\div$ or $/$	Division	Σ	Summation
$:$	Ratio	$\int$	Integral
$<$	Less than	$!$	Factorial, for example, $5! = 5 \times 4 \times 3 \times 2 \times 1$
$>$	Greater than	Δ	Delta indicating difference
$\ll$	Much less than	∂	Partial
$\gg$	Much greater than	π	Pi, its value 3.1415926…
$=$	Equal to	∞	Infinity
$\approx$	Approximately equal to	$\circ$	Degree
$\neq$	Not equal to	$()$	Parentheses
$\equiv$	Identical with	$[]$	Brackets
$\leq$	Equal to or less than	$\{\}$	Braces

The Greek Alphabet and Roman Numerals

As you take more and more mathematics and engineering classes, you will see that the Greek alphabetic characters quite commonly are used to express angles, dimensions, and physical variables in drawings and in mathematical equations and expressions. Take a few moments to learn and memorize these characters. Knowing these symbols will save you time in the long run when communicating with other students or when asking a question of your professor. You don't want to refer to ζ (zeta) as that "curly thing" when speaking to your professor! You may also find Roman numerals in use to some extent in science and engineering. The **Greek alphabet** and the Roman numerals are shown in Tables 18.2 and 18.3, respectively.

> Mathematics is a language that has its own symbol and terminology. You should also memorize Greek alphabetic characters because they are used in engineering in drawings and mathematical equations.

TABLE 18.2   **The Greek Alphabet**

A	α	Alpha	I	ι	Iota	P	ρ	Rho
B	β	Beta	K	κ	Kappa	Σ	σ	Sigma
Γ	γ	Gamma	Λ	λ	Lambda	T	τ	Tau
Δ	δ	Delta	M	μ	Mu	Y	υ	Upsilon
E	ε	Epsilon	N	ν	Nu	Φ	ϕ	Phi
Z	ζ	Zeta	Ξ	ξ	Xi	X	χ	Chi or khi
H	η	Eta	O	o	Omicron	Ψ	ψ	Psi
Θ	θ	Theta	Π	π	Pi	Ω	ω	Omega

TABLE 18.3   **Roman Numerals**

I	= 1	XIV	= 14	XC	= 90
II	= 2	XV	= 15	C	= 100
III	= 3	XVI	= 16	CC	= 200
IIII or IV	= 4	XVII	= 17	CCC	= 300
V	= 5	XVIII	= 18	CCCC or CD	= 400
VI	= 6	XIX	= 19	D	= 500
VII	= 7	XX	= 20	DC	= 600
VIII	= 8	XXX	= 30	DCC	= 700
IX	= 9	XL	= 40	DCCC	= 800
X	= 10	L	= 50	CM	= 900
XI	= 11	LX	= 60	M	= 1000
XII	= 12	LXX	= 70	MM	= 2000
XIII	= 13	LXXX	= 80		

LO² 18.2   Linear Models

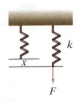

FIGURE 18.1

Linear models are the simplest form of equations commonly used to describe a wide range of engineering situations. In this section, we first discuss some examples of engineering problems where linear mathematical models are found. We then explain the basic characteristics of linear models.

A Linear Spring   In Chapter 10, we discussed Hooke's law, which states that, over the elastic range, the deformation of a spring is directly proportional to the applied force (Figure 18.1) and consequently to the internal force developed in the spring, according to

$$F = kx \qquad \boxed{18.1}$$

where

F = spring force (**N or lb**)
k = spring constant (**N/mm or N/cm or lb/in.**)
x = deformation of the spring (mm or cm or in.)
    (use units that are consistent with k)

It is clear from examining Equation (18.1) that the spring force F depends on how much the spring is stretched or compressed. In mathematics, F is called a **dependent variable**. The spring force is called the dependent variable because its value depends on the deformation of the spring x. Consider the force in a spring with a stiffness of k = 2 N/mm, as shown in Figure 18.2. For the linear model describing the behavior of this spring, constant k = 2 N/mm represents the slope of the line. The value of 2 N/mm tells us that each time the spring is stretched or compressed by an additional 1 mm; as a result, the spring force will be changed by 2 N. Or stated another way, the force required to either compress or extend the spring by and additional 1 mm is 2 N. Moreover, note that for x = 0, the spring force F = 0. Not all springs exhibit linear behaviors. In fact, you find many springs in engineering practice whose behaviors are described by nonlinear models.

x (mm)	F (N)
0	0
5	10
10	20
15	30
20	40

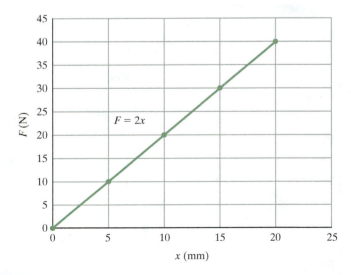

FIGURE 18.2   A linear model for spring force–deflection relationship.

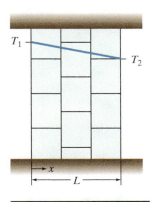

FIGURE 18.3

Temperature Distribution Across a Plane Wall Temperature distribution across a plane wall is another example where a linear mathematical model describes how temperature varies along the wall (Figure 18.3). Under a steady-state assumption, the temperature distribution—how temperature varies across the thickness of the wall—is given by

$$T(x) = (T_2 - T_1)\frac{x}{L} + T_1$$

18.2

where

$$T(x) = \text{temperature distribution} \,(°F \text{ or } °C)$$
$$T_2 = \text{temperature at surface 2} \,(°F \text{ or } °C)$$
$$T_1 = \text{temperature at surface 1} \,(°F \text{ or } °C)$$
$$x = \text{distance from surface 1} \,(ft \text{ or } m)$$
$$L = \text{wall thickness} \,(ft \text{ or } m)$$

For this linear model, T is the dependent variable, and x is the independent variable. The variable x is called an **independent variable**, because the position x is not dependent on temperature. Now, let us consider a situation for which $T_1 = 68°$ F, $T_2 = 38°$ F, and $L = 0.5$ ft. For these conditions, the **slope** of the linear model is given by $(T_2 - T_1)/L = -60°$ F/ft, as shown in Figure 18.4. Note that for the given conditions, the line that describes the relationship between the temperature and position intercepts the temperature axis at the value of 68 (i.e., at $x = 0$, $T = 68°$ F).

We can describe many other engineering situations for which linear relationships exist between dependent and independent variables. For example, as we explained in Chapter 12, *resistivity* is a measure of the resistance of a piece of material to electric current. The resistivity values usually are measured using samples made of a centimeter cube or a cylinder having a diameter of 1 mil and length of 1 ft. The resistance of the sample is then given by

$$R = \frac{\rho\ell}{a}$$

$T_1 = 68°$ F
$T_2 = 38°$ F
$L = 0.5$ ft

x (ft)	T (x)
0	68
0.1	62
0.2	56
0.3	50
0.4	44
0.5	38

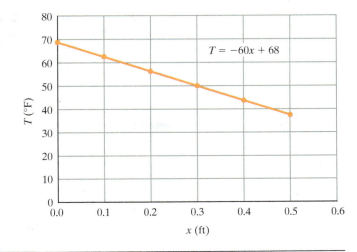

FIGURE 18.4   Temperature distribution along a wall.

$T(°C)$	$T(°F)$	$T(°C)$	$T(°F)$
−40	**−40**	35	95
−35	**−31**	40	104
−30	−22	45	113
−25	−13	50	122
−20	−4	55	131
−15	5	60	140
−10	14	65	149
−5	23	70	158
0	**32**	75	167
5	**41**	80	176
10	50	85	185
15	59	90	194
20	68	**95**	**203**
25	77	**100**	**212**
30	86		

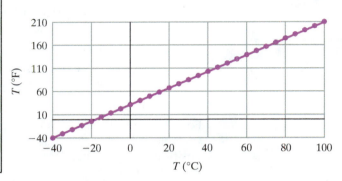

FIGURE 18.5   The relationship between Fahrenheit and Celsius scales.

where ρ is the resistivity, ℓ is the length of the sample, and a is the cross-sectional area of the sample. As you can see, for constant values of ρ and a, there exists a linear relationship between R and the length ℓ.

The relationship among various systems of units is also linear. Let us demonstrate this fact using an example dealing with temperature scales. In Chapter 11, we discussed the relationship between the two temperature scales Fahrenheit and Celsius, which is given by

$$T(°F) = \frac{9}{5}T(°C) + 32$$ **18.3**

We have plotted the relationship between the Fahrenheit and Celsius scales for the temperature range shown in Figure 18.5. Note the slope of the line describing the relationship is $9/5 = 1.8$, and the line intercepts the Fahrenheit axis at 32 (i.e., $T(°C) = 0$, $T(°F) = 32$).

Linear Equations and Slopes

Now that you realize the importance of linear models in describing engineering situations, let us consider some of the basic characteristics of a linear model. As you know, the basic form of a line equation is given by

$$y = ax + b$$ **18.4**

where

$$a = \text{slope} = \frac{\Delta y}{\Delta x} = \frac{\text{change in } y \text{ value}}{\text{change in } x \text{ value}}$$

$$b = y\text{-intercept (the value of } y \text{ at } x = 0)$$

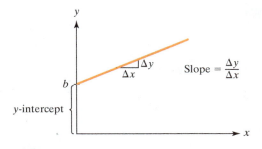

FIGURE 18.6   A linear model.

Equation (18.4) is plotted and shown in Figure 18.6. Note positive values were assumed for the y-intercept and the slope in Figure 18.6. The slope of a linear model shows by how much the dependent variable y changes each time a change in the independent variable x is introduced. Moreover, for a linear model, the value of the slope is always constant.

Comparing our previous models of examples of engineering situations to Equation (18.4): for the spring example, the slope has a value of 2 and the y-intercept is zero. As we mentioned before, the slope value 2 N/mm conveys that each time we stretch the spring by an additional one millimeter, as a result the spring force will increase by 2 N. For the temperature-distribution model, the slope has a value of $-60°$ F/ft, and the y-intercept is given by the value of 68° F. For the temperature-scale example, when comparing Equation (18.3) to Equation (18.4), note that $T(°F)$ corresponds to y, and $T(°C)$ corresponds to x. The slope and y-intercept for this linear model is given by 9/5 and 32, respectively. You easily can see that from the values shown in Figure 18.5. The slope shows that, for any 5° C change, the corresponding Fahrenheit scale change is 9° F, regardless of the position of the change in the temperature scale.

> The slope of a linear model is constant and shows by how much the dependent variable changes each time a change in the independent variable is introduced.

$$a = \text{slope} = \frac{\Delta y}{\Delta x} = \frac{\text{change in } y \text{ value}}{\text{change in } x \text{ value}} = \frac{\overbrace{(-31) - (-40)}^{9}}{\underbrace{(-35) - (-40)}_{5}}$$

$$= \frac{\overbrace{41 - 32}^{9}}{\underbrace{5 - 0}_{5}} = \frac{\overbrace{212 - 203}^{9}}{\underbrace{100 - 95}_{5}} = \frac{9}{5}$$

Linear models could have different forms with different characteristics. We have summarized the characteristics of various linear models in Table 18.4. Make sure to study them carefully.

Linear Interpolation

Occasionally, you need to look up a value from a table that does not have the exact increments to match your need. To shed light on such occasions, let us consider the variation of air density and atmospheric pressure as a function of altitude as shown in Table 10.4 (shown here again as a reference). Now, let us assume you want to estimate the power consumption of a plane that might be flying at an altitude of 7300 m. To carry out this calculation, you would need the density of air at that

TABLE 18.4	A Summary of Linear Models and Their Characteristics

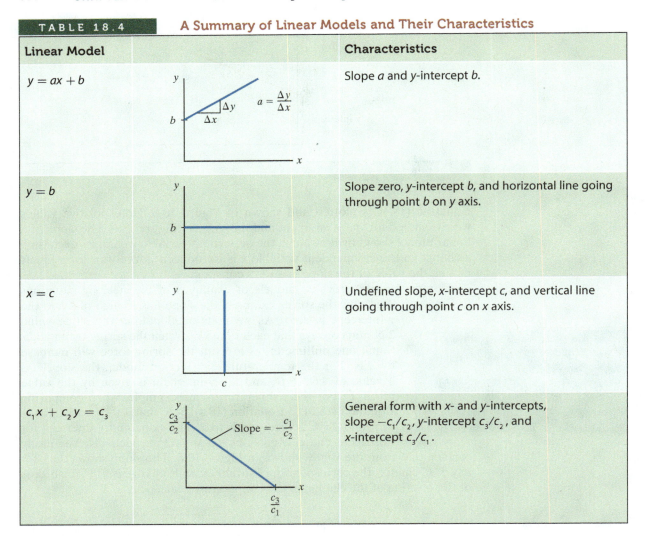

Linear Model		Characteristics
$y = ax + b$		Slope a and y-intercept b.
$y = b$		Slope zero, y-intercept b, and horizontal line going through point b on y axis.
$x = c$		Undefined slope, x-intercept c, and vertical line going through point c on x axis.
$c_1 x + c_2 y = c_3$		General form with x- and y-intercepts, slope $-c_1/c_2$, y-intercept c_3/c_2, and x-intercept c_3/c_1.

altitude. Consequently, you would go to Table 10.4; however, the value of air density corresponding to an altitude of 7300 m is not shown. The altitude increments shown in the table do not match your need, so what do you do?

One approach would be to approximate the air density value at 7300 m using the neighboring values at 7000 m (0.590 kg/m³) and at 8000 m (0.526 kg/m³). We can assume that over an altitude of 7000 m to 8000 m, the air density values change linearly from 0.590 kg/m³ to 0.526 kg/m³. Using the two similar triangles ACE and BCD shown in Figure 18.7 accompanying Table 10.4, we can then approximate the density of air at 7300 m using **linear interpolation** in the manner:

$$\frac{\overline{BC}}{\overline{AC}} = \frac{\overline{BD}}{\overline{AE}}$$

$$\frac{8000 - 7300}{8000 - 7000} = \frac{0.526 - \text{density of air @ 7300 m}}{0.526 - 0.590}$$

and solving for density of air at 7300 m, we get $\rho_{@7300} = 0.578$ kg/m³.

TABLE 10.4	Variation of Standard Atmosphere with Altitude (repeated)

Altitude (m)	Atmospheric Pressure (kPa)	Air Density (kg/m^3)
0 (sea level)	101.325	1.225
500	95.46	1.167
1000	89.87	1.112
1500	84.55	1.058
2000	79.50	1.006
2500	74.70	0.957
3000	70.11	0.909
3500	65.87	0.863
4000	61.66	0.819
4500	57.75	0.777
5000	54.05	0.736
6000	47.22	0.660
7000	41.11	0.590
8000	35.66	0.526
9000	30.80	0.467
10,000	26.50	0.413
11,000	22.70	0.365
12,000	19.40	0.312
13,000	16.58	0.266
14,000	14.17	0.228
15,000	12.11	0.195

Source: Data from U.S. Standard Atmosphere (1962)

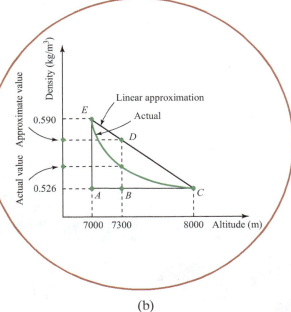

(b)

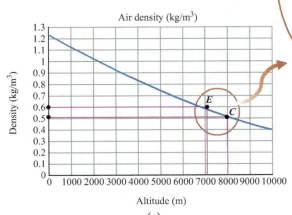

Altitude (m)

(a)

FIGURE 18.7

Systems of Linear Equations

At times, the formulation of an engineering problem leads to a set of linear equations that must be solved simultaneously. In Section 18.5, we will discuss the general form for such problems and the procedure for obtaining a solution. Here, we will discuss a simple graphical method that you can use to obtain the solution for a model that has two equations with two unknowns. For example, consider the following equations with x and y as unknown variables.

$$2x + 4y = 10 \qquad \text{18.5a}$$

$$4x + y = 6 \qquad \text{18.5b}$$

Equations (18.5a) and (18.5b) are plotted and shown in Figure 18.8. The intersection of the two lines represents the $x$solution, which is given by $x = 1$, because, as you can see, at $x = 1$ both equations have the same y value. We then substitute for x into either Equation (18.5a) or Equation (18.5b) and solve for y, which yields a value of $y = 2$. This is also the value you get if you were to draw a perpendicular line to the y-axis from the intersection point.

x	$(10 - 2x)/4$	$6 - 4x$
0	2.5	6
0.5	2.25	4
1	2	2
1.5	1.75	0
2	1.5	-2
2.5	1.25	-4
3	1	-6
3.5	0.75	-8
4	0.5	-10
4.5	0.25	-12
5	0	-14
5.5	-0.25	-16
6	-0.5	-18

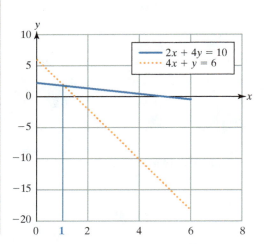

FIGURE 18.8   The plot of Equations (18.5a) and (18.5b).

Answer the following questions to test your understanding of the preceding section.

Before You Go On

1. Why is it important to know the mathematical symbols?

2. Why is it important to know the Greek alphabet?

3. Give an example of a linear model in engineering.

4. What are the basic characteristics of a linear model?

LO³ 18.3 Nonlinear Models

For many engineering situations, **nonlinear models** are used to describe the relationships between dependent and independent variables because they predict the actual relationships more accurately than linear models do. In this section, we first discuss some examples of engineering situations where nonlinear mathematical models are found. We then explain some of the basic characteristics of nonlinear models.

Polynomial Functions

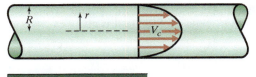

FIGURE 18.9

Laminar Fluid Velocity Inside a Pipe For those of you who are planning to become an aerospace, chemical, civil, or mechanical engineer, later in your studies you will take a fluid mechanics class. In that class, among other topics, you will learn about the flow of fluids in pipes and conduits. For a laminar flow (a well-behaved flow), the velocity distribution—how fluid velocity changes at a given cross-section—inside a pipe (Figure 18.9) is given by

$$u(r) = V_c\left[1 - \left(\frac{r}{R}\right)^2\right]$$

18.6

where

$u(r)$ = fluid velocity at the radial distance r (m/s)
V_c = center line velocity (m/s)
r = radial distance measured from the center of the pipe (m)
R = radius of the pipe (m)

The velocity distribution for a situation where $V_c = 0.5$ m/s and $R = 0.1$ m is shown in Figure 18.10. From Figure 18.10, it is evident that the velocity equation is a second-order polynomial (a nonlinear function) and the slope of this type of model is not constant (it changes with r). For the given example, for any 0.01 m change in r, the dependent variable u changes by different amounts depending on where in the pipe you evaluate the change.

$$\text{slope} = \frac{\text{change in velocity value}}{\text{change in position value}} = \frac{\overbrace{(0.495) - (0.5)}^{-0.005}}{\underbrace{(0.01 - 0)}_{0.01}} \neq \frac{\overbrace{0 - 0.095}^{-0.095}}{\underbrace{0.1 - 0.09}_{0.01}}$$

r	u (r)	r	u (r)
0.1	0	−0.01	0.495
0.09	0.095	−0.02	0.48
0.08	0.18	−0.03	0.455
0.07	0.255	−0.04	0.42
0.06	0.32	−0.05	0.375
0.05	0.375	−0.06	0.32
0.04	0.42	−0.07	0.255
0.03	0.455	−0.08	0.18
0.02	0.48	−0.09	0.095
0.01	0.495	−0.1	0
0	0.5		

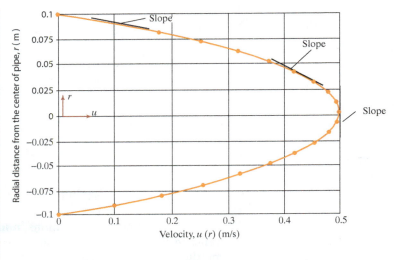

FIGURE 18.10   An example of fluid velocity distribution inside a pipe.

Stopping Sight Distance A model known as *stopping sight distance* is used by civil engineers to design roadways. This simple model estimates the distance a driver needs in order to stop his car while traveling at a certain speed after detecting a hazard (Figure 18.11). The model proposed by the American Association of State Highway and Transportation Officials (AASHTO) is given by

$$S = \frac{V^2}{2g(f \pm G)} + TV \qquad \text{18.7}$$

where

- S = stopping sight distance (ft)
- V = initial speed (ft/s)
- g = acceleration due to gravity, 32.2 ft/s²
- f = coefficient of friction between tires and roadway
- G = grade of road $\left(\dfrac{\%}{100}\right)$
- T = driver reaction time (s)

In the Equation (18.7), the typical value for the coefficient of friction between tires and roadway f is 0.33, the driver reaction time varies between 0.6 to 1.2 seconds; however, when designing roadways, a conservative value of

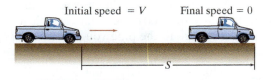

FIGURE 18.11

Speed (mph)	Speed (ft/s)	Stopping Sight Distance (ft)
0	0.0	0
5	7.3	21
10	14.7	47
15	22.0	78
20	29.3	114
25	36.7	155
30	44.0	201
35	51.3	252
40	58.7	309
45	66.0	370
50	73.3	436
55	80.7	508
60	88.0	584
65	95.3	666
70	102.7	753
75	110.0	844
80	117.3	941

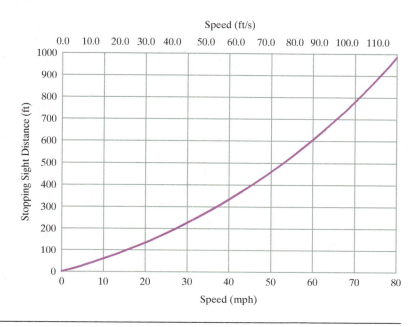

FIGURE 18.12   The stopping sight distance for a car traveling speeds of up to 80 mph.

2.5 seconds commonly is used. In the denominator of Equation (18.7), plus (+) indicates upgrade, whereas minus (–) is for downgrade. A graph showing the stopping sight distance for a flat roadway as a function of initial speed is shown in Figure 18.12 ($G = 0$, $f = 0.33$, and $T = 2.5$ seconds were used to generate this graph). This is another example where a second-order polynomial describes an engineering situation.

Again, note the slope of this model is not constant: that is, for any 5 mph change in speed, the dependent variable S changes by different amounts based on where in the speed range you introduce the change.

$$\text{shown: slope} = \frac{\text{change in stopping distance } S}{\text{change in speed } V} = \frac{\overbrace{(47) - (21)}^{26}}{\underbrace{(10 - 5)}_{5 \text{ mph (7.4 ft/s)}}} \neq \frac{\overbrace{508 - 436}^{72}}{\underbrace{55 - 50}_{5 \text{ mph (7.4 ft/s)}}}$$

We can describe many other engineering situations with second-order polynomials. The trajectory of a projectile under a constant deceleration, power consumption for a resistive element, drag force, or the air resistance to the motion of a vehicle are represented by second-order models.

Deflection of a Beam   The deflection of a cantilever beam is an example of an engineering situation where a higher-order polynomial model is used. For example, the cantilever beam (supported at one end) shown in Figure 18.13 is

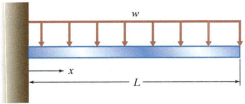

FIGURE 18.13	A cantilever beam.

used to support a load acting on a balcony. The deflection of the centerline of the beam is given by the following fourth-order polynomial equation.

$$y = \frac{-wx^2}{24EI}(x^2 - 4Lx + 6L^2)$$

18.8

where

y = deflection at a given x location (m)
w = distributed load (N/m)
E = modulus of elasticity (N/m²)
I = second moment of area (m⁴)
x = distance from the support as shown (m)
L = length of the beam (m)

The deflection of a beam with a length of 5 m with the modulus of elasticity of $E = 200$ GPa, $I = 99.1 \times 10^6$ mm⁴, and for a load of 10,000 N/m is shown in Figure 18.14.

x (m)	y (m)	x (m)	y (m)
0	0	2.6	−0.01489
0.2	−0.00012	2.8	−0.01678
0.4	−0.00048	3	−0.01873
0.6	−0.00105	3.2	−0.02072
0.8	−0.00181	3.4	−0.02274
1	−0.00275	3.6	−0.02478
1.2	−0.00386	3.8	−0.02685
1.4	−0.00511	4	−0.02893
1.6	−0.00649	4.2	−0.03102
1.8	−0.00799	4.4	−0.03311
2	−0.00959	4.6	−0.03521
2.2	−0.01128	4.8	−0.03732
2.4	−0.01305	5	−0.03942

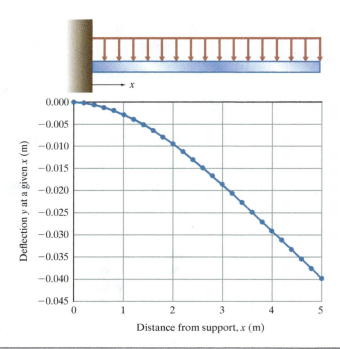

FIGURE 18.14	The deflection of a cantilever beam.

Now that you realize the importance of polynomial models in describing engineering situations, let us consider some of the basic characteristics of polynomial models. The general form of a polynomial function (model) is given by

$$y = f(x) = a_0 + a_1 x + a_2 x^2 + a_3 x^3 + \cdots + a_n x^n \qquad \boxed{18.9}$$

where $a_0, a_1, \ldots, a_n$ are coefficients that could take on different values and n is a positive integer defining the order of the polynomial. For the laminar fluid velocity and the stopping sight distance examples, n is 2, and the deflection of the beam was represented by a fourth-order polynomial.

Unlike linear models, second- and higher-order polynomials have variable slopes, meaning that each time you introduce a change in the value of the independent variable x, the corresponding change in the dependent variable y will depend on where in the x range the change is introduced. To better visualize the slope at a certain x value, draw a tangent line to the curve at the corresponding x value, as shown in Figure 18.15. Another important characteristic of a polynomial function is that the dependent variable y has a zero value at points where it intersects the x axis. For example, for the laminar fluid velocity situation shown in Figure 18.10, the dependent variable, velocity u, has zero values at $r = 0.1$ m and $r = -0.1$ m.

> Unlike linear models, nonlinear models have variable slopes.

As another example, consider the third-order polynomial $y = f(x) = x^3 - 6x^2 + 3x + 10$, as shown in Figure 18.15. This function intersects the x axis at $x = -1$, $x = 2$, and $x = 5$. These points are called the *real roots* of the polynomial function. Not all polynomial functions have real roots. For example, the function $f(x) = x^2 + 4$ does not have a real root. As shown in Figure 18.16, the function does not intersect the x axis. This also should be obvious, because if you were to solve $f(x) = x^2 + 4 = 0$, you would find $x^2 = -4$. Even though this function does not have a real root, it still possesses imaginary roots. You will learn about imaginary roots in your advanced math and engineering classes.

Next, we consider other forms of nonlinear engineering models.

x	f(x) = x³ − 6x² + 3x + 10
−3	−80
−2	−28
−1	0
0	10
1	8
2	0
3	−8
4	−10
5	0
6	28

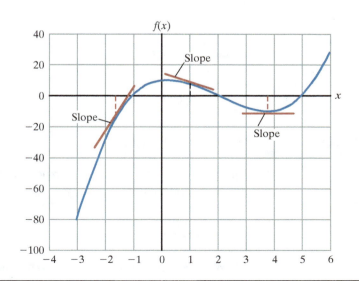

FIGURE 18.15   The real roots of a third-order polynomial function.

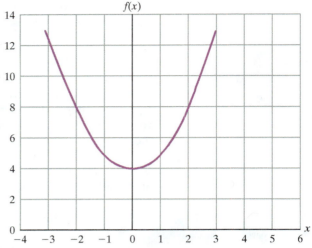

x	$f(x) = x^2 + 4$
−3	13
−2	8
−1	5
0	4
1	5
2	8
3	13

FIGURE 18.16   A function with imaginary roots.

Before You Go On

Answer the following questions to test your understanding of the preceding section.

1. What are the basic characteristics of nonlinear models?

2. What is the difference between the slopes of a nonlinear model and a linear model?

3. What do we mean by the real roots of a polynomial function?

Vocabulary—State the meaning of the following terms:

Laminar Flow _____

Cantilever Beam _____

Polynomial Function _____

LO⁴ 18.4   Exponential and Logarithmic Models

In this section, we will discuss exponential and logarithmic models and their basic characteristics.

The Cooling of Steel Plates In an annealing process—a process wherein materials such as glass and metal are heated to high temperatures and then cooled slowly to toughen them—thin steel plates (k = thermal conductivity = 40 W/m · k, ρ = density = 7800 kg/m³, and c = specific heat = 400 J/kg · K) are heated to temperatures of 900° C and then cooled in an environment with a temperature of 35° C and a heat transfer coefficient of h = 25 W/m² · K. Each plate has a thickness of $L = 5$ cm. We are interested in determining what the temperature of the plate is after one hour.

Those of you who will pursue aerospace, chemical, mechanical, or materials engineering will learn about the underlying concepts that lead to the solution in your heat-transfer class. For now, in order to determine the temperature of a plate after one hour, we use the following exponential equation.

$$\frac{T - T_{environment}}{T_{initial} - T_{environment}} = \exp\left(\frac{-2h}{\rho c L}t\right) \qquad \textbf{18.10}$$

In Equation (18.10), T represents the temperature of the plate at time t. Using Equation (18.10), we have calculated the temperature of the plate after each 12 minute (0.2 hr) interval. The corresponding temperature distribution is shown in Figure 18.17.

Time (hr)	Temperature (°C)
0	900
0.2	722
0.4	580
0.6	468
0.8	379
1	308
1.2	252
1.4	207
1.6	172
1.8	143
2	121
2.2	103
2.4	89
2.6	78
2.8	69
3	62
3.2	57
3.4	52
3.6	49
3.8	46
4	44
4.2	42
4.4	40
4.6	39
4.8	38
5	38

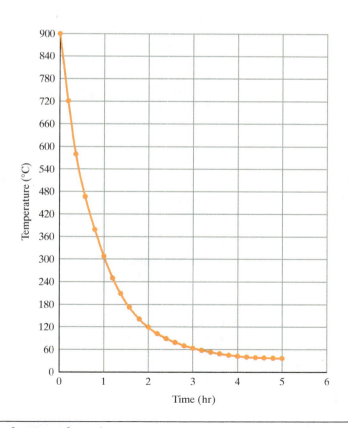

FIGURE 18.17   The cooling of a piece of metal.

From examining Figure 18.17, we note that the temperature of the plate after one hour is 308° C. Moreover, Figure 18.17 shows that, during the first hour, the temperature of the plate drops from 900° C to 308° C (a temperature drop of 592° C). During the second hour, the temperature of the plate drops from 308° C to 121° C (a temperature drop of 187° C), and during the third hour, a temperature drop of 59° C occurs. As you can see, the cooling rate is much higher at the beginning and much lower at the end. You should also note that the temperature begins to level off toward the end of the cooling process (after about 4.6 hours). This is the most important characteristic of an **exponential function**. That is to say, the value of the dependent variable begins to level off as the value of the independent variable gets larger and larger.

> When modeling an engineering problem using an exponential function, the rate of change of the dependent variable is much higher at the beginning and much lower at the end. The change levels off toward the end.

Now, let us look at what is meant by an exponential function. The simplest form of an exponential function is given by $f(x) = e^x$, where e is an irrational number with an approximate value of $e = 2.718281$. To provide you with a visual aid for better understanding of e^x and for comparison, we have plotted functions 2^x, e^x, and 3^x and shown the results in Figure 18.18. Also note that in Figure 18.18, we have shown how the value of e is determined. Spend a few minutes studying Figure 18.18 and note trends and characteristics of the functions.

The exponential functions have important characteristics, as demonstrated by examples shown in Table 18.5. Example 1 shows the changes that occur in a exponential model when the growth rate of an exponential function increases. Example 2 demonstrates similar changes for a decaying exponential function. Note these important effects as you study Table 18.5. A good understanding of these concepts will be beneficial when you take future engineering classes.

Another interesting form of an exponential function is $f(x) = e^{-x^2}$. You will find this type of exponential function in expressing probability distributions. We will discuss probability distributions in more detail in Chapter 19. For comparison, we have plotted the functions $f(x) = 2^{-x^2}$ and $f(x) = e^{-x^2}$ and shown them in Figure 18.19. Note the bell shape of these functions.

Logarithmic Functions

In this section, we will discuss **logarithmic functions**. In order to show the importance of logarithmic functions, we will revisit the cooling of steel plates example, and ask a different question.

The Cooling of Steel Plates (Revisited)   In an annealing process, thin steel plates (k = thermal conductivity = 40 W/m · K, ρ = density = 7800 kg/m³, and c = specific heat = 400 J/kg · K) are heated to temperatures of 900° C and then cooled in an environment with temperature of 35° C and a heat transfer coefficient of $h = 25$ W/m² · K . Each plate has a thickness of $L = 5$ cm. We are now interested in determining how long it would take for a plate to reach a temperature of 50° C. To determine the time that it takes for a plate to reach a temperature of 50° C, we will use the following logarithmic equation.

$$t = \frac{\rho c L}{2h} \ln \frac{T_i - T_f}{T - T_f} = \frac{(7800 \text{ kg/m}^3)(400 \text{ J/kg} \cdot \text{K})(0.05 \text{ m})}{(2)(25 \text{ W/m} \cdot \text{K})} \ln \frac{900 - 35}{50 - 35}$$

$$= 12650 \text{ sec} = 3.5 \text{ hr}$$

n	$\left(1+\dfrac{1}{n}\right)^n$
1	2.000000
2	2.250000
5	2.488320
10	2.593742
20	2.653298
50	2.691588
100	2.704814
200	2.711517
500	2.715569
1000	2.716924
2000	2.717603
5000	2.718010
10000	2.718146

X	$f(x)=2^x$	$f(x)=e^x$	$f(x)=3^x$
0	1	1.00	1.00
0.2	1.15	1.22	1.25
0.4	1.32	1.49	1.55
0.6	1.52	1.82	1.93
0.8	1.74	2.23	2.41
1	2.00	2.72	3.00
1.2	2.30	3.32	3.74
1.4	2.64	4.06	4.66
1.6	3.03	4.95	5.80
1.8	3.48	6.05	7.22
2	4.00	7.39	9.00
2.2	4.59	9.03	11.21
2.4	5.28	11.02	13.97
2.6	6.06	13.46	17.40
2.8	6.96	16.44	21.67
3	8.00	20.09	27.00
3.2	9.19	24.53	33.63
3.4	10.56	29.96	41.90
3.6	12.13	36.60	52.20
3.8	13.93	44.70	65.02
4	16.00	54.60	81.00

As $n \to \infty$
$(1 + 1/n)^n \to 2.7182818285\ldots$

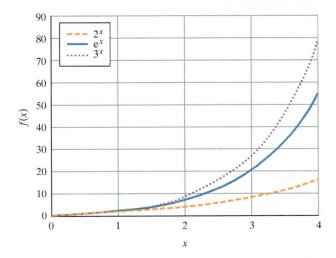

FIGURE 18.18 The comparison of functions 2^x, e^x, and 3^x.

But what is meant by a logarithmic function? The logarithmic functions are defined for the ease of computations. For example, if we let $10^x = y$, then we define $\log y = x$. The symbol "log" reads logarithm to the base-10, or common logarithm. For example, you know any number (other than 0) raised to

TABLE 18.5	Some Important Characteristics of Exponential Functions
Form of the Exponential Function	**Characteristics**
$f(x) = f_0 + a_0 e^{a_1 x}$ (Example 1)	
$f(x) = f_0 - a_0 e^{-a_1 x}$ for $f_0 > a_0$ (Example 2)	

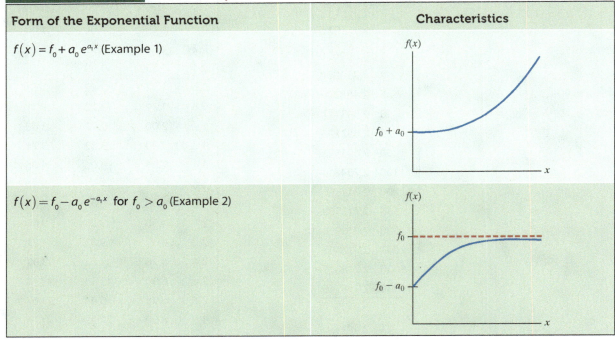

x	$y = 2^{-x^2}$	e^{-x^2}	x	$y = 2^{-x^2}$	e^{-x^2}
−2.6	0.009	0.001	0.2	0.973	0.961
−2.4	0.018	0.003	0.4	0.895	0.852
−2.2	0.035	0.008	0.6	0.779	0.698
−2	0.063	0.018	0.8	0.642	0.527
−1.8	0.106	0.039	1	0.500	0.368
−1.6	0.170	0.077	1.2	0.369	0.237
−1.4	0.257	0.141	1.4	0.257	0.141
−1.2	0.369	0.237	1.6	0.170	0.077
−1	0.500	0.368	1.8	0.106	0.039
−0.8	0.642	0.527	2	0.063	0.018
−0.6	0.779	0.698	2.2	0.035	0.008
−0.4	0.895	0.852	2.4	0.018	0.003
−0.2	0.973	0.961	2.6	0.009	0.001
0	1.000	1.000			

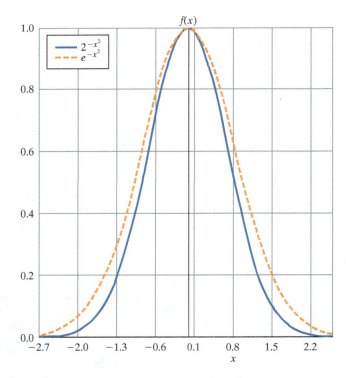

FIGURE 18.19 The plot of $f(x) = 2^{-x^2}$ and $f(x) = e^{-x^2}$.

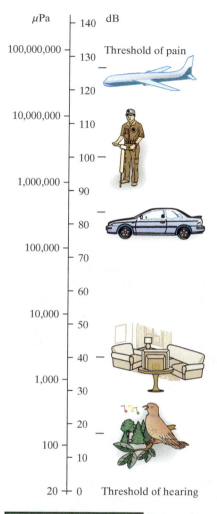

FIGURE 18.20

the power of zero has a value of 1 (e.g., $10^0 = 1$); then using the definition of the common logarithm, log 1 = 0, for $10^1 = 10$, then log 10 = 1, for $10^2 = 100$, then log 100 = 2, and so on. On the other hand, if we let $e^x = y$, then we define *ln* $y = x$, and the symbol *ln* reads the logarithm to the base-e (or natural logarithm). Moreover, the relationship between the natural logarithm and the common logarithm is given by *ln* $x = (ln\ 10)(\log x) = 2.302585 \log x$. Using the logarithmic definitions, we can also prove the following identities, which you will find useful when simplifying engineering relationships.

$$\log xy = \log x + \log y \quad \log \frac{x}{y} = \log x - \log y \quad \log x^n = n\log x$$

Decibel Scale In engineering, the loudness of sound is typically expressed in a unit called decibel (dB) indicated in Figure 18.20. The threshold of hearing (that is, the softest sound that a healthy human can hear) is 20 μPa or 20×10^{-6} Pa. Note this is caused by a very, very small air pressure change. Amazing! At the other end of hearing range lies the threshold of hearing pain, which is caused by approximately 100×10^6 μPa of pressure change. To keep the numbers manageable in the range of hearing (from 20 μPa to 100,000,000 μPa), the decibel scale is defined by dB = $20 \log (I/20)$ μPa, where I represents the pressure change (in μPa) created by the sound source. For example, a sound created by a moving car (creating a pressure change of 200,000 μPa) has a corresponding decibel rating of $20 \log (200{,}000\ \mu Pa/20)\ \mu Pa = 80$ dB. The decibel rating of common sounds is shown in the accompanying figure.

Finally, the following are some additional mathematical relationships that you may find helpful during your engineering education.

$$x^n x^m = x^{n+m} \quad (xy)^n = x^n y^n \quad (x^n)^m = x^{nm} \quad x^0 = 1\ (x \neq 0)$$

$$x^{-n} = \frac{1}{x^n} \quad \frac{x^n}{x^m} = x^{n-m} \quad \left(\frac{x}{y}\right)^n = \frac{x^n}{y^n}$$

Before You Go On

Answer the following questions to test your understanding of the preceding section.

1. What are some of the important characteristics of exponential functions?

2. Explain the difference between the slopes of a nonlinear model and an exponential model.

3. Why do we define logarithmic functions, and how are they defined?

4. How do we define the natural logarithm?

Vocabulary—State the meaning of the following terms:

Exponential Function _____

Logarithmic Function _____

Natural Logarithm _____

Decibel Scale _____

LO⁵ 18.5   Matrix Algebra

As you will learn later during your engineering education, the formulation of many engineering problems, such as the vibration of machines, airplanes, and structures; joint deflections of structural systems; current flow through branches of electrical circuits; and the fluid flow in pipe networks leads to a set of linear algebraic equations that are solved simultaneously. A good understanding of matrix algebra is essential in the formulation and solution of these models. As is the case with any topic, matrix algebra has its own terminology and follows a set of rules. We will provide a brief overview of matrix terminology and matrix algebra in this section.

Basic Definitions

During your engineering education, you will learn about different types of physical variables. There are those that are identifiable by a single value or magnitude. For example, time can be described by a single value such as two hours. These types of physical variables which are identifiable by a single value are called *scalars*. Temperature is another example of a scalar variable. On the other hand, if you were to describe the velocity of a vehicle, you not only have to specify how fast it is moving (speed), but also its direction. The physical variables that possess both magnitude and direction are called *vectors*. There are also other quantities that require specifying more than two pieces of information to describe them accurately. For example, if you were to describe the location of a car parked in a multi-story garage (with respect to the garage entrance), you would need to specify the floor (z coordinate), and then the location of the car on that floor (x and y coordinates). A matrix is often used to describe situations that require many values. A **matrix** is an array of numbers, variables, or mathematical terms. The numbers or the variables that make up the matrix are called the **elements of matrix**. The *size* of a matrix is defined by its number of rows and columns. A matrix may consist of m rows and n columns. For example,

$$[N] = \begin{bmatrix} 6 & 5 & 9 \\ 1 & 26 & 14 \\ -5 & 8 & 0 \end{bmatrix} \qquad \{L\} = \begin{Bmatrix} x \\ y \\ z \end{Bmatrix}$$

matrix $[N]$ is a 3 by 3 (or 3×3) matrix whose elements are numbers, and $\{L\}$ is a 3 by 1 matrix with its elements representing variables x, y, and z. The $[N]$ is called a square matrix. A *square* matrix has the same number of rows and columns. The element of a matrix is denoted by its location. For example, the element in the first row and the third column of a matrix $[N]$ is denoted by n_{13} (it reads *n* sub 13), which has a value of 9. In this book, we denote the matrix by a **boldface letter** in brackets like [] and {}: for example, [**N**], [**T**], {**F**}; and the elements of matrices are represented by regular lowercase letters. The {} is used to distinguish a column matrix. A column matrix is defined as a matrix that has one column but could have many rows. On the other hand, a row matrix is a matrix that has one row but could have many columns. Examples of column and row matrices follow.

$$\{A\} = \begin{Bmatrix} 1 \\ 5 \\ -2 \\ 3 \end{Bmatrix} \quad \text{and} \quad \{X\} = \begin{Bmatrix} x_1 \\ x_2 \\ x_3 \end{Bmatrix}$$

are examples of column matrices, whereas

$$[C] = \begin{bmatrix} 5 & 0 & 2 & -2 \end{bmatrix} \quad \text{and} \quad [Y] = \begin{bmatrix} y_1 & y_2 & y_3 \end{bmatrix}$$

are examples of row matrices.

Diagonal and Unit Matrices A diagonal matrix is one that only has elements along its principal diagonal; the elements are zero everywhere else. An example of a 4×4 diagonal matrix follows.

$$[A] = \begin{bmatrix} 5 & 0 & 0 & 0 \\ 0 & 7 & 0 & 0 \\ 0 & 0 & 4 & 0 \\ 0 & 0 & 0 & 11 \end{bmatrix}$$

The diagonal along which values 5, 7, 4, and 11 lie is called the *principal diagonal*. An *identity* or *unit matrix* is a diagonal matrix whose elements consist of a value of 1. An example of an identity matrix is.

$$[I] = \begin{bmatrix} 1 & 0 & 0 & . & . & 0 & 0 \\ 0 & 1 & 0 & . & . & 0 & 0 \\ 0 & 0 & 1 & . & . & 0 & 0 \\ . & . & . & . & . & . & . \\ . & . & . & . & . & . & . \\ 0 & 0 & 0 & . & . & 1 & 0 \\ 0 & 0 & 0 & . & . & 0 & 1 \end{bmatrix}$$

Matrix Addition or Subtraction

Two matrices can be added together or subtracted from each other provided that they are of the same size—each matrix must have the same number of rows and columns. We can add matrix $[A]_{m \times n}$ of dimension *m* by *n* (having *m* rows and *n* columns) to matrix $[B]_{m \times n}$ of the same dimension by adding the like elements. Matrix subtraction follows a similar rule, as shown.

$$[A] \pm [B] = \begin{bmatrix} 10 & 3 & . & . & 2 \\ 5 & 1 & . & . & 0 \\ . & . & . & . & . \\ . & . & . & . & . \\ 9 & 2 & . & . & 7 \end{bmatrix} \pm \begin{bmatrix} 2 & 12 & . & . & 8 \\ 1 & 7 & . & . & 15 \\ . & . & . & . & . \\ . & . & . & . & . \\ 4 & 55 & . & . & 10 \end{bmatrix}$$

$$= \begin{bmatrix} (10 \pm 2) & (3 \pm 12) & . & . & (2 \pm 8) \\ (5 \pm 1) & (1 \pm 7) & . & . & (0 \pm 15) \\ . & . & . & . & . \\ . & . & . & . & . \\ (9 \pm 4) & (2 \pm 55) & . & . & (7 \pm 10) \end{bmatrix}$$

Matrix Multiplication

In this section, we will discuss the rules for multiplying a matrix by a scalar quantity and by another matrix.

Multiplying a Matrix by a Scalar Quantity When a matrix [A] is multiplied by a scalar quantity with a magnitude such as 5, the operation results in a matrix of the same size whose elements are the product of elements in the original matrix and the scalar quantity. For example, when we multiply matrix [A] of size 3 × 3 by a scalar quantity 5, this operation results in another matrix of size 3 × 3 whose elements are computed by multiplying each element of matrix [A] by 5, as shown here.

$$5[A] = 5 \begin{bmatrix} 4 & 0 & 1 \\ -2 & 9 & 2 \\ 5 & 7 & 10 \end{bmatrix} = \begin{bmatrix} 20 & 0 & 5 \\ -10 & 45 & 10 \\ 25 & 35 & 50 \end{bmatrix} \qquad \textbf{18.11}$$

Multiplying a Matrix by Another Matrix Whereas any size matrix can be multiplied by a scalar quantity, matrix multiplication can be performed only when the number of columns in the *premultiplier* matrix is equal to the number of rows in the *postmultiplier* matrix. For example, matrix [A] of size $m \times n$ can be premultiplied by matrix [B] of size $n \times p$ because the number of columns n in matrix [A] is equal to number of rows n in matrix [B]. Moreover, the multiplication results in another matrix, say [C], of size $m \times p$. Matrix multiplication is carried out according to the following rule. Consider the multiplication of the following 3 by 3 [A] and the 3 by 2 [B] matrices.

$$[A][B] = \begin{bmatrix} 2 & 4 & 1 \\ 1 & 6 & 5 \\ -2 & 3 & 8 \end{bmatrix}_{3 \times 3} \begin{bmatrix} 7 & 23 \\ 12 & 9 \\ 16 & 11 \end{bmatrix}_{3 \times 2} = [C]_{3 \times 2}$$

Note the number of columns in matrix [A] is equal to number of rows in matrix [B], and the multiplication will result in a matrix of size 3 by 2. The elements in the first column of the resulting [C] matrix are computed from

$$[C] = \begin{matrix} 7 & 12 & 16 \leftarrow \end{matrix}$$

$$[C] = \begin{bmatrix} 2 & 4 & 1 \\ 1 & 6 & 5 \\ -2 & 3 & 8 \end{bmatrix} \begin{bmatrix} 23 \\ \uparrow & 9 \\ \uparrow & 11 \end{bmatrix}$$

$$= \begin{vmatrix} (7)(2) + (12)(4) + (16)(1) & c_{12} \\ (7)(1) + (12)(6) + (16)(5) & c_{22} \\ (7)(-2) + (12)(3) + (16)(8) & c_{32} \end{vmatrix} = \begin{vmatrix} 78 & c_{12} \\ 159 & c_{22} \\ 150 & c_{32} \end{vmatrix}$$

and the elements in the second column of the $[C]$ matrix are

$$23 \quad 9 \quad 11 \leftarrow$$

$$[C] = \begin{bmatrix} 2 & 4 & 1 \\ 1 & 6 & 5 \\ -2 & 3 & 8 \end{bmatrix} \begin{bmatrix} 7 \\ 12 & \uparrow \\ 16 & \uparrow \end{bmatrix} = \begin{vmatrix} 78 & c_{12} \\ 159 & c_{22} \\ 150 & c_{32} \end{vmatrix}$$

$$= \begin{vmatrix} 78 & (23)(2) + (9)(4) + (11)(1) \\ 159 & (23)(1) + (9)(6) + (11)(5) \\ 150 & (23)(-2) + (9)(3) + (11)(8) \end{vmatrix} = \begin{vmatrix} 78 & 93 \\ 159 & 132 \\ 150 & 69 \end{vmatrix}$$

If you are dealing with larger matrices, the elements in the other columns are computed in a similar manner. Also, when multiplying matrices, keep in mind that matrix multiplication is not commutative except for very special cases. That is,

$$[A][B] \neq [B][A] \qquad \boxed{18.12}$$

This may be a good place to point out that if $[I]$ is an identity matrix and $[A]$ is a square matrix of matching size, then it can be readily shown that the product of

$$[I][A] = [A][I] = [A] \qquad \boxed{18.13}$$

EXAMPLE 18.1

Given matrices: $[A] = \begin{bmatrix} 0 & 5 & 0 \\ 8 & 3 & 7 \\ 9 & -2 & 9 \end{bmatrix}$, $[B] = \begin{bmatrix} 4 & 6 & -2 \\ 7 & 2 & 3 \\ 1 & 3 & -4 \end{bmatrix}$, and $\{C\} = \begin{Bmatrix} -1 \\ 2 \\ 5 \end{Bmatrix}$,

perform the following operations.

(a) $[A] + [B] = ?$

(b) $[A] - [B] = ?$

(c) $3[A] = ?$

(d) $[A][B] = ?$

(e) $[A]\{C\} = ?$

(f) Show that $[I][A] = [A][I] = [A]$

We will use the operation rules discussed in the preceding sections to answer these questions.

(a) $[A] + [B] = ?$

$$[A] + [B] = \begin{bmatrix} 0 & 5 & 0 \\ 8 & 3 & 7 \\ 9 & -2 & 9 \end{bmatrix} + \begin{bmatrix} 4 & 6 & -2 \\ 7 & 2 & 3 \\ 1 & 3 & -4 \end{bmatrix}$$

$$= \begin{bmatrix} (0+4) & (5+6) & (0+(-2)) \\ (8+7) & (3+2) & (7+3) \\ (9+1) & (-2+3) & (9+(-4)) \end{bmatrix} = \begin{bmatrix} 4 & 11 & -2 \\ 15 & 5 & 10 \\ 10 & 1 & 5 \end{bmatrix}$$

(b) $[A] - [B] = ?$

$$[A] - [B] = \begin{bmatrix} 0 & 5 & 0 \\ 8 & 3 & 7 \\ 9 & -2 & 9 \end{bmatrix} - \begin{bmatrix} 4 & 6 & -2 \\ 7 & 2 & 3 \\ 1 & 3 & -4 \end{bmatrix}$$

$$= \begin{bmatrix} (0-4) & (5-6) & (0-(-2)) \\ (8-7) & (3-2) & (7-3) \\ (9-1) & (-2-3) & (9-(-4)) \end{bmatrix} = \begin{bmatrix} -4 & -1 & 2 \\ 1 & 1 & 4 \\ 8 & -5 & 13 \end{bmatrix}$$

(c) $3[A] = ?$

$$3[A] = 3\begin{bmatrix} 0 & 5 & 0 \\ 8 & 3 & 7 \\ 9 & -2 & 9 \end{bmatrix} = \begin{bmatrix} 0 & (3)(5) & 0 \\ (3)(8) & (3)(3) & (3)(7) \\ (3)(9) & (3)(-2) & (3)(9) \end{bmatrix} = \begin{bmatrix} 0 & 15 & 0 \\ 24 & 9 & 21 \\ 27 & -6 & 27 \end{bmatrix}$$

(d) $[A][B] = ?$

$$[A][B] = \begin{bmatrix} 0 & 5 & 0 \\ 8 & 3 & 7 \\ 9 & -2 & 9 \end{bmatrix}\begin{bmatrix} 4 & 6 & -2 \\ 7 & 2 & 3 \\ 1 & 3 & -4 \end{bmatrix}$$

$$= \begin{bmatrix} (0)(4)+(5)(7)+(0)(1) & (0)(6)+(5)(2)+(0)(3) & (0)(-2)+(5)(3)+(0)(-4) \\ (8)(4)+(3)(7)+(7)(1) & (8)(6)+(3)(2)+(7)(3) & (8)(-2)+(3)(3)+(7)(-4) \\ (9)(4)+(-2)(7)+(9)(1) & (9)(6)+(-2)(2)+(9)(3) & (9)(-2)+(-2)(3)+(9)(-4) \end{bmatrix}$$

$$= \begin{bmatrix} 35 & 10 & 15 \\ 60 & 75 & -35 \\ 31 & 77 & -60 \end{bmatrix}$$

(e) $[A]\{C\} = ?$

$$[A]\{C\} = \begin{bmatrix} 0 & 5 & 0 \\ 8 & 3 & 7 \\ 9 & -2 & 9 \end{bmatrix}\begin{Bmatrix} -1 \\ 2 \\ 5 \end{Bmatrix} = \begin{Bmatrix} (0)(-1)+(5)(2)+(0)(5) \\ (8)(-1)+(3)(2)+(7)(5) \\ (9)(-1)+(-2)(2)+(9)(5) \end{Bmatrix} = \begin{Bmatrix} 10 \\ 33 \\ 32 \end{Bmatrix}$$

(f) Show that $[I][A] = [A][I] = [A]$

$$[I][A] = \begin{bmatrix} 1 & 0 & 0 \\ 0 & 1 & 0 \\ 0 & 0 & 1 \end{bmatrix} \begin{bmatrix} 0 & 5 & 0 \\ 8 & 3 & 7 \\ 9 & -2 & 9 \end{bmatrix} = \begin{bmatrix} 0 & 5 & 0 \\ 8 & 3 & 7 \\ 9 & -2 & 9 \end{bmatrix} \text{ and}$$

$$[A][I] = \begin{bmatrix} 0 & 5 & 0 \\ 8 & 3 & 7 \\ 9 & -2 & 9 \end{bmatrix} \begin{bmatrix} 1 & 0 & 0 \\ 0 & 1 & 0 \\ 0 & 0 & 1 \end{bmatrix} = \begin{bmatrix} 0 & 5 & 0 \\ 8 & 3 & 7 \\ 9 & -2 & 9 \end{bmatrix}$$

Transpose of a Matrix

As you will see in the classes which you will take later, the formulation and solution of engineering problems lend themselves to situations wherein it is desirable to rearrange the rows of a matrix into the columns of another matrix.

In general, to obtain the **transpose** of a matrix $[B]$ of size $m \times n$, the first row of the given matrix becomes the first column of $[B]^T$, the second row of $[B]$ becomes the second column of $[B]^T$, and so on, leading to the mth row of $[B]$ becoming the mth column of $[B]^T$, and resulting in a matrix with the size of $n \times m$. The matrix $[B]^T$ reads as the transpose of the $[B]$ matrix.

Sometimes, in order to save space, we write the solution matrices, which are column matrices, as row matrices using the transpose of the solution which is another use for the transpose of a matrix. For example, we represent the solution given by the U matrix.

$$\{U\} = \begin{Bmatrix} 7 \\ 4 \\ 9 \\ 6 \\ 12 \end{Bmatrix} \text{ by } [U]^T = \begin{bmatrix} 7 & 4 & 9 & 6 & 12 \end{bmatrix}$$

This is a good place to define a symmetric matrix. A *symmetric matrix* is a square matrix whose elements are symmetrical with respect to its principal diagonal. An example of a symmetric matrix follows.

$$[A] = \begin{bmatrix} 1 & 4 & 2 & -5 \\ 4 & 5 & 15 & 20 \\ 2 & 15 & -3 & 8 \\ -5 & 20 & 8 & 0 \end{bmatrix}$$

EXAMPLE 18.2

Given the following matrices:

$$[A] = \begin{bmatrix} 0 & 5 & 0 \\ 8 & 3 & 7 \\ 9 & -2 & 9 \end{bmatrix} \text{ and } [B] = \begin{bmatrix} 4 & 6 & -2 \\ 7 & 2 & 3 \\ 1 & 3 & -4 \end{bmatrix}, \text{ perform the following operations:}$$

(a) $[A]^T = ?$ and (b) $[B]^T = ?$

(a) As explained earlier, the first, second, third, …, and mth rows of a matrix become the first, second, third, …, and mth column of the transpose matrix, respectively.

$$[A]^T = \begin{bmatrix} 0 & 8 & 9 \\ 5 & 3 & -2 \\ 0 & 7 & 9 \end{bmatrix}$$

(b) Similarly,

$$[B]^T = \begin{bmatrix} 4 & 7 & 1 \\ 6 & 2 & 3 \\ -2 & 3 & -4 \end{bmatrix}$$

Determinant of a Matrix

Up to this point, we have defined essential matrix terminology and discussed basic matrix operations. In this section, we will define what is meant by a **determinant** of a matrix. Let us consider the solution to the following set of simultaneous equations:

$$a_{11}x_1 + a_{12}x_2 = b_1 \qquad \text{18.14a}$$

$$a_{21}x_1 + a_{22}x_2 = b_2 \qquad \text{18.14b}$$

Expressing Equations (18.14a) and (18.14b) in a matrix form, we have

$$\overbrace{\begin{bmatrix} a_{11} & a_{12} \\ a_{21} & a_{22} \end{bmatrix}}^{[A]} \begin{Bmatrix} x_1 \\ x_2 \end{Bmatrix} = \begin{Bmatrix} b_1 \\ b_2 \end{Bmatrix}$$

To solve for the unknowns x_1 and x_2, we may first solve for x_2 in terms of x_1, using Equation (18.14 b), and then substitute that relationship into Equation (18.14a). These steps are shown next.

$$x_2 = \frac{b_2 - a_{21}x_1}{a_{22}} \implies a_{11}x_1 + a_{12}\left(\frac{b_2 - a_{21}x_1}{a_{22}}\right) = b_1$$

Solving for x_1:

$$x_1 = \frac{b_1 a_{22} - a_{12} b_2}{a_{11} a_{22} - a_{12} a_{21}} \qquad \text{18.15a}$$

After we substitute for x_1 in either Equation (18.14a) or (18.14b), we get

$$x_2 = \frac{a_{11} b_2 - b_1 a_{21}}{a_{11} a_{22} - a_{12} a_{21}} \qquad \text{18.15b}$$

Referring to the solutions given by Equations (18.15a) and (18.15b), we see that the denominators in these equations represent the product of coefficients in the main diagonal minus the product of the coefficient in the other diagonal of the [A] matrix. The $a_{11}a_{22} - a_{12}a_{21}$ is the determinant of the 2 × 2 [A] matrix and is represented in one of following ways:

$$\mathbf{Det}[A] \text{ or } \mathbf{det}[A] \text{ or } \begin{vmatrix} a_{11} & a_{12} \\ a_{21} & a_{22} \end{vmatrix} = a_{11}a_{22} - a_{12}a_{21}$$

18.16

Only the determinant of a square matrix is defined. Moreover, keep in mind that the determinant of the [A] matrix is a single number. That is, after we substitute for the values of an, a_{11}, a_{12}, and a_{21} into $a_{11}a_{22} - a_{12}a_{21}$, we get a single number.

Let us now consider the determinant of a 3 by 3 matrix such as

$$[C] = \begin{bmatrix} c_{11} & c_{12} & c_{13} \\ c_{21} & c_{22} & c_{23} \\ c_{31} & c_{32} & c_{33} \end{bmatrix}$$

which is computed in the following manner:

$$\begin{bmatrix} c_{11} & c_{12} & c_{13} \\ c_{21} & c_{22} & c_{23} \\ c_{31} & c_{32} & c_{33} \end{bmatrix} = c_{11}c_{22}c_{33} + c_{12}c_{23}c_{31} + c_{13}c_{21}c_{32} - c_{13}c_{22}c_{31} - c_{11}c_{23}c_{32} - c_{12}c_{21}c_{33}$$

18.17

There is a simple procedure called *direct expansion* that you can use to obtain the results given by Equation (18.17). Direct expansion proceeds in the following manner. First we repeat and place the first and the second columns of the matrix [C] next to the third column, as shown in Figure 18.21. Then, we add the products of the diagonal elements lying on the solid arrows and subtract them from the products of the diagonal elements lying on the dashed arrows. This procedure shown in Figure 18.21 results in the determinant value given by Equation (18.17).

The direct-expansion procedure cannot be used to obtain higher-order determinants. Instead, we resort to a method that first reduces the order of the determinant—to what is called a minor—and then evaluates the lower-order determinants. You will learn about minors later in your other classes.

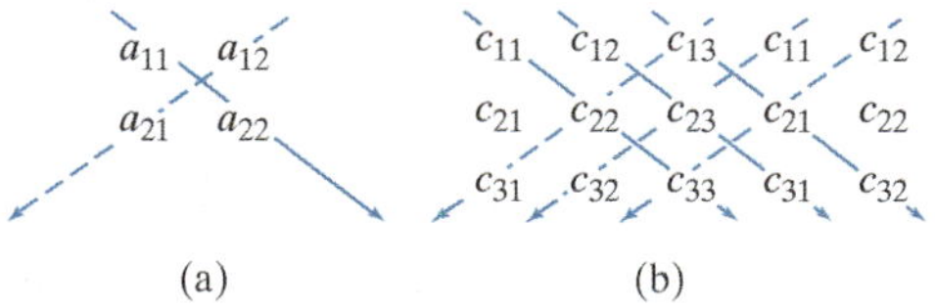

(a)   (b)

FIGURE 18.21 Direct-expansion procedure for computing the determinant of (a) 2 × 2 matrix, and (b) 3 × 3 matrix.

EXAMPLE 18.3

Given the following matrix: $[A] = \begin{bmatrix} 1 & 5 & 0 \\ 8 & 3 & 7 \\ 6 & -2 & 9 \end{bmatrix}$, calculate the determinant of $[A]$.

As explained earlier, using the direct-expansion method, we repeat and place the first and the second column of the matrix next to the third column as shown, compute the products of the elements along the solid arrows, and then subtract them from the products of elements along the dashed arrows, as shown in Figure 18.22. Use of this method results in the following solution.

FIGURE 18.22

The direct expansion method for Example 18.3.

$$\begin{vmatrix} 1 & 5 & 0 \\ 8 & 3 & 7 \\ 6 & -2 & 9 \end{vmatrix} = (1)(3)(9) + (5)(7)(6) + (0)(8)(-2) \\ -(5)(8)(9) - (1)(7)(-2) - (0)(3)(6) = -109$$

When the determinant of a matrix is zero, the matrix is called a *singular*. A singular matrix results when the elements in two or more rows of a given matrix are identical. For example consider the following matrix:

$[A] = \begin{bmatrix} 2 & 1 & 4 \\ 2 & 1 & 4 \\ 1 & 3 & 5 \end{bmatrix}$, whose rows one and two are identical.

As shown next, the determinant of $[A]$ is zero.

$$\begin{vmatrix} 2 & 1 & 4 \\ 2 & 1 & 4 \\ 1 & 3 & 5 \end{vmatrix} = (2)(1)(5) + (1)(4)(1) + (4)(2)(3) \\ -(1)(2)(5) - (2)(4)(3) - (4)(1)(1) = 0$$

Matrix singularity can also occur when the elements in two or more rows of a matrix are linearly dependent. For example, if we multiply the elements of the second row of matrix $[A]$ by a scalar factor such as 7, then the resulting

matrix $[A] = \begin{bmatrix} 2 & 1 & 4 \\ 14 & 7 & 28 \\ 1 & 3 & 5 \end{bmatrix}$ is singular because rows one and two are now

linearly dependent. As shown next, the determinant of the new $[A]$ matrix is zero.

$$\begin{vmatrix} 2 & 1 & 4 \\ 14 & 7 & 28 \\ 1 & 3 & 5 \end{vmatrix} = (2)(7)(5) + (1)(28)(1) + (4)(14)(3) \\ -(1)(14)(5) - (2)(28)(3) - (4)(7)(1) = 0$$

Solutions of Simultaneous Linear Equations

As we discussed earlier, the formulation of many engineering problems leads to a system of algebraic equations. As you will learn later in your math and engineering classes, there are a number ways that we can use to solve a set of linear equations. In the section that follows, we will discuss one of these methods that you can use to obtain solutions to a set of linear equations.

Gauss Elimination Method We will begin our discussion by demonstrating the Gauss elimination method, using an example. Consider the following three linear equations with three unknowns: x_1, x_2, and x_3.

$$2x_1 + x_2 + x_3 = 13 \qquad \text{18.18a}$$

$$3x_1 + 2x_2 + 4x_3 = 32 \qquad \text{18.18b}$$

$$5x_1 - x_2 + 3x_3 = 17 \qquad \text{18.18c}$$

Step 1: We begin by dividing the first equation, Equation (8.18a), by 2: the coefficient of the x_1 term. This operation leads to

$$x_1 + \frac{1}{2}x_2 + \frac{1}{2}x_3 = \frac{13}{2} \qquad \text{18.19}$$

Step 2: We multiply Equation (18.19) by 3: the coefficient of x_1 in Equation (18.18b).

$$3x_1 + \frac{3}{2}x_2 + \frac{3}{2}x_3 = \frac{39}{2} \qquad \text{18.20}$$

We then subtract Equation (18.20) from Equation (18.18b). This step will eliminate x_1 from Equation (18.18b). This operation leads to

$$\begin{aligned} 3x_1 + 2x_2 + 4x_3 &= 32 \\ -\left(3x_1 + \frac{3}{2}x_2 + \frac{3}{2}x_3 = \frac{39}{2}\right) & \\ \hline \frac{1}{2}x_2 + \frac{5}{2}x_3 &= \frac{25}{2} \end{aligned} \qquad \text{18.21}$$

Step 3: Similarly, to eliminate x_1 from Equation (18.18c), we multiply Equation (18.19) by 5: the coefficient of x_1 in Equation (18.18c).

$$5x_1 + \frac{5}{2}x_2 + \frac{5}{2}x_3 = \frac{65}{2} \qquad \text{18.22}$$

We then subtract the above equation from Equation (18.18c), which will eliminate x_1 from Equation (18.18c). This operation leads to

$$\begin{aligned} 5x_1 - x_2 + 3x_3 &= 17 \\ -\left(5x_1 + \frac{5}{2}x_2 + \frac{5}{2}x_3 = \frac{65}{2}\right) & \\ \hline -\frac{7}{2}x_2 + \frac{1}{2}x_3 &= -\frac{31}{2} \end{aligned} \qquad \text{18.23}$$

Let us summarize the results of the operations performed during steps 1 through 3. These operations eliminated x_1 from Equations (18.18b) and (18.18c).

$$x_1 + \frac{1}{2}x_2 + \frac{1}{2}x_3 = \frac{13}{2} \qquad \text{18.24a}$$

$$\frac{1}{2}x_2 + \frac{5}{2}x_3 = \frac{25}{2}$$

18.24b

$$-\frac{7}{2}x_2 + \frac{1}{2}x_3 = -\frac{31}{2}$$

18.24c

Step 4: To eliminate x_2 from Equation (18.24c), first we divide Equation (18.24b) by $\frac{1}{2}$, the coefficient of x_2.

$$x_2 + 5x_3 = 25$$

18.25

Then, we multiply Equation (18.25) by $\frac{-7}{2}$, the coefficient of x_2 in Equation (18.24c), and subtract that equation from Equation (18.24c). These operations lead to

$$-\frac{7}{2}x_2 + \frac{1}{2}x_3 = -\frac{31}{2}$$
$$-\left(-\frac{7}{2}x_2 - \frac{35}{2}x_3 = -\frac{175}{2}\right)$$
$$\overline{\qquad 18x_3 = 72 \qquad}$$

18.26

Dividing both sides of Equation (18.26) by 18, we get

$$x_3 = 4$$

Summarizing the results of the previous steps, we have

$$x_1 + \frac{1}{2}x_2 + \frac{1}{2}x_3 = \frac{13}{2}$$

18.27

$$x_2 + 5x_3 = 25$$

18.28

$$x_3 = 4$$

18.29

Step 5: Now we can use back substitution to compute the values of x_2 and x_3. We substitute for x_3 in Equation (18.28) and solve for x_2.

$$x_2 + 5(4) = 25 \implies x_2 = 5$$

Next, we substitute for x_3 and x_2 in Equation (18.27) and solve for x_1.

$$x_1 + \frac{1}{2}(5) + \frac{1}{2}(4) = \frac{13}{2} \implies x_1 = 2$$

Inverse of a Matrix

In the previous sections, we discussed matrix addition, subtraction, and multiplication, but you may have noticed that we did not say anything about matrix division. That is because such an operation is not defined formally. Instead, we define an **inverse of a matrix** in such a way that when it is multiplied by the original matrix, the identity matrix is obtained.

$$[A]^{-1}[A] = [A][A]^{-1} = [I]$$

18.30

In Equation (18.30), $[A]^{-1}$ is called the inverse of $[A]$. Only a square and nonsingular matrix has an inverse. In the previous section, we explained the Gauss elimination method that you can use to obtain solutions to a set of linear equations. Matrix inversion allows for yet another way of solving for the solutions of a set of linear equations. As you will learn later in your math and engineering classes, there are a number of ways to compute the inverse of a matrix.

Before You Go On

Answer the following questions to test your understanding of the preceding section.

1. What do we mean by the elements of a matrix?

2. What do we mean by the size of a matrix?

3. What do we mean by the transpose of a matrix?

4. What do we mean by the determinant of a matrix?

Vocabulary—State the meaning of the following terms:

Unit Matrix _____

Square Matrix _____

Identity Matrix _____

LO⁶ 18.6   Calculus

Calculus commonly is divided into two broad areas: differential and integral calculus. In the following sections, we will explain some key concepts related to differential and integral calculus.

Differential Calculus

A good understanding of differential calculus is necessary to determine the *rate of change* in engineering problems. The rate of change refers to how a

> Calculus is divided into two broad areas: differential and integral calculus.

dependent variable changes with respect to an independent variable. Let's imagine that on a nice day, you decided to go for a ride. You get into your car and turn the engine on, and you start on your way for a nice drive. Once you are cruising at a constant speed and enjoying the scenery, your engineering curiosity kicks in and you ask yourself, how has the speed of my car been changing? In other words, you are interested in knowing the *time rate of change* of speed, or the tangential acceleration of the car.

As defined above, the rate of change shows how one variable changes with respect to another variable. In this example, speed is the *dependent variable* and time is the *independent variable.* The speed is called the dependent variable, because the speed of the car is a function of time. On the other hand, the time variable is not dependent on the speed, and hence, it is called an independent variable. If you could define a function that closely described the speed in terms of time, then you would *differentiate* the function to obtain the acceleration. Related to the example above, there are many other questions that you could have asked:

What is the time rate of fuel consumption (gallons per hour)?
What is the distance rate of fuel consumption (miles per gallon)?
What is the time rate of change of your position with respect to a known location (i.e., speed of the car)?

Engineers calculate the rate of change of variables to design products and services. The engineers who designed your car had to have a good grasp of the concept of rate of change in order to build a car with a predictable behavior. For example, manufacturers of cars make certain information available, such as miles per gallon for city or highway driving conditions. Additional familiar examples dealing with rates of change of variables include:

How does the temperature of the oven change with time after it is turned on?
How does the temperature of a soft drink change over time after it is placed in a refrigerator?

Again, the engineers who designed the oven and refrigerator understood the rate of change concept to design a product that functions according to established specifications. Traffic flow and product movement on assembly lines are other examples where a detailed knowledge of the rate of change of variables are sought.

During the next two years, as you take your calculus classes, you will learn many new concepts and rules dealing with differential calculus. Make sure you take the time to understand these concepts and rules. In your calculus classes, you may not apply the concepts to actual engineering problems, but be assured that you will use them eventually in your engineering classes. Some of these concepts and rules are summarized in Table 18.6. Examples that demonstrate how to apply these differentiation rules follow. As you study these examples, keep in mind that our intent here is to introduce some rules, not to explain them thoroughly.

EXAMPLE 18.4

Identify the dependent and independent variables for the following situations: water consumption and traffic flow.

For the water consumption situation, the mass or the volume is the dependent variable, with time the independent variable. For the traffic flow problem, the number of cars is the dependent variable and time is the independent variable.

EXAMPLE 18.5

Find the derivative of $f(x) = x^3 - 10x^2 + 8$.

We use rule 3 and 5, $f'(x) = nx^{n-1}$, from Table 18.6 to solve this problem as shown.

$$f'(x) = 3x^2 - 20x$$

TABLE 18.6	Summary of Definitions and Derivative Rules	
	Definitions and Rules	**Explanation**
1	$f'(x) = \dfrac{df}{dx} = \lim\limits_{h \to 0} \dfrac{f(x+h) - f(x)}{h}$	The definition of the derivative of the function $f(x)$.
2	If $f(x) = $ constant then $f'(x) = 0$	The derivative of a constant function is zero.
3	If $f(x) = x^n$ then $f'(x) = nx^{n-1}$	The Power Rule (see Example 18.5).
4	If $f(x) = a \cdot g(x)$ where a is a constant then $f'(x) = a \cdot g'(x)$	The rule for when a constant such as a is multiplied by a function (see Example 18.6).
5	If $f(x) = g(x) \pm h(x)$ then $f'(x) = g'(x) \pm h'(x)$	The rule for when two functions are added or subtracted (see Example 18.7).
6	If $f(x) = g(x) \cdot h(x)$ then $f'(x) = g'(x) \cdot h(x) + g(x) \cdot h'(x)$	The Product Rule (see Example 18.8).
7	If $f(x) = \dfrac{g(x)}{h(x)}$ then $f'(x) = \dfrac{h(x) \cdot g'(x) - g(x) \cdot h'(x)}{\left[h(x)\right]^2}$	The Quotient Rule (see Example 18.9).
8	If $f(x) = f[g(x)] = f(u)$ where $u = g(x)$ then $f'(x) = \dfrac{df(x)}{dx} = \dfrac{df(x)}{du} \cdot \dfrac{du}{dx}$	The Chain Rule.
9	If $f(x) = \left[g(x)\right]^n = u^n$ where $u = g(x)$ then $f'(x) = n \cdot u^{n-1} \cdot \dfrac{du}{dx}$	The Power Rule for a general function such as $g(x)$ (see Example 18.10).
10	If $f(x) = \ln\left|g(x)\right|$ then $f'(x) = \dfrac{g'(x)}{g(x)}$	The rule for natural logarithm functions (see Example 18.11).
11	If $f(x) = \exp(g(x))$ or $f(x) = e^{g(x)}$ then $f'(x) = g'(x) \cdot e^{g(x)}$	The rule for exponential functions (see Example 18.12).

EXAMPLE 18.6

Find the derivative of $f(x) = 5(x^3 - 10x^2 + 8)$.

We use rule 4, $f'(x) = a \cdot g'(x)$, from Table 18.6 to solve this problem. For the given problem, if $a = 5$ and $g(x) = x^3 - 10x^2 + 8$, then the derivative of $f(x)$ is $f'(x) = 5(3x^2 - 20x) = 15x^2 - 100x$.

EXAMPLE 18.7

Find the derivative of $f(x) = (x^3 - 10x^2 + 8) \pm (x^5 + 5x)$.

We use rule 5, $f'(x) = g'(x) \pm h'(x)$, from Table 18.6 to solve this problem as shown.

$$f'(x) = (3x^2 - 20x) \pm (5x^4 + 5)$$

EXAMPLE 18.8

Find the derivative of $f(x) = (x^3 - 10x^2 + 8)(x^5 - 5x)$.

We use rule 6, $f'(x) = g'(x) \cdot h(x) + g(x) \cdot h'(x)$, from Table 18.6 to solve this problem as shown. For this problem, $h(x) = (x^5 - 5x)$ and $g(x) = (x^3 - 10x^2 + 8)$.

$$f'(x) = (3x^2 - 20x)(x^5 + 5x) + (x^3 - 10x^2 + 8)(5x^4 - 5)$$
$$= 8x^7 - 70x^6 + 40x^4 = 20x^3 + 150x^2 - 40$$

EXAMPLE 18.9

Find the derivative of $f(x) = (x^3 - 10x^2 + 8)/(x^5 - 5x)$.

We use rule 7, $f'(x) = \left[h(x) \cdot g'(x) - g(x) \cdot h'(x) \right] / \left[h(x) \right]^2$, from Table 18.6 to solve this problem as shown. For this problem, $h(x) = (x^5 = 5x)$ and $g(x) = (x^3 - 10x^2 + 8)$.

$$f'(x) = \frac{(x^5 - 5x)(3x^2 - 20x) - (x^3 - 10x^2 + 8)(5x^4 - 5)}{(x^5 - 5x)^2}$$

EXAMPLE 18.10

Find the derivative of $f(x) = (x^3 - 10x^2 + 8)^4$.

We use rule 9, $f'(x) = n \cdot u^{n-1} \cdot \dfrac{du}{dx}$, from Table 18.6 to solve this problem as shown. For this problem, $u = (x^3 - 10x^2 + 8)$.

$$f'(x) = 4(x^3 - 10x^2 + 8)^3 (3x^2 - 20x)$$

EXAMPLE 18.11

Find the derivative of $f(x) = ln|x^3 - 10x^2 + 8|$.

We use rule 10, $f'(x) = g'(x)/g(x)$, from Table 18.6 to solve this problem as shown. For $g(x) = x^3 - 10x^2 + 8$.

$$f'(x) = \frac{(3x^2 - 20x)}{x^3 - 10x^2 + 8}$$

EXAMPLE 18.12

Find the derivative of $f(x) = e^{(x^3 - 10x^2 + 8)}$.

We use rule 11, $f'(x) = g'(x) \cdot e^{g(x)}$, from Table 18.6 to solve this problem as shown. For this problem, $g(x) = x^3 - 10x^2 + 8$.

$$f'(x) = (3x^2 - 20x)e^{(x^3 - 10x^2 + 8)}$$

Integral Calculus

Integral calculus plays a vital role in the formulation and solution of engineering problems. To demonstrate the role of integrals, consider the following examples.

EXAMPLE 18.13

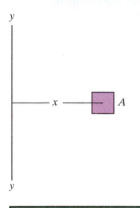

FIGURE 18.23

Small area element located at a distance *x* from the *y–y* axis.

Recall in Chapter 7, we discussed a property of an area known as the second moment of area. The second moment of area, also known as the area moment of inertia, is an important property of an area that provides information on how hard it is to bend something and therefore plays an important role in design of structures. We explained that for a small area element *A* located at a distance *x* from the axis *y–y*, as shown in Figure 18.23, the area moment of inertia is defined by

$$I_{y-y} = x^2 A \qquad \text{18.31}$$

We also included more small area elements, as shown in Figure 18.24. The area moment of inertia for the system of discrete areas shown about the *y–y* axis is now

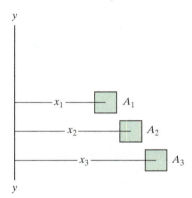

FIGURE 18.24   Second moment of area for three small area elements.

$$I_{y-y} = x_1^2 A_1 + x_2^2 A_2 + x_3^2 A_3 \qquad \text{18.32}$$

Similarly, we can obtain the second moment of area for a cross-sectional area, such as a rectangle or a circle, by summing the area moment of inertia of all the little area elements that makes up the cross-section. However, for a continuous cross-sectional area, we use integrals instead of summing the $x^2 A$ terms to evaluate the area moment of inertia. After all, the integral sign, $\int$, is nothing but a big "S" sign, indicating summation.

$$I_{y-y} = \int x^2 \, dA \qquad \text{18.33}$$

We can obtain the area moment of inertia of any geometric shape by performing the integration given by Equation (18.33). For example, let us derive a formula for a rectangular cross-section about the *y–y* axes.

$$\overbrace{I_{y-y} = \int_{-w/2}^{w/2} x^2 \, dA}^{\text{step 1}} = \overbrace{\int_{-w/2}^{w/2} x^2 \, h dx}^{\text{step 2}} = \overbrace{h \int_{-w/2}^{w/2} x^2 \, dx}^{\text{step 3}} = \overbrace{\frac{1}{12} h w^3}^{\text{step 4}}$$

Step 1: The second moment of the rectangular cross-sectional area is equal to the sum (integral) of little rectangles.

Step 2: We substitute for $dA = hdx$ (see Figure 18.25).

Step 3: We simplify by taking out h (constant) outside the integral.

Step 4: The solution. We will discuss the integration rules later.

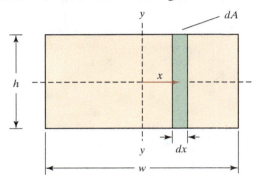

FIGURE 18.25   Differential element used to calculate the second moment of area.

EXAMPLE 18.14

As a civil engineer, you may be assigned the task of determining the force exerted by water that is stored behind a dam. We discussed the concept of hydrostatic pressure in Chapter 10 and stated that, for fluid at rest, the pressure increases with the depth of fluid as shown in Figure 18.26 and according to

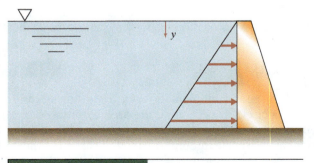

FIGURE 18.26   The variation of pressure with depth.

$$P = \rho g y \qquad \text{18.34}$$

Where

P = fluid pressure at a point located a distance y below the water surface (Pa or lb/ft^2)

ρ = density of the fluid (kg/m^3 or $slugs/ft^3$)

g = acceleration due to gravity ($g = 9.81 \, m/s^2$ or $g = 32.2 \, ft/s^2$)

y = distance of the point below the fluid surface (m or ft)

Since the force due to the water pressure varies with depth, we need to add the pressure exerted on areas at various depths to obtain the net force. Consider the force acting at depth y over a small area dA, as shown in Figure 18.27.

The procedure for computing the total force is demonstrated using the following steps. Note these steps make use of integrals.

$$\text{Net Force} = \overbrace{\int_0^H dF}^{\text{step 1}} = \overbrace{\int_0^H p\,dA}^{\text{step 2}} = \overbrace{\int_0^H \rho gy\,dA}^{\text{step 3}} = \overbrace{\rho g \int_0^H y\,dA}^{\text{step 4}} = \overbrace{\rho gw \int_0^H y\,dy}^{\text{step 5}} = \overbrace{\frac{1}{2}\rho gw\,H^2}^{\text{step 6}}$$

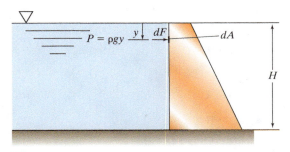

FIGURE 18.27   The forces due to pressure acting on a vertical surface.

Step 1: The net force is equal to the sum (integral) of all the little forces acting at different depths.

Step 2: We substitute for $dF = p\,dA$ (recall, force is equal to pressure times area).

Step 3: We make use of the relationship between the fluid pressure and depth of the fluid, that is, $P = \rho gy$.

Step 4: We simplify by assuming a constant fluid density and constant g.

Step 5: We substitute for $dA = w\,dy$, where w is the width of the dam.

Step 6: The solution. We will discuss the integration rules later.

We could come up with many more examples to emphasize the role of integrals in engineering applications.

During the next few years, as you take your calculus classes, you will learn many new concepts and rules dealing with integral calculus. Make sure you take the time to understand these concepts and rules. Some of these integral concepts and rules are summarized in Table 18.7. Examples that demonstrate how to apply some of these rules follow. As you study the examples, keep in mind again that our intent is to familiarize you with these rules, not to provide a detailed coverage.

EXAMPLE 18.15   Evaluate $\int (3x^2 - 20x)\,dx$.

We use rules 2 and 6 from Table 18.7 to solve this problem, as shown.

$$\int (3x^2 - 20x)\,dx = \int 3x^2 dx + \int -20x\,dx = 3\int x^2 dx - 20 \int x\,dx$$

$$= 3\left[\frac{1}{2+1}x^3\right] - 20\left[\frac{1}{1+1}x^2\right] + C$$

$$= x^3 - 10x^2 + C$$

TABLE 18.7	Summary of Basic Integral Rules	
	Definitions and Rules	**Explanation**
1	$\int a\,dx = ax + C$	The integral of a constant a.
2	$\int x^n\,dx = \dfrac{1}{n+1}x^{n+1} + C$	True for $n \neq -1$ (see Example 18.15).
3	$\int \dfrac{a}{x}\,dx = a\,\ln\lvert x\rvert + C$	True for $x \neq 0$ (see Example 18.18).
4	$\int e^{ax}\,dx = \dfrac{1}{a}e^{ax} + C$	The rule for exponential function.
5	$\int a \cdot f(x)\,dx = a\int f(x)\,dx$	When $a = $ constant (see Example 18.16).
6	$\int [f(x) \pm g(x)]\,dx = \int f(x)\,dx \pm \int g(x)\,dx$	See Example 18.17.
7	$\int [u(x)]^n\,u'(x)\,dx = \dfrac{[u(x)]^{n+1}}{n+1} + C$	The substitution method.
8	$\int e^{u(x)}u'(x)\,dx = e^{u(x)} + C$	The substitution method (see Example 18.20).
9	$\int \dfrac{u'(x)}{u(x)}\,dx = \ln\lvert u(x)\rvert + C$	The substitution method.

EXAMPLE 18.16

Evaluate $\int 5(3x^2 - 20x)\,dx$.

We use rule 5, $\int a \cdot f(x)\,dx = a\int f(x)\,dx$, from Table 18.7 to solve this problem. For the given problem, $a = 5$ and $f(x) = 3x^2 - 20x$, then using the results of Example 18.15, we get $\int 5(3x^2 - 20x)\,dx = 5(x^3 - 10x^2 + C)$.

EXAMPLE 18.17

Evaluate $\int [(3x^2 - 20x) \pm (5x^4 - 5)]\,dx$.

We use rule 6, $\int [f(x) \pm g(x)]\,dx = \int f(x)\,dx \pm \int g(x)\,dx$, from Table 18.7 to solve this problem, as shown.

$$\int [(3x^2 - 20x) \pm (5x^4 - 5)]\,dx = \int (3x^2 - 20x)\,dx \pm \int (5x^4 - 5)\,dx$$
$$= (x^3 - 10x^2 + C_1) \pm (x^5 - 5x + C_2)$$

EXAMPLE 18.18

Evaluate $\int \dfrac{10}{x}\,dx$.

We use rule 3, $\int a/x\,dx = a\ln|x| + C$, from Table 18.7 to solve this problem, as shown.

$$\int \frac{10}{x}\,dx = 10\ln|x| + C$$

EXAMPLE 18.19

Evaluate $\int [(x-1)(x^2 - 2x)]\,dx$.

We use the substitution method (rule 7) from Table 18.7 to solve this problem, as shown. For this problem, $u = x^2 - 2x$ and $du/dx = 2x - 2 = 2(x-1)$, and rearrange the terms as $du = 2(x-1)\,dx$ or $du/2 = (x-1)\,dx$. Making these substitutions, we get

$$\int [(x-1)(x^2 - 2x)]\,dx = \int u\,\frac{du}{2} = \frac{1}{2}\int u\,du = \frac{1}{2}\left(\frac{u^2}{2} + C\right) = \frac{1}{2}\left[\frac{(x^2 - 2x)^2}{2}\right] + C$$

EXAMPLE 18.20

Evaluate $\int \left[(x-1)e^{(x^2 - 2x)}\right]\,dx$.

We use the substitution method (rule 8), $\int e^{u(x)}u'(x)\,dx = e^{u(x)} + C$, from Table 18.7 to solve this problem, as shown. From the previous example, $u = x^2 - 2x$ and $du/2 = (x-1)\,dx$. Making these substitutions, we get

$$\int \left[(x-1)e^{(x^2 - 2x)}\right]\,dx = \frac{1}{2}\int e^u\,du = \frac{1}{2}(e^u) + C = \frac{1}{2}(e^{(x^2 - 2x)}) + C$$

Before You Go On

Answer the following questions to test your understanding of the preceding section.

1. Why is important to know differential calculus?

2. Why is important to know integral calculus?

3. What do we mean by rate of change and give an example?

4. What does the integral sign represent?

Vocabulary—State the meaning of the following terms:

Dependent Variable _____

Independent Variable _____

LO⁷ 18.7   Differential Equations

Many engineering problems are modeled using differential equations with a set of corresponding boundary and/or initial conditions. As the name implies, **differential equations** contain derivatives of functions or differential terms. Moreover, the differential equations are derived by applying the fundamental laws and principles of nature (some of which we described earlier) to a very small volume or a mass. These differential equations represent the balance of mass, force, energy, and so on. **Boundary conditions** provide information about what is happening physically at the boundaries of a problem. **Initial conditions** tell us about the initial conditions of a system (at time $t = 0$), before a disturbance or a change is introduced. When possible, the exact solution of these equations renders detailed behavior of the system under the given set of conditions. Examples of governing equations, boundary conditions, initial conditions, and solutions are shown in Table 18.8.

TABLE 18.8   **Examples of Governing Differential Equations, Boundary Conditions, Initial Conditions, and Exact Solutions for Some Engineering Problems**

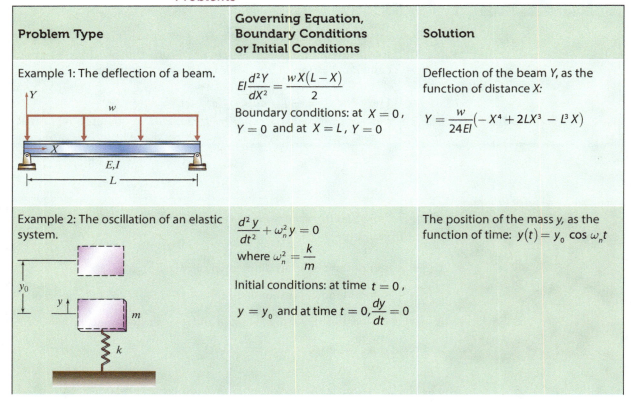

Problem Type	Governing Equation, Boundary Conditions or Initial Conditions	Solution
Example 1: The deflection of a beam.	$EI\dfrac{d^2Y}{dX^2} = \dfrac{wX(L-X)}{2}$ Boundary conditions: at $X = 0$, $Y = 0$ and at $X = L$, $Y = 0$	Deflection of the beam Y, as the function of distance X: $Y = \dfrac{w}{24EI}\left(-X^4 + 2LX^3 - L^3X\right)$
Example 2: The oscillation of an elastic system.	$\dfrac{d^2y}{dt^2} + \omega_n^2 y = 0$ where $\omega_n^2 = \dfrac{k}{m}$ Initial conditions: at time $t = 0$, $y = y_0$ and at time $t = 0, \dfrac{dy}{dt} = 0$	The position of the mass y, as the function of time: $y(t) = y_0 \cos \omega_n t$

Example 3: The temperature distribution along a long fin.

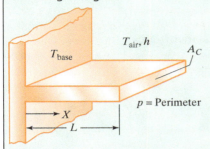

T_{air}, h

A_C

T_{base}

$p = \text{Perimeter}$

X

L

$$\frac{d^2T}{dX^2} - \frac{hp}{kA_c}(T - T_{air}) = 0$$

Boundary conditions:

at $X = 0$, $T = T_{base}$

as $L \to \infty$, $T = T_{air}$

Temperature distribution along the fin as the function of X:

$$T = T_{air} + (T_{base} - T_{air})e^{-\sqrt{\frac{hp}{kA_c}}X}$$

In Example 1, the function Y represents the deflection of the beam at the location denoted by the position variable X. As shown in Table 18.8, the variable X varies from zero to L, the position of the beam. Note X is measured from the left-support point. The load or the force acting on the beam is represented by W. The boundary conditions tell us what is happening at the boundaries of the beam. For Example 1, at supports located at $X = 0$ and $X = L$, the deflection of the beam Y is zero. Later, when you take your differential equation class, you will learn how to obtain the solution for this problem, as given in Table 18.8. The solution shows, for a given load W, how the given beam deflects at any location X. Note that, if you substitute $X = 0$ or $X = L$ in the solution, the value of Y is zero. As expected, the solution satisfies the boundary conditions.

The differential equation for Example 2 is derived by applying Newton's second law to the given mass. Moreover, for this problem, the initial conditions tell us that, at time $t = 0$, we pulled the mass upward by a distance of y_0 and then release it without giving the mass any initial velocity. The solution to Example 2 gives the position of the mass, as denoted by the variable y, with respect to time t. It shows the mass will oscillate according to the given cosinusoidal function.

In Example 3, T represents the temperature of the fin at the location denoted by the position X, which varies from zero to L. Note that X is measured from the base of the fin. The boundary conditions for this problem tell us that the temperature of the fin at its base is T_{base} and the temperature of the tip of the fin will equal the air temperature, provided that the fin is very long. The solution then shows how the temperature of the fin varies along the length of the fin.

Again, please keep in mind that the purpose of this chapter was to focus on important mathematical models and concepts and to point out why mathematics is so important in your engineering education. Detailed coverage will be provided later in your math classes.

Before You Go On

Answer the following questions to test your understanding of the preceding section.

1. What do we mean by a differential equation?

2. In engineering, what does a differential equation represent?

Vocabulary—State the meaning of the following terms:

Governing Differential Equation _____

Boundary Condition _____

Initial Condition _____

SUMMARY

LO¹ Mathematical Symbols and Greek Alphabet

By now, you understand the importance of mathematics in engineering. Mathematics is a language that has its own symbols and terminology, and it is important for you to know what they mean and to use them properly when communicating with others. You should also memorize Greek alphabetic characters because they are used in engineering to express angles, dimensions, and physical variables in drawings and in mathematical equations.

LO² Linear Models

You should understand the importance of linear models in describing engineering problems and their solutions. Linear models are the simplest form of equations used to describe a range of engineering situations. You should also know the defining characteristics of these models and what they represent. For example, the slope of a linear model shows by how much the dependent variable y changes each time a change in the independent variable x is introduced. Moreover, for a linear model, the value of the slope is always constant.

LO³ Nonlinear Models

You should recognize nonlinear equations, their characteristics, and how they are used to describe engineering problems. You should know that, unlike linear models, nonlinear models have variable slopes, meaning each time you introduce a change in the value of the independent variable x, the corresponding change in the dependent variable y will depend on where in the x range the change is introduced.

LO⁴ Exponential and Logarithmic Models

You should be able to identify exponential and logarithmic functions, their important characteristics, and how they are used to model engineering problems. The simplest form of an exponential function is given by $f(x) = e^x$, where e is an irrational number with an approximate value of $e = 2.718281$. Moreover, when modeling an engineering problem using an exponential function, the rate of change of a dependent variable is much higher at the beginning and much lower at the end (it levels off toward the end). You should also know that the logarithmic functions are defined for the ease of computations. For example, if we let $10^x = y$, we define $\log y = x$: then, using the definition of the common logarithm, $\log 1 = 0$ or $\log 100 = 2$. On the other hand, if we let $e^x = y$, we define $ln\, y = x$, and the symbol ln reads the logarithm to the base-e, or natural logarithm. Moreover, the relationship between the natural logarithm and the common logarithm is given by $ln\, x = (ln\, 10)(\log x) = 2.302585 \log x$.

LO⁵ Matrix Algebra

You should know the rules for adding and subtracting matrices and for multiplying a matrix by a scalar quantity or by another matrix. The formulation and solution of engineering problems lend themselves to situations wherein it is desirable to rearrange the rows of a matrix into the columns of another matrix, which leads to the idea to transpose a matrix. You should also realize that the formulation of many engineering problems leads to a set of linear algebraic equations that are solved simultaneously. Therefore, a good understanding of matrix algebra is essential.

LO⁶   Calculus

You should know that calculus is divided into two broad areas: differential and integral calculus. You should also know that differential calculus

deals with understanding the rate of change—how a variable may change with respect to another variable—and that integral calculus is related to the summation or addition of things.

LO⁷   Differential Equations

You should understand that differential equations contain derivatives of functions and represent the balance of mass, force, energy, and so on; and boundary conditions provide information about what is happening physically at the boundaries of a problem. Moreover, you should know that initial conditions provide information about a system before a disturbance or a change is introduced.

KEY TERMS

Boundary Condition 720
Calculus 711
Dependent Variable 682
Determinant 706
Differential Equation 720
Exponential Function 696
Greek Alphabet 681
Independent Variable 683

Initial Condition 720
Inverse of a Matrix 710
Linear Interpolation 686
Linear Model 682
Logarithmic Function 696
Math Symbol 680
Matrix 700
Matrix Addition 701

Matrix Element 700
Matrix Multiplication 702
Matrix Subtraction 701
Nonlinear Model 689
Slope 683
Transpose 705

APPLY WHAT YOU HAVE LEARNED

Waste can be classified into two broad categories of municipal and industrial waste. Municipal waste is basically the trash that we throw away every day. It consists of items such as food scraps, packaging materials, bottles, cans, and so on. On the other hand, as the name implies, industrial waste refers to waste that is produced in industry. This type of waste includes construction, renovation, and demolition materials; medical waste; and waste generated during exploration, development, and production of fossil fuels and rocks and minerals. Visit the EPA website and collect data on total annual municipal solid waste (MSW) generation in the United States from 1960 to recent year. Plot the total municipal solid waste generation data as a function of year. What is the form of this function, linear or nonlinear? Can you identify any characteristics for this graph? Estimate the rate of change of MSW for each decade (1960–1970, 1970–1980, 1980–1990, 1990–2000, and 2000–2010). Write a brief report explaining your findings.

Horiyan/Shutterstock.com
antpkr/Shutterstock.com
koya979/Shutterstock.com
MichaelJayBerlin/Shutterstock.com
mylisa/Shutterstock.com
nito/Shutterstock.com
grynold/Shutterstock.com

PROBLEMS

Problems that promote lifelong learning are denoted by 🔑

18.1 The force–deflection relationships for three springs are shown in the accompanying figure. What is the stiffness (spring constant) of each spring? Which one of the springs is the stiffest?

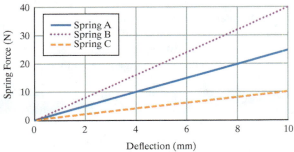

Problem 18.1

18.2 In the accompanying diagram, spring A is a linear spring and spring B is a hard spring, with characteristics that are described by the relationship $F = kx^n$. Determine the stiffness coefficient k for each spring. What is the exponent n for the hard spring? In your own words, also explain the relationship between the spring force and the deflection for the hard spring and how it differs from the behavior of the linear spring.

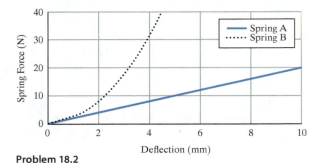

Problem 18.2

18.3 The equations describing the position of a water stream (with respect to time) coming out of the hose, shown in the accompanying figure, are given by

$$x = x_0 + (v_x)_0 t$$
$$y = y_0 + (v_y)_0 t - \frac{1}{2}gt^2$$

In these relationships, x and y are position coordinates, x_0 and y_0 are initial coordinates of the tip of the hose, $(v_x)_0$ and $(v_y)_0$ are the initial velocities of water coming out of the hose in the x and y directions, $g = 9.81$ m/s², and t is time.

Plot the x and the y position of the water stream as a function of time. Also, plot the path the water stream will follow as a function of time. Compute and plot the components of velocity of the water stream as a function of time.

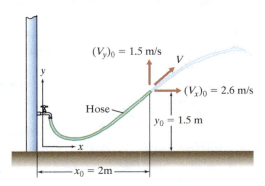

Problem 18.3

18.4 In Chapter 12, we explained that the electric power consumption of various electrical components can be determined using the following power formula: $P = VI = RI^2$ where P is power in watts, V is the voltage, I is the current in amps, and R is the resistance of the component in ohms.

Plot the power consumption of an electrical component with a resistance of 145 ohms. Vary the value of the current from zero to 4 amps. Discuss and plot the change in power consumption as the function of current drawn through the component.

18.5 The deflection of a cantilevered beam supporting the weight of an advertising sign is given by

$$y = \frac{-Wx^2}{6EI}(3L - x)$$

where

y = deflection at a given x location (m)
W = weight of the sign (N)
E = modulus of elasticity (N/m^2)
I = second moment of area (m^4)
x = distance from the support as shown (m)
L = length of the beam (m)

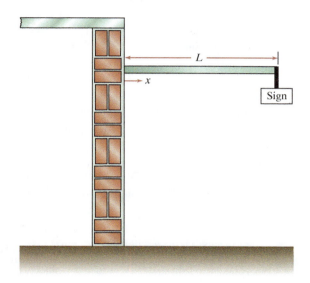

Problem 18.5

Plot the deflection of a beam with a length of 3 m, the modulus of elasticity of $E = 200$ GPa, and $I = 1.2 \times 10^6$ mm^4 and for a sign weighing 1500 N. What is the slope of the deflection of the beam at the wall $(x = 0)$ and at the end of the beam where it supports the sign $(x = L)$.

18.6 As we explained in earlier chapters, the drag force acting on a car is determined experimentally by placing the car in a wind tunnel. The drag force acting on the car is determined from

$$F_d = \frac{1}{2} C_d \rho V^2 A$$

where

F_d = measured drag force $(N \text{ or } lb)$
C_d = drag coefficient (unitless)
ρ = air density $(kg/m^3 \text{ or } slugs/ft^3)$
V = air speed inside the wind tunnel $(m/s \text{ or } ft/s)$
A = frontal area of the car $(m^2 \text{ or } ft^2)$

The power requirement to overcome the air resistance is computed by

$$P = F_d V$$

Plot the power requirement (in hp) to overcome the air resistance for a car with a frontal area of 2800 in^2, a drag coefficient of 0.4, and for an air density of 0.00238 slugs/ft^3. Vary the speed from zero to 110 ft/s (75 mph). Also, plot the rate of change of power requirement as a function of speed.

18.7 The cooling rate for three different materials is shown in the accompanying figure. The mathematical equation describing the cooling rate for each material is of the exponential form $T(t) = T_{initial} e^{-at}$. In this relationship, $T(t)$ is the temperature of material at time t and the coefficient a represents the thermal capacity and resistance of the material. Determine the initial temperature and the a coefficient for each material. Which material cools the fastest, and what is the corresponding a value?

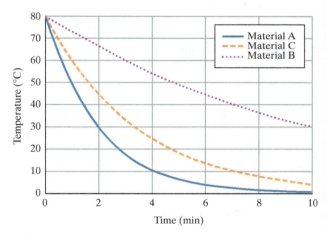

Problem 18.7

18.8 As explained in earlier chapters, fins, or extended surfaces, commonly are used in a variety of engineering applications to enhance cooling. Common examples include a motorcycle engine head, a lawn mower engine head, heat sinks used in electronic equipment, and finned tube heat exchangers in room heating and cooling applications. For long fins, the temperature distribution along the fin is given by:

$$T - T_{ambient} = (T_{base} - T_{ambient})e^{-mx}$$

where

$$m = \sqrt{\frac{hp}{kA}}$$

h = heat transfer coefficient $(W/m^2 \cdot K)$

p = perimeter of the fin $2(a + b)(m)$

A = cross-sectional area of the fin $(a \cdot b)(m^2)$

k = thermal conductivity of the fin material $(W/m \cdot K)$

Problem 18.8

What are the dependent and independent variables?

Next, consider aluminum fins of a rectangular profile shown in the accompanying figure, which are used to remove heat from a surface whose temperature is 100° C. The temperature of the ambient air is 20° C.

Plot the temperature distribution along the fin using the following data: $k = 180$ W/m · K, $h = 15$ W/m² · K, $a = 0.05$ m, and $b = 0.015$ m. Vary x from zero to 0.015 m. What is the temperature of the tip of the fin? Plot the temperature of the tip as a function of k. Vary the k value from 180 to 350 W/m · K.

18.9 Use the graphical method discussed in this chapter to obtain the solution to the following set of linear of equations.

$$x + 3y = 14$$
$$4x + y = 1$$

18.10 Use the graphical method discussed in this chapter to obtain the solution to the following set of linear of equations.

$$-2x_1 + 3x_2 = 5$$
$$x_1 + x_2 = 10$$

18.11 Without using your calculator, answer the following question. If 6^4 is approximately

equal to 1300, then what is the approximate value of 6^8 ?

18.12 Without using your calculator, answer the following questions. If log 8 = 0.9 , then what are the values of log 64, log 80, log 8000, and log 6400?

18.13 Plot the functions, $y = x$, $y = 10^x$, and $y = \log x$. Vary the x value from 1 to 3. Is the function $y = \log x$ a mirror image of $y = 10^x$ with respect to $y = x$, and if so, why? Explain.

18.14 Using a sound meter, the following measurements were made for the following sources: rock band at a concert $(100 \times 10^6 \ \mu Pa)$, a jackhammer $(2 \times 10^6 \ \mu Pa)$, and a whisper $(2000 \ \mu Pa)$. Convert these readings to dB values.

18.15 A jet plane taking off creates a noise with a magnitude of approximately 125 dB. What is the magnitude of the pressure disturbance (in μPa)?

18.16 Identify the size and the type of the given matrices. Denote whether the matrix is a square, a column, a diagonal, a row, or a unit (identity).

a. $\begin{bmatrix} 3 & 2 & 0 \\ 2 & 4 & 5 \\ 0 & 5 & 6 \end{bmatrix}$ b. $\begin{Bmatrix} x \\ x^2 \\ x^3 \\ x^4 \end{Bmatrix}$

c. $\begin{bmatrix} 4 & 0 \\ 0 & 8 \end{bmatrix}$ d. $\begin{bmatrix} 1 & y & y^2 & y^3 \end{bmatrix}$

e. $\begin{bmatrix} 1 & 0 & 0 \\ 0 & 1 & 0 \\ 0 & 0 & 1 \end{bmatrix}$

18.17 Given matrices: $[A] = \begin{bmatrix} 4 & 2 & 1 \\ 7 & 0 & -7 \\ 1 & -5 & 3 \end{bmatrix}$,

$[B] = \begin{bmatrix} 1 & 2 & -1 \\ 5 & 3 & 3 \\ 4 & 5 & -7 \end{bmatrix}$, and $\{C\} = \begin{Bmatrix} 1 \\ -2 \\ 4 \end{Bmatrix}$,

perform the following operations.

a. $[A] + [B] = ?$

b. $[A] - [B] = ?$

c. $3[A] = ?$

d. $[A][B] = ?$

e. $[A]\{C\} = ?$

f. $[A]^2 = ?$

g. Show that $[I][A] = [A][I] = [A]$

18.18 Given the following matrices:

$$[A] = \begin{bmatrix} 2 & 10 & 0 \\ 16 & 6 & 14 \\ 12 & -4 & 18 \end{bmatrix} \text{ and } [B] = \begin{bmatrix} 2 & 10 & 0 \\ 4 & 20 & 0 \\ 12 & -4 & 18 \end{bmatrix},$$

calculate the determinant of $[A]$ and $[B]$ by direct expansion. Which matrix is singular?

18.19 Solve the following set of equations using the Gaussian method.

$$x + 3y = 14$$
$$4x + y = 1$$

18.20 Solve the following set of equations using the Gaussian method.

$$-2x_1 + 3x_2 = 5$$
$$x_1 + x_2 = 10$$

18.21 Solve the following set of equations using the Gaussian method.

$$\begin{bmatrix} 1 & 1 & 1 \\ 2 & 5 & 1 \\ -3 & 1 & 5 \end{bmatrix} \begin{Bmatrix} x_1 \\ x_2 \\ x_3 \end{Bmatrix} = \begin{Bmatrix} 6 \\ 15 \\ 14 \end{Bmatrix}$$

18.22 As we explained in Chapter 13, an object having a mass m and moving with a speed V has a kinetic energy, which is equal to

$$Kinetic\ Energy = \frac{1}{2}mV^2.$$

Plot the kinetic energy of a car with a mass of 1500 kg as the function of its speed. Vary the speed from zero to 35 m/s (126 km/h). Determine the rate of change of kinetic energy of the car as function of speed and plot it. What does this rate of change represent?

18.23 In Chapter 13, we explained that when a spring is stretched or compressed from its unstretched position, elastic energy is stored in the spring and that energy will be released when the spring is allowed to return to its unstretched position. The elastic energy stored in a spring when stretched or compressed is determined from

$$Elastic\ Energy = \int_0^x F dx$$

Obtain expressions for the elastic energy of a linear spring described by $F = kx$ and a hard spring whose behavior is described by $F = kx^2$.

18.24 For Example 1 in Table 18.8, verify that the given solution satisfies the governing differential equation and the boundary conditions.

18.25 For Example 3 in Table 18.8, verify that the given solution satisfies the governing differential equation and the initial conditions.

18.26 We presented Newton's Law of Gravitation in Chapter 10. We also explained the acceleration due to gravity. Create a graph that shows the acceleration due to gravity as a function of distance from the earth's surface. Change the distance from sea level to an altitude of 5000 m.

18.27 For Problem 18.26, plot the weight of a person with a mass of 80 kg as a function of distance from the earth's surface.

18.28 An engineer is considering storing some radioactive material in a container she is creating. As a part of her design, she needs to evaluate the ratio of volume to surface area of two storage containers. Create curves that show the ratio of volume to surface area of a sphere and a square container. Create another graph that shows the difference in the ratios. Vary the radius or the side dimension of a square container from 50 cm to 4 m.

18.29 As we mentioned in Chapter 10, engineers used to use pendulums to measure the value of g at a location. The formula used to measure the acceleration due to gravity is

$$g = \frac{4\pi^2 L}{T^2}$$

where g is acceleration due to gravity (m/s^2), L is the length of pendulum, and T is the period of oscillation of the pendulum (the time that it takes the pendulum to complete one cycle). For a pendulum of 2 m long, create a graph that could be used for locations between an altitude of 0 and 2000 m, and shows g as a function of T.

18.30 The mass moment of inertia I of a disk is given by

$$I = \frac{1}{2}mr^2$$

where m is the mass of the disk and r is the radius. Create a graph that shows I as a function of r for a steel disk with a density of 7800 kg/m^3. Vary the r value from 10 cm to 25 cm. Assume a thickness of 1 cm.

18.31 Use the linear interpolation method discussed in Section 18.2 to estimate the density of air at an altitude 4150 m.

18.32 For the cooling of steel plates discussed in Section 18.4 (Figure 18.17) using linear interpolation, estimate the temperature of the plate at time equal to 1 hr, from the temperature data at 0.8 hr and 1.2 hour. Compare the estimated temperature value to the actual value of 308° C. What is the percentage of error?

18.33 For the stopping sight distance problem of Figure 18.12, estimate the stopping distance for speed of 27 mph, using the 25 mph and 30 mph data. Compare the estimated stopping distance value to the actual value from Equation (18.7). What is the percentage of error?

18.34 The variation of air density at the standard pressure as a function of temperature is given in the accompanying table. Use linear interpolation to estimate the air density at 27° C and 33° C.

Temperature (°C)	Air Density (kg/m³)
0	1.292
5	1.269
10	1.247
15	1.225
20	1.204
25	1.184
30	1.164
35	1.146

18.35 The air temperature and speed of sound for the U.S. standard atmosphere is given in the accompanying table. Using linear interpolation, estimate the air temperatures and the corresponding speeds of sound at altitudes of 1700 m and 11,000 m.

Altitude (m)	Air Temperature (K)	Speed of Sound (m/s)
500	284.9	338
1000	281.7	336
2000	275.2	332
5000	255.7	320
10,000	223.3	299
15,000	216.7	295
20,000	216.7	295

For Problems 18.36 through 18.42 use the data from the accompanying table shown below.

Electricity Generation by Fuel, 1980-2030 (billion kilowatt-hours)—Data from U.S. Department of Energy

Year	Coal	Petroleum	Natural Gas	Nuclear	Renewable/Other	
1980	1161.562	245.9942	346.2399	251.1156	284.6883	actual values
1990	1594.011	126.6211	372.7652	576.8617	357.2381	actual values
2000	1966.265	111.221	601.0382	753.8929	356.4786	actual values
2005	2040.913	115.4264	751.8189	774.0726	375.8663	actual values
2010	2217.555	104.8182	773.8234	808.6948	475.7432	projected values
2020	2504.786	106.6799	1102.762	870.698	515.1523	projected values
2030	3380.674	114.6741	992.7706	870.5909	559.1335	projected values

Data from U.S. Department of Energy

18.36 Estimate the amount of electricity that is projected to be generated from coal in 2017.

18.37 Estimate the amount of electricity that is projected to be generated from petroleum in 2018.

18.38 Estimate the amount of electricity that is projected to be generated from natural gas in 2024.

18.39 Estimate the amount of electricity that is projected to be generated from nuclear fuel in 2022.

18.40 Estimate the amount of electricity that is projected to be generated from renewable and other sources in 2017.

18.41 Using linear interpolation, estimate the percentage change in the amount of electricity that was generated using coal in 2007 compared to 1987.

18.42 Using linear interpolation, estimate the percentage change in the total amount of electricity that was generated in 2007 compared to 1987.

18.43 Investigate what is meant by numerical analysis. Write a brief report explaining your findings, and give examples.

18.44 Investigate Taylor series expansion. Explain how the Taylor series is used in your calculator to compute values of functions such as $\sin(x)$, $\cos(x)$, and e^x, where x represent any value. Write a brief report explaining your findings, and give examples.

18.45 Investigate the Fourier series. Explain how the Fourier series is used in engineering. Write a brief report explaining your findings, and give examples.

Probability and Statistics in Engineering

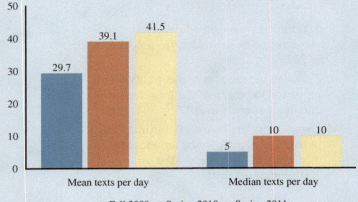

Number of texts sent/received per day, 2009–2011

Based on adults who use text messaging on their cell phones

■ Fall 2009 ■ Spring 2010 ■ Spring 2011

Source: The Pew Research Center's Internet & Americal Life Project, April 26 – May 22, 2011 Spring Tracking Survey. n=2,277 adult internet users ages 18 and older, including 755 cell phone interviews. Interviews were conducted in English and Spanish. *May 2010 data is for English-speaking Hispanics only.

Every day we use probability and statistics to predict future events. We use statistics to forecast weather and prepare for related emergencies, predict the resutlt of a political race, or the side effects of a new drug or a new technology. Statistical models also are used by engineers to address quality control and reliability concerns.

LEARNING OBJECTIVES

LO¹ **Probability—Basic Ideas:** explain the basic ideas of probability and give examples

LO² **Statistics—Basic Ideas:** describe the basic ideas of statistics and give examples

LO³ **Frequency Distribution:** know how to organize data in a way that pertinent information and conclusions can be extracted

LO⁴ **Measures of Central Tendency and Variation—Mean, Median, and Standard Deviation:** explain the means by which we can measure the dispersion of a reported data set

LO⁵ **Normal Distribution:** describe what we mean by a probability distribution and the characteristics of a probability distribution that has a bell-shaped curve

DISCUSSION STARTER

WHAT IS DISTRACTED DRIVING?

Distracted driving is any activity that could divert a person's attention away from the primary task of driving. All distractions endanger driver, passenger, and bystander safety. These types of distractions include:

- Texting

- Using a cell phone or smartphone

- Eating and drinking

- Talking to passengers

- Grooming

- Reading, including maps

- Using a navigation system

- Watching a video

- Adjusting a radio, CD player, or MP3 player

But because text messaging requires visual, manual, and cognitive attention from the driver, it is by far the most alarming distraction.

Key Facts and Statistics

- In 2011, 3,331 people were killed in crashes involving a distracted driver, compared to 3,267 in 2010. An additional 387,000 people were injured in motor vehicle crashes involving a distracted driver, compared to 416,000 injured in 2010.

- 10% of injury crashes in 2011 were reported as distraction-affected crashes.

- As of December 2012, 171.3 billion text messages were sent in the United States (includes Puerto Rico, the Territories, and Guam) every month.

- 11% of all drivers under the age of 20 involved in fatal crashes were reported as distracted at the time of the crash. This age group has the largest proportion of drivers who were distracted.

- For drivers 15–19 years old involved in fatal crashes, 21% of the distracted drivers were distracted by the use of cell phones.

- At any given daylight moment across America, approximately 660,000 drivers are using cell phones or manipulating electronic devices while driving; a number that has held steady since 2010.

- Engaging in visual–manual subtasks (such as reaching for a phone, dialing, and texting) associated with the use of hand-held phones and other portable devices increases the risk of getting into a crash by three times.

- Sending or receiving a text takes a driver's eyes from the road for an average of 4.6 seconds, which is the equivalent—at 55 mph—of driving the length of an entire football field blind.

- Headset cell-phone use is not substantially safer than hand-held use.

- A quarter of teens respond to a text message once or more every time they drive. 20 percent of teens and 10% of parents admit that they have extended, multi-message text conversations while driving.

Source: http://www.distraction.gov/

To the Students: Do you text while driving? How many of your friends or classmates do you think text while driving? How many of you in this class have texted while driving? How should we organize this data so that we can extract useful information?

S tatistical models are being used increasingly more often by practicing engineers to address quality control and reliability issues and to perform failure analyses. Civil engineers use statistical models to study the reliability of construction materials and structures and to design for flood control and water supply management. Electrical engineers use statistical models for signal processing or for developing voice-recognition software. Mechanical engineers use statistics to study the failure of materials and machine parts and to design experiments. Manufacturing engineers use statistics for quality control assurance of the products they produce. These are but a few examples of why an understanding of statistical concepts and models is important in engineering. We will begin by explaining some of the basic ideas in probability and statistics. We will then discuss frequency distributions, measure of central tendency (mean and median), measure of variation within a data set (standard deviation), and normal distributions.

LO¹ 19.1   Probability—Basic Ideas

If you were to ask your instructor how many students are enrolled in your engineering class this semester, she could give you an exact number: say 60. On the other hand, if you were to ask her how many students will be in the class next year, or the year after, she would not be able to give you an exact number. She might have an estimate based on trends or other pieces of information, but she cannot know exactly how many students will be enrolled in the class next year. The number of students in the class next year, or the year after, is *random*. There are many situations in engineering that deal with random phenomena. For example, as a civil engineer, you may design a bridge or a highway. It is impossible for you to predict exactly how many cars will use the highway or go over the bridge on a certain day. As a mechanical engineer, you may design a heating, cooling, and ventilating system to maintain the indoor temperature of a building at a comfortable level. Again, it is impossible to predict exactly how much heating will be required on a future day in January. As a computer engineer, you may design a network for which you cannot predict its future usage exactly. For these types of situations, the best we can do is to predict outcomes using **probability** models.

Probability has its own terminology; therefore, it is a good idea to spend a little time to familiarize yourself with it. In probability, each time you repeat an experiment is called a **trial**. The result of an experiment is called an **outcome**. A **random experiment** is one that has random outcomes—random outcomes cannot be predicted exactly. To gain a better understanding of these terms, imagine a manufacturing setting wherein cell phones are being assembled. You are positioned at the end of the assembly line, and in order to perform a final quality check, you are asked to remove cell phones at random from the assembly line and turn them on and off. Each time you remove a cell phone and turn it on and off, you are conducting a random experiment. Each time you pick up a phone is a *trial*, with a result that can be marked as a good phone or a bad phone. The result of each experiment is called an **outcome**. Now, suppose in one day you check 200 phones, and out of these phones, you find five bad phones. Then, the *relative frequency* of finding bad phones is given by $5/200 = 0.025$. In general, if you were to repeat an experiment n times under the same conditions, with a certain outcome occurring m times, the relative frequency of the outcome is given by m/n. As n gets larger, then the probability p of a specific outcome is given by $p = m/n$.

> Probability is an area of science that deals with predicting (estimating) the likelihood of an event to occur.

EXAMPLE 19.1

Each question on a multiple-choice exam has five answers listed. Knowing that only one of the answers is correct, if you are unprepared for the exam, what is the probability that you pick the correct answer?

$$p = \frac{1}{5} = 0.2$$

For those of you who follow sports, you may have noticed that sometimes the probability of a certain outcome is expressed in terms of odds. For example, the odds in favor of your team winning may be given as 1 to 2. What does "odds in favor of an event" mean? The odds in favor of an event occurring is defined by probability (occuring)/probability (not occuring). Therefore, if the probability of your team winning is given by 0.33, then the odds in favor of your team winning is given by 0.33/0.66 = 1/2 or 1 to 2. On the other hand, if the odds are expressed as x to y, then the probability of a specific outcome is calculated from $x/(x + y)$. For this example, as expected, $p = 1/(1 + 2) = 0.33$.

As you take advanced classes in engineering, you will learn more about the mathematical models that provide probabilities of certain outcomes. Our intent here is to make you aware of the importance of probability and statistics in engineering, not to provide detailed coverage of these topics.

LO² 19.2 Statistics—Basic Ideas

Statistics is that area of science that deals with collection, organization, analysis, and interpretation of data. Statistics also deals with methods and techniques that can be used to draw conclusions about the characteristics of something with a large number of data points—commonly called a **population**—using a smaller portion of the entire data. For example, using statistics, we can predict the outcome of an election in a state, say with two million registered voters, by gathering information only from 1000 people about how they are planning to vote. As this example demonstrates, it is neither feasible nor practical to contact two million people to find out how they are planning to vote. However, the sample selected from a population must represent the characteristics of the population. It is important to note that, in statistics, population does not refer necessarily to people but to all of the data that pertain to a situation or a problem. For example, if a company is producing 15,000 screws a day and they want to examine the quality of the manufactured screws, they may select only 500 screws randomly for a quality test. In this example, 15,000 screws is the population, and the 500 selected screws represents the sample.

Statistical models are becoming common tools in the hands of practicing engineers to address quality control and reliability issues and to perform failure analyses. At this stage of your education, it is important to realize that, in order to use statistical models, you need first to completely understand the underlying concepts. The next sections are devoted to some of these important concepts.

LO³ 19.3   Frequency Distributions

As we have said repeatedly throughout the text, engineers are problem solvers. They apply physical laws, chemical laws, and mathematics to design, develop, test, and supervise the manufacture of millions of products and services. Engineers perform tests to learn how things behave or how well they are made. As they perform experiments, they collect data that can be used to explain certain things better and to reveal information about the quality of products and services they provide. In the previous section, we defined what we mean by population and samples. In general, any statistical analysis starts with identifying the population and the sample. Once we have defined a sample that represents the population and have collected information about the sample, then we need to organize the data in a certain way such that pertinent information and conclusions can be extracted. To shed light on this process, consider the following example.

> Statistics is an area of science that deals with collection, organization, analysis, and interpretation of data.

EXAMPLE 19.2   The scores of a test for an introductory chemistry class of 26 students are shown here. Certainly, the scores of your class would be better than these! We are interested in drawing some conclusions about how good this class is. The scores of a test for Example 19.2:

Scores: 58, 95, 80, 75, 68, 97, 60, 85, 75, 88, 90, 78, 62, 83, 73, 70, 70, 85, 65, 75, 53, 62, 56, 72, 79, 87

As you can see from the way the data (scores) are represented, we cannot easily draw a conclusion about how good this chemistry class is. One simple way of organizing the data better would be to identify the lowest and the highest scores, and then group the data into equal intervals or ranges: say a range of size 10, as shown in Table 19.1. When data is organized in the manner shown in Table 19.1, it is commonly referred to as a **grouped frequency distribution**.

TABLE 19.1   Grouped Frequency Distribution for Example 19.2

Scores	Range	Frequency
58, 53, 56	50–59	3
68, 60, 62, 65, 62	60–69	5
75, 75, 78, 73, 70, 70, 75, 72, 79	70–79	9
80, 85, 88, 83, 85, 87	80–89	6
95, 97, 90	90–99	3

The way the scores are now organized in Table 19.1 reveals some useful information. For example, three students did poorly and three performed admirably. Moreover, nine students received scores that were in the range of 70–79, which is considered an average performance. These average scores also constitute the largest frequency in the given data set. Another useful piece of information, which is clear from examining Table 19.1, is that the frequency (the number of scores in a given range) increases from 3 to 5 to 9 and then decreases from 6 to 3. Another way of showing the range of scores and their frequency is by using a *bar graph* (what is commonly called a **histogram**). The height of the bars shows the frequency of the data within the given ranges. The histogram for Example 19.2 is shown in Figure 19.1.

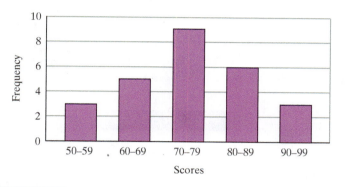

FIGURE 19.1   The histogram for the scores given in Table 19.1.

Cumulative Frequency

The data can be organized further by calculating the **cumulative frequency**. The cumulative frequency shows the cumulative number of students with scores up to and including those in the given range. We have calculated the cumulative frequency for Example 19.2 and shown it in Table 19.2. For Example 19.2, eight scores fall in the range of 50 to 69, and 17 students' scores (the majority of the class) show an average or below-average performance.

The cumulative frequency distribution can also be displayed using a histogram or a *cumulative frequency polygon,* as shown in Figures 19.2 and 19.3, respectively. These figures convey the same information as contained in Table 19.2. However, it might be easier for some people to absorb the information when it is presented graphically. Engineers use graphical communication when it is the clearer, easier, and more convenient way to convey information.

> A histogram is a way to show the range of data and their frequency. The height of the bars shows the frequency of the data within the given ranges.

TABLE 19.2   Cumulative Frequency Distribution for Example 19.2

Range	Frequency	Cumulative Frequency	
50–59	3	3	3
60–69	5	$3 + 5 = 8$	8
70–79	9	$3 + 5 + 9 = 17$ or $8 + 9 = 17$	17
80–89	6	$3 + 5 + 9 + 6 = 23$ or $17 + 6 = 23$	23
90–99	3	$3 + 5 + 9 + 6 + 3 = 26$ or $23 + 3 = 26$	26

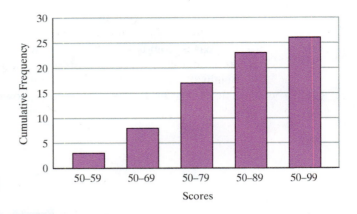

The cumulative-frequency histogram for
Example 19.2.

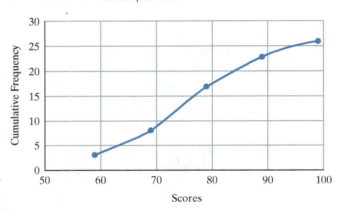

The cumulative-frequency polygon for Example 19.2.

Before You Go On

Answer the following questions to test your understanding of the preceding section.

1. Describe the basic ideas of probability.

2. What does statistics entail?

3. Describe at least two ways to organize data such that useful information could be obtained.

Vocabulary—State the meaning of the following terms:

Outcome _____

A Population _____

Frequency Distribution _____

Histogram _____

LO⁴ 19.4   Measures of Central Tendency and Variation—Mean, Median, and Standard Deviation

In this section, we will discuss some simple ways to examine the central tendency and variations within a given data set. Every engineer should have some understanding of the basic fundamentals of statistics and probability for analyzing experimental data and experimental errors. There are always inaccuracies associated with all experimental observations. If several variables are measured to compute a final result, then we need to know how the inaccuracies associated with these intermediate measurements will influence the accuracy of the final result. There are basically two types of observation errors: systematic errors and random errors. Suppose you were to measure the boiling temperature of pure water at sea level and standard pressure with a thermometer that reads 104° C. But you know from your physics background that the temperature of boiling water at standard conditions is 100° C. And if readings from this thermometer are used in an experiment, it will result in systematic errors. Therefore, *systematic errors*, sometimes called *fixed errors*, are errors associated with using an inaccurate instrument. These errors can be detected and avoided by properly calibrating instruments. On the other hand, *random errors* are generated by a number of unpredictable variations in a given measurement situation. Mechanical vibrations of instruments or variations in line voltage friction or humidity could lead to fluctuations in experimental observations. These are examples of random errors.

Suppose two groups of students in an engineering class measured the density of water at 20° C. Each group consisted of ten students. They reported the results shown in Table 19.3. We would like to know if any of the reported data is in error.

TABLE 19.3	Reported Densities of Water at 20° C
Group A Findings	**Group B Findings**
ρ (kg/m³)	ρ (kg/m³)
1020	950
1015	940
990	890
1060	1080
1030	1120
950	900
975	1040
1020	1150
980	910
960	1020
$\rho_{avg} = 1000$	$\rho_{avg} = 1000$

Let us first consider the **mean** (arithmetic average) for each group's findings. The mean of densities reported by each group is 1000 kg/m³. The mean alone cannot tell us whether any student or which student(s) in each group may have made a mistake. What we need is a way of defining the dispersion of the reported data. There are a number of ways to do this. Let us compute how much each reported density deviates from the mean, add up all the deviations, and then take their average. Table 19.4 shows the deviation from the mean for each reported density. As one can see, the sum of the deviations is zero for both groups. This is not a coincidence. In fact, the sum of deviations from the mean for any given sample is always zero. This can be readily verified by considering the following:

$$\bar{x} = \frac{x_1 + x_2 + x_3 + \cdots + x_{n-1} + x_n}{n} = \frac{1}{n}\sum_{i=1}^{n} x_i \qquad \boxed{19.1}$$

$$d_i = \left(x_i - \bar{x}\right) \qquad \boxed{19.2}$$

where x_i represents data points, $\bar{x}$ is the average, n is the number of data points, and d_i represents the deviation from average.

$$\sum_{i=1}^{n} d_i = \sum_{i=1}^{n}\left(x_i - \bar{x}\right) = \sum_{i=1}^{n} x_i - \sum_{i=1}^{n} \bar{x} \qquad \boxed{19.3}$$

TABLE 19.4   Deviations from the Mean

Group A			Group B		
ρ	$\left(\rho - \rho_{avg}\right)$	$\left\|\left(\rho - \rho_{avg}\right)\right\|$	ρ	$\left(\rho - \rho_{avg}\right)$	$\left\|\left(\rho - \rho_{avg}\right)\right\|$
1020	+ 20	20	950	−50	50
1015	+ 15	15	940	−60	60
990	−10	10	890	−110	110
1060	+60	60	1080	+ 80	80
1030	+30	30	1120	+ 120	120
950	−50	50	900	−100	100
975	−25	25	1040	+ 40	40
1020	+20	20	1150	+ 150	150
980	−20	20	910	−90	90
960	−40	40	1020	+ 20	20
	$\Sigma = 0$	$\Sigma = 290$		$\Sigma = 0$	$\Sigma = 820$

$$\sum_{i=1}^{n} d_i = n\bar{x} - n\bar{x} = 0 \qquad \boxed{19.4}$$

Therefore, the average of the deviations from the mean of the data set cannot be used to measure the spread of a given data set. What if one considers the absolute value of each deviation from the mean? We can then calculate the average of the absolute values of deviations. The result of this approach is shown in the third column of Table 19.4. For group A, the mean deviation is 29, whereas for group B the mean deviation is 82. It is clear that the result provided by group B is more scattered than the group A data. Another common way of measuring the dispersion of data is by calculating the **variance**. Instead of taking the absolute values of each deviation, one may simply square the deviations and compute their averages:

$$v = \frac{\sum_{i=1}^{n} (x_i - \bar{x})^2}{n - 1} \qquad \boxed{19.5}$$

Notice, however, for the given example the variance yields units that are $(kg/m^3)^2$. To remedy this problem, we can take the square root of the variance, which results in a number that is called **standard deviation**.

$$s = \sqrt{\frac{\sum_{i=1}^{n} (x_i - \bar{x})^2}{n - 1}} \qquad \boxed{19.6}$$

This may be an appropriate place to say a few words about why we use $n - 1$ rather then n to obtain the standard deviation. This is done to obtain conservative values because (as we have mentioned) generally the number of experimental trials are few and limited. Let us turn our attention to the standard deviations computed for each group of densities in Table 19.5. Group A has a standard deviation (34.56) that is smaller than group B's (95.22). This shows the densities reported by group A are bunched near the mean ($\rho = 1000 \ kg/m^3$), whereas the results reported by group B are more spread out. The standard deviation can also provide information about the frequency of a given data set. For normal distribution (discussed in Section 19.5) of a data set, we will show that approximately 68% of the data will fall in the interval of (mean − s) to (mean + s), about 95% of the data should fall between (mean − 2s) to (mean + 2s), and almost all data points must lie between (mean − 3s) to (mean + 3s).

> **Median** is the value in the middle of a data. It is that value that separates the higher half of the data from its lower half.

In Section 19.3, we discussed grouped frequency distribution. The mean for a grouped distribution is calculated from

$$\bar{x} = \frac{\Sigma (xf)}{n} \qquad \boxed{19.7}$$

where

x = midpoints of a given range

f = frequency of occurrence of data in the range

$n = \Sigma f$ = total number of data points

TABLE 19.5	Standard Deviation Calculation for Each Group	
Group A		**Group B**
$\left(\rho - \rho_{avg}\right)^2$		$\left(\rho - \rho_{avg}\right)^2$
400		2500
225		3600
100		12,100
3600		6400
900		14,400
2500		10,000
625		1600
400		22,500
400		8100
1600		400
$\Sigma = 10{,}750$		$\Sigma = 81{,}600$
$s = 34.56 \ (\text{kg/m}^3)$		$s = 95.22 \ (\text{kg/m}^3)$

The standard deviation for a grouped distribution is calculated from

$$s = \sqrt{\frac{\Sigma(x - \bar{x})^2 f}{n - 1}}$$

19.8

Next, we demonstrate the use of these formulas.

EXAMPLE 19.3

For Example 19.2, using Equations (19.7) and (19.8), calculate the mean and standard deviation of the class scores.

Consult Tables 19.6, 19.7, and 19.8, respectively, while following the solution. To calculate the mean, first we need to evaluate the midpoints of data for each range and then evaluate the Σxf as shown.

TABLE 19.6	Data for Example 19.3
Range	**Frequency**
50–59	3
60–69	5
70–79	9
80–89	6
90–99	3

TABLE 19.7	Evaluating Midpoints of Data an Σxf		
Range	Frequency f	Midpoint x	xf
50–59	3	54.5	163.5
60–69	5	64.5	322.5
70–79	9	74.5	670.5
80–89	6	84.5	507
90–99	3	94.5	283.5
	$n = \Sigma f = 26$		$\Sigma xf = 1947$

TABLE 19.8	Computing Standard Deviation				
Range	Frequency f	Midpoint x	$\bar{x}$	$x - \bar{x}$	$(x - \bar{x})^2 f$
50–59	3	54.5	74.9	−20.4	1248.5
60–69	5	64.5	74.9	−10.4	540.8
70–79	9	74.5	74.9	−0.4	1.44
80–89	6	84.5	74.9	9.6	552.96
90–99	3	94.5	74.9	19.6	1152.5
					$\Sigma(x - \bar{x})^2 f = 3496$

Using Equation (19.7), the mean of the scores is

$$\bar{x} = \frac{\Sigma(xf)}{n} = \frac{1947}{26} = 74.9$$

Similarly, using Equation (19.8), we calculate the standard deviation, as shown in Table 19.8.

$$n = \Sigma f = 26$$
$$n - 1 = 25$$
$$s = \sqrt{\frac{\Sigma(x - \bar{x})^2 f}{n - 1}} = \sqrt{\frac{3496}{25}} = 11.8$$

Normal distribution is discussed next.

LO⁵ 19.5 Normal Distribution

In Section 19.1, we explained what we mean by a statistical experiment and outcome. Recall that the result of an experiment is called an outcome. In an engineering situation, we often perform experiments that could have many outcomes. To organize the outcomes of an experiment, it is customary to make use of probability distributions. A probability distribution shows the probability values for the occurrence of the outcomes of an experiment. To better understand the concept of probability distribution, let's turn our attention to Example 19.2. If we were to consider the chemistry test as an experiment with outcomes represented by student scores, then we can calculate a probability value for each range of scores by dividing each frequency by 26 (the total number of scores). The probability distribution for Example 19.2 is given in Table 19.9. From examining Table 19.9, you should note that the sum of probabilities is 1, which is true for any probability distribution. The plot of the probability distribution for Example 19.2 is shown in Figure 19.4. Moreover, if this was a typical chemistry test with typical students, then we might be able to use the probability distribution for this class to predict how students might do on a similar test next year. Often, it is difficult to define what we mean by a typical class or a typical test. However, if we had a lot more students take this test and incorporate their scores into

TABLE 19.9		Probability Distribution for Example 19.2	
Range	Frequency	Probability	
50–59	3	$\dfrac{3}{26}$	0.115
60–69	5	$\dfrac{5}{26}$	0.192
70–79	9	$\dfrac{9}{26}$	0.346
80–89	6	$\dfrac{6}{26}$	0.231
90–99	3	$\dfrac{3}{26}$	0.115
		$\Sigma p = 1$	

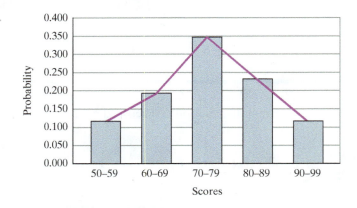

| FIGURE 19.4 | Plot of probability distribution for Example 19.2. |

our analysis, we might be able to use the results of this experiment to predict the outcomes of a similar test to be given later. As the number of students taking the test increases (leading to more scores), the line connecting the midpoint of scores shown in Figure 19.4 becomes smoother and approaches a bell-shaped curve. We use the next example to further explain this concept.

EXAMPLE 19.4

In order to improve the production time, the supervisor of assembly lines in a manufacturing setting of computers has studied the time that it takes to assemble certain parts of a computer at various stations. She measures the time that it takes to assemble a specific part by 100 people at different shifts and on different days. The record of her study is organized and shown in Table 19.10.

Based on data provided, we have calculated the probabilities corresponding to the time intervals that people took to assemble the parts. The probability distribution for Example 19.4 is shown in Table 19.10 and Figure 19.5.

TABLE 19.10   **Data Pertaining to Example 19.4**

Time That It Takes a Person to Assemble the Part (minutes)	Frequency	Probability
5	5	0.05
6	8	0.08
7	11	0.11
8	15	0.15
9	17	0.17
10	14	0.14
11	13	0.13
12	8	0.08
13	6	0.06
14	3	0.03
	$\Sigma = 100$	$\Sigma = 1$

Again, note that the sum of probabilities is equal to 1. Also note that if we were to connect the midpoints of time results (as shown in Figure 19.5), we would have a curve that approximates a bell shape. As the number of data points increases and the intervals decrease, the probability-distribution curve becomes smoother. A probability distribution that has a bell-shaped curve is called a **normal distribution**. The probability distribution for many engineering experiments is approximated by a normal distribution.

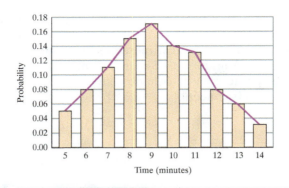

y-axis: Probability (0.00 to 0.18)
x-axis: Time (minutes) (5 to 14)

FIGURE 19.5   Plot of probability distribution for Example 19.4.

The detailed shape of a normal-distribution curve is determined by its mean and standard deviation values. For example, as shown in Figure 19.6, an experiment with a small standard deviation will produce a tall, narrow curve; whereas a large standard deviation will result in a short, wide curve. However, it is important to note that since the normal probability distribution represents all possible outcomes of an experiment (with the total of probabilities equal to 1), the area under any given normal distribution should always be equal to 1. Also, note normal distribution is symmetrical about the mean.

In statistics, it is customary and easier to normalize the mean and the standard deviation values of an experiment and work with what is called the *standard normal distribution,* which has a mean value of zero ($\bar{x} = 0$) and a standard deviation value of 1 ($s = 1$). To do this, we define what commonly is referred to as a **z score** according to

$$z = \frac{x - \bar{x}}{s}$$

19.9

In Equation (19.9), z represents the number of standard deviations from the mean. The mathematical function that describes a normal-distribution curve or a standard normal curve is rather complicated and may be beyond the level of your current understanding. Most of you will learn about it later in your statistics or engineering classes. For now, using Excel, we have generated a table that shows the areas under portions of the standard normal-distribution curve, shown in Table 19:11. At this stage of your education, it is important for you to know how to use the table and solve some problems. A more detailed explanation will be provided in your future classes. We will next demonstrate how to use Table 19.11, using a number of example problems.

> A probability distribution shows the probability values for the occurrence of the outcomes of an experiment. A probability distribution that has a bell-shaped curve is called a normal distribution.

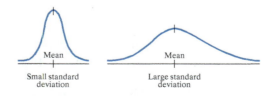

Small standard deviation   Large standard deviation

FIGURE 19.6   The shape of a normal distribution curve as determined by its mean and standard deviation.

TABLE 19.11

Areas Under the Standard Normal Curve—The Values Were Generated Using the Standard Normal Distribution Function of Excel

Note that the standard normal curve is symmetrical about the mean.

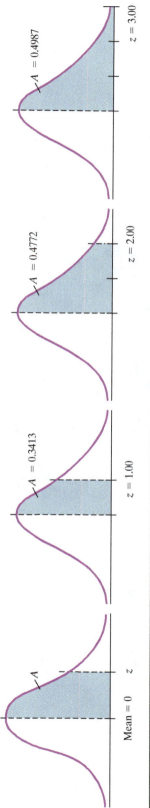

Z	A	Z	A	Z	A	Z	A	Z	A	Z	A	Z	A
0	0.0000	0.13	0.0517	0.26	0.1026	0.39	0.1517	0.52	0.1985	0.65	0.2422	0.78	0.2823
0.01	0.0040	0.14	0.0557	0.27	0.1064	0.4	0.1554	0.53	0.2019	0.66	0.2454	0.79	0.2852
0.02	0.0080	0.15	0.0596	0.28	0.1103	0.41	0.1591	0.54	0.2054	0.67	0.2486	0.8	0.2881
0.03	0.0120	0.16	0.0636	0.29	0.1141	0.42	0.1628	0.55	0.2088	0.68	0.2517	0.81	0.2910
0.04	0.0160	0.17	0.0675	0.3	0.1179	0.43	0.1664	0.56	0.2123	0.69	0.2549	0.82	0.2939
0.05	0.0199	0.18	0.0714	0.31	0.1217	0.44	0.1700	0.57	0.2157	0.7	0.2580	0.83	0.2967
0.06	0.0239	0.19	0.0753	0.32	0.1255	0.45	0.1736	0.58	0.2190	0.71	0.2611	0.84	0.2995
0.07	0.0279	0.2	0.0793	0.33	0.1293	0.46	0.1772	0.59	0.2224	0.72	0.2642	0.85	0.3023
0.08	0.0319	0.21	0.0832	0.34	0.1331	0.47	0.1808	0.6	0.2257	0.73	0.2673	0.86	0.3051
0.09	0.0359	0.22	0.0871	0.35	0.1368	0.48	0.1844	0.61	0.2291	0.74	0.2704	0.87	0.3078
0.1	0.0398	0.23	0.0910	0.36	0.1406	0.49	0.1879	0.62	0.2324	0.75	0.2734	0.88	0.3106
0.11	0.0438	0.24	0.0948	0.37	0.1443	0.5	0.1915	0.63	0.2357	0.76	0.2764	0.89	0.3133
0.12	0.0478	0.25	0.0987	0.38	0.1480	0.51	0.1950	0.64	0.2389	0.77	0.2794	0.9	0.3159

(Continued)

TABLE 19.11 Areas Under the Standard Normal Curve—The Values Were Generated Using the Standard Normal Distribution Function of Excel (*Continued*)

Z	A	Z	A	Z	A	Z	A	Z	A	Z	A	Z	A
0.91	0.3186	1.1	0.3643	1.29	0.4015	1.48	0.4306	1.67	0.4525	1.86	0.4686	2.05	0.4798
0.92	0.3212	1.11	0.3665	1.3	0.4032	1.49	0.4319	1.68	0.4535	1.87	0.4693	2.06	0.4803
0.93	0.3238	1.12	0.3686	1.31	0.4049	1.5	0.4332	1.69	0.4545	1.88	0.4699	2.07	0.4808
0.94	0.3264	1.13	0.3708	1.32	0.4066	1.51	0.4345	1.7	0.4554	1.89	0.4706	2.08	0.4812
0.95	0.3289	1.14	0.3729	1.33	0.4082	1.52	0.4357	1.71	0.4564	1.9	0.4713	2.09	0.4817
0.96	0.3315	1.15	0.3749	1.34	0.4099	1.53	0.4370	1.72	0.4573	1.91	0.4719	2.1	0.4821
0.97	0.3340	1.16	0.3770	1.35	0.4115	1.54	0.4382	1.73	0.4582	1.92	0.4726	2.11	0.4826
0.98	0.3365	1.17	0.3790	1.36	0.4131	1.55	0.4394	1.74	0.4591	1.93	0.4732	2.12	0.4830
0.99	0.3389	1.18	0.3810	1.37	0.4147	1.56	0.4406	1.75	0.4599	1.94	0.4738	2.13	0.4834
1	0.3413	1.19	0.3830	1.38	0.4162	1.57	0.4418	1.76	0.4608	1.95	0.4744	2.14	0.4838
1.01	0.3438	1.2	0.3849	1.39	0.4177	1.58	0.4429	1.77	0.4616	1.96	0.4750	2.15	0.4842
1.02	0.3461	1.21	0.3869	1.4	0.4192	1.59	0.4441	1.78	0.4625	1.97	0.4756	2.16	0.4846
1.03	0.3485	1.22	0.3888	1.41	0.4207	1.6	0.4452	1.79	0.4633	1.98	0.4761	2.17	0.4850
1.04	0.3508	1.23	0.3907	1.42	0.4222	1.61	0.4463	1.8	0.4641	1.99	0.4767	2.18	0.4854
1.05	0.3531	1.24	0.3925	1.43	0.4236	1.62	0.4474	1.81	0.4649	2	0.4772	2.19	0.4857
1.06	0.3554	1.25	0.3944	1.44	0.4251	1.63	0.4484	1.82	0.4656	2.01	0.4778	2.2	0.4861
1.07	0.3577	1.26	0.3962	1.45	0.4265	1.64	0.4495	1.83	0.4664	2.02	0.4783	2.21	0.4864
1.08	0.3599	1.27	0.3980	1.46	0.4279	1.65	0.4505	1.84	0.4671	2.03	0.4788	2.22	0.4868
1.09	0.3621	1.28	0.3997	1.47	0.4292	1.66	0.4515	1.85	0.4678	2.04	0.4793	2.23	0.4871

TABLE 19.11 Areas Under the Standard Normal Curve–The Values Were Generated Using the Standard Normal Distribution Function of Excel (*Continued*)

Z	A	Z	A	Z	A	Z	A	Z	A	Z	A	Z	A
2.24	0.4875	2.43	0.4925	2.62	0.4956	2.81	0.4975	3	0.4987	3.19	0.4993	3.38	0.4996
2.25	0.4878	2.44	0.4927	2.63	0.4957	2.82	0.4976	3.01	0.4987	3.2	0.4993	3.39	0.4997
2.26	0.4881	2.45	0.4929	2.64	0.4959	2.83	0.4977	3.02	0.4987	3.21	0.4993	3.4	0.4997
2.27	0.4884	2.46	0.4931	2.65	0.4960	2.84	0.4977	3.03	0.4988	3.22	0.4994	3.41	0.4997
2.28	0.4887	2.47	0.4932	2.66	0.4961	2.85	0.4978	3.04	0.4988	3.23	0.4994	3.42	0.4997
2.29	0.4890	2.48	0.4934	2.67	0.4962	2.86	0.4979	3.05	0.4989	3.24	0.4994	3.43	0.4997
2.3	0.4893	2.49	0.4936	2.68	0.4963	2.87	0.4979	3.06	0.4989	3.25	0.4994	3.44	0.4997
2.31	0.4896	2.5	0.4938	2.69	0.4964	2.88	0.4980	3.07	0.4989	3.26	0.4994	3.45	0.4997
2.32	0.4898	2.51	0.4940	2.7	0.4965	2.89	0.4981	3.08	0.4990	3.27	0.4995	3.46	0.4997
2.33	0.4901	2.52	0.4941	2.71	0.4966	2.9	0.4981	3.09	0.4990	3.28	0.4995	3.47	0.4997
2.34	0.4904	2.53	0.4943	2.72	0.4967	2.91	0.4982	3.1	0.4990	3.29	0.4995	3.48	0.4997
2.35	0.4906	2.54	0.4945	2.73	0.4968	2.92	0.4982	3.11	0.4991	3.3	0.4995	3.49	0.4998
2.36	0.4909	2.55	0.4946	2.74	0.4969	2.93	0.4983	3.12	0.4991	3.31	0.4995	3.5	0.4998
2.37	0.4911	2.56	0.4948	2.75	0.4970	2.94	0.4984	3.13	0.4991	3.32	0.4995	3.51	0.4998
2.38	0.4913	2.57	0.4949	2.76	0.4971	2.95	0.4984	3.14	0.4992	3.33	0.4996	3.52	0.4998
2.39	0.4916	2.58	0.4951	2.77	0.4972	2.96	0.4985	3.15	0.4992	3.34	0.4996	3.53	0.4998
2.4	0.4918	2.59	0.4952	2.78	0.4973	2.97	0.4985	3.16	0.4992	3.35	0.4996	...	...
2.41	0.4920	2.6	0.4953	2.79	0.4974	2.98	0.4986	3.17	0.4992	3.36	0.4996	...	...
2.42	0.4922	2.61	0.4955	2.8	0.4974	2.99	0.4986	3.18	0.4993	3.37	0.4996	3.9	0.5000

EXAMPLE 19.5

Using Table 19.11, show that for a standard normal distribution of a data set, approximately 68% of the data will fall in the interval of $-s$ to s, about 95% of the data falls between $-2s$ to $2s$, and approximately all of the data points lie between $-3s$ to $3s$.

In Table 19.11, $z = 1$ represents one standard deviation above the mean and 34.13% of the total area under a standard normal curve. On the other hand, $z = -1$ represents one standard deviation below the mean and 34.13% of the total area, as shown in Figure 19.7. Therefore, for a standard normal distribution, 68% of the data fall in the interval of $z = -1$ to $z = 1$ ($-s$ to s). Similarly, $z = -2$ and $z = 2$ (two standard deviations below and above the mean) each represent 0.4772% of the total area under the normal curve. Then, as shown in Figure 19.7, 95% of the data fall in the interval of $-2s$ to $2s$. In the same way, we can show that 99.7% (for $z = -3$ then $A = 0.4987$ and $z = 3$ then $A = 0.4987$) or almost all of the data points lie between $-3s$ to $3s$.

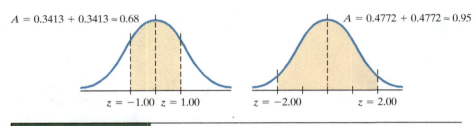

| **FIGURE 19.7** | The area under a normal curve for Example 19.5. |

EXAMPLE 19.6

For Example 19.4, calculate the mean and standard deviation, and determine the probability that it will take a person between 7 and 11 minutes to assemble the computer parts. Refer to Table 19.12 when following solution steps.

TABLE 19.12	Data for Example 19.6			
Time (minutes) x	Frequency f	xf	$x - \bar{x}$	$(x - \bar{x})^2 f$
5	5	25	−4.22	89.04
6	8	48	−3.22	82.95
7	11	77	−2.22	54.21
8	15	120	−1.22	22.33
9	17	153	−0.22	0.82
10	14	140	0.78	8.52
11	13	143	1.78	41.19
12	8	96	2.78	61.83
13	6	78	3.78	85.73
14	3	42	4.78	168.55
		$\Sigma xf = 922$		$\Sigma (x - \bar{x})^2 f = 515.16$

$$\bar{x} = \frac{\Sigma xf}{n} = \frac{922}{100} = 9.22 \text{ minutes}$$

$$s = \sqrt{\frac{\Sigma(x - \bar{x})^2 f}{n - 1}} = \sqrt{\frac{515.16}{99}} = 2.28 \text{ minutes}$$

The value 7 is below the mean value (9.22), and the z value corresponding to 7 is determined from

$$z = \frac{x - \bar{x}}{s} = \frac{7 - 9.22}{2.28} = -0.97$$

From Table 19.11, $A = 0.3340$. Similarly, the value 11 is above the mean value and the z score corresponding to 11 is computed from

$$z = \frac{x - \bar{x}}{s} = \frac{11 - 9.22}{2.28} = 0.78$$

For Table 19.11, $A = 0.2823$. Therefore, the probability that it will take a person between 7 and 11 minutes to assemble the computer part is $0.3340 + 0.2823 = 0.6163$ as shown in Figure 19.8.

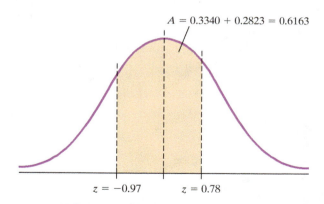

$$A = 0.3340 + 0.2823 = 0.6163$$

$$z = -0.97 \qquad z = 0.78$$

FIGURE 19.8 Area under the probability distribution curve for Example 19.6.

EXAMPLE 19.7

For Example 19.4, determine the probability that it will take a person longer than 10 minutes to assemble the computer parts.

For this problem, the z score is

$$z = \frac{x - \bar{x}}{s} = \frac{10 - 9.22}{2.28} = 0.34$$

From Table 19.11, $A = 0.1331$. Since we wish to determine the probability that it takes longer than 10 minutes to assemble the part, we need to calculate the area, $0.5 - 0.1331 = 0.3669$, as shown in Figure 19.9. The probability that it will take a person longer than 10 minutes to assemble the computer part is approximately 0.37.

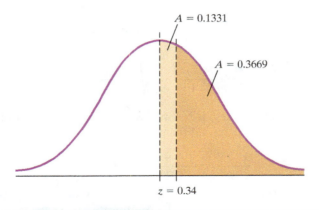

$A = 0.1331$

$A = 0.3669$

$z = 0.34$

| FIGURE 19.9 | Areas under the probability distribution curve for Example 19.7. |

In closing, keep in mind that the purpose of this chapter was to make you aware of the importance of probability and statistics in engineering, not to provide a detailed coverage of statistics. As you take statistics classes and advanced classes in engineering, you will learn much more about statistical concepts and models.

Before You Go On

Answer the following questions to test your understanding of the preceding section.

1. Describe ways by which we can measure the dispersion of a reported data set.

2. Explain why the mean of a data set is not a good way of defining the dispersion of the reported data.

3. In your own words, explain what we mean by standard deviation.

4. What do we mean by probability distribution?

5. What are some of the important characteristics of a normal distribution?

Vocabulary—State the meaning of the following terms:

Median _____

Probability Distribution_____

Normal Distribution_____

SUMMARY

LO¹ Probability—Basic Ideas

By now, you should understand the important role of probability and statistics in various engineering disciplines and be familiar with their terminologies. Probability deals with that branch of science that attempts to predict the likelihood of an event to occur. In probability, each time you repeat an experiment is called a *trial*. The result of an experiment is called an *outcome*, and a *random experiment* is one that has random outcomes—random outcomes cannot be predicted exactly.

LO² Statistics—Basic Ideas

Statistics is that area of science that deals with the collection, organization, analysis, and interpretation of data. Statistics also deals with methods and techniques that can be used to draw conclusions about the characteristics of something with a large number of data points—commonly called a *population*—using a smaller portion of the entire data.

LO³ Frequency Distribution

One simple way of organizing the data (for drawing conclusions) would be to identify the lowest and the highest data points, and then group the data into equal intervals or ranges. When data is organized in this manner, it is commonly referred to as a grouped *frequency distribution*. Another way of showing the range of scores and their frequency is by using a bar graph or a *histogram*. The height of the bars shows the frequency of the data within the given ranges.

LO⁴ Measures of Central Tendency and Variation—Mean, Median, and Standard Deviation

You should have a good grasp of statistical measures of central tendency and variation. You should know how to compute basic statistical information such as mean, variance, and standard deviation for a set of data points. You also should understand that the value of the mean alone does not provide useful information about the dispersion of data; the standard deviation value gives a better idea about how scattered (or spread out) the data is.

LO⁵ Normal Distribution

A probability distribution shows the probability values for the occurrence of the outcomes of an experiment, and a probability distribution that has a bell-shaped curve is called a *normal distribution*. It also is important to know that the detailed bell shape of a normal-distribution curve is determined by its mean and standard deviation values. An experiment with a small standard deviation will produce a tall, narrow curve; whereas a large standard deviation will result in a short, wide curve. You should also know that the area under any given normal distribution should always be equal to 1.

KEY TERMS

Cumulative Frequency 735	Median 739	Random Experiment 732
Grouped Frequency	Normal Distribution 743	Standard Deviation 739
Distribution 734	Outcome 732	Trial 732
Histogram 735	Population 733	Variance 739
Mean 738	Probability 732	Z Score 744

APPLY WHAT YOU HAVE LEARNED

The Body Mass Index (BMI) is a way of determining obesity and whether someone is overweight. It is computed from

$$BMI = \frac{mass(\text{in kg})}{[height(\text{in meter})]^2}.$$

The BMI values in the range of 18.5 to 24.9, 25.0 to 29.9, and > 30.0 are considered healthy, overweight, and obese, respectively.

Height (m)	Mass (kg)						
	50	55	60	65	70	75	80
1.5	22.2	24.4	26.7	28.9	31.1	33.3	35.6
1.6	19.5	21.5	23.4	25.4	27.3	29.3	31.3
1.7	17.3	19.0	20.8	22.5	24.2	26.0	27.7
1.8	15.4	17.0	18.5	20.1	21.6	23.1	24.7
1.9	13.9	15.2	16.6	18.0	19.4	20.8	22.2

Your instructor will pass along a sheet in class wherein you can record your mass and height anonymously. Your instructor will then make the collected data available to the entire class. Use the collected data and perform the following tasks:

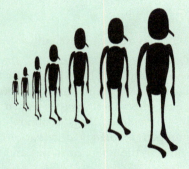

1. Create histograms for height and mass.
2. Calculate the mean and standard deviation of the class height and mass.
3. Calculate the probability distribution for the given height and mass ranges, and plot the probability distribution curves.
4. Calculate the BMI values for the entire class, and group the results into healthy, overweight, and obese.

Discuss your findings in a brief report.

PROBLEMS

Problems that promote lifelong learning are denoted by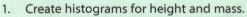

19.1   The scores of a test for an engineering class of 30 students are shown here. Organize the data in a manner similar to Table 19.1 and use Excel to create a histogram.

  Scores: 57, 94, 81, 77, 66, 97, 62, 86, 75, 87, 91, 78, 61, 82, 74, 72, 70, 88, 66, 75, 55, 66, 58, 73, 79, 51, 63, 77, 52, 84

19.2   For Problem 19.1, calculate the cumulative frequency and plot a cumulative-frequency polygon.

19.3   For Problem 19.1, using Equations (19.1) and (19.6), calculate the mean and standard deviation of the class scores.

19.4 For Problem 19.1, using Equations (19.7) and (19.8), calculate the mean and standard deviation of the class scores.

19.5 For Problem 19.1, calculate the probability distribution and plot the probability-distribution curve.

19.6 In order to improve the production time, the supervisor of assembly lines in a manufacturing setting of cellular phones has studied the time that it takes to assemble certain parts of a phone at various stations. She measures the time that it takes to assemble a specific part by 165 people at different shifts and on different days. The record of her study is organized and shown in the accompanying table.

Time That it Takes a Person to Assemble the Part (minutes)	Frequency
4	15
5	20
6	28
7	34
8	28
9	24
10	16

Plot the data and calculate the mean and standard deviation.

19.7 For Problem 19.6, calculate the probability distribution and plot the probability-distribution curve.

Screw Length (cm)	Pipe Diameter (in.)
2.55	1.25
2.45	1.18
2.55	1.22
2.35	1.15
2.60	1.17
2.40	1.19
2.30	1.22
2.40	1.18
2.50	1.17
2.50	1.25

19.8 Determine the average, variance, and standard deviation for the following parts. The measured values are given in the accompanying table.

19.9 Determine the average, variance, and standard deviation for the following parts. The measured values are given in the accompanying table.

2 × 4 Lumber Width (in.)	Steel Spherical Balls (cm)
3.50	1.00
3.55	0.95
3.45	1.05
3.60	1.10
3.55	1.00
3.40	0.90
3.40	0.85
3.65	1.05
3.35	0.95
3.60	0.90

19.10 The next time you make a trip to a supermarket, ask the manager if you can measure the mass of at least 10 cereal boxes of your choice. Choose the same brand and the same size boxes. Tell the manager this is an assignment for a class. Report the average mass, variance, and standard deviation for the cereal boxes. Does the manufacturer's information noted on the box fall within your measurement?

19.11 Repeat Problem 19.10 using three other products, such as cans of soup, tuna, or peanuts.

19.12 Obtain the height, age, and mass of players for your favorite professional basketball team. Determine the average, variance, and standard deviation for the height, age, and mass. Discuss your findings. If you do not like basketball, perform the experiment using data from a soccer team, football team, or a sports team of your choice.

19.13 For Example 19.4, determine the probability that it will take a person between 5 and 10 minutes to assemble the computer parts.

19.14 For Example 19.4, determine the probability that it will take a person longer than 7 minutes to assemble the computer parts.

19.15 For Problem 19.6 (assuming normal distribution), determine the probability that it will take a person between 5 to 8 minutes to assemble the phone.

19.16 Imagine that you and four of your classmates have measured the density of air and recorded the values shown in the accompanying table. Determine the average, variance, and standard deviation for the measured density of air.

Density of Air (kg/m³)
1.27
1.21
1.28
1.25
1.24

19.17 Imagine that you and four of your classmates have measured the viscosity of engine oil and recorded the values shown in the accompanying figure. Determine the average, variance, and standard deviation for the measured viscosity of oil.

Viscosity of Engine Oil (N.s/m²)
0.15
0.10
0.12
0.11
0.14

19.18 Assuming a standard normal distribution (Table 19.11), what percentage of the data falls between $-1.5\ s$ to $1.5\ s$?

19.19 Assuming a standard normal distribution (Table 19.11), what percentage of the data falls between $-0.5\ s$ to $0.5\ s$?

19.20 Typical heating values of coal from various parts of the U.S. are shown in the accompanying table. Calculate the average, variance, and standard deviation for the given data.

Coal from County and State of	Higher Heating Value (Btu/lbm)
Musselshell, Montana	12,075
Emroy, Utah	13,560
Pike, Kentucky	15,040
Cambria, Pennsylvania	15,595
Williamson, Illinois	13,710
McDowell, West Virginia	15,600

Source: Babcock and Wilcox Company, *Steam: Its Generation and Use.*

19.21 Typical heating values of natural gas from various parts of the U.S. are shown in the accompanying table. Calculate the average, variance, and standard deviation for the given data.

Source of Gas	Heating Value (Btu/lbm)
Pennsylvania	23,170
Southern California	22,904
Ohio	22,077
Louisiana	21,824
Oklahoma	20,160

Source: Babcock and Wilcox Company, *Steam: Its Generation and Use.*

19.22 As an electrical engineer, you have designed a new efficient light bulb. In order to predict its life expectancy, you conducted a series of experiments on 135 of these light bulbs and gathered the data shown in the table. Plot the data and calculate the mean and standard deviation.

Number of Hours the Light Bulb Functioned before Failing	Frequency
700	15
800	20
900	34
1000	28
1100	22
1200	16

19.23 For Problem 19.22, calculate the probability distribution and plot the probability distribution curve.

19.24 For Problem 19.22, determine the probability (assuming normal distribution) that a light bulb would have a life expectancy between 800 and 1000 hours.

19.25 For Problem 19.22, determine the probability (assuming normal distribution) that a light bulb would have a life expectancy greater than 1000 hours.

19.26 For Problem 19.22, determine the probability (assuming normal distribution) that a light bulb would have a life expectancy less than 900 hours.

19.27 As a mechanical engineer working for an automobile manufacturer, you conduct a survey and collect the following data in order to study the performance of an engine that was designed many years ago. Plot the data and calculate the mean and standard deviation.

Miles Driven before a Need for an Engine Maintenance	Frequency
70,000	12
80,000	17
90,000	22

100,000	33
110,000	42
120,000	30
130,000	24
140,000	15
150,000	11

19.28 For Problem 19.27, calculate the probability distribution and plot the probability distribution curve.

19.29 For Problem 19.27, determine the probability (assuming normal distribution) that a car would need engine maintenance between 70,000 and 90,000 miles.

19.30 For Problem 19.27, determine the probability (assuming normal distribution) that a car would need engine maintenance after 100,000 miles.

19.31 For Problem 19.27, determine the probability (assuming normal distribution) that a car would need engine maintenance before 85,000 miles.

19.32 For Problem 19.27, determine the probability (assuming normal distribution) that a car would need engine maintenance before 90,000 miles.

19.33 As an engineer working for a water bottling company, you collect the following data in order to test the performance of the bottling systems. Plot the data and calculate the mean and standard deviation.

Milliliters of Water in the Bottle	Frequency
485	13
490	17
495	25
500	40
505	23
510	18
515	15

19.34 For Problem 19.33, calculate the probability distribution and plot the probability distribution curve.

19.35 For Problem 19.33, determine the probability (assuming normal distribution) that a bottle would be filled between 500 and 515 milliliters.

19.36 For Problem 19.33, determine the probability (assuming normal distribution) that a bottle would be filled with more than 495 milliliters.

19.37 For Problem 19.33, determine the probability (assuming normal distribution) that a bottle would be filled with less than 500 milliliters.

19.38 For Problem 19.33, determine the probability (assuming normal distribution) that a bottle would be filled with less than 495 milliliters.

19.39 As a chemical engineer working for a tire manufacturer, you collect the following data in order to test the performance of tires. Plot the data and calculate the mean and standard deviation.

Miles with Acceptable (Reliable) Wear	Frequency
30,000	15
35,000	20
40,000	34
45,000	32
50,000	22
55,000	16

19.40 For Problem 19.39, calculate the probability distribution and plot the probability distribution curve.

19.41 For Problem 19.39, determine the probability (assuming normal distribution) that a tire could be used reliably between 45,000 and 55,000 miles.

19.42 For Problem 19.39, determine the probability (assuming normal distribution) that a tire could be used reliably for more than 50,000 miles.

19.43 For Problem 19.39, determine the probability (assuming normal distribution) that a tire could be used reliably for less than 45,000 miles.

19.44 For Problem 19.39, determine the probability (assuming normal distribution) that a tire could be used reliably for less than 50,000 miles.

Class Experiments—Problems 19.45 through 19.50 are experiments that are performed in class.

19.45 Your instructor will pass along an unopened bag of Hershey's Kisses. You are to estimate the number of Kisses in the bag and write it down on a piece of paper. Your instructor will then collect the data and share the results with the class. Your assignment is to organize the data per your instructor's suggestion and calculate the mean and standard deviation. Compute the probability distribution. Does your data distribution approximate a normal distribution? Answer any additional questions that your instructor might ask.

19.46 Your instructor will ask for a volunteer in class. You are to estimate his or her height in inches (or in cm) and write it down on a piece of paper. Your instructor will then collect the data and share the results with the class. Your assignment is to organize the data per your instructor's suggestion and calculate the mean and standard deviation of the data. Compute the probability distribution. Does your data distribution approximate a normal distribution? Answer any additional questions that your instructor might ask.

19.47 Your instructor will ask for a volunteer in class. You are to estimate his or her mass in lbm (or in kg) and write it down on a piece of paper. Your instructor will then collect the data and share the results with the class. Your assignment is to organize the data per your instructor's suggestion and calculate the mean and standard deviation of the data. Compute the probability distribution. Does your data distribution approximate a normal distribution? Answer any additional questions that your instructor might ask.

19.48 You are to write down on a piece of paper the number of credits you are taking this semester. Your instructor will then collect the data and share the results with the class. Calculate the mean and standard deviation of the data. Assuming a normal distribution, determine the probability that a student is taking between 12 to 15 credits this semester. What is the probability that a student is taking less than 12 credits?

19.49 You are to write down on a piece of paper how much (to the nearest penny) money you have on you. Your instructor will then collect the data and share the results with the class. Your assignment is to organize the data per your instructor's suggestion and calculate the mean and standard deviation of the data. Assuming a normal distribution, determine the probability that a student has between $5 to $10. What is the probability that a student has less than $10?

19.50 You are to write down your waist size on a piece of paper. If you don't know your waist size, ask your instructor for a measuring tape. Your instructor will then collect the data and share the results with the class. Your assignment is to organize the data per your instructor's suggestion and calculate

the mean and standard deviation of the data. Assuming a normal distribution, determine the probability that a student will have a waist size that is less than 34 inches. What is the probability that a student will have a waist size that is between 30 in. to 36 in.?

19.51 As an agricultural engineer you are asked to collect corn and wheat production data for the most recent 10 years. You are to use the principles we discussed in this chapter to organize the data. For example, you can present the data using histogram or could calculate the mean and standard deviation for the mentioned crops for the given period. Use PowerPoint slides to present your findings.

19.52 Collect data on how many Apple iPhones and iPads have been sold in the United States since 2010. Use the principles we discuss in this chapter to organize the data. Can you identify any patterns and draw any conclusions? Discuss your findings in a brief report.

Engineering Economics

Source: © Lightscapes Photography, Inc./CORBIS

Economic considerations play a vital role in product and service development and in the engineering design decision-making process.

LEARNING OBJECTIVES

LO¹ **Cash Flow Diagrams**: explain how the diagram is used in analysis of engineering economics problems and give examples

LO² **Simple and Compound Interest**: explain what they mean and how they differ and give examples

LO³ **Future Worth of a Present Amount and Present Worth of a Future Amount**: know how to compute the future worth of any present amount (principal) and present worth of any future amount

LO⁴ **Effective Interest Rate**: explain what it means and give an example

LO⁵ **Present and Future Worth of Series Payment**: know how to calculate the present and future worth of a series of payments

LO⁶ **Interest–Time Factors**: know how to use interest-time factor tables to set up and solve problems

LO⁷ **Choosing the Best Alternatives—Decision Making**: understand how to use engineering economic principles to select the best alternative from among many choices

LO⁸ **Excel Financial Functions**: know how to use Excel functions to set up and solve engineering economics problems

DISCUSSION STARTER

CONSUMER GUIDE ON CREDIT CARDS

Interest Rates

One of the most important things to understand about your credit card is its interest rate.

An interest rate is the price you pay for borrowing money. For credit cards, the interest rates are stated as a yearly rate, called the annual percentage rate (APR).

One Credit Card may have Several APRs. Here are some common APR terms you should know:

Different APRs for Different Types of Transactions. Your credit card will always have a purchase APR—the amount of interest you will pay on purchases. For many cards, you only have to pay interest on purchases if you carry over a balance. Your card likely will also have a different—often higher—APR for cash advances or balance transfers.

Introductory APR. Your card may have a lower APR during an introductory period and a higher rate after that period ends. Under Federal law, the introductory period must last at least six months, and the credit card company must tell you what your rate will be after the introductory period expires. For example, your introductory rate may be 8.9 percent for six months and then go up to 17.9 percent.

Penalty APR. Your APR may increase if you trigger one of the penalty terms, for example, by paying your bill late or making a payment that is returned.

APR(s) Can Be Fixed or Variable

A **fixed-rate APR** is set at a certain percent and cannot change during the period of time outlined in your credit card agreement. If your company does not specify a time period, the rate cannot change as long as your account is open.

A **variable-rate APR** may change depending upon an index that is outside of the credit card company's control, such as the prime rate (an index that represents the interest rate most banks charge their most credit-worthy customers) or Treasury bill rate (the rate paid by the government on its short-term borrowing). The credit card application and agreement will tell you how often your card's APR may change.

Card issuers may offer combinations of fixed and variable rates—for example, a fixed-rate APR that becomes a variable rate after your introductory period ends. Read your credit card agreement carefully to understand when or if your APR may change.

Source: United States Federal Reserve

To the students: Assuming you have a credit card balance of $2,000 and an APR of 12%, if you were to pay an extra $20 above the minimum payment due each month, how much money do you think you would save?

As we explained in Chapter 3, economic factors always play important roles in engineering design decision making. If you design a product that is too expensive to manufacture, then it cannot be sold at a price that consumers can afford and still be profitable to your company. The fact is that companies design products and provide services not only to make our lives better but also to make money! In this section, we will discuss the basics of engineering economics. The information provided here not only applies to engineering projects but can also be applied to financing a car or a house or borrowing from or investing money in banks. Some of you may want to apply the knowledge gained here to determine your student loan or credit card payments. Therefore, we advise you to develop a good understanding of engineering economics; the information presented here could help you manage your money more wisely.

LO¹ 20.1   Cash Flow Diagrams

Cash flow diagrams are visual aids that show the flow of costs and revenues over a period of time. Cash flow diagrams show *when the cash flow occurs, the cash flow magnitude, and whether the cash flow is out of your pocket (cost) or into your pocket (revenue).* It is an important visual tool that shows the timing, the magnitude, and the direction of cash flow. To shed more light on the concept of the cash flow diagram, imagine that you are interested in purchasing a new car. Being a first-year engineering student, you may not have much money in your savings account at this time; for the sake of this example, let us say that you have $1200 to your name in a savings account. The car that you are interested in buying costs $15,500; let us further assume that including the sales tax and other fees, the total cost of the car would be $16,880. Assuming you can afford to put down $1000 as a down payment for your new shiny car, you ask your bank for a loan. The bank decides to lend you the remainder, which is $15,880 at 8% interest. You will sign a contract that requires you to pay $315.91 every month for the next five years. You will soon learn how to calculate these monthly payments, but for now let us focus on how to draw the cash flow diagram. The cash flow diagram for this activity is shown in Figure 20.1. Note in Figure 20.1 the direction of the arrows representing the money given to you by the bank and the payments that you must make to the bank over the next five years (60 months).

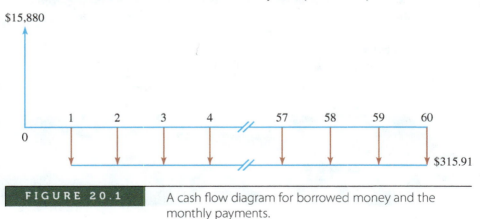

FIGURE 20.1   A cash flow diagram for borrowed money and the monthly payments.

EXAMPLE 20.1

Draw the cash flow diagram for an investment that includes purchasing a machine that costs $50,000 with a maintenance and operating cost of $1000 per year. It is expected that the machine will generate revenues of $15,000 per year for five years. The expected salvage value of the machine at the end of five years is $8000.

The cash flow diagram for the investment is shown in Figure 20.2. Again, note the directions of arrows in the cash flow diagram. We have represented the initial cost of $50,000 and the maintenance cost by arrows pointing down, while the revenue and the salvage value of the machine are shown by arrows pointing up.

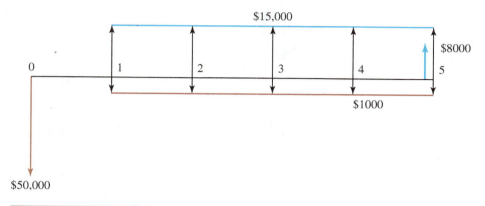

FIGURE 20.2   The cash flow diagram for Example 20.1.

LO² 20.2   Simple and Compound Interest

Interest is the extra money—in addition to the borrowed amount—that is paid for the purpose of having access to the borrowed money. **Simple interest** is the interest that is paid only on the initial borrowed or deposited amount. For simple interest, the interest accumulated on the principal each year will not collect interest itself. Only the initial principal will collect interest. For example, if you deposit $100.00 in a bank at 6% simple interest, after six years you will have $136 in your account. In general, if you deposit the amount P at a rate of i% for a period of n years, then the total future value F of the P at the end of the nth year is given by

$$F = P + (P)(i)(n) = P(1 + ni)$$ **20.1**

EXAMPLE 20.2

Compute the future value of a $1500 deposit, after eight years, in an account that pays a simple interest rate of 7%. How much interest will be paid to this account?

You can determine the future value of the deposited amount using Equation (20.1), which results in

$$F = P(1 + ni) = 1500[1 + 8(0.07)] = \$2340$$

And the total interest to be paid to this account is

$$interest = (P)(n)(i) = (1500)(8)(0.07) = \$840$$

TABLE 20.1	The Effect of Compounding Interest		
Year	Balance at the Beginning of the Year (dollars and cents)	Interest for the Year at 6% (dollars and cents)	Balance at the End of the Year, Including the Interest (dollars and cents)
1	100.00	6.00	106.00
2	106.00	6.36	112.36
3	112.36	6.74	119.10
4	119.10	7.14	126.24
5	126.24	7.57	133.81
6	133.81	8.02	141.83

Simple interests are very rare these days! Almost all interest charged to borrow accounts or interest earned on money deposited in a bank is computed using **compound interest**. The concept of compound interest is discussed next.

Compound Interest

Under the compounding interest scheme, the interest paid on the initial principal will also collect interest.

Under the compounding interest scheme, the interest paid on the initial principal will also collect interest. To better understand how the compound interest earned or paid on a principal works, consider the following example. Imagine that you put $100.00 in a bank that pays you 6% interest compounding annually. At the end of the first year (or the beginning of the second year) you will have $106.00 in your bank account. You have earned interest in the amount of $6.00 during the first year. However, the interest earned during the second year is determined by ($106.00)(0.06) = $6.36. That is because the $6.00 interest of the first year also collects 6% interest, which is 36 cents itself. Thus, the total interest earned during the second year is $6.36, and the total amount available in your account at the end of the second year is $112.36. Computing the interest and the total amount for the third, fourth, fifth and the sixth year in a similar fashion will lead to $141.83 in your account at the end of the sixth year. Refer to Table 20.1 for detailed calculations. Note the difference between $100.00 invested at 6% simple interest and 6% interest compounding annually for a duration of six years. For the simple interest case, the total interest earned, after six years, is $36.00, whereas the total interest accumulated under the annual compounding case is $41.83 for the same duration.

LO³ 20.3 Future Worth of a Present Amount and Present Worth of a Future Amount

Now we will develop a general formula that you can use to compute the **future value F of any present amount** (principal) P, after n years collecting i% interest compounding annually. The cash flow diagram for this situation is shown in Figure 20.3. In order to demonstrate, step-by-step, the compounding effect of the interest each year, Table 20.2 has been developed. As shown in Table 20.2,

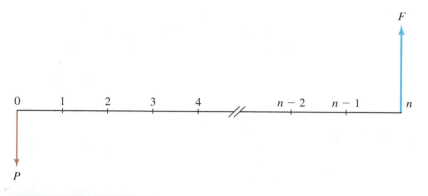

| FIGURE 20.3 | The cash flow diagram for future worth of a deposit made in the bank today. |

starting with the principal P, at the end of the first year we will have $P + Pi$ or $P(1 + i)$. During the second year, the $P(1 + i)$ collects interest in an amount of $P(1 + i)i$, and by adding the interest to the $P(1 + i)$ amount that we started with in the second year, we will have a total amount of $P(1 + i) + P(1 + i)i$. Factoring out the $P(1 + i)$ term, we will have $P(1 + i)^2$ dollars at the end of the second year. Now by following Table 20.2 you can see how the interest earned and the total amount are computed for the third, fourth, fifth,..., and the nth year. Consequently, you can see that the relationship between the present worth P and the future value F of an amount collecting $i\%$ interest compounding annually after n years is given by

$$F = P(1 + i)^n$$

20.2

Many financial institutions pay interest that compounds more than once a year. For example, a bank may pay you an interest rate that compounds semiannually (twice a year), or quarterly (four times a year), or monthly

| TABLE 20.2 | The Relationship Between Present Value P and the Future Value F |

Year	Balance at the Beginning of the Year	Interest for the Year	Balance at the End of the Year, Including the Interest
1	P	$(P)(i)$	$P + (P)(i) = P(1 + i)$
2	$P(1 + i)$	$P(1 + i)(i)$	$P(1 + i) + P(1 + i)(i) = P(1 + i)^2$
3	$P(1 + i)^2$	$P(1 + i)^2(i)$	$P(1 + i)^2 + P(1 + i)^2(i) = P(1 + i)^3$
4	$P(1 + i)^3$	$P(1 + i)^3(i)$	$P(1 + i)^3 + P(1 + i)^3(i) = P(1 + i)^4$
5	$P(1 + i)^4$	$P(1 + i)^4(i)$	$P(1 + i)^4 + P(1 + i)^4(i) = P(1 + i)^5$
.	......	....	....
n	$P(1 + i)^{n-1}$	$P(1 + i)^{n-1}(i)$	$P(1 + i)^{n-1} + P(1 + i)^{n-1}(i) = P(1 + i)^n$

(12 periods a year). If the principal P is deposited for a duration of n years and the interest given is compounded m periods (or m times) per year, then the future value F of the principal P is determined from

$$F = P\left(1 + \frac{i}{m}\right)^{nm}$$

20.3

EXAMPLE 20.3

Compute the future value of a $1500 deposit made today, after eight years, in an account that pays an interest rate of 7% that compounds annually. How much interest will be paid to this account?

The future value of the $1500 deposit is computed by substituting in Equation (20.2) for P, i, and n, which results in the amount that follows:

$$F = P(1 + i)^n = 1500(1 + .07)^8 = \$2577.27$$

The total interest earned during the eight-year life of this account is determined by calculating the difference between the future value and the present deposit value.

$$interest = \$2577.27 - \$1500 = \$1077.27$$

EXAMPLE 20.4

Compute the future value of a $1500 deposit, after eight years, in an account that pays an interest rate of 7% that compounds monthly. How much interest will be paid to this account?

To determine the future value of the $1500 deposit, we substitute in Equation (20.3) for P, i, m and n. The substitution results in the future value shown next.

$$F = 1500\left(1 + \frac{0.07}{12}\right)^{(8)(12)} = 1500\left(1 + \frac{0.07}{12}\right)^{96} = \$2621.73$$

And the total interest is

$$interest = \$2621.73 - \$1500 = \$1121.73$$

The results of Examples 20.2, 20.3, and 20.4 are compared and summarized in Table 20.3. Note the effects of simple interest, interest compounding annually, and interest compounding monthly on the total future value of the $1500 deposit.

Present Worth of a Future Amount

Let us now consider the following situation. You would like to have $2000 available to you for a down payment on a car when you graduate from college in, say, five years. How much money do you need to put in a certificate of deposit (CD) with an interest rate of 6.5% (compounding annually) today? The relationship between the future and present value was developed earlier and is given by Equation (20.2). Rearranging Equation (20.2), we have

TABLE 20.3		Comparison of Results for Examples 20.2, 20.3, and 20.4			
Example Number	Principal (dollars)	Interest Rate	Duration (years)	Future Value (dollars and cents)	Interest Earned (dollars and cents)
Example 20.2	1500	7% simple	8	2340.00	840.00
Example 20.3	1500	7% compounding annually	8	2577.27	1077.27
Example 20.4	1500	7% compounding monthly	8	2621.73	1121.73

$$P = \frac{F}{(1+i)^n} \qquad \boxed{20.4}$$

and substituting in Equation (20.4) for the future value F, the interest rate i, and the period n, we have

$$P = \frac{2000}{(1+0.065)^5} = \$1459.76$$

This may be a relatively large sum to put aside all at once, especially for a first-year engineering student. A more realistic option would be to put aside some money each year. Then the question becomes, how much money do you need to put aside every year for the next five years at the given interest rate to have that $2000 available to you at the end of the fifth year? To answer this question, we need to develop the formula that deals with a series of payments or series of deposits. This situation is discussed in Section 20.5.

LO⁴ 20.4   Effective Interest Rate

If you deposit $100.00 in a savings account, at 6% compounding monthly, then, using Equation (20.3), at the end of one year you will have $106.16 in your account. The $6.16 earned during the first year is higher than the stated 6% interest, which could be understood as $6.00 for a $100.00 deposit over a period of one year. In order to avoid confusion, the stated or the quoted interest rate is called the **nominal interest rate**, and the actual earned interest rate is called the **effective interest rate**. To determine the relationship between the nominal and effective interest rates, let's imagine that we have deposited an arbitrary amount P in an account that pays $i\%$ compounding m times in one year. Then the effective interest rate can be determined from

> In order to avoid confusion, the stated or the quoted interest rate is called the nominal interest rate, and the actual earned interest rate is called the effective interest rate.

$$i_{eff} = \frac{\text{amount available at the end of year 1} - \text{amount started with}}{\text{amount started with}}$$

$$= \frac{P\left(1 + \dfrac{i}{m}\right)^m - P}{P}$$

After factoring out the P's and simplifying, the relationship between the nominal rate, i, and the effective rate, i_{eff}, becomes

$$i_{eff} = \left(1 + \frac{i}{m}\right)^m - 1 \qquad \boxed{20.5}$$

TABLE 20.4		The Effect of the Frequency of Interest Compounding Periods		
Compounding Period	Total Number of Compounding Periods	Total Amount after 1 Year (dollars and cents)	Interest (dollars and cents)	Effective Interest Rate
Annually	1	$100(1 + 0.06) = 106.00$	6.00	6%
Semiannually	2	$100\left(1 + \dfrac{0.06}{2}\right)^2 = 106.09$	6.09	6.09%
Quarterly	4	$100\left(1 + \dfrac{0.06}{4}\right)^4 = 106.13$	6.13	6.13%
Monthly	12	$100\left(1 + \dfrac{0.06}{12}\right)^{12} = 106.16$	6.16	6.16%
Daily	365	$100\left(1 + \dfrac{0.06}{365}\right)^{365} = 106.18$	6.18	6.18%

where *m* represents the number of compounding periods per year. To better understand the compounding effect of interest, let us see what happens if we deposit $100.00 in an account for a year based on one of the following quoted interests: 6% compounding annually, 6% semi-annually, 6% quarterly, 6% monthly, and 6% daily. Table 20.4 shows the difference among these compounding periods, the total amount of money at the end of one year, the interest earned, and the effective interest rates for each case.

When comparing the five different interest compounding frequencies, the difference in the interests earned on a $100.00 investment over a period of a year may not seem much to you, but as the principal and the time of deposit are increased this value becomes significant. To better demonstrate the effect of principal and time of deposit, consider the following example.

EXAMPLE 20.5   Determine the interest earned on $5000 deposited in a savings account, for 10 years, based on one of the following quoted interest rates: 6% compounding annually, semiannually, quarterly, monthly, and daily. The solution to this problem is presented in Table 20.5.

TABLE 20.5		The Solution of Example 20.5	
Compounding Period	Total Number of Compounding Periods	Total Future Amount Using Eq. (16.8) (dollars and cents)	Interest (dollars and cents)
Annually	10	$5000(1 + 0.06)^{10} = 8954.23$	3954.23
Semiannually	20	$5000\left(1 + \dfrac{0.06}{2}\right)^{20} = 9030.55$	4030.55
Quarterly	40	$5000\left(1 + \dfrac{0.06}{4}\right)^{40} = 9070.09$	4070.09

| Monthly | 120 | $5000\left(1+\dfrac{0.06}{12}\right)^{120} = 9096.98$ | 4096.98 |
| Daily | 3650 | $5000\left(1+\dfrac{0.06}{365}\right)^{3650} = 9110.14$ | 4110.14 |

EXAMPLE 20.6

Determine the effective interest rates corresponding to the nominal rates: (a) 7% compounding monthly, (b) 16.5% compounding monthly, (c) 6% compounding semiannually, (d) 9% compounding quarterly.

We can compute the i_{eff} for each case by substituting for i and m in Equation (20.5).

(a) $i_{\text{eff}} = \left(1+\dfrac{i}{m}\right)^{m} - 1 = \left(1+\dfrac{0.07}{12}\right)^{12} - 1 = 0.0722$ or 7.22%

(b) $i_{\text{eff}} = \left(1+\dfrac{0.165}{12}\right)^{12} - 1 = 0.1780$ or 17.80%

(c) $i_{\text{eff}} = \left(1+\dfrac{0.06}{2}\right)^{2} - 1 = 0.0609$ or 6.09%

(d) $i_{\text{eff}} = \left(1+\dfrac{0.09}{4}\right)^{4} - 1 = 0.0930$ or 9.30%

Before You Go On

Answer the following questions to test your understanding of the preceding section.

1. What do we mean by a cash flow diagram?

2. What is a simple interest rate?

3. What is a compounding interest rate?

4. What do we mean by nominal and effective interest rates?

5. What is a future worth of a present amount?

Vocabulary—State the meaning of the following terms:

Future Worth_____

Present Worth_____

Nominal Interest Rate _____

LO⁵ 20.5 Present and Future Worth of Series Payment

In this section, we will first formulate the relationship between a present lump sum, P, and future uniform series payments, A, and then from that relationship we will develop the formula that relates the uniform series of payments A to the future lump sum F. This approach is much easier to follow as you will see. To derive these relationships, let us first consider a situation where we have borrowed some money, denoted by P, at an annual interest rate i from a bank, and we are planning to pay the loan yearly, in equal amounts A, in n years, as shown in Figure 20.4.

To obtain the relationship between P and A, we will treat each future payment separately and relate each payment to its present equivalent value using Equation (20.4); we then add all the resulting terms together. This approach leads to the following relationship:

$$P = \frac{A}{(1+i)} + \frac{A}{(1+i)^2} + \frac{A}{(1+i)^3} + \cdots + \frac{A}{(1+i)^{n-1}} + \frac{A}{(1+i)^n} \qquad \boxed{20.6}$$

As you can see, Equation (20.6) is not very user-friendly, so we need to simplify it somehow. What if we were to multiply both sides of Equation (20.6) by the term $(1+i)$? This operation results in the following relationship:

$$P(1+i) = A + \frac{A}{(1+i)} + \frac{A}{(1+i)^2} + \frac{A}{(1+i)^3} + \cdots + \frac{A}{(1+i)^{n-2}} + \frac{A}{(1+i)^{n-1}} \qquad \boxed{20.7}$$

Now if we subtract Equation (20.6) from Equation (20.7), we have

$$P(1+i) - P = A + \frac{A}{(1+i)} + \frac{A}{(1+i)^2} + \frac{A}{(1+i)^3} + \cdots + \frac{A}{(1+i)^{n-2}} + \frac{A}{(1+i)^{n-1}}$$

$$-\left[\frac{A}{(1+i)} + \frac{A}{(1+i)^2} + \frac{A}{(1+i)^3} + \cdots + \frac{A}{(1+i)^{n-1}} + \frac{A}{(1+i)^n} \right] \qquad \boxed{20.8a}$$

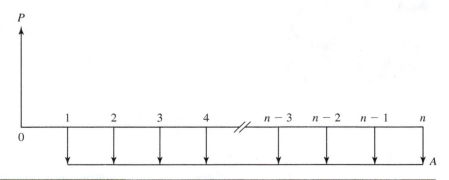

| **FIGURE 20.4** | The cash flow diagram for a borrowed sum of money and its equivalent series payments. |

Simplifying the right-hand side of Equation (20.8a) leads to the following relationship:

$$P(1+i) - P = A - \frac{A}{(1+i)^n} \qquad \text{20.8b}$$

And after simplifying the left-hand side of Equation (20.8a), we have

$$P(i) = \frac{A\left[(1+i)^n - 1\right]}{(1+i)^n} \qquad \text{20.8c}$$

Now if we divide both sides of Equation (20.8c) by i, we have

$$P = A\left[\frac{(1+i)^n - 1}{i(1+i)^n}\right] \qquad \text{20.9}$$

Equation (20.9) establishes the relationship between the present value of a lump sum P and its equivalent uniform series payments A. We can also rearrange Equation (20.9), to represent A in terms of P directly, as given by the following formula:

$$A = \frac{P(i)(1+i)^n}{(1+i)^n - 1} = P\left[\frac{(i)(1+i)^n}{(1+i)^n - 1}\right] \qquad \text{20.10}$$

Future Worth of Series Payment

To develop a formula for computing the future worth of a series of uniform payments, we begin with the relationship between the present worth and the future worth, Equation (20.2), and then we substitute for P in Equation (20.2) in terms of A, using Equation (20.9). This procedure is demonstrated, step-by-step, next. The relation between a present value and a future value is given by Equation (20.2):

$$F = P(1+i)^n \qquad \text{20.2}$$

And the relationship between the present worth and a uniform series is given by Equation (20.9):

$$P = A\left[\frac{(1+i)^n - 1}{i(1+i)^n}\right] \qquad \text{20.9}$$

Substituting into Equation (20.2) for P in terms of A using Equation (20.9), we have

$$F = P(1+i)^n = A\overbrace{\left[\frac{(1+i)^n - 1}{i(1+i)^n}\right]}^{P}(1+i)^n \qquad \text{20.11}$$

Simplifying Equation (20.11) results in the direct relationship between the future worth F and the uniform payments or deposits A, which follows:

$$F = A\left[\frac{(1+i)^n - 1}{i}\right] \qquad \text{20.12}$$

And by rearranging Equation (20.12), we can obtain a formula for A in terms of future worth F:

$$A = F\left[\frac{i}{(1+i)^n - 1}\right] \qquad \boxed{20.13}$$

Now that we have all the necessary tools, we turn our attention to the question we asked earlier about how much money you need to put aside every year for the next five years to have $2000 for the down payment of your car when you graduate. Recall that the interest rate is 6.5% compounding annually. The annual deposits are calculated from Equation (20.13), which leads to the following amount:

$$A = 2000\left[\frac{0.065}{(1 + 0.065)^5 - 1}\right] = \$351.26$$

Putting aside $351.26 in a bank every year for the next five years may be more manageable than depositing a lump sum of $1459.76 today, especially if you don't currently have access to that large a sum!

It is important to note that Equations (20.9), (20.10), (20.12), and (20.13) apply to a situation wherein the uniform series of payments or revenues *occur annually*. Well, the next question is, how do we handle situations where the payments are made monthly? For example, a car or a house loan payment occurs monthly. Let us now modify our findings by considering the relationship between present value P and uniform series payments or revenue A that occur more than once a year at the same frequency as the frequency of compounding interest per year. For this situation, Equation (20.9) is modified to incorporate the frequency of compounding interest per year, m, in the following manner:

$$P = A\left[\frac{\left(1 + \dfrac{i}{m}\right)^{nm} - 1}{\dfrac{i}{m}\left(1 + \dfrac{i}{m}\right)^{nm}}\right] \qquad \boxed{20.14}$$

Note that in order to obtain Equation (20.14), we simply substituted in Equation (20.9) for i, i/m, and for n, nm. Equation (20.14) can be rearranged to solve for A in terms of P according to

$$A = P\left[\frac{\left(\dfrac{i}{m}\right)\left(1 + \dfrac{i}{m}\right)^{nm}}{\left(1 + \dfrac{i}{m}\right)^{nm} - 1}\right] \qquad \boxed{20.15}$$

Similarly, Equations (20.12) and (20.13) can be modified for situations where A occurs more than once a year—at the same frequency as the compounding interest—leading to the following relationship:

$$F = A\left[\frac{\left(1 + \dfrac{i}{m}\right)^{mn} - 1}{\dfrac{i}{m}}\right] \qquad \boxed{20.16}$$

$$A = F\left[\dfrac{\dfrac{i}{m}}{\left(1 + \dfrac{i}{m}\right)^{mn} - 1}\right]$$

20.17

Finally, when the frequency of a uniform series is different from the frequency of compounding interest, i_{eff} must first be calculated to match the frequency of the uniform series.

EXAMPLE 20.7

Let us return to the question we asked earlier about how much money you need to put aside for the next five years to have $2000 for the down payment on your car when you graduate. Now consider the situation where you make your deposits every month, and the interest rate is 6.5% compounding monthly.

The deposits are calculated from Equation (20.17), which leads to the following:

$$A = F\left[\dfrac{\dfrac{i}{m}}{\left(1 + \dfrac{i}{m}\right)^{mn} - 1}\right] = 2000\left[\dfrac{\dfrac{0.065}{12}}{\left(1 + \dfrac{0.065}{12}\right)^{(12)(5)} - 1}\right] = \$28.29$$

Putting aside $28.29 in the bank every month for the next five years is even more manageable than depositing $351.26 in a bank every year for the next five years, and it is certainly more manageable than depositing a lump sum of $1459.76 in the bank today!

EXAMPLE 20.8

Determine the monthly payments for a five-year, $10,000 loan at an interest rate of 8% compounding monthly.

To calculate the monthly payments, we use Equation (20.15).

$$A = P\left[\dfrac{\left(\dfrac{i}{m}\right)\left(1 + \dfrac{i}{m}\right)^{nm}}{\left(1 + \dfrac{i}{m}\right)^{nm} - 1}\right] = 10{,}000\left[\dfrac{\left(\dfrac{0.08}{12}\right)\left(1 + \dfrac{0.08}{12}\right)^{60}}{\left(1 + \dfrac{0.08}{12}\right)^{60} - 1}\right] = \$202.76$$

EXAMPLE 20.9

In this example problem, we show how to deal with situations when the frequency of a uniform series is different from the frequency of compounding interest. As we mentioned previously, you must first calculate an i_{eff} that matches the frequency of the uniform series. Consider the following situations in which you deposit $2000 every three months for one year. (a) The interest is 18% compounding quarterly. (b) The interest is 18% compounding monthly. Compare the future values of the deposits at the end of year one.

When following the solution, note that deposits are made at the end of the current month or the beginning of the next month. In part (a), the frequency of deposits matches the interest compounding frequency. Consequently, the future value is simply calculated from

$$F = 2000\left[\frac{(1 + 0.045)^4 - 1}{0.045}\right] = \$8,556.38$$

In part (b), the interest compounding frequency is 12, whereas the frequency of deposits is 4. In order to understand how part (b) differs from part (a), let us look at the balance at the beginning and the end of each month, as shown in Table 20.6.

As you can see for situation (b), at the end of year 1, the future value of the deposits is \$8,564.99, which is slightly higher than the value for situation (a) at \$8,556.38.

TABLE 20.6   Balance at the Beginning and the End of Each Month for Example 20.9

Month	Balance at the Beginning of Months (dollars and cents)	Interest for the Month at 1.5% (dollars and cents)	Balance at the End of the Month (dollars and cents)
1	0.00	0.00	0.00
2	0.00	0.00	0.00
3	0.00	0.00	0.00
4	2000	30.00	2030.00
5	2030.00	30.45	2060.45
6	2060.45	30.90	2091.35
7	2000 + 2091.35 = 4091.35	61.37	4152.72
8	4152.72	62.29	4215.01
9	4215.01	63.22	4278.23
10	2000 + 4278.23 = 6278.23	94.17	6372.40
11	6372.40	95.58	6467.98
12	6467.98	97.01	6564.99
	2000 + 6564.99 = 8,564.99		

Alternatively, first, we could have computed the i_{eff} that matches the deposits' frequency and then used it to compute the future value. These steps are

$$i_{eff} = \left(1 + \frac{0.045}{3}\right)^3 - 1 = 0.0456$$

$$F = 2000\left[\frac{(1 + 0.0456)^4 - 1}{0.0456}\right] = \$8,564.02$$

Answer the following questions to test your understanding of the preceding section.

1. What do we mean by a uniform series payment?

2. What is the present worth of a uniform series payment?

3. What is the future worth of a uniform series payment?

Vocabulary—State the meaning of the following term:

Uniform Series_____

LO⁶ 20.6   Interest–Time Factors

The engineering economics formulas that we have developed so far are summarized in Tables 20.7 and 20.8. The definitions of the terms in the formulas are given here:

P = present worth, or present cost—lump sum ($)
F = future worth, or future cost—lump sum ($)
A = uniform series payment, or uniform series revenue ($)
i = nominal interest rate
i_{eff} = effective interest rate
n = number of years
m = number of interest compounding periods per year

The interest–time factors shown in the fourth column of Table 20.7 are used as shortcuts to avoid writing long formulas when evaluating equivalent values of various cash flow occurrences.

For example, when evaluating the series payment equivalence of a present principal, instead of writing

$$A = P\left[\frac{(i)(1 + i)^n}{(1 + i)^n - 1}\right]$$

we write $A = P(A/P, i, n)$, where, of course,

$$(A/P, i, n) = \left[\frac{(i)(1 + i)^n}{(1 + i)^n - 1}\right]$$

Interest–time factors are used as shortcuts to avoid writing long formulas when evaluating equivalent values of various cash flow occurrences.

In this example, the $(A/P, i, n)$ term is called the **interest–time factor**, and it reads A given P at $i\%$ interest rate, for a duration of n years. It is used to find A, when the present principal value P is given, by multiplying P by the value of the interest–time factor $(A/P, i, n)$. As an example, the numerical values of interest–time factors for $i = 8\%$ are calculated and shown in Table 20.9.

TABLE 20.7   A Summary of Formulas for Situations when *i* Compounds Annually and the Uniform Series *A* Occurs Annually

To Find	Given	Use This Formula	Interest–Time Factor
F	P	$F = P(1+i)^n$	$(F/P, i, n) = (1+i)^n$
P	F	$P = \dfrac{F}{(1+i)^n}$	$(P/F, i, n) = \dfrac{1}{(1+i)^n}$
P	A	$P = A\left[\dfrac{(1+i)^n - 1}{i(1+i)^n}\right]$	$(P/A, i, n) = \left[\dfrac{(1+i)^n - 1}{i(1+i)^n}\right]$
A	P	$A = P\left[\dfrac{(i)(1+i)^n}{(1+i)^n - 1}\right]$	$(A/P, i, n) = \left[\dfrac{(i)(1+i)^n}{(1+i)^n - 1}\right]$
F	A	$F = A\left[\dfrac{(1+i)^n - 1}{i}\right]$	$(F/A, i, n) = \left[\dfrac{(1+i)^n - 1}{i}\right]$
A	F	$A = F\left[\dfrac{(i)}{(1+i)^n - 1}\right]$	$(A/F, i, n) = \left[\dfrac{(i)}{(1+i)^n - 1}\right]$

TABLE 20.8   A Summary of Formulas for Situations when *i* Compounds *m* Times per Year and the Uniform Series *A* Occurs at the Same Frequency

To Find	Given	Use This Formula
i_{eff}	i	$i_{eff} = \left(1 + \dfrac{i}{m}\right)^m - 1$
F	P	$F = P\left(1 + \dfrac{i}{m}\right)^{nm}$
P	F	$P = \dfrac{F}{\left(1 + \dfrac{i}{m}\right)^{nm}}$
P	A	$P = A\left[\dfrac{\left(1 + \dfrac{i}{m}\right)^{nm} - 1}{\dfrac{i}{m}\left(1 + \dfrac{i}{m}\right)^{nm}}\right]$
A	P	$A = P\left[\dfrac{\left(\dfrac{i}{m}\right)\left(1 + \dfrac{i}{m}\right)^{nm}}{\left(1 + \dfrac{i}{m}\right)^{nm} - 1}\right]$

F	A	$F = A\left[\dfrac{\left(1+\dfrac{i}{m}\right)^{mn}-1}{\dfrac{i}{m}}\right]$
A	F	$A = F\left[\dfrac{\dfrac{i}{m}}{\left(1+\dfrac{i}{m}\right)^{mn}-1}\right]$

TABLE 20.9   The Interest–Time Factors for $i = 8\%$

n	(F/P, i, n)	(P/F, i, n)	(P/A, i, n)	(A/P, i, n)	(F/A, i, n)	(A/F, i, n)
1	1.08000000	0.92592593	0.92592593	1.08000000	1.00000000	1.00000000
2	1.16640000	0.85733882	1.78326475	0.56076923	2.08000000	0.48076923
3	1.25971200	0.79383224	2.57709699	0.38803351	3.24640000	0.30803351
4	1.36048896	0.73502985	3.31212684	0.30192080	4.50611200	0.22192080
5	1.46932808	0.68058320	3.99271004	0.25045645	5.86660096	0.17045645
6	1.58687432	0.63016963	4.62287966	0.21631539	7.33592904	0.13631539
7	1.71382427	0.58349040	5.20637006	0.19207240	8.92280336	0.11207240
8	1.85093021	0.54026888	5.74663894	0.17401476	10.63662763	0.09401476
9	1.99900463	0.50024897	6.24688791	0.16007971	12.48755784	0.08007971
10	2.15892500	0.46319349	6.71008140	0.14902949	14.48656247	0.06902949
11	2.33163900	0.42888286	7.13896426	0.14007634	16.64548746	0.06007634
12	2.51817012	0.39711376	7.53607802	0.13269502	18.97712646	0.05269502
13	2.71962373	0.36769792	7.90377594	0.12652181	21.49529658	0.04652181
14	2.93719362	0.34046104	8.24423698	0.12129685	24.21492030	0.04129685
15	3.17216911	0.31524170	8.55947869	0.11682954	27.15211393	0.03682954
16	3.42594264	0.29189047	8.85136916	0.11297687	30.32428304	0.03297687
17	3.70001805	0.27026895	9.12163811	0.10962943	33.75022569	0.02962943
18	3.99601950	0.25024903	9.37188714	0.10670210	37.45024374	0.02670210
19	4.31570106	0.23171206	9.60359920	0.10412763	41.44626324	0.02412763

(Continued)

TABLE 20.9   The Interest–Time Factors for $i = 8\%$ (Continued)

n	(F/P, i, n)	(P/F, i, n)	(P/A, i, n)	(A/P, i, n)	(F/A, i, n)	(A/F, i, n)
20	4.66095714	0.21454821	9.81814741	0.10185221	45.76196430	0.02185221
21	5.03383372	0.19865575	10.01680316	0.09983225	50.42292144	0.01983225
22	5.43654041	0.18394051	10.20074366	0.09803207	55.45675516	0.01803207
23	5.87146365	0.17031528	10.37105895	0.09642217	60.89329557	0.01642217
24	6.34118074	0.15769934	10.52875828	0.09497796	66.76475922	0.01497796
25	6.84847520	0.14601790	10.67477619	0.09367878	73.10593995	0.01367878
26	7.39635321	0.13520176	10.80997795	0.09250713	79.95441515	0.01250713
27	7.98806147	0.12518682	10.93516477	0.09144810	87.35076836	0.01144810
28	8.62710639	0.11591372	11.05107849	0.09048891	95.33882983	0.01048891
29	9.31727490	0.10732752	11.15840601	0.08961854	103.96593622	0.00961854
30	10.06265689	0.09937733	11.25778334	0.08882743	113.28321111	0.00882743
31	10.86766944	0.09201605	11.34979939	0.08810728	123.34586800	0.00810728
32	11.73708300	0.08520005	11.43499944	0.08745081	134.21353744	0.00745081
33	12.67604964	0.07888893	11.51388837	0.08685163	145.95062044	0.00685163
34	13.69013361	0.07304531	11.58693367	0.08630411	158.62667007	0.00630411
35	14.78534429	0.06763454	11.65456822	0.08580326	172.31680368	0.00580326
36	15.96817184	0.06262458	11.71719279	0.08534467	187.10214797	0.00534467
37	17.24562558	0.05798572	11.77517851	0.08492440	203.07031981	0.00492440
38	18.62527563	0.05369048	11.82886899	0.08453894	220.31594540	0.00453894
39	20.11529768	0.04971341	11.87858240	0.08418513	238.94122103	0.00418513
40	21.72452150	0.04603093	11.92461333	0.08386016	259.05651871	0.00386016
41	23.46248322	0.04262123	11.96723457	0.08356149	280.78104021	0.00356149
42	25.33948187	0.03946411	12.00669867	0.08328684	304.24352342	0.00328684
43	27.36664042	0.03654084	12.04323951	0.08303414	329.58300530	0.00303414
44	29.55597166	0.03383411	12.07707362	0.08280152	356.94964572	0.00280152
45	31.92044939	0.03132788	12.10840150	0.08258728	386.50561738	0.00258728
46	34.47408534	0.02900730	12.13740880	0.08238991	418.42606677	0.00238991
47	37.23201217	0.02685861	12.16426741	0.08220799	452.90015211	0.00220799
48	40.21057314	0.02486908	12.18913649	0.08204027	490.13216428	0.00204027
49	43.42741899	0.02302693	12.21216341	0.08188557	530.34273742	0.00188557
50	46.90161251	0.02132123	12.23348464	0.08174286	573.77015642	0.00174286

Additional values of interest–time factors for other interest rates can be created using Excel. Keep in mind that you can use those tables or other similar tables found in the back of most engineering economics text books to determine interest–time factors for interest rates that compound more frequently than once a year. To do so, however, you must first divide the quoted nominal interest rate i by the number of the compounding frequency m and use the resulting number to pick the appropriate interest table to use. You must then multiply the number of years n by the number of the compounding frequency m and use the outcome of n times m as the period when looking up interest–time factors. For example, if a problem states an interest rate of 18% compounding monthly for four years, you use the 1.5% interest table $(18/12 = 1.5)$, and for the number of periods, you will use $48(4 \times 12 = 48)$.

EXAMPLE 20.10

What is the equivalent present worth of the cash flow given in Figure 20.5? Put another way, how much money do you need to deposit in the bank today in order to be able to make the withdrawals shown? The interest rate is 8% compounding annually.

The present worth (PW) of the given cash flow is determined from

$$PW = 1000(P/A,\ 8\%,\ 4) + 3000(P/F,\ 8\%,5) + 5000(P/F,\ 8\%,7)$$

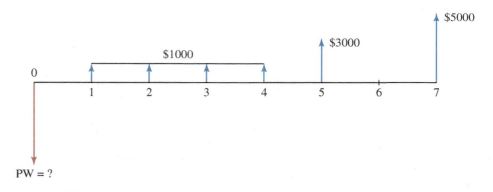

FIGURE 20.5   The cash flow diagram for Example 20.10.

We can use Table 20.8 to look up the interest–time factor values, which leads to

$(P/A, 8\%, 4) = 3.31212684$
$(P/F,\ 8\%,5) = 0.68058320$
$(P/F,\ 8\%,7) = 0.58349040$
$PW = (1000)(3.31212684) + (3000)(0.68058320) + (5000)(0.58349040)$
$PW = \$8271.32$

Therefore, if today, you put aside \$8271.32 in an account that pays 8% interest, you can withdraw \$1000 in the next four years, and \$3000 in five years, and \$5000 in seven years.

LO⁷ 20.7 Choosing the Best Alternatives— Decision Making

Up to this point, we have been discussing general relationships that deal with money, time, and interest rates. Let us now consider the application of these relationships in an engineering setting. Imagine that you are assigned the task of choosing which air-conditioning unit to purchase for your company. After an exhaustive search, you have narrowed your selection to two alternatives, both of which have an anticipated 10 years of working life. Assuming an 8% interest rate, find the best alternative. Additional information is given in Table 20.10. The cash-flow diagrams for each alternative are shown in Figure 20.6.

TABLE 20.10	Data to Be Used in Selection of an Air-Conditioning Unit	
Criteria	**Alternative A**	**Alternative B**
Initial cost	$100,000	$85,000
Salvage value after 10 years	$10,000	$5000
Operating cost per year	$2500	$3400
Maintenance cost per year	$1000	$1200

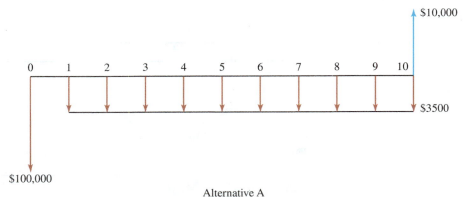

Alternative A

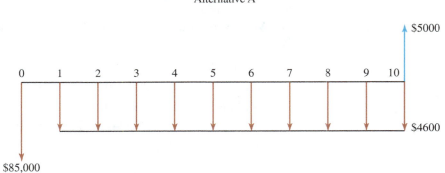

Alternative B

| FIGURE 20.6 | The cash-flow diagrams for the example problem. |

Here we will discuss three different methods that you can use to choose the best economical alternative from many options. The three methods are commonly referred to as (1) present worth (PW) or present cost analysis, (2) annual worth (AW) or annual cost analysis, and (3) future worth (FW) or future cost analysis. When these methods are applied to a problem, they all lead to the same conclusion. So in practice, you need only apply one of these methods to evaluate options; however, in order to show you the details of these procedures, we will apply all of these methods to the preceding problem.

Present Worth or Present Cost Analysis With this approach you compute the total present worth or the present cost of each alternative and then pick the alternative with the lowest present cost or choose the alternative with the highest present worth or profit. To employ this method, you begin by calculating the equivalent present value of all cash flow. For the example problem mentioned, the application of the present worth analysis leads to:

Alternative A:

$$PW = -100,000 - (2500 + 1000)(P/A, \ 8\%, \ 10) + 10,000(P/F, \ 8\%, \ 10)$$

The interest–time factors for $i = 8\%$ are given in Table 20.9.

$$PW = -100,000 - (2500 + 1000)(6.71008140) + (10,000)(0.46319349)$$
$$PW = -118,853.35$$

Alternative B:

$$PW = -85,000 - (3400 + 1200)(P/A, \ 8\%, \ 10) + 5000(P/F, \ 8\%, \ 10)$$
$$PW = -85,000 - (3400 + 1200)(6.71008140) + 5000(0.46319349)$$
$$PW = -113,550.40$$

Note that we have determined the equivalent present worth of all future cash flow, including the yearly maintenance and operating costs and the salvage value of the air-conditioning unit. In the preceding analysis, the negative sign indicates cost, and because alternative B has a lower present cost, we choose alternative B.

Annual Worth or Annual Cost Analysis Using this approach, we compute the equivalent annual worth or annual cost value of each alternative and then pick the alternative with the lowest annual cost or select the alternative with the highest annual worth or revenue. Applying the annual worth analysis to our example problem, we have

Alternative A:

$$AW = -(2500 + 1000) - 100,000(A/P, \ 8\%, \ 10) + 10,000(A/F, \ 8\%, \ 10)$$
$$AW = -(2500 + 1000) - (100,000)(0.14902949) + (10,000)(0.06902949)$$
$$AW = -17,712.65$$

Alternative B:

$$AW = -(3400 + 1200) - 85,000(A/P, \ 8\%, \ 10) + 5000(A/F, \ 8\%, \ 10)$$
$$AW = -(3400 + 1200) - 85,000(0.14902949) + 5000(0.06902949)$$
$$AW = -16,922.35$$

Note that using this method, we have determined the equivalent annual worth of all cash flow, and because alternative B has a lower annual cost, we choose alternative B.

Future Worth or Future Cost Analysis This approach is based on evaluating the future worth or future cost of each alternative. Of course, you will then choose the alternative with the lowest future cost or pick the alternative with the highest future worth or profit. The future worth analysis of our example problem follows.

Alternative A:

$$FW = +10,000 - 100,000\,(F/P,\ 8\%,\ 10) - (2500 + 1000)\,(F/A,\ 8\%,\ 10)$$
$$FW = +10,000 - 100,000\,(2.15892500) - (2500 + 1000)\,(14.48656247)$$
$$FW = -256,595.46$$

Alternative B:

$$FW = +5,000 - 85,000\,(F/P,\ 8\%,\ 10) - (3400 + 1200)\,(F/A,\ 8\%,\ 10)$$
$$FW = +5,000 - 85,000\,(2.15892500) + (3400 + 1200)\,(14.48656247)$$
$$FW = -245,146.81$$

> You can use present worth, annual worth, or future worth analysis to choose the best economical alternative from many options.

Because alternative B has a lower future cost, again we choose alternative B. Note that regardless of which method we decide to use, alternative B is economically the better option. Moreover, for each alternative, all of the approaches discussed here are related to one another through the interest–time relationships (factors). For example,

Alternative A:

$$PW = AW\,(P/A,\ 8\%,\ 10) = (-17,712.65)(6.71008140) = -118,853.32$$

or

$$PW = FW\,(P/F,\ 8\%,\ 10) = (-256,595.46)(0.46319349) = -118,853.34$$

Alternative B:

$$PW = AW\,(P/A,\ 8\%,\ 10) = (-16,922.35)(6.71008140) = -113,550.40$$

or

$$PW = FW\,(P/F,\ 8\%,\ 10) = (-245,146.81)(0.46319349) = -113,550.40$$

LO[8] 20.8 Excel Financial Functions

You also can use Excel **financial functions** to solve engineering economic problems. Examples of Excel's financial functions and how they may be used are given in Table 20.11. Please pay close attention to the terminology used by

Formula	Equation Number	Excel Financial Function	Example/ Section	How to Use Excel to Solve the Example Problem
$F = P(1 + i)^n$ $F = P(1 + i/m)^{nm}$ F Future value (Future worth) i Interest per year (%) m Interest compounding frequency n Period (years) P Present value (Present worth)	(20.2) (20.3)	**= FV (Rate, Nper, Pmt, Pv, Type)** Fv Future value (Future worth) Rate Interest rate per period Nper Total number of payment periods in an annuity Pmt Payment made each period Pv Present value (Present worth) Type Number 0 or 1 and indicates when payments are due. If type is omitted, it is assumed to be 0. (0: at the end of the period, 1: at the beginning of the period) Investment or payment is described as negative number and income is described as positive number.	*Example 20.3* *Example 20.4*	= FV(0.07,8,,−1500) = 2,577.28 = FV(0.07/12,8 * 12,,−1500) = 2,621.74
$i_{eff} = \left(1 + \dfrac{i}{m}\right)^m - 1$ i_{eff} Effective interest rate per year (%) i Normal interest per year (%) m Interest compounding frequency	(20.5)	**= EFFECT (Nominal interest per year, Npery)** Npery Number of compounding periods per year	*Example 20.6a* *Example 20.6c*	= EFFECT(0.07,12) = 7.23% = EFFECT(0.06,2) = 6.09%
$P = \dfrac{F}{(1 + i)^n}$ P Present value (Present worth) i Interest per year (%) n Period (years) F Future value (Future worth)	(20.4)	**= PV (Rate, Nper, Pmt, Fv, Type)** Pv Present value (Present worth) Rate Interest rate per period Nper Total number of payment periods in an annuity Pmt Payment made each period Fv Future value (Future worth) Type Number 0 or 1 indicates when payments are due. If type is omitted, it is assumed to be 0.	*Section 20.3*	= PV(0.0065,5,2000,,) = −1,459.76
$A = F\left[\dfrac{i}{(1 + i)^n - 1}\right]$	(20.13)	**= PMT (Rate, Nper, Pv, Fv, Type)** PMT Payment for a loan based on constant payments and a constant interest rate Rate Interest rate per period Nper Total number of payment periods in an annuity	*Section 20.5* *Example 20.7*	= PMT(0.065,5,0,2000) = −351.27 = PMT(0.065/12,5*12,0,2000) = −28.30

(Continued)

Example of Excel's Financial Functions and How They May be Used (Continued)

Formula	Equation Number	Excel Financial Function	Example/Section	How to Use Excel to Solve the Example Problem
$$A = P\left[\dfrac{\left(\dfrac{i}{m}\right)\left(1+\dfrac{i}{m}\right)^{nm}}{\left(1+\dfrac{i}{m}\right)^{nm}-1}\right]$$	(20.15)	Pv Present value (Present worth) Fv Future value (Future worth) Type Number 0 or 1 and indicates when payments are due. If type is omitted, it is assumed to be 0. (0: at the end of the period, 1: at the beginning of the period)	*Example 20.8*	= PMT(0.08/12,5*12,10000) = −202.76
$$A = F\left[\dfrac{\left(\dfrac{i}{m}\right)}{\left(1+\dfrac{i}{m}\right)^{nm}-1}\right]$$	(20.17)	Investment or payment is described as negative number and income is described as positive number.		
A Annuity, or Payment for a loan based on uniform payments and a constant interest rate i Interest per year (%) n Period (years) m Interest compounding frequency F Future value (Future worth) P Present value (Present worth)		*Data for the Example in Section 20.7* **Alternative A** Initial cost 100,000 Salvage value after 10 years 10,000 Operating cost per year 2,500 Maintenance cost per year 1,000 **Alternative B** Initial cost 85,000 Salvage value after 10 years 5,000 Operating cost per year 3,400 Maintenance cost per year 1200	*Example 20.10* *Section 20.7*	PW = −100000−PV(0.08,10,−(2500+1000))+PV(0.08,10,0,−10000) = −118,853.35 AW = −(2500+1000)−PMT(0.08,10,−100000)−PMT(0.08,10,0,−10000) = −17,712.65 FW = 10000−FV(0.08,10,0,−100000)−FV(0.08,10,−(2500+1000)) = −256,595.47 PW = −85000−PV(0.08,10,−(3400+1200))+PV(0.08,10,0,−5000) = −113,550.41 AW = −(3400+1200)−PMT(0.08,10,−85000)+PMT(0.08,10,0,−5000) = −16,922.36 FW = 5000−FV(0.08,10,0,−85000)−FV(0.08,10,−(3400+1200)) = −245,146.81

= PV(0.08, 4,1000)+PV(0.08,5,0, 3000)+PV(0.08,7,0,5000)
= −8,271.33

Excel, and the sign of variables, while following the solutions to the example problems.

EXAMPLE 20.11

A bank charges you, the credit card holder, 13.24% compounding monthly. Imagine that you have accumulated debt in an amount of $4,000. Your credit card statement shows a minimum monthly payment of $20.00. Assuming you wise up and realize that you better pay off your debt before charging on your card again, how long would it take to pay off the debt completely if you were to make the minimum payments? How long would it take if you were to make a monthly payment of $50.00?

This problem could be solved by trial and error for the value of *n*, using Equations (20.14) or (20.15), or better yet, using Excel's NPER function. This function returns the number of periods for an investment, given the interest rate, the uniform series payments, and the present value are known.

For the minimum payments of $20.00, the `=NPER(0.1324/12, -20, -4000)` will then return the value 106.19 months or 8.85 years.

For monthly payments of $50.00, the `=NPER(0.1324/12, -50, -4000)` will return 57.66 months or 4.8 years.

The moral of the story is "try not to get into debt, but if you do, pay the debt as quickly as possible!"

Finally, it is worth noting that you can take semester-long classes in engineering economics. Some of you will eventually do so. You will learn more in depth about the principles of money–time relationships, including rate-of-return analysis, benefit–cost ratio analysis, general price inflation, bonds, depreciation methods, evaluation of alternatives on an after-tax basis, and risk and uncertainty in engineering economics. For now, our intent has been to introduce you to engineering economics, but keep in mind that we have just scratched the surface! We cannot resist ending this section with definitions of some of these important concepts that you will learn more about later.

Bonds

States, counties, and cities issue bonds to raise money to pay for various projects, such as schools, highways, convention centers, and stadiums. Corporations also issue bonds to raise money to expand or to modernize their facilities. There are many different types of bonds, but basically, they are loans that investors make to government or corporations in return for some gain. When a bond is issued, it will have a *maturity date* (a year or less to 30 years or longer), *par value* (the amount originally paid for the bond and the amount that will be repaid at maturity date), and an *interest rate* (percentage of par value that is paid to bond holder at regular intervals).

Depreciation

Assets (such as machines, cars, and computers) lose their value over a period of time. For example, a computer purchased today by a company for $2000 is not worth as much in three or four years. Companies use this reduction in value of an asset against their before-tax income. There are rules and guidelines that specify what can be depreciated, by how much, and over what period of time. Examples of depreciation methods include the Straight Line and the Modified Accelerated Cost Recovery System (MACRS).

Life-Cycle Cost

In engineering, the term *life-cycle cost* refers to the sum of all the costs that are associated with a structure, a service, or a product during its life span. For example, if you are designing a bridge or a highway, you need to consider the costs that are related to the initial definition and assessment, environmental study, conceptual design, detailed design, planning, construction, operation, maintenance, and disposal of the project at the end of its life span.

Before You Go On

Answer the following questions to test your understanding of the preceding section.

1. What are interest–time factors, and how are they used in the analysis of engineering economics problems?

2. Based on engineering economics principles, explain how would you choose the best alternative from among many choices.

3. Give some examples of Excel financial functions.

SUMMARY

LO¹ Cash Flow Diagrams

Economics plays an important role in engineering decision making. Moreover, a good understanding of the fundamentals of engineering economics could also benefit you in better managing your lifelong financial activities. Cash flow diagrams are visual aids that show the flow of costs and revenues over a period of time. Cash flow diagrams show when the cash flow occurs, the cash flow magnitude, and whether the cash flow is out of your pocket (cost) or into your pocket (revenue).

LO² Simple and Compound Interest

Interest is the extra money (in addition to the borrowed amount) that one must pay for the purpose of having access to the borrowed money. Simple interest is the interest that one pays only on the initial borrowed amount, while under the compounding interest scheme, the interest paid on the initial principal will also collect interest.

LO³ Future Worth of a Present Amount and Present Worth of a Future Amount

You should know the relationship among money, time, and interest rate. You should be familiar with how these relationships were derived. For example, the future value F of any present amount (principal) P, after n years collecting $i\%$ interest compounding annually, is given by $F = P(1 + i)^n$ and vice versa $P = F/(1 + i)^n$.

LO⁴ Effective Interest Rate

If you were to deposit $1000.00 in a savings account, at 4% compounding monthly, then at the end of one year you will have $1040.74 in your account. The $40.74 earned during the first year is higher than the stated 4% interest, which could be understood as $40.00 for a $1000.00 deposit over a period of one year. In order to avoid confusion, the stated or the quoted interest rate is called the *nominal interest rate,* and the actual earned interest rate is called the *effective interest rate.*

LO⁵ Present and Future Worth of Series Payment

Consider a situation where you have borrowed some money for a new car, denoted by P, at an interest rate i from a bank and you are expected to pay the loan monthly, in equal amounts A, in n years. The monthly payments are commonly referred to as uniform series payments, denoted by A. Moreover, there is a relationship between the borrowed money P (present worth) and the series payment A. Consider another situation where you deposit money A each month in a savings account that pays an interest rate i for a period of n years. There is a relationship between the series payment A and the future worth F of the payments.

LO⁶ Interest–Time Factors

The interest–time factors are used as shortcuts to avoid writing long formulas when evaluating equivalent values of various cash flow occurrences. For example, when evaluating the series payment equivalence of a present principal, instead of writing

$$A = P\left[\frac{(i)(1+i)^n}{(1+i)^n - 1}\right]$$

we write $A = P(A/P, i, n)$, where, of course,

$$(A/P, i, n) = \left[\frac{(i)(1+i)^n}{(1+i)^n - 1}\right]$$

In this example, the $(A/P, i, n)$ term is called the *interest–time factor*, and it reads A given P at $i\%$ interest rate, for a duration of n years. It is used to find A, when the present principal value P is given, by multiplying P by the value of the interest–time factor $(A/P, i, n)$.

LO⁷ Choosing the Best Alternatives—Decision Making

You can use any of the following three different methods to choose the best (most economical) alternative from many options: present worth (PW) or present cost analysis, (2) annual worth (AW) or annual cost analysis, and (3) future worth (FW) or future cost analysis.

LO⁸ Excel Financial Functions

You can use Excel financial functions to solve engineering economic problems; however, pay close attention to the terminology used by Excel and the sign of variables.

KEY TERMS

Cash Flow Diagram 760
Compound Interest 762
Effective Interest Rate 765
Excel Financial Functions 780
Future Worth of a Present
  Amount 762

Future Worth of a Series
  Payment 769
Interest-Time Factors 773
Nominal Interest Rate 765
Present Worth of a Future
  Amount 764

Present Worth of a Series
  Payment 768
Series Payment 768
Simple Interest Rate 761

APPLY WHAT YOU HAVE LEARNED

Nearly all of you will start a family and purchase a house in the near future. Obviously, most of us cannot pay for a house in cash, so we borrow money from a bank. Imagine your house loan payment extends for 30 years at 6%, and you make your home mortgage payments monthly. Create a table that shows after how many months 25%, 50%, and 75% of the money you borrowed from the bank is paid off. Do your findings depend on how much money you borrow? What if your loan payment were to extend for 15 years at 5%? How much money would you save? Explain your findings in a brief report.

PROBLEMS

Problems that promote lifelong learning are denoted by 🔑

20.1 Compute the future value of the following deposits made today:

a. $10,000 at 6.75% compounding annually for 10 years

b. $10,000 at 6.75% compounding quarterly for 10 years

c. $10,000 at 6.75% compounding monthly for 10 years

20.2 Compute the interest earned on the deposits made in Problem 20.1.

20.3 How much money do you need to deposit in a bank today if you are planning to have $5000 in four years by the time you get out of college? The bank offers a 6.75% interest rate that compounds monthly.

20.4 How much money do you need to deposit in a bank each month if you are planning to have $5000 in four years by the time you get out of college? The bank offers a 6.75% interest rate that compounds monthly.

20.5 Determine the effective rate corresponding to the following nominal rates:

a. 6.25% compounding monthly

b. 9.25% compounding monthly

c. 16.9% compounding monthly

20.6 Using Excel or a spreadsheet of your choice, create interest–time factor tables, similar to Table 20.9, for $i = 6.5\%$ and $i = 6.75\%$.

20.7 Using Excel or a spreadsheet of your choice, create interest–time factor tables, similar to Table 20.9, for $i = 7.5\%$ and $i = 7.75\%$.

20.8 Using Excel or a spreadsheet of your choice, create interest–time factor tables, similar to Table 20.9, for $i = 8.5\%$ and $i = 9.5\%$.

20.9 Using Excel or a spreadsheet of your choice, create interest–time factor tables, similar to Table 20.9, that can be used for $i = 8.5\%$ compounding monthly.

20.10 Most of you have credit cards, so you already know that if you do not pay the balance on time, the credit card issuer will charge you a certain interest rate each month. Assuming that you are charged 1.25% interest each month on your unpaid balance, what are the nominal and effective interest rates? Also,

determine the effective interest rate that your own credit card issuer charges you.

20.11 You have accepted a loan in the amount of $15,000 for your new car. You have agreed to pay the loan back in four years. What is your monthly payment if you agree to pay an interest rate of 9% compounding monthly? Solve this problem for $i = 6\%$, $i = 7\%$, and $i = 8\%$, each compounding monthly.

20.12 How much money will you have available to you after five years if you put aside $100.00 a month in an account that gives you 6.75% interest compounding monthly?

20.13 How long does it take to double a deposit of $1000

a. at a compound annual interest rate of 6%

b. at a compound annual interest rate of 7%

c. at a compound annual interest rate of 8%

d. If instead of $1000 you deposit $5000, would the time to double your money be different in parts (a)–(c)? In other words, is the initial sum of money a factor in determining how long it takes to double your money?

Now use your answers to verify a rule of thumb that is commonly used by bankers to determine how long it takes to double a sum of money. The rule of thumb commonly used by bankers is given by

$$\text{time period to double a sum of money} \approx \frac{72}{\text{interest rate}}$$

20.14 Imagine that as an engineering intern you have been assigned the task of selecting a motor for a pump. After reviewing motor catalogs, you narrow your choice to two motors that are rated at 1.5 kW. Additional information collected is shown in an accompanying table. The pump is expected to run 4200 hours every year. After checking with your electric utility company, you determine the average cost of electricity is about 11 cents per kWh. Based on the information given here, which one of the motors will you recommend to be purchased?

Criteria	Motor X	Motor Y
Expected useful life	5 years	5 years
Initial cost	$300	$400
Efficiency at the operating point	0.75	0.85
Estimated maintenance cost	$12 per year	$10 per year

20.15 What is the equivalent present worth of the cash flow given in the accompanying figure? Assume $i = 8\%$.

20.16 What is equivalent future worth of the cash flow given in the accompanying figure? Assume $i = 8\%$.

20.17 What is the equivalent annual worth of the cash flow given in the accompanying figure? Assume $i = 8\%$.

20.18 What are the equivalent present worth, annual worth, and future worth of the cash flow given in the accompanying figure? Assume $i = 8\%$.

20.19 You are to consider the following projects. Which project would you approve if each project creates the same income? Assume $i = 8\%$ and a period of 15 years.

	Project X	Project Y
Initial cost	$55,000	$80,000
Annual operating cost	$15,000	$10,000
Annual maintenance cost	$6,000	$4,000
Salvage value at the end of 15 years	$10,000	$15,000

20.20 In order to purchase a new car, imagine that you recently have borrowed $15,000

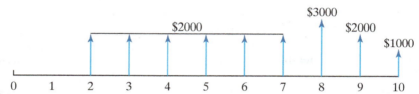

Problem 20.15

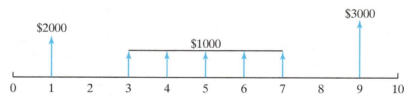

Problem 20.16

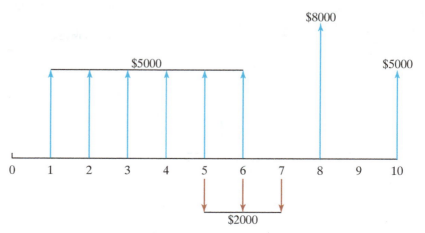

Problem 20.17

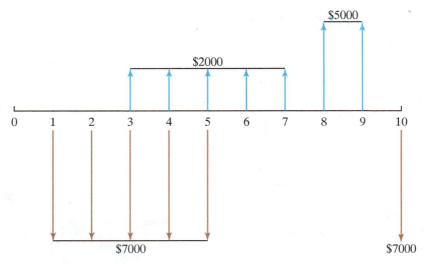

Problem 20.18

from a bank that charges you according to a nominal rate of 8%. The loan is payable in 60 months. (a) Calculate the monthly payments. (b) Assume the bank charges a loan fee of 4.5% of the loan amount payable at the time they give you the loan. What is the effective interest rate that you actually are being charged?

20.21 Imagine the company that you work for borrows $8,000,000 at 8% interest, and the

Year	Amount
1	$1,000,000
2	$1,000,000
3	$1,000,000
4	$1,000,000
5	$1,000,000
6	$1,000,000
7	$?

loan is to be paid in seven years according to the schedule shown. Determine the amount of the last payment.

20.22 You need to borrow $12,000 to buy a car, so you visit two banks and are given two alternatives. The first bank allows you to pay $2595.78 at the end of each year for six years. The first payment is to be made at the end of the first year. The second bank offers equal monthly loan payments of $198.87, starting at the end of first month. What are the interest rates that the banks are charging? Which alternative is more attractive?

20.23 What is the value of X if the given cash flow diagrams are equivalent? Assume $i = 8\%$.

20.24 Your future company has been presented with an opportunity to invest in a project with the following cash flow for ten years. If the company would like to make at least 8% on its investment, would you invest in the project?

Problem 20.23

Initial investment	$10,000,000
Income (annual)	$3,000,000
Labor cost (annual)	$400,000
Material cost (annual)	$150,000
Maintenance cost (annual)	$80,000
Utility cost (annual)	$200,000

20.25 Your future company has purchased a machine and has entered into a contract that requires the company to pay $2000 each year for the upgrade of machine components at the end of years 6, 7, and 8. In anticipation of the upgrade cost, your company has decided to deposit equal amounts (X) at the end of each year for five years in a row in an account that pays $i = 6\%$. The first deposit is made at the end of the first year. What is the value of X?

20.26 Your car loan payment extends for six years at 8% interest compounded monthly. After how many months do you pay off half of your loan?

20.27 What are the equivalent annual worth and future worth of the cash flow given in Problem 20.15? Assume $i = 8\%$.

20.28 What are the equivalent present worth and annual worth of the cash flow given in Problem 20.16? Assume $i = 8\%$.

20.29 What are the equivalent present worth and future worth of the cash flow given in Problem 20.17? Assume $i = 8\%$.

20.30 The cash flow given in Problem 20.18 is to be replaced by an equivalent cash flow with equal amounts (X) at the end of years 6, 7, 8, 9, and 10. What is the value of X?

20.31 The cash flow given in Problem 20.15 is to be replaced by an equivalent cash flow with equal amounts (X) at the end of years 5 through 10. What is the value of X?

20.32 The cash flow given in Problem 20.16 is to be replaced by an equivalent cash flow with equal amounts (X) at the end of years 9 and 10. What is the value of X?

20.33 Solve Problem 20.1 using Excel.

20.34 Solve Problem 20.3 using Excel.

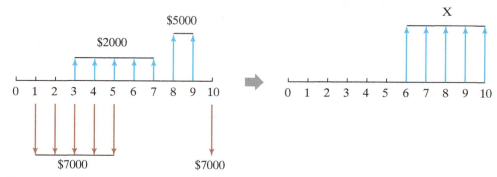

Problem 20.30

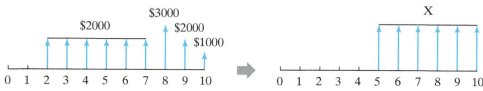

Problem 20.31

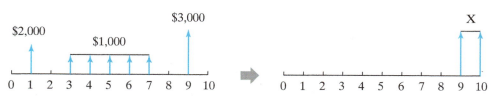

Problem 20.32

20.35 Solve Problem 20.4 using Excel.

20.36 Solve Problem 20.5 using Excel.

20.37 Solve Problem 20.11 using Excel.

20.38 Solve Problem 20.15 using Excel.

20.39 Solve Problem 20.16 using Excel.

20.40 Solve Problem 20.17 using Excel.

20.41 Solve Problem 20.18 using Excel.

20.42 Solve Problem 20.19 using Excel.

20.43 Solve Problem 20.21 using Excel.

20.44 Solve Problem 20.22 using Excel.

20.45 Solve Problem 20.23 using Excel.

20.46 Solve Problem 20.24 using Excel.

20.47 Solve Problem 20.25 using Excel.

20.48 Solve Problem 20.26 using Excel.

20.49 Solve Problem 20.31 using Excel.

20.50 Solve Problem 20.32 using Excel.

20.51 Investigate what type of APRs your credit card company charges you. Assuming your credit card has a balance of $2000.00, if you were to pay an extra $20 above your minimum payment, how much money would you save? Write a brief report explaining your findings.

20.52 The U.S. Treasury Department issues savings bonds that commonly are purchased to pay for a child's future education expenses, plan one's own retirement, or give as a gift. Investigate what is meant by I saving bonds. Write a brief report explaining your findings and give examples.

20.53 The U.S. Bureau of Labor Statistics produces data sets that show how the price of goods and services changes from month to month and from year to year. Investigate the Consumer Price Indexes (CPI), and write a brief report explaining your findings and giving examples.

20.54 As we mentioned in this chapter, assets such as machines, cars, and computers lose their value over a period of time. Companies use this reduction in value of an asset against their before-tax income. An example of a depreciation method is called the straight line. Investigate what is meant by straight line depreciation. Write a brief report explaining your findings and giving examples.

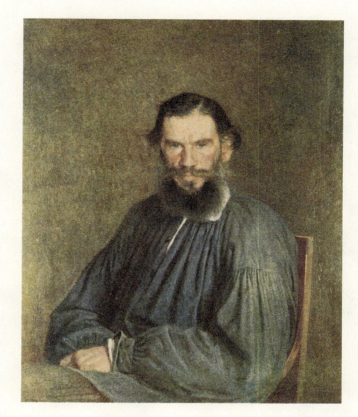

*"Everyone thinks of changing the world,
but no one thinks of changing oneself."*
—Leo Tolstoy (1828–1910)

http://www.biography.com/people/leo-tolstoy-9508518

A Summary of Formulas Discussed in the Book

traffic flow: $q = \dfrac{3600\,n}{T}$

average speed $= \dfrac{\text{distance traveled}}{\text{time}}$

average acceleration $= \dfrac{\text{change in velocity}}{\text{time}}$

volume flow rate $= \dfrac{\text{volume}}{\text{time}}$

angular speed: $\omega = \dfrac{\Delta\theta}{\Delta t}$

the relationship between linear and angular speed: $V = r\omega$

average angular acceleration $= \dfrac{\text{change in angular speed}}{\text{time}}$

density $= \dfrac{\text{mass}}{\text{volume}}$

specific volume $= \dfrac{\text{volume}}{\text{mass}}$

specific gravity $= \dfrac{\text{density of a material}}{\text{density of water@4° C}}$

specific weight $= \dfrac{\text{weight}}{\text{volume}}$

mass flow rate $= \dfrac{\text{mass}}{\text{time}}$

mass flow rate $= (\text{density})(\text{volume flow rate})$

linear momentum: $\overrightarrow{L} = m\overrightarrow{V}$

spring force (Hooke's law): $F = kx$

Newton's second law: $\sum F = ma$

Newton's law of gravitational attraction: $F = \dfrac{Gm_1 m_2}{r^2}$

Weight: $W = mg$

hydrostatic pressure: $P = \rho g h$

buoyancy: $F_B = \rho V g$

stress–strain relation (Hooke's law): $\sigma = E\varepsilon$

Temperature conversion:

$$T(^\circ C) = \frac{5}{9}\left(T(^\circ F) - 32\right) \qquad T(^\circ F) = \frac{9}{5}\left(T(^\circ C)\right) + 32$$

$$T(K) = T(^\circ C) + 273.15 \qquad T(^\circ R) = T(^\circ F) + 459.67$$

Fourier's law: $q = kA\dfrac{T_1 - T_2}{L}$

Newton's law of cooling: $q = hA\left(T_s - T_f\right)$

radiation: $q = \varepsilon \sigma A T_s^4$

coefficient of thermal linear expansion: $\alpha_L = \dfrac{\Delta L}{L\,\Delta T}$

coefficient of thermal volumetric expansion: $\alpha_v = \dfrac{\Delta V}{V\,\Delta T}$

Coulomb's law: $F_{12} = \dfrac{k q_1 q_2}{r^2}$

Ohm's law: $V = RI$

electrical power: $P = VI$

kinetic energy $= \dfrac{1}{2}mV^2$

change in potential energy $= \Delta PE = mg\Delta h$

elastic energy $= \dfrac{1}{2}kx^2$

conservation of mechanical energy: $\Delta KE + \Delta PE + \Delta EE = 0$

conservation of energy—first law of thermodynamics: $Q - W = \Delta E$

$$\text{power} = \frac{\text{work}}{\text{time}} = \frac{(\text{force})(\text{distance})}{\text{time}} \quad \text{or} \quad \text{power} = \frac{\text{energy}}{\text{time}}$$

$$\text{efficiency} = \frac{\text{actual output}}{\text{required input}}$$

standard deviation: $s = \sqrt{\dfrac{\sum\limits_{i=1}^{n}(x_i - \bar{x})^2}{n-1}}$

The Greek Alphabet

A	α	alpha
B	β	beta
Γ	γ	gamma
Δ	δ	delta
E	ε	epsilon
Z	ζ	zeta
H	η	eta
Θ	θ	theta
I	ι	iota
K	κ	kappa
Λ	λ	lambda
M	μ	mu
N	ν	nu
Ξ	ξ	xi
O	o	omicron
Π	π	pi
P	ρ	rho
Σ	σ	sigma
T	τ	tau
Υ	υ	upsilon
Φ	ϕ	phi
X	χ	chi or khi
Ψ	ψ	psi
Ω	ω	omega

Some Useful Trigonometric Relationships

Pythagorean relation:

$$a^2 + b^2 = c^2$$

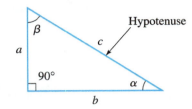

$$\sin \alpha = \frac{\text{opposite}}{\text{hypotenuse}} = \frac{a}{c}$$

$$\sin \beta = \frac{\text{opposite}}{\text{hypotenuse}} = \frac{b}{c}$$

$$\cos \alpha = \frac{\text{adjacent}}{\text{hypotenuse}} = \frac{b}{c}$$

$$\cos \beta = \frac{\text{adjacent}}{\text{hypotenuse}} = \frac{a}{c}$$

$$\tan \alpha = \frac{\sin \alpha}{\cos \alpha} = \frac{\text{opposite}}{\text{adjacent}} = \frac{a}{b}$$

$$\tan \beta = \frac{\sin \beta}{\cos \beta} = \frac{\text{opposite}}{\text{adjacent}} = \frac{b}{a}$$

The sine rule:

$$\frac{a}{\sin \alpha} = \frac{b}{\sin \beta} = \frac{c}{\sin \theta}$$

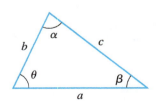

The cosine rule:

$$a^2 = b^2 + c^2 - 2bc(\cos \alpha)$$

$$b^2 = a^2 + c^2 - 2ac(\cos \beta)$$

$$c^2 = a^2 + b^2 - 2ba(\cos \theta)$$

Some other useful trignometry identities:

$$\sin^2\alpha + \cos^2\alpha = 1$$

$$\sin 2\alpha = 2 \sin \alpha \cos \alpha$$

$$\cos 2\alpha = \cos^2\alpha - \sin^2\alpha = 2 \cos^2\alpha - 1 = 1 - 2 \sin^2\alpha$$

$$\sin(-\alpha) = -\sin \alpha$$

$$\cos(-\alpha) = \cos \alpha$$

$$\sin(\alpha + \beta) = \sin \alpha \cos \beta + \sin \beta \cos \alpha$$

$$\sin(\alpha - \beta) = \sin \alpha \cos \beta - \sin \beta \cos \alpha$$

$$\cos(\alpha + \beta) = \cos \alpha \cos \beta - \sin \alpha \sin \beta$$

$$\cos(\alpha - \beta) = \cos \alpha \cos \beta + \sin \alpha \sin \beta$$

$$\theta = \frac{S_1}{R_1} = \frac{S_2}{R_2}$$

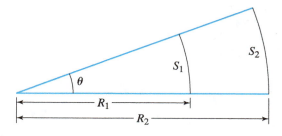

Some Useful Mathematical Relationships

$\pi = 3.14159\ldots$

$2\pi = 360$ degrees

$1 \text{ radian} = \dfrac{180}{\pi} = 57.2958°$

$1 \text{ degree} = \dfrac{\pi}{180} = 0.0174533 \text{ rad}$

$x^n x^m = x^{n+m} \qquad (xy)^n = x^n y^n$

$(x^n)^m = x^{nm} \qquad x^0 = 1 \ (x \neq 0)$

$x^{-n} = \dfrac{1}{x^n} \qquad \dfrac{x^n}{x^m} = x^{n-m} \qquad \left(\dfrac{x}{y}\right)^n = \dfrac{x^n}{y^n}$

$\log = $ logarithm to the base 10 (common logarithm)

$10^x = y \qquad \log y = x \qquad \log 1 = 0$

$\log 10 = 1 \qquad \log 100 = 2 \qquad \log 1000 = 3$

$\log xy = \log x + \log y$

$\log \dfrac{x}{y} = \log x - \log y$

$\log x^n = \ \ n \log x$

$e = 2.71828\ldots$

$\ln = $ logarithm to the base e (natural logarithm)

$e^x = y \qquad \ln y = x$

$\ln x = (\ln 10)(\log x) = 2.302585 \log x$

Some Useful Area Formulas

Triangle $A = \dfrac{1}{2}bh$

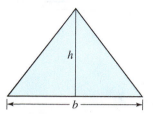

Rectangle $A = bh$

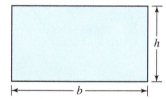

Parallelogram $A = bh$

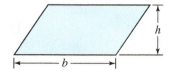

Trapezoid $A = \frac{1}{2}(a+b)h$

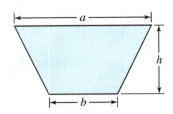

n-sided polygon $A = \left(\frac{n}{4}\right)b^2 \cot\left(\frac{180°}{n}\right)$

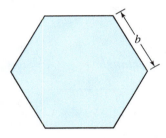

Circle $A = \pi R^2$

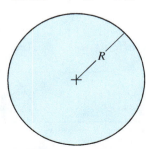

Ellipse $A = \pi ab$

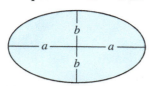

Cylinder $A = 2\pi Rh$

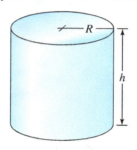

Right circular cone $A = \pi Rs = \pi R\sqrt{R^2 + h^2}$

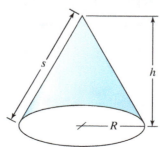

Sphere $A = 4\pi R^2$

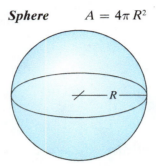

Trapezoidal rule $A \approx h\left(\dfrac{1}{2}y_0 + y_1 + y_2 + \cdots + y_{n-2} + y_{n-1} + \dfrac{1}{2}y_n\right)$

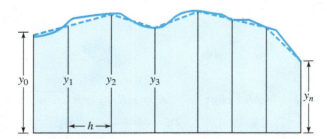

Some Useful Volume Formulas

Cylinder $\qquad V = \pi R^2 h$

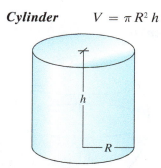

Right circular cone $\qquad V = \dfrac{1}{3} \pi R^2 h$

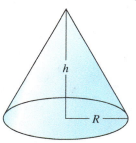

Section of a cone $\qquad V = \dfrac{1}{3} \pi h \left(R_1^2 + R_2^2 + R_1 R_2 \right)$

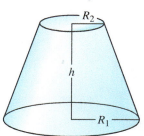

Sphere $V = \dfrac{4}{3}\pi R^3$

Section of a sphere $V = \dfrac{1}{6}\pi h (3a^2 + 3b^2 + h^2)$

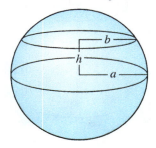

Index

A

Absolute temperature, 359–363
Absolute zero, 358–363
Academic dishonesty, 132
Acceleration
    angular, 256–257
    average, 250, 793
    due to gravity, 251–252
    instantaneous, 250
    linear, 250
    normal, 250
Accreditation board for engineering and
        technology, 16–17
Active solar system, 462–463
AFUE (annual fuel utilization
        efficiency), 453
Air, 662–664
Air standards in the U.S. *See* Water and air
        standards in the U.S.
Alternating current, 402–403
Alternative concepts, evaluation of, 55–56
American Wire Cage (AWC), 408–409
Ampere (unit), 399–400
Annual fuel utilization efficiency (AFUE), 453
Annuity, 773
ANSI (American National Standards Institute),
        81, 594
Area, 203–211
    approximation of planar, 208–211
    calculations and measurement, 207–208
    units of, and equivalent values, 206
    useful formulas for, 207–208, 795–796
ASTM (American Society for Testing and
        Materials), 81–82
Atmosphere, 663–664
AWG, 408–409

B

Bonds, 783
Boundary conditions, 720
Brass, 651
British Gravitational (BG) system
        of units, 157

Bronze, 651
Bulk modulus of compressibility, 72
Buoyancy, 212, 793

C

Calculus, 711–719
Capacitors, 415
Cash flow diagrams, 760–761
Cast iron, 653
Cell references in Excel, 491–493
Cells and their addresses in Excel, 485
Central tendency, measures of, 737–741
Chain rule, 713
Charge, 399
Civil engineering design process, 62–67
Civil engineering, drawings in, 603
Classes, importance of attending, 38
Classmates, know your, 44
Clothing insulation, 379
Coal, 456–457
Code of ethics for engineers,
    fundamental canons, 127
    professional obligations, 129–131
    revision of February 2001, 131
    rules of practice, 127–128
    statement by NSPE executive committee,
        131–132
Coefficient of thermal expansion, 383–384,
        793
Color Rendition Index (CRI), 421–423
Command history in MATLAB,
        543–544
Command window in MATLAB,
        543–544
Communication, engineering,
    engineering problems, basic steps in solution
        of, 104–108
    graphical, 113–115
    oral, 111–113
    skills and presentation of engineering work,
        103–104
    written, 108–111

Communication, written,
  executive summary, 108
  progress report, 108
  short memos, 108–109
  technical reports, detailed, 109–111
Components and systems, 171–173
Composites, 660–661
Compression strength, 71–72, 643
Conceptualization, 54–56
Concrete, 654–655
Conduction, 364–367
Conflict of interest, 132
Conflict resolution, 76
Conservation of Energy, 439–443
Conservation of mass, 175, 280–283
Conservation of mass and energy, 175
Conservation of mechanical energy, 439–440
Contract, 132
Convection, 371–375
Cooling of steel plates, 694–695, 696
Coordinate systems, 192–193
Coulomb's law, 793
Creating formulas in Excel, 487–490
CRI (Color Rendition Index), 421–423
Cumulative frequency, 735–736
Curve fitting in Excel, 505–509
Curve fitting with MATLAB, 572

D

Daylight saving, 243–244
Decibel scale, 699
Decimal multiples and prefixes for SI base
  units, 155
Decision-making and engineering economics,
  778–780
Deflection of a beam, 691–692
Degree-days and energy estimation, 381–382
Density, 71, 272–273, 642, 793
Depreciation, 783–784
Derivative rules, 713
Design considerations, additional, 67–74
Design process, engineering, 52–53
Design, sustainability in, 67–69
Determinant of a matrix, 706–708
Developer tab in Excel, 516–526
Diagonal matrix, 701
Dialog box in Excel, 524–531
Differential equations, 720–722
Dimensional homogeneity. *See* Unit conversion
  and dimensional homogeneity
Dimensioning and tolerancing, 593–596

Dimensions and units, fundamental, 146–187
  British Gravitational (BG) system of units,
    157
  components and systems, 171–173
  decimal multiples and prefixes for SI base
    units, 155
  derived units, 156
  International System (SI) of units,
    151–157
  physical laws and observations, 173–178
  significant digits (figures), 169–171
  systems of units, 151–160
  systems of units and conversion factors, 162
  unit conversion and dimensional
    homogeneity, 160–169
  U.S. Customary system of units, 158–160
  variables, used for defining ,190
Direct current, 402–403
disp command in MATLAB, 545–547
Drawings and symbols,
    586–630
  ANSI, 594
  in civil engineering, 603
  dimensioning and tolerancing, 593–596
  in electrical and electronic engineering, 603
  isometric view, 596–599
  logic gates, 612, 614
  mechanical, 590–603
  orthographic views, 590–593
  sectional views, 599–603
  solid modeling and, 603–611
  symbols, 611–616
Duty cycle, 417

E

Economics, engineering, 70, 758–784
  annuity, 773
  bonds, 783
  cash flow diagrams, 760–761
  decision-making and, 778–780
  depreciation, 783–784
  effective interest rate, 765–767
  Excel financial functions and, 780–784
  future and present worth, 762–765
  interest, simple and compound, 761–762
  interest-time factors, 773–777
  life-cycle cost, 784
  present and future worth of series of
    payments, 768–773
Effective interest rate, 765–767
Efficacy, 420–425

Efficiency of heating, cooling, and refrigeration systems, 450–454
Efficiency, 447–454
Elastic energy, 437–438, 793
Electric charge, 399
Electric current,
    American Wire Cage (AWC) and, 408–409
    capacitors and, 415
    direct and alternating, 402–403
    duty cycle and, 417
    and electrical circuits and components, 407–416
    and electric motors, 416–417
    electric power and, 409–411, 793
    electromotive force and, 400–401
    as a fundamental dimension, 399–400
    Kirchhoff's current law, 403–404
    lighting systems and, 417–425
    Ohm's law and, 408, 793
    parallel circuit and, 413–415
    photoemission and, 401–402
    power plants and, 402
    and related variables, 396–431
    residential power distribution and, 404–406
    series circuit and, 412–413
    voltage and electric power, 399–406
Electrical and electronic engineering, drawings in, 603
Electrical circuits, 407–416
Electrical resistivity, 71, 642
Electric motors, 416–417
Electric power, 409–411
Electric resistance, 407–408, 412–415
Electromotive force, 400–401
Electronic spreadsheets (MS Excel), 482–539
    basics, 484–493
    cell references in, 491–493
    cells and their addresses, 485
    creating formulas in, 487–490
    curve fitting, 505–509
    developer tab, 516–526
    dialog box, 524–531
    exporting data files into MATLAB, 567–569
    for loop, 523–524
    functions, 493–499
    functions in VBA, 518–520
    inserting cells, columns, and rows, 486–487
    logical functions, 497–498
    matrix algebra, 510–515
    matrix computation with, 509–515
    naming worksheets, 485
    now() and today() functions, 495
    plotting with, 499–509
    range of cells, 485–486
    recording macros, 520–523
    VBA's object-oriented syntax, 523
    VBA (Visual Basic for Applications), 516–531
Element-by-element operations in MATLAB, 548–550
Elements, chemical, 270
Emissivity, 375–376
Energy
    coal and, 456–457
    conservation of, 439–443, 794
    conservation of mechanical, 439–440, 793
    efficiency, 447–454
    efficiency of heating, cooling, and refrigeration systems and, 450–454
    elastic, 437–438
    ethanol and biodiesel, 470
    first law of thermodynamics and, 440–442, 794
    hydropower, 460
    and internal combustion engine efficiency, 449
    joule, 435
    kilowatt, 444
    kilowatt-hour, 444
    kinetic, 434–435
    motor and pump efficiency and, 449–450
    natural gas and, 457–458
    nuclear, 459–460
    photovoltaic systems, 465
    potential, 435–436
    power and, 443–447, 794
    and power plant efficiency, 448–449
    solar, 460–465
    sources, generation, and consumption, 455–470
    thermal, 438–439
    watts and horsepower, 444–447
    wind, 465–470
    work, mechanical energy, and thermal energy, 434–439
Energy estimation and degree-days, 381–382
Engineering as a profession, 9–13
Engineering career, preparing for 32–49
    classes, importance of attending, 38
    classmates, know your, 44
    engineering organization, getting involved with an 42–43
    examinations, preparation for, 41–42
    graduation plan, 43–44
    help, getting, 38–39

Engineering career, preparing for, (continued)
　notes, taking good, 39–40
　study groups, 41
　study habits and strategies, 36–42
　study place, 40–41
　time budgeting, 34–36
　transition from high school to college, 34
　upper-division engineering students, get to
　　know, 44
　volunteer work, 44
　voting in elections, 44
Engineering design, introduction to, 50–101
　alternative concepts, evaluation of, 55–56
　electrical resistivity, 71
　bulk modulus of compressibility, 72
　civil engineering design process, 62–67
　compression strength, 71–72
　conceptualization, 54–56
　conflict resolution, 76
　density, 71
　design considerations, additional, 67–74
　design process, engineering, 52–53
　evaluation, 57
　heat capacity, 72
　modulus of elasticity (Young's modulus), 71
　modulus of resilience, 71
　modulus of rigidity (shear modulus), 71
　modulus of toughness, 72
　need for product or service, recognizing need
　　for, 53
　optimization, 57–59
　patent, trademark, and copyright, 73–74
　presentation, 59–60
　problem definition and understanding, 53–54
　project scheduling and task chart, 76–78
　research and preparation, 54
　standards and codes, engineering, 78–83
　strength-to-weight ratio, 72
　synthesis, 56–57
　teams, common traits of good, 75–76
　teamwork, 74–76
　tensile strength, 71
　thermal conductivity, 72
　thermal expansion, 72
　vapor pressure, 72
　viscosity, 72
　water and air standards in the U.S., 84–89
Engineering disciplines, 14–25
Engineering organization, getting involved with
　an, 42–43
Engineering problems,
　defining, 104

　simplifying, 104
　solution or analysis, 105
　verifying the results, 105
Engineering technology, 24
Engineering work is all around you, 6–9
Engineer's creed
　academic dishonesty, 132
　cases 133–137
　conflict of interest, 132
　contract, 132
　plagiarism, 132
　professional responsibility, 133
EPA (Environmental Protection Agency),
　86–89
Ethanol and biodiesel, 470
Ethics, engineering, 124–143
　code of, of the NSPE, 126–132
　engineer's creed, 132–137
Evaluation, 57
Examinations, preparation for, 41–42
Excel financial functions, 780–784
Excel. See Electronic spreadsheets
Executive summary, 108
Exponential and logarithmic models, 694–700

F

Fahrenheit and Celsius scales, 684
First law of thermodynamics, 440–442
Flow rate, 274–275
Footcandle, 420
for loop (VBA), 523–524
for loop in MATLAB, 555–556
Force(s), 148, 296–302
　boundary and initial conditions and, 311
　external, 310–311
　friction and, 301–302
　and force-related variables in engineering,
　　294–348
　internal, 310–311
　linear impulse and, 336–338
　moment, torque and, 307–311
　and Newton's laws in mechanics, 303–306
　normal, 301–302
　pressure and stress and, 314–335
　reaction, 310–311
　spring, and Hooke's law, 299–301, 793
　tendencies of a, 298
　units of, 299
　viscous, 299
　work and, 312–313
Format command in MATLAB, 545–546

Formulas in MATLAB, 547–548
Fourier's Law, 365–367, 793
fprintf command in MATLAB, 545–547
Frequency. *See* Periods and frequencies
Frequency distributions, 734–736
Functions in Excel, 493–499
Functions in VBA, 518–520
Functions, loop control, and conditional statements (MATLAB), 552–560
Fundamentals of Engineering Exam, 17
Future and present worth, 762–765

G

Gauss elimination method, 709–710
Glass fibers, 659–660
Glass, 658–660
Graduation plan, 43–44
Gravity, specific, 272–273, 793
Greek alphabet, 681, 794
Groundwater, 664
Grouped frequency distribution, 734–735

H

Hardwood, 656
Heat capacity, 72, 643
Heating value, 379–380
Heating values of fuels, 379–381
Heat transfer, 363–376
Heat transfer coefficient, 372–375
Help, getting, 38–39
Histogram, 735
Homework presentation, 105–108
Hooke's law, 299–301, 793
Humidity, 664–665
Hydropower, 460

I

if and *if-else* statements in MATLAB, 557–558
Importing data files into MATLAB, 567–569
Independent variable, 683
Initial conditions, 720
Inserting cells, columns, and rows in Excel, 486–487
Integral rules, 718
Interest, simple and compound, 761–762
Interest-time factors, 773–777
Internal combustion engine efficiency, 449
International System (SI) of units, 151–157

Introduction to the engineering profession, 4–30
   accreditation board for engineering and technology, 16–17
   engineering as a profession, 9–13
   engineering disciplines, 14–25
   engineering technology, 24
   engineering work is all around you, 6–9
   professional engineer, 17–24
   sustainability concerns, 7–9
   traits of good engineers, 13–14
Inverse of a matrix, 710–711
ISO (International Organization of Standards), 80, 83
Isometric view, 596–599

J

Joule (unit), 435

K

Kilowatt, 444
Kilowatt-hour, 444
Kinetic energy, 434–435, 793

L

Kirchhoff's Current Law, 403–404
Laminar fluid velocity, 689–690
Length and time, engineering variables involving, 248–257
Length(s),
   area and, 203–211
   area, approximation of planar, 208–211
   area calculations and measurement, 207–208
   area, useful formulas for, 207–208
   coordinate systems, 192–193
   as a fundamental dimension, 191–201
   and length-related variables, 188–235
   measurement and calculation of, 194–198
   nominal sizes versus actual sizes and, 198–201
   ratio of two, radians and strain, 201–202
   second moment of area and, 217–222
   units of, and equivalent values, 194
   units of area and equivalent values, 206
   volume and, 212–217
Life-cycle cost, 784
Lighting system audit, 425
Lighting systems, 417–425
Linear equations and slopes, 684–685

Linear interpolation, 685–684
Linear models, 682–689
Linear spring, 682
Logical functions in Excel, 497–498
Logic gates, 612, 614
Lumen, 420

M

Macros (Excel), 520–523
Mass, 148
    conservation of, 175, 280–283
    density, specific weight, specific gravity,
      specific volume, and 272–273
    flow rate, 274–275, 793
    as a fundamental dimension, 268–272
    and mass-related variables in engineering,
      266–292
    moment of inertia, 275–278
    momentum and, 278–279
Mass-related variables in engineering, 266–292
Material properties related to temperature,
    383–386
Material selection, 70–72
    bulk modulus of compressibility, 72
    compression strength, 71–72
    density, 71
    electrical resistivity, 71
    heat capacity, 72
    modulus of elasticity (Young's modulus), 71
    modulus of resilience, 71
    modulus of rigidity (shear modulus), 71
    modulus of toughness, 72
    strength-to-weight ratio, 72
    tensile strength, 71
    thermal conductivity, 72
    thermal expansion, 72
    vapor pressure, 72
    viscosity, 72
Materials
    air, 662–664
    composites, 660–661
    concrete, 654–655
    glass, 658–660
    glass fibers, 659–660
    metals, 648–653
    plastics, 656–658
    polymers, 657
    properties of, 641–648
    selection and origin, 634–641
    silicon, 658
    thermoplastics, 657

    thermosets, 657
    water, 664–665
    wood, 656
Mathematics
    boundary conditions, 720
    calculus, 711–719
    chain rule, 713
    cooling of steel plates, 694–695, 696
    decibel scale, 699
    deflection of a beam, 691–692
    derivative rules, 713
    determinant of a matrix, 706–708
    diagonal matrix, 701
    differential equations, 720–722
    exponential and logarithmic models, 694–700
    Fahrenheit and Celsius scales, 684
    formulas, 795–797
    Gauss elimination method, 709–710
    Greek alphabet, 681, 794
    independent variable, 683
    initial conditions, 720
    integral rules, 718
    inverse of a matrix, 710–711
    laminar fluid velocity, 689–690
    linear equations and slopes, 684–685
    linear interpolation, 685–684
    linear models, 682–689
    linear spring, 682
    matrix algebra, 700–711
    nonlinear models, 689–694
    polynomials functions, 689–694
    relationships, 795
    Roman numerals, 681
    scalar, 700
    slope, 683, 684
    stopping sight distance, 690–691
    symbols, 680
    systems of linear equations, 688
    temperature distribution across a plane wall, 683
    transpose of a matrix, 705
    trigonometric identities, 795
    trigonometric relationships, 794–795
    unit matrix, 701
MATLAB, 540–585
    basics, 542–552
    command history, 543–544
    command window, 543–544
    curve fitting with, 572
    disp command, 545–547
    element-by-element operations, 548–550
    for loop, 555–556
    format command, 545–546

formulas in, 547–548
fprintf command, 545–547
functions, loop control, and conditional
     statements, 552–560
if and *if-else* statements, 557–558
importing data files into, 567–569
matrix computations with, 569–572
matrix operations in, 550–552
M-file in, 559–560
plotting with, 561–569
range of values, generating a, 547
set of linear equations, solving in, 571,
     574–575
symbolic mathematics with, 573–575
while loop, 556
workspace, 543
workspace, saving your, 547
Matrix algebra, 700–711
Matrix algebra in Excel, 510–515
Matrix computations with Excel, 509–515
Matrix computations with MATLAB, 569–572
Matrix operations in MATLAB, 550–552
Matter, states of, 268
Mean, 738
Mechanical engineering, drawings in, 590–603
Mechanics, 33–306
Median, 739
Mesosphere, 664
Metabolic rate, 377–378
Metals, 648–653
M-file in MATLAB, 559–560
Modulus
   bulk, of compressibility, 318
   of elasticity (Young's modulus), 71, 328–330,
     642
   of resilience, 71
   of rigidity (shear modulus), 71, 328–330
   of toughness, 72
Modulus of elasticity, 71, 328–330, 642
Modulus of resilience, 71, 643
Modulus of rigidity, 71, 328–330, 642
Moment, 307
Moment of inertia, 275–278
Momentum, 278–279, 793
Motion, angular, 255–257
Motor and pump efficiency, 449–450
MS Excel. *See* Electronic spreadsheets

N

Naming worksheets in Excel, 485
Natural gas, 457–458

NCEES (National Council of Examiners for
     Engineering and Surveying), 17–18
Need for product or service, recognizing need
     for, 53
Newton, Isaac, 174
Newton's
   first law, 303
   law of cooling, 793
   law of universal gravitation, 250–251,
     304–306, 793
   second law, 303–304, 793
   third law, 304
NFPA (National Fire Protection Association),
     82–83
Nonlinear models, 689–694
Normal distribution, 741–750
Notes, taking good, 39–40
now() and today() functions in Excel, 495
NSPE (National Society of Professional
     Engineers), 5
Nuclear, 459–460
Numerical versus symbolic solutions, 167–169

O

Ohm (unit), 408
Ohm's law, 408, 793
Optimization, 57–59
Orthographic views, 590–593
Outcome (probability and statistics), 732

P

Parallel circuit, 413–415
Pascal's law, 316–318
Passive solar (energy) system, 463–465
Patent, trademark, and copyright, 73–74
Photoemission, 401–402
Photovoltaic systems, 465
Physical laws and observations, 173–178
Plasma, 268
Plastics, 656–658
Plotting
   with MATLAB, 561–569
   with Excel, 499–509
Polymers, 657
Polynomials functions, 689–694
Population (probability and statistics), 732
Potential energy, 435–436, 793
Power, 443–447
Power plant efficiency, 448–449
Power plants, 402

Present and future worth of series of payments, 768–773
Presentation, 59–60
Pressure
    absolute and gauge, 319–322
    atmospheric, 318–319
    blood, 322
    common units of, 315
    gauge, 320–322
    hydraulic systems and, 323–325
    hydrostatic, 793
    modulus of elasticity, rigidity, and bulk, 326–335
    Pascal's Law and, 316–318
    and stress, 314–335
    vapor, 322
Probability and statistics, 730–757
    central tendency, measures of, 737–741
    cumulative frequency, 735–736
    frequency distributions, 734–736
    grouped frequency distribution, 734–735
    histogram, 735
    mean, 738
    median, 739
    normal distribution, 741–750
    outcome, 732
    population, 732
    probability basic ideas, 732–733
    random experiment, 732
    standard deviation, 739–741
    statistics basic ideas, 733
    trial, 732
    variance, 739
    variation, measures of, 737–741
    Z score, 744
Probability basic ideas, 732–733
Problem definition and understanding, 53–54
Professional engineer, 17–24
Progress report, 108
Project scheduling and task chart, 76–78
Properties of, 641–648

Q

Queuing, 281

R

Radians, 201
Radiation, 375–381
Random experiment, 732
Range of cells (Excel), 485–486

Range of values in MATLAB, 547
Recording macros (Excel), 520–523
Research and preparation, 54
Residential power distribution, 404–406
Resistor, 412–415
Roman numerals, 681
R-value, 367–371

S

Scalar, 700
Seasonal energy efficiency ratio (SEER), 453–454
Second moment of area and, 217–222
Sectional views, 599–603
SEER (seasonal energy efficiency ratio), 453–454
Selection and origin, 634–641
Series circuit, 412–413
Set of linear equations, solving in MATLAB, 571, 574–575
Short memos, 108–109
SI (International System of units). *See* International System of units
Significant digits (figures), 169–171
Silicon, 658
Silicone, 658
Slope, 683, 684
Softwood, 656
Solar energy, 460–465
Solar energy system
    active, 462–463
    passive, 463–465
Solid modeling, 603–611
Specific heat, 385–386
Specific weight, 272–273, 793
Speed
    angular, 255–256, 793
    average, 248–249, 793
    instantaneous, 249
    relationship between angular and linear, 793
    traffic average, 246–248
Stainless steel, 653
Standard deviation, 739–741
Standards and codes, engineering, 78–83
Statistics basic ideas, 733
Stefan–Boltzmann constant, 375–376
Stopping sight distance, 690–691
Strain, 202
Stratosphere, 663
Strength-to-weight ratio, 72, 643
Stress, shear, 326

Stress–strain relation (Hooke's law), 793
Study groups, 41
Study habits and strategies, 36–42
Study place, 40–41
Surface water, 664
Sustainability concerns, 7–9
Symbolic mathematics with MATLAB, 573–575
Symbols, 680
Symbols in engineering, 611–616
Synthesis, 56–57
Systems of linear equations, 688
Systems of units, 151–160
Systems of units and conversion factors, 162

T

Task chart. *See* Project scheduling and task chart
Teams, common traits of good, 75–76
Teamwork, 74–76
Technical reports, detailed,
   abstract, 109
   apparatus and experimental procedures, 109
   appendix, 111
   conclusions and recommendations, 111
   data and results, 110–111
   discussion of results, 111
   objectives, 109
   references, 111
   theory and analysis, 109
   title, 109
Temperature, 355
   absolute zero, 358–363
   conduction and, 364–367
   convection and, 371–375
   conversion, 793
   degree-days and energy estimation, and, 381–382
   difference and heat transfer, 363–376
   distribution across a plane wall, 683
   as a fundamental dimension, 353–363
   heating values of fuels and, 379–381
   material properties related to, 383–386
   measurement of, and its units, 356–358
   radiation and, 375–381
   specific heat and, 385–386
   and temperature-related variables, 350–395
   thermal comfort and, 377–379
   thermal expansion and, 383–384
   thermal resistance and, 367–371
Tensile strength, 71, 642

Thermal comfort, 377–379
Thermal conductivity, 72, 365–367, 643
Thermal energy, units of, 438–439
Thermal expansion, 72, 383–384
Thermal resistance, 367–371
Thermistor, 358–359
Thermocouple wire, 358–359
Thermoplastics, 657
Thermosets, 657
Time, 148–149
   daylight saving, 243–244
   engineering variables involving length and, 248–257
   flow of traffic, 246–248, 793
   as a fundamental dimension, 238–240
   measurement of, 241–244
   periods and frequencies, 244–245
   and time-related variables in engineering, 236–265
Time budgeting, 34–36
Time zones, 242–243
Ton of cooling, 444
Torque, 307–311
Traffic
   average speed, 246–247
   density, 246–247
   flow, 246–248, 793
Traits of good engineers, 13–14
Transient (or unsteady) problem, 239
Transition from high school to college, 34
Transpose of a matrix, 705
Trapezoidal rule, 208
Trial (probability and statistics), 732
Troposphere, 664

U

Upper-division engineering students, get to know, 44
U.S. Customary System of Units, 158–160
Underwriters Laboratories (UL), 83
Unit conversion and dimensional homogeneity, 160–169
Unit matrix, 701
Unsteady (or transient) problem, 239

V

Vapor pressure, 72, 643
Variables, dimensions and units used for defining, 190
Variance, 739

Variation, measures of, 737–741
VBA
    for loop, 523–524
    functions in, 518
    object-oriented syntax of, 523
Velocity
    changes in, 250
    linear, 248–249
Viscosity, 72, 643
Voltage, 400–401
Volume,
    calculations, 213–217
    flow rate, 253–254, 793
    formulas, 797
    specific, 272–273, 793
Volunteer work, 44
Voting in elections, 44

W

Water, 664–665
Water and air standards in the U.S., 84–89
Watts and horsepower, 444–447
Weight, 148, 793
    specific, 272–273, 793
while loop in MATLAB, 556
Wind, 465–470
Wood, 656
Work, 312–313, 434–439
Workspace in MATLAB, 543
Workspace, saving your, in MATLAB, 547

Z

Z score, 744